SILFS

Volume 1

New Essays in Logic and Philosophy of Science

Volume 1
New Essays in Logic and Philosophy of Science
Marcello D'Agostino, Giulio Giorello, Federico Ladisa, Telmo Pievani and Corrado Sinigaglia, eds

SILFS Series Editor
Marcello D'Agostino   marcello.dagostino@unife.it

New Essays in Logic and Philosophy of Science

Edited by
Marcello D'Agostino,
Giulio Giorello,
Federico Laudisa,
Telmo Pievani,
and
Corrado Sinigaglia

© Individual author and College Publications 2010. All rights reserved.

ISBN 978-1-84890-003-5

College Publications
Scientific Director: Dov Gabbay
Managing Director: Jane Spurr
Department of Computer Science
King's College London, Strand, London WC2R 2LS, UK

http://www.collegepublications.co.uk

Cover designed and created by Laraine Welch
Printed by Lightning Source, Milton Keynes, UK

All rights reserved. No part of this publication may be reproduced, stored in a retrieval system or transmitted in any form, or by any means, electronic, mechanical, photocopying, recording or otherwise without prior permission, in writing, from the publisher.

Table of contents

Editors' preface	ix
List of contributors	xi
PART I LOGIC AND COMPUTING	1
UMBERTO RIVIECCIO A bilattice for contextual reasoning	3
FRANCESCA POGGIOLESI Reflecting the semantic features of S5 at the syntactic level	13
GISÈLE FISCHER SERVI Non monotonic conditionals and the concept *I believe only A*	27
CARLO PENCO, DANIELE PORELLO Sense and Proof	37
ANDREA PEDEFERRI Some reflections on plurals and second order logic	47
G. CASINI, H. HOSNI Default-assumption consequence relations in a preferential setting: reasoning about normality	53
BIANCA BORETTI, SARA NEGRI On the finitization of Priorean linear time	67
RICCARDO BRUNI Proof-theoretic aspects of quasi-inductive definitions	81
GIACOMO CALAMAI Remarks on a proof-theoretic characterization of polynomial space functions	95
PART II PHYSICS AND MATHEMATICS	115
VINCENZO FANO, GIOVANNI MACCHIA How contemporary cosmology bypasses Kantian prohibition against a science of the universe	117
GIULIA GIANNINI Poincaré and the electromagnetic world picture. For a revaluation of his conventionalism	131
MARCO TOSCANO "Besides quantity": the epistemological meaning of Poincaré's qualitative analysis	141

LAURA FELLINE
Structural explanation from special relativity to quantum
mechanics   153
MIRIAM COMETTO
When the structure is not a limit. On continuity through
theory-change   163
GIANLUCA INTROZZI
Approaches to wave/particle duality: historical analysis and
critical remarks   173
MARCO PEDICINI, MARIO PIAZZA
An application of von Neumann algebras to computational
complexity   183
MIRIAM FRANCHELLA
Phenomenology and intuitionism: the pros and cons of a
research program   195
LUCA BELLOTTI
A note on the circularity of set-theoretic semantics for set theory   207
VALERIA GIARDINO
The use of figures and diagrams in mathematics   217
PAOLA CANTÙ
The role of epistemological models in Veronese's and Bettazzi's
theory of magnitudes   229

PART III LIFE SCIENCES   243

PIETRO OMODEO
Evolution by increasing complexity in the framework of Darwin's
theory   245
STEFANO GIAIMO, GIUSEPPE TESTA
Gene: an entity in search of concepts   257
ELENA CASETTA
Categories, taxa, and chimeras   265
FLAVIO D'ABRAMO
Final, efficient and complex causes in biology   279
LUDOVICA LORUSSO
The concept of race and its justification in biology   289

PART IV ECONOMICS AND SOCIAL SCIENCES   301

MARCO NOVARESE, ALESSANDRO LANTERI, CESARE
TIBALDESCHI
Learning, generalization, and the perception of information: an
experimental study   303
ANDREA POZZALI
Tacit knowledge and economics: recent findings and perspectives
of research   319

VIVIANA DI GIOVINAZZO
From individual well-being to economic welfare. Tibor Scitovsky
explains why (consumers') dissatisfaction leads to a joyless
economy 327

FEDERICA RUSSO
Explaining causal modelling. Or, what a causal model ought to
explain 347

ENZO DI NUOSCIO
The epistemological statute of the rationality principle.
Comparing Mises and Popper 363

ALBERTINA OLIVERIO
Evolution, cooperation and rationality: some remarks 377

FRANCESCO DI IORIO
Self-organization of the mind and methodological individualism
in Hayek's thought 389

SIMONA MORINI
Can ethics be naturalized? 403

STEFANO VASELLI
Searle's collective intentionality and the "invisible hand"
explanations 409

SERGIO LEVI
The naturalness of religion and the action representation system 423

ANDREA ZHOK
On value judgement and the ethical nature of economic
optimality 433

DARIO ANTISERI
Carl Menger, Ludwig von Mises, Friedrich A. von Hayek and
Karl Popper: four Viennese in defense of methodological
individualism 447

PART V NEUROSCIENCE AND PHILOSOPHY OF MIND 463

FABIO BACCHINI
Jaegwon Kim and the threat of epiphenomenalism of mental
states 465

WOLFGANG HUEMER
Philosophy of mind between reduction, elimination and
enrichment 481

LAURA SPARACI
Discourse and action: analyzing the possibility of a structural
similarity 493

ALESSANDRO DELL'ANNA
Visuomotor representations: Jacob and Jeannerod between
enaction and the two visual systems hypothesis 505

DANIELA TAGLIAFICO
Mirror neurons and the "radical view" on simulation ... 515
VINCENZO G. FIORE
Multiple realizations of the mental states: hunting for plausible chimeras ... 529
ARTURO CARSETTI
The embodied meaning and the "unfolding" of the mind's eyes ... 539
KATJA CRONE
Consciousness and the problem of different viewpoints ... 547
GIULIA PIREDDA
The whys and hows of extended mind ... 559
CARMELA MORABITO
Movement in the philosophy of mind: traces of the motor model of mind in the history of science ... 571
JEAN-LUC PETIT
The brain, the person and the world ... 585

PART VI   GENERAL PHILOSOPHY OF SCIENCE ... 601

GUSTAVO CEVOLANI, VINCENZO CRUPI, ROBERTO FESTA
The whole truth about Linda: probability, verisimilitude, and a paradox of conjunction ... 603
ANTONINO FRENO
Probabilistic Graphical Models and logic of scientific discovery ... 617
MASSIMILIANO CARRARA, DAVIDE FASSIO
Perfected science and the knowability paradox ... 629
LUCA TAMBOLO
Two problems for normative naturalism ... 637
SILVANO ZIPOLI CAIANI
Explaining the scientific success. A critique of an abductive defence of scientific realism ... 649
MARIO ALAI
Van Fraassen, observability and belief ... 663
MARCO GIUNTI
Reduction in dynamical systems: a representational view ... 677
ALEXANDER AFRIAT
Duhem, Quine and the other dogma ... 695
EDOARDO DATTERI
Bionic simulation of biological systems: a methodological analysis ... 711
LUCA GUZZARDI
Some remarks on a heuristic point of view about the role of experiment in the physical sciences ... 725

Editors' preface

The papers collected in this volume stem from the contributions delivered to the conference of the Italian Society for Logic and Philosophy of Science (SILFS) that took place in Milan on 8-10 October 2007. The aim of the Society, since its foundation in 1952, has always been that of bringing together scholars — working in the broad areas of Logic, Philosophy of Science and History of Science — who share an open-minded approach to their disciplines and regard them as essentially requiring continuous confrontation and bridge-building to avoid the vanity of over-specialism. In this perspective, logicians and philosophers of science should not indulge in inventing and cherishing their own "internal problems" — although these may occasionally be an opportunity for conceptual clarification — but should primarily look at the challenging conceptual and methodological questions that arise in any genuine attempt to extend our "objective" knowledge of the physical, biological, social and intellectual environment in which we are embedded. As Ludovico Geymonat used to put it: "[good] philosophy should be sought in the folds of science itself".

Accordingly, the accepted contributions were distributed into six sections, five of which — "Logic and Computing", "Physics and Mathematics", "Life Sciences", "Economics and Social Sciences", "Neuroscience and Philosophy of Mind" — were devoted to the discussion of cutting-edge problems that arise from current-day scientific research, while the remaining section on "General Philosophy of Science" was focused on foundational and methodological questions that are common to all areas.

The very good response to the call for abstracts and the high quality of the accepted presentations persuaded the SILFS President, Giulio Giorello, to launch the idea of a refereed volume based on the best contributions. Authors were therefore invited to submit papers, inspired by their talks, which were anonimously refereed and subsequently revised before being finally accepted for publication in this volume. So, what we are presenting here is by no means the proceedings of the conference, but rather the result of a long (and, alas, time-consuming) process that started after the conference, although being inspired by it. We do hope that readers may enjoy the result of this collective effort and appreciate the strong inderdisciplinary spirit that pervades it.

We wish to thank all the authors and, especially, Luca Guzzardi who generously offered us his competent support in editing this volume.

M.D'A, G.G., F.L, T.P. C.S.

Milano, 31 July 2010

List of contributors

ALEXANDER AFRIAT, Department of Philosophy, University of West Brittany
MARIO ALAI, Department of Philosophy, University of Urbino
DARIO ANTISERI, Department of Political Sciences, LUISS, Rome
FABIO BACCHINI, Faculty of Architecture of Alghero, University of Sassari
LUCA BELLOTTI, Department of Philosophy, University of Pisa
BIANCA BORETTI, Department of Philosophy, University of Helsinki
RICCARDO BRUNI, Department of Philosophy, University of Florence
GIACOMO CALAMAI, Department of Mathematics and Information Sciences "Roberto Magari", University of Siena
PAOLA CANTÙ, Centre d'Epistémologie et Ergologie Comparatives, Université de Provence, Aix-en-Provence
MASSIMILIANO CARRARA, Department of Philosophy, University of Padua
ARTURO CARSETTI, Department of Philosophy, University of Rome "Tor Vergata"
ELENA CASETTA, Department of Philosophy, University of Turin
GIOVANNI CASINI, Scuola Normale Superiore di Pisa
GUSTAVO CEVOLANI, Department of Philosophy, University of Bologna
MIRIAM COMETTO, Department of Philosophy, University of Rome III
KATJA CRONE, Department of Philosophy, Humboldt University, Berlin
VINCENZO CRUPI, Department of Critical Care, University of Florence
FLAVIO D'ABRAMO, Department of Philosophical and Epistemological Studies, University of Rome "La Sapienza"
EDOARDO DATTERI, Department of Human Sciences, University of Milan-Bicocca
ALESSANDRO DELL'ANNA, Department of Philosophy, University of Genoa
VIVIANA DI GIOVINAZZO, Department of Sociology and Social Research, University of Milan-Bicocca
FRANCESCO DI IORIO, LUISS University, Rome and EHESS-CREA, Ecole Polytechnique, Paris
ENZO DI NUOSCIO, Faculty of Human Sciences, University of Molise
ENZO FANO, Department of Philosophy, University of Urbino
DAVIDE FASSIO, Department of Philosophy, University of Padua
LAURA FELLINE, Department of Education Science and Philosophy, University of Cagliari
ROBERTO FESTA, Department of Philosophy, Languages, and Literatures, University of Trieste
VINCENZO G. FIORE, Laboratory Of Computational Embodied Neuroscience, Institute of Cognitive Sciences and Technologies, CNR
GISÉLE FISCHER SERVI, Department of Philosophy, Univeristy of Parma
MIRIAM FRANCHELLA, Department of Philosophy, University of Milan
ANTONINO FRENO, Information Engineering Department, University of Siena

STEFANO GIAIMO, European School of Molecular Medicine and FIRC Institute of Molecular Oncology, Milan
GIULIA GIANNINI, Department of Human Sciences, University of Bergamo
VALERIA GIARDINO, Institute Jean Nicod (CNRS-EHESS-ENS), Paris
MARCO GIUNTI, Department of the Science of Education and Philosophy, University of Cagliari
LUCA GUZZARDI, Astronomical Observatory of Brera, Milan
HYKEL HOSNI, Scuola Normale Superiore di Pisa
WOLFGANG HUEMER, Department of Philosophy, University of Parma
GIANLUCA INTROZZI, Department of Nuclear and Theoretical Physics, University of Pavia
ALESSANDRO LANTERI, European School of Economics, Milan
SERGIO LEVI, Department of Philosophy, University of Milan
LUDOVICA LORUSSO, Department of Economics, Business and Regulation, University of Sassari
GIOVANNI MACCHIA, Department of Philosophy, University of Urbino
CARMELA MORABITO, Department of Philosophy, University of Rome "Tor Vergata"
SIMONA MORINI, Faculty of Design and Arts, IUAV, Venice
SARA NEGRI, Department of Philosophy, University of Helsinki
MARCO NOVARESE, Centre for Cognitive Economics, Department of Legal Sciences and Economics, University of West Piedmont, Vercelli
ALBERTINA OLIVERIO, Faculty of Human Sciences, University of Chieti and Pescara
PIETRO OMODEO, Department of Evolutionary Biology, University of Siena
ANDREA PEDEFERRI, Department of Philosophy, George Washingotn University, Washington DC
MARCO PEDICINI, CNR — Istituto per le Applicazioni del Calcolo M. Picone, Roma
CARLO PENCO, Department of Philosophy, University of Genoa
JEAN-LUC PETIT, Faculty of Philosophy, Linguistics and the Science of Education, University of Strasbourg and College de France, Paris
MARIO PIAZZA, Department of Philosophy, University of Chieti and Pescara
GIULIA PIREDDA, Institute Jean Nicod (CNRS-EHESS-ENS), Paris
FRANCESCA POGGIOLESI, Department of Philosophy and Moral Sciences, Vrije Universiteit Brussel
DANIELE PORELLO, Department of Philosophy, Univeristy of Genoa
ANDREA POZZALI, European University of Rome
UMBERTO RIVIECCIO, Department of Philosophy, University of Genoa
FEDERICA RUSSO, Department of Philosophy, University of Kent
LAURA SPARACI, Neuroscience Department, Child Neuropsychiatry Unit, Children's Hospital "Bambino Gesù", Rome
DANIELA TAGLIAFICO, Laboratory of Ontology, University of Turin
LUCA TAMBOLO, Department of Philosophy, Languages, and Literatures, University of Trieste
GIUSEPPE TESTA, Laboratory of Stem Cell Epigenetics, European Institute of Oncology and European School of Molecular Medicine, Milan

CESARE TIBALDESCHI, Department of Legal Sciences and Economics, University of West-Piedmont, Vercelli
MARCO TOSCANO, Department of Human Sciences, University of Bergamo
STEFANO VASELLI, Department of Philosophical and Epistemological Studies, University of Rome "La Sapienza"
SILVANO ZIPOLI CAIANI, Department of Philosophy, University of Milan
ANDREA ZHOK, Department of Philosophy, University of Milan

PART I

LOGIC AND COMPUTING

A bilattice for contextual reasoning
UMBERTO RIVIECCIO

1 Introduction

Bilattices are algebraic structures introduced by Ginsberg [4] as a uniform framework for inference in Artificial Intelligence. In the last two decades the bilattice formalism proved useful in many fields: however it has never been applied to contextual reasoning so far. My aim here is to sketch one such possible application.

The basic idea is to treat contexts as "truth values" that form a bilattice, that is a lattice equipped with two partial orders. We shall employ a bilattice construction introduced in [4] for the "justifications" of a *Truth Maintenance System*, i.e. sets of premises used for derivations, but we will apply it to sets of formulas representing not premises in the usual logical sense but cognitive contexts.

The usual logical connectives will then be defined as lattice operators on the set of truth values, for instance conjunction and disjunction correspond respectively to the *meet* and *join* with respect to the so called "truth ordering". Non classical connectives may also be defined, such as those corresponding to the *meet* and *join* w.r.t. the second partial order, usually called the "knowledge ordering".

The next step would be to construct a suitable inference mechanism for contextual bilattices. This has been done for the general case (see for instance [4] and [1]), but it remains to show that such mechanisms may be successfully applied to contextual reasoning. In section 4 we shall see an example of a possible application.

2 Bilattices

Given a set B, a bilattice may be defined as a structure $\langle B, \leq_t, \leq_k \rangle$ where $\langle B, \leq_t \rangle$ and $\langle B, \leq_k \rangle$ are both complete lattices.

The elements of B are intended to represent truth values ordered according to the degree of truth and the degree of knowledge (or information): the relation $\leq_t$ corresponds to the truth ordering and $\leq_k$ to the knowledge ordering. Intuitively, given two sentences p and q, $v(p) \leq_t v(q)$ means that the agent has stronger evidence for the truth of q than for the truth of p and weaker evidence for the falsity of q than for that of p, while $v(p) \leq_k v(q)$ means that the agent has stronger evidence for *both* the truth and falsity of q than for the truth and falsity of p (thus allowing for inconsistency).

The *meet* and *join* operations on the two lattices correspond to propositional connectives in the bilattice-based logics. Conjunction and disjunction are defined respectively as the greatest lower bound and least upper bound with respect to the truth ordering. Given two sentences p, q and a valuation v, we have:

$$v(p \wedge q) = \mathrm{glb}_t \{v(p), v(q)\}$$
$$v(p \vee q) = \mathrm{lub}_t \{v(p), v(q)\}$$

The corresponding connectives relative to the knowledge ordering have been called *consensus* ($\otimes$) and *gullability* ($\oplus$) operator (see [3]). They are defined as follows:

$$v(p \otimes q) = \mathrm{glb}_k \{v(p), v(q)\}$$
$$v(p \oplus q) = \mathrm{lub}_k \{v(p), v(q)\}$$

Intuitively, we may interpret $v(p) \otimes v(q)$ as the most information that $v(p)$ and $v(q)$ agree on, while the gullability operator $\oplus$ simply accepts any information from both $v(p)$ and $v(q)$.

We have a bilattice negation if there is a function $\neg : B \to B$ such that:

1. if $v(p) \leq_t v(q)$ then $\neg v(q) \leq_t \neg v(p)$,

2. if $v(p) \leq_k v(q)$ then $\neg v(p) \leq_k \neg v(q)$,

3. $v(p) = \neg \neg v(p)$.

In other words, we require that the negation be an involutive operator that reverses the truth ordering while leaving the knowledge ordering unchanged: this corresponds to the intuition that the amount of information one has concerning some sentence p should not be altered when considering its negation $\neg p$. The existence of such a negation operator is a minimal requirement for logical bilattices (indeed some authors consider it part of the basic definition of bilattice); for the kind of bilattices we will construct there is a straightforward way to define it (see below, section 3).

The smallest non-trivial bilattice is the one corresponding to Belnap's logic (see [2]), which has four elements, that is exactly the least and greatest elements with respect to the two orderings. This bilattice can be constructed from the cartesian product of the classical two-point truth set with itself: $\{0, 1\} \times \{0, 1\} = \{(0, 1), (1, 0), (0, 0), (1, 1)\}$. We may interpret the first element of each ordered pair as representing the evidence for the truth of a sentence p, while the second element represents the evidence for the falsity of p. In this way we can understand Belnap's values in terms of the classical ones: $(1, 0)$ corresponds to "at least true", $(0, 1)$ to "at least false", $(0, 0)$ to "unknown" (i. e. not known to be either true or false) and $(1, 1)$ to "contradictory" (i. e. known to be both true and false).

Several more complex bilattices have been introduced in the literature for a variety of applications (for instance to deal with default reasoning, with modal operators etc.), many of them built like Belnap's using two copies of some lattice, the first for the positive evidence and the second for the negative. In the next section we shall employ this procedure to construct a bilattice for contextual reasoning.

3 Contexts as truth values

In cognitive processes, the notion of context may be defined as a part of the epistemic state of an agent, i.e. as a set of implicit assumptions. These assumptions enable us to assign a reference to indexical expressions such as "this", "here", "now" etc., and so to determine the truth value of the sentences involving them. The simplest way to formalize this is to identify contexts with subsets of the knowledge base, i.e. sets of formulas.

Let F be the set of all formulas in the knowledge base and let $C_1, \ldots, C_n \subseteq F$ be sets of formulas intended to represent contexts. To each sentence p we may associate the set $C^+ = \{C_1, \ldots, C_n\}$ of all contexts in which p holds. We assume each C_i to be a set of sentences, possibly cointaining contextual "axioms" such as "Speaker = ...", " Time = ..." etc., that logically imply p. We shall denote this writing $v(p) = [C^+]$. This is the basic idea that provides a link with the multi-valued setting of bilattices, that is the idea to treat contexts as truth values.

If we want to handle inconsistent beliefs, we may also consider the set C^- of all contexts in which $\neg p$ holds, without requiring that C^+ and C^- be disjoint, so that we may have some context in which both p and $\neg p$ hold. Therefore, instead of writing only $v(p) = [C^+]$, meaning that the value of p is given by the contexts in which p holds, we shall write $v(p) = [C^+, C^-]$, meaning that the value of p is given by the contexts in which p holds together with the contexts in which $\neg p$ holds.

Now we proceed to define an order relation on these "truth values". We may order contexts in a natural way by set inclusion. For instance, $C_2 \subseteq C_1$ intuitively means that C_2 is more general than C_1, since C_2 requires fewer assumptions than C_1. (The most general context is thus the empty context, corresponding to sentences that are completely context-independent, while the least general context is simply the set of all formulas.) This intuition may be extended to sets of contexts as follows.

Given two sets of contexts $C = \{C_1, \ldots, C_m\}$ and $D = \{D_1, \ldots, D_n\}$, we set $C \leq D$ iff for all $C_i \in C$ there is some $D_j \in D$ such that $D_j \subseteq C_i$. This means that, for every context in C, there is some context in D which is more general: so if we know that p holds in D and q holds in C, we can conclude that p is less context-dependent than q.

So far the relation we have defined is in fact only a preorder, since we may have $C \leq D$ and $D \leq C$ but $C \neq D$. This happens for instance if $C = \{\{p\}\}$ and $D = \{\{p\}, \{p, q\}\}$. However, to obtain an order we just need to consider the equivalence classes under this preorder; or, equivalently, we might require that each set of contexts be minimal, in the sense that for every $C = \{C_1, \ldots, C_m\}$ there should be no $C_i, C_j \in C$ such that $C_i \subseteq C_j$. In other words, what we are assuming is that all sets are free of redundant contexts (such as C_j in our example).

Of course this is not the only possible way to define an order relation on (sets of) contexts. For instance one might consider the logical (instead of just the set inclusion) relationship between the propositions representing contexts. For suppose we have two contexts C and D such that $C \not\subseteq D$ and $D \not\subseteq C$ but $D \subseteq \text{th}(C)$, that is $C \vDash p$ for each sentence $p \in D$. Since D is contained in the logical consequences of C, from a deductive point of view we might expect to have $C \leq D$. According to this intuition, we should replace the previous definition with the more general one: $C \leq D$ iff $D \subseteq \text{th}(C)$.

However, we shall not employ this definition here because it would not allow to construct a lattice of contexts in an effective way. In fact, in order to determine if $C \leq D$ we would have to check if $D \subseteq \text{th}(C)$, which is notoriously a complex task from a computational point of view.

Instead, we prefer to adopt the simpler set inclusion definition, delaying the hard part of the job to a later stage of the inference process. So if p holds in C and q holds in D, with $D \subseteq \text{th}(C)$, this relation will not be reflected in the values assigned to p and q by our initial valuation until we have applied some suitable closure operator (such as the one introduced in [4]).

Adopting the set inclusion order relation we are now able to define a lattice of sets of contexts. Let F be the set of all formulas in the knowledge base and $P(F)$ its power set, that is the set of all possible contexts. If we denote the set of all sets of contexts by $L = P(P(F))$, then the structure $\langle L, \leq \rangle$ is the lattice of sets of contexts.

As we have said, in order to consider inconsistent beliefs we employ two copies of L, one for the contexts in which a sentence p holds and the other for those in which $\neg p$ holds. In this way we obtain a structure that may be called a "contextual bilattice" $\langle L \times L, \leq_t, \leq_k \rangle$, the underlying set being formed by the ordered pairs $[C^+, C^-]$ of elements of the lattice of sets of contexts. In this structure the truth and knowledge order relations may be defined as follows. For any two elements $[C^+, C^-], [D^+, D^-] \in L \times L$:

$$[C^+, C^-] \leq_t [D^+, D^-] \quad \text{iff} \quad C^+ \leq D^+ \quad \text{and} \quad C^- \geq D^-$$

$$[C^+, C^-] \leq_k [D^+, D^-] \quad \text{iff} \quad C^+ \leq D^+ \quad \text{and} \quad C^- \leq D^-.$$

It can be verified that our definition reflects the previous considerations on the two orderings. The logical connectives on the contextual bilattice may then be easily defined as lattice operators. Let $\langle L, \leq, \cdot, + \rangle$ be the lattice of sets of contexts, with $\cdot$ and $+$ denoting respectively the meet and join operations. It is easy to see that the propositional connectives on $\langle L \times L, \leq_t, \leq_k \rangle$ result as follows. For any two elements $[C^+, C^-], [D^+, D^-] \in L \times L$:

$$[C^+, C^-] \wedge [D^+, D^-] = [(C^+ \cdot D^+), (C^- + D^-)]$$

$$[C^+, C^-] \vee [D^+, D^-] = [(C^+ + D^+), (C^- \cdot D^-)]$$

$$[C^+, C^-] \otimes [D^+, D^-] = [(C^+ \cdot D^+), (C^- \cdot D^-)]$$

$$[C^+, C^-] \oplus [D^+, D^-] = [(C^+ + D^+), (C^- + D^-)].$$

Negation is simply defined as a function swapping the "truth" and "falsity degree", that is we have $\neg [C^+, C^-] = [C^-, C^+]$ for all $[C^+, C^-] \in L \times L$. It can be easily verified that this operation meets all the requirements stated in the previous section.

We may also note that our contextual bilattice has nice structural properties, for instance that it is *interlaced*, that is for any $x, y, z \in L \times L$ we have:

(i) $x \leq_t y$ implies $x \otimes z \leq_t y \otimes z$ and $x \oplus z \leq_t y \oplus z$

(ii) $x \leq_k y$ implies $x \wedge z \leq_t y \wedge z$ and $x \vee z \leq_t y \vee z$.

This means that each operation associated with one of the lattice orderings is monotonic with respect to the other ordering. As noted by Fitting [3], this is another kind of connection between the two orderings besides that provided by the negation operator.

4 Reasoning with contexts

The next step would be to construct a suitable inference mechanism for "contextual bilattices". As we have anticipated, there exist general inference procedures for bilattices (see [4] and [1]), but it remains to show that they may be successfully applied to contextual reasoning. As a preliminary result, we will see an application to our setting of the closure operator defined by Ginsberg [4].

As we have said in the previous section, we begin with an arbitrary valuation and then apply to it a closure operator to derive the logical consequences of the set of initial beliefs we are interested in.

First of all we need to define what it means for a valuation to be closed in the bilattice framework. Let $v : F \to B$ be a valuation from the set of all formulas into a bilattice B. We say that v is *closed* iff, for any formulas $p, q \in F$, the following conditions hold:

(i) if $p \vDash q$ then $v(p) \leq_t v(q)$

(ii) $v(p) \wedge v(q) \leq_k v(p \wedge q)$

(iii) $v(\neg p) = \neg v(p)$.

Item (i) says that if p implies (in classical logic) q, then the value of q should be at least as true as that of p.

Item (ii) says that one should know at least as much about a conjunction as about each one of the conjuncts, and in some cases one may know more. For instance, consider the four-point bilattice corresponding to Belnap's logic, and recall that $(0, 0)$ and $(0, 1)$ may be intuitively interpreted as (respectively) "unknown" and "at least false". We may have $v(p) = v(\neg p) = (0,0)$ for some sentence p, but still we would expect to have $v(p \wedge \neg p) = (0, 1)$ instead of $v(p \wedge \neg p) = v(p) \wedge v(\neg p) = (0, 0)$. So we see that in some cases it is reasonable to have $v(p \wedge q) >_k v(p) \wedge v(q)$.

Item (iii) just states the intuitive requirement that the negation should map the truth value assigned to p to that assigned to $\neg p$.

Given two valuations v and w, we will say that w is an *extension* of v in case w is "more informed" than v, that is in case for each formula $p \in F$ we have $v(p) \leq_k w(p)$. We shall write $v \leq_k w$ to indicate that w is an extension of v.

Now we define $\text{cl}(v)$, the *closure* of a valuation $v : F \to B$, as follows:

$$\text{cl}(v) = \bigotimes \{w : v \leq_k w \text{ and } w \text{ is closed}\}$$

where the symbol $\bigotimes$ denotes the infinitary version of the consensus operator, i.e. the meet with respect to the knowledge lattice. As Ginsberg [4] has shown, it is possible to construct an effective procedure that computes $\text{cl}(v)$ for a given valuation v.

Now we shall see how this closure operator works with the contextual bilattice. Consider a set A of assumptions that may represent the agent's initial beliefs. We define an initial valuation $v : F \to L \times L$ as follows. For each sentence $p \in F$:

$$v(p) = \begin{cases} [(\{p\}), (F)] & \text{if } p \in A \\ [(F), (F)] & \text{otherwise.} \end{cases}$$

In this way we are labeling each sentence in A as self-justified, i.e associated with a context consisting only of itself (recall that F, the set of all formulas, is the least element in the lattice of contexts L, since it amounts to having no information at all about in which contexts a sentence holds). Then we apply the closure operator to v. We now have the following result:

THEOREM 1. *Let v be a valuation defined as above. Then, for each $\{p_1, \ldots, p_n, q\} \subseteq A$:*

$$\text{cl}(v(q)) \geq_k [(\{p_1, \ldots, p_n\}), (F)] \quad \text{iff} \quad p_1, \ldots, p_n \vDash q$$

$$\text{cl}(v(q)) \geq_k [(F), (\{p_1, \ldots, p_n\})] \quad \text{iff} \quad p_1, \ldots, p_n \vDash \neg q$$

Proof. Let $\bigoplus$ denote the infinitary version of the gullability operator $\oplus$ and let $\bar{v}$ be a valuation defined as follows. For any $q \in F$:

$$\bar{v}(q) = \bigoplus_{X \subseteq A} \{[(X), (F)] : X \vDash q\} \oplus \bigoplus_{Y \subseteq A} \{[(F), (Y)] : Y \vDash \neg q\}.$$

Clearly, to prove the theorem it is sufficient to show that $\bar{v}(q) = \text{cl}(v(q))$ for any $q \in F$. First we show that for every closed valuation w, if $v(q) \leq_k w(q)$ for any $q \in F$, then

$$\bigoplus_{X \subseteq A} \{[(X), (F)] : X \vDash q\} \leq_k w(q).$$

Since w is closed, for any set $\{p_1, \ldots, p_n\} \subset F$ such that $\{p_1, \ldots, p_n\} \vDash q$ we have $\wedge_i w(p_i) \leq_k w(\wedge_i p_i) \leq_t w(q)$ $(1 \leq i \leq n)$. By hypothesis $v(p_i) \leq_k w(p_i)$: since we are in an interlaced bilattice, this implies $\wedge_i v(p_i) \leq_k \wedge_i w(p_i)$ for all $p_i \in F$. By the definition of v we have $\wedge_i v(p_i) = \wedge_i [(\{p_i\}), (F)] = [(\{p_1, \ldots, p_n\}), (F)]$. So we also have

$$[(\{p_1, \ldots, p_n\}), (F)] = \wedge_i v(p_i) \leq_k \wedge_i w(p_i) \leq_k w(\wedge_i p_i) \leq_t w(q).$$

Since F is the minimal element in the lattice of contexts L, this implies that $[(\{p_1,\ldots,p_n\}),(F)] \leq_k w(q)$. Hence

$$\bigoplus_{X \subseteq A} \{[(X),(F)] : X \vDash q\} \leq_k w(q).$$

Recalling that $w(\neg q) = \neg w(q)$ for all $q \in F$, we may show in the same way that

$$\bigoplus_{Y \subseteq A} \{[(F),(Y)] : Y \vDash \neg q\} \leq_k w(q).$$

So we have that $\bar{v}(q) \leq_k w(q)$ for any closed extension w of v.

Now it remains only to show that $\bar{v}$ is closed, for then it will clearly coincide with the greatest lower bound of all the closed extensions of v. We consider each item of the definition of closed valuation.

(i) Let $p,q \in F$ such that $p \vDash q$. We have to show that $\bar{v}(p) \leq_t \bar{v}(q)$. Note that $Z \vDash p$ implies $Z \vDash q$ for any $Z \subset F$, so any set of formulas appearing in

$$\bigoplus_{S \subseteq A} \{[(S),(F)] : S \vDash p\}$$

will also be in

$$\bigoplus_{X \subseteq A} \{[(X),(F)] : X \vDash q\}.$$

Hence

$$\bigoplus_{S \subseteq A} \{[(S),(F)] : S \vDash p\} \leq_k \bigoplus_{X \subseteq A} \{[(X),(F)] : X \vDash q\}$$

and since the falsity component is the same in both elements (F), this implies that

$$\bigoplus_{S \subseteq A} \{[(S),(F)] : S \vDash p\} \leq_t \bigoplus_{X \subseteq A} \{[(X),(F)] : X \vDash q\}.$$

Similarly, for all $Z \subset F$, if $Z \vDash \neg q$ then by contraposition $Z \vDash \neg p$. Hence we have

$$\bigoplus_{T \subseteq A} \{[(F),(T)] : T \vDash \neg p\} \leq_t \bigoplus_{Y \subseteq A} \{[(F),(Y)] : Y \vDash \neg q\}.$$

Recalling again that the bilattice is interlaced, we may conclude that $\bar{v}(p) \leq_t \bar{v}(q)$.

(ii) We have to show that $\bar{v}(p) \wedge \bar{v}(q) \leq_k \bar{v}(p \wedge q)$ for all $p,q \in F$. For this it is sufficient so show that in the lattice of contexts we have

$$\sum_{X \subseteq A} \{[(X)] : X \vDash p\} \cdot \sum_{Y \subseteq A} \{[(Y)] : Y \vDash q\} \leq \sum_{Z \subseteq A} \{[(Z)] : Z \vDash p \wedge q\}$$

where $\cdot$ and $\sum$ denote respectively the meet and the infinitary join in this lattice.

Note that $X \vDash p$ and $Y \vDash q$ imply $X \cup Y \vDash p \wedge q$, so for any X and Y appearing in

$$\sum_{X \subseteq A} \{[(X)] : X \vDash p\} \cdot \sum_{Y \subseteq A} \{[(Y)] : Y \vDash q\}$$

there will be some Z in

$$\sum_{Z \subseteq A} \{[(Z)] : Z \vDash p \wedge q\}$$

such that $Z = X \cup Y$. Therefore we have

$$\sum_{X,Y \subseteq A} \{[(X \cup Y)] : X \vDash p \text{ and } Y \vDash q\} \leq \sum_{Z \subseteq A} \{[(Z)] : Z \vDash p \wedge q\}.$$

So we just need to observe that

$$\sum_{X,Y \subseteq A} \{[(X) \cdot (Y)] : X \vDash p \text{ and } Y \vDash q\} =$$

$$= \sum_{X,Y \subseteq A} \{[(X \cup Y)] : X \vDash p \text{ and } Y \vDash q\}$$

and we are done.

(iii) We have to show that for all $q \in F$ we have $\neg \bar{v}(q) = \bar{v}(\neg q)$. This is straightforward, since in any bilattice B we have $\neg(a \oplus b) = \neg a \oplus \neg b$ for all $a, b \in B$. Hence:

$$\neg \left(\bigoplus_{Y \subseteq A} \{[(Y),(F)] : Y \vDash \neg q\} \oplus \bigoplus_{X \subseteq A} \{[(F),(X)] : X \vDash q\} \right) =$$

$$= \neg \bigoplus_{Y \subseteq A} \{[(Y),(F)] : Y \vDash \neg q\} \oplus \neg \bigoplus_{X \subseteq A} \{[(F),(X)] : X \vDash q\} =$$

$$= \bigoplus_{Y \subseteq A} \{[(F),(Y)] : Y \vDash \neg q\} \oplus \bigoplus_{X \subseteq A} \{[(X),(F)] : X \vDash q\}.$$

∎

We see therefore that, whenever a sentence q (or its negation) holds in some context $C = \{p_1, \ldots, p_n\}$, this information is punctually reflected in the value assigned to q once we have applied the closure operator.

5 Future work

What we have sketched in the present paper is of course just a proposal: it is clear that most of the work has still to be done.

The closure operator we have seen in the previous paragraph is just one of the possible definitions for a consequence relation on the contextual bilattice. Arieli and Avron [1] introduced several other consequence relations for bilattices, with corresponding Gentzen-style deduction systems, that could be employed as well. All these inference mechanism, however, have been introduced with other applications in mind: it could be desirable to investigate the possibility of developing inference systems specifically designed for contextual bilattices.

Another related issue would be to incorporate inference rules that are *local* in the sense of [5], that is relative to a given context. I believe these may be interesting topics for future investigation.

BIBLIOGRAPHY

[1] O. Arieli and A. Avron. Reasoning with logical bilattices. *Journal of Logic, Language and Information*, 5 (1): 25–63, 1996.

[2] N.D. Belnap Jr. A useful four-valued logic. In J.M. Dunn and G. Epstein (eds.), *Modern Uses of Multiple-Valued Logic*, Reidel, Dordrecht 1977, pages 8–37.

[3] M. Fitting. Bilattices in logic programming. In G. Epstein (ed.), *Proceedings of the Twentieth International Symposium on Multiple-Valued Logic*, IEEE, 1990, pages 238–246.

[4] M.L. Ginsberg. Multivalued logics: a uniform approach to inference in artificial intelligence. *Computational Intelligence*, 4 (3): 265–316, 1988.

[5] F. Giunchiglia. Contextual Reasoning. *Epistemologia* (special issue on *I Linguaggi e le Macchine*), 16: 345–364, 1993.

Reflecting the semantic features of S5 at the syntactic level

FRANCESCA POGGIOLESI

1 Introduction

Modal logic is one of the non-classical logics that has most flourished in recent years. There are many interesting systems of modal logic, but usually attention is focussed on the ones that are called *normal* modal systems. These systems, that can be easily presented in Hilbert-style, enjoy simple and interesting semantic properties. Moreover, they can be set out in a cube known as the cube of modal logic. In this cube each system extends and is extended by another system, except the weakest one **K** and the strongest one **S5**. This last system represents the main concern of this paper.

S5 is an important modal system, not only for being the most powerful of the cube of modal logic and for having deep philosophical issues, but also for enjoying peculiar Kripke semantics features. It is a well-known fact that there are two different kinds of Kripke frames for **S5**: Kripke frames where the accessibility relation is an equivalence relation, i.e. it enjoys the properties of reflexivity, transitivity and symmetry (or, equivalently, the properties of reflexivity and euclideaness), and Kripke frames where the accessibility relation is the universal relation, i.e. it can simply be omitted.

Unluckily, so much cannot be said of the syntactic level. There are many attempts at finding a Gentzen calculus for this system, but each of them is unsatisfactory for one of the following two reasons: either it presents several defects, e.g. it not cut-free [5, 2] or it does not not enjoy the subformula property [8], or it does not fully reflect the semantic richness of **S5**, i.e it can only treat **S5** as a system whose accessibility relation satisfies several conditions [3, 4], or it can only treat **S5** as a system where there is no accessibility relation [1].

Our goal in this paper is threefold. (i) We want to exploit the tree-hypersequent method, introduced in [6], to present a new calculus for **S5** that reflects the more complex way of describing this system semantically. (ii) We want to show how the tree-hypersequent method leads us to the construction of a second sequent calculus for **S5** (introduced in [7]) that reflects the second way of describing this system semantically. (iii) We want to emphasize analogies and differences between the two calculi mentioned above. This goal will be realized by exposing the several results obtainable in these calculi in a parallel.

We start our task by informally introducing the tree-hypersequent method. More precisely, we explain what a tree-hypersequent is by constructing this object step-by-step. Let us, then, refresh the simple notion of empty hypersequent. An empty hypersequent is a syntactic object of the form

$$\overbrace{-/-/-}^{n}$$

which is to say: n slashes that separate $n+1$ dashes. If the order of the dashes is taken into account (as it is not standardly done), we can look to this structure as a tree-frame of Kripke semantics, where the dashes are meant to be the worlds of the tree-frame and the slashes the relations between worlds in the tree-frame. Following this analogy the dash that is at the extreme left of the empty hypersequent denotes a world at distance one in the corresponding tree-frame, the dash after denotes a world a distance two in the corresponding tree-frame, and so on.

In a tree-frame, at every distance, except the first one, we may find n different possible worlds: how do we express this fact in our syntactic object? We separate different dashes that are at the same distance with a semi-colon and get, this way, an empty tree-hypersequent. An example of empty tree-hypersequent is an object of the following form (see figure on the left):

$-/-;-$ ⤳

that corresponds to a tree-frame (see figure on the right) with a world at distance one related with two different worlds at distance two. Another example of empty tree-hypersequent is the following (see figure on the left):

$-/(-/-);(-/-)$ ⤳

that corresponds to a tree-frame (see figure on the right) with a world at distance one related with two different worlds at distance two, each of which is, in its turn, related with another world at distance three.

In order to obtain a tree-hypersequent we fill the dashes with sequents which are objects of the form $M \Rightarrow N$, where M and N are multisets of formulas.

Next section will be used to present the first calculus for the system **S5**. Third section will be dedicated to the explanation of the passage from the first calculus for **S5** to the second one. The last sections will be exploited to briefly present the mains results that are obtainable with these calculi.

2 The first sequent calculus for S5

We define the modal propositional language $\mathcal{L}^{\Box}$ in the following way:

atoms: $p^0, p^1, \ldots$

logical constant: $\square$

connectives: $\neg, \wedge$

The other classic connectives can be defined as usual, as well as the constant $\diamond$ and the formulas of the modal language $\mathcal{L}^{\square}$.

Syntactic Conventions

$\alpha, \beta, \ldots$: formulas,

$M, N, \ldots$: finite multisets of formulas,

$\Gamma, \Delta, \ldots$: sequents(SEQ). The empty sequent ($\Rightarrow$) is included.

$G, H, \ldots$: tree-hypersequents (THS),

$\underline{X}, \underline{Y}, \ldots$: finite multisets of tree-hypersequents (MTHS), $\emptyset$ included.

We point out that for the sake of brevity we might use the following notation: given $\Gamma \equiv M \Rightarrow N$ and $\Pi \equiv P \Rightarrow Q$, we will write $\Gamma.\Pi$ instead of $M, P \Rightarrow N, Q$.

DEFINITION 1. The notion of tree-hypersequent is inductively defined in the following way:

- if $\Gamma \in$ SEQ, then $\Gamma \in$ THS,

- if $\Gamma \in$ SEQ and $\underline{X} \in$ MTHS, then $\Gamma/\underline{X} \in$ THS.

DEFINITION 2. The intended interpretation of a tree-hypersequent is:

- $(M \Rightarrow N)^\delta := \bigwedge M \to \bigvee N$,

- $(\Gamma/G_1; \ldots; G_n)^\delta := \Gamma^\delta \vee \square G_1^\delta \vee \ldots \vee \square G_n^\delta$.

In order to display the rules of the calculi, we will use the notation $G[*]$ to refer to a tree-hypersequent G together with one hole $[*]$, where the hole should be understood, metaphorically, as a zoom that allows us to focus attention on a particular point, $*$, of G. Such an object becomes a real tree-hypersequent whenever the symbol "$*$" is *appropriately* replaced by (i) a sequent Γ, and in this case we will write $G[\Gamma]$ to denote the tree-hypersequent G together with a specific occurrence of a sequent Γ in it; (ii) two sequents, Γ/Σ, one after another and separated by a slash, and in this case we will write $G[\Gamma/\Sigma]$ to denote the tree-hypersequent G together with a specific occurrence of a sequent Γ immediately followed by a specific occurrence of a sequent Σ; (iii) a tree-hypersequent H, and in this case we will write $G[H]$ to denote the tree-hypersequent G together with a specific occurrence of a tree-hypersequent H in it.

The calculus *CSS5* is composed of:

Initial Tree-hypersequents

$G[p, M \Rightarrow N, p]$

Propositional Rules

$$\dfrac{G[M \Rightarrow N, \alpha]}{G[\neg\alpha, M \Rightarrow N]} \neg A \qquad \dfrac{G[\alpha, M \Rightarrow N]}{G[M \Rightarrow N, \neg\alpha]} \neg K$$

$$\dfrac{G[\alpha, \beta, M \Rightarrow N]}{G[\alpha \wedge \beta, M \Rightarrow N]} \wedge A \qquad \dfrac{G[M \Rightarrow N, \alpha] \quad G[M \Rightarrow N, \beta]}{G[M \Rightarrow N, \alpha \wedge \beta]} \wedge K$$

Modal Rules

$$\dfrac{G[\Box\alpha, M \Rightarrow N/\alpha, S \Rightarrow T]}{G[\Box\alpha, M \Rightarrow N/S \Rightarrow T]} \Box A \qquad \dfrac{G[M \Rightarrow N/ \Rightarrow \alpha; \underline{X}]}{G[M \Rightarrow N, \Box\alpha/\underline{X}]} \Box K$$

Special Logical Rules

$$\dfrac{G[\Box\alpha, \alpha, M \Rightarrow N]}{G[\Box\alpha, M \Rightarrow N]} \, t \qquad \dfrac{G[\Box\alpha, M \Rightarrow N/\Box\alpha, S \Rightarrow T]}{G[\Box\alpha, M \Rightarrow N/S \Rightarrow T]} \, 4$$

$$\dfrac{G[\alpha, M \Rightarrow N/\Box\alpha, S \Rightarrow T]}{G[M \Rightarrow N/\Box\alpha, S \Rightarrow T]} \, b \qquad \dfrac{G[\Box\alpha, M \Rightarrow N/\Box\alpha, S \Rightarrow T]}{G[M \Rightarrow N/\Box\alpha, S \Rightarrow T]} \, 5$$

As the reader can easily observe, the calculus $CSS5$ reflects at the syntactic level the first way of describing **S5** semantically. Indeed the four special logical rules, t, 4, b and 5, are meant to capture the frame properties of reflexivity, transitivity, symmetry and euclideaness, respectively. It is interesting to note that each of these special logical rules have a (admissible) structural counterpart:

Special Structural Rules

$$\dfrac{G[\Gamma/(\Sigma/\underline{X}); \underline{X}']}{G[\Gamma \bullet \Sigma/\underline{X}; \underline{X}']} \, \tilde{t} \qquad \dfrac{G[\Gamma/(\Sigma/\underline{X}); \underline{X}']}{G[\Gamma/(\Rightarrow /\Sigma/\underline{X}); \underline{X}']} \, \tilde{4}$$

$$\dfrac{G[\Gamma/(\Sigma/(\Delta/\underline{X}); \underline{X}'); \underline{X}'']}{G[\Gamma \bullet \Delta/(\Sigma/\underline{X}'); \underline{X}; \underline{X}'']} \, \tilde{b} \qquad \dfrac{G[\Gamma/(\Sigma/(\Delta/\underline{X}); \underline{X}'); \underline{X}'']}{G[\Gamma/(\Delta/\underline{X}); (\Sigma/\underline{X}'); \underline{X}'']} \, \tilde{5}$$

LEMMA 3. *The rules $\tilde{t}$ and $\tilde{b}$ are height-preserving admissible in the calculus $CSS5$. The rule $\tilde{4}$ and $\tilde{5}$ are admissible in the calculus $CSS5$.*

Proof. By induction on the height of the derivation of the premiss. In case the last applied rule is the modal rule $\Box A$, we exploit one of the special logical rules to solve the case. ∎

As we will see in the last section the existence of these special structural rules is crucial for the proof of cut-admissibility.

3 The second sequent calculus for S5

Let us concentrate on the special structural rule $\tilde{5}$. Roughly speaking, this rule allows us to move from the symbol "/" to the symbol ";". In more intuitive terms: this rule allows us to move from the presence of the accessibility relation of Kripke semantics to its absence.[1] Given this fact, an idea seems to naturally arise: we could construct an alternative sequent calculus for **S5** where we still have n different sequents a time, but there is no longer an order on these sequents, (there is no longer an accessibility relation over the set of worlds), i.e. a sequent calculus where we no longer need to deal with the two symbols "/" and ";", but with just one of them.

This section will be dedicated to the realization of such an idea by means of the development of another Gentzen system for modal logic **S5**, which, by contrast with $CSS5$, reflects, at the syntactic level, the simplicity of the Kripke frames where the accessibility relation is absent. In this new sequent calculus we will use hypersequents where precisely we only have the meta-linguistic symbol "|".[2] Let us emphasize that the return to hypersequents is motivated by the work with tree-hypersequents. In other words, hypersequents stand to tree-hypersequents, as Kripke frames with universal relation stand to Kripke frames.

DEFINITION 4. An hypersequent is a syntactic object of the form:

$$M_1 \Rightarrow N_1 | M_2 \Rightarrow N_2 | \ldots | M_n \Rightarrow N_n$$

where $M_i \Rightarrow N_i$ $(i = 1, \ldots, n)$ is a classical sequent.

DEFINITION 5. The intended interpretation of a hypersequent is definable in the following inductive way:

- $(M \Rightarrow N)^\tau := \bigwedge M \to \bigvee N$,

- $(\Gamma_1 | \Gamma_2 | \ldots | \Gamma_n)^\tau := \Box \Gamma_1^\tau \vee \Box \Gamma_2^\tau \vee \ldots \vee \Box \Gamma_n^\tau$

A hypersequent is then just a *multiset* of classical sequents, which is to say the order of the sequents in a hypersequents does not count.

The postulates of the calculus $CSS5_s$ are:

---

[1]We underline that it is an easy (but quite tedious) work to show that the rule $\tilde{5}$ is invertible, which is to say, that the following rule:

$$\frac{G[\Gamma/(\Delta/\underline{X});(\Sigma/\underline{X}');\underline{X}'']}{G[\Gamma/(\Sigma/(\Delta/\underline{X});\underline{X}');\underline{X}'']}$$

is also admissible in the calculus $CSS5$.

[2]We follow the tradition in choosing the symbol "|". If, on the contrary, we had decided to follow the notation adopted in the previous section, we should have chosen the symbol ";".

Initial Hypersequents

$$G \mid p, M \Rightarrow N, p$$

Propositional Rules

$$\frac{G \mid M \Rightarrow N, \alpha}{G \mid \neg\alpha, M \Rightarrow N} \neg A \qquad \frac{G \mid \alpha, M \Rightarrow N}{G \mid M \Rightarrow N, \neg\alpha} \neg K$$

$$\frac{G \mid \alpha, \beta, M \Rightarrow N}{G \mid \alpha \wedge \beta, M \Rightarrow N} \wedge A \qquad \frac{G \mid M \Rightarrow N, \alpha \quad G \mid M \Rightarrow N, \beta}{G \mid M \Rightarrow N, \alpha \wedge \beta} \wedge K$$

Modal Rules

$$\frac{G \mid \alpha, \Box\alpha, M \Rightarrow N}{G \mid \Box\alpha, M \Rightarrow N} \Box A_1 \qquad \frac{G \mid M \Rightarrow N \mid \Rightarrow \alpha}{G \mid M \Rightarrow N, \Box\alpha} \Box K$$

$$\frac{G \mid \Box\alpha, M \Rightarrow N \mid \alpha, S \Rightarrow T}{G \mid \Box\alpha, M \Rightarrow N \mid S \Rightarrow T} \Box A_2$$

4 Admissibility of the structural rules

In this section we will show which rules are height-preserving admissible in the calculi $CSS5$ and $CSS5_s$ (the proofs of height-preserving admissibility are developed by straightforward induction on the height of the derivation of the premise); moreover we will prove that the logical and modal rules are height-preserving invertible.

LEMMA 6. *The rules of internal weakening:*

$$\frac{G[M \Rightarrow N]}{G[M, P \Rightarrow N, Q]} W \qquad \frac{G \mid M \Rightarrow N}{G \mid M, P \Rightarrow N, Q} W_s$$

are height-preserving admissible in, respectively, $CSS5$ and $CSS5_s$.

LEMMA 7. *The rules of external weakening:*

$$\frac{G[\Gamma/\underline{X}]}{G[\Gamma/\underline{X}; \Sigma]} EW \qquad \frac{G}{G \mid M \Rightarrow N} EW_s$$

are height-preserving admissible in, respectively, $CSS5$ and $CSS5_s$.

LEMMA 8. *The rules of merge:*

$$\frac{G[\Delta/(\Gamma/\underline{X}); (\Pi/\underline{X'}); Y]}{G[\Delta/(\Gamma \cdot \Pi/\underline{X}; \underline{X'}); Y]} merge \qquad \frac{G \mid M \Rightarrow N \mid P \Rightarrow Q}{G \mid M, P \Rightarrow N, Q} merge_s$$

are height-preserving admissible in, respectively, $CSS5$ and $CSS5_s$.

LEMMA 9. *The rule of necessitation:*

$$\frac{G}{\Rightarrow /G}\ rn$$

is height-preserving admissible in $CSS5$.

LEMMA 10. *All the logical and modal rules of* $CSS5$ *and* $CSS5_s$ *are height-preserving invertible.*

Proof. The proof proceeds by induction on the derivation of the premise of the rule considered. The cases of logical rules of $CSS5$ and $CSS5_s$ are dealt with in the classical way. The only differences — the fact that we are dealing with tree-hypersequents, and hypersequent, respectively, and the cases where last rule applied is one of the modal rules or one of the special logical rules — are dealt with easily.

The rule $\Box A$ and the special logical rules of $CSS5$, as well as the two modal rules $\Box A_1$ and $\Box A_2$ of $CSS5_s$, are all trivially height-preserving invertible since the premise is concluded by weakening from the conclusion, and weakening is height-preserving admissible.

We show in detail the invertibility of the rule $\Box K$ of the calculus $CSS5$ (the proof of the invertibility of the rule $\Box K$ of the calculus $CSS5_s$ is analogous). If $G[M \Rightarrow N, \Box\alpha/\underline{X}]$ is an initial tree-hypersequent, then so is $G[M \Rightarrow N/ \Rightarrow \alpha; \underline{X}]$. If $G[M \Rightarrow N, \Box\alpha/\underline{X}]$ is obtained by a logical rule $\mathcal{R}$, we apply the inductive hypothesis on the premise(s), $G[M' \Rightarrow N', \Box\alpha/\underline{X}]$ ($G[M'' \Rightarrow N'', \Box\alpha/\underline{X}]$) and we obtain derivation(s), of height $n-1$, of $G[M' \Rightarrow N'/ \Rightarrow \alpha; \underline{X}]$ ($G[M'' \Rightarrow N''/ \Rightarrow \alpha; \underline{X}]$). By applying the rule $\mathcal{R}$, we obtain a derivation of height n of $G[M \Rightarrow N/ \Rightarrow \alpha; \underline{X}]$. If $G[M \Rightarrow N, \Box\alpha/\underline{X}]$ is of the form $G[\Box\beta, M' \Rightarrow N', \Box\alpha/S \Rightarrow T]$ and is obtained by the modal rule $\Box A$, we apply the inductive hypothesis to $G[\Box\beta, M' \Rightarrow N', \Box\alpha/\beta, S \Rightarrow T]$ and we obtain a derivation of height $n-1$ of $G[\Box\beta, M' \Rightarrow N'/ \Rightarrow \alpha; \beta, S \Rightarrow T]$. By applying the rule $\Box A$, we obtain a derivation of height n of $G[\Box\beta, M' \Rightarrow N'/ \Rightarrow \alpha; S \Rightarrow T]$. If $G[M \Rightarrow N, \Box\alpha/\underline{X}]$ is obtained by one of the special logical rules or by the modal rule $\Box K$ in which $\Box\alpha$ is not the principal formula, then the case can be dealt with analogously to the one of the rule $\Box A$. Finally, if $G[M \Rightarrow N, \Box\alpha/\underline{X}]$ is preceded by the modal rule $\Box K$ and $\Box\alpha$ is the principal formula, the premise of the last step gives the conclusion. ∎

LEMMA 11. *The rules of contraction:*

$$\frac{G[\alpha, \alpha, M \Rightarrow N]}{G[\alpha, M \Rightarrow N]}\ CA \qquad \frac{G \mid \alpha, \alpha, M \Rightarrow N}{G \mid \alpha, M \Rightarrow N}\ CA_s$$

and

$$\frac{G[M \Rightarrow N, \alpha, \alpha]}{G[M \Rightarrow N, \alpha]}\ CK \qquad \frac{G \mid M \Rightarrow N, \alpha, \alpha}{G \mid M \Rightarrow N, \alpha}\ CK_s$$

are height-preserving admissible in, respectively, $CSS5$ *and* $CSS5_s$.

Finally we remind the reader that in the calculus $CSS5$ we have shown the (height-preserving) admissibility of four special structural rules.

5 Adequacy of the calculi

In this section we prove that the sequent calculi $CSS5$ and $CSS5_s$ prove exactly the same formulas as the corresponding Hilbert system **S5**.

THEOREM 12.

(i) If $\vdash \alpha$ in **S5**, then $\vdash \Rightarrow \alpha$ in $CSS5$.

(ii) If $\vdash \alpha$ in **S5**, then $\vdash \Rightarrow \alpha$ in $CSS5_s$.

Proof. By induction on the height of proofs in $CSS5$ and $CSS5_s$, respectively. The classical axioms and the modus ponens are proved as usual; the axioms T, the axiom 4, the axiom B are proved by exploiting the corresponding special logical rules in $CSS5$, and the modal rules in $CSS5_s$. We present the proof of the axiom 5.[3]

$CSS5 \vdash \Rightarrow \neg\Box\neg\alpha \to \Box\neg\Box\neg\alpha$

$$
\cfrac{\cfrac{\cfrac{\cfrac{\cfrac{\cfrac{\cfrac{\cfrac{\Box\neg\alpha \Rightarrow /\alpha \Rightarrow \alpha; \Box\neg\alpha \Rightarrow}{\Box\neg\alpha \Rightarrow /\neg\alpha \Rightarrow \neg\alpha; \Box\neg\alpha \Rightarrow} \neg^*}{\Box\neg\alpha \Rightarrow / \Rightarrow \neg\alpha; \Box\neg\alpha \Rightarrow} \Box A}{\Rightarrow / \Rightarrow \neg\alpha; \Box\neg\alpha \Rightarrow} 5}{\Rightarrow / \Rightarrow \neg\alpha; \Rightarrow \neg\Box\neg\alpha} \neg K}{\Rightarrow \Box\neg\alpha / \Rightarrow \neg\Box\neg\alpha} \Box K}{\Rightarrow \Box\neg\alpha, \Box\neg\Box\neg\alpha} \Box K}{\neg\Box\neg\alpha \Rightarrow \Box\neg\Box\neg\alpha} \neg A}{\Rightarrow \neg\Box\neg\alpha \to \Box\neg\Box\neg\alpha} \to K
$$

$CSS5_s \vdash \Rightarrow \neg\Box\neg\alpha \to \Box\neg\Box\neg\alpha$

$$
\cfrac{\cfrac{\cfrac{\cfrac{\cfrac{\cfrac{\cfrac{\cfrac{\Rightarrow |\Box\neg\alpha \Rightarrow |\alpha \Rightarrow \alpha}{\Rightarrow |\Box\neg\alpha \Rightarrow | \Rightarrow \neg\alpha, \alpha} \neg K}{\Rightarrow |\Box\neg\alpha \Rightarrow |\neg\alpha \Rightarrow \neg\alpha} \neg A}{\Rightarrow |\Box\neg\alpha \Rightarrow | \Rightarrow \neg\alpha} \Box A_2}{\Rightarrow | \Rightarrow \neg\Box\neg\alpha| \Rightarrow \neg\alpha} \neg K}{\Rightarrow \Box\neg\alpha| \Rightarrow \neg\Box\neg\alpha} \Box K}{\Rightarrow \Box\neg\alpha, \Box\neg\Box\neg\alpha} \Box K}{\neg\Box\neg\alpha \Rightarrow \Box\neg\Box\neg\alpha} \neg A}{\Rightarrow \neg\Box\neg\alpha \to \Box\neg\Box\neg\alpha} \to K
$$

∎

---

[3]Notice that we use the derived rules for the connective $\to$. Moreover in the case where repeated running applications of a same rule $\mathcal{R}$ take place, we write the rule $\mathcal{R}$ with the symbol "*" as index.

THEOREM 13.

(i) If $\vdash G$ in $CSS5$, then $\vdash (G)^\tau$ in **S5**.

(ii) If $\vdash G$ in $CSS5_s$, then $\vdash (G)^\tau$ in **S5**.

Proof. By induction on the height of proofs in **S5**. (i) The technique to develop this proof consists of the following two steps: first of all, the sequent(s) affected by the rule should be isolated and the corresponding implication proved, then the implication should be transported up all along the tree so that, by modus ponens, the desired result is immediately achieved. (ii) The case of the axioms is trivial, while for the inductive steps with the propositional rules all we need is classical logic and the fact that if **S5**$\vdash \alpha_1 \to (\alpha_2 \to \ldots \to (\alpha_n \to \beta)\ldots)$, then **S5** $\vdash \Box\alpha_1 \to (\Box\alpha_2 \to \ldots \to (\Box\alpha_n \to \Box\beta)\ldots)$. As for the inductive steps for modal rules, we again exploit the fact that, if **S5** $\vdash \alpha_1 \to (\alpha_2 \to \ldots \to (\alpha_n \to \beta)\ldots)$, then **S5** $\vdash \Box\alpha_1 \to (\Box\alpha_2 \to \ldots \to (\Box\alpha_n \to \Box\beta)\ldots)$ and the axioms T and 4. ■

6 Cut-admissibility

We dedicate this last section to the proof of the admissibility of the cut-rule in $CSS5$ and $CSS5_s$. We then have to present the two cut-rules. In order to introduce the cut-rule for the calculus $CSS5$ we firstly need the following two definitions.

DEFINITION 14. Given two tree-hypersequents, $G[\Gamma]$ and $G'[\Gamma']$, the relation of *equivalent position* between two of their sequents, in this case Γ and Γ', $G[\Gamma] \sim G'[\Gamma']$, is defined inductively in the following way:

- $\Gamma \sim \Gamma'$

- $\Gamma/\underline{X} \sim \Gamma'/\underline{X}'$

- If $H[\Gamma] \sim H'[\Gamma']$, then $\Delta/H[\Gamma]; \underline{X} \sim \Delta'/H'[\Gamma']; \underline{X}'$

DEFINITION 15. Given two tree-hypersequents $G[\Gamma]$ and $G'[\Gamma']$ such that $G[\Gamma] \sim G'[\Gamma']$, the operation of *product*, $G[\Gamma] \otimes G'[\Gamma']$, is defined inductively in the following way:

- $\Gamma \otimes \Gamma' := \Gamma \cdot \Gamma'$

- $(\Gamma/\underline{X}) \otimes (\Gamma'/\underline{X}') := \Gamma \cdot \Gamma'/\underline{X}; \underline{X}'$

- $(\Delta/H[\Gamma]; \underline{X}) \otimes (\Delta'/H'[\Gamma']; \underline{X}') :=$
  $\Delta \cdot \Delta'/(H[\Gamma] \otimes H'[\Gamma']); \underline{X}; \underline{X}'$

Cut-rule of the calculus $CSS5$. Given two tree-hypersequents $G[M \Rightarrow N, \alpha]$ and $G'[\alpha, P \Rightarrow Q]$ such that $G[M \Rightarrow N, \alpha] \sim G'[\alpha, P \Rightarrow Q]$, the cut-rule is:

$$\frac{G[M \Rightarrow N, \alpha] \quad G'[\alpha, P \Rightarrow Q]}{G \otimes G'[M, P \Rightarrow N, Q]} \; cut_\alpha$$

Cut-rule of the calculus $CSS5_s$. The cut-rule of the calculus $CSS5_s$ is simpler than the previous one and it is the following:

$$\frac{G \mid M \Rightarrow N, \alpha \quad G' \mid \alpha, P \Rightarrow Q}{G \mid G' \mid M, P \Rightarrow N, Q} \; cut_\alpha^s$$

Contrary to the other structural rules (that we were given a chance to observe in section 4), the cut-rules of the calculi $CSS5$ and $CSS5_s$ are not so similar between each other. The reason for this is quite simple. In the calculus $CSS5$ we deal with tree-hypersequents and then, when we have to fuse two tree-hypersequents by means of an application of a cut-rule, we should ensure that the tree-shape is kept. In the calculus $CSS5_s$, instead, we deal with hypersequents, which are just multisets of sequents, therefore when we fuse two hypersequents by means of a cut-rule, we can do it arbitrarily since there is no particular structure to keep.

Each cut-rule is admissible in the corresponding calculus, as the following theorems states.

THEOREM 16. *The rule cut_α is admissible in the calculus $CSS5$.*

Proof. The proof is developed by induction on the complexity of the cut-formula, which is the number (≥ 0) of the occurrences of logical symbols in the cut-formula α, with subinduction on the sum of the heights of the derivations of the premises of the cut-rule. The proof has the same structure as the proof of admissibility of cut for the sequent calculus of first-order logic, (see for example [9]). However, for the sake of clarity, we consider in detail two cases: (i) the case of a cut where both premises have been introduced by a modal rule — $\Box K$ and $\Box A$, respectively — and the cut-formula is the principal formula of these rules; (ii) the case of a cut where the left premise has been introduced by the modal rule $\Box K$, while the right premise has been introduced by the rule t. Moreover, the cut-formula is the principal formula of both these rules.

With this second case we want to underline the indispensability of the special structural rules introduced in Section 3.

(i):[4]

$$\frac{G[M \Rightarrow N/ \Rightarrow \beta]}{G[M \Rightarrow N, \Box\beta]} \Box K \quad \frac{G'[\Box\beta, \Pi/\beta, \Psi]}{G'[\Box\beta, \Pi/\Psi]} \Box A$$
$$\overline{G \otimes G'[M \Rightarrow N . \Pi/\Psi]} \quad cut_{\Box\beta}$$

We reduce to:

$$\frac{G[M \Rightarrow N/ \Rightarrow \beta] \quad \frac{G[M \Rightarrow N, \Box\beta] \quad G'[\Box\beta, P \Rightarrow Q/\beta, Z \Rightarrow W]}{G \otimes G'[M, P \Rightarrow N, Q/\beta, Z \Rightarrow W]} cut_{\Box\beta}}{\frac{G \otimes G \otimes G'[M, M, P \Rightarrow N, N, Q/Z \Rightarrow W]}{G \otimes G'[M, P \Rightarrow N, Q/Z \Rightarrow W]} C^* + merge^*} cut_\beta$$

where the first cut is eliminable by induction on the sum of the heights of the derivations of the premises of cut and the second cut is eliminable by induction on the complexity of the cut formula.

(ii):

$$\frac{G[M \Rightarrow N/ \Rightarrow \beta]}{G[M \Rightarrow N, \Box\beta]} \Box K \quad \frac{G'[\Box\beta, \beta, P \Rightarrow Q]}{G'[\Box\beta, P \Rightarrow Q]} t$$
$$\overline{G \otimes G'[M, P \Rightarrow N, Q]} \quad cut_{\Box\beta}$$

We reduce to:

$$\frac{G[M \Rightarrow N/ \Rightarrow \beta]}{G[M \Rightarrow N, \beta]} \tilde{t} \quad \frac{G[M \Rightarrow N, \Box\beta] \quad G'[\Box\beta, \beta, P \Rightarrow Q]}{G \otimes G'[\beta, M, P \Rightarrow N, Q]} cut_{\Box\beta}$$
$$\frac{G \otimes G \otimes G'[M, M, P \Rightarrow N, N, Q]}{G \otimes G'[M, P \Rightarrow N, Q]} C^* + merge^* \quad cut_\beta$$

where the first cut is eliminable by induction on the sum of the heights of the derivations of the premises of cut and the second cut is eliminable by induction on the complexity of the cut formula.

In those cases where the last applied rule on the left premise is the rule $\Box K$ and the last applied rule on the right premise is the the rule 4 or the rule 5, and the cut-formula is principal in both the left and the right premises, the situation is a little bit more complicated but can be dealt with by adopting the technique showed in [6]. ∎

THEOREM 17. *The rule cut^s_α is admissible in the calculus $CSS5_s$.*

---

[4]In the cases (i) and (ii), we assume to write, for the sake of clarity, the rule $\Box K$ without the aid of the multiset of tree-hypersequents $\underline{X}$. We rely on the reader for understanding the rule correctly, anyway.

Proof. The proof is developed by induction on the complexity of the cut-formula, with subinduction on the sum of the heights of the derivations of the premises of the cut-rule. The proof has been fully developed in [7]. However, for the sake of clarity, we consider in detail two cases: (i) the case of a cut where the left premise has been introduced by the modal rule $\Box K$, while the right premise has been introduced by the rule $\Box A_2$. Moreover, the cut-formula is the principal formula of both these rules; (ii) the case of a cut where the left premise has been introduced by the modal rule $\Box K$, while the right premise has been introduced by the rule $\Box A_1$. Moreover, the cut-formula is the principal formula of both these rules.

(i):

$$\dfrac{\dfrac{G \mid M \Rightarrow N \mid\, \Rightarrow \beta}{G \mid M \Rightarrow N, \Box\beta}\,\Box K \quad \dfrac{G' \mid \Box\beta, P \Rightarrow Q \mid \beta, Z \Rightarrow W}{G' \mid \Box\beta, P \Rightarrow Q \mid Z \Rightarrow W}\,\Box A_2}{G \mid G' \mid M, P \Rightarrow N, Q \mid Z \Rightarrow W}\,cut^s_{\Box\beta}$$

We reduce to:

$$\dfrac{G \mid M \Rightarrow N \mid\, \Rightarrow \beta \quad \dfrac{\dfrac{G \mid M \Rightarrow N, \Box\beta \quad G' \mid \Box\beta, P \Rightarrow Q \mid \beta, Z \Rightarrow W}{G \mid G' \mid M, P \Rightarrow N, Q \mid \beta, Z \Rightarrow W}\,cut^s_{\Box\beta}}{G \mid G \mid G' \mid M \Rightarrow N \mid M, P \Rightarrow N, Q \mid Z \Rightarrow W}\,cut^s_\beta}{G \mid G' \mid M, P \Rightarrow N, Q \mid Z \Rightarrow W}\,merge^*+C^*$$

where the first cut is eliminable by induction on the sum of the heights of the derivations of the premises of cut and the second cut is eliminable by induction on the complexity of the cut formula.

(ii):

$$\dfrac{\dfrac{G \mid M \Rightarrow N \mid\, \Rightarrow \beta}{G \mid M \Rightarrow N, \Box\beta}\,\Box K \quad \dfrac{G' \mid \Box\beta, \beta, P \Rightarrow Q}{G' \mid \Box\beta, P \Rightarrow Q}\,\Box A_1}{G \mid G' \mid M, P \Rightarrow N, Q}\,cut^s_{\Box\beta}$$

We reduce to:

$$\dfrac{G \mid M \Rightarrow N \mid\, \Rightarrow \beta \quad \dfrac{\dfrac{G \mid M \Rightarrow N, \Box\beta \quad G' \mid \Box\beta, \beta, P \Rightarrow Q}{G \mid G' \mid \beta, M, P \Rightarrow N, Q}\,cut^s_{\Box\beta}}{G \mid G \mid G' \mid M \Rightarrow N \mid M, P \Rightarrow N, Q}\,cut^s_\beta}{G \mid G' \mid M, P \Rightarrow N, Q}\,merge^*+C^*$$

where the first cut is eliminable by induction on the sum of the heights of the derivations of the premises of cut and the second cut is eliminable by induction on the complexity of the cut formula. ■

BIBLIOGRAPHY

[1] A. Avron. The method of hypersequents in the proof theory of propositional non-classical logic. In C. Steinhorn and J. Strauss W. Hodges and M. Hyland (ed.) *Logic: from foundations to applications*, Oxford University Press, Oxford 1996, pages 1–32.

[2] S. Blumey and L. Humberstone. A perspective on modal sequent logic. *Publications of the Research Institute for Mathematical Sciences, Kyoto University*, 27: 763–782, 1991.

[3] K. Došen. Sequent systems for modal logic. *Journal of Symbolic Logic*, 50: 149–159, 1985.

[4] A. Indrezejczak. Generalised sequent calculus for propositional modal logics. *Logica Trianguli*, 1: 15–31, 1997.

[5] K. Matsumoto and M. Ohnishi. Gentzen method in modal calculi. *Osaka Mathematical Journal*, 11: 115–120, 1959.

[6] F. Poggiolesi. The method of tree-hypersequent for modal propositional logic. In David Makinson, Jacek Malinowski and Heinrich Wansing (eds.), *Trends in Logic IV: Towards Mathematical Philsophy*, Springer, Berlin 2009, pages 31–51.

[7] F. Poggiolesi. A cut-free simple sequent calculus for modal logic S5. *Review of Symbolic Logic*, 1: 3–15, 2008.

[8] M. Sato. A cut-free gentzen-type system for the modal logic s5. *Journal of Symbolic Logic*, 45: 67–8, 1980.

[9] A. Troelestra and H. Schwichtenberg. *Basic Proof Theory*. Cambridge University Press, Cambridge 1996.

[10] H. Wansing. Sequent calculi for normal modal propositional logics. *Journal of Logic and Computation*, 4: 125–142, 1994.

Non monotonic conditionals and the concept *I believe only A*

GISÈLE FISCHER SERVI

1 Overview and related work

In ordinary reasoning all assertions are taken to be assertions of beliefs based on the best available information rather than on absolute truth. Even though natural languages are seldom explicitely epistemic (it is not supposed that in practice we can distinguish truth from rationally supported belief), much of our ordinary reasoning functions because we implicitly recognize the epistemic character of our convictions. Thus, representing beliefs in a language that is concise in the way that natural language is and chosing their logic accordingly could be a way to attack the common sense reasoning challenge.

The goal of this paper is to adopt such a strategy in relation to the problem of revisable reasoning. The first and most fundamental factor is the use of (First Order) Intuitionistic Logic (IL) since, appropriately, it is based on the idea that the reasoner aims not at truth but at warranted belief. Intuitionistic assertions, much as natural language assertions, are understood to be *implicitly epistemic*: to assert A means to be committed to the belief in A. The choice of IL as the basic logical machinery makes the present approach different from other epistemic analyses of defeasible reasoning such as those developed by Moore, Halpern and Levesque ([11], [5], [10]) where epistemic commitment is made explicit and modal belief statements are in themselves taken to be *true or false*. Moreover, in using IL there is the additional advantage of working within a timeless logic developed with an eye to an underlying temporal framework. Again we note that the *implicit* consideration of the flow of beliefs through time is part and parcel of any analysis of how revisable reasoning works.

Ultimately, we shall define the logic φ, a "normality" conditional extension of IL. The idea to use this kind of approach to deal with the topic "exceptions to the rules" was initially suggested by Gabbay in [4] and developed by Fischer Servi in [2]. Lately, the conditionally based approach has enjoyed some popularity, particularly through the work of Lehman and his colleagues ([8], [9]) whose rational has been to provide a good paradigm for the concept of non monotonic consequence. The systems presented below are somewhat related to the Lehman approach but, aside from being intuitionistic rather than classical, they are conceptually different. First, their aim is not to provide a list of basic desiderata on a non monotonic consequence relation so as to predict when a non-deductive proposed conception deviates from standard norms. Rather, they describe formally how a subject may be committed to general cognitive policies and *at the same time* have actual beliefs that contradict them. Second, and not unrelatedly, the strength of the primitive conditional operator may vary up to satisfying the property of Rational Monotony without losing semantical completeness (similarly to Boutilier in [1], although he features a conditional operator *defined* within a classical conditional logic combined with an alethic modal system). Last, the model-theoretic interpretation of *normality* does not follow the trend which has been to understand it as a comparative notion requiring a measure on states of affairs. Normal, then, is what happens in "most normal worlds". In this paper, we investigate a sort of absolute concept of normality so that "not flying birds" count as being plainly exceptional and not merely "less normal than flying birds". Since, under our interpretation, normality is a black and white affair, our semantics does not require access to the subtleties of a normality order on belief states.

As other conditionally based non monotonic logics, the logic φ does not determine the kind of flexible concept of inference we are looking for: there are no rules that say, given some *facts*, which beliefs we may adopt on the basis of those very facts; indeed, the conclusion of a conditional assertion cannot be separated from its premises. For in order to preserve consistency, premises and conditional consequences form an indivisible whole. To solve this problem we introduce a sentence operator "*Only* A", meant to represent the notion "*all* the subject *actually* believes is A". The addition of this operator leads to the definition of an extension of φ, the logic Φ, which yields a genuine concept of non monotonic inference, one in which conclusions can be detached from their premises without risking monotonicity or contradiction.

Logics of "Only knowing" are not new, but these differ significantly from the system Φ defined below. For instance, the systems introduced and studied in Levesque [10], Halpern [5] and Halpern and Lakemayer [6], [7] are not conditionally-based and the operator *Only* is not primitive but defined by means of a classical bimodal logic. Moreover, the intended meanings of "*Only A*" diverge: for these authors, it is "the subject believes that at least A is true and at most $\sim A$ is false" as opposed to our "the subject believes at least A and at most what is compatible with A-normality". Conditions imposed on the latter concept imply that asserting "*Only A*" amounts to making the decision to apply some *closure* conditions with respect to normality; briefly: assume the actual circumstances to be the normal circumstances so that what is not or cannot be endorsed on the basis of this assumption is denied.

The paper is organized as follows: the next paragraph is devoted to the logic φ; following it is a paragraph that explores the logic of believing only A. Sketches of proofs can be found in the appendix.

2 The logic φ

First, let us define the language $\mathcal{L}$, a first order language based on the propositional constants: $\sim, \vee, \wedge, \rightarrow, >$ (with $\leftrightarrow$ defined as usual). The letters A, B, C,... denote well formed formulas in $\mathcal{L}$. Associated with the first four constants are the usual meanings that IL confers on negation, disjunction, conjunction and material implication (resp.); $>$ is a binary modal connective denoting the intentional link expressed in natural language by such adverbs as "normally" or "typically"; hence, $>$ connects a sentence which represents a condition or a context for assuming another sentence. Formulas having the form $A>B$ will be called N-*conditionals* and are thought to refer to big numbers, the way things are in general. Sentences are said to be *basic* if they do not contain $>$. Understand the N-conditional $A(x) > B(x)$ where x is a free variable as a shorthand for the set of its ground instances. From now on we will omit the metavariable x and simply write $A > B$. Hence, each N-conditional stands for a set of sentences each of which eventually can be propositionally combined with other sentences in $\mathcal{L}$. The concept of negation deserves a special mention: $\sim(A>B)$ is to be distinguished from $A>\sim B$; in the first case, given A, $\sim B$ is not necessarily an exception, but in the second it happens to be the rule.

Let us now define the logic φ: it extends IL with the following axioms and rules: for all sentences A, B, C,

(1) $(A>B \land A>C) \to A>B \land C$;

(2) $(A \land B)>C \to A>(B \to C)$;

(3) $(A>B) \land (C>B) \to (A \lor C)>B$;

(4) $(A>B) \land (A>C) \to (A \land B)>C$;

(5) $A>B \land \sim(A>C) \to A \land \sim C>B$.

(6) $(A>B) \lor \sim(A>B)$;

(7) $\dfrac{A \to B}{A > B}$;

(8) $\dfrac{B \leftrightarrow C}{B>A \leftrightarrow C>A}$;

(9) $\dfrac{A > \bot}{\sim A}$.

Axioms (1)–(4) and rules (7) and (8) are familiar in conditionally inspired non monotonic logics. Note that (6) is meaningful only because φ is intuitionistically based; it says that something counts as normal or it doesn't, hence normality is decided once and for all and is independent of actual evidence. According to (8) the conditional is weaker than material implication and with (9) we have that absurdity is never normal. Last, note that by dropping (3) and (5) we obtain a system equivalent to the logic Γ_1 in [2].

So far we have separated the question of inference from the issue of defining the pragmatic predictive policies that are relevant in a particular domain. The latter go together with ordinary (basic) assertions and determine an extralogical *theory*, while inferencing remains strictly deductive. The logic φ; is not fully equipped to deliver a concept of non monotonic inference within a theory. To solve this problem, we put

DEFINITION 1. Let $\mathcal{T}$ be a theory (as described above) and A and B two sentences Then, the expression "*given a background theory $\mathcal{T}$, if A, then expect B*" (in symbols: $\mathcal{T}; A \mathrel{\mid\!\sim} B$) denotes the following *non monotonic inference relation*:

there is a formula X such that

$$\mathcal{T} \vdash A > X \quad \text{and} \quad \mathcal{T}; A \vdash X \to B,$$

where $\mathcal{T} \vdash A$ means there are formulas $A_1, \ldots, A_n \in \mathcal{T}$ such that $A_1 \land \cdots \land A_n \to B$ is a theorem of φ.

According to this definition, an *expectation* (with respect to a given theory) is a metadescription of two monotonic derivations which require that when a subject makes a supposition, he is taken to be responsible for all its implications. Expectations (as given by Definition 1) are similar to the rules of Default Logic, but while the latter are usually primitive, the former are derived. Obviously in our case, it is normality and not coherence that is involved in the rules. It is immediate to see that expectations are monotonic with respect to theories, i.e. if $\mathcal{T} \subseteq \mathcal{T}'$, then $\mathcal{T}; A \hspace{-0.3em}\mid\hspace{-0.5em}\sim\hspace{-0.3em} B$ implies $\mathcal{T}'; A \hspace{-0.3em}\mid\hspace{-0.5em}\sim\hspace{-0.3em} B$.

Many properties of $>$ carry over to $\hspace{-0.3em}\mid\hspace{-0.5em}\sim\hspace{-0.3em}$ (reflexivity, reasoning by cases, restricted transitivity etc.). Notable exceptions are the properties of cautious and rational monotony (i.e. (4) and (5)) which not unsurprisingly are also a problem for Default Logic. However, if we take theories whose axioms are restricted to N-conditionals (as Lehman does in [9] with conditional assertions, the analogues of atomic N-conditionals), then expectations do satisfy these two properties.

Moreover, if we take the classical version of φ, we can show that $\hspace{-0.3em}\mid\hspace{-0.5em}\sim\hspace{-0.3em}$ has, at least, the inferential power of [9]'s Preferential Entailment. So given the above restriction on the base logic, if each conditional assertion $A \mathrel{\mid\!\sim} B$ belonging to a knowledge base $\mathcal{K}$ is translated into the N-conditional $A > B$ thus yielding the background theory $\mathcal{K}'$, by induction it is easy to prove that for any sentences A and B,

THEOREM 2. *If $\mathcal{K} \vdash (A \mathrel{\mid\!\sim} B)$ (in P), then $\mathcal{K}'; A \hspace{-0.3em}\mid\hspace{-0.5em}\sim\hspace{-0.3em} B$*

Let us now define the semantics of φ. To interpret the intuitionistic base, we have a set M of states of knowledge and a partial order R_i on M. To interpret N-conditionals, we add a three place relation R between states and particular sets of states, "belief sets", where each belief set consists of all states in which a sentence is believed. To be precise,

DEFINITION 3. $\mathcal{M} = (M, R_i, R, [\,])$ is a φ-*model* if R_i is a partial order on M, $R \subseteq M \times \mathcal{P}(M) \times M$ and $[\,]$ is a function which assigns to each sentence A a set $[A]$ of elements of M. Moreover, for every $m, n, p \in M$:

(M1) if $(m, [A], n) \in R$, then $n \in [A]$;

(M2) if $[A] \neq \emptyset$, then $(m, [A], n) \in R$, for some m and n;

(M3) if $(m, [A], n) \in R$, nR_ip and $p \in [B]$, then $(m, [A] \cap [B], p) \in R$;

(M4) if $(m, [A] \cup [B], n) \in R$, then $(m, [A], n) \in R$ or $(m, [B], n) \in R$;

(M5) when $(m, [A], n) \in R$ implies $n \in [B]$ for every n, then
$(m, [A] \cap [B], n) \in R$ implies $(m, [A], n) \in R$ for every n;

(M6) if mR_in, then $(m, [A], p) \in R$ if and only if $(n, [A], p) \in R$;

(M7) if $(m, [A] \cap [\sim C], n) \in R$, and if there is p such that $(m, [A], p) \in R$ and $p \notin [C]$, then $(m, [A], n) \in R$.

(M8) the function $[\,]$ satisfies the usual intuitionistic constraints on sentential connectives; furthermore,
$[A > B] = \{m \in M : \text{if } (m, [A], n) \in R, \text{ then } n \in [B]\}$.

Given a φ-model $\mathcal{M}$ and a sentence A, we say that a state $m \in M$ accepts A, if $m \in [A]$. Furthermore, A is said to be *accepted* in $\mathcal{M}$, if $M \subseteq [A]$. Last, A is φ-*valid* if A is accepted in all φ-models.

Definition 3 converts into an informal account of reasoning with N-conditionals. The set M is a collection of cognitive states, i.e. they are rational idealizations of psychological states of belief. Take R_i to be a forward directed temporal relation on M such that a later state may contain more knowledge than an earlier state. In each state, the subject asserts both basic assertions and N-conditionals. Given $m \in M$, the set $\{n \in M : (m, [A], n) \in R\}$ stands for the set of *normal* states that concern A, from the point of view of the state m. These states, call them A-*normal* states, are not necessarily *complete* states of knowledge, but they contain enough information to evaluate N-conditionals having context A. Thus, $[A > B]$ is accepted in a state m, if B is accepted in all states that m considers to be A-*normal* (viz. (M8)). Constraints are put on the selection of A-normal states that correspond to axioms and rules (1)-(9). Condition (M6) deserves a brief comment: it is in fact a constraint on the relationship between normality and time. It literally says that normality persists through time: either we know that condition A is normally accompanied by B or we know that it isn't, independently of the knowledge that may be gained as time flows forward. In other words, actual knowledge has no effect on established normality claims. Also it is important to note that the subject may believe $A > B$ although he may ignore A or he may believe A and ignore B; moreover, he might eventually learn $A \wedge \sim B$ without giving up the knowledge of $A > B$. Note that given a sentence A, what is accepted in a state does not affect the choice of the set of A-normal states associated with that state; hence in the actual state of knowledge, the N-conditionals may remain together with the knowledge that falsifies them. This is a very plausible feature of φ-models. After all, if we know that normally birds fly and we find out that Tweety the bird is a penguin and thus it does not fly, we don't think that we have generated a contradiction: learning that Tweety is a bird which does not fly need not make us change our mind about the fact that normally birds fly. Bo assertions can be accepted since one expresses definite actual knowledge while the other refers to statistical evidence. Now, it can be proved

THEOREM 4. *(special completeness) The monotonic logic φ is complete with respect to φ-models i.e. A is a theorem of φ if and only if it is φ-valid.*

From Definition 1 and the previous theorem, it follows that

COROLLARY 5. *(general completeness) $\mathcal{T}; A \vdash B$ if and only if there is a formula X such that for every φ-model $\mathcal{M}$ and for every $m \in M$, if m accepts $\mathcal{T}$, then m also accepts $A > X$ and $A \wedge X \to B$.*

Corollary 5 does not give a good semantical characterization of $\mathcal{T}$-expectations under φ since it is expressed in terms of an *existential* condition. But we can do better; for this purpose consider:

DEFINITION 6. Given a theory $\mathcal{T}$, a sentence A and a φ-model $\mathcal{M}$, an A-*orthodox state (according to $\mathcal{T}$)* in $\mathcal{M}$ is a state in M which accepts the axioms in T and also accepts any formula X such that $T \vdash A > X$.

Thus, given a φ-model $\mathcal{M}$, any A-orthodox state according to T accepts both the theory T and whatever that theory considers to be normal with respect to A. In other words, orthodox states are states in which what is known to be normal (with respect to some T and some A) does turn out to be actual. Using this terminology, we can prove another completeness result. First,

THEOREM 7. $T; A \mathrel{\mid\!\sim} \bot$ if and only if there are no φ-models that have A-orthodox states according to T.

Then, we obtain the following completeness theorem for expectations under φ.

THEOREM 8. $T; A \mathrel{\mid\!\sim} B$ if and only if there are no φ-models that have A-orthodox states (according to T) which accept $A > \sim B$, i.e. in all φ-models all A-orthodox states (according to T) accept $\sim(A > \sim B)$.

Hence, $T; A \mathrel{\mid\!\sim} B$ iff in all φ-models and in all their A-orthodox states (according to T), B cannot be exceptional. Note that when one wants to use these theorems, it is advisable to construct special kinds of φ-models which we call *full* φ-models. Full models satisfy conditions (M1)–(M7) (of Definition 3) except for the fact that any reference to belief sets (for instance $[A]$) is replaced by arbitrary subsets of the domain M (say $Q \subseteq M$); furthermore they satisfy condition (M8). Thus, in full φ-models, the model *structure* is independent of the (belief) assignments given to the formulas. It can be shown that the φ *canonical* model cannot be a full model (see [2]).

3 Believing *Only A*

The relational concept $T; A \mathrel{\mid\!\sim} B$ may be thought as defining a non-deductive inference rule, where the premise A lends only partial support to the conclusion B (under T). Strictly speaking, one cannot really term B to be a conclusion, since B cannot be asserted on its own, i.e. detached from its premise A. Thus unlike deductive rules, expectations do not provide the means by which we acquire new statements that can be added to the ones we already possess.

However, when premises are qualified as being the *only beliefs* that a subject feels entitled to hold, we may allow him to effect a transition from beliefs to other beliefs that acquire the same status. To see this, consider a subject who believes that "Tweety is a bird" and that "birds normally fly". If all he believes about Tweety is that he is a bird, he will not think "well, maybe he's dead or he is a penguin" but he'll simply believe that Tweety is a normal bird. The point is that an evidential base is usually given with the understanding that anything that is not explicitly stated as a part of the base or implicitly in it is to be ignored. And this is the same as saying that things are as normal as that evidential base allows. Hence, from our point of view, the assertion "*Only A*" calls for the *decision* to consider normality induced beliefs to be full beliefs. Obviously, the operator *Only* is context-sensitive: when it is no longer *Only A* because it is *Only* $(A \wedge B)$, there may be changes in beliefs. It is important to realize that if such is the case, it is not thought "an *error* was made and it ought to be corrected" for given what was originally believed, the previous conclusions still stand. Now the question arises whether a logic can be devised to describe how changes in beliefs may occur under the responsibility of the operator *Only*. Our answer is the Φ-logic in the language $\mathcal{L}'$ which is the language $\mathcal{L}$ extended with the unary operator *Only* applicable to *objective* formulas. Let us say that a formula is *plain* if it does not contain the operator *Only*. Then, an *objective* formula is a formula that is both basic and plain.

The following are the axioms and rules that determine the logic Φ: axioms and rules of IL, axioms and rules (1)–(9) restricted to plain formulas and if A is an objective sentence,

(O1) $Only\, A \rightarrow B \vee \sim B$

(O2) $\sim\sim Only\, A \vee \sim Only\, A$

(O3) $Only\, A \rightarrow (A > B \rightarrow B)$

(O4) $\dfrac{B \leftrightarrow C}{Only\, A \leftrightarrow Only\, B}$

According to (O1), when the subject only believes A (for any objective A), he closes off his deliberations and considers his actual beliefs to be complete, i.e. although the subject is ready to revise his judgment, he believes things in an all or nothing way. Another kind of closure condition is conveyed by (O2): it says that, given any objective sentence A, if the subject thinks that *possibly* all his information does not amount to precisely A, then he thinks that it is *impossible* to believe that all he believes is A; alternatively, if things *could* be as normal as A allows, then it is *impossible* for the subject to conceive that things could be otherwise. Here closure amounts to transforming an existential condition (possibility) into a universal one (impossibility to believe the negation). It is interesting to note that axiom (O2) describes a characteristic feature of the process of jumping to conclusions i.e. the intention of elevating tentative beliefs (it is consistent to hold that things are normal) into "use-beliefs" (it is impossible to hold that things are abnormal). Note, furthermore, that axioms (O1) and (O2) are meaningful only because we are using IL. Last, read (O3) as saying that detachment is forced by the circumstance of deciding that there is a piece of evidence and no more. Thus, we have

THEOREM 9. *If* $\mathcal{T} \vdash A > B$, *then* $\mathcal{T}; Only\, A \vdash B$. *Moreover, if* $\mathcal{T}; A \hspace{-0.3em}\mid\hspace{-0.5em}\sim B$, *then* $\mathcal{T}; Only\, A \vdash B$.

So the basic picture is this: all the subject believes is a finite set of objective statements. Given this set, he treats tentative conclusions as assertions of beliefs without the qualification of tentativity. When new objective statements are added to this set, the subject must recompute with the Φ-logic what are the expected consequences.

Now, let us provide for a semantical analysis of the operator *Only*. Consider first,

DEFINITION 10. Given a φ-model $\mathcal{M}$, $n \in M$ is an *endpoint* in $\mathcal{M}$ iff for every $m \in M$, $(mR_i n)$ implies $nR_i m$.

DEFINITION 11. Given a φ-model $\mathcal{M}$, $m \in M$ is a *closed A-orthodox state* in $\mathcal{M}$ iff m is an endpoint and $(m, [A], m) \in R$.

Then, put

DEFINITION 12. A φ-model $\mathcal{M}$ is a Φ-*model* if

(MO1) for every A and for every $m \in M$, either,

- for every $n \in M$ such that $mR_i n$, n is not a closed A-orthodox state

or

- for all $n \in M$ such that $mR_i n$, there is $p \in M$ such that $nR_i p$ and p is a closed A-orthodox state

and

(MO2) $m \in [Only A]$ iff m is a closed A-orthodox state.

Hence, every Φ-model is a φ-model where, for all objective sentences A, each state either has no closed A-orthodox states "below" it or on the contrary every state from that state onwards has a closed A-orthodox state "below" it. Moreover, $Only\ A$ is asserted at a state if and only if that state is a closed A-orthodox state. We can prove

THEOREM 13. *The monotonic logic Φ is complete with respect to Φ-models, i.e. A is a theorem of Φ if and only if it is Φ-valid.*

At this point, we can express what it means to infer on the basis of $Only\ A$ and a theory $\mathcal{T}$ simply in terms of the belief in A and $\mathcal{T}$. It is easy to check

THEOREM 14. $\mathcal{T}, Only\ A \vdash B$ *iff in all Φ-models, all closed A-orthodox states that accept $\mathcal{T}$ accept B.*

In order to get an intuitive picture of the above semantics, consider the class of all *generated models* by some element in the domain of each Φ-model. Roughly speaking, the models generated with respect to a Φ-model are obtained by fixing an element in the domain of the model, by considering all elements R-related to it and last by taking all elements that are in the ancestral of R_i whith respect to those elements; the "generated" relations are the appropriate restrictions of the relations in the original model. It is easy to see that validity in a model is preserved in all models that can be generated from that model. Now, given an objective sentence A, among all possible generated models there are some that have an A-orthodox state below every state as there are those who have no A-orthodox states. This suggests the following

DEFINITION 15. An *A-orthodox model* is a Φ-model $\mathcal{M} = (M, R_i, R, [\ ])$ such that for every $m \in M$, there is $n \in M$ such that $mR_i n$ and n is a closed A-orthodox state.

Then it is obvious:

THEOREM 16.

If $T; Only\, A \vdash B$, then $\sim\sim B$ is accepted in all states of all A-orthodox models which are models of T.

We close with the following remarks. According to the previous theorem, the process of jumping to conclusions under the assumption that beliefs should include no more and no less than what is considered to be normal in the face of a certain piece of evidence, involves drawing a *classical* conclusion in a special kind of intuitionistic setting concerning that evidence. Moreover, according to our analysis, normality inferences are non monotonic not because their logic is but because, in certain circumstances, we decide to consider high plausibility sufficient for full belief.

Appendix: sketches of proofs

Proof of THEOREM 4.

See Theorems 6, 8 in [2] for part of the proof. In addition, consider axiom (5). Let $m \in [A>B]$, $m \in [\sim(A>C)]$ and suppose that both

$(m, [A \wedge \sim C], n) \in R$ and there is p such that $(m, [A], p) \in R$ and $p \notin [C]$.

Then, by condition (M9), we have $(m, [A], n) \in R$. Hence $n \in [B]$ and $m \in [A \wedge \sim C) > B]$. Analogously for axiom (3).

As for completeness, define a φ-*canonical model* as in Def. 4.1 in [2]; briefly, take M the collection of all sets that are *nice* with respect to φ, define the *canonical relation* R_i as set inclusion and the *canonical relation* R as follows: given $[A]=\{m \in M : A \in m\}$, put $(m, [A], n) \in R$ iff for every B, if $A > B \in m$, then $B \in n$. As an example of how to proceed, suppose that $(m, [A \wedge \sim C], n) \in R$ and that there is p such that $(m, [A], p) \in R$ but $p \notin [C]$. We show that $(m, [A], n) \in R$, i.e. for every B, if $A > B \in m$, then $B \in n$. From hypothesis, $A > C \notin m$. By (6) and the fact that M is nice in φ, $\sim(A>C) \in m$. Let $A > B \in m$; given (9), we have $B \in n$.

Proof of THEOREM 7

If $T; A \mid\!\!\sim \bot$, then there is Y such that $T \vdash A > Y$ and $T \vdash Y \to \sim A$. If m is an A-orthodox state according to T, $m \in [Y \wedge \sim A]$, impossible. On the other hand, suppose that $T; A \not\mid\!\!\sim \bot$; Consider the set $\Gamma = T \cup \{X : T \vdash A > X\}$. It is easy to see that Γ is consistent. Extend Γ to s, a nice set with respect to T; by construction, s is an A-orthodox state according to T.

Proof of THEOREM 8

Let $T; A \mid\!\!\sim B$. Given a φ-model $\mathcal{M}$, any A-orthodox state m in $\mathcal{M}$ will not accept $A > \sim B$, since by hypothesis $m \in [B]$. If $T; A \not\mid\!\!\sim B$, then $T' = T \cup (A > \sim B)$ is consistent. Apply Theorem 7 to find a φ-model and an A-orthodox state according to T'.

Proof of THEOREM 13

To prove validity, consider a Φ-*model* $\mathcal{M} = (M, R_i, R, [\,])$. We first try out axiom (O1); just bear in mind that *Only A* holds at all endpoints. As for axiom (O2), if $m \notin [\sim Only\,A]$, then there is n such that $mR_i n$ and $n \in [Only\,A]$; hence by (MO1) it can only be that for all p such that $mR_i p$, there is a closed A-rthodox state q, with $pR_i q$. The intuitionistic interpretation of double negation yields $m \in [\sim\sim Only\,A]$. That (O3) is valid is a consequence of (MO2).

To prove completeness, reconsider the canonical model for the φ-logic except that now put $(m, [A], n) \in R$ iff for every B, if $A > B \in m$, then $B \in n$ and when $m = n$ for a maximal consistent nice set m, $Only\,A \in m$. To prove that (MO2) holds in the new canonical model use axioms (O1) and (O3). Before proving that condition (MO1) holds in the canonical model, note that if m is a maximal consistent nice set and $(m, [A], m) \in R$, then m is a closed A-orthodox state. Now to proceed with the proof, deny that for all n such that $mR_i n$, n is not a closed orthodox state, i.e there is p such that $mR_i p$ and p is a closed A-orthodox state. Hence, $m \notin \sim Only\,A$. Using axiom (O2), it is easy to prove the other half of condition (MO1).

BIBLIOGRAPHY

[1] C. Boutilier. Conditional logics of normality: a modal approach. *Artificial Intelligence* 68: 87–154, 1994.
[2] G. Fischer Servi. Nonmonotonic consequence based on intuitionistic logic. *Journal of Symbolic Logic* 57 (4): 1176–1197, 1992.
[3] G. Fischer Servi. *Quando l'eccezione è la regola*. McGraw-Hill, Milan 2001.
[4] D.M. Gabbay. Intuitionistic basis for nonmonotonic logic. In *Proceedings of the Conference on Automated Deduction*. Lecture Notes on Computer Science 138: 260–273, Springer Verlag, 1982
[5] J.H. Halpern. A Theory of Knowledge and Ignorance for Many Agents. *Journal of Logic and Computation* 7 (1): 79–108, 1997.
[6] J.H. Halpern, G. Lakemeyer. Levesque's axiomatization of only knowing is Incomplete. *Artificial Intelligence* 74: 381–387, 1995.
[7] J.H. Halpern, G. Lakemeyer. Multiagent only knowing. *Journal of Logic and Computation* 11 (1): 41–70, 2001.
[8] S. Kraus, D. Lehman and M. Magidor. Nonmonotonic reasoning, preferential models and cumulative logics. *Artificial Intelligence* 44 (1-2): 167–207, 1990.
[9] D. Lehman and M. Magidor. What does a conditional knowledge base entail? *Artificial Intelligence* 55: 1–60, 1992.
[10] H.J. Levesque. All I know: a dtudy of autoepistemic logic. *Artificial Intelligence* 42: 263–309, 1990.
[11] R.C. Moore. Semantical considerations on non monotonic logic. *Artificial Intelligence*, 25: 75–94, 1985.

Sense and Proof

CARLO PENCO, DANIELE PORELLO

1 Frege, Carnap and the contrast between cognitive and semantic sense

In a paper given at the 1997 SILFS Conference [18], C. Penco claims that a strong tension between a semantic and a cognitive notion of sense was already present in Frege's writings.

Many authors have discussed recently this tension in Frege, but few of them remark that Carnap was probably the first to attempt to compose the tension.

Carnap ([1]) soon realized that his conception of intension, although suitable to treat a semantic notion of sense as truth condition on the lines developed by Wittgenstein, was not enough to solve the problem posed by Frege on belief contexts. He devised therefore the notion of intensional structure and intensional isomorphism, as a more fine-grained notion that the semantic notion o intension and intensional equivalence. For many reasons however his notion did not appear successful. In this short paper we try to develop a new perspective which tries to compose the tension between a semantic and a cognitive notion of sense, suggesting the use of proof theory as a mean to interpret the cognitive notion of sense.

In the Fregean view, sense is something which is there to be grasped from a speaker. The concept of grasping a sense or understanding is co-original with the notion of sense. However it is not clear in Frege what is the object of understanding. Frege himself made interesting remarks on the limitations of understanding complex mathematical formula with all their connections with other parts of mathematics. In his famous argument on the "intuitive difference of thought" (as it has been named by Gareth Evans) Frege claims that if it is possible to understand two sentences, and coherently believe one and disbelieve the other, then those sentences express different senses or thoughts. The Fregean example is the belief that Hesperus is a planet and Phosphorus is not a planet, given by the ignorance of the identity Hesperus=Phosphorus. A lack of knowledge permits a person who hold erroneous beliefs to be considered rational, because the erroneous beliefs express two different thoughts, whose truth is not evident to the believer.

The criterion links sense identity to what a subject believes, to her limited accessibility to the information, to what she grasps, with a limited understanding. Here the limitation is given by empirical information, but there are many passages where Frege accepts the idea of limited computational capacity as a way to explain why two expressions expresses different senses.

The semantic notion of sense is linked to the criterion of logical equivalence: in an example proposed by Frege (letter to Husserl 1906, see [10].) two sentences which express a logical equivalence like $A \to B$ and $\neg(A \wedge \neg B)$ express the same thought. Frege probably connects this idea with the criterion of immediate recognizability, given that most of his examples — given in his last writings on logic — deal with elementary logical equivalences between connectives. The idea of sameness of sense given by logical equivalence is coherent with the concept of sense as truth conditions made by Wittgenstein in his Tractatus.

The second, semantic notion clashes with the first if we try to apply also to the semantic notion the criterion of the intuitive difference of thoughts; if we may believe two sentences as having different truth value because of lack of information, why shouldn't we accept the same uncertain attitude towards two sentences with the same truth conditions, but such that we accept one of them while remaining uncertain on the truth condition of the other? Actually we may lack computational capacity, while understanding the basic notions used, the meaning of the connectives, the working of the symbolism. We may imagine a person who is so slow that she cannot realize that two logical equivalent sentences produce the same truth tables. Therefore she may believe that $A \to B$ is true while $\neg(A \wedge \neg B)$ is false; therefore the two sentences would express — contrary to what Frege says — two different (cognitive) senses. This possibility is enforced by what Frege says in a later paper on negation, when he assert that A and not not A express different thoughts.

We claim that the ambiguity in Frege's concept of sense depends on the lack of logical instruments that has been developed later with proof theory. And we think that proof theory may help to define different levels of understanding; in this way we may distinguish the semantic and the cognitive aspect of sense, composing the original tension, in a way which is different from the original suggestions given by Dummett with the choice of a verificationistic theory of meaning.

2 Understanding sense as truth conditions

Let us begin with an analysis of what it means "to understand sense as truth conditions". When we consider complex sentences, the understanding of truth conditions amounts to knowing the corresponding truth tables.

The sense of a complex sentence A is then known, when we know for which values of the atomic propositions occurring in A, the sentence is true.

Consider the example considered above, we have two logically equivalent sentences (1) $A \to B$ (2) $\neg(A \wedge \neg B)$.

If a subject understands $A \to B$ then, by definition of material implication, she knows that $A \to B$ is true when either A is false or B is true.

If the subject understands $\neg(A \wedge \neg B)$, then she knows that the sentence is true when it is false that A and B is true; which entails, if the meaning of the conjunction and the negation is known, that he knows that $\neg(A \wedge \neg B)$ is true when either A is false or B is true.

So the truth conditions of the sentences involved are the same. If we do not consider how subjects grasp those truth conditions, then we face the following problems.

First, for example, we cannot concede the possibility that a subject might believe (1) true and (2) false. Suppose for example that a subject, by mistake or logical confusion, believes that $A \to B$ is true and believes that $\neg(A \land \neg B)$ is false. Then, she believes that it is the case that A is false or B is true, since she believes (1) true. But if she believes that $\neg(A \land \neg B)$ is false, then she believes that $(A \land \neg B)$ is true; so she believes that A is true and B is false. Since the information the subject should manage is not coherent (A is false or B is true, A is true and B is false), the only way in which we may claim that she believes (1) true and (2) false is claiming that she is accepting a contradiction. We have no means to say that she doesn't realize that she is contradicting. We may however ask whether it is possible that a subject believes that (1) is true while having no opinion about the truth value of (2).

Even in this case, if we keep the hypotheses at issue, we cannot represent such situation. If a subject believes that $A \to B$ is true, then she believes that A is false or B is true, and since these are precisely the truth-conditions of $\neg(A \land \neg B)$ and the subject knows that if these condition are satisfied then $\neg(A \land \neg B)$ is true (since she understands both sentences), then we are forced to say that the subject believes that (2) is also true. But why does she believe it? Simply because of our definitions, since we cannot deny it without contradicting our definitions.

From this argument, it follows that if a subject believes a sentence A, then he must believe all the sentences that are logically equivalent to A, no matter how complex they are. Moreover, the subjects *immediately* believes all the logically equivalent sentences, since we just proved it as a fact simply entailed by our hypotheses. With our notion of understanding sense as truth conditions we are compelled to make our speaker logically omniscient.

3   A weaker notion of understanding

The logical omniscience of a subject described by the assumptions we made concerning the notion of understanding a sentence is completely useless from a cognitive, or computational, point of view.

It seems that if we assume the definition of understanding a sentence as mere grasping truth conditions, even if we try to employ a cognitive notion of sense (e. g. the one presented by the immediate recognizability criterion) we are not allowed to define a cognitive difference between logically equivalent sentences.

If however we look closer at the same argument we mentioned to find out the truth conditions of the sentences involved, we note that the formula $\neg(A \land \neg B)$ requires more calculation than $A \to B$.

The notion of sense of a sentences in mere terms of truth conditions fails to capture all the information concerning the complexity of the process of grasping the truth condition, which seem to be the relevant aspect in a cognitive notion of sense.

In the following sections, we will advance a proposal for defining a cognitive notion of sense which is coherent and compatible with an objective one.

A proper notion of cognitive sense should be grounded not on a strong notion of understanding requiring full grasping of truth conditions, but on a weaker notion of understanding based on the idea of limited knowledge (or even on bounded rationality) [20]. We will therefore need to enrich the notion of sense considering another aspect of sense that seems more suitable to deal with cognitive aspects, namely the notion of sense as computing procedure hinted at by Frege (see [9]).

4 Limited knowledge and procedures

Assuming that understanding a formula is not necessarily understanding directly its truth conditions, but understanding its mode of composition and the meaning of the connectives, we can state the problem at issue with the following question:

(Q) Assuming that a subject understands (1) $A \to B$ and (2) $\neg(A \wedge \neg B)$, and moreover accepts (1) $A \to B$, what does she need in order to accept also (2) $\neg(A \wedge \neg B)$?

The point is that if we aim to describe a cognitive notion of sense, we cannot consider the process of understanding of the two sentence as immediate [6]. We need to consider the process of understanding, say (2) given (1), as a process mediated by a procedure, or a computation. Otherwise we would lose the information concerning the complexity of the process of understanding sentences which is essential for a cognitive notion of sense. Therefore we reformulate the truth conditional approach considering the procedure of grasping truth conditions as a constitutive feature of understanding sentences.

Our basic definition is then that a subject understands a sentence when she can perform a *procedure* of grasping truth conditions of the sentence.

A good way to represent procedures, as we shall see in more detail in the next section, is the notion of proof which may be defined within some suitable logical calculus. The notion of proof, or more generally the notion of justification, has been applied for example by Michael Dummett [5]. to define his justification semantics.

However, our proposal is to keep the truth conditional approach — since it allows to state clearly the relationship between a cognitive notion of sense and an objective, or semantic, notion of sense — enriching it by means of a notion of procedure, rather than proposing an alternative semantic theory based on different key concepts.[1] It is useful to remark that the approach we are suggesting is not to be intended as a representation of the explicit knowledge speakers have. We are not claiming that it is always the case that someone who accepts a sentence is able to justify it showing a proof. We are rather suggesting that the notion of proof is a useful tool for representing implicit knowledge [6, p. 139] speakers show when they understand sentences. The problem (Q) therefore may be generally solved in our setting saying that a subject can understand (1) and (2) since she can manage a procedure to grasp their truth conditions, but — in case she believes one true and the other

---

[1] The justification semantics proposed by Dummett uses the notion of proof as semantic value, and then it is committed with intuitionistc logic. Here we are presenting an approach which could be applied in different logical calculi, provided they have a proof theory and a truth values semantics, since it is not decided yet which formal calculus is adequate to represent semantic understanding.

false — she may fail to realize she reaches a contradiction, since the subject may fail to manage the procedure of detecting the logical relationship between (1) and (2).

5 Proof theory for representing procedures

We sketch a formal setting for representing the definitions we proposed which allows us to keep both a classical truth conditional style semantics and to account for the complexity of the procedure of grasping truth conditions.

We propose to use proof theory to represent the procedure of grasping truth conditions; in this way we can describe the failure of detecting logically equivalent sentence as a lack of logical competence due to the complexity of the sentences involved.

Here we will not refer precisely to a particular logical calculus, rather we will propose some general idea which may be applied in different logical frameworks.[2] We consider two notions of sense: a *semantic* one and a *cognitive* one. Remark that in this way we may keep a notion of *co-tenability* of thoughts, which states some important intuition about the relationship between the meaning of a sentence and a subject who understands it.

(S) The semantic sense of a sentence A is the whole class of rules defining a proof of A, which lead to the truth conditions of A.

(C) The cognitive sense of a sentence A is a class of rules a subject manages, which lead to grasp (partial) truth conditions of A.

So, the relationship between (C) and (S) can be stated in terms of an inclusion, namely the cognitive competence of a subject amounts to manage a subclass of the rules of inference which are defined in a logical calculus.

It is important to remark that the partial understanding we are defining depends on subjects just in the sense that subjects manage some of the rules required to build a proof of a given sentence. It doesn't mean that a subject has individual or private rules for building proofs.[3] Moreover, the partial comprehension can also be stated in terms of complexity bounds on the application of those rules. For example, if we assume that a subject is able to perform a *modus ponens*, we don't want to assume that he is able to draw the

---

[2]An interesting choice would be to state our definitions within linear logic (see [13]). Linear logic may be considered more general than intuitionistic or classical logic, in the sense that both can be embedded in linear logic, and it may also be considered as an analysis of the properties of classical and intuitionistic proofs. Briefly, linear logic allows to define where resources are actually needed to be bound and where we can assume an unbounded number of tokens. This aspect is crucial in order to go into the relationship between a cognitive and a semantic notion of sense. Moreover, linear logic has been applied to define formal grammars for natural languages working both for syntactical aspect of sentence understanding and for composition of meanings (see for example [15] and [2]). However, it is not clear if we can consider the semantics of linear logic, based on the algebraic structure of phase space (see [11]), as a truth value semantics: we would need in particular to investigate which notion of truth is formalized by that structure. We leave a deeper examination of this approach for further work.

[3]The fact that we don't allow individual or private strategies aims to keep some features of the Fregean anti-psychologism: the sense of a sentence doesn't depend on the representation nor depends on private aspects of the comprehension of meaning.

conclusion at any degree of complexity: we are not assuming she can perform a proof consisting in an unbounded iterated or nested applications of *modus ponens*. Therefore, the class defining the cognitive notion will be a sub-class of the class involved in the objective notion of sense.[4]

We consider some example using natural deduction for classical logic. In order to get the truth conditions for (1) $A \to B$, a subject may be able to manage the following procedure, represented by the logical inference:

(π):

$$\frac{B}{A \to B}$$

So, the subject knows that she is accepting $A \to B$, since she is accepting B.

In this way we may represent a partial comprehension of truth conditions, in the sense that we do not need to assume subjects are able to grasp all the possible case described by a truth table (in this example, that $A \to B$ may be true also if A is false). Moreover, in order to get truth conditions for (2), a subject may be able to perform the following procedure:

(π'):

$$\frac{\neg A}{\neg(A \land \neg B)}$$

which doesn't entail a complete knowledge of truth conditions of (2). So we can assume that a subject can understand (1) and (2) and she can accept (1) but she can fail to accept (2) simply because she can fail to manage the procedure of detecting the connection between (1) and (2). Consider again Frege example. Suppose a subject understands (1) and (2), by means of the procedures we mentioned. Moreover, she considers (1) true. We can represent what the subject needs in order to accept also (2), for example by means of a proof like:

(π''):

$$\cfrac{\cfrac{\cfrac{[A \land \neg B]^1}{A} \land E \quad \cfrac{\pi}{A \to B}}{B} \to E \quad \cfrac{[A \land \neg B]^1}{\neg B} \land E}{\cfrac{\bot}{\neg(A \land \neg B)} \neg I, 1} \neg E$$

Here π represents the procedure the subject can perform in order to understand and then accept (1). We used a natural deduction proof just to show an example, we could have chosen other proof theoretical calculi.[5]

---

[4]Remark that it is difficult to speak of partial understanding using mere truth-conditional definition of sense since, as we saw, even if we try to define a partial understanding by means of the immediate recognizability criterion, we are led either to make the subject contradict himself or to make the subject logically omniscient.

[5]We can read the proof in the following way. A subject can manage the procedure π that leads her to accept (1); in order to accept (2) a proof is required. Assume by contradiction, $A \land \neg B$. Eliminating conjunction, we obtain A and we obtain $\neg B$. From hypothesis (1) and

So we can use the proof π'' to represent the process of accepting (2) given (1). In this way it is possible to argue about complexity bounds to put on the process itself. Of course our approach should take into account data describing subjects effective performances.

Consider now an example showing a kind of same level complexity and ask again the same question concerning what a subject needs in order to detect logically equivalent sentences. We consider commutative use of conjunction. May a subject accept $A \wedge B$ while not accepting $B \wedge A$? If she understands both sentences, then she grasps in a certain way a procedure represented by the rules for the conjunction.

If she accepts $A \wedge B$, then she should accept both A and B. But if this metalinguistic use of "and" is commutative, then she has to accept also $B \wedge A$. In this case, if we are in a commutative framework, we have two formulas which share a same level of understanding complexity, besides a common logical content; therefore we may say that a subject who accepts one of the two sentences is not rational if he doesn't accept the other, since she can manage both procedures.[6]

We conclude mentioning an example taken from the literature on the application of proof theoretical notions in formal semantics which show how proof theory can represent subjects' different performances also in case of quantified sentences [16]. Consider the problem of quantifier-scope ambiguity. The sentence "Someone loves everyone" allows two different readings, depending on the narrow ($\forall \exists$) or wide ($\exists \forall$) scope. A careful examination of proofs representing the meanings of those sentences[7] it is possible to express the fact that the preferred reading ($\exists \forall$) has a lower complexity degree than the other. In this way it is possible to develop a quite precise notion of cognitive relationship between subject performances and meaning of a sentence.

Summing up, in case of two logically connected sentences, we can claim that it is rational, or better it is possible without contradiction, to accept a sentence while having no opinion concerning the other, when the complexity of the logically connected sentences is different.

So we provided the theoretical possibility for suspending judgement until, by reflection, calculation or other means, a subject can access a procedure for grasping the truth conditions of the sentence.

This approach points at a more sophisticated notion of rationality which

---

from A, we obtain by eliminating conditional B. But B and $\neg B$ entail a contradiction ($\bot$), so we can apply the rule of introducing negation ($\neg I$) and discharge hypotheses marked by 1. We chose this example, and we used the intuitionistic rule for negation, just not to be committed a priori with classical logic. Actually the distinction between partial and full understanding stated in terms of proof can be reformulated for semantic theory that insist on other notion of semantic value.

[6] The aim of this proof theoretic approach to the complexity of understanding meaning would be a sort of normal form for the proof representing procedures of grasping truth conditions. Actually we cannot define a normal form procedure without considering empirical data concerning subjects effective performance. The notion of normalization of proof, which is a central issue in proof theory, may be a staring point in order to define classes of meanings sharing a same measure of such complexity, so to give a proof-theoretical account of the criterion of immediate recognizability.

[7] Actually, the representation at issue employs *proof nets*, which is the peculiar proof theory for linear logic.

can include the process of learning new procedures (for example, by means of interaction) rather than considering rationality as static set of features subjects have, or should have.

6 Conclusion

The proposal we presented allows to consider both a cognitive and a semantic notion of sense and, this is the most interesting point, we can see how the two notions interact: they are not distinct features, as it happens for example in many attempts to conciliate those two aspects of Fregean notion of sense. We stated the relationship between cognitive sense and semantic or objective sense in terms of a partial understanding speakers have of meaning.

The semantic notion of sense is given by the whole class of procedures, or proofs, of the given sentence that give the truth conditions of the sentence, while the cognitive notion of sense is defined as a partial access to semantic sense. Moreover we used a general notion of procedure, represented by the notion of proof in some suitable logical system. So it seems that the opposition between a truth-conditional semantics and a justification-semantics (Dummett) may be weakened considering the proof as a way to grasp truth conditions of a given sentence.

BIBLIOGRAPHY

[1] R. Carnap. *Meaning and necessity*. University of Chicago Press, Chicago 1947.
[2] B. Carpenter. *Type Logical semantics*. MIT Press, Cambridge, MA 1997.
[3] M. Dummett. What is a theory of meaning (ii). In Evans and McDowell (eds.), *Truth and Meaning*, Oxford University Press, Oxford 1976.
[4] M. Dummett. *Elements of Intuitionism*. Oxford University Press, Oxford 1977.
[5] M. Dummett. What does the appeal to use do for the theory of meaning? In A. Margalit (ed.), *Meaning and Use*. Reidel, Dordrecht 1979.
[6] M. Dummett. *The Logical Basis of Metaphysics*. Harvard University Press, Cambridge, MA 1991.
[7] M. Dummett. Meaning in terms of justification. *Topoi*, 21: 11–19, 2002.
[8] G. Frege. Über Sinn und Bedeutung. *Zeitschrifft für Philosophie und Philosophische Kritik*. 100: 25–50, 1892.
[9] G. Frege. *Grundgesetze der Arithmetik, vol I*. Pohle, Jena, 1893. Rist. Olms, Hildesheim, 1962; Engl. tr. M. Furth (ed.) *Frege. The Basic law of Arithmetic*, University of California Press, Berkeley 1964.
[10] G. Frege. Wissenshaftliches Briefwechsel. Felix Meiner, Hamburg, Hamburg 1976. Engl. Tr. *Philosophical and Mathematical Correspondence*, ed. by B. Mc Guinness and translated by H. Kaal, Blackwell, Oxford 1980.
[11] J.-Y. Girard. Linear logic. *Theoretical Computer Science*, 50: 1–102, 1987.
[12] J.-Y. Girard, Y. Lafont and P. Taylor. *Proofs and Types*. Cambridge Tracts in Theoretical Computer Science 7. Cambridge University Press, Cambridge 1989.
[13] J.-Y. Girard. *Le Point Aveugle. Cours de Logique I. Vers la Perfection*. Hermann, Paris 2006.
[14] A. Heyting. *Intuitionism*. North-Holland, Amsterdam 1956.
[15] G. Morrill. *Type logical Grammars*. Reidel, Dordrecht 1994.
[16] G. Morrill. Incremental processing and acceptability. *Computational Linguistics*, 26 (3): 319–338, 2000.
[17] Y. Moschovakis. Sense and denotation as algorithm and value. In *Proceedings of the 1993 ASL Summer Meeting in Helsinki*. Lecture Notes in Logic, Springer-Verlag, 1993.
[18] C. Penco. Frege e carnap: verso una teoria integrata del senso. In R. Cordeschi, V. Fano, V. M. Abrusci and C. Cellucci (eds.), *Prospettive della Logica e della Filosofia della scienza: Atti del Convegno SILFS*, ETS, Pisa 1998, pages 345–360.
[19] C. Penco. Frege, sense and limited rationality. *Review of Modern Logic*, 9: 53–65, 2003.

[20] C. Penco. Frege: Two theses, two senses. *History and Philosophy of Logics*, 24 (2): 87–109, 2003.
[21] G. Sundholm. Proof theory and meaning. In D. Gabbay and F. Guenthner (eds.), *Handbook of Philosophical Logic: Volume III: Alternatives to Classical Logic*, Reidel, Dordrecht 1986, pages 471–506.
[22] J. van Benthem. *Language in Action. Categories, Lambdas and Dynamic Logic*. MIT Press, Cambridge, MA 1995.

Some reflections on plurals and second order logic

ANDREA PEDEFERRI

1 Logic or set theory?

The role of second order logic is a controversial issue both for technical reasons and for its metalogical entailments. Following the famous Quine's thesis that second order logic is "set theory in sheep's clothing", it has been widely mantained that by using second order logic we enter in the realm of mathematics and consequently second order logic can no longer be called a proper "logic". The point is particularly focused on quantifiers that directly imply sets along with their "staggering" ontology. This is even more true, for instance, in the frameworks like the so-called neo logicist (or neo fregean) project, where the use of second order logic is somewhat natural. Historically speaking, the sharp distinction between first and second (higher) order logic arose only in the first decades of the last century, in particular with th 1915 Lownheim theorem, when, as Stewart Shapiro notes, "first order languages [became] *de facto* standard in logic".

However it has always been tried to find a way to handle second order logic. Two are, basically, the paths to do it: starting form second order logc and "impoverish" it and its properties towards first order, or, on the opposite side, starting from first order and "enrich" it coming closer and closer to second order. In both cases the crestion of "hybrids" didn't solve the problem since the contrast between lack of completeness from on hand and lack of compactness from the other hand syntetized by the result of Lindström seems unsolvable.

In this paper I'd like to scketch: first a rough account of one of many attempts to settle the problem: a solution that allows to use a second order quantification that does not refer to set (the approach of Boolos of a monadic second order with plural quantifiers). I will then point out the flaws of this attempt. I will finally argue on the philosophical nature of the problem

2 Boolos' way

The question whether second order logic can be labbelled as "logic" has become central in many logical debates especially after the arguments used by Quine in his *Philosophy of Logic* [3]. According to Quine we can understand second order quantification only if we make use of quantifiers varying over subsets of the domain: this leads to a direct entailment with sets and with their problemlatic commitments. It seems, therefore, that second order logic holds a very strong "mathematical character", and this derives from its ability to give cathegorical characterizations of infinite structures. The Lindström Theorem, moreover, appears to be a definitive proof in favor of the "pure logicality" only of first order logic, since it shows that the only logic for which completeness, compactness and Löwenheim-Skolem Theorem hold is first order logic.

In this sense it can possibly be a solution to use a second order quantification that does not refer to sets, and the approach of the late George Boolos of a monadic second order with plural quantifiers could fill the bill. This system has been introduced by Boolos in *To Be is To Be a Value of a Variable (or To Be Some Values of Some Variables)* and in *Nominalist Platonism* (both reprinted in [1]). He argued that it is legitimate introduce two new quantifiers to account for the plural locutions we find in the natural language such as, for instance, "there are some apples on the table".

Roughly, we call first order monadic a language $L1$ such that the set of non-logical symbols does not have function symbols or n-ary relation variables for $n > 1$. The restriction for non-logical symbols disappears in second order languages, but there aren't variables ranging on n-ary function or relations (for $n > 1$). In the standard semantic of second order the quantifiers must be defined only on unbounded first order variables. So in the comprehension schema

$$\exists X \forall x (Xx \leftrightarrow Ax)$$

the second order monadic formula Ax must have x free and not have X free. Boolos develops for second order monadic languages a model-theoretic semantic that allows to not entail the existence of proper classes. Monadic second-order logic is a subsystem of second-order logic which admits only quantification over properties, not over (polyadic) relations. The focus is on quantifiers and to avoid quantifiers involving concept such as class or property, Boolos introduce *plural* quantifiers (usually now indicated by $\forall xx$ and $\exists yy$) which can account for expression in naturar language such as "there are some objects".

In this way we can account for sentences which can not be paraphrased in first order logic (a practice that is also not so natural), most famously the Geach-Kaplan sentence:

some critics admire only one another

If we consider as domain the class of critics the sentence con be formalized at second order as:

$$\exists X (\exists x Xx \wedge \forall x \forall y ((Xx \wedge Axy) \rightarrow (x \neq y \wedge Xy)))[1]$$

The formula can be now read as: "there are some critics each of whom admires a person only if that person is one of them and non of whom admires

---

[1]Kaplan proves as interpreting in the formula Axy with $(x = 0 \vee x = y + 1)$ we obtain

$$\exists X (\exists x Xx \wedge \forall x \forall y ((Xx \wedge (x = 0 \vee x = y + 1)) \rightarrow x \neq y \wedge Xy))$$

which is true in any non standard models of arithmetics but is false in the standard one, namely the model for natural numbers.

himself"[1, p. 74]. We can then introduce in the informal metalanguage used in the construction of formal semantics the plural quantifiers, like in natural languages, and then use them for interpreting the existential quantifiers of monadic second order logic.

Boolos concerns about higher order logic in general seem to be oriented on its ontological enatailmens and its topic neutrality (on this topic see Linnebo03). Since second order's quantifiers range over sets, it seems that second order logic, as opposite to first order, is not topic neutral; in this case we must remember that Quine's criterion is formulated in terms of first order languages, and it can not be automatically assumed that it can be transfered to second order. Boolos suggests we have to separate what a theory is ranging on from its argument: arithmetics, for instance, can be built not only to concern numbers, but also the addition operation. Is it possible to give a clarification of the distinction between range ad argument? The range can be considered as a thecnical notion connected to quantification, while the argument recalls the intuitive notion of "concernin something". We can also regard the argument as built from theory's operations (in arithmetics, for instance, the addition and the multiplication) and from the range of the quantifiers if it is required an esplicit limitation. In a theory where the quantification is absolutely free, the range will not be a part of the argument.

3 Questions and answers

Now, let's see how to use plurals in contexts, like neo-logicism, requiring second order logic. Monadic second-order logic is the subsystem of second-order logic which admits only quantification over properties, not over (polyadic) relations. The problem with the abstraction principles used in the neologicism is that they uses dyadic second-order logic, so it seems that this kind of plural logic alone does not have sufficient expressive power to accommodate the needs of neo-logicism. The neo-logicist may attempt to solve this problem by regarding equinumerosity as a primitive logical quantifier or by simulating dyadic second-order quantification in some suitable extension of plural first order logic. The "trick", in these extension, is to introduce a pairing function on the domain such that it is possible to quantify over dyadic relations by using plural quantification over ordered pairs. To do this we define a pairing function on a given domain B as a $1 \rightarrow 1$ function from $B \times B$ to B. A theory admits the pairing if there exists a pairing function definable within it: that is, if there exists a formula $A(x, y, z)$ such that in every model M of the theory there exists a pairing function F on the domain of M such that for every assignment s on M, $M, s \models A(x, y, z)$ if and only if $f(s(x), s(y)) = s(z)$.

However, does the use of ordere pairs solve the problem? If it technically works, even though could be questioning about the legitimacy of the extension of the logical sistem to account for the pairs, many concerns arise about the real meaning and the ontological commitmens of these new entities. In fact, as Peter Simons pointed out:

> ordered pairs are in the eyes of some, myself included, weird entities and under as much suspicion as classes or other dubious characters. Certainly the idea that we can use ordered pairs to explicate the idea of order seems to me completely back to front, but that is not at issue here. Rather we may doubt whether the fourth term of the comparison

monadic predicate : plurality of individuals : : dyadic predicate : plurality of ordered pairs

is indeed as transparent as it needs to be when we are weighing carefully the metaphysical implications of the interpretations of a logical theory [5, p. 261].

If these technical and philosophical issues seem to cut the possibility to use plurals in the neofregean context, they nonetheless bring our attention to the greater problem about the logical status of second order. Remember that, for instance, Nelson Goodman used first order by treating classes as individuals; this seems like a trick. Plurals, on the other hand, play an important dual role. First, If we use them we honestly mantain that we don't want to commmit ourself to classes by remaining at first order. Second, we also realize that there exists a level of complexity we must account for. The concern with second order could be ontological — I don't want to be compormised with classes — or meta-logical. However, second order is able to describe complex states of affairs: to say with a joke, if the world is messed up it's not second order's fault. We said above that Lindström Theorem draws a line between the "pure logicality" of first order logic and the "mathematicality" of second order logic. Is the validity of completeness, compactness and Löwenheim-Skolem Theorem the only qualification we need to call a formal system "logic"? They could well be considered only strong *desiderata*. We know that limitative theorems like incompleteness clearly point out how higher order systems lack of the adequacy between sintax and semantics. Löwenheim theorem, however, shows that even at the first order there is a gap between the twos. So the lacking of expressive power of first order rapresented by the lacking of categoricity, could well be considered an important fault too. Technical solutions like the one we saw before or all the works on the semantical aspects (Henkin semantics etc.), the use of infinitary logics, etc. at the end lead to the same a basic point: a sort of incommensurability of the two logics. At this point I think it is necessary a wider and more philosophical reflection. To be provocative it could be asked what is the real meaaing of the question wehater oe not second order logic is to be called logic. Following an insight of Nelson Goodman we can shift from the question "what is logic" to the question "when is logic". If this argument is sound so it could be sound also a metatheorical shifting: from a uniquely deductive orientiation, to a orientation more semantic and interpretative. In this sense the relationship between language and world gains importance. What about the fact that logc should be decontextualized and free of ontological entailments? This is true, but in the act of contextualizing it is unavoidable the intervention of semantics. In this case second order is a formidable tool since it provides better models for important aspect of the world, like mathematics. If second order does not belong to logic it belongs at least to mathematical logic.

I think that a reflection on the nature of the distinction between what is logic and what is not logic is a necessary outcome of the study of second order logic. I think this reflection transcends technical issues and entails a careful philosphical reconsideration of key concepts like validity, provability, truth, completeness and cathegoricity. I think that the results on the status and on the differences among different orders of logic are solid and well grounded. I think, however, that these very results push us to overcame the sharp boundaries and to hold a different philosphical position towards logic. It's not only a problem of names. If I want to call scond order a proper logic is because I don't fully understand whay limitative theorems and their logical and philosphical entailments are to be considered more logical than the ones of second order. For this reason I think the problem is philosophical. I dont' want to argue on the results. I want to argue on the meaning of these results. Why is "more logical" completeness compared with cathegoricity? What is the meanig of beeing logical? Maybe if we start from a less narrow point of view on this question, we can try to find a better and new understanding of what logic is and what it is for.

BIBLIOGRAPHY

[1] G. Boolos. *Logic, Logic and Logic*. Harvard University Press, Cambridge, MA 1998.
[2] O. Linnebo. Plural quantification exposed. *Noûs*, 37 (1): 71–92, 2003.
[3] W.V.O. Quine. *Philosophy of Logic*. Harvard University Press, Cambridge, MA 1970
[4] S. Shapiro. *Foundations without Foundationalism*. Clarendon Press, Oxford 1991.
[5] P. Simons. Higher order quantification and ontological commitments. *Dialectica*, 5 (4), 1997.
[6] C. Wright. *Frege's Conception of Numbers as Objects*. Aberdeen University Press. Aberdeen 1983.
[7] C. Wright e B. Hale. *The Reason's Proper Study*. Oxford University Press, Oxford 2001.

Default-assumption consequence relations in a preferential setting: reasoning about normality

G. CASINI, H. HOSNI

1 Introduction

The study of non monotonic (or defeasible) inference is dotted with considerations about normality. Non monotonic conditionals (or defaults) are in fact often attached an intuitive semantics to the effect that "in normal situations if θ then ϕ". This intuitive reading can be represented semantically in terms of an ordering over the set of (propositional) valuations leading to the so-called *preferential semantics* for non monotonic reasoning, see e.g. [9, 5, 6]. The idea here is that an agent performs non monotonic inferences under a two-fold assumption on its knowledge base: on the one hand the agent behaves as if the information at its disposal is "complete", that is to say, *all the relevant information available to the agent at that particular time is explicitly represented in the knowledge base*; on the other hand the agent reasons under the assumption that the current situation is in fact a *normal situation*. In a realistic scenario, however, both assumption could be violated: new or more refined information might become available to the agent to the effect that either some previously held beliefs turns out to be contradicted by new evidence (thus violating "completeness"), or the newly acquired evidence leads the agent to believe that the situation at hand is in fact not normal. In both cases some of the previously inferred conclusions might need to be abandoned. This makes the agent's reasoning essentially *defeasible*.

Thus, under the assumption that an agent is identified with a consequence relation, there is a very tight connection between the non-monotonicity of an intelligent agent's reasoning and its reasoning about normality. This paper intends to shed further light on this connection by characterizing normality in terms of preferential semantics. Given the amplitude of the field we shall restrict ourselves to a particularly relevant class of non monotonic logics, namely those based on Default-assumption consequence relations (Dacr) [7]. The key idea underlying the Dacr approach consists in defining a consequence relation in such a way that the conclusions of a given set of premises are established *modulo* a maxiconsistent subset of a given set of background assumptions. Indeed the role of this latter set is to represent the situation or context K under which a set of sentences Γ can be said to *normally entail* a sentence θ. Since a Dacr is defined relative to a set of background assumptions K, each such set determines a distinct consequence relation.

The purpose of this paper is to identify a principled set of epistemic operations on a set of background assumptions under which a given consequence relation stays fixed. We shall identify a *fixed* set of default assumptions arising in this way with a *normal situation*. Thus our approach builds on well-known results about preferential reasoning to introduce a new characterization of normality in terms of the stability of a given preferential consequence relation.

The paper is organized as follows. We shall begin by reviewing the key notions leading to the correspondence between finite default assumption consequence and preferential reasoning. This will provide us with an adequate setting to move on to the central topic of the paper, namely a preferential characterization of normality. The key step in our formalization consists in characterizing normality in terms of *stable ordering relations*. Building on this intuitions we shall define a normality operator and investigate its formal properties. The final part of the paper is devoted to a discussion of our normality operator in the light of the epistemic operations of Expansion, Contraction and Revision, as defined within the standard AGM approach [1].

2 Preferential Frameworks

We begin by briefly recalling some important features of Default-assumption consequence relations and to point out the correspondence, in the finite case, between those and the family of preferential consequence relations. Since this correspondence has already appeared in the literature [2], we will only outline the key facts, omitting proofs and examples.

To fix the notation, let $L = \{p_1, \ldots, p_n\}$ be a finite set of propositional letters, and let ℓ be the propositional language generated from L in the usual, recursive way. The sentences of ℓ will be denoted by lowercase Greek letters $\alpha, \beta, \gamma, \ldots$, and subsets of ℓ will be denoted by capital Roman letters $A, B, C, \ldots$. Let W be the set of classical (two-valued) valuations on ℓ. Since we shall be interested in *injective models* only, we assume that every element of W is a distinct valuation on L (see e.g. Sec. 3.3. of [7] for a discussion of this assumption). As usual $\vDash$ will denote the classical (Tarskian) consequence relation. For $A \subseteq \ell$, we shall write $[A]_W$ for set valuations in W which satisfy every sentence in A:

$$[A]_W = \{w \in W \mid w \vDash \phi \text{ for every } \phi \in A\}.$$

Note that since L is finite, W is finite too with $|W| = 2^n$, hence given any sentence α we can identify a particular set of valuations (all the valuations which satisfy that sentence), and, conversely, given a set of valuations $V \subseteq W$, we can find a sentence β which is satisfied precisely by those valuations. Thus, there is a bijection between the sentences in ℓ (modulo logical equivalence) and the subsets of W. In the light of this, given a set of sentences A, we denote by A_w the subset of A which is satisfied by the valuation w, that is to say, $A_w = \{\phi \in A \mid w \vDash \phi\}$. Note that most of the results of this paper depend on this finiteness assumption.

2.1 Default-assumption consequence relations

In logic-based AI it is customary to assume that the key features of an intelligent agent's reasoning can be modeled using a consequence relation. In this perspective it is natural to interpret a given set of premises A as the relevant information which is actually available to the agent at a specific time. For example, in the case of a robotic agent performing a navigation task, the set A might encode its sensory data, the map of its current working space and so on. We shall refer to this kind of information as to the *agent's local premisses*. Yet there is another kind of information which plays a fundamental role in modeling the behaviour of intelligent agents, information that is about what is normally the case. Thus, for example, the above mentioned robot might be given information to the effect that "slippery surfaces normally impede correct motion", "colliding with obstacles usually result into failure" and so on. The key feature of the Default-assumption consequence relation approach consists in enabling the agent to take into account such *default information* when drawing conclusion from its local premisses. Of course local premises and default information can interact (and in particular they might conflict) within any reasoning task performed by the agent. This is why such an interaction is subject to the constraint of maxi-consistency:

DEFINITION 1 (Maximally A-consistent sets). Given two sets of sentences, K and A, we say that a set K' is a *maximally A-consistent* subset of K iff K' is consistent with A and for no K'' s.t. $K' \subset K'' \subseteq K$, K'' is consistent with A.

Given a set of local premises A and a set of default information K, the maximally A-consistent subsets of K represent *all the default-information which is compatible with the agent's knowledge*. Thus it is natural to think of those sets as representing what the agent might expect or presume to be true in a situation in which A holds. This motivates the following definition.

DEFINITION 2 (Default-assumption consequence relation). ϕ is a default-assumption consequence of the set of premises A, given a set of default-assumptions K, (written $A \mathrel{\vert\!\sim}_K \phi$) if and only if ϕ is a classical consequence of the union of A and every maximally A-consistent subset of K:

$$A \mathrel{\vert\!\sim}_K \phi \Leftrightarrow A \cup K' \vDash \phi, \text{ for all maximally } A\text{-consistent } K' \subseteq K.$$

2.2 Preferential semantics

We now put forward the correspondence between Default-assumption consequence relations and the so-called preferential semantics for non monotonic reasoning. The central idea involved in this latter is that of generalizing classical (tarskian) semantics by allowing an ordering over the set of valuations (models) of the language. Thus, given a set of premises A, the set of its preferential consequences is defined as the set of the classical consequences of the *preferred* models classically satisfying A. To put this more precisely, let $\delta \subseteq W \times W$ be an irreflexive and transitive binary relation (indeed a strict order). As usual we shall write $w <_\delta v$ for $(w, v) \in \delta$. The intuitive interpretation of $w <_\delta v$ is that the situation represented by w is *more normal* than (preferred to) the one represented by v. This leads to the following definition.

DEFINITION 3 (Preferential consequence relation). $A \mathrel{\vert\!\sim}_\delta \phi$ if and only if ϕ

is classically satisfied by every δ-minimal valuation in $[A]_W$:

$$A \mathrel{\mid\!\sim}_\delta \phi \Leftrightarrow w \vDash \phi, \forall w \in min_\delta([A]_W).$$

where $min_\delta(U)$ is the set of *minimal valuations* in U with respect to δ, that is:

$$min_\delta(U) = \{w \in U \mid \not\exists v \in U, \text{s.t. } v <_\delta w\}.$$

Note that the finiteness and the ordering of our structure guarantees the existence of such a set.

In order to recall the key fact of this section we need to define the notion of *generated strict order*.

DEFINITION 4 (Generated strict order). Given a set K of sentences, we say that δ_K is the strict order generated by δ if and only if $\delta = \{(w,v) \in W \times W \mid K_w \supset K_v\}$.

As already pointed out above, the correspondence between preferential injective models and (what amounts to) default-assumption consequence relations is known in the literature at least since [2], where the following representation result is proved.

THEOREM 5. *(Makinson, Freund)[6, 2]*

1. *Given a default-assumption consequence relation $\mathrel{\mid\!\sim}_K$, we can define an injective preferential consequence relation $\mathrel{\mid\!\sim}_{\delta_K}$ such that $A \mathrel{\mid\!\sim}_{\delta_K} \phi$ holds just if $A \mathrel{\mid\!\sim}_K \phi$ and conversely,*

2. *given an arbitrary injective preferential consequence relation $\mathrel{\mid\!\sim}_\delta$, we can define a default-assumption consequence relation $\mathrel{\mid\!\sim}_K$ such that $A \mathrel{\mid\!\sim}_K \phi$ holds just if $A \mathrel{\mid\!\sim}_\delta \phi$.*

Note that Theorem 5 can be immediately extended to preorders (i.e. reflexive, transitive relations) $\varepsilon \subseteq W^2$ via the notion of *generated preorder*:

$$\varepsilon_K = \{(w,v) \in W \times W \mid K_w \supseteq K_v\}$$

(or equally, $\varepsilon = \{(w,v) \in W \times W \mid v \vDash \psi \Rightarrow w \vDash \psi$ for every $\psi \in K\}$). Indeed ε_K extends δ_K since

$$\varepsilon_K = \delta_K \cup \{(w,v) \mid K_w = K_v\}.$$

Summing up, we started off by noting that when modeling "intelligent" agents, we can purposefully distinguish between *local premises* and *default information*. The Default-assumption consequence relation approach formalizes this intuition by means of maxi-consistent reasoning, as we saw in definition 2. We then introduced preferential consequence relations in definition 3 and recalled the representation result linking the two.

We complete the set up for our characterization of normality by showing that each preferential order δ determines uniquely the preferential consequence relation $\mathrel{\mid\!\sim}_\delta$.

PROPOSITION 6. *Given two strict orders $\delta, \delta' \subseteq W^2$*

$$\delta = \delta' \iff \hspace{0.1cm}\vdash_\delta = \vdash_{\delta'}$$

Proof. The direction from left to right follows directly from Definition 3. As to the converse, assume that $\vdash_\delta = \vdash_{\delta'}$ but $\delta \neq \delta'$. Then there is at least a pair (w, v) such that either $(w, v) \in \delta$ and $(w, v) \notin \delta'$, or $(w, v) \in \delta'$ and $(w, v) \notin \delta$. Assume, without loss of generality, that the first case holds. Let γ be a sentence satisfied only by w and v. Since $(w, v) \in \delta$ and $(w, v) \notin \delta'$ we have that $min_\delta(\{w, v\}) = \{w\}$, while

$$min_{\delta'}(\{w, v\}) = \begin{cases} \{w, v\}, & \text{if } (v, w) \notin \delta' \\ \{v\}, & \text{if}(v, w) \in \delta'. \end{cases}$$

Either way, by the injectivity of the preferential model, we have $\{\phi \mid \gamma \vdash_\delta \phi\} \neq \{\phi \mid \gamma \vdash_{\delta'} \phi\}$, contradicting the hypothesis that $\vdash_\delta = \vdash_{\delta'}$. ∎

Note that by Theorem 5 this extends immediately to default-assumption consequence relations, that is to say, each ordering determines uniquely the corrisponding default-assumption consequence relation.

3 Characterizing normality

It is immediate from definition 2 that distinct choices of a default information set K might give rise to indistinguishable consequence relations $\vdash_K$. If we think of a consequence relation as an agent, this can be intuitively interpreted as saying that given a set of default information K there is a certain amount of "change" that we can operate on a set K itself *while keeping its generated ordering fixed*, that is to say, according to our discussion of preferential reasoning, without altering the *normality* of the situation at hand. Roughly speaking then, our characterization of normality could be viewed as identifying the "epistemic changes" that a default-assumption consequence relation is capable of tolerating before "disgregating".

Given a default assumption set K and a sentence ϕ, we shall say that the generated strict ordering δ_K is *stable with respect to* ϕ just if $\delta_K = \delta_{K \cup \{\phi\}}$.

It so happens that the statement and the proof of many results is greatly simplified if we take reflexive orders as primitives instead of strict orders. This however does not make any conceptual difference, as theorem 9 below shows.

LEMMA 7.

$\delta_K = \delta_{K \cup \{\phi\}}$ *if and only if for every $(w, v) \in \delta_K$ and every w, v s.t. $K_w = K_v$, $v \vDash \phi$ implies $w \vDash \phi$.*

Proof. (Sketch) We shall omit the details of the simple yet rather tedious proof. The key step consists in showing that if $\delta_K \neq \delta_{K \cup \phi}$, then either $\exists (w, v) \in \delta_K$ such that $v \vDash \phi$ does not imply $w \vDash \phi$ or $\exists w, v \in W$ with $K_w = K_v$ such that $v \vDash \phi$ but $w \nvDash \phi$. ∎

LEMMA 8. $\varepsilon_K = \varepsilon_{K \cup \{\phi\}}$ *if and only if $v \vDash \phi$ implies $w \vDash \phi$ for every $(w, v) \in \varepsilon_K$*

Proof. The implication from left to right follows directly from the definition of $\varepsilon_{K\cup\{\phi\}}$.

As to the other direction note that

$$\varepsilon_K = \{(w,v) \mid v \vDash \psi \Rightarrow w \vDash \psi \text{ for every } \psi \in K\}. \tag{1}$$

$$\varepsilon_{K\cup\{\phi\}} = \{(w,v) \mid v \vDash \psi \Rightarrow w \vDash \psi \text{ for every } \psi \in K \cup \{\phi\}\}. \tag{2}$$

Now, since $v \vDash \phi$ implies $w \vDash \phi$ for every $(w,v) \in \varepsilon_K$, then equations (1) and (2) define exactly the same pairs. Thus $\varepsilon_K = \varepsilon_{K\cup\phi}$. ∎

The upshot of Lemma 7 and Lemma 8 is the following:

THEOREM 9. $\delta_K = \delta_{K\cup\{\phi\}}$ *if and only if* $\varepsilon_K = \varepsilon_{K\cup\{\phi\}}$, *that is, if and only if* $v \vDash \phi$ *implies* $w \vDash \phi$, *for every* $(w,v) \in \varepsilon_K$.

As a consequence of Theorem 9 we shall be freely swapping between δ_K and ε_K in what follows.

3.1 The normality operator ▷

Recall from Proposition 6 that every distinct default-assumption consequence relation is semantically represented by a distinct strict preferential order. We now define a preferential model and a corresponding notion of satisfiability, with the desideratum that only those sentences which, if added to K keep $\vdash_K$ fixed, should be satisfied. This satisfiability relation gives us the building block to construct our normality operator.

Let $\mathcal{C}$ be the class of models of the form $\mathfrak{M} = (W, \varepsilon)$, with W and ε as above. We say that $\mathfrak{M}$ satisfies ϕ, written $\mathfrak{M} \Vdash \phi$, just if ϕ is *compatible* with ε, that is

$$\mathfrak{M} \Vdash \phi \quad \text{iff} \quad v \vDash \phi \Rightarrow w \vDash \phi, \forall (w,v) \in \varepsilon. \tag{3}$$

We can now define our normality operator ▷ by putting $K \triangleright \phi$ just if ϕ is satisfied by every model $\mathfrak{M} \in \mathcal{C}$ that satisfies K, in the sense of formula (3):

DEFINITION 10.

$$K \triangleright \phi \quad \text{iff} \quad \forall \mathfrak{M} \in \mathcal{C}, \text{ if } \mathfrak{M} \Vdash \psi, \forall \psi \in K, \text{ then } \mathfrak{M} \Vdash \phi.$$

The next Proposition justifies the intuitive reading of ▷ as a *normality operator* in the light of the above discussion.

PROPOSITION 11.

$$K \triangleright \phi \quad \text{iff} \quad \varepsilon_K = \varepsilon_{K\cup\{\phi\}}.$$

Proof. ($\Rightarrow$): suppose that $K \triangleright \phi$. This amounts to say that $v \vDash \phi \Rightarrow w \vDash \phi$ holds in every preorder ε such that $v \vDash \psi \Rightarrow w \vDash \psi$, for every $\psi \in K$ and every $(w,v) \in \varepsilon$. Since ε_K is one of those preorders, we have $v \vDash \phi \Rightarrow w \vDash \phi$ for every $(w,v) \in \varepsilon_K$. So $\varepsilon_K = \varepsilon_{K\cup\{\phi\}}$.

($\Leftarrow$): Suppose that $\varepsilon_K = \{(w,v) \in W \times W \mid v \vDash \psi \Rightarrow w \vDash \psi$ for every $\psi \in K\} = \varepsilon_{K\cup\{\phi\}}$. Let $\mathfrak{M} = (W, \varepsilon)$ be an arbitrary model in $\mathcal{C}$. If $\mathfrak{M} \Vdash \psi$ for every $\psi \in K$, then every pair $(w,v) \in \varepsilon$ satisfies $v \vDash \psi \Rightarrow w \vDash \psi$ for every $\psi \in K$. But since all those pairs of valuations are in ε_K, it follows that

$\varepsilon \subseteq \varepsilon_K$. Since $\varepsilon_K = \varepsilon_{K \cup \{\phi\}}$, then $v \vDash \phi \Rightarrow w \vDash \phi$ for every $(w,v) \in \varepsilon_K$. It therefore holds that $v \vDash \phi \Rightarrow w \vDash \phi$ for every $(w,v) \in \varepsilon$, that is to say, $\mathfrak{M} \Vdash \phi$. But $\mathfrak{M}$ was an arbitrary model, so we conclude that $K \triangleright \phi$, as required. ∎

It is natural to ask, at this point, which kind of object is the operator $\triangleright$. We shall begin by observing that $\triangleright$ is a Tarskian operator.

PROPOSITION 12. $\triangleright$ *satisfies*

REF: $K \triangleright \phi, \forall \phi \in K$ (*Reflexivity*)

MON: *if* $K \triangleright \phi$ *then* $K \cup \{\psi\} \triangleright \phi$ (*Monotonicity*)

CT: *if* $K \cup \{\psi\} \triangleright \phi$ *and* $K \triangleright \psi$ *then* $K \triangleright \phi$ (*Cut*)

Proof. The proof is straightforward and is omitted. ∎

The following proposition relates $\triangleright$ to the classical consequence relation $\vDash$.

PROPOSITION 13.

⊤: $K \triangleright \top$, *for any tautology* $\top$ *of* ℓ (Tautology).

⊥: $K \triangleright \bot$, *for any contradiction* $\bot$ *of* ℓ (Contradiction).

sLLE: *If* $\vDash \phi \leftrightarrow \psi$ *and* $K \cup \{\phi\} \triangleright \gamma$ *then* $K \cup \{\psi\} \triangleright \gamma$ (Singleton Left Logical Equivalence)

RLE: *If* $\vDash \phi \leftrightarrow \psi$ *and* $K \triangleright \phi$ *then* $K \triangleright \psi$ (Right Logical Equivalence)

Proof. (Sketch) (⊤) is straightforward. Both (sLLE) and (RLE) follow from the fact that $\vDash \phi \leftrightarrow \psi$ implies $\mathfrak{M} \Vdash \phi$ if and only if $\mathfrak{M} \Vdash \psi$. As to (⊥) it is proven by noting that adding a contradiction to K does not affect the maximally A-consistent subsets of K and therefore leaves the generated order unchanged. ∎

Note that since $\triangleright$ aims at characterizing invariance under any "normal refinement" of a default information set, it is only sensitive to *contingent facts* and therefore disregards as uninformative both tautologies (as we remarked above) and contradictions. This latter case can be illustrated by taking $K = \{p,q\}$ and $A = \{\neg q\}$. Clearly there is only one maximally A-consistent subset of K, namely $K_1 = \{p\}$. Let us now add a contradiction to K, so $K' = \{p,q,\alpha \wedge \neg \alpha\}$. Again, there is only one maximally A-consistent subset of K', which is still $K_1 = \{p\}$.

Note also that although default-assumption consequence relations are not closed under substitution of logically equivalent default information sets (see below, with respect to Right Weakening), (sLLE) ensures that $\triangleright$ is closed under singleton substitution.

We now move on to consider the behaviour of $\triangleright$ with respect to the standard propositional connectives. Let us begin with the properties which $\triangleright$ does satisfy.

$$\frac{\Delta \cup \{\phi\} \vartriangleright \gamma \quad \Delta \cup \{\psi\} \vartriangleright \gamma}{\Delta \cup \{\phi \vee \psi\} \vartriangleright \gamma} \quad \text{(Disjunction in the premises (OR))}$$

Let $\mathfrak{M} = (W, \varepsilon)$ be a model. Assume $\Delta \cup \{\phi\} \vartriangleright \gamma$, $\Delta \cup \{\psi\} \vartriangleright \gamma$ and $\mathfrak{M} \Vdash \rho$ for every $\rho \in \Delta \cup \{\phi \vee \psi\}$, which means that for every $(w, v) \in \varepsilon$, if $v \vDash \rho$, then $w \vDash \rho$. Take one of those pairs $(w, v) \in \varepsilon$. We need to check three cases:

1) $v \vDash \phi \vee \psi$ and $w \vDash \phi \vee \psi$.
   Since $w \vDash \phi \vee \psi$, then either $w \vDash \phi$ or $w \vDash \psi$. Hence at least one of $v \vDash \phi \Rightarrow w \vDash \phi$ and $v \vDash \psi \Rightarrow w \vDash \psi$ is satisfied. Either way, $v \vDash \gamma$ implies $w \vDash \gamma$

2) $v \nvDash \phi \vee \psi$ and $w \vDash \phi \vee \psi$.
   The same argument as (1) applies.

3) $v \nvDash \phi \vee \psi$ and $w \nvDash \phi \vee \psi$.
   We have $v \nvDash \phi$, $v \nvDash \psi$, $w \nvDash \phi$ and $w \nvDash \psi$. Then $v \vDash \phi$ implies $w \vDash \phi$ and $v \vDash \psi$ implies $w \vDash \psi$. Hence $v \vDash \gamma$ implies $w \vDash \gamma$.

Summing up, if we assume that $\Delta \cup \{\phi\} \vartriangleright \gamma$, $\Delta \cup \{\psi\} \vartriangleright \gamma$, and $\mathfrak{M} \Vdash \rho$ for every $\rho \in \Delta \cup \{\phi \vee \psi\}$, then, for every pair $(w, v) \in \varepsilon$, $v \vDash \gamma \Rightarrow w \vDash \gamma$ holds, that is $\mathfrak{M} \Vdash \gamma$. So $\Delta \cup \{\phi \vee \psi\} \vartriangleright \gamma$, as required.

Introduction of conjunction $(I\wedge)$: $\{\phi\} \cup \{\psi\} \vartriangleright \phi \wedge \psi$.

If an ordering ε satisfies ϕ and ψ, then for every $(w, v) \in \varepsilon$, if $v \vDash \phi$, then $w \vDash \phi$, and if $v \vDash \psi$, then $w \vDash \psi$. Therefore, for such (w, v), if $v \vDash \phi \wedge \psi$, we have that $v \vDash \phi$ and $v \vDash \psi$, so also $w \vDash \phi$ and $w \vDash \psi$, i.e. $w \vDash \phi \wedge \psi$. This property, together with MON and CT, gives us the AND rule:

$$\frac{K \vartriangleright \phi \quad K \vartriangleright \psi}{K \vartriangleright \phi \wedge \psi}$$

Cautious Introduction of disjunction $(CI\vee)$: $\{\phi\} \cup \{\psi\} \vartriangleright \phi \vee \psi$.

If an ordering ε satisfies ϕ and ψ, that means that for every $(w, v) \in \varepsilon$, if $v \vDash \phi$, then $w \vDash \phi$, and if $v \vDash \psi$, then $w \vDash \psi$. Then, for such (w, v), if $v \vDash \phi \vee \psi$, we have that $v \vDash \phi$ or $v \vDash \psi$, so also $w \vDash \phi$ or $w \vDash \psi$, i.e. $w \vDash \phi \vee \psi$. Note that we need both premises to derive the disjunction. In particular the classical Introduction of disjunction ($\phi \vartriangleright \phi \vee \psi$) is not valid.
To see this, take $(w, v) \in \varepsilon$ s.t. $v \nvDash \phi$, $v \vDash \psi$, $w \nvDash \phi$ and $w \nvDash \psi$, so $v \vDash \phi \vee \psi$ and $w \nvDash \phi \vee \psi$. For this pair $v \vDash \phi \Rightarrow w \vDash \phi$, but $v \vDash \phi \vee \psi \nRightarrow w \vDash \phi \vee \psi$.

Finally, let us look at some of the properties which $\vartriangleright$ does not satisfy.

$$\frac{\vDash \phi \rightarrow \psi \quad K \vartriangleright \phi}{K \vartriangleright \psi} \quad (Right\,Weakening - RW)$$

$$\frac{K \vartriangleright \phi \rightarrow \psi \quad K \vartriangleright \phi}{K \vartriangleright \psi} \quad (Modus\,Ponens - MP)$$

To see that $\triangleright$ satisfies neither of the above, let (w,v) be $v \nvDash \phi$, $v \vDash \psi$, $w \nvDash \phi$, and $w \nvDash \psi$. This pair satisfies $v \vDash \phi \Rightarrow w \vDash \phi$, both v and w satisfy $\phi \rightarrow \psi$, but $v \vDash \psi \nRightarrow w \vDash \psi$.

Let us briefly comment on these two negative results. The failure of (MP) boils down to the fact that the normality closure of a sentence given a set of default information does not obey the laws of material implication. Given the *default* nature of reasoning based on normality, this is hardly surprising. As to the failure of (RW) it implies that $\triangleright$ does not satisfy Supraclassicality (that is, $K \vDash \phi$ implies $K \triangleright \phi$). This is a real drawback only if we require the information in K to be *undefeasible* (since in this case, the agent would fail to expect as normal what is logically entailed by its information). However, K is meant to represent *defeasible information*, that is to say, information which the agent has no reason to take for granted.

4 Default-revision

In the last section we have seen how, given a default-assumption consequence relation $\vdash_K$ characterized by the generated preorder ε_K, an operator $\triangleright$ can be introduced to characterize those sentences whose addition to the default-assumption set K can be "absorbed" or "tolerated" by the consequence relation itself. This naturally leads to consider the kinds of "epistemic change" allowed by $\triangleright$. To do so, we shall rely on the standard AGM approach to theory change, [1, 8].

The AGM model, which aims at characterizing the epistemic behaviour of ideally rational agents, is centered around two key constraints: Logical closure and consistency. The former imposes that, given a set of sentences K, an agent should behave as if it accepted not only the information contained in K but also all its (classical) logical consequences. The latter amounts to the requirement that no logical inconsistency should arise after the correct instantiation of any of the three epistemic operations of expansion, contraction and revision.

It is well known, however, that a set of default-assumptions K is closed under classical consequence, any default-assumption consequence relation built up from K collapses into classical consequence (see, e.g. [7], Theor. 2.7.). In order to avoid this, we shall weaken the requirement of logical closure to *closure under the normality operator*:

$$C_\triangleright(A) = \{\alpha | A \triangleright \alpha\}.$$

In what follows, it will be useful to make the following terminological distinction. We shall refer to the finite set of default-assumptions K which determines a default-assumption consequence relation as the *default base*, while we shall call *default sets* those default bases D which are closed under the normality operator, that is to say such that $D = C_\triangleright(D)$.

Besides the logical constraints of closure and consistency, the standard approach to theory revision adopts, as a heuristic constraint, the principle of so-called *informational economy*. Roughly speaking, this amounts to requiring that any epistemic operation performed by an agent should result in the smallest possible "loss of information". This heuristic principle guides both the formalization of the normative postulates for expansion, contraction and revision, as well as the explicit constructions, such as the Epistemic Entrenchment approach [4]. In the reminder we shall focus on the normality operators of Expansion, Contraction and Revision, leaving other constructions to future work.

4.1 Expansion

Expansion formalizes the epistemic operation of simply adding a sentence to a default set D. That is, if an agent acquires the information that normally α holds, then it will simply add α to D and close this set under $C_\rhd$:

$$D^+_\alpha := C_\rhd(D \cup \{\alpha\})$$

It is straightforward to see that normality-expansion thus defined, satisfies the basic AGM postulates for expansion (+1)-(+5) (see [3]):

(+ 1) D^+_α is a default set.
(+ 2) $\alpha \in D^+_\alpha$.
(+ 3) $D \subseteq D^+_\alpha$.
(+ 4)   If $\alpha \in D$, then $D^+_\alpha = D$.
(+ 5)   If $D \subseteq H$, then $D^+_\alpha \subseteq H^+_\alpha$.

The last postulate

(+ 6) For all default sets D and all sentences α, D^+_α is the smallest default set that satisfies (+1) − −(+5)

deserves a little more attention. (+ 6) is known as the *minimality postulate*. It imposes that the new default set does not contain any extra information with respect to the addition of α to D. To see that (+ 6) holds, let H be such that $H \subset D^+_\alpha$. Assume H satisfies (+ 2) and (+ 3), that is $D \cup \{\alpha\} \subseteq H$. If H satisfies also (+ 1), then $H = C_\rhd(H)$; given $D^+_\alpha = C_\rhd(D \cup \{\alpha\})$, we have $D^+_\alpha \subseteq H$, by the monotonicity of $\rhd$, contradiction.

4.2 Contraction

Normality-contraction amounts to removing a sentence from a given default set. As in the AGM case, the problem of contraction is two-fold: not only a specific sentence needs removing from a default set, we also need to make sure that its deduction is blocked within the resulting contracted set. Of course there can be many ways to achieve this. In the spirit of the AGM approach we shall not describe here any particular procedure for normality-contraction, but only the desiderata that any such function should reasonably satisfy.

Given a default set D and a sentence α let *contr* be defined as follows:

$$contr_D(\alpha) = \begin{cases} \emptyset & \text{if } \alpha \notin D \text{ or } \alpha = \top \text{ or } \alpha = \bot, \\ B \subseteq D \text{ s.t. } \alpha \notin C_\rhd(D - B) & \text{otherwise} \end{cases}$$

Note that the clause relative to $\alpha = \top$ and $\alpha = \bot$ is motivated by the fact that neither tautologies nor contradictions are relevant to the normality closure. We can now define normality-contraction by letting

$$D^-_\alpha := C_\rhd(D - contr_D(\alpha)).$$

As in the case of normality expansion, it is straightforward to prove that normality contraction satisfies the basic AGM postulates for contraction (-1)–(- 4), which in the present setting read as follows:

(- 1) D_α^- is a default set.
(- 2) $D_\alpha^- \subseteq D$
(- 3) If $\alpha \notin D$, then $D_\alpha^- = D$
(- 4) If $\not\models \alpha$, then $\alpha \notin D_\alpha^-$

A word of caution is due in the case of

(- 5) If $\alpha \in D$, then $D \subseteq (D_\alpha^-)_\alpha^+$

This postulate, usually referred to as "recovery", is by far the most controversial principle of the AGM model and in general it is not satisfied by the normality contraction operator. It will therefore be pleasing to note, in the next section, that its failure does not affect the desired properties of the normality revision operator.

Also the last basic AGM postulate for contraction requires some attention:

(- 6) If $\models \alpha \leftrightarrow \beta$, then $D_\alpha^- = D_\beta^-$

The intuition is clearly that normality contraction should be well-behaved with respect to classical equivalence. Indeed it follows from (RLE) that $\beta \notin D_\alpha^-$ and $\alpha \notin D_\beta^-$, but this alone does not guarantee that $D_\alpha^- = D_\beta^-$. However this can be ensured in various ways depending on the particular construction at hand. The details of such constructions will be discussed in a follow-up paper.

4.3 Revision

The normality-revision of a default set D consists in adding to D a sentence which is potentially inconsistent with D itself. Given that we require any epistemic change to preserve consistency, the problem amounts to defining a function *, which takes a default set D and a sentence α as inputs and returns a new consistent default set containing α. Here, as in the AGM approach, consistency means classical consistency.

First of all we need to make sure that our closure operator is *consistency preserving*. The fact that $K \triangleright \bot$ for any contradiction $\bot$ clearly implies that the normality closure $C_\triangleright$ is not consistency preserving. Thus, in order to meet our desideratum, we need to suitably constrain $\triangleright$. This motivates the definition of

$$K \triangleright' \phi \Leftrightarrow K \triangleright \phi \text{ and } \exists w \in W \text{ s.t. } w \models \sigma, \forall \sigma \in K \cup \{\phi\}$$

that is:

$$K \triangleright' \phi \Leftrightarrow \forall \mathfrak{M} \in \mathcal{C}, \text{ if } \mathfrak{M} \Vdash \psi, \forall \psi \in K, \text{ then } \mathfrak{M} \Vdash \phi$$
$$\text{and } \exists w \in W \text{ s.t. } w \models \sigma, \forall \sigma \in K \cup \{\phi\}$$

The following result shows that, in the case of a consistent default base, the constrained closure operator does preserve consistency, but it does so at the price of eliminating only contradictions, which are irrelevant to the construction of maximally consistent sets.

PROPOSITION 14.

$$C_{\rhd'}(K) = \begin{cases} C_{\rhd}(K) - \bot & \text{if } K \text{ is consistent} \\ \ell & \text{otherwise} \end{cases}$$

Proof. If K is inconsistent, then there is no valuation satisfying it and therefore it vacuously entails ℓ. Thus $\rhd'$ is an explosive operator.

Otherwise, if K is consistent, there is a valuation w such that $w \vDash \psi$ for any $\psi \in K$. Since $K_w = K$, then $w \leq_{\varepsilon_K} v$ holds for every $v \in W$. We have to show that $C_{\rhd'}(K) = C_{\rhd}(K) - \bot$. $C_{\rhd'}(K) \subseteq C_{\rhd}(K) - \bot$ is immediate, since $C_{\rhd'}(K) \subseteq C_{\rhd}(K)$ and $\bot \notin C_{\rhd'}(K)$ hold. To show that $C_{\rhd}(K) - \bot \subseteq C_{\rhd'}(K)$, assume that there is a $\psi \neq \bot$ s.t. $\psi \in C_{\rhd}(K)$. Since $\psi \neq \bot$, there must be a valuation $v \in W$ such that $v \vDash \psi$. Moreover, the fact that $\psi \in C_{\rhd}(K)$ forces $t \vDash \psi$ for every $t \leq_{\varepsilon_K} v$. Now it follows from $w \leq_{\varepsilon_K} v$, that $w \vDash \psi$, i.e. $w \vDash \sigma, \forall \sigma \in K \cup \{\phi\}$. Hence we get $\psi \in C_{\rhd'}(K)$. ∎

Consistency preservation and explosion make $\rhd'$ an intuitively appealing operator in the characterization of normality: it would surely be counterintuitive if an ideally rational agent could hold a sentence as a *normal contradiction*.

It is immediate to note that $\rhd'$ satisfies exactly the same properties satisfied by $\rhd$, apart, obviously, from $(\bot)$. It is likewise easy to see that $\rhd'$ behaves exactly as $\rhd$ insofar as the properties of expansion and contraction are concerned. As a consequence we can use $C_{\rhd'}$ to characterize *normality revision* via the so-called Levi Identity (see e.g. [3] p.69), which defines revision by means of a combination of expansion and contraction:

$$D_\alpha^* = (D_{\neg\alpha}^-)_\alpha^+.$$

The Levi Identity formalizes the two-step operation of revision, where the initial contraction guarantees the consistency of the result while the final expansion guarantees the success of the revision.

A major result in the field points out that the revision operator defined via the Levi Identity satisfies the AGM postulates. Of course, in the AGM model this result is obtained by taking classical consequence as closure operator. However, given our goal of defining a revision operation which preserves the consistency of the default set D (i.e. $(*5) : \bot \in D_\alpha^*$ iff $\vDash \neg\alpha$), we cannot rely on the shortcut provided by the Levi Identity. This is an immediate consequence of the fact that $\rhd$ fails to satisfy (RW). More specifically, let us take a singleton default set $K = \{\beta\}$, and assume that $\vDash \beta \to \neg\alpha$ and $\nvDash \neg\alpha \to \beta$ (i.e. $\neg\alpha$ is a consequence of β, but they are not logically equivalent). Since (RW) is not satisfied, we have that $\neg\alpha \notin C_{\rhd'}(\beta)$. If we add α to K, since $\neg\alpha \notin C_{\rhd'}(\beta)$, we obtain $C_{\rhd'}(K)_{\neg\alpha}^- = C_{\rhd'}(K)$, i.e. $C_{\rhd'}(K)_\alpha^* = C_{\rhd'}(K)_\alpha^+$. Hence we get a default base $K' = \{\beta, \alpha\}$ such that $\alpha \wedge \neg\alpha \notin C_{\rhd'}(K')$, but, by $(I\wedge)$, $\alpha \wedge \beta \in C_{\rhd'}(K')$. But since $\vDash \beta \to \neg\alpha$, the former cannot be consistent.

We conclude that, in order to define a revision operation which preserves the consistency of default sets, we can indeed contract with respect to the classical relation $\vDash$, but we need to expand with respect to $\triangleright'$, since this latter is required to ensure the success of expansion of a default base Δ.

Let us now define the function $\div$ which, given a default set D and a formula α to be contracted as arguments, returns a default set $D_\alpha^\div$ such that $D_\alpha^\div \nvDash \alpha$ (instead of $D_\alpha^- \ntriangleright' \alpha$ as above).

The desired properties of this contraction function are then restated as follows:
- ($\div$ 1) $D_\alpha^\div$ is a default set.
- ($\div$ 2) $D_\alpha^\div \subseteq D$
- ($\div$ 3) If $D \nvDash \alpha$, then $D_\alpha^\div = D$
- ($\div$ 4) If $\nvDash \alpha$, then $\alpha \notin D_\alpha^\div$
- ($\div$ 5) If $\alpha \in D$, then $D \subseteq (D_\alpha^\div)_\alpha^+$
- ($\div$ 6) If $\vDash \alpha \leftrightarrow \beta$, then $D_\alpha^\div = D_\beta^\div$

The function $\div$ can also be defined in terms of a classical contraction function $\dot{-}$ and the the classical operator Cl (see [3]):

$$D_\alpha^\div := Cl(D)_\alpha^{\dot{-}} \cap D.$$

The following Proposition ensures that we characterized indeed the intended notion of contraction. The proof is straightforward and will therefore be omitted.

PROPOSITION 15. *If the classical contraction function $\dot{-}$ satisfies $(K\dot{-}1)$-$(K\dot{-}4)$ and $(K\dot{-}6)$, then the default contraction function $\div$ satisfies $(\div 1)$-$(\div 4)$ and $(\div 6)$*

On the other hand we can take a default contraction function $\div$, built on top of the classical contraction operation $\dot{-}$ and recover the default-revision operation by means of the Levy-identity:

$$D_\alpha^* := (D_{\neg\alpha}^\div)_\alpha^+$$

The new revision operation satisfies the AGM logical postulates.

THEOREM 16. *If the default contraction function $\div$ satisfies $(\div 1)$-$(\div 4)$ and $(\div 6)$, and the normality expansion function $+$ satisfies $(+ 1)$-$(+ 6)$, then the normality revision function $*$, defined via the Levi identity, satisfies $(* 1)$-$(* 6)$.*

Proof. The proof is a straightforward adaptation from the original (see, e.g. [3] theor.3.2). (Note that the satisfaction of ($\div$ 5) is not required for the representation theorem.) ∎

5 Conclusions

In this paper we proposed a characterization based on Default-assumption consequence relations of the sorts of changes that an idealized agent can be expected to perform on its default information. By relying on the correspondence between Dacr and preferential semantics we have described some particularly important operations of "change" leading to the notion of normality closure of a sentence with respect to a set of default information $\triangleright$. Finally we have formulated the problem of revising a set of background assumptions — default revision — and highlighted its close correspondence to the standard AGM account of theory revision. The normality operator seems also promising as a tool for of providing a computationally efficient characterization of default revision, as we aim to show in future work.

BIBLIOGRAPHY

[1] C. E. Alchourròn, P. Gärdenfors and D. Makinson. On the logic of theory change: Partial meet contraction and revision functions. *Journal of Symbolic Logic*, 50: 510–530, 1985
[2] M. Freund. Preferential Reasoning in the Perspective of Poole Default Logic. *Artificial Intelligence* 98 (1–2): 209–235, 1998
[3] P. Gärdenfors. *Knowledge in Flux*. MIT Press, Cambridge, MA 1988.
[4] P. Gärdenfors, and D. Makinson. Revisions of knowledge systems using epistemic entrenchment. In M. Vardi (ed.), *Proceedings of the Second Conference on Theoretical Aspects of Reasoning about Knowledge*, Morgan Kaufmann, 1988, pages 83-95.
[5] S. Kraus, D. Lehmann and M. Magidor. Nonmonotonic reasoning, preferential models and cumulative logic. *Artificial Intelligence* 44: 167–207, 1990.
[6] D. Makinson. General patterns in nonmonotonic reasoning. In D. M. Gabbay, C. J. Hogger and J. A. Robinson (eds.), *Handbook of logic in artificial intelligence and logic programming* (vol. 3), Oxford University Press, Oxford 1994, pages 35-110.
[7] D. Makinson, *Bridges from Classical to Nonmonotonic Logic*. King's College Text in Computing, volume 5, King's College Publications, London 2005.
[8] H. Rott. *Change, Choice and Inference: A Study of Belief Revision and Nonmonotonic Reasoning*. Oxford University Press, Oxford 2001
[9] Y. Shoham. A semantical approach to nonmonotonic logics. In M. Ginsberg (ed.), *Readings in Nonmonotonic Reasoning*, Morgan Kaufmann, 1987, pages 227–250.

On the finitization of Priorean linear time

BIANCA BORETTI, SARA NEGRI

1 Introduction

The birth of temporal logic is closely connected with the name of Arthur Prior and his interest in classical philosophical problems, such as the conflict between fatalism and free will. The study of the answers given to this question by ancient philosophers including Aristotle and Diodorus Cronus, and medieval ones such as Ockham and Peter de Rivo gave him the idea to develop a logic of time on the model of the then nascent modal logic: Temporal operators for future and for past were to be formulated in analogy to the modalities $\Box$ and $\Diamond$ of necessity and possibility. Further operators were later introduced to denote the next and the previous moment (von Wright [22], Scott [20]). The introduction of the "until" and "since" operators into linear-time logic by Kamp [11] allowed the formulation of a more expressive temporal logic.

The importance of temporal logic increased greatly as a consequence of its application to computer science. Several versions of temporal logic have been considered, each reflecting the properties of the intended frames (linear, branching, circular, ...) or the presence or absence of past operators. In particular, Linear Time Logic (LTL) is a temporal logic without past operators that corresponds to discrete frames isomorphic to the natural numbers.

Propositional linear time logic is decidable, as shown for instance by Kesten et al. [12] with tableau methods, but the inherent presence of induction makes the development of a finitary proof system problematic. Decidability has not been, so far, established through terminating proof search for the whole logic, but only for fragments, as in the tableau system proposed by Schmitt and Goubalt-Larrecq [19]. Whereas tableau systems involve non-local rules, that is, global correctness conditions, systems of natural deduction or sequent calculus for full LTL typically either require a rule with an infinite number of premisses or are not normalizable/cut free ([5], [8]). Several attempts have been made in order to obtain a finitary cut-free calculus for LTL . A significant indirect contribution is found in [10], where the finite model property is used to give an upper bound on the number of premisses of an infinitary rule, formally similar to the one used in temporal logic, for the logic of common knowledge. The semantic method allows to prove completeness for the calculus but not cut elimination.

We have a different goal in this work: Instead of trying to finitize the calculus for linear time, we identify a finitary fragment of the system. We use the method of internalization of the possible world semantics within the syntax of sequent calculi, as developed by Negri in [14, 15]. A labelled system G3LT for Priorean linear time is introduced, in Section 2.1, by adding to the basic calculus for temporal logic the mathematical rules that correspond to the properties of the intended class of frames. In particular, discreteness is given by an infinitary rule that states: If x is less than y, then x is the predecessor of y, or it is the predecessor of the predecessor of y, or ... and so on. Structural properties, such as the admissibility of weakening, contraction, and cut, are proved syntactically in Section 2.2, along the guidelines of the general method by Negri and von Plato [16].

A weaker system $G3LT_f$ is formulated in Section 3 by replacing the infinitary rule with two finitary counterparts that permit the splitting of an interval $[x, y]$ with an immediate successor of x and an immediate predecessor of y, respectively. Every sequent derivable in the finitary system is derivable in the infinitary one. The converse fails, but we identify a fragment of G3LT for which conservativity with respect to $G3LT_f$ is proved.

We conclude with a discussion of related literature.

2 A sequent calculus for Priorean linear time

Among the various versions of linear time logic, we consider here the calculus proposed by Prior in [18] (system 7.2, p. 178), which is characterized by the presence of both future and past operators: In addition to the traditional **G**, "it is and always will be", and **H**, "it is and always has been", also the next and last instant **T**, "tomorrow", and **Y**, "yesterday", are considered. If past operators are dropped, we obtain a system corresponding to the one commonly called unary LTL.

The view, developed after Prior's work, of temporal logic as a special modal logic, makes the use of Kripke semantics very natural. Kripke frames are interpreted as ordered sets of instants in the flow of time, with the accessibility relation being the order of temporal precedence. The syntax of temporal logic can thus be developed within the *method of internalization of Kripke semantics* for modal and non-classical logics: Semantic elements, such as possible worlds and accessibility relations, appear on a par with logical constants in systems of inference and the rules are directly generated from the semantic explanation of the logical constants. The systems of inference that result from this internalization are called *labelled systems*. From the extensive literature on labelled and hybrid systems (cf. [7] and the references discussed in [15]), we shall follow the method introduced by the second author in [14]. The treatment of temporal logic requires nontrivial extensions of the basic method and we shall therefore proceed with a self-contained presentation rather than relying on a general background (that can however be found in section 1 of [15]).

2.1 Logical and mathematical rules

Our sequent calculus for linear time is obtained as follows: The starting point is the cut- and contraction-free sequent calculus G3 that was introduced by Ketonen in the 1940's and recently systematically presented in [21]. In [16, 17] and in [13] a general method was presented for extending the basic logical sequent calculus without losing the structural properties such as admissibility of cut: Axioms for specific theories are suitably converted into inference rules to be added to the logical sequent calculus while preserving all the structural properties of the basic sequent system. For systems with internalized Kripke semantics the syntax of the calculus has to be enriched with labels and relations: Every formula in a sequent $\Gamma \Rightarrow \Delta$ is either a relational atomic formula $x \leq y$, $x \prec y$, $x = y$, or a labelled formula $x : A$. Intuitively, relational atoms and labelled formulas are the counterpart of the accessibility or equality relations and of the forcing relation $x \Vdash A$ of Kripke models, respectively.

The rules for the propositional connectives are analogous to the standard rules, with the active and principal formulas all marked by the same label x. For temporal operators, the rules are obtained from the meaning explanations in terms of their relational semantics:

$x \Vdash \mathbf{G}A$ (resp. $x \Vdash \mathbf{H}A$) iff for all y, $x \leq y$ (resp. $y \leq x$) implies $y \Vdash A$

$x \Vdash \mathbf{F}A$ (resp. $x \Vdash \mathbf{P}A$) iff for some y, $x \leq y$ (resp. $y \leq x$) and $y \Vdash A$

$x \Vdash \mathbf{T}A$ (resp. $x \Vdash \mathbf{Y}A$) iff for all y, $x \prec y$ (resp. $y \prec x$) implies $y \Vdash A$

The left-to-right direction in the explanation above justifies the left rules, the right-to-left direction the right rules. The rôle of the quantifiers is reflected in the variable conditions for rules $R\mathbf{G}$, $L\mathbf{F}$, $R\mathbf{T}$, $R\mathbf{H}$, $L\mathbf{P}$ and $R\mathbf{Y}$ below.

The logical rules for the calculus are given in Table 1. Observe that initial sequents are restricted to labelled atomic formulas $x : P$ or relational atoms At. This feature, common to all G3 systems of sequent calculus, is needed to ensure invertibility of the rules (Lemma 5) and other structural properties. In addition to the logical rules of Table 1, we have mathematical rules that correspond to the frame properties of accessibility relations.

Rules for Equality

$$\frac{x = x, \Gamma \Rightarrow \Delta}{\Gamma \Rightarrow \Delta}\ EqRef$$

$$\frac{y : P, x = y, x : P, \Gamma \Rightarrow \Delta}{x = y, x : P, \Gamma \Rightarrow \Delta}\ EqSubst \qquad \frac{At(y), x = y, At(x), \Gamma \Rightarrow \Delta}{x = y, At(x), \Gamma \Rightarrow \Delta}\ EqSubst_{At}$$

Rules for the Order Relation

$$\frac{x \leq z, x \leq y, y \leq z, \Gamma \Rightarrow \Delta}{x \leq y, y \leq z, \Gamma \Rightarrow \Delta}\ Trans \qquad \frac{x \leq x, \Gamma \Rightarrow \Delta}{\Gamma \Rightarrow \Delta}\ Ref$$

Rules for the Successor Relation

$$\frac{y = z, y \prec x, z \prec x, \Gamma \Rightarrow \Delta}{y \prec x, z \prec x, \Gamma \Rightarrow \Delta}\ UnPred \qquad \frac{y = z, x \prec y, x \prec z, \Gamma \Rightarrow \Delta}{x \prec y, x \prec z, \Gamma \Rightarrow \Delta}\ UnSucc$$

$$\frac{y \prec x, \Gamma \Rightarrow \Delta}{\Gamma \Rightarrow \Delta}\ L\text{-}Ser \qquad \frac{x \prec y, \Gamma \Rightarrow \Delta}{\Gamma \Rightarrow \Delta}\ R\text{-}Ser \qquad \frac{x \leq y, x \prec y, \Gamma \Rightarrow \Delta}{x \prec y, \Gamma \Rightarrow \Delta}\ Inc$$

Rules L-Ser and R-Ser have the condition that y is not in the conclusion.

The order relation $x \leq y$ is defined as the transitive and reflexive closure of the immediate successor relation $x \prec y$, that is,

Initial sequents:

$x : P, \Gamma \Rightarrow \Delta, x : P$ $\qquad\qquad At, \Gamma \Rightarrow \Delta, At$

Propositional rules:

$$\frac{x : A, x : B, \Gamma \Rightarrow \Delta}{x : A\&B, \Gamma \Rightarrow \Delta} L\& \qquad \frac{\Gamma \Rightarrow \Delta, x : A \quad \Gamma \Rightarrow \Delta, x : B}{\Gamma \Rightarrow \Delta, x : A\&B} R\&$$

$$\frac{x : A, \Gamma \Rightarrow \Delta \quad x : B, \Gamma \Rightarrow \Delta}{x : A \vee B, \Gamma \Rightarrow \Delta} L\vee \qquad \frac{\Gamma \Rightarrow \Delta, x : A, x : B}{\Gamma \Rightarrow \Delta, x : A \vee B} R\vee$$

$$\frac{\Gamma \Rightarrow \Delta, x : A \quad x : B, \Gamma \Rightarrow \Delta}{x : A \supset B, \Gamma \Rightarrow \Delta} L\supset \qquad \frac{x : A, \Gamma \Rightarrow \Delta, x : B}{\Gamma \Rightarrow \Delta, x : A \supset B} R\supset$$

$$\frac{}{x : \bot, \Gamma \Rightarrow \Delta} L\bot$$

Temporal rules

$$\frac{y : A, x : \mathbf{G}A, x \leq y, \Gamma \Rightarrow \Delta}{x : \mathbf{G}A, x \leq y, \Gamma \Rightarrow \Delta} LG \qquad \frac{x \leq y, \Gamma \Rightarrow \Delta, y : A}{\Gamma \Rightarrow \Delta, x : \mathbf{G}A} RG$$

$$\frac{x \leq y, y : A, \Gamma \Rightarrow \Delta}{x : \mathbf{F}A, \Gamma \Rightarrow \Delta} LF \qquad \frac{x \leq y, \Gamma \Rightarrow \Delta, x : \mathbf{F}A, y : A}{x \leq y, \Gamma \Rightarrow \Delta, x : \mathbf{F}A} RF$$

$$\frac{y : A, x : \mathbf{T}A, x \prec y, \Gamma \Rightarrow \Delta}{x : \mathbf{T}A, x \prec y, \Gamma \Rightarrow \Delta} LT \qquad \frac{x \prec y, \Gamma \Rightarrow \Delta, y : A}{\Gamma \Rightarrow \Delta, x : \mathbf{T}A} RT$$

$$\frac{y : A, x : \mathbf{H}A, y \leq x, \Gamma \Rightarrow \Delta}{x : \mathbf{H}A, y \leq x, \Gamma \Rightarrow \Delta} LH \qquad \frac{y \leq x, \Gamma \Rightarrow \Delta, y : A}{\Gamma \Rightarrow \Delta, x : \mathbf{H}A} RH$$

$$\frac{y \leq x, y : A, \Gamma \Rightarrow \Delta}{x : \mathbf{P}A, \Gamma \Rightarrow \Delta} LP \qquad \frac{y \leq x, \Gamma \Rightarrow \Delta, x : \mathbf{P}A, y : A}{y \leq x, \Gamma \Rightarrow \Delta, x : \mathbf{P}A} RP$$

$$\frac{y : A, x : \mathbf{Y}A, y \prec x, \Gamma \Rightarrow \Delta}{x : \mathbf{Y}A, y \prec x, \Gamma \Rightarrow \Delta} LY \qquad \frac{y \prec x, \Gamma \Rightarrow \Delta, y : A}{\Gamma \Rightarrow \Delta, x : \mathbf{Y}A} RY$$

Rules RG, LF, RT, RH, LP and RY have the condition that y is not in the conclusion.

Table 1. Logical rules for the system G3LT

$$x \leq y \equiv \exists n \in \mathbb{N}\,(x \prec^n y)$$

This means that if $x \leq y$, then y is reachable from x by iterating finitely many times the immediate successor relation.

The iterated successor relation is defined inductively by the following clauses, that result in the mathematical rules below:

$x \prec^0 y \equiv x = y$,
$x \prec^1 y \equiv x \prec y$,
$x \prec^{n+1} y \equiv \exists z (x \prec^n z \,\&\, z \prec y)$ for $n > 0$.

Rules for the Iterated Successor Relation

$$\frac{x \prec^n y, y \prec z, \Gamma \Rightarrow \Delta}{x \prec^{n+1} z, \Gamma \Rightarrow \Delta} LDef \qquad \frac{\Gamma \Rightarrow \Delta, x \prec^{n+1} z, x \prec^n y \quad \Gamma \Rightarrow \Delta, x \prec^{n+1} z, y \prec z}{\Gamma \Rightarrow \Delta, x \prec^{n+1} z} RDef$$

Rule $LDef$ has the condition that y is not in the conclusion.

Infinitary Rule

The left-to-right direction of the definition of $x \leq y$ as the transitive closure of $x \prec y$ gives the following infinitary rule

$$\frac{\{x \prec^n y, x \leq y, \Gamma \Rightarrow \Delta\}_{n \in \mathbb{N}}}{x \leq y, \Gamma \Rightarrow \Delta} T^\omega$$

The right-to-left direction gives, for every $n \in \mathbb{N}$, the following generalized form of rule *Inc*

$$\frac{x \leq y, x \prec^n y, \Gamma \Rightarrow \Delta}{x \prec^n y, \Gamma \Rightarrow \Delta} Inc_n$$

This rule is admissible in our system by induction on n. The proof uses equality rules for $n = 0$, and *Trans* for the inductive case.

Finally, we observe that the *closure condition* required for admissibility of contraction (see e.g. [14] p. 510) does not bring to new rules in the system above since the contracted instances of *Trans*, *UnPred*, and *UnSucc* are special cases of *Ref* and *EqRef*.

2.2 Structural properties

Next we prove the structural properties of the system G3LT.

LEMMA 1. *Sequents of the form* $x : A, \Gamma \Rightarrow \Delta, x : A$, *with A an arbitrary modal formula, are derivable in* G3LT.

Proof. By induction on the length of the formula A. ∎

In order to guarantee invertibility of all the rules, initial sequents with relational atoms as principal cannot be of the form $x \prec^n y, \Gamma \Rightarrow \Delta, x \prec^n y$ for $n > 1$. However, sequents of this form are easily derivable:

LEMMA 2. *Sequents of the form* $x \prec^n y, \Gamma \Rightarrow \Delta, x \prec^n y$ *are derivable in* G3LT *for all $n \in \mathbb{N}$*.

Proof. By induction on n. For $n = 0, 1$, observe that $x = y, \Gamma \Rightarrow \Delta, x = y$ and $x \prec y, \Gamma \Rightarrow \Delta, x \prec y$ are initial sequents. For the inductive case, assume a derivation of $x \prec^n z, z \prec y, \Gamma \Rightarrow \Delta, x \prec^{n+1} y, x \prec^n z$ with z different from x, y and not in Γ, Δ, and derive the claim for $n+1$ by applying *RDef* with right premiss $x \prec^n z, z \prec y, \Gamma \Rightarrow \Delta, x \prec^{n+1} y, z \prec y$ and then *LDef*. ∎

Substitution of labels is defined in the obvious way for relational atoms and labelled formulas, and extended to multisets componentwise. We have:

LEMMA 3. *If $\Gamma \Rightarrow \Delta$ is derivable in* G3LT, *then also $\Gamma(y/x) \Rightarrow \Delta(y/x)$ is derivable, with the same derivation height.*

Proof. By induction on the height h of the derivation. If $h = 0$, then $\Gamma \Rightarrow \Delta$ is either an initial sequent or a conclusion of $L\bot$. In both cases, the sequent $\Gamma(y/x) \Rightarrow \Delta(y/x)$ is also an initial sequent or a conclusion of $L\bot$.

Suppose that the claim holds for $h = n$, and consider the last rule applied in the derivation. If it is a propositional rule or a temporal or mathematical rule without a variable condition, apply the inductive hypothesis to the premiss(es) and then the rule. If the last rule is a rule with a variable condition, we need to avoid a clash with the eigenvariable: In that case, we apply twice the inductive hypothesis to the premiss(es) first to replace the eigenvariable with a fresh variable not appearing in the derivation, and then to perform the desired substitution. ∎

In what follows, Greek lower case letters are used for denoting labelled and relational formulas.

THEOREM 4. *The rules of left and right weakening*

$$\frac{\Gamma \Rightarrow \Delta}{\varphi, \Gamma \Rightarrow \Delta} LWk \qquad \frac{\Gamma \Rightarrow \Delta}{\Gamma \Rightarrow \Delta, \varphi} RWk$$

are height-preserving admissible in G3LT.

Proof. By induction on the height of the derivation. If $\Gamma \Rightarrow \Delta$ is an initial sequent or a conclusion of $L\bot$, also $\varphi, \Gamma \Rightarrow \Delta$ and $\Gamma \Rightarrow \Delta, \varphi$ are. The cases of rules without variable condition are straightforward. If the last step is a rule with a variable condition, we first apply Lemma 3 to avoid a clash of variables and then the inductive hypothesis and the rule in question. ∎

LEMMA 5. *All rules of* G3LT *are height-preserving invertible.*

Proof. The proof of height-preserving invertibility for propositional rules, for rule $LDef$, and for temporal rules with a variable condition is by induction on the height of derivation (clash of variables is avoided through the substitution lemma). The condition that $x \prec^n y, \Gamma \Rightarrow \Delta, x \prec^n y$ is not initial for $n > 1$ is essential for the invertibility of rule $LDef$. ∎

THEOREM 6. *The rules of left and right contraction*

$$\frac{\varphi, \varphi, \Gamma \Rightarrow \Delta}{\varphi, \Gamma \Rightarrow \Delta} LCtr \qquad \frac{\Gamma \Rightarrow \Delta, \varphi, \varphi}{\Gamma \Rightarrow \Delta, \varphi} RCtr$$

are height-preserving admissible in G3LT.

Proof. By simultaneous induction on the height of derivation for left and right contractions. For $h = 0$, note that if $\varphi, \varphi, \Gamma \Rightarrow \Delta$ (resp. $\Gamma \Rightarrow \Delta, \varphi, \varphi$) is an initial sequent or a conclusion of $L\bot$, so is $\varphi, \Gamma \Rightarrow \Delta$ (resp. $\Gamma \Rightarrow \Delta, \varphi$). For $h = n+1$, we distinguish two cases: If none of the contraction formulas is principal in the last rule, we apply the inductive hypothesis to the premiss(es) and then the rule; If one of the contraction formulas is principal, we first apply height-preserving inversion to the premiss(es), then inductive hypothesis, and last the rule; If both are principal, necessarily in a mathematical rule, by the closure condition contraction is absorbed into the contracted instance of the rule. ∎

The system G3LT has mathematical rules that act on both the left- and the right-hand sides of sequents, and a measure of complexity for relational atoms is needed in the proof of cut elimination, as in [6].

DEFINITION 7. The length of a labelled formula $x : A$ is defined as the length of A. The length of relational and equality formulas is defined as follows: $l(x \prec y) = l(x \leq y) = l(x = y) = 1$ and $l(x \prec^n y) = n$ for $n \geq 1$.

THEOREM 8. *The rule of cut*

$$\frac{\Gamma \Rightarrow \Delta, \varphi \quad \varphi, \Gamma' \Rightarrow \Delta'}{\Gamma, \Gamma' \Rightarrow \Delta, \Delta'} Cut$$

is admissible in G3LT.

Proof. By induction on the length of the cut formula and a subinduction on the sum of the heights of the derivations of the premises of cut. The proof has the structure of the proof of cut elimination for modal logics (see [14], Theorem 4.13). However, we have to consider here an essentially new case, because of the simultaneous presence of mathematical rules that act on both the left- and the right-hand sides of sequents. This is the case with the cut formula $x \prec^{n+1} y$ principal in both premises of cut:

$$\cfrac{\cfrac{\Gamma \Rightarrow \Delta, x \prec^{n+1} y, x \prec^n z \quad \Gamma \Rightarrow \Delta, x \prec^{n+1} y, z \prec y}{\Gamma \Rightarrow \Delta, x \prec^{n+1} y} RDef \quad \cfrac{x \prec^n w, w \prec y, \Gamma' \Rightarrow \Delta'}{x \prec^{n+1} y, \Gamma' \Rightarrow \Delta'} LDef}{\Gamma, \Gamma' \Rightarrow \Delta, \Delta'} Cut$$

This derivation is transformed as follows:

We first cut the left premiss of *RDef* with the conclusion of *LDef*

$$1. \quad \cfrac{\Gamma \Rightarrow \Delta, x \prec^{n+1} y, x \prec^n z \quad \cfrac{x \prec^n w, w \prec y, \Gamma' \Rightarrow \Delta'}{x \prec^{n+1} y, \Gamma' \Rightarrow \Delta'} LDef}{\Gamma, \Gamma' \Rightarrow \Delta, \Delta', x \prec^n z} Cut$$

thus obtaining a cut of shorter height. Then we cut the right premiss of *RDef* with the conclusion of *LDef*

$$2. \quad \cfrac{\Gamma \Rightarrow \Delta, x \prec^{n+1} y, z \prec y \quad \cfrac{x \prec^n w, w \prec y, \Gamma' \Rightarrow \Delta'}{x \prec^{n+1} y, \Gamma' \Rightarrow \Delta'} LDef}{\Gamma, \Gamma' \Rightarrow \Delta, \Delta', z \prec y} Cut$$

thus obtaining another cut of shorter height. Finally, we use the sequents thus obtained and the premiss of *LDef* as follows:

$$\cfrac{\Gamma, \Gamma' \stackrel{2}{\Rightarrow} \Delta, \Delta', z \prec y \quad \cfrac{\Gamma, \Gamma' \stackrel{1}{\Rightarrow} \Delta, \Delta', x \prec^n z \quad \cfrac{x \prec^n w, w \prec y, \Gamma' \Rightarrow \Delta'}{x \prec^n z, z \prec y, \Gamma' \Rightarrow \Delta'} Hp\text{-}Subst}{z \prec y, \Gamma, \Gamma', \Gamma' \Rightarrow \Delta, \Delta', \Delta'} Cut}{\cfrac{\Gamma, \Gamma, \Gamma', \Gamma', \Gamma' \Rightarrow \Delta, \Delta, \Delta', \Delta', \Delta'}{\Gamma, \Gamma' \Rightarrow \Delta, \Delta'} Ctr^*}$$

Here the two cuts are on formulas of smaller length and *Hp-Subst* denotes a height-preserving substitution. ∎

COROLLARY 9. *The following generalized rules of substitution of equals*

$$\cfrac{y \prec^n z, x = y, x \prec^n z, \Gamma \Rightarrow \Delta}{x = y, x \prec^n z, \Gamma \Rightarrow \Delta} \qquad \cfrac{z \prec^n y, x = y, z \prec^n x, \Gamma \Rightarrow \Delta}{x = y, z \prec^n x, \Gamma \Rightarrow \Delta} \qquad \cfrac{y : A, x = y, x : A, \Gamma \Rightarrow \Delta}{x = y, x : A, \Gamma \Rightarrow \Delta}$$

are admissible in G3LT.

Proof. Similar to the proof of admissibility of the replacement rule in predicate logic with equality (Theorem 6.5.3 in [17]). Using a cut of the premisses of the rules with the derivable sequents $x \prec^n z, x = y \Rightarrow y \prec^n z$ and $z \prec^n x, x = y \Rightarrow z \prec^n y$ and $x : A, x = y \Rightarrow y : A$, respectively, and admissibility of the rules of cut and contraction. ∎

Because of the internalization of the semantics, most labelled sequents cannot be directly interpreted as temporal formulas. However, we can single out a class of sequents with a plain correspondence to their associated formulas:

DEFINITION 10. A *purely logical* sequent is a sequent that contains no relational atoms and in which every formula is labelled by the same variable.

The Hilbert-style system for Priorean linear time logic can be embedded into our calculus: We show that the purely logical sequents corresponding to the temporal axioms and the modal rules are derivable/admissible in G3LT. Admissibility of *modus ponens* follows by cut elimination.

PROPOSITION 11. *The following purely logical sequents*

$x : \mathbf{G}(A \supset B), x : \mathbf{G}A \Rightarrow x : \mathbf{G}B$ $\quad x : \mathbf{T}(A \supset B), x : \mathbf{T}A \Rightarrow x : \mathbf{T}B$
$x : \mathbf{T}\neg A \Rightarrow x : \neg \mathbf{T}A$ $\quad x : \neg \mathbf{T}A \Rightarrow x : \mathbf{T}\neg A$
$x : \mathbf{G}A \Rightarrow x : A \,\&\, \mathbf{TG}A$ $\quad x : A, x : \mathbf{G}(A \supset \mathbf{T}A) \Rightarrow x : \mathbf{G}A$
$x : \mathbf{TG}A \Rightarrow x : \mathbf{GT}A$ $\quad x : \mathbf{GT}A \Rightarrow x : \mathbf{TG}A$
$x : \mathbf{TY}A \Rightarrow x : A$ $\quad x : A \Rightarrow x : \mathbf{TY}A$

and their temporal mirror images.[1] *are derivable in* G3LT

Proof. By root-first proof search from the sequent to be derived. Note that derivability of $x : A, x : \mathbf{G}(A \supset \mathbf{T}A) \Rightarrow x : \mathbf{G}A$ and of its temporal mirror image require an application of T^ω. ∎

PROPOSITION 12. *The necessitation rules for* **G**, **H**, **T** *and* **Y**

$$\frac{\Rightarrow x : A}{\Rightarrow x : \mathbf{G}A}\,GNec \quad \frac{\Rightarrow x : A}{\Rightarrow x : \mathbf{H}A}\,HNec \quad \frac{\Rightarrow x : A}{\Rightarrow x : \mathbf{T}A}\,TNec \quad \frac{\Rightarrow x : A}{\Rightarrow x : \mathbf{Y}A}\,YNec$$

are admissible in G3LT.

Proof. Let us suppose that we have a derivation of $\Rightarrow x : A$. By Lemma 3 we obtain a derivation of $\Rightarrow y : A$ and by admissibility of weakening we obtain the sequents $x \leq y \Rightarrow y : A$, $y \leq x \Rightarrow y : A$, $x \prec y \Rightarrow y : A$, and $y \prec x \Rightarrow y : A$. We finally conclude $\Rightarrow x : \mathbf{G}A$, $\Rightarrow x : \mathbf{H}A$, $\Rightarrow x : \mathbf{T}A$, and $\Rightarrow x : \mathbf{Y}A$ by a single step of $R\mathbf{G}$, $R\mathbf{H}$, $R\mathbf{T}$ and $R\mathbf{Y}$, respectively. ∎

COROLLARY 13. *The calculus* G3LT *is complete with respect to Priorean linear time logic.*

The equivalences $\mathbf{G}A \supset\subset (A \,\&\, \mathbf{TG}A)$ and $\mathbf{H}A \supset\subset (A \,\&\, \mathbf{YH}A)$ define recursively the operator **G** in terms of **T** and the operator **H** in terms of **Y**, respectively. The left-to-right directions are axioms (see Proposition 11); their converses, $(A \,\&\, \mathbf{TG}A) \supset \mathbf{G}A$ and $(A \,\&\, \mathbf{YH}A) \supset \mathbf{H}A$, are easily derivable by means of the following admissible rules:

$$\frac{x = y, x \leq y, \Gamma \Rightarrow \Delta \quad x \prec z, z \leq y, x \leq y, \Gamma \Rightarrow \Delta}{x \leq y, \Gamma \Rightarrow \Delta}\,Mix_1$$

$$\frac{x = y, x \leq y, \Gamma \Rightarrow \Delta \quad x \leq z, z \prec y, x \leq y, \Gamma \Rightarrow \Delta}{x \leq y, \Gamma \Rightarrow \Delta}\,Mix_2$$

---

[1]The *temporal mirror image* of a purely logical sequent is obtained by replacing each occurrence of a future (resp. past) operator by its past (resp. future) analogue. For example, the temporal mirror image of $x : \mathbf{G}P \Rightarrow x : P \,\&\, \mathbf{TG}P$ is $x : \mathbf{H}P \Rightarrow x : P \,\&\, \mathbf{YH}P$.

both with the condition that z is not in the conclusion. Rules Mix_1 and Mix_2 correspond to the frame properties $x \leq y \supset (x = y \vee \exists z(x \prec z \mathbin{\&} z \leq y))$ and $x \leq y \supset (x = y \vee \exists z(x \leq z \mathbin{\&} z \prec y))$ that permit the splitting of an interval $[x, y]$ with an immediate successor of x and an immediate predecessor of y respectively.

PROPOSITION 14. *Rules Mix_1 and Mix_2 are admissible in* G3LT.

Proof. Whenever the premises of rules Mix_1 or Mix_2 are derivable, so are the sequents $x \prec^n y, x \leq y, \Gamma \Rightarrow \Delta$ for all n. An application of rule T^ω gives the desired conclusion. ∎

Finally, completeness of G3LT implies admissibility of the rules of left and right linearity:

PROPOSITION 15. *The rules of left and right linearity*

$$\frac{y \leq z, z \leq x, y \leq x, \Gamma \Rightarrow \Delta \quad z \leq y, z \leq x, y \leq x, \Gamma \Rightarrow \Delta}{z \leq x, y \leq x, \Gamma \Rightarrow \Delta} \, L\text{-}Lin$$

$$\frac{y \leq z, x \leq z, x \leq y, \Gamma \Rightarrow \Delta \quad z \leq y, x \leq z, x \leq y, \Gamma \Rightarrow \Delta}{x \leq z, x \leq y, \Gamma \Rightarrow \Delta} \, R\text{-}Lin$$

are admissible in G3LT.

Proof. By means of two applications of T^ω, with principal formulas $x \leq z$ and $x \leq y$, and derivability of $x \prec^m z, x \leq z, x \prec^n y, x \leq y, \Gamma \Rightarrow \Delta$ for every $m, n \in \mathbb{N}$, whenever the premises of R-Lin are derivable. ∎

3 A non-standard system for linear time

We define the system G3LT$_f$ by substituting, in the calculus G3LT, the rules T^ω, *LDef* and *RDef*, with the rules Mix_1, Mix_2, L-Lin and R-Lin as primitive.

The standard frame for linear time logic corresponds to the set of the integers $\mathbb{Z}$: Every instant greater (smaller) than x can be reached from x by finitely many iterations of the immediate successor (predecessor) relation. This condition corresponds to the infinitary rule T^ω of the calculus G3LT.

Because of the absence of T^ω, the system G3LT$_f$ allows non-standard frames that consist of several, possibly infinite, consecutive copies of the integers, $\mathbb{Z} \oplus \cdots \oplus \mathbb{Z}$: Even though every point is the unique immediate successor of its unique immediate predecessor (and viceversa), it is not always true that between any two points x, y such that $x \leq y$, there are finitely many other points.

It is easy to verify that the system G3LT$_f$ can be embedded in G3LT: Every sequent derivable in G3LT$_f$ is derivable in G3LT.

THEOREM 16. *If* $\vdash_{G3LT_f} \Gamma \Rightarrow \Delta$*, then* $\vdash_{G3LT} \Gamma \Rightarrow \Delta$.

Proof. Every rule of G3LT$_f$ except Mix_1, Mix_2, L-Lin and R-Lin is a rule of G3LT, and Mix_1, Mix_2, L-Lin and R-Lin are admissible in G3LT, by Proposition 14 and Proposition 15, respectively. ∎

The converse fails because of the infinitary rule: For instance, any proof search for the induction principle $x : A, x : \mathbf{G}(A \supset \mathbf{T}A) \Rightarrow x : \mathbf{G}A$ would require infinitely many applications of rule Mix_1. Nevertheless, we identify a conservative fragment for which derivability in G3LT implies derivability in G3LT$_f$. Our result is confined to purely logical sequents, but this condition is not restrictive, since, as we noticed before, only purely logical sequents can be interpreted as corresponding modal formulas.

THEOREM 17. *If a purely logical sequent $\Gamma \Rightarrow \Delta$ is derivable in G3LT and the operators $\mathbf{G}, \mathbf{H}$ do not appear in its positive part, nor $\mathbf{F}, \mathbf{P}$ in its negative part, then $\Gamma \Rightarrow \Delta$ is derivable without the infinitary rule.*

Proof. We show that all the applications of the infinitary rule can be dispensed with. Without loss of generality, we assume that the given derivation is minimal in the sense that no shortenings, such as those arising from applications of height-preserving contraction, are possible: This excludes rule instances such as transitivity with a reflexitity atom as principal. Observe that all relational atoms $x \leq y$, in particular those concluded by T^ω, have to disappear before the conclusion. We consider one such downmost atom and the rule that makes it disappear: Rules $R\mathbf{G}$, $R\mathbf{H}$, $L\mathbf{F}$ and $L\mathbf{P}$ are excluded because they would introduce $\mathbf{G}, \mathbf{H}$ in the positive part or $\mathbf{F}, \mathbf{P}$ in the negative part. Thus, the atom can disappear by means of *Inc* or *Ref*.

If the atom concluded by T^ω is removed by *Ref*, we have

$$\dfrac{\{x \prec^n x, x \leq x, \Gamma' \Rightarrow \Delta'\}_{n \in \mathbb{N}}}{x \leq x, \Gamma' \Rightarrow \Delta'} T^\omega$$
$$\vdots$$
$$\dfrac{x \leq x, \Gamma'' \Rightarrow \Delta''}{\Gamma'' \Rightarrow \Delta''} Ref$$

We take the leftmost premiss of T^ω and transform the derivation into the following

$$\dfrac{x = x, x \leq x, \Gamma' \Rightarrow \Delta'}{x \leq x, \Gamma' \Rightarrow \Delta'} EqRef$$
$$\vdots$$
$$\dfrac{x \leq x, \Gamma'' \Rightarrow \Delta''}{\Gamma'' \Rightarrow \Delta''} Ref$$

The application of T^ω is removed from the derivation.

If the atom concluded by T^ω is removed by *Inc*, we have

$$\dfrac{\{x \prec^n y, x \leq y, \Gamma' \Rightarrow \Delta'\}_{n \in \mathbb{N}}}{x \leq y, \Gamma' \Rightarrow \Delta'} T^\omega$$
$$\vdots$$
$$\dfrac{x \leq y, x \prec y, \Gamma'' \Rightarrow \Delta''}{x \prec y, \Gamma'' \Rightarrow \Delta''} Inc$$

The second premiss of T^ω has the form $x \prec y, x \leq y, x \prec y, \Gamma''' \Rightarrow \Delta'$, with $\Gamma' \equiv x \prec y, \Gamma'''$. By height-preserving contraction we obtain $x \leq y, \Gamma' \Rightarrow \Delta'$ and proceed with the derivation until we reach $x \prec y, x \leq y, \Gamma'' \Rightarrow \Delta''$. Then we conclude $x \prec y, \Gamma'' \Rightarrow \Delta''$ by an application of *Inc*. Note that the derivation is shortened, contrary to the assumption of minimality.

If the atom concluded by T^ω is removed by applications of *Trans* followed by applications of *Inc*, we have the derivation

$$\cfrac{\cfrac{\cfrac{\{x \prec^n y, x \leq y, z_1 \leq y, \ldots, z_{m-1} \leq y, x \leq z_1, \ldots, z_m \leq y, x \prec z_1, \ldots, z_m \prec y, \Gamma' \Rightarrow \Delta'\}_{n \in \mathbb{N}}}{x \leq y, z_1 \leq y, \ldots, z_{m-1} \leq y, x \leq z_1, \ldots, z_m \leq y, x \prec z_1, \ldots, z_m \prec y, \Gamma' \Rightarrow \Delta'} T^\omega}{\vdots} }{\cfrac{\cfrac{x \leq y, z_1 \leq y, \ldots, z_{m-1} \leq y, x \leq z_1, \ldots, z_m \leq y, x \prec z_1, \ldots, z_m \prec y, \Gamma'' \Rightarrow \Delta''}{x \leq z_1, \ldots, z_m \leq y, x \prec z_1, \ldots, z_m \prec y, \Gamma'' \Rightarrow \Delta''} \text{Trans} \times m}{\cfrac{x \leq z_1, \ldots, z_m \leq y, x \prec z_1, \ldots, z_m \prec y, \Gamma''' \Rightarrow \Delta'''}{x \prec z_1, \ldots, z_m \prec y, \Gamma''' \Rightarrow \Delta'''} \text{Inc} \times (m+1)}}$$

These can be transformed into the following derivation

$$\cfrac{\cfrac{\cfrac{x \prec^{m+1} y, x \leq y, z_1 \leq y, \ldots, z_{m-1} \leq y, x \leq z_1, \ldots, z_m \leq y, x \prec z_1, \ldots, z_m \prec y, \Gamma' \Rightarrow \Delta'}{\cfrac{x \prec z_1, \ldots, z_m \prec y, x \leq y, z_1 \leq y, \ldots, z_{m-1} \leq y, x \leq z_1, \ldots, z_m \leq y, x \prec z_1, \ldots, z_m \prec y, \Gamma' \Rightarrow \Delta'}{x \leq y, z_1 \leq y, \ldots, z_{m-1} \leq y, x \leq z_1, \ldots, z_m \leq y, x \prec z_1, \ldots, z_m \prec y, \Gamma' \Rightarrow \Delta'} \mathcal{C}} \mathcal{I}}{\vdots}}{\cfrac{\cfrac{x \leq y, z_1 \leq y, \ldots, z_{m-1} \leq y, x \leq z_1, \ldots, z_m \leq y, x \prec z_1, \ldots, z_m \prec y, \Gamma'' \Rightarrow \Delta''}{x \leq z_1, \ldots, z_m \leq y, x \prec z_1, \ldots, z_m \prec y, \Gamma'' \Rightarrow \Delta''} \text{Trans} \times m}{\cfrac{x \leq z_1, \ldots, z_m \leq y, x \prec z_1, \ldots, z_m \prec y, \Gamma''' \Rightarrow \Delta'''}{x \prec z_1, \ldots, z_m \prec y, \Gamma''' \Rightarrow \Delta'''} \text{Inc} \times (m+1)}}$$

Here $\mathcal{I}$ stands for m applications of height-preserving invertibility of rule *LDef* and $\mathcal{C}$ for several application of height-preserving contraction. Again, the derivation is shortened, contrary to the assumption.

Note that if the atom concluded by T^ω is removed by an application of $EqSubst_{At}$, we have the following derivation:

$$\cfrac{\cfrac{\{x \prec^n y, z = y, x \leq y, x \leq z, \Gamma' \Rightarrow \Delta'\}_{n \in \mathbb{N}}}{z = y, x \leq y, x \leq z, \Gamma' \Rightarrow \Delta'} T^\omega}{\cfrac{\vdots}{\cfrac{z = y, x \leq y, x \leq z, \Gamma'' \Rightarrow \Delta''}{z = y, x \leq z, \Gamma'' \Rightarrow \Delta''} EqSubst_{At}}}$$

It is possible to permute up rule $EqSubst_{At}$ with respect to rule T^ω. We modify each premiss of T^ω as follows:

$$\cfrac{\cfrac{x \prec^n y, x \leq y, z = y, x \leq z, \Gamma' \Rightarrow \Delta'}{x \prec^n y, x \leq y, x \prec^n z, z = y, x \leq z, \Gamma' \Rightarrow \Delta'} LWk}{\cfrac{\vdots}{\cfrac{\cfrac{x \prec^n y, x \leq y, x \prec^n z, z = y, x \leq z, \Gamma'' \Rightarrow \Delta''}{x \prec^n y, x \prec^n z, z = y, x \leq z, \Gamma'' \Rightarrow \Delta''} Inc_n}{x \prec^n z, z = y, x \leq z, \Gamma'' \Rightarrow \Delta''} EqSubst_n}}$$

We can now apply the modifications previously considered. The case of $EqSubst_{At}$ with active formulas $z = x, x \leq y, z \leq y$ is analogous. ∎

COROLLARY 18. *If* $\vdash_{G3LT} \Gamma \Rightarrow \Delta$ *and* $\Gamma \Rightarrow \Delta$ *is as in the previous theorem, then* $\vdash_{G3LT_f} \Gamma \Rightarrow \Delta$.

Proof. By Theorem 17, $\Gamma \Rightarrow \Delta$ is derivable without using rule T^ω. ∎

4 Related work

From the extensive literature on labelled and hybrid systems, we have followed the method introduced by the second author in [14]. The latter is, compared with the development within Gabbay's labelled deductive systems (see, e.g., chapter 4 in [9]), more explicitly proof-theoretic.

In Baaz *et al.* [4] first-order linear time temporal logic, with future operators □ and ○, that correspond to our **G** and **T**, is compared to the logic for branching time gaps, the frames of which are well-founded trees of copies of ℕ: Whereas in the former an infinitary rule is needed, the latter is formulated as a cut-free extension of Gentzen's system for classical predicate logic with finitary rules for temporal operators. A conservativity result is then obtained for the □-free fragment of the system.

If we drop the rule of right linearity from G3LT$_f$, we obtain a labelled sequent calculus that generalizes the propositional fragment of the system in [4]: It is easy to verify that for every propositional formula derivable in the latter system, the corresponding purely logical sequent is derivable in G3LT$_f$ and if a purely logical sequent does not contain past operators and is derivable in G3LT$_f$ without using *R-Lin*, then the corresponding formula is derivable by means of the rules in [4]. However, we can prove a stronger conservativity result: Our theorem has only the condition that endsequents do not contain **G** in the positive part (nor **F** in the negative part), whereas in [4] the modality □, corresponding to **G**, cannot appear at all in the formula to be derived.

We have identified in our work a finitary fragment of Priorean linear time logic by substituting the rule that corresponds to the reflexive and transitive closure with two weaker finitary counterparts. A somewhat related result is presented by Antonakos and Artemov [2, 3] for the different, but qualitatively similar, logic of common knowledge.

BIBLIOGRAPHY

[1] R. Alonderis. Proof-theoretical investigations of temporal logic with time gaps. *Lithuanian Mathematical Journal* 40: 197–212, 2000.
[2] E. Antonakos. Justified Knowledge is Sufficient. Technical Report TR-2006004, CUNY Ph.D. Program in Computer Science, 2006.
[3] S. Artemov. Proofs, Evidence, Knowledge, *Lecture Notes for ESSLLI*, 2006.
[4] M. Baaz, A. Leitsch, and R. Zach. Completeness of a first-order temporal logic with time gaps, *Theoretical Computer Science*, 160: 241–270, 1996.
[5] A. Bolotov, A. Basukoski, O. Grigoriev, and V. Shangin. Natural deduction calculus for linear-time temporal logic. In M. Fisher, W. van der Hoek, B. Konev and A. Lisitsa (eds.), *Logics in artificial intelligence*. Lecture notes in Artificial Intelligence 4160, pages 56–68, 2006.
[6] B. Boretti and S. Negri. Equality in the presence of apartness: An application of structural proof analysis to intuitionistic axiomatics. In *Constructivism: Mathematics, Logic, Philosophy and Linguistics*, Philosophia Scientiae, Cahier spécial 6, pages 61–79, 2006.

[7] K. Broda, D. M. Gabbay, L. Lamb, and A. Russo. *Compiled Labelled Deductive Systems: A Uniform Presentation of Non-Classical Logics*. Research Study Press, 2004.
[8] J. Brotherston and A. Simpson. Complete sequent calculi for induction and infinite descent, in *LICS '07: Proceedings of the 22nd Annual IEEE Symposium on Logic in Computer Science*, IEEE Computer Society, Washington, 51–62, 2007.
[9] D.M. Gabbay, M. Finger, and M. Reynolds. *Temporal Logic: Mathematical Foundations and Computational Aspects* vol. 2. Oxford University Press, 2000.
[10] G. Jäger, M. Kretz, and T. Studer. Cut-free common knowledge, *Journal of Applied Logic*, 5: 681–689, 2007.
[11] J.A.W. Kamp. *Tense Logic and the Theory of Order*, PhD thesis, UCLA, 1968.
[12] Y. Kesten, Z. Manna, H. McGuire, and A. Pnueli. *Decision Algorithm for Full Propositional Temporal Logic*. Lecture Notes In Computer Science 697, pages 97–109, 1993.
[13] S. Negri. Contraction-free sequent calculi for geometric theories, with an application to Barr's theorem. *Archive for Mathematical Logic*, 42: 389–401, 2003.
[14] S. Negri. Proof analysis in modal logic. *Journal of Philosophical Logic*, 34: 507–544, 2005.
[15] S. Negri. Proof analysis in non-classical logics. In C. Dimitracopoulos, L. Newelski, D. Normann and J. Steel (eds.), *Logic Colloquium '05: Proceedings of the Annual European Summer Meeting of the Association for Symbolic Logic*. ASL Lecture Notes in Logic 28, pages 107–128, Cambridge University Press, Cambridge 2007.
[16] S. Negri and J. von Plato. Cut elimination in the presence of axioms. *The Bulletin of Symbolic Logic*, 4: 418–435, 1998.
[17] S. Negri and J. von Plato. *Structural Proof Theory*. Cambridge University Press, Cambridge 2001.
[18] A.N. Prior. *Past, Present, and Future*. Clarendon Press, Oxford, 1967.
[19] P.H. Schmitt and J. Goubault-Larrecq. A Tableau System for Linear-TIME Temporal Logic. *Lecture Notes In Computer Science* 1217: 130–144, 1997.
[20] D. Scott. *The Logic of Tenses*. Stanford University, 1965.
[21] A. Troelstra and H. Schwichtenberg, *Basic Proof Theory*, 2nd edition. Cambridge University Press, Cambridge 2000.
[22] G.H. von Wright. And Next. *Acta Philosophica Fennica* 18: 293–304, 1965.

Proof theoretic aspects of quasi-inductive definitions

RICCARDO BRUNI

1 Introduction

The *Revision Theory of Truth* (see [6]) is a proposal originally conceived in such a way to provide a solution to the problem, in the philosophy of logic, of defining languages with their own truth predicate. In this direction, it can be presented as exploiting the idea that truth is the result of a process based on *reaching consensus by interaction*, featuring *exchange of information* and *stages of approximations*. From the point of view of formal logic, there are thus many reasons why a precise account of this approach would be a notable project to pursue. Here I make some proposals in this direction, by favouring, so to say, an abstract and extensional view of the Revision Theory, which is regarded as a mean for transifinitely defining sets of natural numbers.

The crucial notion in this sense, as defined by J. Burgess [3], is that of *quasi-inductive definition* which, for a set-theoretical operator $f : \mathcal{P}(\mathbb{N}) \to \mathcal{P}(\mathbb{N})$ whatsoever (i.e., one that is not necessarily monotone), is a construction satisfying the following clauses:

$$\begin{aligned} f_0 &= \emptyset \\ f_{\alpha+1} &= f(f_\alpha) \\ f_\lambda &= \liminf\nolimits_{\beta < \lambda} f_\beta = \bigcup\nolimits_{\alpha < \lambda} \bigcap\nolimits_{\alpha \leq \beta < \lambda} f_\beta, \quad \lambda \text{ limit} \end{aligned}$$

It is true that, owing to its mixing features from both ordinary inductive definitions based on monotone operators (i.e., those constructions which are mostly known in mathematics, logic and computer science), and from their generalization to operators whatsoever, the usual closure arguments showing the existence of fixed points for constructions of the inductive sorts, fail in the quasi-inductive case. However, quasi-inductive definitions prove to have an interesting behaviour with respect to a natural modification of the notions involved therein.

That is, if we let $f_\infty := \bigcup_\alpha \bigcap_{\alpha \leq \beta} f_\beta$ indicate the *stability set* for the operator f, the construction admits levels f_γ such that $f_\gamma = f_\infty$ (in fact, levels of this sort for countable γ's).

As to the logical analysis, here I stick to problems of *axiomatization* and *proof theory*. Though I offer a solution in the first direction by presenting an accordingly defined family of theories in Section 2, as well as I give some initial indication in the second one by providing proof-theoretic results in Sections 3,4, the main character of the paper resides in setting some long-term research objectives. This will turn out perhaps more clearly from the tenor of my closing comments, which also connect the present contribution with some pre-existing literature.

2 The theories QID(K)

The language $\mathcal{L}_0$ I start from extends the language $\mathcal{L}_{Ar}$ of Peano arithmetic, by means of a *new sort* of individual variables for ordinal numbers. Furthermore, it contains *individual constants* $0, 0_\Omega, \omega$, symbols for *number-theoretic primitive recursive functions*, as well as symbols for all *primitive recursive ordinal functions*.[1] Among the *predicate constants* we have $=, <_\Omega$, where the latter is intended to apply to ordinal terms (see below), and is thus marked by Ω so to distinguish it from the primitive recursive number-theoretic order relation which is used, e.g., in the definition of bounded numerical quantification.

The collection $TERM^N$ of *arithmetical terms* is defined as usual. The collection of *ordinal terms* $TERM^\Omega$, instead, is the least containing $0_\Omega, \omega$, individual variables for ordinals and which is closed under primitive recursive ordinal functions. Formulas of $\mathcal{L}_0$ are built up as usual by closing the collection of *atomic formulas* $s = t, p <_\Omega q$ (s, t being terms whatsoever, $p, q \in TERM^\Omega$) under logical connectives and quantification on *both sorts* of variables.

For the sake of readability, we use different symbols for terms of the two aforementioned sorts. In particular, $x, y, z, ..., m, n, ...$ will be used in the following as metavariables for arithmetical variables and terms respectively, while lower-case greek letters $\alpha, \beta, \gamma, ...$ will do the same for ordinal ones.

Now, let $\mathcal{L}^+$ indicate $\mathcal{L}_{Ar} \cup \{X^1\}$, X being a fresh unary predicate variable. We call a formula $A(x, X)$ of $\mathcal{L}^+$ which contains at most x as a free (individual) variable, an *(arithmetical) operator form*. Notice that there is no constraint as to how the higher-order variable occurs in A. Operator forms are then grouped into complexity classes **K**'s according to their logical complexity (that is, $\mathbf{K} = \Sigma_n, \Pi_n$ or Δ_n, for some $n \in \mathbb{N}$). Consequently, I shall speak in the following of **K**-operator forms, avoiding sometimes to explicitly indicate their free variables which, is intended, are as in the informal definition above.

The language $\mathcal{L}(\mathbf{K})$ of QID(**K**) is obtained from $\mathcal{L}_0$ by adding binary predicate constants $\mathcal{H}^A$, yielding formulas $\mathcal{H}^A(n, \alpha)$, for each **K**-operator form $A(x, X)$ in $\mathcal{L}^+$. The notion of formula of this expanded language must be then re-defined as expected. Finally, bounded quantification is introduced as usual on arithmetical variables, and *via* $<_\Omega$ for ordinal ones.[2]

In the following, I shall make use of some notational conventions:

$$n \in \mathcal{H}_\alpha^A \ :\equiv \ \mathcal{H}^A(n, \alpha)$$
$$(\mathcal{H}_\alpha^A \equiv \mathcal{H}_\beta^A) \ :\equiv \ \forall x(x \in \mathcal{H}_\alpha^A \leftrightarrow x \in \mathcal{H}_\beta^A)$$

---

[1] For a characterization of which the reader is referred to, e.g., [10]. However, see also footnote 4.

[2] For the sake of readability, the subscript Ω will be dropped whenever the context allows it.

Furthermore, put

$$x \in \mathcal{H}_+^A(\alpha) :\equiv \exists \beta < \alpha \forall \delta < \alpha(\beta \leq \delta \to x \in \mathcal{H}_\delta^A)$$
$$x \in \mathcal{H}_-^A(\alpha) :\equiv \exists \beta < \alpha \forall \delta < \alpha(\beta \leq \delta \to x \notin \mathcal{H}_\delta^A)$$
$$x \in \mathcal{H}_+^A(\infty) :\equiv \exists \beta \forall \delta(\beta \leq \delta \to x \in \mathcal{H}_\delta^A)$$
$$x \in \mathcal{H}_-^A(\infty) :\equiv \exists \beta \forall \delta(\beta \leq \delta \to x \notin \mathcal{H}_\delta^A)$$

For a fixed **K**, the axioms of the theory QID(**K**) are then grouped as follows:

I. *Logical axioms*, comprising a complete formalization of classical *first-order predicate logic with equality*;[3]

II. *Arithmetical axioms*, with the axioms of Peano arithmetic and the full schema of complete induction;

III. *Ordinal-theoretic axioms*, with standard assumptions on the ordering, on ordinal individual constants, and with the list of defining equations on the stock of primitive recursive functions on the ordinals;[4]

IV. *Transfinite induction*, for all formulas of the language $\mathcal{L}(\mathbf{K})$;

V. A *QID-group* of axioms, featuring, for every **K**-operator form $A(x, X)$, the universal closure of

$(QID.1)$ $\quad x \in \mathcal{H}_{0_\Omega}^A \to x \neq x$
$(QID.2)$ $\quad x \in \mathcal{H}_{\alpha+1}^A \leftrightarrow A(x, \mathcal{H}_\alpha^A)$
$(QID.3)$ $\quad Lim(\lambda) \to [x \in \mathcal{H}_\lambda^A \leftrightarrow \exists \alpha < \lambda \forall \beta < \lambda(\alpha \leq \beta \to x \in \mathcal{H}_\beta^A)]$
$(QID.4)$ $\quad \exists \lambda [Lim(\lambda) \wedge \alpha < \lambda \wedge (\mathcal{H}_+^A(\lambda) \equiv \mathcal{H}_+^A(\infty)) \wedge (\mathcal{H}_-^A(\lambda) \equiv \mathcal{H}_-^A(\infty))]$

(where, $Lim(\lambda) :\equiv (0 < \lambda \wedge \forall \beta < \lambda(\beta + 1 < \lambda))$, and $\alpha \leq \beta :\equiv (\alpha < \beta \vee \alpha = \beta)$).

As usual, for most of the arguments concerning quasi-inductive processes it is required to state how to deal with generalized ordinal additions. In this sense it can be proved

LEMMA 1. *(a) For every **K**-operator form A, QID(**K**) proves:*

(a.1) $x \in \mathcal{H}_{\alpha+0}^A \leftrightarrow x \in \mathcal{H}_\alpha^A$

(a.2) $x \in \mathcal{H}_{\alpha+\beta'}^A \leftrightarrow x \in \mathcal{H}_{(\alpha+\beta)'}^A$

(a.3) $Lim(\lambda) \to \forall x(x \in \mathcal{H}_{\alpha+\lambda}^A \leftrightarrow \exists \beta < \lambda \forall \delta < \lambda(\beta \leq \delta \to x \in \mathcal{H}_{\alpha+\delta}^A))$

*(b) For every **K**-operator form A, QID(**K**) proves*

$$\forall \alpha \beta \gamma (\mathcal{H}_\alpha^A \equiv \mathcal{H}_\beta^A \to \mathcal{H}_{\alpha+\gamma}^A \equiv \mathcal{H}_{\beta+\gamma}^A)$$

---
[3] Notice then, that our theories are of first-order in both sorts of individual variables.
[4] The reader should notice that, for the sake of the results which will be quoted below, only a finite stock of axioms on ordinal functions are in fact required (in particular, those for ordinal addition, multiplication and their inverses). For an exact list see [2].

Proof. (a.1) and (a.2) are trivial application of the ordinal sum defining equations and the identity axioms.

The proof of (a.3) is, for the most part, a matter of ordinal arithmetic. As a matter of facts:

$$
\begin{aligned}
Lim(\lambda) &\rightarrow Lim(\alpha + \lambda) \\
&\rightarrow x \in \mathcal{H}^A_{\alpha+\lambda} \\
&\leftrightarrow \exists \beta < \alpha + \lambda \forall \delta < \alpha + \lambda (\beta \leq \delta \rightarrow x \in \mathcal{H}^A_\delta)
\end{aligned}
$$

Hence one proves

$$\exists \beta < \alpha + \lambda \forall \delta < \alpha + \lambda (\beta \leq \delta \rightarrow x \in \mathcal{H}^A_\delta) \leftrightarrow \exists \beta < \lambda \forall \delta (\beta \leq \delta \rightarrow x \in \mathcal{H}^A_{\alpha+\delta})$$

by making use of the defining properties of ordinal addition.

Having proved that, (b) then comes from an easy induction on γ, using (a.1–3). ∎

Then, by standard arguments (see, e.g., [4, pp. 390–394]), it is possible, within QID(**K**), to prove those results which help describing the structure of revision processes.

First, let

$$
\begin{aligned}
STAB_A(\lambda) &:= [Lim(\lambda) \wedge (\mathcal{H}^A_+(\lambda) \equiv \mathcal{H}^A_+(\infty)) \wedge (\mathcal{H}^A_-(\lambda) \equiv \mathcal{H}^A_-(\infty))] \\
PER_A(\alpha) &:= STAB_A(\alpha) \wedge \exists \beta[0 < \beta \wedge \forall \gamma (\mathcal{H}^A_\alpha \equiv \mathcal{H}^A_{\alpha+\beta\gamma})]
\end{aligned}
$$

By the first line, one wants to concisely refer to those limit ordinal, which exist in arbitrary amount by $(QID.4)$, and which "stabilize" the quasi-inductive construction. The second formula is a way to express the property of those "stable" ordinals (in the previous sense of the expression) which happen to occur again and again in the quasi-inductive iteration of A, according to a certain "period".

Then, one has:

PROPOSITION 2. *For every* **K**-*operator form* A

$$QID(\mathbf{K}) \vdash \exists \alpha PER_A(\alpha)$$

Proof. Take

$$
\begin{aligned}
\alpha &= \min \xi.(STAB_A(\xi)) \\
\delta &= \min \xi.(\alpha < \xi \wedge STAB_A(\xi)) \\
\beta &= \delta - \alpha
\end{aligned}
$$

Then, the theorem follows by induction on the γ occurring in the formula defining $PER_A(\alpha)$. ∎

For α such that $PER_A(\alpha)$ holds, let $p(\alpha)$ indicate its period. Moreover, let[5]

$$\pi_0 := \min \xi.(PER_A(\xi))$$

Then, more detailed information as to the structure of the quasi-inductive iterations can be obtained by putting

$$\begin{aligned}
BOUND_A(\alpha) &:= \exists \beta \forall \gamma (\beta \leq \gamma \to \mathcal{H}_\alpha^A \not\equiv \mathcal{H}_\gamma^A) \\
CONF_A(\alpha) &:= \forall \beta \exists \gamma (\beta \leq \gamma \wedge \mathcal{H}_\alpha^A \equiv \mathcal{H}_\gamma^A) \\
CYCL_A(\alpha) &:= \exists \beta < \pi_0 + p(\pi_0)(\pi_0 \leq \beta \wedge \mathcal{H}_\alpha^A \equiv \mathcal{H}_\beta^A)
\end{aligned}$$

LEMMA 3. *In* **QID(K)** *we have, for every* **K**-*operator form A:*
(i) $CONF_A(\alpha) \wedge \alpha \leq \beta \to CONF_A(\beta)$
(ii) $BOUND_A(\alpha) \wedge \beta \leq \alpha \to BOUND_A(\beta)$
(iii) $CONF_A(\alpha) \leftrightarrow CYCL_A(\alpha)$
(iv) $x \in \mathcal{H}_{\pi_0}^A \wedge \pi_0 < \alpha \to x \in \mathcal{H}_\alpha^A$

Proof. (i) By contradiction, using ordinal subtraction, Lemma 1.(b) and right-monotonicity of ordinal addition.

(ii) A similar argument to the one used for (i) applies.

(iii) The direction from right to left comes from (i), and $CONF_A(\pi_0)$ which in turn is an easy consequence of $PER_A(\pi_0)$.

As to the other direction, for some $\beta > \pi_0$ it is, by hypothesis, $\mathcal{H}_\beta^A \equiv \mathcal{H}_\alpha^A$. But, $\pi_0 < \beta$ implies, by ordinal division $\beta = \pi_0 + p(\pi_0)\gamma + \delta$ for some $\delta < p(\pi_0)$. But then, by $PER_A(\pi_0)$ and, again, Lemma 1.(b) we have

$$\mathcal{H}_{\pi_0}^A \equiv \mathcal{H}_{\pi_0+p(\pi_0)\gamma}^A \to \mathcal{H}_{\pi_0+\delta}^A \equiv \mathcal{H}_{\pi_0+p(\pi_0)\gamma+\delta}^A \equiv \mathcal{H}_\beta^A$$

(iv) An immediate consequence of the stabilization property of π_0 and (iii). ∎

To summarize, **QID(K)** proves that, for every given $A(x, X)$ operator form, there exists an ordinal π_0 which: (1) is the least stable point of the corresponding quasi-inductive definition; (2) it is then such that, below it, there exist only 'bounded' points, i.e. levels in the iteration which do not appear again after π_0 itself; (3) it is followed only by 'cyclic' points, which, on the contrary, do always reappear, and which coincides with the 'confinal' ones; (4) it is such that elements of $\mathcal{H}_{\pi_0}^A$, which by the way are the stable ones, feature a specific persistency property.

Next sections deal with proof-theoretic results concerning an instance of the family of theories I have just introduced.

---

[5]It is clear that this π_0 should carry with it a trace of the connection with the operator $A(x, X)$, to the iteration of which it refers to. However, as far as the results below are concerned this dependence can be dropped for the sake of readability. Hence, it will explicitly appear in formulas involving $\mathcal{H}_{\pi_0}^A$ only.

3 A result on lower bound

The results I present in this section are worth emphasizing in at least two senses. First, because they go in the direction of obtaining proof-theoretic lower bounds for theories of quasi-inductive definitions as I introduced them, by means of a comparison with formal frameworks for the ordinary (monotone) inductive case. In a second place, because they throw some light on the syntactic characterization of that subclass of quasi-inductive constructions which, beside stable points, admit also fixed points in the usual sense of the expression (namely, levels H_α^A such that $H_\alpha^A \equiv H_{\alpha+1}^A$)[6].

Recall that an operator form $A(x, X)$ of $\mathcal{L}_{Ar} \cup \{X^1\}$, is $(X\text{-})positive$ if and only if X has only positive occurrences in $A(x, X)$ (i.e., it occurs in the scope of an even number of negations). The collection of positive (negative) operator forms is inductively defined from formulas of $\mathcal{L}_{Ar}$, and formulas $X(t)$ ($\neg X(t)$) for t term, by closing it under boolean connectives and quantifiers applied to positive and negative operator forms.

It is well known that positive operator forms are related to *monotone* operators. In order to encompass theories for *iterated* inductive definitions one is required to deal with *parametric* operator forms, namely formulas $A(x, X, Y)$ in the language $\mathcal{L}_{Ar} \cup \{X^1, Y^1\}$ which are X-positive, whereas Y is free to occurr positively and/or negatively.

On operator forms of this sort it is possible to prove the following:[7]

LEMMA 4. *(i) For every $\boldsymbol{K}$ and for every X-positive $\boldsymbol{K}$-operator form $A(x, X, Y)$, QID($\boldsymbol{K}$) proves*

$$A(x, B, D) \land \forall x(B(x) \to C(x)) \to A(x, C, D)$$

where $B(x), C(x), D(z)$ are arbitrary formulas of $\mathcal{L}(\boldsymbol{K})$ with displayed free variables.

(ii) For π_0 as defined in the previous section, and for a given X-positive $\boldsymbol{K}$-operator form $A(x, X, Y)$, we have

$$\vdash_{\mathsf{QID}(\boldsymbol{K})} x \in \mathcal{H}_{\pi_0}^A \leftrightarrow A(x, \mathcal{H}_{\pi_0}^A, B)$$

where $B(x)$ is an arbitrary formula of $\mathcal{L}(\boldsymbol{K})$.

(iii) Given an X-positive $\boldsymbol{K}$-operator form $A(x, X, Y)$, the following is provable in QID($\boldsymbol{K}$) for every formula $B(x), C(x)$ from $\mathcal{L}(\boldsymbol{K})$

$$\forall \alpha[(A(x, B, C) \to B(x)) \to (x \in \mathcal{H}_\alpha^A \to B(x))].$$

Proof. (i) By an easy induction on $A(x, X, Y)$, with a secondary induction so to cope with the case $A(x, X, Y) \equiv \neg B(x, X, Y)$ with $B(x, X, Y)$ X-negative.

---

[6] A more general result on this latter topic, the proof of which requires the machinery of Infinite Time Turing Machines, can be found in [15, Prop. 8].

[7] I implicitly assume to work with a slight modification of the syntax of our theories as I defined it above, so to let QID($\boldsymbol{K}$) have predicates $\mathcal{H}^A$ also for every $\boldsymbol{K}$-operator form A of this parametric sort. It is immediate to verify that the results from Section 2 goes through for the accordingly adjusted theories.

(ii) The direction from left to right is an application of Lemma 3.(iv) and $(QID.2)$. As to the other direction, using again Lemma 3.(iv), (i), $(QID.2)$ one shows

$$A(x, \mathcal{H}^A_{\pi_0}, B) \to \forall \delta(\pi_0 < \delta \to x \in \mathcal{H}^A_{\delta+1}).$$

This can be easily expanded by induction to an argument showing $x \in \mathcal{H}^A_{\pi_0+1} \to x \in \mathcal{H}^A_\delta$ for ordinals δ's greater than π_0, which, in turn, yields, $x \in \mathcal{H}^A_{\pi_0+p(\pi_0)}$ under the assumption. The theorem then follows by the periodicity property of π_0.

(iii) By induction on α ordinal, where both non-trivial cases of the induction argument follow from (i), IH, and axioms $(QID.2-3)$. ∎

This result can be used to define an interpretation for theories ID_α of *transfinitely iterated* inductive definitions into suitably chosen instances of theories $\mathsf{QID}(\mathbf{K})$. Theories ID_α were conceived in order to axiomatically characterize those inductively defined collections of natural numbers which are built by referring to previously constructed ones, along recursive well-orderings α's.

The syntax of these theories comes from a natural modification of the one for theories of 'plain' inductive definitions. Then, for a fixed (primitive recursive) well-ordering $\prec$ up to a countable ordinal α, the language $\mathcal{L}_\alpha$ of ID_α is obtained by adding to the language of Peano arithmetic unary predicate constants $\mathcal{P}^A_y$ for every X-positive operator form $A(x,y,X,Y)$.[8]

As made explicit by the axioms of theories of this sort, for a given operator form $A(x,y,X,Y)$, x has the role of representing elements of the constructed sets along the iteration, while y does the same for the levels of the iteration itself.

As a matter of facts, the axioms of ID_α are those of Peano Arithmetic plus:

$(ID.1)_\alpha$ $\quad \forall \beta \prec \alpha \forall x(A_\beta(x, \mathcal{P}^A_\beta, \mathcal{P}^A_{\prec\beta}) \to \mathcal{P}^A_\beta(x))$
$(ID.2)_\alpha$ $\quad \forall \beta \prec \alpha \forall x((A_\beta(x, B, \mathcal{P}^A_{\prec\beta}) \to B(x)) \to \forall z(\mathcal{P}^A_\beta(z) \to B(z)))$
$(TI)_\alpha$ $\quad \forall \beta \prec \alpha(\forall \gamma \prec \beta(B(\gamma) \to B(\beta))) \to \forall \beta \prec \alpha B(\beta)$

Here, $B(x)$ is an arbitrary formula of $\mathcal{L}_\alpha$, while $A_y(x,X,Y)$, $\mathcal{P}^A_y(x)$ and $\mathcal{P}^A_{\prec y}(x)$ abbreviate respectively $A(x,y,X,Y)$, $\mathcal{P}^A(\langle x,y\rangle)$, and $((x)_2 \prec y \land \mathcal{P}^A(x))$, for $\langle \cdot \rangle$ primitive recursive pairing function with projections $(\cdot)_i$ ($i = 1, 2$).

So, we have:

THEOREM 5. *For every fixed α and $\prec$, there exists a translation $(\cdot)^\circ_\alpha$ of $\mathcal{L}_\alpha$ into $\mathcal{L}(\Pi^0_\infty)$ such that*

$$\mathsf{ID}_\alpha \vdash A \Rightarrow \mathsf{QID}(\Pi^0_\infty) \vdash (A)^\circ_\alpha$$

Proof. For a given operator form $A(x,y,X,Y)$, let $A'(z,X,Y)$ indicate $A((z)_1, (z)_2, X, Y)$.

---

[8] As customarily, lower case greek letters will be used for arithmetical terms which are in the field of the well-ordering relation $\prec$.

Then, the relevant clauses for $(\cdot)_\alpha^\circ$ reads

$$(\mathcal{P}_y^A(x))_\alpha^\circ := \langle x,y \rangle \in \mathcal{H}_{\pi_0}^{A'}$$
$$(\mathcal{P}_{\prec y}^A(x))_\alpha^\circ := ((x)_2 \prec y \wedge x \in \mathcal{H}_{\pi_0}^{A'})$$

The definition of the translating function is further completed by the usual clauses for the commutation of it with respect to logical connectives and quantification.

The theorem then follows by an easy induction on the length of the proof of A in ID_α. ∎

4 Kripke-Platek with ordinals and existence of Σ_3-substructures

The set-theoretical setting I will work with is based on a version of Kripke-Platek theory for admissible sets with urelemente and ordinals, featuring a strong reflection assumption which is tailored so to cope with the stabilization axiom from the QID-group above. As in the previous section, we stick here to the instance $\mathsf{QID}(\Pi_\infty)$ of our family of theories.[9]

The language $\mathcal{L}_{KP}^\Omega(\Sigma_3)$ of the theory which I will refer to as KPu_{S3}^Ω, extends the language of Peano arithmetic by the binary relation symbol $\in$ for membership, a set constant N for natural numbers and the unary relation symbol S^1, Ad^1 and Ord^1 for sets, admissible sets and ordinals respectively.

Equality is defined by the formula

$$x = y := (x \in \mathsf{N} \wedge y \in \mathsf{N} \wedge x =_N y) \vee (\mathsf{S}(x) \wedge \mathsf{S}(y) \wedge \forall z \in x(z \in y) \wedge \forall z \in y(z \in x))$$

where $=_N$ is the equality relation for natural numbers.

For its standard part, the theory KPu_{S3}^Ω is formulated à la Jäger (see [7]). This means that its axioms are grouped as follows:

I. *Ontological Axioms.*

    I.1 $x \in \mathsf{N} \vee \mathsf{S}(x)$

    I.2 $\vec{x} \in \mathsf{N} \to f(\vec{x}) \in \mathsf{N}$

    I.3 $R(\vec{x}) \to \vec{x} \in \mathsf{N}$

    I.4 $x \in y \to \mathsf{S}(y)$

    I.5 $\mathsf{Ad}(x) \to Tran(x) \wedge \mathsf{N} \in x$

    I.6 $\mathsf{Ad}(x) \wedge \mathsf{Ad}(y) \to x \in y \vee x = y \vee y \in x$

    I.7 $\mathsf{Ad}(x) \to A^{(x)}$

    I.8 $\mathsf{Ord}(x) \to Tran(x) \wedge (\forall z \in x)Tran(z)$

---

[9] The reader should compare the main result of this section with the corresponding one in [2]. Further, the remark there concerning quasi-inductively iterated functions to be provably unique in a set-theoretical setting, allows to prove in fact the result of this section with respect to KPu_{S2}^Ω (which is defined as expected).

[where f and R are function and relation symbols from $\mathcal{L}_{Ar}$; $Tran(x)$ abbreviates the formula $(\mathsf{S}(x) \land \forall z \in x \forall w \in z(w \in x))$; and, in I.7, A is any of the statements from group III below, with $A^{(x)}$ indicating the formula which is obtained from A by relativizing all of its unbounded quantifiers to x.]

II. *Number-theoretic axioms.* To this group belong the axioms for Peano arithmetic except the induction schema.

III. *Kripke-Platek axioms.* Here we have the (universal closure of the) following set-theoretical assumptions

$$(\mathsf{PAIR}) \quad \exists z(x \in z \land y \in z)$$
$$(\mathsf{T-HULL}) \quad \exists z(x \subseteq z \land Tran(z))$$
$$(\Delta_0 - \mathsf{SEP}) \quad \exists z \forall x [x \in z \leftrightarrow x \in y \land A(x)]$$
$$(\Delta_0 - \mathsf{COLL}) \quad \forall x \in w \exists y B(x,y) \to \exists z \forall x \in w \exists y \in z B(x,y)$$

[where the latter two schemas are restricted to Δ_0-formulas $A(z)$ and $B(z,w)$.]

IV. *Induction principles.* I assume to have both full complete induction on the natural numbers, and full $\in$-induction:

$$(\mathsf{IND}_N) \quad \forall x \in \mathsf{N}(\forall y \in \mathsf{N}(y < x \to A(y))) \to A(x)) \to \forall x \in \mathsf{N}(A(x))$$
$$(\mathsf{IND}_\in) \quad \forall x(\forall y \in x(A(y) \to A(x)) \to \forall x A(x)$$

V. Σ_3-*substructurality.* Let:

$$x \prec_3 V :\equiv \mathsf{Ord}(x) \land \forall z_1, ..., z_n \in x(\varphi(z_1, ..., z_n) \leftrightarrow \varphi^{(x)}(z_1, ..., z_n))$$

for every Σ_3-formula $\varphi(x_1, ..., x_n)$ with displayed free variables.

Then, I admit as an axiom schema

$$(\Sigma_3 - \mathsf{SUB}) \quad \forall y \exists x(y \in x \land x \prec_3 V)$$

One has of course:

FACT 6. $\mathsf{KPu}_{S3}^{\Omega}$ *is a conservative extension of* KPu_{S3}, *which is obtained from the former theory by deleting axiom* I.8.

Further, the KP-part of the axioms suffices to prove:

PROPOSITION 7. *The following statements are provable in* KPu *(even restricting $\in$-induction to Δ_0-formulas):*

1. The (UNION) axiom

$$\exists z [\mathsf{S}(z) \land \forall x(x \in z \leftrightarrow \exists y \in w(x \in y))]$$

2. $\exists z[\mathsf{S}(z) \land \forall x(x \in z \leftrightarrow (x = y \lor x = w))]$.

3. *The schema of separation up to Δ formulas.*

4. The schema of collection for Σ formulas.[10]

For a proof of 1, the reader is referred to, e.g., [13]. 2 is an easy consequence of Δ_0 separation and (PAIR). For 3 and 4, see [5, pp. 50–52].

The same set of standard assumptions allows one to prove everything that is required for the basic properties of the ordinals, and ordinal arithmetic (see, e.g, [1]).

As to the novel assumption (Σ_3-SUB), a quick way to get into the strength of it is to consider the next two propositions.

Before going into that, let

$$s(x) = \min \xi . (x \in \xi \wedge \xi \prec_3 V)$$

the existence of which, for every x, is ensured by (Σ_3-SUB) and (IND$_\in$). Then:

PROPOSITION 8. *The following statements are provable in* $\mathsf{KPu}_{S3}^{\Omega}$:

1. (Σ_3 − SEP).

2. (Σ_3 − COLL).

3. $y \prec_3 V \to Lim(y)$.[11]

Proof. 1. For every a set and for every Σ_3-formula $A(x)$, argue straightforwardly, using (Σ_3-SUB), with respect to:

$$b = \{x \in a \mid A^{(s(a))}(x)\}$$

which exists by (Δ_0 − SEP).

2. The result is immediate, by (Σ_3-SUB), with respect to $s(a)$ for every given set a. Alternatively, one can even find, by applying (Δ_0 − COLL) and (Δ_0 − SEP)

$$b = \{y \mid (\exists x \in a) A^{s(a)}(x, y)\}$$

which, by (Σ_3-SUB) again, gives exactly the range of $A(x, y)$.

3. Apply (Σ_3 − SUB) to $\exists x(x = a + 1)$, for $a \in y$ set. ∎

Inspection of the argument usually yielding Σ-Recursion in plain Kripke-Platek set theory, shows that this essentially follows from Σ-Collection which is provable in KPu (see, e.g, [5, pp. 54–55], [1, pp. 26–28]). Then, by straightforwardly modifying it in the light of PROP. 8 above, one gets:

PROPOSITION 9 (Σ_3 Recursion). *Let g be a $(n+2)$-ary Σ_3 function.*[12] *Then a Σ_3 $(n+1)$-ary function f can be defined in such a way that*

$$\vdash_{\mathsf{KPu}_{S2}^{\Omega}} f(x, \vec{w}) = g(x, \vec{w}, f \upharpoonright x)$$

---

[10] Owing to the Σ reflection principle (see [1, p. 16]), this class of formulas is, provably in KPu, equivalent to Σ_1 formulas.

[11] Where $Lim(x) := (\emptyset < x \wedge \forall y < x(y + 1 < x))$, with $y + 1$ abbreviating $y \cup \{y\}$.

[12] If Φ is a class of formulas complexity, then a function f is said to be a k-ary Φ *function* over the theory S iff its graph is provably of complexity Φ in S. Since we are interested in function which are Φ over KPu only, reference to the formal framework is omitted here.

(where $f \upharpoonright x := \{\langle w, \vec{z}, f(w, \vec{z})\rangle \mid w \in x\}$).

From this, by applying the usual strategy for converting the above formulation of the recursion result into the one 'by clauses' (with the liminf operation substituting the usual limit one), it is possible to obtain the following corollary:

COROLLARY 10 (Quasi-recursion). *Given a unary function g which associates subsets of N to subsets of N, there exists a function h on the ordinals satisfying, provably in* $\mathsf{KPu}_{S2}^{\Omega}$

$$h(0) = \emptyset$$
$$h(\alpha+1) = g(h(\alpha))$$
$$h(\lambda) = \liminf_{\beta<\lambda} h(\beta), \; \lambda \; limit$$

For an h solving the equations of COR. 10 given a function g, we write h_g. Then, for the sake of the result below, we shall actually be working in the extension of our theory $\mathsf{KPu}_{S3}^{\Omega}$ that contains *terms* for all functions h_g of this sort. It is possible to prove that this theory, let us call it $\mathsf{H} - \mathsf{KPu}_{S3}^{\Omega}$, is a conservative extension of the $\mathsf{KPu}_{S3}^{\Omega}$ system of axioms for what is needed in order to prove the subsequent theorem. Owing to that, we will omit any reference to the extended theory in the next result.

Now, all of the basic ingredients needed for the announced translation result have been displayed. It remains to state the theorem, which is as follows:

THEOREM 11. *There exist a function $(\cdot)^{\bullet}$ yielding formulas of $\mathcal{L}_{KP}^{\Omega}(\Sigma_3)$ out of formulas of $\mathcal{L}(\Pi_{\infty})$, such that, for every formula A*

$$\mathsf{QID}(\Pi_{\infty}) \vdash A \Rightarrow \mathsf{KPu}_{S3}^{\Omega} \vdash (A)^{\bullet}$$

Proof. First, notice that on terms translation the only thing that one is required to mention are the counterparts of 0_{Ω} and ω, which are obvious. Having said that, the defining clauses for $(\cdot)^{\bullet}$ go as follows:[13]

$$\begin{aligned}
(s = t)^{\bullet} &:= s = t \\
(s <_{\Omega} t)^{\bullet} &:= s \in t \\
(s \in \mathcal{H}_{\alpha}^{A})^{\bullet} &:= s \in h_{g_A}(\alpha), \; [\text{where } g_A := \{\langle x,y\rangle \mid A^{\mathsf{N}}(y,x)\}] \\
(\neg A)^{\bullet} &:= \neg(A^{\bullet}) \\
(A \wedge B)^{\bullet} &:= (A^{\bullet} \wedge B^{\bullet}) \\
(\forall x A(x))^{\bullet} &:= \forall x(x \in \mathsf{N} \to A^{\bullet}(x)) \\
(\forall \alpha A(\alpha))^{\bullet} &:= \forall x(\mathrm{Ord}(x) \to A^{\bullet}(x)).
\end{aligned}$$

The theorem is then proved by an easy induction on the length of the proof of A in $\mathsf{QID}(\Sigma_2)$, using COR. 10 for axioms $(QID.1-3)$ and $(\Sigma_3 - \mathsf{SUB})$ for $(QID.4)$. ∎

---

[13] By abuse of notation, I have retained the same symbol for terms on both sides of the translation clauses for formulas.

5 Comments and further work

In this final section, I would like to hint at some perspective goal as it was said in the introduction. To be very concrete, I will mention something on this in the course of listing below the issues that, in my opinion, should guide the future developments of the investigation.

1. First item in the list is of course the need for refining the shape of the analysis I began to pursue in the present contribution. This requires to make it clear whether or not the interpretation theorems I presented can be turned into results establishing *sharp* bounds. As it is customary in a proof-theoretic sort of a logical investigation, this might be done by providing a full ordinal analysis for the theories here at stake. It is easy to predict that such an outcome will not be easily achieved, though the intermediate steps in the pursuing of it, which usually requires developing *ad hoc* tecniques, might be of independent interest.

2. In view of this goal, it could be useful to pursue in the direction of subsystems of second-order arithmetic, a similar investigation to what I started to do here in the direction of set theory. The most interesting aspect in doing this is related with a conjecture made by P. Welch, which, in turn, generalizes a result by M. Rathjen [12]. Welch's claim concerns the fact that towers of Σ_2-*extendibles* (i.e., levels of the Gödel's hierarchy of constructible sets admitting Σ_2-end extensions), which can be naturally related to quasi-inductive definitions, might be intrinsically connected with the proof theory of $(\Pi_3^1\text{-CA})$. Further, it is suggested that such a connection might prove to be generic (i.e., to hold unchanged for Σ_n-extendibles and $(\Pi_n^1\text{-CA})$, for every n). It should be noticed that the full version of our theory for arithmetical quasi-inductive definition seems to go beyond that. It is not impossible, however, that something which is worth working out in this respect may turn out by combining and comparing my approach with some recent suggestions of Welch's (see [15]).

3. As a final item in the list, I would like to mention something which goes in the direction of exploiting the revision-theoretic pattern directly in the costruction of mathematical structures. The core assumption we have used here for the 'upper' embedding result, (Σ_3-SUB), is obtained as a modification of a similar assumption (though restricted to Σ_1 formulas), that M. Rathjen used in order to provide an interpretation of certain non-monotone inductive construction of universes within Martin-Löf type theory (see Rathjen [11], and the related work by A. Setzer [14]). Rathjen's result is part of an interesting discussion on the limits of constructive methods in metamathematics (the use of which, in turn, can be connected with a reformulation of an Hilbert-style program in the foundations of mathematics, as explained by Rathjen in

that paper). At the same time, theories for non-monotone inductive definitions by G. Jä [8] have been proved to be related to the model construction of a theory based on S. Feferman's T_0 for Explicit Mathematics (see [9]), which is conceived in order to capture some form of a predicative approach to mathematics. In terms of our presentation of quasi-inductive definitions as a generalization of the inductive cases, these results suggest that it might be worth trying to devise some structure directly conveying the properties of quasi-inductive constructions, into an appropriately chosen framework for constructive mathematics. As a result, one might have a discussion on the limits of constructive mathematics to be re-assessed accordingly. Though the goal is ambitious and, to be honest, not within our grasp in the short time, we think it to be a sensible aim to think of for the far-future development of this investigation.

BIBLIOGRAPHY

[1] J. Barwise. *Admissible Sets and Structures*. Springer-Verlag, Berlin 1975.
[2] R. Bruni. A note on theories for quasi-inductive definitions. *The Review of Symbolic Logic*, 2: 684–699, 2009.
[3] J. Burgess. The truth is never simple. *The Journal of Symbolic Logic*, 51: 663–681, 1986.
[4] A. Cantini. *Logical Frameworks for Truth and Abstraction. An Axiomatic Study*. Elsevier, Amsterdam 1996.
[5] K. Devlin. *Constructibility*. Springer-Verlag, Berlin 1984.
[6] A. Gupta and N. Belnap. *The Revision Theory of Truth*. MIT Press, Cambridge MA 1993.
[7] G. Jäger. The strength of admissibility without foundation. *The Journal of Symbolic Logic*, 49: 867–879, 1984.
[8] G. Jäger. First-order theories for non-monotone inductive denitions: recursively inaccessible and Mahlo. *Journal of Symbolic Logic*, 66: 1073–1089, 2001.
[9] G. Jäger and T. Studer. Extending the system T_0 of Explicit Mathematics: the limit and Mahlo axioms. *Annals of Pure and Applied Logic*, 114: 79–101, 2002.
[10] R. B. Jensen and C. Karp. Primitive recursive set functions. In D. Scott (ed.) *Axiomatic Set Theory*. Part 1, pages 143–176. Proceedings of Symposia in Pure Mathematics, vol. 13, part 1, American Mathematical Society, Providence, Rhode Island 1971.
[11] M. Rathjen. The constructive Hilbert program and the limits of Martin-Löf type theory. *Synthese*, 147: 81–120, 2005.
[12] M. Rathjen. An ordinal analysis of parameter free Π_2^1-comprehension. *Archive for Mathematical Logic*, 44: 263–362, 2005.
[13] V. Salipante. *On the Consistency Strength of the Strict Π_1^1 Reflection Principle*. Ph.D. Thesis, Institut für Informatik und Angewandte Mathematik, Universität Bern, 2005.
[14] A. Setzer. Extending Martin-Löf type theory by one Mahlo-universe. *Archiv for Mathematical Logic*, 39: 155–181, 2000.
[15] P. Welch. Weak systems of determinacy and arithmetical quasi-inductive denitions. Revised version, to appear in *the Journal of Symbolic Logic*, 2009.

Remarks on a proof-theoretic characterization of polynomial space functions

GIACOMO CALAMAI

1 Introduction

Parallel computations have been studied for long as a tool for classifying natural collections of sub-recursive functions: it is well known, for instance, that the class of functions computable in deterministic polynomial space coincides with the set of languages decidable in (parallel) alternating polynomial time.

Several machine-independent characterizations of complexity classes have been developed in the field of the so called *Implicit Computational Complexity Theory*, by introducing concepts like *ramification* and *data tiering* (see, for instance, [6, 1, 2]). The use of ramified data links computational complexity to levels of definitional abstraction and clarifies the correspondence between subrecursion and complexity, by requiring that recurrence principles respect the separation between data objects which are used computationally in different guises.

Ramified recurrence with parameter substitution was introduced by Leivant and Marion in 1995 [7] as a quite general variant of ramified recurrence, where the parameters of a recursive call may be altered at each iteration using previously defined functions, thereby enabling the simulation of parallel alternating computing. It follows that the functions definable by ramified recurrence with parameter substitution, using only standard constructors in the algebra of binary strings, are precisely the polynomial space computable functions.

We give here a proof-theoretic characterization of poly-space operations, generalizing the results of Cantini [5], which were proven for polynomial time functions. To this aim we propose a classical ramified sequent calculus with extensionality on the set of binary strings, which comprises full untyped combinatory logic together with a principle of ramified Π_i^s-induction with parameter substitution.

Applicative systems [4, 5, 8] provide a natural framework for a proof-theoretic approach to computational complexity: all objects may be regarded as operations or rules, in the sense of combinatory logic, together with binary strings. We also assume a many-sorted structure with copies $\mathbb{W}_0, \mathbb{W}_1, \mathbb{W}_2, ...$ of the algebra $\mathbb{W} = \{0, 1\}^*$ of binary strings as our tiered universes. These ramification conditions impose a strictly predicative regime, which distinguishes between different uses of variables in induction schemas. In addition, parameter substitution allows the representation of parallel computing, by giving a tree-structure to the usual recursion on notation scheme and leading to a branching of the computation flow.

The axiom of positive ramified i-safe induction with parameter substitution has the following general aspect. It includes, for each positive i-safe formula $C(x,v)$ such that all tiers in C are strictly bounded by i (i.e. **tier**(C) $< i$, see the definition below),

$$(\forall v \in W_k)(\vec{\sigma}_0(\vec{v}) \in W_k)...(\forall v \in W_k)(\vec{\sigma}_1(\vec{v}) \in W_k) \wedge$$
$$(\forall v \in W_k)C(\epsilon, v) \wedge$$
$$(\forall x \in W_i)(\forall v \in W_k)((C(x, \vec{\sigma}_0 v) \rightarrow C(x0, v)) \wedge$$
$$(\forall x \in W_i)(\forall v \in W_k)((C(x, \vec{\sigma}_1 v) \rightarrow C(x1, v)) \rightarrow$$
$$(\forall x \in W_i)(\forall v \in W_k)C(x, v)$$

where $\vec{\sigma}_0 \equiv \sigma_{0_1},...,\sigma_{0_l}$, $\vec{\sigma}_1 \equiv \sigma_{1_1},...,\sigma_{1_l}$ and $i, k, l \in \{0, 1, 2, ...\}$.

Intuitively, the above schema is a ramified counterpart of a principle of *positive induction with substitution* for primitive recursion (see [3]), where substitutions of recursive parameters have to respect the main tiering proviso of i-safeness. Nevertheless, the ramified sequent calculus which will be introduced below comprehends a slight strengthening of this induction principle, where the crucial substitution functions can be effectively proved to be provably total (see §4 below).

We prove that the word algebra introduced by Leivant and Marion can be naturally embedded in our calculus, proving a completeness result. Analogously, soundness is carried out by adopting a realizability interpretation inside our sequents system: as realizing operations we adopt directly vector-valued functions in Leivant-Marion's algebra, in order to realize positive sequents of our system.

Hence it turns out that the ramified sequent calculus characterizes exactly those terms (i.e. programs) defining polynomial space operations on the algebra of binary words.

2 Syntactical framework

Our basic ramified language $\mathcal{L}_\mathbf{r}$ contains

- countably many individual variables $x_1, x_2, x_3, ...$;
- logical constants $\rightarrow, \wedge, \vee, \exists, \forall$;
- predicate symbols W_i (with $i \in \{0, 1, 2, ...\}$) for copies $\mathbb{W}_0, \mathbb{W}_1, \mathbb{W}_2, ...$ of the many-sorted structure of the algebra $\mathbb{W}$; the symbol = for equality;
- individual constants $K, S, PAIR, L, R, \epsilon, \mathbf{s}_0, \mathbf{s}_1, pr, D$;
- binary function symbol Ap (application operation).

Terms are inductively defined from variables and constants via application Ap. x, y, z, u, v, w, f, g stand for metavariables, while t, t', t'', s, s', r, r', etc. are metavariables for terms. We write (ts) instead of $Ap(ts)$, and outer brackets are usually omitted, while the missing ones are restored by associating to the left: for instance, xyz stands for $((xy)z)$.

We adopt familiar shorthands for special terms: $\langle t,s \rangle :=$ PAIRts (= the ordered pair composed by t and s); $(t)_1 :=$ L(t) (= the left projection of t) and $(t)_1 :=$ R(t) (= the right projection of t). Also, we have $\mathbf{s}_0(\epsilon) := 0$, $\mathbf{s}_1(\epsilon) := 1$ and $t^- :=$ prt, $t0 := \mathbf{s}_0 t$, $t1 := \mathbf{s}_1 t$.

Finally, we have that D$pqrs$ converges to p (resp. q) whenever r,s denote binary strings a,b (respectively), and $a = b$ (resp. $a \neq b$): so D represents a conditional.

As usual, if α is a binary word, $\bar{\alpha}$ stands for its corresponding *numeral*: a numeral is any term obtained from the constant zero (empty sequence ϵ) by means of a finite number of successors applications.

Formulas are inductively generated by means of the logical operations and quantifiers from atomic formulas (atoms) of the form $t = s$ and W$_i t$, with $i \in \{0,1,2,...\}$. If A is an expression (term or formula), A(x) means that x may occur free in A, while A$[x := t]$ stands for the result of substituting t for the free occurrence of x. FV(A) means that "x occurs free in A". We also have the standard definition of λ-abstraction in combinatory logic.

DEFINITION 1. *If t is an arbitrary term of $\mathcal{L}_\mathbf{r}$, $\lambda x.t$ is introduced by induction on the notion of $\mathcal{L}_\mathbf{r}$-term:*

(*i.*) $\lambda x.x :=$ SKK

(*ii.*) $\lambda x.t :=$ Kt if $x \notin$ FV(t)

(*iii.*) $\lambda x.(ts) :=$ S$(\lambda x.t)(\lambda x.s)$, if $x \in$ FV(ts).

Of course, $\lambda x.t$ has exactly the same free variables of t, minus x. We use $|t|$ as the length of the term t. We also use the notation $t \in$ W$_i :=$ W$_i t$, with $i \in \{0,1,...,n\}$. We write

$$g : \mathrm{W}_i \to \mathrm{W}_i$$
$$g : \mathrm{W}_i^{n+1} \to \mathrm{W}_i$$

as abbreviation for the formulas

$$\forall x(x \in \mathrm{W}_i \to gx \in \mathrm{W}_i)$$
$$\forall x(x \in \mathrm{W}_i \to (gx : \mathrm{W}_i^n \to \mathrm{W}_i)).$$

where $i \in \{0,1,2,...\}$.

As to negation, we let $\neg$A := A $\to \bot$, where $\bot :=$ K = S.

DEFINITION 2. *A formula A is* positive *iff A is $\to$-free, i.e. iff A is inductively generated from atoms of the form $t = s$, $t \in$ W$_i$ ($i \in \{0,1,2,...\}$) by means of $\wedge, \vee, \exists, \forall$.*

Finally, we introduce the notion of *rank*.

DEFINITION 3. *Inductive definition of* **rk**(A) *(A arbitrary $\mathcal{L}_\mathbf{r}$-formulas)*

- **rk**(A) = 0 if A is positive; else

- **rk**($\neg$A) = **rk**(A) + 1;

- $\mathbf{rk}(A \circ B) = \max\{\mathbf{rk}(A), \mathbf{rk}(B)\} + 1$ where $\circ$ is a binary connective;
- $\mathbf{rk}(QxA) = \mathbf{rk}(A) + 1$, if $Q = \forall, \exists$.

$\mathbf{rk}(A)$ is called the *rank* of A.

3 The Ramified Sequent Calculus

RC^s_{ext} (*Ramified Calculus with Extensionality and induction with parameter substitution*) is a classical sequent calculus, based upon the rules of G3c of [9].

The language of RC^s_{ext} is $\mathcal{L}_\mathbf{r}$; in RC^s_{ext} sequents (finite multisets of $\mathcal{L}_\mathbf{r}$-formulas) are derived of the form $\Gamma \Rightarrow \Delta$, where Γ, Δ are (possibly empty) multisets of formulas.

The intended meaning of a sequent $A_1, ..., A_n \Rightarrow B_1, ..., B_k$ is $A_1 \wedge ... \wedge A_n$ imply $B_1 \vee ... \vee B_k$. The expression "Γ, Δ" stands for the (multi)-set-theoretic union of Γ with Δ.

A. Identity Axioms

1. $\Gamma, A \Rightarrow A, \Delta$, where $A := (t = s), t \in W_i$, and where $i \in \{0, 1, 2, ...\}$
2. $\Gamma \Rightarrow t = t, \Delta$
3. $\Gamma, t = s, t \in W_i \Rightarrow s \in W_i, \Delta$, where $i \in \{0, 1, 2, ...\}$
4. $\Gamma, t = s, r = p \Rightarrow tr = sp, \Delta$
5. $\Gamma, t = s, s = r \Rightarrow t = r, \Delta$

B. Combinatory Logic and Pairing

6. $\Gamma \Rightarrow \mathrm{K}ts = t, \Delta$
7. $\Gamma \Rightarrow \mathrm{S}tsr = tr(sr), \Delta$
8. $\Gamma \Rightarrow \mathrm{L}(\mathrm{PAIR}ts) = t, \Delta$
9. $\Gamma \Rightarrow \mathrm{R}(\mathrm{PAIR}ts) = s, \Delta$

C. Definition by Cases

10. $\Gamma, t \in W_i, s \in W_i, t = s \Rightarrow \mathrm{D}rpts = r, \Delta$, where $i \in \{0, 1, 2, ...\}$
11. $\Gamma, t \in W_i, s \in W_i \Rightarrow t = s, \mathrm{D}rpts = p, \Delta$, where $i \in \{0, 1, 2, ...\}$

D. Binary Successor and Predecessor

12. $\Gamma \Rightarrow \epsilon \in W_i, \Delta$, where $i \in \{0, 1, 2, ...\}$

13. $\Gamma, t \in W_i \Rightarrow s_0 t \in W_i, \Delta$, where $i \in \{0, 1, 2, ...\}$
    $\Gamma, t \in W_i \Rightarrow s_1 t \in W_i, \Delta$, where $i \in \{0, 1, 2, ...\}$

14. $\Gamma, t \in W_i \Rightarrow \text{pr} t \in W_i, \Delta$, where $i \in \{0, 1, 2, ...\}$

15. $\Gamma \Rightarrow \text{pr}\epsilon = \epsilon, \Delta$

16. $\Gamma, t \in W_i \Rightarrow t = \epsilon, \text{pr}(t0) = t, \text{pr}(t1) = t, \Delta$, where $i \in \{0, 1, 2, ...\}$

17. $\Gamma, s_i t = \epsilon \Rightarrow \Delta$, where $i \in \{0, 1\}$

18. $\Gamma, t0 = r1 \Rightarrow \Delta$

19. $\Gamma, t \in W_i \Rightarrow \text{pr}(s_0 t) = t, \Delta$, where $i \in \{0, 1, 2, ...\}$
    $\Gamma, t \in W_i \Rightarrow \text{pr}(s_1 t) = t, \Delta$, where $i \in \{0, 1, 2, ...\}$

Extensionality Axiom for Operations

$$\Gamma, \forall x (tx = sx) \Rightarrow t = s, \Delta$$

where $x \notin \text{FV}(t = s)$.

We also have in RC^s_{ext} the standard inferences for the classical connectives and quantifiers (the well-known logical rules for introducing $\wedge, \vee, \neg, \rightarrow, \forall, \exists$ on the right-hand side and on the left-hand side), and (context-sharing[1]) cut.

Recall that we are working in a ramified framework. The use of data objects are now classified into *tiers*. We have a generic concept of *data tiering* when the use of an object α is of *higher* tier if and only if it is *global*, i.e. if and only if α is used as an iterator for functions over lower tiers.

Inside our framework the notion of tier assumes the following quite natural aspect.

DEFINITION 4.

(i.) Let α be a binary string; then

$$\textbf{tier}(\alpha) = l \quad if \quad \alpha \in W_l$$

where $l \in \{0, 1, 2, ...\}$. Let also $\bar{\rho}$ be a finite sequence $\bar{\rho} := \alpha_1, ..., \alpha_k$, where $\alpha_1 \in W_{j_1}, ..., \alpha_k \in W_{j_k}$, we have

$$\textbf{tier}(\bar{\rho}) = \max\{\textbf{tier}(\alpha_1), ..., \textbf{tier}(\alpha_k)\}.$$

(ii.) Let A be an arbitrary $\mathcal{L}_\textbf{r}$-formula; we have

- $\textbf{tier}(A) = 0$ if $A := (t = s)$
- $\textbf{tier}(t \in W_j) = j$, with $j \in \{0, 1, 2, ...\}$
- $\textbf{tier}(\neg A) = \textbf{tier}(A)$

---

[1] Rules of inference with several premises are using the same context.

- **tier**$(A \circ B) = \max\{$**tier**$(A),$ **tier**$(B)\}$, where $\circ$ is a binary connective
- **tier**$(\mathcal{Q}xA) =$ **tier**(A), with $\mathcal{Q} = \exists, \forall$.

In RC^s_{ext} we have a principle of Π^s_i-*ramified positive i-safe induction with parameter substitution*.

Assume that $C(x, v)$ is a j-positive formula, t is an arbitrary term, $\vec{v}$ are substitution eigen-parameters not free in Γ; also assume that all tiers in C are strictly bounded by j, i.e. **tier**$(C) < j$; then:

$$\frac{\begin{cases} \begin{cases} \vec{v} \in W_k \Rightarrow \vec{\sigma}_0(\vec{v}) \in W_k \\ \vdots \\ \vec{v} \in W_k \Rightarrow \vec{\sigma}_1(\vec{v}) \in W_k \end{cases} \\ \Gamma, \vec{v} \in W_k \Rightarrow C(\epsilon, \vec{v}), \Delta \\ \Gamma, a \in W_j, \vec{v} \in W_k, C(a, \vec{\sigma}_0\vec{v}) \Rightarrow C(a0, \vec{v}), \Delta \\ \Gamma, a \in W_j, \vec{v} \in W_k, C(a, \vec{\sigma}_1\vec{v}) \Rightarrow C(a1, \vec{v}), \Delta \end{cases}}{\Gamma, t \in W_j, \vec{v} \in W_k \Rightarrow C(t, \vec{v}), \Delta}$$

We observe that all the main formulas of non-logical axioms and rules of RC^s_{ext}, as well as the main induction formulas in the ramified rule above, are all positive.

4 Elementary steps

4.1 Preparatory normal forms

We turn to a preparatory weak cut elimination argument for RC^s_{ext}: firstly, we define the derivability relation $\mathrm{RC}^s_{ext} \vdash^m_n \Gamma \Rightarrow \Delta$.

DEFINITION 5.

- $\mathcal{D} \vdash^m_n \Gamma \Rightarrow \Delta$ $(m, n \in \omega)$ holds iff $\mathcal{D}$ is a locally correct tree *(modulo axioms and rules of RC^s_{ext})* of depth $\leq m$ with root $\Gamma \Rightarrow \Delta$, such that each cut occurring in $\mathcal{D}$ applies to formulas of rank $< n$;

- $\mathrm{RC}^s_{ext} \vdash^m_n \Gamma \Rightarrow \Delta$ *(or, simply, $\vdash^m_n \Gamma \Rightarrow \Delta$)* means that there exists some derivation $\mathcal{D} \vdash^m_n \Gamma \Rightarrow \Delta$.
  $\mathrm{RC}^s_{ext} \vdash^m_n \Gamma \Rightarrow \Delta$ is read as "$\Gamma \Rightarrow \Delta$ is RC^s_{ext}-derivable with length $\leq m$ and rank $< n$"; $\mathcal{D}$ is usually called "derivation" of the given sequent;

- if $\mathcal{D} \vdash^m_1 \Gamma \Rightarrow \Delta$, the derivation $\mathcal{D}$ is called "quasi-normal"; a sequent $\Gamma \Rightarrow \Delta$ is "positive" iff each $C \in \Gamma \cup \Delta$ is positive.

As expected, we have that RC^s_{ext} satisfies a weak-cut elimination property.

THEOREM 6. (*Weak Cut Elimination*)
For all sequents Γ, Δ, $\mathcal{D} \vdash^m_n \Gamma \Rightarrow \Delta$ $(m, n \in \omega)$ entails $\mathcal{D} \vdash^m_1 \Gamma \Rightarrow \Delta$:

i.e. every derivation can be transformed into a quasi-normal derivation of the same conclusion.

Due to the fact that all the main formulas of non-logical axioms and rules of RC^s_{ext} are positive, now we obtain the following corollary, which directly follows from the weak cut-elimination theorem.

COROLLARY 7. *Assume that* $\Gamma \Rightarrow \Delta$ *is a positive sequent such that* $\vdash^m_1 \Gamma \Rightarrow \Delta$; *then* $\Gamma \Rightarrow \Delta$ *has a* RC^s_{ext}-*derivation which only contains positive formulas.*

Then we have a standard

THEOREM 8. (*Normal Form*)
If $\Gamma \Rightarrow \Delta$ *is positive, i.e. if each* $C \in \Gamma \cup \Delta$ *is positive, and* RC^s_{ext} *proves* $\Gamma \Rightarrow \Delta$, *then there exists a* RC^s_{ext}-*derivation of* $\Gamma \Rightarrow \Delta$ *which only contains positive formulas.*

4.2 The open term model for RC^s_{ext}

We can define a standard *open term model* $\mathcal{M}(\lambda\eta)$ which is based on a straightforward extension of the usual $\lambda\eta$-reduction, using the well-known equivalence of $\lambda\eta$ and standard combinatory logic with extensionality: in order to deal with the new constants of RC^s_{ext} (such as L, R, D and pr) one extends $\lambda\eta$-reduction by the obvious reduction clauses for these new constants, and checks that the so-obtained new reduction relation enjoys the usual Church-Rosser property. So we can interpret the language of RC^s_{ext} as follows:

(i.) the universe of the model $\mathcal{M}(\lambda\eta)$ now consists of the set of all $\mathcal{L}_\mathbf{r}$-terms;

(ii.) the standard equality relation is reduction to a common reduct;

(iii.) the many-sorted structure composed by j-copies $\mathbb{W}_0, \mathbb{W}_1, \mathbb{W}_2, ...$ (with $j \in \{0, 1, 2, ...\}$) of $\mathbb{W}$ is interpreted as the set of all $\mathcal{L}_\mathbf{r}$-terms t such that t reduces to the standard term canonically designating some binary string $\alpha \in \mathbb{W}$;

(iv.) finally, basic constants are interpreted onto themselves, and application of t to s is simply syntactical application (namely, the term ts).

As usual, we write $\mathcal{M}(\lambda\eta) \models A$ in order to express that the formula A is true in the open term model $\mathcal{M}(\lambda\eta)$ of the untyped lambda calculus with extensionality.

From the purely complex-theoretical point of view, it is clear that the interpretation above is useless, because it trivializes the distinction among the copies of $\mathbb{W}$. However, we will see that a simple witnessing method in a realizability interpretation is sufficient to exploit the distinction among the ramified copies of the algebra $\mathbb{W}$.

5 The class LM

We recall here the Leivant-Marion characterization of the class of polynomial space functions as the set of functions over the algebra $\mathbb{W} = \{0,1\}^*$ defined by ramified $\mathbb{W}$-recurrence with parameter substitution [7].

As to the word constructors of the algebra, we call *sources* the 0-ary constructors, and *successors* the unary ones: we consider algebras with several successors, such as the word algebra $\mathbb{W}$ isomorphic to $\{0,1\}^*$ with one source ϵ (empty word) and two successors S_0, S_1 (binary successors).

As usual, we have destructor and conditional functions, which are defined by (unramified flat) recurrence:

$$P(\epsilon) = \epsilon$$
$$P(S_i(\alpha)) = \alpha \quad i \in \{0,1\}$$

and

$$C(\alpha, \beta, \gamma, \delta) = \alpha \quad if \quad \gamma = \delta \quad \gamma, \delta \in \mathbb{W}$$
$$C(\alpha, \beta, \gamma, \delta) = \beta \quad if \quad \gamma \neq \delta \quad \gamma, \delta \in \mathbb{W}$$

Also, we consider standard projection functions, defined as

$$\Pi_i^{n,m}(\alpha_1, ..., \alpha_n, \alpha_{n+1}, ..., \alpha_{n+m}) = \alpha_i$$

where $1 \leq i \leq n+m$.

The main concepts of *data tiering* and *ramified recurrence* are essential for introducing functions definition schemata. We have a many sorted structure $\mathcal{S}(\mathbb{W})$ with copies $\mathbb{W}_0, \mathbb{W}_1, \mathbb{W}_2, ...$ of the algebra $\mathbb{W}$ as universes: these copies are our *tiers*.

We say that a function φ is defined by *composition* iff

$$\varphi(\bar{\alpha}, \bar{\beta}, \bar{\gamma}) = G(\bar{\alpha}, \bar{\beta}, \bar{\gamma}, H(\bar{\alpha}, \bar{\beta}, \bar{\gamma}))$$

where

$$H : \mathbb{W}_h \times \mathbb{W}_j \times \mathbb{W}_l \to \mathbb{W}_k$$
$$G : \mathbb{W}_h \times \mathbb{W}_j \times \mathbb{W}_l \times \mathbb{W}_k \to \mathbb{W}_m$$
$$\varphi : \mathbb{W}_h \times \mathbb{W}_j \times \mathbb{W}_l \to \mathbb{W}_m$$

Finally, if φ is an r-ary function over $\mathbb{W}$, we say that φ is defined by *ramified recurrence with parameter substitution* iff

$$\varphi(\epsilon, \bar{\gamma}, \bar{\alpha}, \bar{\beta}) = \varphi_\epsilon(\bar{\gamma}, \bar{\alpha}, \bar{\beta})$$
$$\varphi(\delta i, \bar{\gamma}, \bar{\alpha}, \bar{\beta}) = \Psi_i(\varphi(\delta, \bar{\sigma}(\bar{\gamma}), \bar{\alpha}, \bar{\beta}), \bar{\alpha}, \bar{\beta})$$

where

$$\varphi : \mathbb{W}_j \times \mathbb{W}_k \times \mathbb{W}_u \times \mathbb{W}_m \to \mathbb{W}_p$$
$$\varphi_\epsilon : \mathbb{W}_k \times \mathbb{W}_u \times \mathbb{W}_m \to \mathbb{W}_p$$
$$\Psi_i : \mathbb{W}_p \times \mathbb{W}_u \times \mathbb{W}_m \to \mathbb{W}_p$$
$$\sigma : \mathbb{W}_k \to \mathbb{W}_k$$

and with the main proviso that $i > p$; so all γ_i and σ_i are in a common tier. Recall that, as usual, the argument of φ displayed first is called the *recurrence argument*, while the argument of Ψ_i displayed first it its *critical argument*. Then we have that in the schema of ramified recurrence with parameter substitution above the tiers of the recurrence argument must be larger than the tier of critical arguments; finally, substitution parameters must have a common tier.

The following examples clarifies the necessity of these restrictions. Consider the exponential function E such that $E(0) = 1$ and $E(n+1) = E(n) + E(n)$. Clearly, addition (over $\mathbb{N}$) is definable by simple ramified recurrence (without substitution), since for each j, k, with $j > k$, we can define a copy of addition $+_{jk} : \mathbb{N}_j \times \mathbb{N}_k \to \mathbb{N}_k$.

Nevertheless, the above definition of exponentiation cannot be ramified, since the first input of ramified addition must be at a tier higher then the output: similar arguments show that the definition of E via multiplication cannot be ramified either.

Consider now the ramification conditions in case of recurrence with substitution. If we look at the definition of

$$exp(0, u) = u$$
$$exp(n+1, u) = exp(n, 2u)$$

then $exp(n, 1) = 2^n$ and this is obtained by simple (unramified) recurrence with substitution. Nevertheless, the above definition of exp cannot be ramified, since the tier of $2x$ is lower than the tier of x, with respect to any ramified definition of $\lambda x.2x$. Hence the tier of the second argument of exp is not well-defined.

We should also consider

$$f(0, x, y) = y$$
$$f(n+1, x, y) = f(n, x, (y+x)+x)$$

with $f : \mathbb{N}_1 \times \mathbb{N}_1 \times \mathbb{N}_0 \to \mathbb{N}_0$, observing that here the tiering conditions on recurrence argument are satisfied. However, the main condition that the substitution parameters must have a common tier is now violated, and by induction on n we have $f(n, x, 0) = 2^n \cdot x$, for all $n > 0$.

The examples above clarify the necessity of tiering constraints on ramified recurrence with substitution definitions.

DEFINITION 9. LM (Leivant-Marion sorted system with ramified recurrence and parameter substitution) is the smallest class of input-sorted functions which contains the constructors *(source ϵ and binary successors S_0 and S_1)*, the destructor, conditional and projections functions *(P, C and Π, respectively)*, all over each input-sorts, and which is closed under the schemes of composition and ramified recurrence with parameter substitution.

The many-sorted structure of the algebra $\mathbb{W}$ implies the existence of infinite copies of the same constructor function, one for each level, or tier. Nevertheless, the infinite copies are related by means of a so-called *coercion* function, which is definable by ramified $\mathbb{W}$-recurrence in the following manner. Given $\mathbb{W} = \mathbb{W}_0, \mathbb{W}_1, \mathbb{W}_2, \ldots$ and j, m with $j > m$, a coercion function

$k_{jm}^W : \mathbb{W}_j \to \mathbb{W}_m$ is defined as

$$k_{jm}^W(\mathbf{c}_i^j(\alpha_1...\alpha_{r_i})) = \mathbf{c}_i^m(k_{jm}^W(\alpha_1)...k_{jm}^W(\alpha_{r_i}))$$

where $\mathbf{c}_i$ is any constructor of tier i in the algebra.

The crucial result is the following.

THEOREM 10. [7]
A function φ is computable in polynomial space if and only if φ is LM-definable by ramified recurrence with parameter substitution. In fact, that definition can be obtained using three tiers only ($\mathbb{W}_0, \mathbb{W}_1$ and $\mathbb{W}_2$).

We extend the notion of LM-computability to the case of *vector-valued* functions.

We need functions defined on k-tuples of vectors of binary strings, assuming vectors of binary strings as values. Here *vector* stands for *finite sequence* of binary strings.

More explicitly, we want that, if a function φ is LM-computable, then

$$\varphi : \mathbb{W}_{i_1}^{n_1} \times ... \times \mathbb{W}_{i_p}^{n_k} \to \mathbb{W}_{h_r}^{m_1} \times ... \times \mathbb{W}_{h_l}^{m_o}$$

where $1 \leq n_1, ..., n_k, m_1, ..., m_o$.

In the sequel, ρ, σ, τ range over finite sequences of binary strings, $\bar{\rho}, \bar{\sigma}, \bar{\tau}$ range over vectors of vectors of binary strings (namely, $\bar{\rho} := \rho_1...\rho_n$), while $\alpha, \beta, \gamma, \delta$ stand for binary strings.

We give here a formal definition of the notion of LM-computability for vector-valued functions.

DEFINITION 11. Assume that $(-,-)$ is the standard set-theoretic pairing constructor. If $\varphi : \mathbb{W}_{i_1}^{n_1} \times ... \times \mathbb{W}_{i_p}^{n_k} \to \mathbb{W}_{h_u}^m$, we say that $\varphi \in$ LM iff there exist $\varphi_1, ..., \varphi_m \in$ LM such that, if $q = n_1 + ... + n_k$, $1 \leq j \leq m$, then

$$\varphi_j : \mathbb{W}_{i_s}^q \to \mathbb{W}_{h_t};$$
$$\varphi(\rho_1, ..., \rho_k) = (\sigma_1, ..., \sigma_m), \text{ where}:$$
$$\sigma_j = \varphi_j(\alpha_1, ..., \alpha_{n_1}, ..., \alpha_{n_1+...+n_{k-1}+1}, ..., \alpha_q)$$
$$\rho_1 = (\alpha_1, ..., \alpha_{n_1})$$
$$\vdots$$
$$\rho_k = (\alpha_{n_1+...+n_{k-1}+1}, ..., \alpha_q).$$

Also, if $I^{n_1,...,n_k}$ is a bijection between $\mathbb{W}_{i_s}^q$ and $\mathbb{W}_{i_1}^{n_1} \times ... \times \mathbb{W}_{i_p}^{n_k}$, where $q = n_1 + ... + n_k$, i.e.

$$I^{n_1,...,n_k}(\alpha_1, ..., \alpha_q) = ((\alpha_1, ..., \alpha_{n_1}), ..., (\alpha_{n_1+...+n_{k-1}+1}, ..., \alpha_q))$$

and if $\varphi : \mathbb{W}_{i_1}^{n_1} \times ... \times \mathbb{W}_{i_p}^{n_k} \to \mathbb{W}_{h_r}^{m_1} \times ... \times \mathbb{W}_{h_l}^{m_o}$, we have that $\varphi \in$ LM iff there exists $\overline{\varphi}$ of arity

$$\mathbb{W}_{i_1}^{n_1} \times ... \times \mathbb{W}_{i_p}^{n_k} \to \mathbb{W}^v$$

(with $v = m_1 + ... + m_o$) such that

$$\varphi(\rho_1, ..., \rho_k) = I^{m_1,...,m_o}(\overline{\varphi}(\rho_1, ..., \rho_k))$$

where $\rho_1 \in \mathbb{W}_{i_1}^{n_1}, ..., \rho_k \in \mathbb{W}_{i_p}^{n_k}$.

Observe that it follows from the above formalization that in the definition of a vector-valued function in the class LM the tiers of the vectors can be different: nevertheless the ramification conditions has to be met as well.[2]

6 Π_i^s-induction yields ramified recurrence with parameter substitution

That our system captures at least the polynomial space operations is easily implied by a straight interpretability argument.

DEFINITION 12. (*Total Provability*)
$F : \mathbb{W}_{i_1} \times ... \times \mathbb{W}_{i_k} \to \mathbb{W}_n$ is *provably total* in the system RC_{ext}^s iff there exists a closed term f_F such that:

(i.) RC_{ext}^s proves the sequent $f_F : \mathbb{W}_{i_1} \times ... \times \mathbb{W}_{i_k} \to \mathbb{W}_n$ which has the form $\Gamma \Rightarrow \Delta$, where

$$\begin{aligned}\Gamma &:= \{x_1 \in \mathbb{W}_{i_1}, ..., x_k \in \mathbb{W}_{i_k}\} \\ \Delta &:= \{f_F x_1...x_k \in \mathbb{W}_n\};\end{aligned}$$

(ii.) $\mathcal{M}(\lambda\eta) \models (f_F \overline{\alpha_1}...\overline{\alpha_k}) = \overline{F(\alpha_1, ..., \alpha_k)}$, for every $\alpha_1, ..., \alpha_k \in \mathbb{W}$.

THEOREM 13. (*Lower Bound*)
If $\varphi \in \mathrm{LM}$, then φ is provably total in RC_{ext}^s: hence every polynomial space function on $\mathbb{W}$ is provably total in RC_{ext}^s.

Proof. Consider the *source* operator in the algebra LM, say $Z(\alpha) = \epsilon$, where $\alpha \in \mathbb{W}_j$, with $j \in \{0, 1, 2, ...\}$. If $\varphi(\vec{\alpha}, \gamma, \vec{\beta}) = Z(\vec{\beta}) = \epsilon$, simply choose $f_\varphi = \lambda x.\epsilon$: then $\epsilon \in \mathbb{W}_j$, for $j \in \{0, 1, 2, ...\}$.

As to the *binary successors* functions $\mathbf{S}_i(\beta) = \beta i$ (where $i \in \{0, 1\}$) and $\beta \in \mathbb{W}_j$, for $j \in \{0, 1, 2, ...\}$, if we have $\varphi(\vec{\alpha}, \gamma, \vec{\beta}) = \beta i$, choose $f_\varphi := \lambda x.\mathbf{S}_i x$. Then $f_\varphi : \mathbb{W}_j \to \mathbb{W}_j$, for $j \in \{0, 1, 2, ...\}$.

Similar arguments hold for the *binary predecessor* function P in the class LM: of $\varphi(\vec{\alpha}, \gamma, \vec{\beta}) = P(\beta)$, simply choose $f_\varphi := \lambda x.\mathrm{pr}x$; then $f_\varphi : \mathbb{W}_j \to \mathbb{W}_j$, for axioms on $\mathbb{W}_j$-predecessor operations in RC_{ext}^s, with $j \in \{0, 1, 2, ...\}$.

In the case of the usual *projection functions*

$$\Pi_i^{n,m}(\alpha_1, ..., \alpha_n, \alpha_{n+1}, ..., \alpha_{n+m}) = \alpha_i$$

where $1 \leq i \leq n + m$, we can apply the standard properties of combinatory logic, letting

$$\Pi_i^{n,m} = \lambda x_1...x_{n+m}.x_i.$$

---

[2]It is clear that LM is closed under the schema of ramified recurrence with parameter substitution, where the ramification conditions have to be extended to the case of vector-valued functions. Assume that $\varphi : \mathbb{W}_j \times \mathbb{W}_k \to \mathbb{W}_i \times \mathbb{W}_m$ where, for instance, $\varphi(\alpha, \beta) = (\varphi_1(\alpha, \beta), \varphi_2(\alpha, \beta))$ and each φ_1, φ_2 is well defined. Hence we have that if φ is defined by ramified recurrence with parameter substitution, then $j > \max\{i, m\}$.

If we consider the *conditional axioms* in the function algebra LM,

$$C(\alpha, \beta, \gamma, \delta) = \alpha \quad if \quad \gamma = \delta \quad \gamma, \delta \in W_j$$
$$C(\alpha, \beta, \gamma, \delta) = \beta \quad if \quad \gamma \neq \delta \quad \gamma, \delta \in W_j$$

then the test for equality on W_j is trivially represented by means of the conditional operators in the calculus RC^s_{ext}; so we can apply the corresponding conditional axioms about definition by cases on W_j. For instance, assume $\varphi(\alpha, \beta, \gamma, \delta) = C(\alpha, \beta, \gamma, \delta) = \alpha$ if $\gamma = \delta$ and $\gamma, \delta \in W_j$. Then $f_\varphi := \lambda\alpha\beta\gamma\delta.D\alpha\beta\gamma\delta = \alpha$ and $f_\varphi := W_j \to W_j$, with $j \in \{0, 1, 2, ...\}$.

Assume that φ is defined by means of the schema of *composition* in the algebra LM:

$$\varphi(\alpha_1, ..., \alpha_p, \beta_1, ..., \beta_q, \gamma_1, ..., \gamma_v) = G(\vec{\alpha}, \vec{\beta}, \vec{\gamma}, H(\vec{\alpha}, \vec{\beta}, \vec{\gamma})),$$

where every $\alpha \in W_h$, every $\beta \in W_j$ and every $\gamma \in W_l$, with $h, j, l \in \{0, 1, 2, ...\}$. Then by induction hypothesis G and H are already provably total, and so representable by means of terms f_G and f_H such that

$$f_G : W_h \times W_j \times W_l \times W_k \to W_m$$
$$f_H : W_h \times W_j \times W_l \to W_k$$

Then we can finally find

$$f_\varphi : \lambda\vec{\alpha}\vec{\beta}\vec{\gamma}.f_G(\vec{\alpha}\vec{\beta}\vec{\gamma}(f_H(\vec{\alpha}\vec{\beta}\vec{\gamma}))).$$

Assume that φ is defined in LM by *ramified recurrence with parameter substitution* on $\mathbb{W}$:[3]

$$\varphi(\epsilon, \vec{\gamma}, \vec{\alpha}, \vec{\beta}) = \varphi_\epsilon(\vec{\gamma}, \vec{\alpha}, \vec{\beta})$$
$$\varphi(\delta i, \vec{\gamma}, \vec{\alpha}, \vec{\beta}) = \varphi_i(\varphi(\delta, \vec{\sigma}_i(\vec{\gamma}), \vec{\alpha}, \vec{\beta}), \vec{\alpha}, \vec{\beta})$$

where $i \in \{0, 1\}$ and

$$\varphi : W_j \times W_k \times W_u \times W_m \to W_p$$
$$\varphi_\epsilon : W_k \times W_u \times W_m \to W_p$$
$$\varphi_i : W_p \times W_u \times W_m \to W_p$$
$$\vec{\sigma}_i : W_k \to W_k$$

Recall that the tier of the recurrence argument is larger than the tier of critical arguments, while substitution parameters have a common tier.

Then by induction hypothesis there exist closed terms $f_\epsilon, f_i, f_\sigma$ ($i \in \{0, 1\}$) such that

$$f_\epsilon : W_k \times W_u \times W_m \to W_p$$
$$f_i : W_p \times W_u \times W_m \to W_p$$
$$f_\sigma : W_k \to W_k$$

---

[3] For the sake of simplicity we use in the proof only one substitution function.

Then we apply the conditional axioms and the fixed point theorem in order to find an f such that

$$f\epsilon\gamma\alpha\beta \simeq f_\epsilon\gamma\alpha\beta$$
$$f(\xi i)\gamma\alpha\beta \simeq f_i(f\xi(f_\sigma\gamma)\alpha\beta)\alpha\beta.$$

We want

$$f: W_j \times W_k \times W_u \times W_m \to W_p$$

with $j > p$. Then we want to prove, by positive i-safe Π_i^s-induction,

$$(\forall \xi \in W_j. \forall \gamma \in W_k. \forall \alpha \in W_u. \forall \beta \in W_m)(f\xi\gamma\alpha\beta \in W_p).$$

If $\xi = \epsilon$, then

$$(\forall \gamma \in W_k. \forall \alpha \in W_u. \forall \beta \in W_m)((f\epsilon\gamma\alpha\beta) = (f_\epsilon\gamma\alpha\beta \in W_m))$$

by assumption on f_ϵ. Let us now consider the induction step $\xi \mapsto \xi i$. We assume

$$f\xi(f_\sigma\gamma)\alpha\beta \in W_p$$

for all $\gamma \in W_k$, $\alpha \in W_u$ and $\beta \in W_m$.
But the induction hypothesis guarantee

$$f_i : W_p \times W_u \times W_m \to W_p$$

and hence we can conclude

$$f_i(f\xi(f_\sigma\gamma)\alpha\beta)\alpha\beta \in W_m.$$

∎

7 Realizability for positive sequents

We reach the conclusive soundness estimate on the polynomial space recursive content of RC_{ext}^s by adopting a suitable realizability interpretation inside our sequent-calculus: to this aim we use as realizing operations directly vector-valued functions in Leivant-Marion algebra, in order to realize only positive sequents of our system.

Hence we introduce a notion of *realizability* for positive formulas in the standard open term model $\mathcal{M}(\lambda\eta)$ of RC_{ext}^s, following techniques firstly introduced in [8] and [6].

We have a set $\mathcal{R}$ of *realizers*, which is inductively generated by means of the usually set-theoretic pairing constructor $(-, -)$: thus we have $\alpha \in \mathcal{R}$, for each binary string α and, if $\rho \in \mathcal{R}$ and $\sigma \in \mathcal{R}$, then $(\rho, \sigma) \in \mathcal{R}$.

DEFINITION 14. The realizability relation $\rho \triangleright A$ (ρ realizer such that $\rho \in \mathbb{W}$, A positive formula) is defined by induction:

(i.) $\rho \triangleright t = s \Leftrightarrow \rho = \epsilon$, $\mathcal{M}(\lambda\eta) \models t = s$ and $\epsilon \in W_j$, with $j \in \{0, 1, 2, ...\}$;

(ii.) $\rho \triangleright t \in W_j \Leftrightarrow \rho$ is a binary string such that $\mathcal{M}(\lambda\eta) \models t = \rho$ (where $j \in \{0, 1, 2, ...\}$) and $\rho \in W_j$;

(iii.) $\rho \triangleright (A \vee B) \Leftrightarrow \rho = (i, \rho')$ and either $i = 0$ and $\rho' \triangleright A$, or $i = 1$ and $\rho' \triangleright B$,

(iv.) $\rho \triangleright (A \wedge B) \Leftrightarrow \rho = (\rho_1, \rho_2)$ with $\rho_1 \triangleright A$ and $\rho_2 \triangleright B$;

(v.) $\rho \triangleright \forall x A \Leftrightarrow \rho \triangleright A[x := a]$, where $a \notin FV(A)$;

(vi.) $\rho \triangleright \exists x A \Leftrightarrow \rho \triangleright A[x := t]$, for some term t, which is free for x in A.

The proof of the following lemma is immediate from the definition of realizability and will therefore be omitted.

LEMMA 15. (*Substitution*) *We have, for all positive formula* A:

(i.) $\rho \triangleright A[a := t]$ *and* $\mathcal{M}(\lambda \eta) \models t = s \Rightarrow \rho \triangleright A[a := s]$;

(ii.) $\rho \triangleright A(a) \Leftrightarrow \rho \triangleright A[a := t]$, *for all terms* t. .

If Δ denotes the sequence $A_1, ..., A_m$ of positive formulas, we say that $\vec{\rho}_i$ (with $\vec{\rho} := \rho^1, ..., \rho^m$) realizes the sequence Δ ($\vec{\rho}_i \triangleright \Delta$) iff $\vec{\rho}_i = (\rho_0^1, \rho_1^2, ..., \rho_j^m)$ and

$$(\rho_0^1 = 1 \quad \text{and} \quad \rho_0^1 \triangleright A_1) \vee ... \vee (\rho_0^1 = m \quad \text{and} \quad \rho_j^m \triangleright A_m)$$

with $i \in \{0, 1, 2, ...\}$.

Here the subscript notation clearly indicates the tiers of realizers (but we omit both superscripts and subscripts though, when in no danger of confusion).

Hence, according to the notion $\vec{\rho}_i \triangleright \Delta$, the sequence Δ is understood disjunctively: then $\vec{\rho}_i \triangleright \Delta$ stands for $\vec{\rho}_i = (\rho_i^1, \sharp A)$ and $\rho_i^1 \triangleright A_i$, for some $A_i \in \Delta$: here $\sharp$ is an assumed encoding via bitstring.

Also, $\Gamma_{\vec{a}} \Rightarrow \Delta_{\vec{a}}$ means that the free variables of $\Gamma \Rightarrow \Delta$ occur in the list $\vec{a}$: if $\mathcal{D} \vdash \Gamma_{\vec{a}} \Rightarrow \Delta_{\vec{a}}$, the term list $\vec{r}$ is *suitable* for $\vec{a}$ in $\mathcal{D}$ iff no free variable of $\vec{r}$ occurs as eigenvariables in $\mathcal{D}$.

Finally, if $\vec{r}$ is suitable for $\vec{a}$ in $\mathcal{D} \vdash \Gamma_{\vec{a}} \Rightarrow \Delta_{\vec{a}}$, $\Gamma_{\vec{r}} \Rightarrow \Delta_{\vec{r}}$ stand for the result of substituting each variable in the list $\vec{a}$ by the corresponding terms of the list $\vec{r}$.

THEOREM 16. (*Main Realizability*)
Let $\mathcal{D}$ be a quasi-normal derivation such that $\mathcal{D} \vdash \Gamma_{\vec{a}} \Rightarrow \Delta_{\vec{a}}$, where $\Gamma_{\vec{a}} \Rightarrow \Delta_{\vec{a}}$ only contains positive formulas.
Then there exists a polynomial space function $\varphi_{\mathcal{D}} \in LM$ such that, if $\Gamma_{\vec{r}} \equiv A_1, ..., A_k$ and $\vec{\rho}_i \equiv \rho_0^1...\rho_j^k$, for all $\vec{\rho}_i$, and $\vec{r}$ suitable for $\vec{a}$ in $\mathcal{D}$,

$$\text{if} \quad \rho_0^1 \triangleright A_1 ... \rho_j^k \triangleright A_k \quad \text{then} \quad \varphi_{\mathcal{D}}(\vec{\rho}_i) \triangleright \Delta_{\vec{r}}.$$

Proof. The proof is by induction on the height of a given derivation $\mathcal{D}$. We assume that we have a fixed list $\vec{r}$ of terms, which is suitable for the parameters of the final sequents.

A. *Identity Axioms*
Consider the case when the final sequent has the form $\Gamma, A \Rightarrow A, \Delta$ (A multiset), where $A := (t = s)$ is atomic. Assume μ_j is the realizer of A; then choose $\varphi_{\mathcal{D}}(\vec{\rho}_i, \mu_j) = (\mu_j, \sharp A)$, with $\varphi_{\mathcal{D}} : W_j \to W_j$, with $i, j \in \{0, 1, 2, ...\}$.

Let us now consider the case when the final sequent has the form $\Gamma, t = s, t \in W_j \Rightarrow s \in W_j, \Delta$. So assume $\vec{\rho_i} \triangleright \Gamma$, $\mu_k \triangleright t = s$ and $\theta_j \triangleright t \in W_j$. Then by definition of realizability, $\mu_k = \epsilon$ and $\mathcal{M}(\lambda\eta) \models (t = s) \wedge (t = \theta_j)$: hence we choose $\varphi_\mathcal{D}(\vec{\rho_i}, \mu_k, \theta_j) = (\theta_j, \sharp s \in W_j)$, with $i, j, k \in \{0, 1, 2, ...\}$ and $\varphi_\mathcal{D} : W_j \to W_j$. Similar arguments hold in the case of the remaining identity axioms, namely identity and transitivity.

B. *Combinatory logic and pairing*
If the final sequent of the derivation has the form of one of the four axioms about K, S, left and right projections on the pairing operations, the verification is immediate: we simply choose $\varphi_\mathcal{D}(\vec{\rho_i}) = (\epsilon, \sharp A)$, A being the active formula of the given axiom.

C. *Definition by cases*
As to the W_j-conditional axioms, if we have

$$\Gamma, t \in W_j, s \in W_j, t = s \Rightarrow Drpts = r, \Delta$$

then we have realizers $\vec{\rho_i}, \mu_j, \theta_j$ and ζ_k such that $\vec{\rho_i} \triangleright \Gamma$, $\mu_j \triangleright t \in W_j$, $\theta_j \triangleright s \in W_j$, $\zeta_k \triangleright t = s$. Hence we choose $\varphi_\mathcal{D}(\vec{\rho_i}, \mu_j, \theta_j, \zeta_k) = (\epsilon, \sharp Drpts = r)$, with $i, j, k \in \{0, 1, 2, ...\}$.

Analogously, consider the case when

$$\Gamma, t \in W_j, s \in W_j \Rightarrow t = s, Drpts = p, \Delta.$$

Thus we have realizers such that $\vec{\rho_i} \triangleright \Gamma$, $\mu_j \triangleright t \in W_j$, $\theta_j \triangleright s \in W_j$; then we define

$$\varphi_\mathcal{D}(\vec{\rho_i}, \mu_j, \theta_j) = (\epsilon, \sharp t = s) \quad \text{if} \quad \mu_j = \theta_j$$
$$\varphi_\mathcal{D}(\vec{\rho_i}, \mu_j, \theta_j) = (\epsilon, \sharp Drpts = p) \quad \text{if} \quad \mu_j \neq \theta_j$$

$i, j, \in \{0, 1, 2, ...\}$.

D. *Binary successors and predecessor*
The verification is immediate; we only discuss three cases. Assume the final sequent has the form $\Gamma, t \in W_j \Rightarrow ti \in W_j, \Delta$, with $i \in \{0, 1\}$ and $j \in \{0, 1, 2, ...\}$. Also assume $\vec{\rho_i} \triangleright \Gamma$, $\mu_j \triangleright t \in W_j$. Then by definition of realizability, $\mathcal{M}(\lambda\eta) \models (t = \mu_j)$, and hence we can choose $\varphi_\mathcal{D}(\vec{\rho_i}, \mu_j) = (\mu_j i, \sharp ti \in W_j)$.

Consider $\Gamma, ti = \epsilon \Rightarrow \Delta$; since $\mu_j \triangleright ti = \epsilon$ never holds, we simply choose $\varphi_\mathcal{D}(\vec{\rho_i}, \mu_j) = (\epsilon, \sharp A)$, with A as any fixed formula of Δ. Clearly the i-subscript in $\vec{\rho_i}$ indicates the vector of realizer, as above, such that $i \in \{0, 1, 2, ...\}$.

Finally, assume the last sequent of $\mathcal{D}$ has the form

$$\Gamma, t \in W_j \Rightarrow \text{pr}(t0) = t, \Delta.$$

If $\vec{\rho_i} \triangleright \Gamma$, $\mu_j \triangleright t \in W_j$, hence $\mathcal{M}(\lambda\eta) \models (t = \mu_j)$ and we can choose $\varphi_\mathcal{D}(\vec{\rho_i}, \mu_j) = (\mu_j \epsilon, \sharp \text{pr}(t0) = t)$, with $i, j \in \{0, 1, 2, ...\}$.

E. *Extensionality*
Consider the extensionality axiom

$$\Gamma, \forall x(tx = sx) \Rightarrow t = s, \Delta$$

where $x \notin \mathrm{FV}(t=x)$. We can assume $\vec{\rho_i} \triangleright \Gamma, \mu_j \triangleright \forall x(tx=sx)$; by the definition of the realizability relation, $\mu_j = \epsilon$ and $\mu_j \triangleright ta = sa$, where a is not free in t, s. Then we have that $ta = sa$ is true in the open term model: thus, by extensionality, $t = s$ is also true in $\mathcal{M}(\lambda\eta)$. Hence we can finally choose $\varphi_D(\vec{\rho_i}, \mu_j) = (\epsilon, \sharp t = s)$, with $i, j \in \{0, 1, 2, ...\}$.

F. Ramified positive i-safe induction with parameter substitution
Assume we are in the case of Π_i^s-induction on $\mathbb{W}$, which includes, for each j-positive formula $C(x, \vec{v})$ such that all tiers in C are strictly bounded by j, i.e. **tier**$(C) < j$,

$$\begin{cases} \begin{cases} \vec{v} \in W_k \Rightarrow \vec{\sigma}_0(\vec{v}) \in W_k \\ \vdots \\ \vec{v} \in W_k \Rightarrow \vec{\sigma}_1(\vec{v}) \in W_k \end{cases} \\ \Gamma, \vec{v} \in W_k \Rightarrow C(\epsilon, \vec{v}), \Delta \\ \Gamma, a \in W_j, \vec{v} \in W_k, C(a, \vec{\sigma}_0 \vec{v}) \Rightarrow C(a0, \vec{v}), \Delta \\ \Gamma, a \in W_j, \vec{v} \in W_k, C(a, \vec{\sigma}_1 \vec{v}) \Rightarrow C(a1, \vec{v}), \Delta \end{cases}$$

$$\overline{\Gamma, t \in W_j, \vec{r} \in W_k \Rightarrow C(t, \vec{r}), \Delta}$$

Assume, by induction hypothesis, that there are functions I, $\vec{F} \equiv F_1, ..., F_n$ and H_i in the class LM such that

(i.) $\vec{\rho_i} \triangleright \Gamma, \vec{v_k} \triangleright \vec{v} \in W_k \Rightarrow I(\vec{\rho_i}, \vec{v_k}) \triangleright C(\epsilon, \vec{v}), \Delta$

(ii.) $\vec{\tau_k} \triangleright \vec{v} \in W_k \Rightarrow F_1(\vec{\tau_k}) \triangleright \sigma_1(\vec{v}) \in W_k, ..., \vec{\tau_k}' \triangleright \vec{v} \in W_k \Rightarrow F_n(\vec{\tau_k}') \triangleright \sigma_n(\vec{v}) \in W_k$

(iii.) $\vec{\rho_i} \triangleright \Gamma, \vec{\zeta_j} \triangleright a \in W_j, \vec{\xi_k} \triangleright \vec{v} \in W_k, \vec{\delta_l} \triangleright C(a, \sigma_1(\vec{v}), ..., \sigma_n(\vec{v})) \Rightarrow H_i(\vec{\zeta_j}, \vec{\rho_i}, \vec{\xi_k}, \vec{\delta_l}) \triangleright C(ai, \vec{v}), \Delta$

where $i \in \{0, 1\}$, $k \leq j$ and $i, k, j, l \in \{0, 1, 2, ...\}$.

Hence we have to find a suitable realizing function for the conclusion of the positive ramified induction rule, namely a function Z such that

$$\vec{\rho_i} \triangleright \Gamma, \vec{\tau_j} \triangleright t \in W_j, \vec{\xi_k} \triangleright \vec{v} \in W_k \Rightarrow Z(\vec{\rho_i}, \vec{\tau_j}, \vec{\xi_k}) \triangleright C(t, \vec{v}), \Delta.$$

Then we define, by conditionals and ramified recurrence with parameter substitution,

$$Z(\epsilon, \vec{\rho_i}, \vec{v_k}) = I(\vec{\rho_i}, \vec{v_k})$$
$$Z(\vec{\zeta_j}i, \vec{\rho_i}, \vec{\xi_k}) = \begin{cases} Z(\vec{\zeta_j}, \vec{\rho_i}, \vec{\xi_k}) & if \ Z(\vec{\zeta_j}, \vec{\rho_i}, \vec{\xi_k})_0 \neq 1 \\ \vec{F}(\vec{\zeta_j}, \vec{\rho_i}, \vec{\xi_k}) & if \ \vec{F}(\vec{\zeta_j}, \vec{\rho_i}, \vec{\xi_k})_1 \neq 1 \\ H_i(\vec{\zeta_j}, \vec{\rho_i}, \vec{\xi_k}, Z(\vec{\zeta_j}, \vec{\rho_i}, \vec{F}(\vec{\zeta_j}, \vec{\rho_i}, \vec{\xi_k})))_1 & else \end{cases}$$

where $\vec{\zeta_j} \in W_j$, $\vec{\rho_i} \in W_i$, $\vec{\xi_k} \in W_k$ and $F_1(\vec{\xi_k}) \in W_k, ..., F_n(\vec{\xi_k}) \in W_k$, where $k < j$ must be strict.

By the substitution lemma we have to verify, by secondary induction on realizers $\vec{\zeta_j}, \vec{\xi_k}$,

$$\vec{\rho_i} \triangleright \Gamma \Rightarrow Z(\vec{\rho_i}, \vec{\xi_k}, \vec{\zeta_j}) \triangleright C(\vec{\zeta_j}, \vec{\xi_k}), \Delta$$

for every $\vec{\rho_i}, \vec{\zeta_j}, \vec{\xi_k}$ such that $\vec{\rho_i} \triangleright \Gamma$, $\vec{\zeta_j} \triangleright t \in W_j$ and $\vec{\xi_k} \triangleright \vec{v} \in W_k$.

Now assume $\vec{\zeta_j} = \epsilon$; if $\vec{\rho_i} \triangleright \Gamma$ and $\vec{\nu_k} \triangleright \vec{v} \in W_k$, then

$$Z(\epsilon, \vec{\rho_i}, \vec{\nu_k}) = I(\vec{\rho_i}, \vec{\nu_k}) \triangleright C(\epsilon, \vec{\nu_k}), \Delta$$

by i.) above.

Now consider the case $\vec{\zeta_j} \mapsto \vec{\zeta_j}i$. Assume, by secondary induction hypothesis,

$$Z(\vec{\zeta_j}, \vec{\rho_i}, \vec{\xi_k}) \triangleright C(\vec{\zeta_j}, \vec{\xi_k}), \Delta$$

where $\vec{\zeta_j} \in W_j$, $\vec{\xi_k} \in W_k$, for arbitrary $\vec{\rho_i}, \vec{\xi_k}$.

Case 1. $Z(\vec{\zeta_j}, \vec{\rho_i}, \vec{\xi_k})_0 \neq 1$. This means

$$Z(\vec{\zeta_j}, \vec{\rho_i}, \vec{\xi_k})_1 \triangleright \Delta.$$

Hence

$$Z(\vec{\zeta_j}, \vec{\rho_i}, \vec{\xi_k}) = Z(\vec{\zeta_j}i, \vec{\rho_i}, \vec{\xi_k}) \triangleright C(\vec{\zeta_j}i, \vec{\xi_k}), \Delta$$

Case 2. $Z(\vec{\zeta_j}, \vec{\rho_i}, \vec{\xi_k})_0 = 1$. This means

$$(\star\star) \qquad Z(\vec{\zeta_j}, \vec{\rho_i}, \vec{\xi_k})_1 \triangleright C(\vec{\zeta_j}, \vec{\xi_k}).$$

Subcase 2.1 $\vec{F}(\vec{\zeta_j}, \vec{\rho_i}, \vec{\xi_k})_0 \neq 1$. Then

$$\vec{F}(\vec{\zeta_j}, \vec{\rho_i}, \vec{\xi_k})_1 \triangleright \Delta$$

and so

$$Z(\vec{\zeta_j}, \vec{\rho_i}, \vec{\xi_k}) = \vec{F}(\vec{\zeta_j}, \vec{\rho_i}, \vec{\xi_k}) \triangleright C(\vec{\zeta_j}i, \vec{\xi_k}), \Delta.$$

Subcase 2.2 $\vec{F}(\vec{\zeta_j}, \vec{\rho_i}, \vec{\xi_k})_0 = 1$. Then

$$\vec{F}(\vec{\zeta_j}, \vec{\rho_i}, \vec{\xi_k})_1 \triangleright \vec{\sigma}(\vec{\zeta_j}, \vec{\xi_k}).$$

Then, by secondary induction hypothesis, using the substitution property in $(\star\star)$,

$$Z(\vec{\zeta_j}, \vec{\rho_i}, \vec{F}(\vec{\zeta_j}, \vec{\rho_i}, \vec{\xi_k}))_1 \triangleright C(\vec{\zeta_j}, \vec{\sigma}(\vec{\zeta_j}, \vec{\xi_k})).$$

Hence by the main induction hypothesis, we have that

$$Z(\vec{\zeta_j}i, \vec{\rho_i}, \vec{\xi_k}) = H_i(\vec{\zeta_j}, \vec{\rho_i}, \vec{\xi_k}, Z(\vec{\zeta_j}, \vec{\rho_i}, \vec{F}(\vec{\zeta_j}, \vec{\rho_i}, \vec{\xi_k}))_1)$$

realizes
$$C(\vec{\zeta_j}i, \vec{\xi_k}), \Delta.$$

G. *Cut rule*
By assumption, there exists a positive formula A of a given tier, say j, so that our derivation ends by an application of the rule
$$\frac{\Gamma, A \Rightarrow \Delta \quad \Gamma \Rightarrow A, \Delta}{\Gamma \Rightarrow \Delta}$$
By induction hypothesis we are given realizing functions $\varphi_0, \varphi_1 \in$ LM such that
$$\vec{\rho_i} \triangleright \Gamma, \vec{\mu_k} \triangleright A \;\Rightarrow\; \varphi_0(\vec{\rho_i}, \mu_k) \triangleright \Delta$$
$$\vec{\rho_i} \triangleright \Gamma \;\Rightarrow\; \varphi_1(\vec{\rho_i}) \triangleright A, \Delta$$
where $\varphi_0 : W_j \to W_j$ and $\varphi_1 : W_j \to W_j$. Now we obtain a realizing function $\varphi_\mathcal{D}$ for $\Gamma \Rightarrow \Delta$ by setting
$$\varphi_\mathcal{D}(\vec{\rho_i}) = \varphi_0(\vec{\rho_i}) \quad if \quad R(\varphi_0(\vec{\rho_i})) \neq A$$
$$\varphi_\mathcal{D}(\vec{\rho_i}) = \varphi_1(\vec{\rho_i}, L(\varphi_0(\vec{\rho_i}))) \quad otherwise,$$
where $\varphi_\mathcal{D} : W_j \to W_j$ and $\varphi_\mathcal{D} \in$ LM, with $i, j, k \in \{0, 1, 2, ..\}$.

H. *Logical rules*
As to the logical rules of RC_{ext}^s, we have to deal with the standard inferences for the classical connectives and quantifiers (the well-known logical rules for introducing $\wedge, \vee, \to, \neg, \forall, \exists$ on the right-hand side and on the left-hand side.[4]) Since in these cases the proof on the length of quasi cut-free derivations of sequents of positive formulas in RC_{ext}^s is completely standard, we remind the reader to a similar realizability argument that can be found in [5, pp. 183–185]. ■

COROLLARY 17. *Let f be a closed term; if*
$$RC_{ext}^s \vdash f : W_{j_1} \times ... \times W_{j_k} \to W_j$$
then f defines a polynomial space computable function.

Proof. By cut-elimination we get a quasi-normal derivation of the separated sequent
$$\vec{a} \in W_{j_1}, ..., \vec{b} \in W_{j_k} \Rightarrow f\vec{a}\vec{b} \in W_j.$$
Thus we apply the main realizability theorem: then there exists a function $\varphi_\mathcal{D}$ such that, if $\vec{\rho} \triangleright \vec{a} \in W_{j_1}$ and $\vec{\sigma} \triangleright \vec{b} \in W_{j_k}$, then we have $\varphi_\mathcal{D}(\vec{\rho}, \vec{\sigma}) \triangleright f\vec{a}\vec{b} \in W_j$. By definition of realizability, we have $\mathcal{M}(\lambda\eta) \models \overline{f\vec{\rho}\vec{\sigma}} = \varphi_\mathcal{D}(\vec{\rho}, \vec{\sigma})$, where $\vec{\rho}$ and $\vec{\sigma}$ are arbitrary sequences of binary strings such that $\vec{\rho} \in \mathbb{W}_{j_1}$ and $\vec{\sigma} \in \mathbb{W}_{j_n}$. So f defines a LM-operation $\varphi_\mathcal{D}$ with values in $\mathbb{W}$: hence, by Leivant-Marion's theorem, $\varphi_\mathcal{D}$ is computable in polynomial space. ■

---
[4] More precisely, since $\Gamma \Rightarrow \Delta$ is positive and the derivation is quasi-normal, we should only consider the introduction rules for $\wedge, \vee, \forall, \exists$.

Acknowledgements

I would like to express my gratitude to Professor Andrea Cantini for many inspiring discussions and helpful suggestions.

BIBLIOGRAPHY

[1] S. Bellantoni. *Predicative Recursion and Computational Complexity*. PhD thesis, University of Toronto 1992.
[2] S. Bellantoni and S. Cook. A new recursion-theoretic characterization of the poly-time functions. *Computational Complexity*, 2: 97–110, 1992.
[3] G. Calamai. *Proof-theoretic contributions to computational complexity*. PhD thesis, University of Siena 2008.
[4] A. Cantini. Feasible operations and applicative theories based on $\lambda\eta$, *Mathematical Logic Quarterly*, 46: 291–312, 2000.
[5] A. Cantini. Polytime, combinatory logic and positive safe induction. *Archive for Mathematical Logic*, 41: 169–189, 2002.
[6] D. Leivant. A foundational delineation of computational feasibility. In *Proc. Sixth Annual IEEE Symposium on Logic in Computer Science*, 2–11, IEEE, 1991.
[7] D. Leivant and J. Marion. Ramified recurrence and computational complexity II: substitution and polyspace. In *8th Proceedings of CSL*, LNCS pages 486–500, Springer, 1995.
[8] T. Strahm. Theories with self application and computational complexity. *Information and Computation*, 185: 263–297, 2003.
[9] A.S. Troelstra and H. Schwichtenberg. *Basic Proof Theory*. Cambridge University Press, Cambridge 1997.

PART II

PHYSICS AND MATHEMATICS

How contemporary cosmology bypasses Kantian prohibition against a science of the universe

VINCENZO FANO, GIOVANNI MACCHIA

> *If someone in my laboratory begins to talk of the Universe, I tell him that it is time to leave.* Ernest Rutherford [42, p. 110].

1 Introduction

It is well-known that Kant, in the *Critique of Pure Reason*,[1] discusses explicitly the conceptual limitations to which all our cosmological speculations are subjected. He suggests that the universe as a whole, the object of cosmological explanation, is not an object of possible experience, and, consequently, of knowledge; whence his renunciation of a cosmology as a science, i.e. of a rational cosmology that can explain the cosmos in its totality. His meditations on cosmology claim to prove that the ultimate basic questions potentially pertaining to cosmology about the universe — its finiteness in space, its origin in time — do not stand up to critical examination and are rationally insoluble: it is possible to give apparently cogent reasons to support opposite views in both questions, but then one arrives at what seems to be a contradiction (the so-called cosmological antinomies), and neither answer can be definitely accepted.

But how does Kant reach this result and, above all, what is its epistemological effect on modern cosmology, on its scientific status, in particular in the light of the cosmological principle, generally considered a cornerstone in history and in the foundations of cosmology, and one of the fundamental tenets of its modern standard relativistic models?

2 Kant's attitude

Kant distinguishes between *understanding* and *reason*: the former applies to experience, but not the latter. Reason is the faculty that guides a priori deductions from premises to consequences. Understanding works through *concepts*, which apply to *intuition* — either sensible or pure — whereas reason involves *ideas*. The latter cannot be applied to experience, since they have no empirical correspondent. They are called *transcendental* when they determine the use of understanding in the whole realm of experience. Hence transcendental ideas refer to the notion of *totality*.

---

[1]Second book of *Transcendental Dialectics*. Here we are not concerned with what Kant actually said; on the contrary we are discussing cosmological problems in a generic Kantian style.

Since ideas have no empirical application, they seem useless from a cognitive point of view; nonetheless they provide a guide for understanding. Therefore it is true that in process of knowing, ideas play no *constitutive* role, but they play a *regulative* one.

In particular, reason can push understanding either from premises to consequences, or vice versa from consequences to premises; the former is the *progressive* use of reason, the latter the *regressive* one. The progressive use goes from *condition* to *conditioned*, vice versa it holds for the regressive use. It subsists a similarity between the progressive use and the development of a mathematical *succession*, of which one knows the general term that is the condition; for instance:

$$\frac{1}{2^n} \quad n = 1, 2, 3 \ldots \rightarrow \frac{1}{2}, \frac{1}{4}, \frac{1}{8} \ldots,$$

whereas the regressive case is analogous to a series, of which one does not know if it converges; for instance:

$$\sum_{n=1}^{\infty} \frac{1}{2^n} = \frac{1}{2} + \frac{1}{4} + \frac{1}{8} + \ldots = 1 \quad \text{converging series}$$

$$\sum_{n=1}^{\infty} n = 1 + 2 + 3 + \ldots = \infty \quad \text{diverging series}$$

Cosmological ideas concern the *unconditioned* totality of *phenomena*.[2] The *composition* of all phenomena, that is the *universe*, deserves particular attention. Kant aims to show that the notion of the universe is an idea, so that it has a regulative scientific value and not a constitutive one. On this matter he develops the following proof:

K1. In the progressive use of reason the condition determines each conditioned member, as occurs in the case of successions. On the other hand, in the regressive use the conditioned determines the condition only if the conditioned is *intuited* in its totality. That is, in our mathematical metaphor, only if the series *converges*. For instance, all parts, into which a segment is divisible, are intuitable in their totality.[3]

K2. The notion of universe is not intuitable in its totality; therefore it is like a divergent series. That is, it is an idea: it concerns the unconditioned.

If the universe is an idea, then a *rational cosmology*, i.e. a partly a priori science of the universe, considered as a whole, is not possible. Given this impossibility, one might ask if a completely *empirical* cosmology is not possible as well. According to Kant, though enriched by empirical data, a science must have a synthetic a priori foundation. This is impossible, because the universe does not belong to the realm of *possible experience*. Even though in modern science we do not believe that a totally a priori part is necessary, we introduce many theoretical terms with a definition substantially independent

---

[2] Contrary to Kant, we avoid ascribing a cognitively devaluating character to the term "phenomenon" (*Erscheinung*).

[3] It should be noted that here the problem is not infinity, but *boundlessness*.

from experience.[4] It follows that today's ripe physical theories have an a priori part as well, although it is not necessarily a priori, that is this part could be changed in future research. Hence the Kantian problem is in a certain sense still alive. We will see that the solution of the problem lies precisely in the fact that the a priori part of cosmology must be completely revisable.

Nevertheless the universe is not a mere fancy (*Hirngespinste*),[5] indeed it plays a regulative role for science. That is the universe is not removed from experience in the sense that it is a schema formulated without the guide of intuition, but it is something unconditioned, namely it is beyond the limits of experience. In our mathematical analogy, it is a diverging series. Hence when reason asks if the universe has a beginning in time or an end in space, it finds arguments favouring both a positive (*thesis*) and a negative answer (*antithesis*), so that reason is not able to decide. The common-sense necessity of finding a limit to space and time favours a thesis, whereas the impossibility of intuiting such a limit favours an antithesis. Moreover Kant underlines that the thesis is more reassuring, albeit less likely, whereas the antithesis is less popular but more likely.

The solution of the two *antinomies* is dialectic. To understand Kant's proposal, one can reflect on the fact that there is a difference between the following two pairs of statements:

x smells x does not smell

x smells good x does not smell good

The first pair, in fact, is *contradictory*, that is one of the sentences must be false and the other true, whereas the second is a dialectic *contrariness*, that is the statements could be both false, if, for instance, x is odourless. The cosmological antinomies are similar to the latter.

3 The scientific character of cosmology

We can reformulate the Kantian problem in the following manner.[6] Let o be an object belonging to a collection identified by the set of predicates P. Then, if one observes a certain number of objects of the category P, for which Ux holds, one can suppose that:

G. For every x, if Px, then Ux.

G. concerns a potentially *infinite* set of objects, which could stay in different parts of space and time. Can we suppose that G. holds good for each P-object

---

[4]For example, quantum mechanics assumes that microphysical states are represented as unit vectors in a Hilbert space. This assumption cannot be directly tested; in fact it is accepted by the majority of physicists only because so far predictions of the theory have been confirmed.

[5]*Critique of Pure Reason* (A 222, B 269).

[6]In the following we develop the distinction outlined by Bergia [3, p. 12] between *cosmological observations* and *observations of cosmological relevance*, which, in our opinion, is a capital clarification.

in the universe? If the number of investigated P-objects is sufficiently large and diversified, the answer could be "yes". The reason is that G. could be part of a theoretical background with a certain capacity to represent reality. In our argument, we presuppose an at least partial justification of induction.[7]

Now, we can ask: if G. holds for every P-object, does it give information about the universe? No, because we have no information about the universe as an unitary object. To understand this point, let us consider the following example: a big box is filled with objects of different kinds. Although through observations one can generalise about one or more kinds of objects, one cannot affirm anything about the box as a whole. In order to transform generalisations about kinds of objects into generalisations about the box, one has *already* to know something about the box as a whole. But we know nothing about the whole, i.e. concerning either the box or the universe. The whole is indeed the problem.[8] Inductive reasoning, employed as a form of argument-by-experience, is not useful in cosmology:

> In so far as inductive reasoning is the attempt to provide good reasons for inferring something about unobserved instances, on the basis of observed instances, it has to do with the inference to further *instances* of some regularity or *law*. However, the problem faced in cosmology is not that of finding a warrant in experience for establishing *laws*. This is the task of ordinary physics. (Munitz [31, p. 62])

The aim of cosmology, rather, is to say something about what, in our example, is the box, its dimensions and its global properties, that is its being that self-contained and singular "thing" to which our objects (the observed universe) belong. For this purpose, we do not need a law, but — as we will see later — a model of the universe. Consequently, if "nothing certain is known of what the properties of the spacetime continuum may be as a whole", as Einstein said in 1929,[9] the universe as a whole could have properties different from those discoverable in its local parts.

On the contrary, to justify cosmological induction, Sciama [40] looks at the so-called *interconnectedness of the universe*: each part interacts with each other part; these interconnections have the same importance both between local and very distant regions, so that, at least in principle, it would be possible to obtain some understanding (or even completely deduce the nature) of the whole from any of its parts. Sciama's argument seems strong, but actually it presupposes what it wishes to prove. Indeed we know with certainty only that the *visible* universe is interconnected, but in order to extrapolate this issue to the whole universe, the cosmological induction that the interconnectedness alleges to show is necessary.

Moreover, the expression "the universe as a whole" is susceptible to different interpretations. According to Munitz [32, pp. 60–69], there are two basic ways to understand it: realist and pragmatist. For the first interpretation the "universe" is the name of an existing entity, with its own inherent properties and an intelligible structure; "the universe as a whole" designates just this entity, whose existence is independent of, and antecedent to, every cosmological investigation. The second interpretation, on the contrary, does not presuppose

---
[7] See for instance Glymour [21, p. 110]: "Confirmation or support is a relation among a body of evidence, a hypothesis and a theory".
[8] For instance, Agazzi [1].
[9] Quoted in Kragh [29, p. 142].

as necessary the existence of an independent entity; the concept "the universe as a whole" is only a theoretical construction, that plays a merely pragmatical role, introduced implicitly or explicitly by every cosmological model.[10]

We adhere to the first interpretation both for personal philosophical convictions and for a simple "a priori" reason: if cosmology were considered, from an instrumentalist perspective, only as a mere account of observational data, without presupposing a *real* existence of the cosmos, then our Kantian problem of cosmology as a science would simply disappear, because one limits the epistemological import of cosmology only to the visible universe.

With respect to this cognitive situation, many scholars maintain that cosmology, as a science of the whole universe, is not possible. For instance Agazzi [1], but also Barrow [2], a cosmologist, affirms that cosmology is not a science of the universe, but only a science of the *visible* universe:[11] whether the universe is literally a complete, unique, and intelligibly structured whole is an irresolvable problem, since we can refer, by direct experience, only to the observed universe.[12]

4 The cosmological principles

However, to overcome, at least partially, the *impasse*, it is first of all necessary to remember — with Kanitscheider [27], Harrison [23] and van Fraassen [18] — that cosmology does not produce a complete description of the Universe, with a capital "U", but it formulates possible models of a few important physical features of the Universe.[13] If at every age, a culture constructs its *own* model — religious, artistic, philosophical or scientific — of the Universe,[14] and each model is only one of many possible representations, "a different cosmic picture that is like a mask fitted on the face of the unknown Universe" (Harrison [23, p. 13]), then each of our contemporary cosmological models, too, really explains only the model *u*niverse and not the actual *U*niverse. "The universe is what a cosmological model says it is", Munitz [32, p. 62] declares lapidary.

Secondly, Einstein already in his 1917 paper, introduced, but without elevating it into a general principle, what would be called, and better formalised, by the English astrophysicist Edward Milne in 1933, the "cosmological principle", according to which:[15]

> *The universe*, at a given cosmic time,[16] *is spatially homogeneous and isotropic on sufficiently large scales.*[17]

---

[10] See also Munitz [31, pp. 58–63].

[11] We can define the *visible* universe — that could be very different (smaller, but also bigger) from the *actual* one — as the space inner to a sphere, centered on the observer, whose radius is that of the cosmological horizon (the so-called *surface of last scattering*, when the universe became transparent).

[12] For a recent survey on cosmology and its philosophical problems see Ellis [16].

[13] This does not mean that we endorse what Munitz calls the pragmatic approach, but only that in order to accomplish a scientific cosmology, one has to renounce a science of everything in a literal sense.

[14] In the history of science we can distinguish, for example, between the Pythagorean model, the Atomist model, the Aristotelian model, and so on.

[15] We do not consider the perspective of the inflationary universe, see for instance [41].

[16] It is beyond the limits of this paper to discuss the important problem of the definition of a cosmic time. It is enough to say that it can be defined if homogeneity and isotropy themselves hold.

[17] See Weinberg [43], Peebles [34].

The expression "large scales" relates to the fact that the actual universe possesses local irregularities, so it is evidently inhomogeneous at the scales of galaxies. To give a precise definition of what is meant by "sufficiently" is not easy: in any case, to be meaningful, this term must be small compared to the universe as a whole, or to that portion of it, which, at a particular moment, is theoretically observable.[18] Strictly speaking, homogeneity and isotropy mean, respectively, an absence of both privileged points and directions. However, the cosmological principle is not the unique assumption: the universe is not only homogeneous and isotropic, but also it looks the same to all observers. The cosmological principle is a development of this last observer-equivalence (from a certain point of view more fundamental but less powerful) statement, usually called the *Copernican principle*.[19] It formally states:

> *No place, in the universe, is in a central or specially favoured position.*

Together, these two principles affirm that the general picture of the universe, as seen by an observer from different locations, is essentially the same, that is, the universe, as seen from Earth, is the same as seen by other observers at other points and from whichever perspective in the sky. Furthermore, the Copernican principle makes it possible to derive the homogeneity by extrapolating it directly from the isotropy, utilising the observational evidence that the universe seems to exhibit spherical symmetry around us.[20]

By means of cosmological principle, it is possible to transform statements about objects contained in the universe into statements about the universe itself: the cosmological principle "unites the universe into a homogeneous whole", Harrison [22, p. 175] summarises. But what about the evolution in time? The cosmological principle can be generalised in the following way:

> *The universe*, at all cosmic times, *is spatially homogeneous and isotropic on sufficiently large scales.*

The last principle must not be confused with what defenders either of the steady-state[21] or of the inflationary universe call the *perfect cosmological principle*:

> *Universe, considered at large scale, is* immutable, *that is it is temporally homogeneous as well.*[22]

---

[18] For instance Weinberg [44, p. 24] speaks about scales at least as large as the distance between clusters of galaxies, about 100 million light years.

[19] The reason for this name is evident: Copernicus reminds us that Earth has no special status in the solar system. A similar consideration could be extended to our solar system, our galaxy and so on.

[20] See Bondi [6, p. 13] and Weinberg [44, pp. 23–24]. It should be remembered that homogeneity by no means implies isotropy: in Gödel's model, for instance, the universe is homogeneous but not isotropic; the same holds for a universe with a large-scale magnetic field pointing everywhere in the same direction and having the same magnitude at every point.

[21] For instance Jaakkola [26].

[22] Einstein's formulation was a form of perfect cosmological principle as well, but his static model of the universe was highly unstable under perturbations, and it was empirically falsified by subsequent expansion observations.

Thus, if cosmological principle requires only homogeneity and isotropy in space at each cosmic time, perfect cosmological principle adds the requirement that the physical distribution is the *same* at every cosmic time.[23] This principle is at the heart of the aforementioned steady-state cosmological model, an alternative to the hot big bang model developed in 1948 by Hermann Bondi and Thomas Gold [7], and, separately, by Fred Hoyle [25]. However, today this theory is no longer justified on empirical grounds.

5 How is a scientific cosmology possible?

Now we have generalisations about the content of the box and a principle about the global structure of the box, but by means of which theory we should connect the former with the latter? Again it was Einstein in his 1917 paper who gives the answer: *general relativity*, the theory that in almost complete solitude he had just established.

To sum up, contemporary cosmology is made possible by the following three issues:

1. astrophysical generalisations, which come from observations. In our previous metaphor, the former are analogous to the generalisations on the content of the box. Thus cosmology is fed by astrophysics.

2. the cosmological principle.

3. general relativity, our best theory of gravity, the dominant force on the cosmic scale (obviously, also electromagnetism, thermodynamics and particle physics play a secondary but important role).

Now, before entering into the crucial *philosophical* aspect concerning the epistemological status of the cosmological principle, it is useful to set it in its effective *scientific* background, casting a quick glance at the main aspects of modern cosmology, in order to understand why this principle is so determinant in the last century cosmology, and why it is accepted today by virtually all cosmologists, whatever their theoretical persuasions are.

We can briefly divide cosmology topics into three separate sections — *cosmography*, *theoretical cosmology*, and *cosmogony* —, bearing in mind, however, the importance and richness of their interconnections.[24]

The first aspect, *cosmography*, concerns cosmic objects, our cataloguing them, and charting their positions and motions. From our observation point (the Earth), our only information about them is contained in the directional distribution, and spectral composition, of their electromagnetic radiation.

The second aspect, *theoretical cosmology*, is the research for a theoretical framework where information from cosmography can be organised and comprehended. Physical laws, established on and near Earth, are employed and — "outrageous extrapolation" ([5, p. 1])! — are applied throughout the universe. The latter is an important assumption, because we need really "to escape" from the narrow limits of our observation point.[25]

---

[23] For instance, a universe homogeneously filled by a magnetic field of magnitude M at time t_1 and homogeneously filled with a magnetic field of magnitude $2M$ at time t_2 is always spatially homogeneous, but not temporally homogeneous.

[24] See Berry [5, p. 1].

[25] We note that even though laws of physics are valid *throughout* the universe, this does

The third and last aspect, *cosmogony*, is the most hazardous part of cosmology, where laws of physics are extrapolated to the most distant times and places, in order to study both the remote past (and the origin, supposing that it exists) and remote future of the universe.

In this rough picture of modern cosmology, the physical-mathematical power of the cosmological principle resides in the fact that we are able to determine, by its apparently simple assumptions, some remarkable features concerning the metric which describes cosmological spacetime. Indeed, a direct consequence of homogeneity is the existence, as already suggested, of a universal cosmic time; moreover, the cosmological principle implies that the three-dimensional physical space of the universe must either be static, or expanding, or contracting, uniformly (to our observational knowledge it is expanding).[26] Another immediate mathematical consequence of the homogeneity is that the relative speed of any two galaxies must be proportional to the distance between them,[27] and this is precisely the famous empirical law, at the foundations of big bang cosmologies, found by Hubble.[28]

Finally, we briefly remind the cosmological principle's observational supports: the highly uniform distribution of intensity of cosmic microwave background radiation; the number counts of distant radio sources (their average distribution is the same in all directions); the extremely uniform distribution of distant optical galaxies and of intensity of the cosmic X-ray background radiation. Therefore, nowadays astronomers observe isotropy, and cosmologists postulate homogeneity.[29]

6 The epistemological status of the cosmological principle

In order to look for an answer to the question concerning the philosophical status of the cosmological principle, and its being a non directly verifiable assumption,[30] one can scan the epistemologically most careful handbooks and treatises of cosmology and relativity: for instance Penrose [35], Rindler [37],

---

not mean that they are valid *for* the universe as a whole, an extrapolation even more problematic than the preceding one, which, as we have just seen, is based on the cosmological principle. That is, one thing is to extend inductively the laws of physics tested only on a limited number of objects, to all objects in the universe, another thing is to pose questions and provide answers about the universe as a whole. The former is a problem of induction, the latter is one of cosmology as a science.

[26] The simple assumptions of homogeneity and isotropy make it possible to deduce *kinematically* the spacetime metric of the universe (the so-called *Robertson-Walker metric*), that is to say, before involving a *dynamical* approach constituted by Einstein's equations!

[27] See Weinberg [44, p. 22], Ohanian and Ruffini [33, p. 453].

[28] We note — with Weinberg ([44, p. 23]) — that we can read this fact *both* in the sense that Hubble's observations are an indirect confirmation of the truth of the cosmological principle, *and*, contrariwise, in the sense that the cosmological principle, taking it for granted on a priori grounds, confirms, or implies, Hubble's law. The meaning of our paper is precisely a methodological argumentation favouring the former approach.

[29] However, the empirical "pedigree" of the cosmological principle has been sometimes challenged: Ellis [15], for instance, sustains that it is guaranteed more by philosophical commitments than by empirical evidences. On the contrary, Collins-Hawking [10] even argue that if the universe were not isotropic life might not exist.

[30] See Ellis [14].

D'Inverno [11], Harrison [23] etc.[31] Even if agreement as to its validity and its utility is very remarkable, there is wide divergence of view as to its necessity, significance and logical position: it is not always clear if the cosmological principle is adopted either as an idealisation dictated by the lack of more precise information, or if it is proposed as normative, that is as a restriction on all possible models of the universe. However, in the literature there are essentially three kinds of justifications:[32]

A. Leibnizian. One can generalise the Copernican viewpoint: there is no reason favouring the fact that we find ourselves in a special part of the universe, so if the visible universe is homogeneous and isotropic, then the whole universe should be the same. Weinberg [44, p. 23] indeed says: "Since Copernicus we have learned to beware of supposing that there is anything special about mankind's location in the universe. So if the universe is isotropic around us, it ought to be isotropic about every typical galaxy".

B. Baconian. The visible universe is homogeneous and isotropic, therefore one can generalise to the whole universe the same properties, as Schutz says:

> The simplest approach to applying general relativity is to use the remarkable large-scale uniformity we observe. We see, on scales of 10^3 Mpc, not only a uniform average density but uniformity in other properties: types of galaxies, their clustering densities, their chemical composition and stellar composition. We therefore conclude that, on the large scale, the universe is homogeneous. What is more, on this scale the universe seems to be isotropic about every point [39, p. 319].

C. Kantian. The cosmological principle is an a priori postulate that makes scientific cosmology possible. Coles-Lucchin [9, p. 6] indeed affirm: "It would be very difficult for us to understand the universe if physical conditions, or even the laws of physics themselves, were to vary dramatically from place to place". This Kantian justification often contains a further working hypothesis needed to overcome our present (only present?) ignorance of the universe as a whole and useful to obtain more progress in cosmological investigations; Bondi [6, p. 13] calls it a *simplicity postulate*: the universe should be as simple as possible (also mathematically: [38, p. 63]), i.e. uniform.[33]

Against the Baconian point of view, the argument proposed by Barrow holds, according to which, whether the universe is finite or infinite in space, the visible universe is an insignificant part of the whole, thence this generalisation is not well-founded.

---

[31]To our knowledge the best cosmological book from an epistemological point of view is Bergia [3], which unfortunately has not been translated into English.

[32]On this topic see also Raychaudhuri [36, pp. 2–7], who distinguishes three possible justifications: mathematical, deductive, and empirical.

[33]It is clear that these three kinds of justifications often go together in cosmological literature. In this Gamow's [20, p. 390] thought, the Copernican generalisation is evident, the Baconian viewpoint resides in the middle, and the necessary apriority is a little hidden, but really important, being repeated in those lapidary verbs (our italics) that do not concede alternatives to the possibility of the cosmological enterprise: "In studying the structure [of the universe] we *must* accept the Copernican point of view and deny to man the honour of a privileged position in the universe; in other words, we *must* assume that the structure of space is very much the same in distant regions as it is in the part we can observe. We *cannot* suppose that our particular neighbourhood is specially adorned with beautiful spiral galaxies for the enjoyment of professional and amateur astronomers".

Against the Kantian point of view, one should observe that, following the quantum and relativistic revolutions, it is clear to everyone that no part of science could be completely a priori.

The Leibnizian point of view is more sophisticated than the others, but in our opinion it too is mistaken. Indeed it is based on a sort of Laplacean principle of insufficient reason, or Keynesian[34] principle of indifference, that is based on an a priori probabilistic argument of the kind:

> *There are no reasons to believe that various parts of the universe are different, therefore they are similar.*

Harrison [22, p. 174] calls this argument "the location principle", that states: "It is improbable that human beings have privileged location in the physical universe".[35] Many epistemologists have correctly contested these sorts of arguments, since they are based on ignorance. At best, in our opinion,[36] one could apply a sort of *principle of reasonable similarity*, that is:

> *We have good reasons for believing that the universe is similar in all its parts.*

In other words, the form of these kinds of principles must be a positive, not a negative one. But till now we have no reasons favouring such similarity independently of the application of cosmology. Then, which is the correct justification of the cosmological principle? The answer is much simpler. To begin with we can read again the brief proposition, with which Einstein introduced a sort of forerunner of the cosmological principle in 1917:

> Wenn es uns aber nur auf die Struktur im grossen ankommt, dürfen wir uns die Materie als über ungeheure Räume gleichmässig ausgebreitet vorstellen, so dass deren Verteilungsdichte eine ungeheuer langsam veränderliche Funktion wird.[37] [12, p. 148]

Einstein places no particular emphasis on this hypothesis, that is he introduces it with absolute naturalness, because without any special reflection he follows the *hypothetico-deductive* method, which he himself, Carnap and Hempel were to elaborate a few decades later. Indeed in his celebrated 1934 paper [13, p. 163] he begins with these famous words:

> If you wish to learn from the theoretical physicist anything about the methods which he uses, I would give you the following piece of advice: Don't listen to his words, examine his achievements.

And in the following he explains how theoretical physics works:

---

[34] See [28, chap. IV], where one can also find a very useful historical retrospective.

[35] In Harrison's [23, p. 140] opinion, the Copernican principle, similar to his location principle, asserts too much: it appears "to perpetuate the belief that a center, somewhere or other, exists", whereas "we may say with certainty only that a central location in the cosmos is improbable".

[36] See Fano [17].

[37] "But if we are concerned with the structure only on a large scale, we may represent matter to ourselves as being uniformly distributed over enormous spaces, so that its density is a variable function which varies extremely slowly" [4, p. 21]. In our opinion, with the expression "eine ungeheuer langsam veränderliche Funktion" Einstein does not mean that the density distribution is smooth, but that, if it is considered with a sufficiently coarse grain it is practically constant.

> We have now assigned to reason and experience their place within the system of theoretical physics. Reason gives the structure to the system; the data of experience and their mutual relations are to correspond exactly to consequences in the theory. On the possibility alone of such a correspondence rests the value and the justification of the whole system, and especially of its fundamental concepts and basic laws. But for this, these latter would simply be free inventions of the human mind which admit of no *a priori* justification either through the nature of the human mind or in any other way at all [13, p. 165].

Thus, contrary to most scientists of the third millennium, Einstein is completely aware that in theoretical physics a part of the theory not only does not have a direct justification, but it does not need it at all. For this part receives empirical meaning only indirectly through the whole theoretical network, as definitely clarified by Hempel [24], and Carnap [8] in the fifties. This point of view will become notorious as the "received view" and in spite of its low capacity to describe real science, in our opinion it remains the best way of giving empirical meaning to theoretical terms of science.

Moreover Einstein is conscious as well that he himself is one of the creators (maybe the most important) of this new methodology, not completely understood by Galileo, Newton and the scientists of nineteenth century:

> On the contrary the scientists of those times were for the most part convinced that the basic concepts and laws of physics were not in a logical sense free inventions of the human mind, but rather that they were derivable by abstraction, i.e. by a logical process, from experiments. It was the general Theory of Relativity which showed in a convincing manner the incorrectness of this view. For this theory revealed that it was possible for us, using basic principles very far removed from those of Newton, to do justice to the entire range of the data of experience in a manner even more complete and satisfactory than was possible with Newton's principles [13, p. 166].

In the same years cosmologists reached a full comprehension of the hypothetico-deductive method of their science during the lively polemic, which burst out between the cosmologists Dingle and Milne in the thirties, and which concluded with the victory of the latter, who was one of the most prominent promoters, as already mentioned, of the role of the cosmological principle in contemporary cosmology.[38] It is noteworthy that today cosmologists seem to have forgotten this. Nevertheless, recent cosmology can be counted as one of the fields in which the hypothetico-deductive method, with its combined way of validation and explanation, has reached its most remarkable success.[39]

7 Concluding remarks

We shall now return to Kant's problem. The most recent relativistic cosmology — based on the above mentioned epistemology — solves the Kantian

---

[38] On this subject see Gale [19].

[39] This reflection by McMullin [30, p. 35] clearly epitomises how work on cosmology proceeds incessantly: "The success of hypothetico-deductive methods when applied to the most distant regions of the universe as well as to the universe taken as a whole testifies quite strongly to its fundamental unity. So far as one can see, it might *not* have worked out this way. When the spectra of distant stars, or the velocities of distant galaxies, continue to be interpretable by schemes derived from terrestrial processes, confidence quite properly grows in the assumption that these schemes are not just conventions imposed for convenience's sake or because our minds cannot operate otherwise, but that all parts of the universe are united in a web of physical process which is accessible, through coherent and ever-widening theoretical constructs created and continually modified by us".

antinomy in an odd way: the thesis holds for both space and time, that is time has a beginning and space, though unlimited, is finite. But these conclusions are highly *hypothetical* and not without problems.

To sum up: how is it possible that contemporary cosmology bypassed Kant's prohibition? The answer is very simple: modern science does not necessitate of any a priori foundations. Relativistic cosmology is founded on the cosmological principle, which makes the *convergence* possible of the regressive use of reason relatively to the totality of phenomena. So that the universe becomes a concept, i.e. it is no longer an idea. On the other hand, the cosmological principle is partially confirmed indirectly, as already seen, by the numerous correct predictions of the whole theory.

Acknowledgements

We thank Michel Ghins, Gino Tarozzi, Jos Uffink and an anonymous referee for their meaningful comments to a preceding presentation of this argument.

BIBLIOGRAPHY

[1] E. Agazzi. The universe as a scientific and philosophical problem. In E. Agazzi and A. Cordero (eds.), *Philosophy and the Origin and Evolution of the Universe*, Kluwer, Dordrecht 1991.
[2] J.D. Barrow. *The World within the World*. Oxford University Press, Oxford 1988.
[3] S. Bergia. *Dialogo sul sistema dell'universo*. McGraw Hill, Milano 2002.
[4] J. Bernstein and G. Feinberg (eds.) *Cosmological Constants*. Columbia University Press, New York 1986.
[5] M. Berry. *Principles of Cosmology and Gravitation*. Cambridge University Press, Cambridge 1976.
[6] H. Bondi. *Cosmology*. Cambridge University Press, Cambridge 1952.
[7] H. Bondi and T. Gold. The steady-state theory of the expanding universe. *Monthly Notices of the Royal Astronomical Society*, 108: 252–70, 1948.
[8] R. Carnap. The methodological character of theoretical concepts. In H. Feigl and M. Scriven (eds.), *Minnesota Studies in the Philosophy of Science* **I**, *The Foundations of Science and the Concepts of Psychology and Psychoanalysis*, 1956, pages 38–76.
[9] P. Coles and F. Lucchin.*Cosmology*. John Wiley & Sons, Chichester 1995.
[10] C.B. Collins and S.W. Hawking. Why is the universe isotropic?. *The Astrophysical Journal*, 180: 317–34, 1973.
[11] R. D'Inverno. *Introducing Einstein's Relativity*. Clarendon Press, Oxford 1992.
[12] A. Einstein. Kosmologische betrachtungen zur allgemeine Relativitätstheorie. In *Sitzungsberichte der königlich Preuss. Akad. der Wissensch.*, 1917, pages 142–52.
[13] A. Einstein. On the method of theoretical physics. *Philosophy of Science*, 1: 163–9, 1934
[14] G.F.R. Ellis. Cosmology and verifiability. *Quarterly Journal of the Royal Astronomical Society*, 16: 245–64, 1975.
[15] G.F.R. Ellis. Is the universe expanding? *Gen. Rel. and Grav.*, 9: 87–94, 1978
[16] G.F.R. Ellis. Issues in the philosophy of cosmology. In J. Butterfield and J. Earman (eds.), *Philosophy of Physics*, Elsevier, Amsterdam 2007, pages 1183-285.
[17] V. Fano. A critical evaluation of comparative probability. In *Probability in Keynes: Rationality, Pragmatism and Economic Decision*, 2004; http://philsci-archive.pitt.edu/archive/00003019/01/scazzieriingl.doc.
[18] B. van Fraassen. *The Empirical Stance*. Yale University Press, 2002.
[19] G. Gale. Cosmology: methodological debates in the 1930's and 1940's. http://plato.stanford.edu/entries/cosmology-30s/, 2007.
[20] G. Gamow. Modern cosmology. In M. Munitz (ed.), *Theories of the Universe*, Free Press, Illinois 1957, pages 390–404.
[21] C. Glymour. *Theory and Evidence*. Princeton University Press, Princeton 1980.
[22] E. Harrison. *Masks of the Universe*. Macmillan Publ. Comp., New York 1985.
[23] E. Harrison.*Cosmology*. Cambridge University Press, Cambridge 2000.

[24] C.G. Hempel. *Fundamentals of Concept Formation in Empirical Science.* University of Chicago Press, Chicago 1952.
[25] F. Hoyle. A new model for the expanding universe. *Monthly Notices of the Royal Astronomical Society,* 108: 372–82, 1948
[26] T. Jaakkola. The cosmological principle: theoretical and empirical foundations. *Apeiron,* 4: 14–49, 1989.
[27] B. Kanitscheider. Does physical cosmology transcend the limits of naturalistic reasoning? in P. Weingartner and G.J.W. Dorn (eds.), *Studies in Mario Bunge's Treatise,* Rodopi, Amsterdam 1990.
[28] J.M. Keynes.*A Treatise on Probability* (1921). Mcmillan, London 1988.
[29] H. Kragh. *Conceptions of Cosmos,* Oxford University Press, Oxford 2007.
[30] E. McMullin. Is philosophy relevant to cosmology? In J. Leslie (ed.), *Physical Cosmology and Philosophy,* Macmillan, New York 1990.
[31] M.K. Munitz. *The Mystery of Existence.* Meredith, New York 1965.
[32] M.K. Munitz. *Cosmic Understanding.* Princeton University Press, Princeton 1986.
[33] H. Ohanian and R. Ruffini. *Gravitazione e Spazio-Tempo.* Zanichelli, Bologna 1997.
[34] P.J.E. Peebles. *Principles of Physical Cosmology.* Princeton University Press, Princeton 1993.
[35] R. Penrose. *The Road to Reality.* Alfred Knopf, New York 2005.
[36] A.K. Raychaudhuri. *Theoretical Cosmology.* Clarendon Press, Oxford 1979.
[37] W. Rindler. *Relativity.* Oxford University Press, Oxford 2006.
[38] M. Rowan-Robinson. *Cosmology.* Clarendon Press, Oxford 1996.
[39] B. Schutz.*A First Course in General Relativity.* Cambridge University Press, Cambridge 1985.
[40] D.W.S. Sciama. *The Unity of the Universe.* Doubleday, New York 1961.
[41] M.S. Turner. A sober assessment of cosmology and the new millennium. Astro-ph/0102057 v1, 2001.
[42] W.G. Unruh. Is the physical universe natural? In J. Robson (ed.), *Origin and Evolution of the Universe,* McGill-Queen's University Press, Montreal 1987, pages 109-18.
[43] S. Weinberg. *Gravitation and Cosmology.* Wiley & Sons, New York 1972.
[44] S. Weinberg. *The First Three Minutes.* Basic Books, New York 1993.

Poincaré and the electromagnetic world picture. For a revaluation of his conventionalism

GIULIA GIANNINI

This paper is a part of a broader research on the origin and development of a new relativistic dynamics in Poincaré's work. My aim is to show how the approach of Poincaré to an electromagnetic world picture, joint with the importance of experimental confirmations in his works and the operational origin of physical concepts and theories, can lead to a revaluation of Poincaré's epistemological perspective. The first goal of this paper is, therefore, to offer a rereading of Poincaré's epistemological position that takes into account the context in which such a position was developed and the totality of Poincaré's work. The interpretations given up to now are, in fact, very controversial and they lead to a misunderstanding of Poincaré's philosophy of sciences thus causing it to be undervalued.

It is only by extension that the critical literature usually refers to Poincaré's philosophy using the word "conventionalism". This term is inspired by the conventions Poincaré diffusely wrote about in his epistemological papers and never appears in his production. The appellation, coined by his critics, is at the same time a symptom and cause of the misunderstandings to which Poincaré's thought has been subjected. Different, radical and opposing attitudes seem to be hidden behind the term "conventionalism": we see on the one hand the impossibility of defining reality and on the other the ensuing positions of logic empiricism.[1]

Indeed, Poincaré's work, wide and heterogeneous, gave rise to several controversial interpretations. The most technical essays dedicated to it,[2] are based only on some Poincaré's physical or mathematical papers and they seldom take in account his epistemological work. Moreover, from a philosophical point of view, these technical essays are often devoid of any interpretational attempt. On the contrary, the most celebrated readings of "conventionalism" — notwithstanding the fact that their authors come primarily from the scientific field [3] — are often founded in his most famous epistemological works. Furthermore, a great part of these interpretations ignores the historical context in which Poincaré's thought was developed. In particular, it is neces-

---

[1] On this subject see [6] and [45]; see also the interpretation given by Parrini in [26].

[2] See Logunov's essay on Poincaré's papers *Sur la dynamique de l'électron* [21] and June Barrow-Green's essay on the three bodies problem [2].

[3] From the mathematical field came for example Roberto Torretti [44] and Jan Mooij [23]. From the physical field Adolf Grünbaum [14] and Jerzy Giedymin [9, 11, 10, 12]. From the logical field Gerhard Heinzmann [16, 17, 18, 19] and Elie Zahar [46, 47, 48].

sary to consider the relationships maintained by Poincaré not only with the other mathematicians, but also and especially, with the physicists and the philosophers of his time.[4] Thus, the discordances among the different readings originate mainly in the lack of attention to both the historical context in which the "conventionalism" grew and also the wholeness of Poincaré's work. The different interpretations — in disagreement about the role assigned by Poincaré to the experience — tried to reduce the complex relationship between geometry and the physical world to a sort of moderate empiricism,[5] or to a kind of nominalism.[6] So, they simplified Poincaré's position, extracting some isolated statements from his epistemological papers.

The problems concerning "conventionalism" interpretations are also related to the intrinsic difficulties of Poincaré's work. His work includes different sorts of texts addressed to various publics and is does not contain strictly philosophical terminology. Indeed, Poincaré wrote not only physical or mathematical papers, but also epistemological essays, popular works and lectures. In particular, his well-known popular works are based on the readaptation of previous philosophical papers written in different periods. Therefore, they do not represent a methodical exposition of his epistemology but rather a heterogeneous collection: chapters are autonomous, independent from one another and the terminology is not always uniform. This aspect has often led critics to perceive contradictions in Poincaré's texts and to force the meaning of certain statements to coincide "conventionalism" with different kinds of empiricist or nominalist positions.

In the present paper I will show the originality and the relevance of Poincaré's philosophy of science and the necessity of rereading his epistemological thought. This aim will be pursued through a historical analysis of his works which will focus on his criticism of Mechanics and his approach to an electromagnetic perspective on Nature.

By 1880, Poincaré participated in the criticism of Mechanism that characterised the origin of different world pictures in the second part of the 19th century. Indeed, at this time it was possible to witness at the clash of different theoretical positions concerning distinct world views.[7] Mechanism had been the dominant paradigm over two centuries. Now, however, there were new attempts to formulate unified world pictures, based on rising physical disciplines, such as Thermodynamics (its first two principles date back to the 1860s) and Maxwell's electromagnetism (1873). The electromagnetic world picture, in opposition to the mechanistic one, tried to explain all the natural phenomena, not through a reduction of them to matter and motion, but rather through the laws of the electromagnetic field. The aim of such a point of view on Nature was to base all physics on Electromagnetism, which was

---

[4]The first edition of Poincaré's correspondence with the mathematicians, edited by Pierre Dugac, date to the second half of the eighties [3, 4]. A project of edition of Poincaré's correspondence only began since 1994 at Poincaré's archives and it led in 1999 to the publication of the correspondence between Poincaré and Gösta Mittag-Leffler ([31]; see also [24]). Whereas the publication of the correspondence with the physicists, the chemists and the engineers only appeared in July 2007 [33].

[5]See, among others [1] and [14].

[6]See, among others, [5, 43, 41, 42, 25].

[7]On this aspect see, among others [22, 20, 13, 15].

conceived as the basic discipline to which all the others had to be reduced.

Poincaré's criticism of Mechanism, already present in the afterword of Leibniz's *Monadologie* (1880, [27]) edited by Emile Boutroux, appears as a constant element of his production. At the beginning, it was based on the impossibility of finding a unified mechanical explanation of phenomena and on the problems caused by the relationship between Mechanism and the new experimental evidences. Poincaré claimed that the existence of one mechanical model implied the existence of an infinite number of them. Therefore, it was impossible to determine, among the infinite possible mechanical models, which would be most suitable for describing natural phenomena. Neither the experience nor the convenience (used, for example, in geometry) could help in the choice among the different mechanical models: such a choice was founded then on purely subjective and metaphysical considerations. Due to the impossibility of defining a single mechanical model, Poincaré argued for their insubstantiality. The infinity of such models was, in fact, the first step towards their loss of meaning. Moreover, Poincaré stated, as early as 1894, the uselessness of research into a mechanical model. For him, it was not necessary to find a mechanical explication, but rather to look for Unity of Nature, namely for the common features of all the theories.

Since his lessons in 1887–1888, such a unity appeared as the only aim of scientific researches. In his paper *Les relations entre la physique expérimentale et la physique mathématique* (1900, [34]) Poincaré declared that the attempt of finding a unitary view of Nature clashed with the difficulties linked to the mechanistic interpretation of electrical phenomena. Then, in 1893 [28], he showed that the mechanical effort of giving an unitary explanation of all phenomena by means of mass and motion met with several obstacles. The physicists had difficulties reconciling mechanical description with experimental data. In particular, such an attempt proved to be incompatible with phenomenal irreversibility. The experience showed an amount of irreversible events whereas mechanist hypothesis presupposed the reversibility of all phenomena. Thus, the aim of finding Unity of Nature, while essential, could not be pursued in a mechanist way.

Poincaré's analysis of Mechanism defines itself through criticism of some distinctive concepts of Mechanics. By 1895, in *A propos de la théorie de Larmor* [29], Poincaré affirmed the impossibility of observing the absolute motion:

> L'expérience a révélé une foule de faits qui peuvent se résumer dans la formule suivante: il est impossible de rendre manifeste le mouvement absolu de la matière, ou mieux le mouvement relatif de la matière pondérable par rapport à l'éther; tout ce qu'on peut mettre en évidence, c'est le mouvement de la matière pondérable par rapport à la matière pondérable.[8]

Later, in *La mesure du temps* (1898) [30], he showed the conventional character of measuring procedures of temporal intervals and, more generally, Time's conventionality:

> Nous n'avons pas l'intuition directe de l'égalité de deux intervalles de temps. Les personnes qui croient posséder cette intuition sont dupes d'une illusion [...]. La simultanéité de deux événements, ou l'ordre de leur succession, l'égalité de deux durées,

---

[8] Poincaré [29, page 412].

doivent être définies de telle sorte que l'énoncé des lois naturelles soit aussi simple que possible. En d'autre termes, toutes ces règles, toutes ces définitions ne sont que le fruit d'un opportunisme inconscient.[9]

In another paper (1900, [32]), published in honour of the jubilee of Lorentz's doctoral thesis, Poincaré introduced his method of clock synchronisation by light signals. Then, through a physical interpretation of Lorentz's local time, he reaffirmed the inexistence of Absolute Time. Starting from this text Poincaré also introduced the "principle of relative motion". Furthermore, still in 1900, in a lecture on the principles of mechanics [35], he claimed:

> 1. Il n'y a pas d'espace absolu et nous ne concevons que des mouvements relatifs; cependant on énonce le plus souvent les faits mécaniques comme s'il y avait un espace absolu auquel on pourrait les rapporter;
> 2. Il n'y a pas de temps absolu; dire que deux durées sont égales, c'est une assertion qui n'a par elle-même aucun sens et qui n'en peut acquérir un que par convention;
> 3. Non seulement nous n'avons pas l'intuition directe de l'égalité de deux durées, mais nous n'avons même pas celle de la simultanéité de deux événements qui se produisent sur des théâtres différents; [...]
> 4. Enfin notre géométrie euclidienne n'est elle-même qu'un sorte de convention de langage; nous pourrions énoncer les faits mécaniques en les rapportant à un espace non euclidien qui serait un repère moins commode, mais tout aussi légitime que notre espace ordinaire; l'énoncé deviendrait ainsi beaucoup plus compliqué; mais il resterait possible.
> Ainsi l'espace absolu, le temps absolu, la géométrie même ne sont pas des conditions qui s'imposent à la mécanique; toutes ces choses ne préexistent pas plus à la mécanique que la langue française ne préexiste logiquement aux vérités que l'on exprime en français [35, pp. 142–144].

For Poincaré the concepts of Absolute Space, Absolute Motion and Absolute Time were meaningless, already within Classical Mechanics [7]. The impossibility of determining them in an experimental way showed, in Poincaré's opinion, that they were empty notions, alien to physical processes.

Poincaré's criticism also took into account the concept of Mass. Associated to the electromagnetic field, Mass depended on direction and velocity of body motion and makes sense only for slower than light velocities. In several papers, notably the 1904 Saint-Louis lecture [36] and *La fin de la matière* ([37], published in 1906 and since 1907 included in *La Science et l'Hypothèse* [38]), Poincaré showed that the mechanical concept of a constant mass had to be replaced by the idea of mass depending on velocity and linked to the electromagnetic field (or, at least, acting as if it was related to the field).

At the Saint-Louis conference, Poincaré underlined the crisis of Lavoisier's principle. He affirmed that the total electron's mass (or apparent mass) was composed of two parts: the mechanical mass and the electromagnetic mass. Poincaré explained that the electron was submitted not only to the mechanical inertia but also to an electromagnetic force, which he later defined as *self-induction*. In *La fin de la matière*, he clarified:

> [...] nous savons que les courants électriques présentent une sorte d'inertie spéciale appelée *self-induction*. Un courant une fois établi tend à se maintenir, et c'est pour cela que quand on veut rompre un courant, en coupant le conducteur qu'il traverse, on voit jaillir une étincelle au point de rupture. Ainsi le courant tend à conserver son intensité de même qu'un corps en mouvement tend à conserver sa vitesse ([33, page 246], author's italics).

---

[9]Poincaré [30, pages 2–3].

Thus, there are two different reasons which incite the electron's resistance towards any possible alteration of velocity: its mechanical inertia and its *self-induction*. The latter is derived from the fact that any kind of velocity's alteration correspondeds to an alteration of current. The electromagnetic mass is dependent on velocity and direction, hence it is not constant.

In addition, Poincaré emphasized that Kaufmann's experiments showed the inexistence of the mechanical mass. These experiments revealed rather the existence of only electromagnetic mass, dependent on the electromagnetic field. Moreover, in his 1904 lecture, he claimed that even if Kaufmann's experiments were not confirmed, it would be necessary in any case to consider the mass as variable. Lorentz was obliged to suppose that, in a uniform translated medium, every force was reduced by the same proportion independently of its origin. He did so to preserve the principle of relativity as well as the "indubitable" results of Michelson's experiment. Such a reduction did not deal only with "real" forces but also with force of inertia: the masses of every particle would be influenced by a translation, behaving in the same way as electromagnetic masses of electrons. Thus, mechanical masses, even if they existed, could be constant.

Looking for a unitary view of nature, Poincaré distanced himself from Mechanics' world view and he approached Lorentz's theory as well as Kaufmann's and Abraham's works. On several occasions he refered to Lorentz's theory initially calling it the "less defective" [29] among all electrodynamic theories of moving bodies. Later in 1900 [34], he dubbed it the "most satisfactory", "[...] celle dont on trouvera le plus de traces dans la construction définitive" [34, p. 1172]. Poincaré attentively followed the developments of Lorentz's theory. He actively participated in its elaboration both with remarks on its compatibility with the experience and the principles, and also with personal changes. Following the goal of a unitary world view, able to resolve mechanist contradictions, Poincaré moved towards Lorentz's theory and towards an electromagnetic perspective on Nature.

Despite the fact that a great number of critics considered Poincaré to be linked to a classical idea of science, he instead developed a deep criticism to Mechanics which led him to a completely new conception of Nature. Moreover, from his afterword to *Monadology* in 1880 [27] until the Saint-Louis conference in 1904 [36], Poincaré was increasingly persuaded by the necessity of developing a new physical theory, able to solve the intrinsic contradiction of mechanics and also to recompose the Unity of Nature.[10]

As underlined by Giannetto [7, p. 180], Poincaré's criticism of Mechanics is characterized by a "deconstruction" of its main concepts. Absolute Space, Absolute Time and Absolute Motion became artificial notions for Poincaré, devoid of meaning in classical Mechanics too. The impossibility of defining them through the use of experimental operations revealed that they were empty parameters, external to all physical processes. The same notion of mechanical Mass lost its meaning of basic concept and it was redefined by Poincaré in an electromagnetic way. As previously stated, even if Poincaré did not exclude the possibility of conceiving of a mechanical mass, he recognized

---

[10]Such a theory will find its fulfilment in the subsequent years. See [39, 40].

that Mass, like electromagnetic self-induction, was dependent on velocity. Hence, the Mass was deprived of its mechanical characteristics.

A sort of "deconstruction" was also applied by Poincaré to the concept of Ether. His use of such a term was often interpreted as the evidence of his adhesion to a classical idea of science and as an epistemological impediment for elaborating a real theory of relativity. On the contrary, the term "Ether" was in fact deprived of any previous meaning. Since 1899 [8], Poincaré referred to it as a "metaphysical hypothesis" destined to disappear. Furthermore, when he described Ether's physical properties, he said:

> L'expérience a révélé une foule de faits qui peuvent se résumer dans la formule suivante: il est impossible de rendre manifeste le mouvement absolu de la matière, ou mieux le mouvement relatif de la matière pondérable par rapport à l'éther; tout ce qu'on peut mettre en évidence, c'est le mouvement de la matière pondérable par rapport à la matière pondérable [29, p. 412].

Only the "matière pondérable" could represent a reference frame for Poincaré. The Ether was not considered a material substratum to which phenomena had to be referred. This aspect is confirmed by the fact that in his two papers on electron dynamics, Poincaré only uses the term "Ether" just in the Introduction. He does so in order to explain the impossibility of measuring the motion of matter with respect to Ether. In the other parts of these papers there is no reference to Ether and it has no role in the development of either the calculus or the reasoning (see[49]).

Through an examination of Poincaré's criticism of Mechanics it is possible to understand the importance he gave to experience and experimental data. His reflections frequently arose from experiments (e.g. the experiment of Michelson-Morley and the works of Kaufmann and Abraham). Poincaré often considered the possibility of experimental confirmation to be decisive. Several times, in his scientific works, Poincaré considered experiments capable of condemning the scientific principles and essential to identify the correct theory among a moltitude of possibilities. This aspect is not in contradiction with what he affirmed in his epistemological texts. The role of physical principles and conventions, usually compared to that of geometrical conventions, appeared in Poincaré's work as very complex.

As mentioned at the Saint-Louis conference Poincaré spoke about a "principles' crisis". The use of new measuring instruments allowed new experiments and measures to be carried out which led to results and to conditions of experience that were incompatible with the previous data. Two statement went hand in hand: the generalisation of principles involved conventional elements, and, it was necessary to abandon old principles. For Poincaré, there were contexts in which the introduction of *ad hoc* hypothesis was not sufficient to save the principles. Even if they were not directly falsified by experience, they lost their meaning: the experimental proofs attributed them only a formal value. Thus, the principles did not express anything about physical phenomena anymore. Even if they were not "falsified", they became useless and meaningless.

The experiments acquired a fundamental role in Poincaré's epistemology as starting points for operational definitions. They became the basis upon which it was possible to found a theory that also took into account measuring instruments, namely a theory about the conditions of knowledge [7].

The undervaluation of empirical data and the misunderstanding of Poincaré's statements about the possibility of constructing different theoretical frames often led to study Poincaré's epistemology through the interpretational categories of his geometrical works. Poincaré's new dynamics originated in experiments and in an electromagnetical world view. Nature could only be understood through measuring instruments, which were, for Poincaré, indissolubly linked to the assumption of a specific theory on the structure of the world. Human knowledge was impossible without such instruments and was related to the particular world picture on which the theories were founded. Hence, the awareness that there was no dynamics without a physical world view was the real basis of Poincaré's epistemology in physics.

Such an epistemology, at the same time relevant and original, cannot be reduced to its main interpretations. About the readings which consider Poincaré as an empiricist,[11] it is sufficient to mention that he continually and explicitly criticized empirical positions. The inadequacy of such interpretations is stressed by the fact that Reichenbach, far from seeing Poincaré as one of his forerunners,[12] criticized him for assigning a sort of "subjective arbitrariness" [43] to conventions. The charges of "antirealism" related to nominalist readings[13] are not justified. For Poincaré, Geometry was nothing but linguistic convention, namely a convenient language among the others. His view did not involve an antirealism, but only a rejection of geometrical realism. Even if Geometry indicated a physical reality, for Poincaré it did not coincide with such a reality. Poincaré's statements about the presence of conventional elements in principles could not be seen as "antirealist". Such affirmations did not conceal a reality denial, but rather the awareness of the limits of theories and principles. The experience could not determine them with certainty; consequently they could be true only within certain limits. Laws and principles were nothing but mathematical forms through which it was possible to describe the world. They were contingent and they changed with the shift of theories in the history of science. Such an evolution, even though it revealed the impossibility of a sure and total knowledge of phenomena, allowed reality to gradually show itself.

Even Giedymin's interpretation [9, 11, 10, 12] cannot be considered completely satisfactory even though it is the most accurate and faithful to Poincaré's texts. In order to direct Poincaré's physical epistemology to the geometrical one, Giedymin found the rise of physical "generalised conventionalism" of Poincaré in the work of Hamilton and Hertz. Thus, he reduced Poincaré's physical thought to what he defined a "Pluralist Conception of Theories". In his opinion, the base of Poincaré's whole epistemology was constituted by a rejection of uniqueness according to which experimental data could lead to different possible theoretical constructions. So, with the aim of founding the totality of Poincaré's thought in his geometrical conventionalism, Giedymin focused his attention on Poincaré's works on geometry and mathematical physics, ignoring Poincaré's physical papers.

---

[11] For this aspect see footnote 5.
[12] As Grünbaum maintained on the contrary [14].
[13] See note 6.

In a subsequent paper, Donald Gillies [50] found contradictions in Poincaré's work. In particular, he maintained that Poincaré's scientific practice contradicts his epistemological methodology. Even though Poincaré made a revolutionary advance in his 1905 and 1906 papers [39, 40], such an innovatory step was not followed by his methodological views. On the contrary, in Gillies opinion, such an advance was only made possible by the fact that Poincaré ignored and broke whith his own conservative methodology.

Poincaré's scientific and epistemological activities were never separated. While his scientific works showed the results of an *in fieri* science, his epistemological writings represented rather a philosophical analysis of classical science. This does not imply the presence of a contradiction between his physical and mathematical researches and his philosophical and popular works. While we must consider Poincaré's writings as a cohesive whole, we should not try to impose one part of Poincaré's though to the entirety of his philosophy. Poincaré never had the intention of systematically exposing his philosophy. Therefore makes no sense to look for such an exposition in his writings or to realize an *a posteriori* synthesis of Poincaré's thought. In order to understand his thought it is necessary to avoid any kind of synthesis and, on the contrary, to try to understand all its aspects in the context in which they were formulated.

BIBLIOGRAPHY

[1] K. Ajdukievicz. *Zagadnienia i kierunki filozofii*, Czytelnik, Warszawa, 1948. English translation by H. Skolimowski e A. Quinton, *Problems and Theories of Philosophy*, Cambridge University Press, Cambridge 1973.

[2] J. Barrow-Green. *Poincaré and the Three Body Problem*. American Mathematical Society/London Mathematical Society, Providence 1997.

[3] P. Dugac. La correspondance de Henri Poincaré avec des mathématiciens de A à H, *Cahiers du séminaire d'histoire des mathématiques*, 7: 59–219, 1886.

[4] P. Dugac. Henri Poincaré. La correspondance avec des mathématiciens de J à Z. *Cahiers du séminaire d'histoire des mathématiques*, 10: 83–229, 1889.

[5] F. Enriques. *I problemi della scienza* (1906). Anastatic reprint of 2nd edition (1909), Zanichelli, Bologna 1989.

[6] M. Friedman. Poincaré's conventionalism and the logical positivists. *Foundations of Science*, 1 (2): 299–314, 1995.

[7] E. Giannetto. The rise of special relativity: Henri Poincaré's works before Einstein. *Atti del XVIII congresso di storia della fisica e dell'astronomia*, 1998, pages 171–207..

[8] H. Poincaré. *La théorie de Maxwell et les oscillations hertziennes*, La télégraphie sans fil, G. Carré and C. Naud, Paris 1899.

[9] J. Giedymin. On the origin and significance of Poincaré's conventionalism, *Studies in History and Philosophy of Science*, 8: 271–301, 1977.

[10] J. Giedymin. Geometrical and physical conventionalism of Henri Poincaré in epistemological formulation. *Studies in History and Philosophy of Science*, 22: 1–22, 1981.

[11] J. Giedymin. *Science and Convention. Essay on Henri Poincaré's Philosophy of science and the Conventionalist Tradition*. Pergamon Press, Oxford 1982.

[12] J. Giedymin. Conventionalism, the pluralist conception of theories and the nature of interpretation. *Studies in History and Philosophy of Science*, 23: 423–443, 1992.

[13] B. Giusti Doran. Origins and consolidation of field theory in nineteenth-century Britain: From the mechanical to the electromagnetic view of nature. In R. McCormmach, *Historical Studies in the Physical Sciences*, vol. VI, Princeton University Press, Princeton 1975.

[14] A. Grünbaum. *Philosophical Problems of Space and Time*, Alfred Knopf, New York 1963. 2nd edition (enlarged): Dordrecht, Reidel 1973.

[15] P.M. Harman. *Energy, Force, and Matter. The Conceptual Development of Nineteenth-Century Physics*. Cambridge University Press, Cambridge, 1982.

[16] G. Heinzmann. *Poincaré, Russell, Zermelo et Peano*. Blanchard, Paris 1986.
[17] G. Heinzmann. Poincaré et la philosophie des mathématiques. *Cahiers du séminaire d'histoire des mathématiques*, 9: 99–121, 1988.
[18] G. Heinzmann. Helmholtz and Poincaré's Considerations on the Genesis of Geometry. In L. Boi, D. Flament, J.-M. Salanskis (eds.), *1830-1930: A Century of Geometry. Epistemology, History and Mathematics*, Springer, Berlin/Heidelberg/New York 1992, pages 245–249.
[19] G. Heinzmann. La Philosophie des sciences de Henri Poincaré. In J. Gayon (ed.), *L'épistémologie française de 1850 à 1950*, PUF, Paris 2002.
[20] C. Jungnickel and R. McCormmach. *Intellectual mastery of nature: theoretical physics from Ohm to Einstein*, 2 vols., University of Chicago Press, Chicago 1986; in particular chap. 24 (vol. 2): New foundations for theoretical physics at the turn of the twentieth century.
[21] A.A. Logunov. *On the articles by Henri Poincaré "On the dynamics of the Electron"*, (original edition in russian), Moscow University Press, Moscow, 1988. English translation by G. Pontecorvo: *On the articles by Henri Poincaré "On the dynamics of the Electron"*, Publishing Department of the Joint Institute for Nuclear Research, Dubna 1995.
[22] R. McCormmach. H. A. Lorentz and the electromagnetic view of nature. *Isis*, 61 (4): 459–497, 1970.
[23] J.J.A. Mooij. *La philosophie des mathématiques de Henri Poincaré*. Gauthier-Villars, Paris 1966.
[24] P. Nabonnand. The Correspondance Between Poincaré and Mittag-Leffler. *The Mathematical Intelligencer*, 21: 58–64 1999.
[25] E. Nagel. *The Structure of Science: Problems in the Logic of Scientific Explanation*, Harcourt, Brace and World, New York 1961.
[26] P. Parrini. *Empirismo logico e convenzionalismo*. Franco Angeli, Milano 1983.
[27] J.-H. Poincaré. Note sur les principes de la mécanique dans Descartes et dans Leibnitz. In G.W. Leibniz, *Monadologie*,édition annotée par Émile Boutroux, Delagrave, Paris 1880, pages 225-231.
[28] J.-H. Poincaré. Mécanisme et expérience. *Revue de métaphysique et de morale*, 1: 534–537, 1893.
[29] J.-H. Poincaré. A propos de la théorie de M. Larmor. Eclairage Electrique, t. III and IV (1895). Also published in *Œuvres*, XI vols., Gauthier-Villars, Paris 1916-1956, vol. IX, pages 369-426.
[30] J.-H. Poincaré. La mesure du temps. *Revue de métaphysique et de morale*, VI: 1–13, 1897.
[31] J.-H. Poincaré. *La correspondance entre Mittag-Leffler et Henri Poincaré*. Présentée et annotée par Philippe Nabonnand, Birkhäuser, Basel 1899.
[32] J.-H. Poincaré. La théorie de Lorentz et le principe de réaction. *Archives néerlandaises des sciences exactes et naturelles*, 5: 252–278, 1900.
[33] J.-H. Poincaré. *La correspondance entre Henri Poincaré et les physiciens, chimistes et ingénieurs*. Présentée et annotée par Scott Walter en collaboration avec Étienne Bolmont et André Coret, Birkhäuser, Basel, 2007.
[34] J.-H. Poincaré. Les relations entre la physique expérimentale et la physique mathématique. *Revue générale des sciences pures et appliquées* 11: 1163–1175, 1900.
[35] J.-H. Poincaré. Sur les principes de la mécanique. *Bibliothèque du Congrès international de philosophie*, tome III, pages 457-494, Paris, 1900; or in [38], cap. VI and VII, pages 142-187.
[36] J.-H. Poincaré. L'état actuel et l'avenir de la Physique mathématique. *Bulletin des sciences mathématiques*, 28: 302-324, 1904
[37] J.-H. Poincaré. La fin de la matière. *Athenaeum*, 4086 (Feb 17): 201–202, 1906. Also published in [38], pages 245–250.
[38] H. Poincaré. *La science et l'hypothèse*. Flammarion, Paris 1902 (2nd enlarged edition: 1907).
[39] J.-H. Poincaré. Sur la dynamique de l'électron. *Comptes rendus de l'Académie des sciences*, 140: 1504–1508, 1905. Also published in *Œuvres*, XI vols., Gauthier-Villars, Paris 1916-1956, vol. IX, pages 489–493; and in *La mécanique nouvelle. Conférence, mémoire et notes sur la théorie de la relativité*, edited by E. Guillaume, Gauthier-Villars, Paris 1924, pages 77–81.
[40] J.-H. Poincaré. Sur la dynamique de l'électron. *Rendiconti del Circolo matematico di Palermo*, 21: 129–176, 1906. Also published in *Œuvres*, XI vols., Gauthier-Villars, Paris 1916-1956, vol. IX, pages 494–550.

[41] K. Popper. *Logik der Forschung*, Springer, Vienna 1934. English translation by the author: *The Logic of Scientific Discovery*, Springer, Berlin 1959.
[42] K. Popper. *Conjectures and Refutations*, Routledge and Kegan Paul, London 1969.
[43] H. Reichenbach. *Philosophie der Raum-Zeit-Lehre*, de Gruyter, Berlin and Leipzig 1928. Also published in *Gesammelte Werke*, Bd. 2, Vieweg Verlag, Braunschweig 1977; English translation by M. Reichenbach, *Philosophy of Space and Time*, Dover Pubblcation, New York (1928) 1958.
[44] R. Torretti. *Philosophy of Geometry from Riemann to Poincaré*, Reidel, Dordrecht-Boston, MA 1984
[45] T. Uebel. Transformations of "Conventionalism" in the Vienna Circle. *Philosophia Scientiae*, 3 (2): 75–94, 1999.
[46] E. Zahar. Les fondements des mathématiques d'après Poincaré et Russell. *Fundamenta Scientiae*, VIII: 31–56, 1987.
[47] E. Zahar. Paradoxes in Poincaré's philosophy. *Cahiers d'histoire et de philosophie des sciences*, 23: 109–130, 1987.
[48] E. Zahar. *Poincaré's Philosophy from Conventionalism to Phenomenology*. Open Court, Chicago 2001.
[49] J.P. Provost and C. Bracco. Poincaré et l'éther relativiste. *Bulletin de l'Union des Professeurs de Spéciales*, 211: 11–36, 2005.
[50] D. Gillies. Poincaré: Conservative Methodologist but Revolutionary Scientist. *Philosophia Scientiae*, I: 60–69, D. 1996,.

"Besides quantity": the epistemological meaning of Poincaré's qualitative analysis

MARCO TOSCANO

In one of his most important works, *Stabilité structurelle et Morphogénèse* [56], René Thom introduced the idea of a "qualitative analysis" and he synthetized, in few words, its poor consideration in the scientific thought's tradition:

> L'usage du terme qualitatif a en Science – en Physique surtout – un aspect péjoratif; et un physicien m'a rappelé, non sans véhémence, le mot du Rutherford: "l'accord qualitatif d'une théorie et de l'expérience n'exprime qu'un accord grossier (*Qualitative is nothing but poor quantitative*)" [56, p. 4].

Despite this "bad reputation", the importance of a qualitative-geometrical approach and the revaluation of the concept of "form" appeared, at least, provable. Thom wrote about a: "tendance naturelle de l'esprit à donner à la forme d'une courbe une valeur intrinsèque" [56, p. 4].

My aim, in this brief essay, is to summarize the epistemological aspects concerning the development, by Poincaré (1854–1912), of a qualitative analysis. So, I will consider the early works of Poincaré on the differential equations and their link with the three body problem. On the other hand, I will show the epistemological relations between Poincaré and Leibniz and I will try to explain in which sense the idea of qualtitative analysis (and "Analysis Situs") is a philosophical link between them. At the end I will introduce the epistemological meaning that Poincaré confered to the Analysis Situs. The present essay does not pretend to be a complete treatment of these subjects; it rather constitutes a summing up of different starting points for new researches on Poincaré's philosophy.

Poincaré's studies on differential equations, between 1881 and 1886, are usally considered the basis for his following work on the three body problem [39, 40, 41, 42].[1] Indeed, in his four parts of *Sur les courbes définies par une équation différentielle* Poincaré introduced many new geometrical tools which were later used in his innovative approach to celestial mechanics. In the first part of *Sur les courbes* (1881) Poincaré divided the functional analysis into two parts: the first one, "qualitative analysis", had the aim of studing the integral curves from a geometrical point of view, stressing on the properties linked with their form. The second one, called "quantitative analysis" concerned instead the numerical calculus of the function's values. These two different approaches were considered complementary and, quoting Sturm,[2]

---

[1] By now, every reference to these texts will be taken from *Œuvres* [39].

[2] Poincaré specifically referred to Sturm's theorem (1829) which established the possibility of calculating the number of unique real roots of a polynomial function in a fixed

Poincaré said that qualitative analysis could be considered as a preliminary step towards quantitative analysis. On the other hand, Poincaré emphazised also the specific role of qualitative approach:

> D'ailleurs, cette étude qualitative aura par elle-même un intérêt du premier ordre. Diverses questions fort importantes d'Analyse et de Mécanique peuvent en effet s'y ramener. Prenons pour exemple le problème des trois corps: ne peut-on pas se demander si l'un des corps restera toujours dans une certaine région du ciel ou bien s'il pourra s'éloigner indéfiniment; si la distance des corps augmentera, ou dimimuera à l'infini, ou bien si elle restera comprise entre certaines limites? [...]. Tel est le vast champ de découvertes qui s'ouvre devant les géomètres [39, pp. 4–5].

Poincaré supported with precise arguments the claim that qualitative study was interesting in itself: it could be used to approach different problems, in particular, he quoted the three body problem. His article on the three body problem was proposed to the Oscar II's Prize in 1888. It appeared on *Acta Mathematica* in 1890 [43],[3] but from the passage quoted above, it is possible to argue that by 1881, Poincaré's development of qualitative analysis was adressed to create a new way for solving such a problem: the qualitative-topological approach. As noted by Ivar Ekeland, Poincaré understood that the orbits of planets were not phenomena describable through the universe of calculus, but in spite of this, they belonged to the field of mathematics. From this point of view it was necessary to change the classical quantitative approach in order to use a qualitative one: "Il faut donc sur cette frontière de la connaissance, un changement d'optique. Aux méthodes quantitatives, précises mais limitées, on essaie de suppléer par des méthodes qualitatives, qui portent plus loin mais donnent une image moins distincte" [14, pp. 48–49]. Poincaré seemed to be genuinely persuaded that such a change of perspective did not involve the failure of scientific thought. On the contrary, he considered it necessary to introduce a new way — epistemological more than scientific — for explaining the problems of celestial mechanics. The illusion of the perfect integrability of the planets' orbits was a proof of the reductivism of classical-modern science. In this sense the complex motions of celestial bodies were explained starting from their decomposition in simple-separated parts [15]. Poincaré was aware of the inadequacy of such an epistemological point of view.[4] So he introduced the qualitative approach as means of opening a new geometrical way for understanding celestial motions.

Poincaré's analysis of integral curves was, indeed, of topological nature and it was usally divided in "local qualitative analysis" and "global qualitative

---

interval. As underlined by Christian Gilain [22], Sturm, in his "Mémoire sur les équations différentielles linéaires du second ordre" (1836) [55], emphasized the role of qualitative analysis. Indeed, Sturm underlined the importance of studing the behaviour ("la marche") of an integral curve in order to understand many physical and dynamical phenomena.

[3] On the three body problem in Poincaré and its relationship with qualitative approach, see: Barrow-Green [3, 4, 5]; Chabert, Dalmedico [12]; Laskar [30]; Ekeland [14]; Peterson [35]; Galison [17]; Bartocci [6].

[4] Several years later, in *La Science et l'Hypothèse*, Poincaré wrote: "Si la simplicité était réelle et profonde, elle résisterait à la précision croissante de nos moyens de mesure; si donc nous croyons la nature profondément simple, nous devrions conclure d'une simplicité approchée à une simplicité rigoureuse. C'est ce qu'on faisait autrefois; c'est ce que nous n'avons plus le droit de faire. La simplicité des lois de Képler, par exemple, n'est qu'apparente." [48, p. 165].

analysis" [4, pp. 30–41], [12, 25], [26, pp. 236–252], [24, 6]. The first one, close to Poincaré's doctoral thesis and to the previous works of Briot and Bouquet [37, 9] concerned the study of the curves' behaviour in proximity to a singular point: a *noeud*, a *centre*, a *col* or a *foyer*.[5] On the contrary, the global qualitative analysis studied the behaviour of the integral curves on all the extension of their phase protrait (in Poincaré's case it was a two dimensional phase space); it was considered as new ground, entirely created by Poincaré, in which he introduced many interesting concepts like "conseguent points" and "surface without contact". The main topological result was the demonstration that every curve that did not end in a singular point could be a limit cycle or a curve that wraps itself asymptotically around a limit cycle. Therefore, it is possible to conclude that Poincaré's analysis of integral curves was essentially of topological nature.

Despite this interest in topology, an explicit work on such a discipline, which Poincaré called "Analysis Situs", was published only between 1895 and 1904,[6] several years after his works on differential equations. Nevertheless Poincaré in his *Analyse de ses travaux scientifiques faite par H. Poincaré* [51] (1901 but published in 1921) asserted, about his interest in topology:

> Quant à moi, toutes les voies diverses où je m'étais engagé successivement me conduisaient à l'Analysis Sitûs. J'avais besoin des données de cette science pour poursuivre mes études sur les courbes définies par les équations différentielles et pour les étendre aux équations différentielles d'ordre supérieur et en particuliern à celles du problème des trois corps [51, p. 101].

The different researches that Poincaré was carrying out led him to the Analysis Situs. This discipline was neccessary for him, in order to continue his works on differential equations and for broadering them to encompass the three body problem. The physical interest towards such a problem appeared immediately as one of the aims of Poincaré's development of qualitative approach. His interest in topology was, in a certain sense, dependent on his interest in the three body problem. Topology and qualitative analysis connected to it, were not only theoretical and mathematical tools. In Poincaré they were above all new epistemological structures — alternatives to the classical ones — through which one understands the complexity of celestial motions. It is possible to argue that Poincaré's early interest in topology was founded on his attempt to outline a new way for solving the three body problem and the related question of the stability of the solar system. Poincaré, as stressed by Ekeland [15], did not believe in a "complete integrable world" and even if he was very skilled in the use of classical science's methods, he was also aware of their technical and epistemological limits. He created, indeed, a new perspective and, as Weierstrass[7] wrote, he opened a "new era" in celestial mechanics.

---

[5]In his "local analysis" Poincaré introduced also the calculus of Index for an arbitrary surface $g(S - F - N) = 2g - 2$.

[6]It was a monumental work [44, 45, 46, 47, 49] considered to be the basis of subsequent development in topology. For a clear and concise description of the main results obtained by Poincaré in topology, see: Sarkaria [53]; Aleksandrov [2].

[7]Here I refer to the remark on Poincaré's article on the three body problem written by Karl Weierstrass (1815–1897) and presented to King Oscar II. An integral version of such a remark is traceable in: Barrow-Green [4, pp. 237–239].

From a philosophical point of view the role of the qualitative approach and of topology can be understood in a broader sense in which Analysis Situs is an epistemological link between Poincaré and Leibniz. In a letter that Leibniz (1646–1716) wrote to his friend Christiaan Huygens (1629–1695) in 1679 he introduced, for the first time, the idea of a qualitative analysis. He declared the neccessity of going over the algebrical analysis of "magnitudo":

> apres tous les progres que j'ay faits en ces matieres, je ne fuis pas ancore content de l'Algebre, en ce qu'elle ne donne ny les plus courtes voyes, ny les plus belles constructions de Geometrie. C'est pourquoy lors qu'il s'agit de cela, je croy qu'il nous faut encor une autre Analyse proprement geometrique ou lineare, qui nous esprime directement, situm, comme l'Algebre esprime magnitudem. Et je croy d'en voir le moyen, et qu'on pourroit representer des figures et mesme des machines et mouvemens en caracteres, comme l'Algebre represente les nomebres ou grandeurs.[8]

Leibniz recognized the necessity of finding a mathematical way to expressing the properties that did not concern quantity but rather quality. A new discipline of "situm" was considered as a branch of mathematics that could study the qualitative and "formal" properties.

Moreover, in an essay entitled "De Analysis Situs", Leibniz wrote: "Figura in universum praeter quantitatem continet qualitatem seu formam"[9] and about a new qualitative discipline he said: "Itaque Analis situs appellare placet, quod ea situm recta et immediate explicat, ita ut figurae etiam non delineatae per notas in animo depingantur...".[10] Leibniz also stressed the importance, in this new geometrical analysis, of the "similarity" as an alternative to "equality". Two figures, said Leibniz, were equal when their "magnitudo" was the same (consequently, equality is connected with quantity); on the contrary, differents figures were similar when their "form" was the same. Hence, the concept of form was understood in a deeper sense and it reppresented, in a figure, the mutal position of the parts or, in other words, their "situs". Following this idea Leibniz recognized the existence of mathematical properties not linked with magnitudo or, in general, with quantity. He specified, indeed, that quality or form came before quantity; they expressed fundamental properties of figures. Laurence Bouquiaux observed:

> Leibniz entend refuser la dictature du quantitatif, sans pour cela abandoner la physique mathématique [...] La mathématique que projette — et, dans une certaine mesure, construit — Leibniz, c'est quelque chose comme la Mathesis Unversalis dont parle Descartes, quelque chose qui déborde la mathématique cartésienne. C'est aussi, déjà, notre mathématique. Cette mathesis ne se reduit pas à l'algébre, qui traite de la quantité en général. Elle concerne tout ce qui tombe sous l'immagination, pour autant que cela soit conu distinctement. Elle ne traite pas seulement de la quantité, mais aussi de la disposition des choses. La notion d'ordre, pour être qualitative, n'en pas

---

[8] [27, p. 216]. For the English translation of the letter, see: Leibniz [31, pp. 248–258]. For a historical comment on this letter and the successive writings on the "geometry of position", see: Aiton [1].

[9] "Beside quantity, figure in general include also quality or form". For the original version see: Leibniz [32, p. 179, vol.V]. For the English translation, see: Leibniz [31, p. 254]. As Loemker said [31, p. 258] "De Analysis Situs" is not dated, but it is strictly linked with the geometrical studies of those years.

[10] "I like to call it Analysis Situs, because it explains situation directly and immediatly, so that, even if the figures are not drawn, they are portrayed to the mind trough symbols...". Leibniz [32, pp. 182–183, vol.V]. For the English translation, see: [31, p. 257].

moins, chez Leibniz, mathématique.[...] La vérité d'une figure réside dans son aspect qualitatif plus que dans son aspect quantitatif.[11]

According to this perspective, quality (or form), far from being a superficial aspect of things, became their essence. Also the criticism of Leibniz towards Descartes' dynamics can be understood in this way.[12] Leibniz underlined the importance of a global physics[13] in which every part was considered in its relation to the others; so, through the consideration of position (situs), every part distinguished itself in a universe conceived as an harmonious unity.

The Leibnizian criticism towards Cartesian mechanics is the link between Leibniz and Poincaré. In 1880 Poincaré wrote an afterword to the *Monadologie* edited by Emile Boutroux, in which he explained the difference between the conservation of Descartes' quantity of motion and the conservation of kinetic energy [38].

Poincaré briefly showed how the mechanics systems of Leibniz were distinguished from those of Descartes by their intrinsic unity and by the relationship among their parts. Every change in some part corresponded to a consequent change in the others[14] and every part was defined by its relative position. As Poincaré noted: "Il faut donc qu'il y ait une certaine harmonie dans les phénomènes mécaniques qui affectent les différentes parties d'un système" [38, p. 230]. It is possible to understand such a harmony starting from a mechanical view that is very different from the Cartesian (and Newtonian) one. The role of the "relation" in Leibniz and the importance of the properties connected to it were clear in Poincaré. We do not know if he knew of the letter to Huygens, and if it is possible to conclude a direct influence of Leibniz on him. From a historical point of view this assumption would be a mistake. The rise of Poincaré's interest towards analysis situs was in fact ascribable both to his attempt to find a new way to approaching the three body problem and also to the influence of the works of Riemann and Betti on his mathematical training (but also of the reflection on the notion of "group" caused by the knowledge of Klein's *Erlanger Programm* and Lie's works). Nevertheless, Poincaré dedicated his first epistemological written to Leibniz, showing

---

[11] [10, p. 160]. In the note 88 of the quoted page, Bouquiaux claimed, according with Mates, that the Analysis Situs of Leibniz should not be confused with the 20th century topology. I agree completely with this statement from a mathematical point of view but, on the other hand, I am quite persuaded that there is an epistemological and philosophical link between Leibnizian Analysis Situs and the birth of modern topology in terms of the importance attributed to the qualitative properties. About the writing of Mates, see: Mates [33, p. 240].

[12] On the Leibnizian critics of Descartes' dynamics see, for example: Belaval [7, pp. 494–496]; Duchensneau [13, pp. 133–146]; Bouquiaux [10, pp. 143–144].

[13] Cfr. Giannetto [18, pp. 235–247]. In these pages the author explains the historical and epistemological differences between Leibnizian and Newtonian physics.

[14] "Dans l'hypothèse cartésienne, une molecule quelconque peut prouver dans son mouvement une perturbation sans exercer aucune influences sur les molécules voisines. Avec les lois de Leibnitz, au contraire, dès que la vitesse d'un point quelconque varie, soi en grandeur, soi en direction, la quantité des progrès serait augmentée ou diminuée si il n'y avait aucune autre modification dans le système. Pour que cette quantité ne soit pas altérée, ainsi que l'éxige la loi leibnitienne, il faut que tout changement dans le mouvement d'un atome soit accompagné d'un changement contraire dans le mouvement d'un ou plusieurs autres atomes", [38, p. 230].

a well-founded knowledge of his philosophy of nature (and of its difference from the Cartesian one). Such a philosophy was a part of Poincaré's epistemological education and contributed to the construction of his thought. In this way it is possible to draw, from an epistemological point of view, a link between Poincaré and Leibniz, setting them in a common current in which the importance of "relation" and "form" was made evident.

Now, it is relevant to try to undestand the kind of epistemological meaning given by Poincaré to topology. He defined topology as: "la science que nous fait connaître les propriétés qualitatives des figures géométriques..." [51, p. 100]. He emphazised the importance of a mathematical treatment of the qualitative aspects. Later, in the posthumous work "Pourquoi l'espace a trois dimensions" [50], Poincaré claimed that the Analysis Situs was a third kind of geometry (in addition to metric geometry and projective geometry) in which:

> la quantité est complètement bannie et qui est purement qualitative [...]. Dans cette discipline, deux figures sont équivalentes toutes les fois qu'on peut passer de l'une à l'autre par une déformation continue, quelle que soit d'ailleurs la loi de cette déformation pourvu qu'elle respecte la continuité [...]. Du point de vu de la géométrie métrique, de celui même de la géométrie projective, les deux figures ne sont pas équivalentes; elles le sont au contraire du point de vue de l'Analysis Situs [50, p. 158].

Poincaré, in this passage, stressed the qualitative essence of topology; its objects of study were continuous trasformations groups.[15] From topological point of view, two figures are the same if it is possible to pass from one to the other through a continuous transformation in which the order of the parts is preserved. The notion of "continuum" was considered by Poincaré to be the basis of topology and the aim of such a discipline was to offer a mathematical treatement of this continuum. As Gregory Nowak noted [34], topology was, for Poincaré, "the mathematics of the continuum" [34, p. 373], where the "continuum" had to be understood in its intuitive wealth; it was precisely in this wealth that it was possible to find the qualitative properties that the quantitative approach tended to ignored. For Nowak, the intuitive continuum, in opposition to the "Zahlenmannigfaltigkeit" of Sophus Lie, reppresented for Poincaré the example of a mathematical object misunderstood by its quantitative definition. About the analytical definition of three dimensional continuum,[16] Poincaré wrote:

> cette définition fait bon marché de l'origine intuitive de la notion de continu, et de toutes les richesses que recèle cette notion. Elle rentre dans le type de ces définitons qui sont devenues si fréquentes dans la Mathématique, depuis qu'on tend à "arithmétiser" cette science. Ces définitons, irréprochabiles, nous l'avons dit, au point de vue

---

[15]The groups of trasformations were, for Poincaré, the essence of geometry. Following the main idea of Felix Klein's *Erlanger Programm*, Poincaré indetified every kind of geometry whith a specific group of transformations. The geometrical properties of a figure are defined as the invariants of a characteristic group. For example, in projective geometry, the geometrical properties are those which remained invariant for any kind of projection (or section). The group of topology is a very large group that includes every kind of continuous transformation. The only geometrical properties that it admits are those related to the order of the parts. About the notion of group of transformation in Poincaré, see: Giedymin [19]; Giedymin [20]; Giedymin [21]; Boi [8]; Gray [23]; Sinigaglia [54].

[16]Poincaré referred here, more generally, to the analytical defintion of continuum based on the introduction of n variables. In these terms an n-dimensional continuum was defined by the use of n independent coordinates.

mathématique, ne sauraient satisfaire le philosophe. Elles remplacent l'object à définir et la notion intuitive de cet object par une construction faite avec des matériaux plus simplex; on voit bien alors qu'on peut effectivement faire cette construction avec ces matériaux, mais on voit en même temps qu'on pourrait en faire tout aussi bien beaucoup d'autres; ce qu'elle ne lasse pas voir c'est la raison profonde pour la quelle on a assemblé ces matériaux de cette façon et ne pas d'une autre. Je ne veux pas dire que cette "arithmétisation" des mathématiques soit une mauvaise chose, je dis qu'elle n'est pas tout [50, p. 65].

Poincaré underlined also that Analysis Situs was the mathematical discipline in which the geometrical intuition was actually employed and he specified the intrinsic difference between this kind of intuition and the arithmetical or algebrical ones.[17] Only through geometrical intuition was it indeed possible to understand "la raison profonde pour la quelle on a assemblé ces matriaux de cette façon et ne pas d'une autre" [50, p. 65]. The essence of continuum had to be found in an internal connection among the parts that the analitical definitions did not express in suitable way. Instead the geometrical intuition was an intuition of "order".[18] In the topological amorphous space, the only properties were the qualitative ones and they were dependent upon the conservation of the order.[19] Poincaré considered this qualitative aspect to the basis of every kind of geometry and, moreover, he thought that it had to be considered as the intuitive (and essential) content of continuum. Metric and projective geometry dealt with properties which depended on the introduction of measure instruments in an amorphous space. Such a space included qualitative properties, intellegible through the use of the authentic geometrical intuition, that quantitative approach did not grasp. So, as Nowak underlined, it is possible to understand Poincaré's criticism towards Lie's "Zahlenmannigfaltigkeit" [34, pp. 369–370]. It reppresented, for Poincaré, a "superstructure" built upon the topological continuum. Considering the "Zahlenmannigfaltigkeit" equivalent to the intuitive continuum was both a mathematical and philosophical error. Mathematical because it did not recognize the qualitative properties of continuum. Philosophical because it did not understand the role of geometrical intuition. The intuitive richness of continuum could be grasped in a suitable way only by stressing the order of the parts and on their relationship.

Topology — or Analysis Situs — was a discipline, in Poincaré as in Leibniz before, which dealt with the properties connected to the position, (situs) considered as the reciprocal relation of the parts. "Form" reppresented such

---

[17]Poincaré's geometrical intuition was very similar to the "intution" that the Italian mathematician Federigo Enriques introduced in his *Lezioni di Geometria Proiettiva* (seconda edizione 1904) [16]. He explained the existence of a particular intuition of geometry through which the mathematician could "see" the object of his demonstration. On Poincaré and Enriques, but from a different point of view, see: Israel [28]; Israel, Menghini [29].

[18]Poincaré also criticized the Hilbert's definition of axiom of order: "M. Hilbert a cherché à fonder une géométrie qu'on apellée reationnelle parce qu'elle est affranchie de tout appel à l'intuition. Elle reponse sur un certain nombre d'axiomes ou de postulats qui sont regardés, non comme des vérités intuitives, mais comme des définitions déguisées. Ces axiomes sont répartis en cinq groupes. Pour quatre des ces groupes, j'ai eu l'occasion de dire dans quelle mesure il est légitime de les regarder comme ne renfermant que des définitions déguisées. Je voudrais insister ici sur un de ces groupes, le duexième, celui des "axiomes de l'ordre" [...] pour les axiomes de l'ordre, il me semble qu'il y a quelque chose de plus, que ce sont de véritables propositions intuitives, se rattachant à l'Analysis situs" [50, p. 93–95].

[19]These properties are the only ones conserved by the continuous transformations group.

a relation and it held particular properties which were impossible to grasp by means of a quantitative approach.[20]

Poincaré, in his study on integral curves was interested in their form because he understood that it contained specific qualitative properties. In the same way, he realized that this new point of view could be suitable also for the three body problem. He developed the tools of qualitative approach in order to solve the ticklish question of solar system's stability. He also recognized both the technical and epistemological limits of the analitical approach. On the epistemological side, Analisys Situs was, for Poincaré, the science which focused itself on the relations among the parts, namely on their order. The topological space was an amorphus continuum, it had no metrical or projective properties, of any kind. Nevertheless it held those properties which depended on axioms of order. Poincaré noted that even if the geometrical axioms could be considered as justified conventions, the axioms of order had to be considered different: "pour les axiomes de l'ordre, il me semble qu'il y a quelque chose de plus, que ce sont de vér itables propositions intuitives, se rattachant l'Analisys situs" [50, pp. 94–95]. The qualitative properties of order came before any kind of convention and they concerned the intuition of continuum. Therefore, in Poincaré, the introduction of a qualitative approach was also a rehabilitation of geometrical intuition.

In conclusion, it is possible to make some remarks:

1. Poincaré's studies on integral curves contributed to opening a new mathematical branch of analysis, the qualitative one. Here the attention was focused on the topological properties of the curves and the concept of "form" became essential. This new geometrical way of studing differential equations seemed motivated, in Poincaré, by his physical interest in the three body problem. Several times, in the articles between 1881 and 1886, and later in "Analyse de ses travaux scientifique", Poincaré underlined the applicability of his new topological results to the three body problem. Specifically, he stressed the possibilty of solving the problem of the stability of the solar system. Moreover, he recognized the inadequacy of the classical analytical approach. Poincaré was aware of the intrisic limits of quantitative analysis and contributed, with the qualitative approach, to broadening the boundaries of mathematics, surpassing the scientific and epistemological perspectives of quantitative methods.

2. From a historical point of view there are not, at the present time, explicit evidence of a direct influence of Leibniz on Poincaré; his approach to the Analysis Situs has to be explained considering the mathematical-historical context (i.e. the influence of Riemann and Betti). Nevertheless, the epistemological reasons at the basis of the Analysis Situs (the importance of qualitative properties, namely the importance of "relation"), seem to be very similar in Leibniz and in Poincaré; for example,

---

[20] "Form" was not simply considered to be an outward appearance but instead, it was thought to be the expression of the deeper essence of the phenomena. In this way, there was a consideration of the form similar to that of scolasticism. On this aspect see, for example: Boutot [11].

by the afterword of the *Monadologie*, it is possible to argue that Leibniz played a relevant role in Poincaré's philosophical interests. Poincaré certainly recognized the main role, in Leibnizian thought, of "relation" (essential also for explaining the concept of "harmony") and of the unity of every part of the universe. These aspects returned in Poincaré's later philosophical writtings. The necessity of founding a new mathematical branch dedicated to qualitative aspects came out in both Leibniz and Poincaré's work. It would be quite incorrect to try to consider Leibniz as Poincaré's "forerunner" but, on the other hand, it is clear that Leibnizian philosophy of nature was a part of Poincaré's training and had an influence on him.

3. As Jean Petitot [36] noted, the interest of the twentieth-century science in the "form" corresponded to an attempt to develop "an objective theory of the forms" connected with a new "qualitative ontology". Classical science did not recognize any kind of objectivity of "form"; it was considered only from the subjective point of view by the psychological or phenomenological approaches. Poincaré's qualitative approach seems to be in contrast with the classical tradition; the form was not excluded by mathematics; on the contrary, it became a new mathematical (or in a broader sense "scientific") object. A stimulating epistimological reading of the qualitative approach was delivered by René Thom [56, 57]. He retained that classical science reserved, at its birth, a special place to "efficient cause" while it refused the "formal cause" entirely [57, p. 112]. This aspect corresponded, in Thom's view, to a sort of anthropological essence of classical science which finds its expression in the introduction of force. Thom, also considered that the concept of form was more subtle and fertile than that of force. The ontological meaning of form corresponded to its attitude of offering a global perspective on physical phenomena. Instead, the relationship among parts was excluded by the construction of a "natura automata".[21] The importance that Petitot and Thom gave to "form" may not find real succes in contemporary science and so, such a science is far from being founded on a "qualitative ontology". Nevertheless, science, intended as a product of human culture, cannot be considered in only one way. Instead there are different currents that place themselves side by side and behind which it possible to find differents epistemological issues. The qualitative approach appears as a new perspective through which it is possible to enlarge the boundaries of quantitative methods, both in a mathematical and in an epistemological sense. Following this idea, I believe it is possible to understand the epistemological value of Poincaré's qualitative approach, considering it as a part (if not the most succesful one) of scientific knowledge.

---

[21] This expression was used by Ilya Prigogine and Isabelle Stengers in their most famous work, *La Nouvelle Alliance* [52], and it described the image of nature built up by rational mechanics.

BIBLIOGRAPHY

[1] E. Aiton. *Leibniz: A Biography*, Hilger, Bristol 1985.
[2] P.S. Aleksandrov. Poincaré and topology. *Russian Mathematical Surveys*, XXVII: 157–168, 1972. Also in F. Browder (ed.), *The Mathematical Heritage of Henry Poincaré*, in *Proceedings of Symposia in Pure Mathematics*, XXXIX, 1983, vol. II, pages 245–255.
[3] J. Barrow-Green. Oscar II's prize competition and the error in Poincaré's Memoir on the three body problem. *Archive for history of exact sciences*, 48: 107–131.
[4] J. Barrow-Green. *Poincaré and the Three Body Problem*. American Mathematical Society-London American Society, Providence 1997.
[5] J. Barrow-Green. Henri Poincaré, memoir on the three body problem. In I. Grattan-Guinnes and R. Cooke (eds.), *Landmark Writings in Western Mathematics 1640-1940*, Elsevier, Amsterdam 2005, pages 627–638.
[6] C. Bartocci. Equazioni e orbite celesti: gli albori della dinamica topologica. In J.H. Poincaré, *Geometria e Caso*, Bollati Boringhieri, Torino 1995, pages VII-L.
[7] Y. Belaval. *Leibniz critique de Descartes*. Gallimard, Paris 1960.
[8] L. Boi. "La conception qualitative des mathématiques et le statut épistémologique du concept de groupe. In J.L. Greffe, G. Heinzmann and K. Lorenz (eds.), *Heri Poincaré. Science et phlosophie*, Akademie-Blanchard, Berlin and Paris 1996, pages 315–329.
[9] J.C. Bouquet and C. Briot. Note sur les propriétés des fonctions définies par les équations différentielles. *Journal de l'École Polytechnique*, cahier XLV: 13–26, 1878.
[10] L. Bouquiaux. *L'harmonie et le chaos. Le rationalisme leibnizien et la "nouvelle science"*. Editions de l'institut supérieur de philosophie Louvain-La-Neuve, Paris 1994.
[11] A. Boutot. *L'invention des formes*, Editions Odile Jacob, Paris 1993.
[12] J.L. Chabert and A. Dahan Dalmedico. Les idées nouvelles de Poincaré. In J.L. Chabert and A. Dahan Dalmedico, *Chaos et déterminisme*, Seuil, Paris 1992, pages 274–305.
[13] F. Duchensneau.*La dynamique de Leibniz*. Vrin, Paris 1994.
[14] I. Ekeland. *Le Calcul, l'Imprevu, les figures du temps de Kepler à Thom*. Seuil, Paris 1984.
[15] I. Ekeland. *Le meilleur des mondes possibles*. Seuil, Paris 2000.
[16] F. Enriques. *Lezioni di Geometria Proiettiva*, Zanichelli, Bologna 1904 (seconda edizione ampliata).
[17] P. Galison. *Einstein's clocks, Poincaré's maps: Empires of Time*. Norton and Company, New York 2003.
[18] E.R.A. Giannetto. *Saggi di Storie del Pensiero Scientifico*. Bergamo University Press, Bergamo 2004.
[19] J. Giedymin. On the origin and significance of Poincaré's conventonalism. *Studies in History and Philosophy of science*, 8 (4): 271–301, 1977.
[20] J. Giedymin. *Science and Convention: Essays on Henry Poincar's Philosophy of Science and the Conventionalist Tradition*. Pergamon Press, Oxford 1982.
[21] J. Giedymin. Geometrical and physical conventionalism of Henri Poincar epistemological formulation. *Studies in History and Philosophy of science*, 22 (1): 1–22, 1991.
[22] C. Gilain. La théorie qualitative de Poincaré et le problème de l'integration des équations différentielles. *Cahiers d'histoire et de philosophie des sciences*, XXXIV: 215–242, 1991.
[23] J. Gray. Poincaré and Klein — Groups and Geometry. In L. Boi, D. Flament and J.M. Salanskins (eds.), *1830–1930: A Century of Geometry. Epistemology, History and Mathematics*, Springer, Berlin 1992, pags 35–44.
[24] J. Gray. Poincaré, topological dynamics and the stability of the solar system. In P. Harman and A. Shapiro, *The investigations of difficult things, essays on Newton and the history of exact sciences in honour of D.T. Whiteside*, Cambridge University Press, Cambridge 1992, pages 503–524.
[25] J. Hadamard. Le problème des trois corps. In P. Boutroux, V. Volterra, J. Hadamard and P. Langevin, *Henri Poincaré l'œuvre scientifique, l'œuvre philosophique*, Alcan, Paris 1914, pages 51–114.
[26] J. J. Hadamard. L' Œuvre mathématique de Poincaré. *Acta Mathematica*, XXXVII: 203–287, 1921. Also in in J.H. Poincaré, *Œuvres*, Gauthier-Villars, Paris 1916–56, vol. 11, pages 187–204.
[27] C. Huygens. *Œuvres complètes de Christiaan Huygens*, 22 vols., Société Hollandaise des sciences, La Haye, vol. 8, pages 214–218, 1888.
[28] G. Israel. Poincaré et Enriques: deux points de vue différents sur les relations entre géométrie, méchanique et physique. In L. Boi, D. Flament and J.M. Salanskins (eds.), *1830–1930: A Century of Geometry. Epistemology, History and Mathematics*, Springer, Berlin 1992, pages 107–126.

[29] G. Israel and M. Menghini. The essential tension at work in qualitative analysis: a case study of the opposite points of view of Poincaré and Enriques on the relationship between analysis and geometry. *Historia mathematica*, 25: 379–411, 1998.

[30] J. Laskar. La stabilité du système solaire. In Chabert J.L. and A. Dahan Dalmedico, *Chaos et déterminisme*, Seuil, Paris 1992, pages 170–211.

[31] G.W. Leibniz. Studies in a geometry of situation with a letter to Chrisitan Huygens. In G.W. Leibniz, *Philosophical papers and letters*, ed. by L.E. Loemker, Reidel Publishing Company, Dordrecht and Boston 1969, pages 248–258.

[32] G.W. Leibniz. De Analysis Situs. In G.W. Leibniz, *Mathematische Schriften*, 7 vols., Georg Olms verlag, Hildesheim, Zurich and New York 2004, vol. 5, pages 178–183.

[33] B. Mates. *The philosophy of Leibniz*. Oxford University Press, Oxford 1986.

[34] G. Nowak. The concept of Space and Continuum in Poincaré's Analysis Situs. In J.L. Greffe, G. Heinzmann and K. Lorenz (eds.), *Heri Poincaré. Science et phlosophie*, Akademie-Blanchard, Berlin and Paris 1996, pages 365–377.

[35] I. Peterson. *Newton Clock's, chaos in the solar system*. Freeman, New York 1993.

[36] J. Petitot. Forme. In *Encyclopædia Universalis*, 9: 712–728, 1990.

[37] J.-H. Poincaré. *Sur les propriétés des fonctions définies par les équations aux différences partialles*, Gauthiers-Villars, Paris 1879. Also in J.H. Poincaré, *Œuvres*, Gauthier-Villars, Paris 1916–56, vol. I, pages LX–CIXXX.

[38] J.-H. Poincaré. Note sur les principes de la mécanique dans Descartes et dans Leibniz. In G.W. Leibniz,*Monadologie*, accompagne d'eclarissements par Emile Boutroux, Delagrave, Paris, 1880, pages 225–231.

[39] J.-H. Poincaré. Mémoire sur le courbes définies par une équation différentielle (première partie), *Journal des mathématiques pures et appliquées*, VII: 375–422, 1881. Also in J.H. Poincaré, *Œuvres*, Gauthier-Villars, Paris 1916–56, vol. I, pages 3–44.

[40] J.-H. Poincaré. Mémoire sur le courbes définies par une équation différentielle (deuxième partie). *Journal des mathématiques pures et appliquées*, VIII:251-296, 1882. Also in J.H. Poincaré, *Œuvres*, Gauthier-Villars, Paris 1916–56, vol. I, pages 44–84.

[41] J.-H. Poincaré. Sur le courbes définies par une équation différentielle (troisième, partie). *Journal des mathématiques pures et appliquées*, I: 167–244, 1885. Also in J.H. Poincaré, *Œuvres*, Gauthier-Villars, Paris 1916–56, vol. I, pages 90–161.

[42] J.-H. Poincaré. Sur le courbes définies par une équation différentielle (quatrième, partie), *Journal des mathématiques pures et appliquées*, II: 151–217, 1886. Also in J.H. Poincaré, *Œuvres*, Gauthier-Villars, Paris 1916–56, vol. I, pages167–222.

[43] J.-H. Poincaré. Sur le problème des trois corps et les équation de la dynamique. *Acta Mathematica*, 13:1-270, 1890. Also in J.H. Poincaré, *Œuvres*, Gauthier-Villars, Paris 1916–56, vol. VII, pages 262–479.

[44] J.-H. Poincaré. Analysis Situs. *Journal de l'École Polytechnique*, ser.2 (1): 1–121, 1895. Also in J.H. Poincaré,*Œuvres*, Gauthier-Villars, Paris 1916–56, vol. VI, pages 193–288, 1895.

[45] J.-H. Poincaré. Complément á l'Analysis Situs. *Rendiconti del Circolo matematico di Palermo*, 13:285–343, 1899. Also in J.H. Poincaré, *Œuvres*, Gauthier-Villars, Paris 1916–56, vol VI, pages 290–337.

[46] J.-H. Poincaré. Sur certaines surfaces algébriques; troisième complément à l'Analysis Situs. *Bulletin de la Société mathématique de France*, 1902. Also in J.H. Poincaré, *Œuvres* Gauthier-Villars, Paris 1916–56, vol. VI, pages 373–392.

[47] J.-H. Poincaré. Sur les cycles des surfaces algébriques; quatrième complément à l'Analysis Situs. *Journal de mathématiques pures et appliquées*, 8: 169–214, 1902. Also in J.H. Poincaré, *Œuvres*, Gauthier-Villars, Paris 1916–56, vol. VI, pages 397–434.

[48] J.-H. Poincaré. *La Science et l'Hypothèse*, Flammarion, Paris 1902 (second edition 1907), 1968.

[49] J.-H. Poincaré. Cinquième complément à l'Analysis Situs. *Rendiconti del Circolo matematico di Palermo*, 18: 45–110, 1904. Also in Poincaré, *Œuvres*, Gauthier-Villars, Paris 1916–56, vol. VI, pages 435–498.

[50] J.-H. Poincaré. Pourquoi, l'espace a trois dimensions. In J.H. Poincaré, *Dernières Pensées*, Flammarion, Paris 1913, pages 55–97.

[51] J.-H. Poincaré. Analysis Situs. In: Analyse de ses travaux scientifiques faite par H. Poincaré, *Acta Matematica*, 38: 36–135, 1921.

[52] I. Prigogine and I. Stengers. *La Nouvelle Alliance*. Gallimard, Paris 1979; (2nd edition 1986).

[53] K. Sarkaria. A Look back at Poincaré's analysis situs. In J.L. Greffe, G. Heinzmann and K. Lorenz (eds.), *Heri Poincaré. Science et philosophie*, Akademie-Blanchard, Berlin-Paris 1996, pages 251–258.
[54] C. Sinigaglia. Introduzione. In J.H. Poincaré, *La Scienza e l'Ipotesi*, ed. italiana con originale a fronte a cura di Corrado Sinigaglia, Bompiani, Milano 2003, pages V–XXIV.
[55] C.F. Sturm. Mémorire sur les équations differentielles linéaires du second ordre. *Journal des mathématiques pures et appliquées*, I: 106–186, 1838.
[56] R. Thom. *Stabilité structurelle et Morphogénèse*. Intereditions, Paris 1977.
[57] R. Thom. *Paraboles et Catastrophes*. Flammarion, Paris 1983.

Structural explanation from special relativity to quantum mechanics

LAURA FELLINE

1 Explanation from Relativity to Quantum Theory.

It is often argued against Jeffrey Bub's analysis of the philosophical meaning of the Clifton Bub Halvorson Characterization Theorem (henceforth CBH) [6, 4, 5] that as a mere axiomatisation of the formalism of quantum mechanics, the CBH theorem cannot represent more than a heuristically convenient tool for the research on the foundations of quantum mechanics [10, 17]. As a consequence, this kind of criticisms leads to the claim that Bub's formulation of quantum theory as a principle theory, based on the three information theoretic constraints of the CBH theorem, cannot genuinely explain quantum phenomena.

On the other hand, by exploiting the parallel between Quantum Information Theory (QIT) and Einstein's formulation of Special Relativity (SR), Bub argues that as SR made Lorentz's theory's explanations superfuous, so does QIT with respect to other constructive theories. Some times implicitly, others more explicitly, Bub's defence of QIT often involves some appeal to the greater explanatory capacity of QIT with respect to Bohm's interpretation of quantum mechanics. In other instances he more modestly claims that "the lesson of modern physics is that a principle theory is the best one can hope to achieve as an explanatory account of quantum phenomena" [5, p. 15].

While often challenging the explanatory power of constructive theories *à la* Bohm, Bub is never explicit about *what kind* of explanations or understanding of quantum phenomena we can hope to gain from his principle reconstruction of quantum theory.

In this paper we will analyse two questions: the first is what kind of explanation, if any, QIT is supposed to provide of quantum phenomena. The second, related, question is to which extent Bub's parallel between the explanatory capacity of SR and that of QIT is justified.

The paper is structured as follows. In Section 2 we introduce the theory of structural explanation. In Section 3 we illustrate Bub's version of QIT, based on the CBH theorem, and we discuss the (partial) derivation it provides of entangled states and of the role of the no-bit commitment in such a derivation. In Section 4 it is shown that traditional accounts of explanation (causal or deductive-nomological) are inapplicable to QIT and in Section 5 we finally analyse how structural explanation applies to the CBH account of entanglement, showing how the basic incompleteness of such an account undermines the sufficiency of QIT as a genuine scientific explanation, as in the case of SR.

2 Structural Explanation

Firstly formulated by R.I.G Hughes [12, 13] and by Robert Clifton [7], the theory of structural explanation is meant to provide an account of the explanatory power of some formal accounts of phenomena provided by highly abstract physical theories.

The best available definition of a structural explanation is provided by Robert Clifton [7, p. 7]:

> We explain some feature B of the physical world by displaying a mathematical model of part of the world and demonstrating that there is a feature A of the model that corresponds to B, and is not explicit in the definition of the model.

The importance of structural explanation has been related from its very beginning to the problem of explanation within quantum theory. The discouraging failures experienced so far in finding an account of how the world must be made in order to behave the way quantum mechanics predicts have given to quantum phenomena the bad reputation of "unexplained phenomena". However, there are at least two reasons for which a philosopher of science should not be happy with this oversimplified picture of the situation. The first, dictated by the naturalistic request of keeping philosophy of science as close as possible to real scientific practice, is that "working physicists" would arguably consider the label of "unexplained phenomena" as unwarranted for quantum phenomena. The second reason, of definitely more normative order, is that this analysis of our most fundamental theory in physics would (in fact, already does) represent a serious embarrassment for those (we among them) who maintain that explaining and understanding the world are two central aims of science.

The theory of structural explanation is therefore meant to defend the idea that notwithstanding the lack of an uncontroversial physical interpretation, quantum theory is still capable of providing genuine explanations and understanding of phenomena.

In [8] it is argued that it is the very process of *making explicit the place of the explanandum* within the model that provides understanding in the context of structural explanation. A typical example of how this explanation works can be found in SR. Suppose that we were asked to explain why, according to the SR, there is one velocity which is invariant across all inertial frames.[1] According to Hughes [13, pp. 256–257]:

> A structural explanation of the invariance would display the models of space-time that SR uses, and the admissible coordinate systems for space-time that SR allows; it would then show that there were pairs of events, ε_1, ε_2, such that, under all admissible transformations of coordinates, their spatial separation X bore a constant ratio to their temporal separation T, and hence that the velocity X/T of anything moving from ε_1 to ε_2 would be the same in all coordinate systems. It would also show that only when this ratio had a particular value (call it c) was it invariant under these transformations.

In the rest of this paper we will explore if this kind of explanation can also be found in QIT.

---

[1] The fact that this happens to be the speed of light is, according to Hughes, irrelevant, given that he considers SR a theory about space-time.

3 Quantum Information theory

Drawing from Einstein's well known distinction between *principle* and *constructive theories* [9], Bub's interpretation takes quantum mechanics as a *principle theory* about the possibility and impossibility of information transfer. Here information is meant in the physical sense, i.e. as Shannon entropy: a measure of the uncertainty associated with a random variable, defined as the amount of classical information we gain, on average, when we learn the value of a random variable.

Therefore, according to Bub, just like SR was born in opposition to Lorentz's constructive theory of the electromagnetic field and with the aim to avoid all the problems it presented, quantum theory is similarly opposed to an interpretation of quantum theory as a constructive theory about the behaviour of non-classical waves or particles. Bub's QIT is based on the following three information-theoretic principles:

1. *No superluminal information transfer via measurement.* This principle states that merely performing a local (non-selective) operation on a system A cannot convey any information to a physically distinct system. This constraint corresponds to the no-signalling via entanglement featuring in ordinary quantum mechanics.

2. *No broadcasting.* This principle states the impossibility of perfectly broadcasting the information contained in an unknown physical state. Broadcasting is a generalization of the process of cloning.

3. *No-bit commitment.* This principle states the impossibility of an inconditionally sure bit commitment.[2]

The CBH Characterization Theorem, therefore, demonstrates that the basic kinematic features of a quantum-theoretic description of physical systems (i.e. noncommutativity and entanglement) can be derived from the three information-theoretic constraints.

The formal model utilized by QIT in order to derive such a result is the C^*-algebra. In relation to quantum mechanics, the algebra $\mathcal{B}(\mathcal{H})$ of all bounded operators on a Hilbert $\mathcal{H}$ space is a C^*-algebra, with $*$ the adjoint operation and $\|\cdot\|$ the standard operator norm. A state on a C^*-algebra $\mathcal{C}$ is defined as any positive normalized linear functional $'\rho : C \to \mathbf{C}$ on the algebra. A state is pure iff when $'\rho = \lambda'\rho_1 + (1-\lambda)'\rho_2$ with $\lambda \in (0,1)$, then $'\rho =' \rho_1 =' \rho_2$. Pure states of $\mathcal{B}(\mathcal{H})$ are admitted, that are not representable by vectors in $\mathcal{H}$ (nor by density operators in $\mathcal{H}$). A representation of a C^*-algebra $\mathcal{C}$ is any mapping $\pi : \mathcal{C} \to \mathcal{B}(\mathcal{H})$ that preserves the linear product and the linear $*$ structure of $\mathcal{C}$.

---

[2] The bit commitment is a cryptographic protocol in which one party, Alice, supplies an encoded bit to a second party, Bob, as a warrant for her commitment to the value 0 or 1. The information available in the encoding should be insuffcient for Bob to ascertain the value of the bit at the initial commitment stage, but suffcient, together with further information supplied by Alice at a later stage (the "revelation stage") when she is supposed to open the commitment by revealing the value of the bit, for Bob to be convinced that the protocol does not allow Alice to cheat by encoding the bit in a way that leaves her free to reveal either 0 or 1 at will.

A quantum system A is represented by a C*-algebra $\mathcal{A}$ and a composite system $A + B$ is represented by the C*-algebra $\mathcal{A} \vee \mathcal{B}$. Observables are represented by self-adjoint elements of the algebra. A quantum state is an expectation-valued functional over these observables. The constraint is added that two systems A and B are physically distinct when any state of $\mathcal{A}$ is compatible with any state of $\mathcal{B}$ (C*-independence), that is, for any state ρ_1 of $\mathcal{A}$ and for any state ρ_2 of $\mathcal{B}$, there is some joint state ρ of the joint algebra $\mathcal{A} \vee \mathcal{B}$ such that $\rho|_A = \rho_1$ and $\rho|_B = \rho_2$.

It is important to note that the choice of a C*-algebra is not necessary nor obvious, for weaker algebras (for instance Segal algebras) could also be apt to characterize a quantum theory [17]. Having said this, the CBH theorem demonstrates how quantum theory (which, again, they take to be a theory formulated in C*-algebraic terms for which the algebras of observables pertaining to distinct systems commute, for which the algebra of observables on an individual system is noncommutative, and which allows space-like separated systems to be in entangled states) can be derived from the assumption of the three information-theoretic constraints.

More specifically, it is demonstrated that: 1) From the first constraint (no superluminal information transfer via measurement) it follows that commutativity of distinct algebras is guaranteed (the converse result is proven in [11]: if the observables of distinct algebras commute, then the no superluminal information transfer via measurement constraint holds). Commutativity of distinct algebras is meant to represent no-signalling; 2) CBH demonstrates both that cloning is always allowed by classical (i.e. commutative) theories and that, if any two states can be (perfectly) broadcast, then the algebra is commutative. Therefore, from the second constraint (no broadcasting) follows the noncommutativity of individual algebras. Noncommutativity of individual algebras is the formal representative of the physical phenomenon of interference; 3) If $\mathcal{A}$ and $\mathcal{B}$ represent two quantum systems (i.e., if they are noncommutative and mutually commuting), there are nonlocal entangled states on the C*-algebra $(A) \vee \mathcal{B}$ they generate. This result has been reached in works by Landau, Summers and Bacciagaluppi [14, 16, 1].

However, Bub argues, we still cannot identify quantum theories with the class of noncommutative C*-algebras. It is at this point that the third information-theoretic constraint, the no unconditionally secure bit commitment, is introduced, "to guarantee entanglement maintenance over distance".

The first suggested motivation for the need of the no-bit commitment is the following: the arising of nonlocal entangled states in the account so far provided, follows directly from the choice of the C*-algebra and from its formal properties. On the other hand, "in an information-theoretic characterization of quantum theory, the fact that entangled states can be instantiated nonlocally, should be shown to follow from some information-theoretic principle." [4, p. 6]. It seems, in other words, that the role of the no-bit commitment is to provide an information theoretical ground *in the context of C*-algebra* to the origin of entanglement, which, otherwise, would be a consequence of the sole mathematical machinery used by the theory.

However, if the mathematical structure of reference is a C*-algebra, it would seem that the function of the third principle would be to merely reassess the occurrence of entangled states, which are already part of the theory. But the idea of positing a principle in order to "rule in" something which is already part of the theory is quite peculiar: "ruling states in rather than out by axiom seems a funny game. Indeed, once we start thinking that some states may need to be ruled in by axiom then where would it all end? Perhaps we would ultimately need a separate axiom to rule in every state, and that can't be right" [17, , p. 206]. On the other hand, given that the problem seems to rise from the existence of other weaker algebras where entanglement could not follow from the first two principles, the no-bit commitment could be seen as a constraint on this more general context. But in this case, it is still to be proven that the no-bit commitment would succeed, given that so far there is no proof that it would guarantee the stability of non-local entanglement in this more general context.

Elsewhere Bub suggests that the function of the no-bit commitment is slightly different. The no-bit commitment is incompatible with a set of possible theories that, although not in violation of the no information via measurement and no broadcasting principles, eliminate non local entanglement by assuming, for instance, its decay with distance. Timpson argues that this argument is also dubious, since such kind of theory "is only an option in the sense that we could arrive at such a theory by imposing further requirements to eliminate the entangled states that would otherwise occur naturally in the theory's state space" [17, p. 207].

In [10] the no-bit commitment is interpreted as a dynamical constraint, meant to rule out dynamical theories (such as GRW), which, still coherent with the first two principles, imply a decay of entanglement at the macroscopic level. Timpson also considers this option [17, Ch. 9] but, rightly in our view, quickly rejects it as in evident contrast with Bub's theory's manifested ambitions of being concerned on the "kinematic features of a quantum-theoretic description of physical systems" [4, p. 1].

In summation, with respect to the effectiveness of the no-bit principle in providing an information-theoretic ground to entanglement, we reached the conclusion that the no-bit commitment has a dubious role: either it is redundant (in the context of the C*-algebra); or it is unconvincing (in the case of Segal algebra).

4 The Problem of Explanation in QIT

Bub often contrasts QIT's explanations of quantum phenomena to those provided by constructive interpretations of quantum mechanics — suggesting some times that QIT's explanations are as satisfactory as any other constructive quantum theory, other times advancing the more modest claim that "the lesson of modern physics is that a principle theory is the best one can hope to achieve as an explanatory account of quantum phenomena" [5, p. 19].

Against Bub's claims, however, it could be argued that the point is not whether QIT's explanations are more or less acceptable than, say, explanations in Bohm's theory, but whether QIT can provide any explanation at all.

This problem obviously rises if one is committed to a "constructive" view of scientific explanation. More precisely, someone endorsing Harvey Brown's analysis of explanation within SR would most likely question the parallel proposed by Bub between SR and QIT. Following Brown, within the current "orthodox" conception of SR, spacetime is considered as an entity "of a special kind" which, as it were, "shapes" lengths and time [3, p. 11], [2, p. 14]. This, Brown's argument continues, provides a "constructive" dimension to Minkowski's formulation of special relativity and, at least in the intentions of the proponent of the orthodox view, makes the geometry of spacetime explicative with respect to the relativistic effects.

Now, this kind of analysis (which is dubious for a supporter of structural explanation) is obviously not applicable to QIT. Even if Bub does acknowledge the status of a primitive physical quantity to information, he clearly rejects a view of QIT as providing a "constructive explanation" of quantum phenomena, with the structure of information acting as a sort of 'cause' of the occurrence of quantum phenomena.

But if not constructive, what kind of explanations or understanding of quantum phenomena can we hope to gain from QIT?

Bub claims that, given the CBH theorem, QIT makes Bohm's theory explanatorily irrelevant. On the other hand, in order to say this, he must assume that what is explained by Bohm's theory is already explained by QIT. Bub's argument is that:

> If the information-theoretic constraints apply at the phenomenal level, then, according to Bohm's theory the universe must be in the equilibrium state, and in that case there can be no phenomena that are not part of the empirical content of a quantum theory (i.e., the statistics of quantum superpositions and entangled states).

From which it follows that [5, p. 12]:

> the additional non-quantum structural elements that [no collapse hidden variable] theories postulate cannot be doing any work in providing a physical explanation of quantum phenomena that is not already provided by an empirically equivalent quantum theory.

However, this argument is dubious. In order to be consistent, it must presuppose that empirical prediction is a sufficent condition for explanation. Moreover, it also seems to imply that no other factor contributes to the explanatory power of a theory. However, with no further assumption on what it is to be counted as an explanation, there seems to be no reason to consider the prediction of, say, entanglement as a sufficent condition for its explanation (let alone as the only criterion for its explanation). If this is true, therefore, Bub's argument can only be defended within the context of a Deductive-Nomological (DN) view of explanation.[3]

As a final, but crucial, remark notice that, considering our earlier argument on the role of the no-bit commitment and entanglement, a DN explanation of entanglement is not realizable in QIT. The obvious candidates for acting as

---

[3]If for Bub, as it seems, the special character of SR's explanations lies in the different, "principle", method for the inference of the explanandum, then, given the "natural" place that the DN model occupies within an "axiomatic" view of scientific theories, the DN model seems to be the natural candidate for accounting for the explanatory power of STR.

the laws of nature in QIT, in fact, are the three information-theoretic principles, but we have already seen how the CBH theorem does not provide a full information-theoretic account of entanglement, given a) the arbitrary choice of C*-algebra over other admissible algebras, b) the fact that there is no proof of the derivability of entanglement from the first two information-theoretic principles in the context of weaker algebras, and that c) the introduction of the no-bit commitment is useless as a solution of the problem.

5 Structural Explanation in QIT.

A more convincing way to go, therefore, could be offered by structural explanation: we explain entanglement with QIT by showing how it is part and parcel of the formal model displayed by QIT (i.e. noncommutative C*-algebra), and what its role is in the formalism and its relations with other explicit features of such a formalism. The fact that entanglement rises in QIT as a consequence of the mathematical properties of C*-algebra should not represent a problem here, since exploiting the mathematical resources of the theory is in the very nature of structural explanation [8].

Moreover, this hypothesis is especially suggested by Bub's parallel between QIT and SR, which explicitly applies to the explanatory level as well (see for instance [4]). We have already argued that SR's explanations of relativistic phenomena are structural explanations — from Bub's account, therefore, it should follow that the same applies to QIT.

In this section we want to put forward this hypothesis and discuss to which extent Bub's parallel between SR's and QIT's explanations actually holds.

Let's see in more detail how this kind of account should explain entanglement. We have seen that the CBH theorem starts with the choice of C*-algebra as the background mathematical structure, and how this algebra covers various different physical theories, both classical and quantum. We must therefore notice, first, how within the framework of C*-algebras, classical theories are different from quantum theories in that while the former are characterized by commutative C*-algebras, the C*-algebras representing the latter are non-commutative. This difference is crucial for a structural understanding of entanglement within the context of the CBH theorem, given how, as we have seen above, it can be shown that if $\mathcal{A}$ and $\mathcal{B}$ are two noncommutative and mutually commuting C*-algebras, there exist non local entangled states on the C*-algebra $\mathcal{A} \vee \mathcal{B}$ they generate: "[s]o it seems that entanglement — what Schrödinger [15] called 'the characteristic trait of quantum mechanics, the one that enforces its entire departure from classical lines of thought' — follows automatically in any theory with a noncommutative algebra of observables. That is, it seems that once we assume 'no superluminal information transfer via measurement', and 'no broadcasting', the class of allowable physical theories is restricted to those theories in which physical systems manifest both interference and nonlocal entanglement" [4, p. 6].

At first sight, this would already constitute a structural explanation of the entanglement: Bub's argument has shown how entanglement is part and parcel of the formal structure of any quantum theory, i.e. any theory characterized by a noncommutative C*-algebra. Moreover, the effectiveness of this structural explanation comes from the fact that it highlights how entanglement raises within quantum theories from the noncommutative character of their structure, and therefore, why it does not occur in classical (viz. commutative) theories. In other words, this structural explanation highlights the necessary relation of entanglement with other explicit elements of the formal structure of QIT.

The question, however, is not so simple. We have said that the fact of exploiting the mathematical properties of C*-algebra does not represent a problem for structural explanation. However, the problem still remains of the availability of other algebras where entanglement would possibly not follow. While the use of structures that are purely mathematical is admitted within structural explanation, this does not imply that any mathematical model can be used. More specifically, in a situation like the present, where two different models (C*-algebra and Segal algebra) seems to be acceptable and the explanandum is not an element of one of them, a structural account of the explanandum in terms of only one of the models is obviously to be considered at least partial. We are not asking here for an information-theoretic derivation of entanglement, just for a straight one.

6 Conclusions

Summing up, here's what our discussion came up with. Following Bub, QIT is not meant to provide a "constructive" or causal explanation of quantum phenomena. Against arguments à la Brown, we argued that this alone does not imply that QIT lacks explanatory power. On the other hand, not even a DN explanation of entanglement seems realizable within Bub's theory: a satisfactory information-theoretic DN explanation of entanglement needs either an argument (relying on information-theoretic bases) which compels toward the adoption of a C*-algebra, or an argument which can assure the rising of entangled states also in weaker algebras than the C*-algebra. Following Timpson's analysis and against Bub's suggestion, we have argued that it is still uncertain if the no-bit commitment could effectively work as the needed information theoretic basis for these two options.

We therefore advanced the idea that QIT is aimed to provide structural explanations: a quantum theory is characterized by a non-commutative C*-algebra, and we understand entanglement in the context of the CBH theorem as a basic feature of any non-commutative C*-algebra. But we have seen how also as a structural account, QIT can provide at best a partial explanation.

We are now in the condition to reconsider Bub's claim that as SR makes Lorentz's theory explanatorily irrelevant, so does QIT with respect to the constructive interpretations of quantum mechanics. Based on what we have argued so far there is a big difference between the SR's and Bub's QIT's structural explanations. In the case of SR, a constructive dynamical explanation of relativistic effects is not needed in order to fully understand relativistic effects, for they are completely understood once they are described as four-dimensional objects, as in Minkowski's formalism. The same, we have argued, cannot be said about QIT, which is far from providing a complete and satisfactory structural account of entanglement.

As long as such an account is not completed, QIT will not be able (borrowing the term from Wesley Salmon) to "screen-off" constructive interpretations of quantum mechanics, as SR does with Lorentz's constructive theory.

Acknowledgements

I would like to thank Mauro Dorato and Cosimo Felline for their helpful comments on previous versions of this article.

BIBLIOGRAPHY

[1] G. Bacciagaluppi. Quantum measurement, irreversibility and the physics of information. In P. Busch, P. Lahti and P. Mittelstaedt (eds.), *Symposium on the foundations of modern physics*, World Scientific, Singapore 1993, pages 29–37.
[2] H. Brown and O. Pooley. Minkowski space-time: a glorious non-entity. In D. Dieks (ed.), *The Ontology of Spacetime*, Elsevier, Amsterdam 2006, pages 67–89.
[3] H. Brown and C. Timpson. Why Special Relativity Should Not Be a Template for a Fundamental Reformu- lation of Quantum Mechanics. http://xxx.lanl.gov/abs/quant-ph/0601182, 2006.
[4] J. Bub. Why the quantum. *Studies in the History and Philosophy of Modern Physics*, 35B: 241–266, arXiv:quant-ph/0402149v1, 2004.
[5] J. Bub. Quantum theory is about quantum information. *Foundations of Physics*, 35(4): 541–560, arXiv:quant-ph/0408020v2, 2005.
[6] R. Clifton, J. Bub and H. Halvorson. Characterizing quantum theory in terms of information-theoretic constraints. *Foundations of Physics*, 33(11): 1561–1591, 2003.
[7] R. Clifton. Structural Explanation in Quantum Theory. http://philsci-archive.pitt.edu/archive/00000091/00/explanation-in-QT.pdf, 2001.
[8] M. Dorato and L. Felline. Structural explanations, or how mathematics contributes to our understanding of the physical world. Forthcoming.
[9] A. Einstein. What is the theory of relativity. First published in *The Times*, London, November 28, 13, 1919. Also in A. Einstein, *Ideas and Opinions*, Bonanza Books, New York 1954, pages 227–232.
[10] A. Hagar, M. Hemmo. Explaining the Unobserved: Why Quantum Mechanics Ain't Only About Information. *Foundations of Physics*, 36(9): 1295–1324, 2006.
[11] H. Halvorson. A note on information-theoretic characterizations of physical theories. Quant-ph/0310101, 2003.
[12] R.I.G. Hughes. Bell's theorem, ideology, and structural explanation. In J. Cushing and J. McMullin (eds.), *Philosophical Consequences of Quantum Theory*, Notre Dame, 1989.
[13] R.I.G. Hughes. *The Structure and Interpretation of Quantum Mechanics*. Harvard University Press, 1989.
[14] L.J. Landau. On the violation of Bell's inequality in quantum theory. *Physics Letters A*, 120: 54–56, 1987.
[15] E. Schrödinger. Probability relations between separated systems. *Proc. Camb. Phil. Soc.*, 32: 446–452, 1936.
[16] S. Summers. On the independence of local algebras in quantum field theory. *Reviews in Mathematical Physics*, 2: 201–247, 1990.
[17] C.G. Timpson. Quantum information theory and the Foundations of Quantum Mechanic. http://philsci-archive.pitt.edu/archive/00000091/00/explanation-in-QT.pdf, 2004.

When the structure is not a limit.
On continuity through theory-change

MIRIAM COMETTO

1 Introduction

Within the landscape of scientific realism, a structuralist tendency has been developed in the last 15 years, characterized by the attention for the "structural features" of our knowledge, mainly along two lines:

- On the one hand we have an epistemic tendency, focusing on structures as "all that we can know" [26, 27, 28].

- On the other hand, an ontic tendency, focusing on structure as "all that there is" [14, 7, 5, 9].

Our concern is the epistemic line developed as *Epistemic Structural Realism*. As a position about scientific knowledge and theory change, it claims to be genuinely *realist* and focuses on the notion of *structure* as the one which both allows to save some continuity through theory change and poses some constraints on what we actually know.
In the following we will suggest:

1. that epistemic structural realism (ESR), in its standard formulation, fails its target (namely, being a genuine realist position), turning out as an eviscerated form of realism;

2. how the notion of structure could be employed as a weapon for the realist rather than as a limit (i.e. how the positive insight of ESR concerning theory change could be developed in a genuine realist view).

To accomplish these tasks we will briefly present ESR and some controversial issues; in the second section we will show some alternative understandings of the problem; finally, in the third section, we will outline some conclusive remarks.

2 ESR and its target

Epistemic structural realism (ESR), as mentioned, was explicitly formulated by John Worrall [26, 27] as a weak form of realism, which basically tries to solve two challenging arguments coming respectively from scientific revolutions and from success of science. The first argument is the "pessimistic meta-induction" (PMI), the famous antirealist argument formulated by Larry Laudan in 1981 in his *A confutation of convergent realism*: the history of science is full of theories that have been abandoned, because of being *false*, despite their predictive success; hence there's nothing that leads us to trust that our present theories will not be abandoned in the future for the same reason; furthermore nothing leads us to believe that they are true (or approximately true), nor that their theoretical terms refer to unobservable entities existing in the world. Typical materials for such induction are theories involving entities like phlogiston, caloric and ether. The second argument is the "No-Miracle Argument", famously waved by Hilary Putnam, but also generally supported, at least as an intuition, by the whole realist group. It states that the incredible predictive success of mature scientific theories would be a miracle, if we did not consider their theoretical content true or approximately true.

In such respect, NMA is linked to what might be called the "Success-to-Truth rule" [12], namely the idea that:

1. the success of scientific theories requires an explanation;

2. the best explanation for that success is truth (with all its ontological consequences).

On the contrary PMI is precisely built against such inference and against the metaphysical thesis concerning the ontological status of unobservable entities involved into the scientific theories.

Epistemic Structural Realism, giving some credits to both those arguments, was developed to have "the best of both worlds", coming out as a weaker position, according to which: on the one hand, there is an undeniable radical change at the ontological level (paying heed to PMI); on the other hand there is an important element of continuity through theory change, which motivates an optimistic attitude with respect to the history of science and its progress(paying heed to NMA). Indeed mathematical equations which carry the structure of reality are preserved through theory-change and such a *structural understanding* is all we must attain to as regard to our knowledge of the world: the *structure* is all we can know about it, we know nothing about the *nature* (the ontological features) of our world.

Summarizing the key elements of the position:

1. A notion of "structure" is defined, as basically a net of relations between set of elements: we mathematically know the net of relations existing in the physical world, without knowing the elements.

2. The whole cognitive content of a theory is accounted, according to ESR, via his Ramsey sentence. The Ramsey sentence of a theory T, as is well known, is the sentence T^* formed from T by replacing all the theoretical predicates with variable predicates and quantifying them existentially.[1]

3. To support the proposed schema, epistemic structural realists put forward a case study from history of optics, namely the Fresnel-Maxwell's

---

[1] The Ramsey sentence of a theory $T(T_1 \ldots T_n; O_1 \ldots O_n)$, with $T_1 \ldots T_n$ as the theoretical predicates and $O_1 \ldots O_n$ as the observational predicates, is T^* such that $\exists \tau_1 \ldots \exists \tau_n (\tau_1 \ldots \tau_n; O_1 \ldots O_n)$.

transition, where some empirical success of the previous theory (about the derivations of the amplitudes of reflected and refracted light in various circumstances) and new successful predictions[2] belong to a theory whose theoretical claims have been radically abandoned. Indeed Fresnel's equations were derived within the frame of luminiferous-ether theories, where light is a vibration of molecules of ether.[3] We know that going to Maxwell's theory and beyond, till now, light was discovered as an electromagnetic wave, a perturbation in the disembodied electromagnetic field (and the role of ether changed completely).

4. The crucial point of such an understanding relies on the fundamental dichotomy between the knowable *structure* and the unknowable *nature* of the world. In this sense the notion of structure represents the insurmountable limit of our knowledge.

Item 4) clarifies that this is an epistemic point of view: it does not deny the existence of entities, just denying their knowability.

> [...] if Fresnel was as wrong as he could have been about what oscillates, he was right, not just about optical phenomena, but right also that those phenomena depend on the oscillations of something or other at right angles to the light. His theory was more than empirically adequate, but less than true; instead it was *structurally correct*. There is an important "carry-over" from Fresnel to Maxwell, one at the "higher" level than the merely empirical, but it is a carry over of *structure* rather than content forms obeying the same mathematics [27, p. 340].

Such a view on the cognitive status of scientific theories expressively takes suggestions from philosophers as H.Poincarè, explicitly quoted by Worrall, P.Duhem, R.Carnap and B.Russell and indeed it often faces some common issues, as we will see in the next section.

---

[2] One of the successfull predictions of Fresnel's theory was the "white spot" at the centre of the shadow of an opaque disc held in light diverging from a single slit.

[3] Fresnel's equations signed a fundamental chapter of the history of optics (and of optical ether theories) in which light is conceived as a transverse oscillatory mechanism. According to Louis de Broglie, Fresnel's work between the 1815 and the 1820 seemed to have definitely established the whole light phenomena in the form of a wave theory. To understand the conceptual framework in which he was working we could briefly remember that just before him, Benjamin Thomson, Count Rumford, Humphry Davy and Thomas Young started the abandonment of the "imponderable fluids" as explanation for light's phenomena and heat's phenomena. Fresnel became familiar with Young's work as a result of meeting Arago in 1815. Fresnel's wave theory of light made a major contribution and lay in opposition to the conceptual scheme of imponderable fluids and to the Laplacian corpuscular theory of light and the caloric theory of heat [10, p. 21]. In his interesting study on the conceptual development of Nineteenth-century physics, P.M.Harman states that the concept of light as a form of motion of a medium "was basic to his[Fresnel's] optical theory" and "By 1821 he had reformulated the science of optics in terms of the dynamics of a wave propagating medium, the luminiferous ether... in his theory the vibrations of the ether explained the phenomena of optics" [10, p. 21]. Fresnel envisaged the possibility of "a unified physic based on the mechanical properties of the ether, conceived as a form of ordinary matter" [10, p. 21]. Nevertheless, Harman also notes that the elaboration of a model for such ether was not the primary intention for Fresnel and, moreover, was undertaken only in support of his undulatory theory of light [10, p. 24]. All the vicissitudes of Fresnel's theory are put into a larger context of development of what is called a mechanical explanation.

3   Controversial Issues

Many criticisms have been developed, with respect to ESR, focusing mainly on:

- the vagueness of the very notion of structure.

- the implications of the involved russellian notion of "similarity of structure" which leads to the famous Newman's problem. Basically Newman [17], and also McLendon ([15]), noted that Russell's notion of structure seems to bound structural knowledge to a matter of cardinality. In his papers on ESR, Worrall speaks of mathematical similarities (structural or syntactical continuity) between Fresnel's and Maxwell's theories (and between the theories and the world)and stresses a mathematical continuity between the two theories. In some sense ESR seems to feel the russellian lesson[4] on similarity of structure, thus it is forced to face Newman's problem, as well as Russell's causal theory of perception.[5]

- the implications of "ramseyfication", which apparently bound our structural knowledge to matter of cardinality and empirical knowledge, therefore collapsing on empiricism.[6]

Our concern in this paper is slightly different. We will basically deal with the specific point of the dichotomy *knowable structure/unknowable nature* and the issue whether such distinction is useful vis-à-vis the realist agenda. Since we could characterize what they probably means by nature only *via* their criticism concerning the ontological reference, therefore the mentioned dichotomy induces to consider the preserved equations as uninterpreted and the structure as given by the mere formal equipment. In the final account, whether they like it or not, uninterpreted mathematical equations leads the way.[7] With this sharp distinction, ESR risks to lose its realistic aim, cutting away the link between our theories and the world they were(are) supposed to describe or, moreover, explain.

Indeed, in order to define a genuine realist position we require the following [8]:

1. it must explain the theoretical success — namely motivate the formalism, answer the "why?" question;

---

[4] It ought to be noted that Russell's aim in the *Analysis of Matter* does not directly concern the theory change. He uses the notion of *similarity of structure* in order to account for the relationship between perceptions and stimuli. Among the motivations of his causal theory of perception he poses the need to conciliate the growing abstractness of science with the domain of perceptions. On the other hand, Russell formulates a specific thesis that physical knowledge is structural in character since it does not refer to the 'qualities' (intrinsic properties) rather to the structure of the world.

[5] For further debates see also [4, 13, 20, 25].

[6] A clear formulation of the issues which ramseyfication supposedly brings to the structuralist position is expressed in cite9. On the other hand he also makes some assumptions which are not to be shared by the structural realist. Such assumptions are discussed in [13], which provide some viable way, for the epistemic positions,to develop their claim.

[7] See also [19].

2. it must show the required level of continuity, or correspondence in theory-change, as Worrall himself recognizes:

> [...] the realist needs to show that, from the point of view of the later theory, the fundamental claims of the earlier theory [...] were — though false — nonetheless in some clear sense "approximately correct". He needs to show that, from the point of view of the later theory, we can still explain the success enjoyed by the earlier one [26].

The difficulties with ESR belong to the fact that it seems to be able to satisfy 2, since it shows a level of continuity, though debatable. Not the same for 1.

How is ESR's schema supposed to give an explanation? If we want to mantain the dichotomy, our mathematical equations will explain empirical success just because of NMA: the very compresence of formal continuity by the side of empirical continuity would justify the explicative link. The issue is that if the formalism itself is all we can attain, it is not clear to what extend we would be able to motivate it. Moreover as realists we have reasons to ask (expecting answers, ndr): what does our mathematics say about the world? What is its physical understanding? The questions ask for an understanding of the way we *do* think of nature (reality) and of the reason why we think of nature (reality) in *that* way.

Whatever the scope of Fresnel, whether to predict optical phenomena as Poincarè stated, quite criticized by Worrall, or to understand the nature of light, he had to write the *right* equations in order to successfully fulfill his scope. In order to understand those right equations, the formal description must be meaningful to us. In this situation the issue concerning the representative ability of theories becomes even more pressing [22, 23]: it is debatable whether the very obtained continuity would be enough to satisfy the realist requirements of explanation and to justify a belief that the relations placed by the structure succeed in mirroring the world's relations. The existence of such a correspondence seems to be something about which ESR wants to be realist. The avoidance of the instrumentalist consequences of such a view (which in the end make it collapse into empiricism or into some kind of Platonism)requires that those equations would be considered as interpreted. Yet, as soon as the realist reintroduces interpretation, the dichotomy structure/ nature vanishes, unless he gives some additional definitions of both the concepts.

The property-move

The debate on structuralist positions got rich in the last years, due to many criticisms developed both by realist and antirealist side, so that now we may find some intermediate approaches which, though sympathetic with ESR concerning the epistemic limits our knowledge and the focus on some structural continuity, dispense from the original strong dichotomy. In this section we want to turn to such approaches, since our idea is that the "solution" for the original radicalism concerning structure/nature issue relies on an alternative way of considering the continuity through theory-change.

Let's go back for a while to the previous quotation in section 2:

> If Fresnel was as wrong as he could have been about *what* oscillates, he was right, not just about optical phenomena, but right also that those phenomena *depend on the oscillations of something or other at right angles to the light.* ([27], my emphasis.)

Worrall [27] wants the dependence of light phenomena on the oscillation at right angles to the direction of propagation to be a structural truth, hence a kind of formal, mathematical truth. But it is not! "It is a truth about some *properties* instantiated where there is phenomenon of light" [21, p. 517]. It is something pertaining to the nature (in terms of ontological features) of light, rather than the structure as an absolutely abstruse net between unknown relations.

This point anticipates our idea: the alternative way of considering the continuity, and maybe the very notion of structure, has not need of opposing it to nature. In such line we mainly refer to some papers of Chakravartty [1], Saatsi [21] and Psillos [18, 19]. They focus on the principal properties involved in the shift between Fresnel and Maxwell, looking for the crucial elements that "play the game", having a fundamental role in leading the preserved equations (and hence the successful predictions). I call this way of considering the problem the "property-move", since indeed it put the analysis, at least at the first step, at the level of properties, rather than of the entities.

This analysis is independent from the possible understanding of the entity-reference-problem (see Psillos, Chakravartty, Saatsi: they develop different forms of realism). The point is that it explains the continuity, via some fundamental properties: with respect to such properties, the dichotomy nature/structure has not any reason to hold. Moreover, there is a fundamental difference between Psillos' view and Chakravartty/Saatsi. Psillos in fact is looking for the possibility of restoring a kind of referential continuity between ether and electromagnetic field. On the contrary, Chakravartty and Saatsi, being sympathetic to the structuralists' need of answering PMI, accept the structural continuity of ESR and match it with some properties of light (we will follow this second line). The interesting feature is that they all closely analyse the properties involved in the shift.

The starting point is: let's face the equations asking for what those mathematical relations require, what is their minimal interpretation (not possible, but essential). We will discover in such a way the properties which are necessarily involved in the causal regularities and which lead our inferences concerning the existing of entities (Chakravartty). In other words we are trying to build a kind of *functional-description of involved properties* (Psillos speaks of core-causal-description. As mentioned, he has a slightly different target).

Fresnel's equations express relations between amplitudes or intensities, and angles and direction of propagation. In order to satisfy (and induce) Fresnel's equations, light does not "need" to be a vibration in an elastic, solid, medium called ether, that is it does not need to fully cover the ontological frame/image. Light obeys such equations even within Maxwell's frame, as an electromagnetic wave. According to the mentioned relations, we minimally require that:[8]

1. Oscillations proceed at right angles to the direction of light.[9]

2. It is made of two components, oscillating at right angles one to the

---

[8] We follow an analysis developed with a great accuracy by Saatsi [21].
[9] We commonly say that light *is* a transverse oscillation.

other and such that they obey Huygens principle of superposition. Each component has amplitude and intensity, such that $A^2 \propto I$.

3. They are also linked to other observable parameters (light intensity and refractive index, depending on the media according to Snell's law) for which we have some *continuity conditions*.

4. Finally they are matched with geometrical considerations.

Hence there are few quantifiable attributes of light which are theoretical. The property of amplitude is a transverse vectorial property which satisfies the mentioned principle of superposition. This constraint is not a formal logical-mathematical feature concerning the description of the system, rather it is an higher-order theoretical property of what we call *light*. In such a sense it is a fundamental constraint:*whatever* light might be, it must satisfy that constraint. For the definition of *higher* order properties we still follows Saatsi in saying that they are properties*instantiated by virtue of having some other lower-order property (or properties) meeting certain specifications, and the higher-order property does not uniquely fix the lower-order one(s)* [21, p. 533]. In such a respect they are by definition multiple realisable properties as the lower order properties which let instantiate them are not uniquely fixed. Due to such multiple realisability the higher order properties appear as the best candidates in order to develop a notion of continuity through theory change. On the other hand they are not self-standing properties, as they need to be plugged into some lower-order one which suggest some possible representation. Thus the property of having a transverse vectorial property which satisfies the superposition principle is supported in Fresnel's frame from a set of properties which make light a vibration in an elastic, solid, medium called ether. In Maxwell's frame it is an electromagnetic wave. But the property which explain the reflection and refraction laws is the same in both the cases.

4 Conclusions

Just in light of such a kind of analysis we may endorse the interpretative move, which Worrall refers to,[10] and affirm that if we interpret the amplitudes A as "amplitudes of the 'vibration' of the relevant electric vectors, then Fresnel's

---

[10]In his paper of 1994 Worrall draws a kind of strategy to clarify his idea of the *structural correctness* and he says that Fresnel's theory and Maxwell's theory individuated the structure of light and the former appears as a "genuine sub-theory" (p. 340) of the latter, since indeed you can replace each occurrence of a talk of "molecule of ether being forced away" by a talk of a forced change in the electromagnetic field strength. This replacing strategy sounds highly odd as it come after the assumptio of the criticized dichotomy. As a standard realist, you might say that if you replace one with the other they are sharing something after all, for example their role, and this replaceability could point to a kind of referential continuity (as Psillos does). On the contrary, in ESR point of view it is not clear how you should be allowed to do the substitution. If it is not really important "who" and "why" is doing the work, those "entities" are placeholders with no role: the ontological consequences of such idea make it very difficult to justify the possibility of the substitution. This issue finds no solution (at least temptative) at all if we are only interested in the naked equations just for their form! Everything collapse on the mathematical structure without any link with the physical structure.

equations are fully entailed by Maxwell's theory" [26, p. 159]. Only due to such analysis, pointed on some relevant theoretical properties of light, we may understand the continuity, even structural, in a realist view, which ultimately imply focusing on properties, as a more steady footstep, rather than entities.

Through its structural continuity, ESR aims to satisfy the intuitive realist claim that older successful theories must find an explanation in light of the newer ones. The property-move is basically a way to understand both minimal requirements for the interpretation of the formal apparatus of past theories, and causal roles of the elements involved in theory-changes. Finally it may ground some notion of "approximation" for past theories.

Furthermore, according to Worrall what is preserved would give a description of observable effects through uninterpreted equations. But the "property-move" analysis of the same case study shows that the derivation of those supposed uninterpreted equations relies on theoretical premises and boundary conditions, linked to some properties (also theoretical) of *light*: those equations are "already" interpreted (even minimally)and this interpretation is at least partially carried over. What we know about the structure provides us with fundamental elements which both:

- enable continuity;

- let us understand the "truth- content" of a past theory (with different degrees[11] according to the level of enquiry and to the available knowledge.

Our previous treatment suggests that a declension of the notion of structure, worthwhile for the realist, could not — and should not — leave out of consideration such ontological significance of its elements. Whether or not this would bring to ontic structural positions is a matter for further debates. What matters is that knowing the structure of the world does not represent any more a limit for our knowledge, rather a great weapon (especially in understanding the continuity through theory-change).

Of course the needed idea of structure is different from Worrall's one: it is a net obtainable only with the contribution of fundamental properties overlapping in the core causal descriptions. With such a frame, we can agree with Poincaré that: "if the equations remain true, it is because the relations preserve their reality" without any need for an opposition between Structure and Nature. The dichotomy is not only untenable for the realist, but it obstructs the very comprehension of the continuity through the shift.

The frame is general enough not to commit us yet to sustain a metaphysical identity between ether and electromagnetic field. It is only an environment to account for continuity: we can attain to a "functional" realism pointing on the relevant properties and relations which play fundamental roles in driving derivations and leading to successful predictions. These are our bricks to build and understand the structure. The very formal equipment provides us with what Chakravartty and Saatsi call a *minimal interpretation*. It turns out to

---

[11] For more details on issues concerning the idea of partial/total truth of scienctific theories see also [3].

be built on multiple realizable properties individuating causal roles, on the basis of which we can reconstruct our image of the physical world, which is our constant demand.

Open Issues

Finally we briefly sketch some open issues:

1. How much are Saatsi's higher order properties committed to the lower-order ones?

2. Which position should we adopt with respect to entities? I.e. where does the property-move lead? Different accounts are possible. This is not the place to endorse a discussion about them, we just note that the functional approach to properties is the ground both for Psillos' object-oriented form of realism (in terms of core-causal-descriptions which enable referential continuity) and for multiple realizability of higher level properties chosen by Saatsi on the footstep of Chakravartty minimal interpretation move. The latter two accounts are quite sympathetic to Worrall agnosticism, as mentioned, with respect to the possibility of restoring some kind of referential continuity of entities.

3. The "properties-move" seems to answer PMI, through enabling the continuity. But is it a still satisfactory account when it comes to the analysis of current theories, within which the possibility of separating higher-order/lower order properties, acting and/or abandoned, seems more difficult? Perhaps there is a problem with the kind of explanation that you need for the two dimension of analysis — past theories' success and current theories' success. In the first case we can pick carefully out within the main theoretical framework, looking for the relevant elements and avoiding the other, thanks to the advantage of the successor perspective. This does not seem possible in the second case: coming to the current theory we can't use the same explanatory strategy. It just seems that we need two different strategies to explain the success of science. Maybe it's just that the same principle should be articulated in two different ways, one in which we can clearly separate elements and one in which this move is not so clearly possible.

4. The relevant role that properties and relations play in all the pictures, strengthens somehow the structuralist motivations, as we may see in French's objection to Psillos' strategy: "In the cases he considers [phlogiston, caloric, ether] we have no object-oriented metaphysics, it simply does not figure. The focus is rather on properties and relations" [9]. Structuralism hence focuses on structures "as both that which is carried over through theory change and that in terms of which physical objects can be reconceptualized" [9].

The road is open to further investigations, we just have to keep walking.

BIBLIOGRAPHY

[1] A. Chakravartty. Semirealism. *Studies in History and Philosophy of Science* 29: 391–408, 1998.

[2] A. Chakravartty. Structuralism as a form of scientific realism. *International Studies in the Philosophy of Science*, 18: 151–171, 2004.

[3] N.C.A. da Costa and S. French. *Science and Partial Truth: A Unitary Approach to Models and Scientific Reasoning*. Oxford University Press, Oxford 2003.

[4] W. Demopoulos and M. Friedman. Bertrand Russell's "The analysis of matter": its historical context and contemporary interest. *Philosophy of Science* 52(4): 621–639, 1985.

[5] S. French. Symmetry, structure and the constitution of Objects. *Conference on Symmetries in Physics: New Reflections*, University of Oxford, January 2001. Available at the *PhilSci Archive* at http://philsci-archive.pitt.edu/archive/00000327/.

[6] S. French and J. Ladyman. Remodelling structural realism: quantum physics and the metaphysics of structure. *Synthese* 136: 31–56, 2003.

[7] S. French and J. Ladyman. The dissolution of objects: between Platonism and phenomenalism. *Synthese* 136: 73–77, 2003.

[8] S. French and J.T. Saatsi. Realism about structure: the semantic view and non-linguistic representation. *Philosophy of Science*, 73: 548–559, 2006.

[9] S. French. Structure as a weapon for the realist. *Proceedings of the Aristotelian Society*, pages 1–19, 2006.

[10] P.M. Harman. *Energy, Force, and Matter — The conceptual development of nineteenth-century physics*. Cambridge History of Science Series, Cambridge University Press, Cambridge 1982.

[11] J. Ketland. Empirical adequacy and Ramseyfication. *British Journal for the Philosophy of Science* 55: 287–300, 2004.

[12] P. Kitcher. On the explanatory role of correspondence truth. *PPR*, 66: 346–364, 2002.

[13] J. Meelia and J. Saatsi. Ramseyfication and theoretical content. *British Journal for the Philosophy of Science*, 57: 561–585, 2006.

[14] J. Ladyman. What is structural realism? *Studies in History and Philosophy of Science* 29: 409–424, 1998.

[15] H.J. McLendon. Uses of similarity of structure in contemporary philosophy. *Mind*, 64: 79–95, 1955.

[16] M. Morganti. On the preferability of epistemic structural realism. *Synthese* 142: 81–107, 2004.

[17] M.H. Newman. Mr. Russell's "causal theory of perception". *Mind*, 37: 137–148, 1928.

[18] S. Psillos. Is structural realism the best of both worlds? *Dialectica*, 49(1): 15–46, 1995.

[19] S. Psillos. *Scientific Realism. How Science tracks truth*. Routledge, London 1999.

[20] S. Psillos. Is structural realism possible? *Philosophy of Science*, 68 (Prooceedings), S13–S24, 2001.

[21] J. Saatsi. Reconsidering the Fresnel-Maxwell theory shift: how the realist can have her cake and EAT it too. *Studies in History and Philosophy of Science* 36: 509–538, 2005.

[22] B. van Fraassen. Structure: its shadow and substance. *The British Journal for the Philosophy of Science* 57: 275–307, 2006.

[23] B. van Fraassen. Structuralism and science: some common problems. *Proceedings of the Aristotelian Society*, 2007.

[24] I. Vostis. Is structure not enough. *Philosophy of Science*, 70: 879–890, 2003.

[25] I. Vostis. The upward path to structural realism. *Philosophy of Science* 72: 1361–1372, 2005.

[26] J. Worrall. Structural realism: the best of both worlds? In D. Papineau (ed.), *The Philosophy of Science*, Oxford University Press, Oxford 1996, pages 139–165; originally published in *Dialectica*, 43: 99–124, 1989.

[27] J. Worrall. How to remain (reasonably) optimistic: scientific realism and luminiferous ether. *PSA 1994*, pages 334–342, 1994.

[28] J. Worrall J. and E. Zahar. Ramseyfication and structural realism. In E. Zahar, *Poincar'e's Philosophy: From Conventionalism to Phenomenology*, Open Court, Chicago and La Salle, 2001, pages 236–251.

Approaches to wave/particle duality: historical analysis and critical remarks
GIANLUCA INTROZZI

Introduction
Wave/particle duality was introduced by Einstein [11, 12] in 1909 to justify Planck formula for the energy distribution of the black-body radiation. Since Einstein's seminal papers, different ideas have been suggested about the meaning and interpretation of wave/particle duality.

At least eight different alternatives have been proposed:

1. Just waves, no particles (Schrödinger)

2. Just particles, no waves (Born)

3. Neither waves, nor particles (Heisenberg and Jordan)

4. Waves and particles (Bohm)

5. Waves or particles (Bohr and Pauli)

6. Neither waves, nor particles (Greenberger/Yasin and Englert)

7. Quantons (Lévy-Leblond)

8. Bosons and fermions (Lévy-Leblond)

These alternatives will be discussed in a historical and critical contest, in order to outline the evolution of the duality concept, and the emergence of new descriptions of quantum phenomena.

1 Schrödinger: Just waves, no particles
In two papers published in March [26] and April [27] 1926 Erwin Schrödinger introduced a new equation for non-relativistic electrons, described by means of a wavefunction ψ. The following May [28] he showed the equivalence between the matrix formulation of quantum mechanics (proposed by Heisenberg, Born and Jordan in 1925) and the wave mechanics he had formulated shortly before. Nevertheless, he considered his model better suited for the description of micro-physics phenomena, like discontinuous atomic transitions.

He was aware of the difficulties — like the dispersion of the wave packet — linked to a realistic interpretation of the wavefunction ψ. Indeed, he was hoping to be able to demonstrate the wavelike nature of quantum events.

This idea was presented again in his late works [29], during the 1950s. He suggested a physical description of reality in terms of a field theory without particles. Usual field theories, on the contrary, are characterized by the contemporary presence of fields *and* particles (like the electrons in classical electromagnetic theory, that are considered to be the sources of the field).

In his view, particles are nothing more than an illusory appearance of the wavelike structure of reality: "What we observe as material bodies and forces are nothing but shapes and variations in the structure of space. Particles are just 'schaumkommen' (appearances)." (Erwin Schrödinger) [30]. This idea was never seriously considered by the scientific community, and generated bitter controversies between Schrödinger and other physicists.

2 Born: Just particles, no waves

According to the probabilistic interpretation of quantum mechanics — proposed by Max Born in 1926 — the wavefunction ψ is *not* a physical wave propagating in space, and carrying energy and momentum. It is instead a "probability wave", whose squared module $|\psi|^2$ corresponds to the probability density ϱ of finding a particle in a specified region of space. Therefore, it is a mere mathematical object, defined in the Hilbert space $\mathcal{H}$ associated to the quantum system.

As outlined by Born during his 1954 Nobel Lecture, particles are the only physical entities required by this description of quantum mechanics: "Schrödinger thought that his wave theory made it possible to return to deterministic classical physics. He proposed (and he has recently emphasized his proposal anew's), to dispense with the particle representation entirely, and instead of speaking of electrons as particles, to consider them as a continuous density distribution $|\psi|^2$ (or electric density $e|\psi|^2$). To us in Göttingen this interpretation seemed unacceptable in face of well established experimental facts. At that time it was already possible to count particles by means of scintillations or with a Geiger counter, and to photograph their tracks with the aid of a Wilson cloud chamber." (Max Born) [5, p. 261].

Considering wave/particle duality, Born probabilistic interpretation presents at least one problem: How to explain the interference effect seen when electrons are sent — one by one — through a double slit ? The single electron crossing the apparatus will impinge on the detector in a specific position. When combined with the signals from many other electrons that have crossed or will cross the double slit at different times, it produces an interference pattern. If only one slit is open, the recorded collective image is a diffraction pattern instead. The behavior of each electron is different, according to the different experimental set-up (single or double slit). If particles are the only physical entities, how could the single particle crossing the apparatus be physically effected in such a way to cooperate to different (interference or diffraction) collective results?

3 Heisenberg, Jordan: Neither waves, nor particles

According to this radically anti-realistic view, quantum physics would require the relinquishment of any attempt to visualize the micro-physical world, and a retreat into mathematical formalism. In the mid 1920s, it was endorsed — among others — by Werner Heisenberg and Pascual Jordan.

Within this interpretation, the ontological definition of quantum entities is a pseudo-problem, a relic of the obsolete categories (*objects, waves, particles*...) of classical physics: "Heisenberg [...] cut the Gordian knot by means of a philosophical principle and replaced guess-work by mathematical rule. The principle states that concepts and representations that do not correspond to physically observable facts are not to be used in theoretical description." (Max Born) [5, p. 258].

The supporters of this position were — not by chance — also the authors of the matrix formulation of quantum mechanics. The mathematical tools required in this case (matrices) are different from the theory of differential equations, that represents a common ground between classical mechanics and Schrödinger wave mechanics. Observable physical quantities (*intensities, frequencies*...) are the only entities of the matrix formulation of quantum theory, and are calculated using the matrix formalism without any need for a physical model of the quantum system.

This minimal interpretation resumes attitudes and views of the relation between scientific theories and reality typical of 19th century positivist philosophy. Being "minimal" by choice, its characteristics coincide with its limits: nothing more than the mere comparison between predictions and experimental results is allowed. It is therefore forbidden to build a "Weltanschauung" (worldview), to infer from the physical theory an ontology, an explanation or a descriptive model of the physical reality. Epistemology and praxeology are the only possible philosophical outcomes of such a limited interpretation of the quantum theory.

It is noteworthy that Heisenberg changed his philosophical viewpoint over time: from his initial positivism to operationalism (arguing that it is the theory which decides what can be observed) with his paper on uncertainty [16], then a neo-Kantian interpretation in the 1930s, and finally a "linguistic approach" in the 1940s and 1950s [8].

4 Bohm: Waves and particles

In 1927 Louis de Broglie proposed the *pilot wave theory* [10] assuming that there is a physical wave (carrying energy and momentum) "guiding" each quantum particle. The theory was presented at the Fifth Solvay Conference in Bruxelles, but it was strongly opposed by Wolfgang Pauli with a wrong argument, that was indeed considered to be correct at the time[1]. As a result, pilot wave theory was abandoned even by de Broglie.

The concept was proposed again and further developed by David Bohm [3] and Jean-Pierre Vigier, de Broglie's pupil, from the early 1950s on. The objective coexistence of a physical wave *and* the guided particle is the fundamental assumption of this formulation, known as *causal interpretation* [17] of quantum mechanics.

---

[1] The confutation of Pauli's argument came only twenty-five years later [2] and was sent from the author to Pauli. But he never replied.

Bohm model presents an asymmetry between waves and particles: While the wave does influence the particle, the particle has no influence on the wave. In fact, in Bohm interpretation, the wave has a nomological status: it determines both the trajectory of the associated particle, and the probability density to find the particle in a specified region of space. The single particle wave function, written in polar form, is

$$\psi(\vec{r}, t) = R(\vec{r}, t)\, e^{i\, S(\vec{r}, t)/\hbar} \qquad (1)$$

where both the amplitude $R(\vec{r}, t)$ and the phase $S(\vec{r}, t)$ are real functions. The velocity (and therefore the trajectory) of the particle depends on the phase S:

$$\vec{v}(\vec{r}, t) = \frac{1}{m} \nabla S(\vec{r}, t) \qquad (2)$$

which is de Broglie guidance condition.

The probability density ϱ is given by the amplitude R of the wavefunction:

$$\varrho(\vec{r}, t) = \psi^* \psi = |\psi|^2 = R^2(\vec{r}, t) \qquad (3)$$

The quantum features of a system are also described by means of the amplitude R: the presence of the *quantum potential*

$$Q(\vec{r}, t) = -\frac{\hbar^2}{2m} \frac{\nabla^2 R(\vec{r}, t)}{R(\vec{r}, t)} \qquad (4)$$

differentiates quantum systems from classical ones. According to Bohm interpretation, the non-local aspects of quantum mechanics are direct consequences of the non-local character of the quantum potential Q.

Furthermore, a particle is always accompanied by a wave, but the converse is not always true: There exist *empty waves*. Empty waves do carry energy and momentum, but there is no associated particle. There are cases of physical interest, where the wave splits up into a set of parts which have no appreciable spatial overlap. One of them remains associated with the guided particle, while the other waves result to be empty. Empty waves "may be effected by external potentials, and if recombined (superposed) with the wave containing the particle will [...] influence the subsequent particle motion." (Peter R. Holland) [17, p. 86]. Nevertheless, the experimental detection of empty waves has been unsuccessful [23, 24, 18]. These results make it difficult to believe to the physical existence of pilot waves.

The analysis of Bohm model [14] shows the similarities between this interpretation and the classical description or reality (*realism, causality, determinism*). But it also elucidates the new properties (*non-locality, olism, contextuality*) introduced by quantum mechanics in our representation of the physical world.

Bohm formulation of quantum mechanics is deeply at variance with the orthodox interpretation (also known as "Copenhagen interpretation") from an epistemological point of view. But it is equivalent to the usual interpretation as far as physical predictions are concerned. Considering the strong and controversial epistemological implications of the orthodox interpretation, it is difficult to understand the marginal attention paid by the physicists' community to the causal interpretation [9]. On the contrary, the picture of reality suggested by Bohm formulation is clear and precise, due to its analogy with classical mechanics.

A further developement of the causal interpretation, known as *Bohmian mechanics*, has been proposed [15] during the 1990s. Instead of the the quantum potential Q introduced by Bohm, Madelung guidance condition

$$\vec{v}(\vec{r},t) = \frac{1}{m} \nabla S(\vec{r},t) = \frac{\hbar}{m} \operatorname{Im} \frac{\nabla \psi(\vec{r},t)}{\psi(\vec{r},t)} \qquad (5)$$

is required by Bohmian mechanics, in addition to Schrödinger equation, to describe a single particle quantum system. For a system composed by N identical particles there are N Madelung guidance equations, each describing the dependence of velocity $\vec{v}_k$ ($k=1,2,\ldots N$) on the instantaneous positions of *all* the N particles belonging to the system.

Since (5) is a first order differential equation, in Bohmian mechanics positions and velocities are *not* independent: it would be sufficient to know the initial positions of the N particles of a quantum system, to complete determine its dynamical evolution. On the contrary, classical mechanics is based on Newton equation, that is a second order differential equation. Therefore both the initial positions and velocities have to be specified, in order to calculate particle trajectories for a classical multiparticle system. This example shows the radical departure of Bohmian mechanics from Newtonian dynamics in the description of physical reality. Even if Bohmian interpretation retains features typical of a classical worldview (realism, causality, determinism), its ontology is completely different.

In conclusion, Bohmian mechanics represents both an ontology different from the classical one, and an epistemology alternative to the orthodox interpretation of quantum mechanics.

5 Bohr, Pauli: Waves or particles

The so called "complementarity principle" — formulated in 1927 by Niels Bohr [4] — states that every quantum system presents at least one pair of properties needed to describe it, that cannot be simultaneously known. They are mutually exclusive, in the sense that the observation of one property prevents from the observation of the other one. The pair unavoidably present is wave/particle: A quantum system displays either a wavelike or a corpuscular behavior, but it will never manifest wave *and* particle properties at the same time.

A formulation of the wave/particle duality due to Wolfgang Pauli [25] — and widely accepted among physicists — attributes to the experimental apparatus the actual determination of the system as a wave or as a particle. If the quantum system is observed using a detector apt to reveal particles (like a counter), it will show a corpuscular behavior. If it is analyzed with an instrument predisposed to detect waves (like an interferometer), the same system will display a wavelike behavior instead. According to Pauli interpretation, quantum systems do not have a defined ontological status, but it would be determined by the interaction with a macroscopic apparatus.

For instruments with two alternative paths (like a double-slit or a Mach-Zehnder interferometer), wave/particle duality could be reformulated introducing two new quantities: The *visibility* $\mathcal{V}$ and the *predictability* $\mathcal{P}$ related respectively to wavelike and corpuscular characteristics. The *visibility* $\mathcal{V}$ is simply the relative contrast of the interference fringes, given in terms of the maximum and minimum intensities:

$$\mathcal{V} = \frac{I_{max} - I_{min}}{I_{max} + I_{min}} \qquad (6)$$

The *predictability* $\mathcal{P}$ measures the probability related to the trajectory of the particle, expressed in terms of the relative probabilities $\mathcal{P}_A$ and $\mathcal{P}_B$ for the two possible paths (A or B), as determined by the wavefunction ψ_T:

$$\psi_T = c_A\,\psi_A + c_B\,\psi_B \qquad (7)$$

$$\mathcal{P} = |\mathcal{P}_A - \mathcal{P}_B| = \left| \frac{|c_A|^2 - |c_B|^2}{|c_A|^2 + |c_B|^2} \right| \qquad (8)$$

Both $\mathcal{V}$ and $\mathcal{P}$ are bounded:

$$0 \leq \mathcal{V} \leq 1 \qquad (9)$$

$$0 \leq \mathcal{P} \leq 1 \qquad (10)$$

and the quantitative formulation of Bohr complementarity is expressed by one of the two following conditions, that are mutually exclusive:

$$\mathcal{V} = 1 \qquad \mathcal{P} = 0 \qquad (11)$$

$$\mathcal{V} = 0 \qquad \mathcal{P} = 1 \qquad (12)$$

6 Greenberger, Englert: Neither waves, nor particles

A deeper understanding of wave/particle duality has been provided by D.M. Greenberger and A. Yasin [20] in 1988. They introduced a generalization of quantitative Bohr complementarity, expressed by the inequality

$$\mathcal{V}^2 + \mathcal{P}^2 \leq 1 \qquad (13)$$

that becomes an equality for a *pure quantum state*:

$$\mathcal{V}^2 + \mathcal{P}^2 = 1 \qquad (14)$$

Greenberger/Yasin duality shows that a quantum system can display both wavelike ($\mathcal{V} \neq 0$) *and* corpuscular ($\mathcal{P} \neq 0$) properties at once, but the enhancement of one feature implies the fading of the other one.

Greenberger/Yasin inequality (13) represents a generalization of Bohr complementarity — (11) or (12) — for values of $\mathcal{V}$ and $\mathcal{P}$ different from zero or one, and a departure from the classical concepts of "wave" and "particle" that are crucial for the definition of Bohr complementarity. A minimal quantum system having non integer values for $\mathcal{V}^2$ and $\mathcal{P}^2$ ($\mathcal{V}^2 = 0.4$ and $\mathcal{P}^2 = 0.6$, for instance) is *neither* a wave *nor* a particle.

B.-G. Englert [13] introduced in 1996 another inequality, apparently similar to (13) but utterly different. While (13) refers to the initial state of the *quantum system*, Englert inequality originates from the quantum properties of the *detector*. He defines the *observed* visibility $\mathcal{V}_o$ and the *distinguishability* $\mathcal{D}$, related to the probability of correctly guessing the path — A or B — *after* the interaction between the quantum system and the detector.
Englert inequality

$$\mathcal{V}_o^2 + \mathcal{D}^2 \leq 1 \qquad (15)$$

becomes the equality

$$\mathcal{V}_o^2 + \mathcal{D}^2 = 1 \qquad (16)$$

if the *detector* is prepared in a pure quantum state.

The specific non integer values measured for $\mathcal{V}_o^2$ and $\mathcal{D}^2$ ($\mathcal{V}_o^2 = 0.4$ and $\mathcal{D}^2 = 0.6$, for instance) are *not* a property of the quantum system, but the result of the interaction between the detector and the quantum system, as suggested by W. Pauli. Using a Mach-Zehnder interferometer and a polarimeter with single polarized photons, Paul Kwiat [19] has been able to vary $\mathcal{D}$ between $0 \leq \mathcal{D} \leq 1$, and to verify to a high degree of accuracy relation (16) for pure quantum states, and (15) for mixed states.

Greenberger/Yasin and Englert duality, empirically corroborated by Kwiat experiment, represent a definitive overcoming of the classical concepts of wave or particle. A minimal quantum system, with properties that are *partially* wavelike and *partially* corpuscular, can not be described either as a wave or as a particle: It is an intrinsically non classical system.

7 Lévy-Leblond: Quantons

A further departure from classical concepts is to attribute ontological meaning to the non-classical states described in the previous Section: "The wave-like and the corpuscolar behaviour of microentities are two extreme form of being of the same ontological entity, which is governed by the Greenberger/Yasin inequality." (Gennaro Auletta) [1, p. 526].

Jean-Marc Lévy-Leblond has been an eminent advocate of this position. According to him, the entities of quantum theory should not be identified as waves or particles (characterizing classical, not quantum physics), but as *quantons* [22] instead: "We must [...] abandon the idea that every physical object is either a wave or a particle. Neither it is possible to say, as is sometimes done that particles 'become' waves in the quantum domain and conversely, that waves are 'transformed' into particles. [...] It is, therefore, necessary to acknowledge that we have here a different kind of an entity, one that is specifically quantum. For this reason we name them *quantons*, even though this nomenclature is not yet universally adopted. These quantons behave in a very specific manner [...]" (Jean-Marc Lévy-Leblond) [21, p. 69].

Indeed it should be stressed that Lévy-Leblond proposal could be a purely semantic "solution": Minting a new term[2] ("quanton") does not necessarily correspond to the existence of a new physical entity. For instance, the noun "quanton" could identify a new level of reality (with respect to classical physics), as well as the noun "platypus" identified (at the beginning of 19th century) a new animal that has similarities both with a beaver and with a duck, but is neither a beaver nor a duck. But it could be as well that the zoological analogy for the term "quanton" is not the platypus but the unicorn, a fabulous creature existing just in myths and legends...

8 Lévy-Leblond: Fermions and Bosons

The statistical behavior of quantum systems — according to Bose-Einstein or Fermi-Dirac statistics — is obviously defined only for an ensemble of quantons. But each individual quanton is characterized — as confirmed by contless experimental results — either as a *boson*[3] or as a *fermion*[4].

Even if fermions and bosons are both quantons, according to J.-M. Lévy-Leblond, the fermion/boson dichotomy is the proper connotation for quantum systems: "[...] fermion-boson dichotomy is of enormous importance and [...] it leads to two very different types of quantum behaviour." [21, p. 491], while the wave/particle duality is appropriated just for classical systems. In fact, even if the existence of quantons — as discussed in the previous Section — could be debatable, the physical evidence for two distinct kinds of microentities (fermions and bosons) is unquestionable. The ontological structure of quantum reality would therefore be constituted by fermions and bosons that appear — in the classical limit — as waves (in case of massless bosons) or particles (for massive bosons or fermions) [21, p. 490].

J.-M. Lévy-Leblond concludes: "Thus, even after having insisted on the universality of the concept of the quanton, which seemed to undermine the classical wave-particle duality, [...] we see a new duality appearing on the quantum level — related, for sure in some complex way, to this classical duality, [...] but yet more profound. We gladly leave the question of determining whether this dialectic of one and of two, and its successive incarnations within physical theory, relates to the object of science or to its subject — assuming that *this* dichotomy makes sense." [21, p. 493].

BIBLIOGRAPHY

[1] G. Auletta. *Foundations and Interpretation of Quantum Mechanics*. World Scientific, Singapore 2000.

[2] D. Bohm. Reply to a criticism of a causal re-interpretation of the quantum theory. *Physical Review*, 87: 389–390, 1952.

[3] D. Bohm. A suggested interpretation of the quantum theory in terms of "hidden variables". *Physical Review*, 85: 166–179, 85: 180–193, 1952.

[4] N. Bohr. The quantum postulate and the recent development of atomic theory. In *Atti del Convegno Internazionale dei Fisici [Proceedings of the International Congress of Physicists]*. Zanichelli, Bologna 1928.

[5] M. Born. The statistical interpretation of quantum mechanics — Nobel Lecture, December 11, 1954. In *Nobel Lectures in Physics 1942-1962*, Elsevier, Amsterdam 1964. On-line http://nobelprize.org/nobel_prizes/physics/laureates/1954/born-lecture.pdf.

---

[2]The word "quanton" was introduced by Mario Bunge [6, 7] in 1967.
[3]A quanton with an integer spin quantum number: $s = 0, 1, 2 \ldots$
[4]A quanton with a half-odd spin quantum number: $s = 1/2, 3/2, 5/2 \ldots$

[6] M. Bunge. *Foundations of Physics*. Springer, New York 1967.
[7] M. Bunge. *Quantum Theory and Reality*. Springer, New York 1967.
[8] K. Camilleri. *Heisenberg and the Interpretation of Quantum Mechanics*. Cambridge University Press, Cambridge 2009.
[9] J.T. Cushing. *Quantum Mechanics: Historical Contingency and the Copenhagen Hegemony*. University of Chicago Press, Chicago 1994.
[10] L. de Broglie. Nouvelle dynamique des quanta [The new dynamics of quanta]. In *Electrons et Photons [Electrons and Photons]*. Eds. Gauthier-Villars et c^{ie}, Paris 1928.
[11] A. Einstein. Zum gegenwärtigen Stand des Strahlungsproblems [On the present status of the radiation problem]. *Physikalische Zeitschrift*, 10: 185–193, 1909.
[12] A. Einstein. Über die Entwicklung unserer Anschauungen über das Wesen und die Konstitution der Strahlung [On the development of our views concerning the nature and constitution of radiation]. *Physikalische Zeitschrift*, 10: 817–825, 1909.
[13] G.-B. Englert. Fringe visibility and which-way information: An inequality. *Physical Review Letters*, 77: 2154-2157, 1996.
[14] A. Fine. On the Interpretation of Bohmian Mechanics. In J.T. Cushing, A. Fine and S. Goldstein (eds.), *Bohmian Mechanics and Quantum Theory: An Appraisal*. Kluwer, Dordrecht 1996.
[15] D. Dürr, S. Goldstein and N. Zanghì. On a realistic Theory of Quantum Physics. In S. Albeverio, G. Casati, U. Cattaneo, D. Merlini (eds.), *Stochastic Processes, Physics and Geometry*. World Scientific, Singapore 1990.
[16] H. Heisenberg. Über den anschaulichen Inhalt der quantentheoretischen Kinematik und Mechanik [On the intuitive content of quantum theoretical kinematics and mechanics]. *Zeitschrift für Physik*, 43: 172–198, 1927.
[17] P.R. Holland. *The Quantum Theory of Motion*. Cambridge University Press, Cambridge 1993.
[18] S. Jeffers and J. Sloan. An experiment to detect "empty" waves. *Foundations of Physics Letters*, 7: 333-341, 1994.
[19] P.D.D. Schwindt, P.G. Kwiat and B.-G. Englert. Quantitative wave-particle duality and non-erasing quantum erasure. *Physical Review*, A60: 4285-4290, 1999.
[20] D.M. Greenberger and A. Yasin. Simultaneous wave and particle knowledge in a neutron interferometer. *Physics Letters*, A128: 391–394, 1988.
[21] J.-M. Lévy-Leblond. *Quantics. Rudiments of Quantum Physics*. North Holland, Amsterdam 1990.
[22] J.-M. Lévy-Leblond. On the Nature of Quantons. *Science and Education*, 12: 495–502, 2003.
[23] L.J. Wang, X.Y. Zou and L. Mandel. Experimental test of the de Broglie guided-wave theory for photons. *Physical Review Letters*, 66: 1111-1114, 1991.
[24] X.Y. Zou, T. Grayson and L.J. Wang, L. Mandel. Can an "empty" de Broglie pilot wave induce coherence ? *Physical Review Letters*, 68: 3667-3669, 1992.
[25] W. Pauli. Die philosophische Bedeutung der Idee der Komplementarität [The philosophical significance of the idea of complementarity]. *Experientia*, 6: 72–76, 195.0
[26] E. Schrödinger. Quantisierung als Eigenwertproblem (Erste Mitteilung) [Quantization as an eigenvalue problem (first communication)]. *Annalen der Physik*, 79: 361–376, 1926.
[27] E. Schrödinger. Quantisierung als Eigenwertproblem (Zweite Mitteilung) [Quantization as an eigenvalue problem (second communication)]. *Annalen der Physik*, 79: 489–527, 1926.
[28] E. Schrödinger. Über das Verhältnis der Heisenberg-Born-Jordanschen Quantenmechanik zu der meinen [On the relation between Heisenberg-Born-Jordan's quantum mechanics and mine]. *Annalen der Physik*, 79: 734-756, 1926.
[29] E. Schrödinger. Are there quantum jumps? *The British Journal for the Philosophy of Science*, 3: 109-123, 3: 233-242, 1952.
[30] E. Schrödinger, *The Interpretation of Quantum Mechanics*, Ox Bow Press, Woodbridge 1995.

An application of von Neumann algebras to computational complexity

MARCO PEDICINI, MARIO PIAZZA

1 Introduction

A von Neumann algebra is an algebra of bounded linear operators on a Hilbert space which is closed under the topology of pointwise convergence. *Factors* are von Neumann algebras whose center consists of scalar multiples of the identity and they can be viewed as the elementary constituents from which all the von Neumann algebras are built: every von Neumann algebra is a direct integral (a generalization of direct sum) of factors [9, 10, 11, 16].

The aim of this paper is to show how a von Neumann algebras setting can be used in a natural way to obtain an implicit complexity model for classical Turing machines (TM). In particular, we focus on the *unique* von Neumann algebra which admits finite dimensional approximations and it is a factor II_1: the so called hyperfinite II_1 factor $\mathcal{R}$. More precisely, the focal point is the construction of an ascending sequence (G_i) of finite cyclic groups such that the cardinality of G_i doubles at every step, and whose infinite union is a discrete group G.

Then, we embed the configurations of a TM in the Hilbert space $\ell^2(G)$ of the square summable formal series indexed by elements of G with complex coefficients. Since the number of configurations in a finite part of the tape is bounded, there is a mapping which sends a configuration to a finite dimensional subspace. In this way, we model the transition of the machine by means of endomorphisms on the subspace. The set of endomorphisms constitutes a finite dimensional subalgebra of a von Neumann algebra. Namely, we build an element of $\ell^2(G)$, by superimposition, of all configurations appearing in the computation of the machine. Since DSPACE($T(n)$) is contained in DTIME($2^{cT(n)}$), we obtain that any machine embedded in the von Neumann group algebra of G_n, denoted $\mathcal{N}(G_n)$, implicitly belongs to the computational class DSPACE(N) and so to DTIME(2^{cN}).

Our investigation intersects J.-Y. Girard's recent proposal, conceptually ambitious, to reshape the semantics of computation called *geometry of interaction* (GoI) in the realm of von Neumann algebras [6]. In previous works, [2, 3, 4, 5], GoI is built in a C^*-algebra: proofs correspond to bounded operators of the infinite dimension Hilbert space, and the execution formula corresponds to the power series of the operator itself. Since von Neumann algebras are automatically C^*-algebras, in this hyperfinite setting we maintain all the advantages of the standard view of GoI and at the same time we obtain an universal mathematical object: the *unique* $\mathcal{R}$, contained in any II_1 factor.

2 Preliminaries and fundamental lemma

Let $\mathcal{H}$ be a Hilbert space and $\mathcal{B}(\mathcal{H})$ the algebra of bounded operators on $\mathcal{H}$. For a subset S of $\mathcal{B}(\mathcal{H})$ we define

$$S' = \{x \in \mathcal{B}(\mathcal{H}) | xy = yx \text{ for all } y \in S\}$$

this set is called the *commutant* of S. One of the first and fundamental results in the theory of von Neumann algebras is the double commutant theorem:

THEOREM 1 (Murray-von Neumann). *Let M be a self-adjoint subalgebra of $\mathcal{B}(\mathcal{H})$ containing the identity operator 1. Then the following conditions are equivalent:*

1. *$M'' = M$;*

2. *M is σ-strongly closed;*

3. *M is σ-weakly closed;*

4. *M is strongly closed;*

5. *M is weakly closed.*

This holds for all unital $*$-subalgebra of $\mathcal{B}(\mathcal{H})$ and in particular for von Neumann algebras:

DEFINITION 2. A von Neumann algebra is a $*$-subalgebra M of $\mathcal{B}(\mathcal{H})$ coinciding with its bicommutant $M = M''$.

Clearly, the definition of von Neumann algebra leads to equivalent definitions under the replacement of the bicommutant condition with equivalent conditions as it is stated in Theorem 1.

DEFINITION 3. A von Neumann algebra M is called a *factor* if it has trivial centre i.e. $Z(M) = M \cap M' = \mathbb{C}1$.

A *projection* is any element $p \in M$ such that $p = p^* = p^2$. Projections play a crucial role in the study (and the classification) of factors.

In particular, if M is a factor any two projections are comparable in the sense that one is equivalent to a sub-projection of the other. The equivalence $e \simeq f$ is given in the sense of Murray-von Neumann if (and only if) there exists a partial isometry u such that $e = u^*u$ and $f = uu^*$.

A *sub-projection* of a projection e is any projection f such that $ef = f$.

A projection is *finite* when it is not equivalent to any of its proper sub-projections.

Moreover, a factor M is said to be of type II if M contains non-zero finite projections and there exists no non-zero minimal projections in M. A type II factor is said to be of type II$_1$ if it does not contain any non unitary isometry (i.e. it is a finite factor).

The rest of this section is devoted to introduce the notion of *von Neumann group algebra of G*. Let G be a (discrete) group fixed throughout the article. Let $\mathbb{C}[[G]]$ denote the set of all functions from G to $\mathbb{C}$ expressed as formal sums, that is, a function $a : G \to \mathbb{C}$, $g \mapsto a(g)$, will be written as $\sum_{g \in G} a(g)g$. For each $a \in \mathbb{C}[[G]]$, we define

$$||a|| := (\sum_{g \in G} |a(g)|^2)^{1/2} \in [0, \infty],$$

and $tr(a) := a(1) \in C$.

Now, let us define

$$\ell^2(G) := \{a \in \mathbb{C}[[G]] : ||a|| < \infty\}.$$

Let $\mathbb{C}G$ denote the complex group ring of formal sums with *finite* support. Then, we view $\mathbb{C} \subset \mathbb{C}G \subset \ell^2(G) \subset \mathbb{C}[[G]]$. There is a well-defined external multiplication map

$$\ell^2(G) \times \ell^2(G) \to \mathbb{C}[[G]], \quad (a, b) \mapsto a \cdot b,$$

where, for each $g \in G$, $(a \cdot b)(g) := \sum_{h \in G} a(h)b(h^{-1}g)$; this series converges in $\mathbb{C}$, and, moreover, $|(a \cdot b)(g)| \leq ||a|| \, ||b||$, by the Cauchy-Schwarz inequality. The external multiplication extends the multiplication of $\mathbb{C}G$.

Then $\ell^2(G)$ is a separable Hilbert space with the scalar product defined by

$$\langle \sum_{g \in G} \lambda_g g, \sum_{g \in G} \mu_g g \rangle = \sum_{g \in G} \lambda_g \overline{\mu}_g$$

where $\overline{\lambda}$ denotes the complex conjugate of λ.

Let us denote the space of the bounded operators in the Hilbert space $\ell^2(G)$ by $\mathcal{B}(\ell^2(G))$. Let $\mathcal{U}(\ell^2(G))$ the sub-algebra of unitary operators (operators such that $uu^* = u^*u = 1$) from $\ell^2(G)$ to $\ell^2(G)$.

Let us define the left regular representation λ' of G: it is a function $\lambda' : G \to \mathcal{U}(\ell^2(G))$ such that the image of an element of the group G is an unitary operator $\lambda'(g) : \ell^2(G) \to \ell^2(G)$ so that

$$\lambda'(g) \sum a_h h := \sum a_h gh.$$

Finally, let us denote with $\lambda : \mathbb{C}G \to \mathcal{B}(\ell^2(G))$ the extension of the left regular representation λ' of G in

$$\lambda : \mathbb{C}G \to \mathcal{B}(\ell^2(G))$$

where

$$\lambda(\sum_{g \in G} a(g)g) := \sum_{g \in G} a(g) \lambda'(g).$$

We recall that the *von Neumann group algebra of G*, denoted by $\mathcal{N}(G)$, is the closure of $\lambda(\mathbb{C}G)$ in the strong operator topology on $\mathcal{B}(\ell^2(G))$.

This object can be easily described as the algebra of (right) G-equivariant bounded operators from $\ell^2(G)$ to $\ell^2(G)$:

$$\mathcal{N}(G) = \{\alpha : \ell^2(G) \to \ell^2(G) \, | \, ||\alpha|| < +\infty,$$
$$\text{for any } h \in \ell^2(G) \text{ and } g \in G, \, \alpha(h)g = \alpha(hg)\}.$$

In other words, α is equivariant by the group action. This definition can be also rephrased as: $\mathcal{N}(G)$ is the ring of bounded $\mathbb{C}G$-endomorphisms of the right $\mathbb{C}G$-module $\ell^2(G)$; see [8, 1.1]. We view $\mathcal{N}(G)$ as a subset of $\ell^2(G)$ by the map $\alpha \mapsto \alpha(1)$, where 1 denotes the identity element of $\mathbb{C}G \subset \ell^2(G)$ and moreover,

$$\mathcal{N}(G) = \{a \in \ell^2(G) | a \cdot \ell^2(G) \subset \ell^2(G)\}. \tag{1}$$

The action of $\mathcal{N}(G)$ on $\ell^2(G)$ is given by the external multiplication. Notice that $\mathcal{N}(G)$ contains $\mathbb{C}G$ as a subring and that there exists an induced *trace map* $tr : \mathcal{N}(G) \to \mathbb{C}$, i.e., $tr(\alpha) := tr(\alpha(1)) = \alpha(1)(1)$. Of course, $\mathcal{N}(G)$ is a von Neumann algebra. For more details on von Neumann algebras we refer the reader to standard monographs, for example [8, 1].

2.1 General construction of the Hyperfinite II_1 factor

Let us recall that the fundamental result of Murray and von Neumann ensures that the hyperfinite II_1 factor is essentially unique (precisely, that there is one isomorphism class, [11]). This means that its construction can be achieved in many different ways.

In particular, the discrete group G is obtained by an approximation procedure, i.e., by giving an ascending sequence of finite subgroups

$$G_0 \subset G_1 \subset G_2 \subset \cdots \subset G_n \subset \ldots$$

such that $G = \bigcup_{i=0}^{\infty} G_i$. In this case we say G is *locally finite*. This condition is not sufficient to make $\mathcal{N}(G)$ a factor.

Now, we need to establish the following lemma.

LEMMA 4. $\bigcup_{i=0}^{\infty} \mathcal{N}(G_i)$ *is weakly dense in the von Neumann group algebra of* $G = \bigcup_{i=0}^{\infty} G_i$.

Proof. First we prove that any $a \in \mathcal{N}(G_i)$ is indeed in $\mathcal{N}(G)$. In fact, if $a \in \ell^2(G_i)$, then, by Equation 1, $a \cdot \ell^2(G_i) \in \ell^2(G_i)$. To prove also that $a \in \mathcal{N}(G)$, we must show that $a \cdot b \in \mathcal{N}(G)$, for any $b \in \mathcal{N}(G)$; since $a \cdot b \in \mathbb{C}[G]$, we show that $a \cdot b$ is bounded. In fact, by definition

$$||a \cdot b|| = \left(\sum_{g \in G} |(a \cdot b)(g)|^2\right)^{1/2} = \left(\sum_{g \in G} \left|\sum_{h \in G_i} a(h)b(h^{-1}g)\right|^2\right)^{1/2},$$

and in order to prove $||a \cdot b|| \leq c$ for some c, we have

$$||a \cdot b||^2 \leq ||a||^2 \left(\sum_{h \in G_i} \sum_{g \in G} |b(h^{-1}g)|^2 + (\sum_{h_1 \neq h_2 \in G_i} \sum_{g \in G} |b(h_1^{-1}g)||b(h_2^{-1}g)|\right) \leq$$

$$\leq ||a||^2 (|G_i|||b||^2 + \sum_{h_1 \neq h_2 \in G_i} 2||b||^2) = \tag{2}$$

$$= ||a||^2 (|G_i| + 2|G_i|(|G_i| - 1))||b||^2.$$

Specifically, in order to prove inequality (2), we consider

$$\sum_{g \in G} |b(h_1^{-1}g)||b(h_2^{-1}g)| = \sum_{g \in G} |b(g)||b(h_2^{-1}h_1g)| \leq 2\sum_{g \in G} |b(g)|^2 = 2||b||^2$$

in fact, for any $g \in G$ and for any fixed pair $h_1, h_2 \in G_i$, we define

$$c(g) = \begin{cases} |b(g)|^2 & \text{if } |b(g)| > |b(h_2^{-1}h_1 g)|, \\ |b(h_2^{-1}h_1 g)|^2 & \text{if } |b(g)| \leq |b(h_2^{-1}h_1 g)|. \end{cases}$$

Since any term of $\sum_{g \in G} |b(g)|^2$ appears in $\sum_{g \in G} c(g)$ at most twice we have the following inequality

$$\sum_{g \in G} |b(g)||b(h_2^{-1}h_1 g)| \leq \sum_{g \in G} c(g) \leq 2 \sum_{g \in G} |b(g)|^2 = 2||b||^2.$$

Since $\mathcal{N}(G)$ is a von Neumann algebra, any pointwise convergent sequence (α_k) of operators in $\mathcal{N}(G_i)$ converges to an element in $\mathcal{N}(G)$. The reason is that if $\alpha_k \in \mathcal{N}(G_i)$, then $\alpha_k \in \mathcal{N}(G)$ for any k, which is dense by definition and so $\phi \in \mathcal{N}(G)$. ∎

An immediate consequence of the Lemma 4, is that $\mathcal{N}(G)$ is an AFD (approximately finite dimensional) von Neumann algebra, that is an algebra A such that there is a family A_i of finite dimensional subalgebras whose union $\bigcup_{i=0}^{\infty} A_i$ is σ-weakly dense in A.

The next proposition establishes an important property of any von Neumann group algebra built on an abelian discrete group. This result relies on the remark that by the Pontryagin duality this algebra can be mapped on the space of bounded operators on $L^2([0,1], \sigma)$ (where σ is the Lebesgue measure). Moreover, this mapping is obtained by considering the dual $\hat{G}$ of the group G equipped with the Haar measure μ, and by building the space $L^2(\hat{G}, \mu)$. This mapping has effects on the projections thanks to the finiteness of the Haar measure on the elements of the dual, namely this algebra contains no minimal projections. Since any *maximal self-adjoint abelian subalgebra* (MASA) of $\mathcal{R}$ has no mimimal projections, this fact implies that these algebras are all isomorphic and in particular they are isomorphic to the von Neumann group algebra on G.

THEOREM 5. *The von Neumann group algebra of any discrete abelian group G is isomorphic to a MASA of $\mathcal{R}$.*

This means that there exists an embedding $J : \mathcal{N}(G) \to \mathcal{R}$. Moreover, this embedding can be projected on the subalgebras $\mathcal{N}(G_i)$, so that $J_i : \mathcal{N}(G_i) \to \mathcal{R}$.

3 A construction of an AFD von Neumann algebra

In this section, we present an instance of the construction of the G_i such that the Lagrange index, i.e., the relative dimension of successive groups, is $[G_i : G_{i+1}] = 2$. Moreover, we define $G_{i+1} = G_i + \mu_i G_i$ where μ_i is a new generator. The sequence starts at G_0, the trivial group containing the sole identity, $G_1 := C_2$ is the cyclic group of 2 elements, and in general $G_i := C_{2^i}$. Cyclic groups are abelian with one group generator. We give a construction of the multiplication table of G_i in such a way that its group generator coincides with the element $\mu_i = 2^i$. In fact, this construction is recursive and uses an auxiliary permutation matrix $T(i)$, which generalises the twist and gives an isomorphic presentation of C_{2^i}:

$$T(i) := \begin{cases} \begin{pmatrix} 1 \end{pmatrix} & \text{if } i = 0 \\ \begin{pmatrix} 0 & I_{2^{i-1}} \\ T(i-1) & 0 \end{pmatrix} & \text{otherwise.} \end{cases} \tag{3}$$

In a similar way we give the construction of the multiplication table of G_i:

$$G_i := \begin{cases} \begin{pmatrix} 0 \end{pmatrix} & \text{if } i = 0, \\ \begin{pmatrix} G_{i-1} & 2^{i-1} + G_{i-1} \\ 2^{i-1} + G_{i-1} & T(i-1).G_{i-1} \end{pmatrix} & \text{otherwise.} \end{cases} \tag{4}$$

The main feature of this construction is that any G_i is subgroup of G_{i+1} and it appears directly in its multiplication table as in the top-left corner. Another important property is that the generator of the group is the element 2^i.

In Figure 1, we have depicted with grey tones the multiplication tables of several G_i's, in order to graphically point out the structure of these groups. The first groups are:

$$G_1 = \begin{pmatrix} 0 & 1 \\ 1 & 0 \end{pmatrix} \qquad G_2 = \begin{pmatrix} 0 & 1 & 2 & 3 \\ 1 & 0 & 3 & 2 \\ 2 & 3 & 1 & 0 \\ 3 & 2 & 0 & 1 \end{pmatrix}$$

$$G_3 = \begin{pmatrix} 0 & 1 & 2 & 3 & 4 & 5 & 6 & 7 \\ 1 & 0 & 3 & 2 & 5 & 4 & 7 & 6 \\ 2 & 3 & 1 & 0 & 6 & 7 & 5 & 4 \\ 3 & 2 & 0 & 1 & 7 & 6 & 4 & 5 \\ \underline{4} & 5 & 6 & 7 & 2 & 3 & 1 & 0 \\ 5 & 4 & 7 & 6 & 3 & 2 & 0 & 1 \\ 6 & 7 & 5 & 4 & 1 & 0 & 3 & 2 \\ 7 & 6 & 4 & 5 & 0 & 1 & 2 & 3 \end{pmatrix}$$

The underlined element of the table is μ_i and let us consider $\mu_3 = 2^2 = 4$ and its row in the multiplication table. With the aid of this table one may compute the different orbits for the iterated action of μ_i over elements of G_i,

$$O_i(x) := (x, \mu_i \cdot x, \mu_i \cdot \mu_i \cdot x, \ldots, \mu_i^{2^i-1} \cdot x), \qquad \text{for all } x \in G_i$$

for instance $O_3(1) = (1, 5, 3, 7, 0, 4, 2, 6)$

$$(O_3(x))_{0 \le x \le 7} = \begin{pmatrix} 0 & 4 & 2 & 6 & 1 & 5 & 3 & 7 \\ 1 & 5 & 3 & 7 & 0 & 4 & 2 & 6 \\ 2 & 6 & 1 & 5 & 3 & 7 & 0 & 4 \\ 3 & 7 & 0 & 4 & 2 & 6 & 1 & 5 \\ 4 & 2 & 6 & 1 & 5 & 3 & 7 & 0 \\ 5 & 3 & 7 & 0 & 4 & 2 & 6 & 1 \\ 6 & 1 & 5 & 3 & 7 & 0 & 4 & 2 \\ 7 & 0 & 4 & 2 & 6 & 1 & 5 & 3 \end{pmatrix}.$$

Furthermore, we have also a formula defining the generator of G_i:

$$\mu_i = \bigotimes_{k=0}^{i} I_{2^k} \oplus 2^k. \tag{5}$$

EXAMPLE 6. Let us compute the generators of the G_i's:

$\mu_0 = (0)$
$\mu_1 = ((0) + 1) \otimes (0) = (10)$
$\mu_2 = ((01) + 2) \otimes (10) = (23) \otimes (10) = (2310)$
$\mu_3 = ((0123) + 4) \otimes (2310) = (4567) \otimes (2310) = (45672310)$

we note that for any i we have

$$\mu_{i+1} = (I_{2^i} + 2^i) \otimes \mu_i$$

and by iteratively expanding the above formula we get

$$\mu_{i+1} = (I_{2^i} + 2^i) \otimes \mu_i = (I_{2^i} + 2^i) \otimes (I_{2^{i-1}} + 2^{i-1}) \otimes \mu_{i-1}.$$

4 Deterministic TMs encoded in a von Neumann Algebra

In this section, we present an encoding of TM's in operators acting on von Neumann group algebra $\mathcal{N}(G)$ of the group G, that we have introduced in the previous section. Leaving aside the obvious differences, our construction is similar to the one given by Nishimura and Ozawa [13, 12]. Without loss of generality, we are concerned with one-way infinite tape machines whose alphabet is reduced to a unique symbol $A = \{1\}$. For any deterministic, one-way infinite tape TM with set of states Q and alphabet A,

$$\mu : Q \times A \to Q' \times A' \times \{-1, +1\},$$

where $A' = A \cup \{\Box\}$ and Q' is $Q' = Q \cup \{q_F^1, q_F^0\}$, we consider an encoding of the transition function acting on the configuration space:

$$C = \{(q, p, f) \mid q \in Q, p \in \mathbb{N}, f : \mathbb{N} \to A'\}.$$

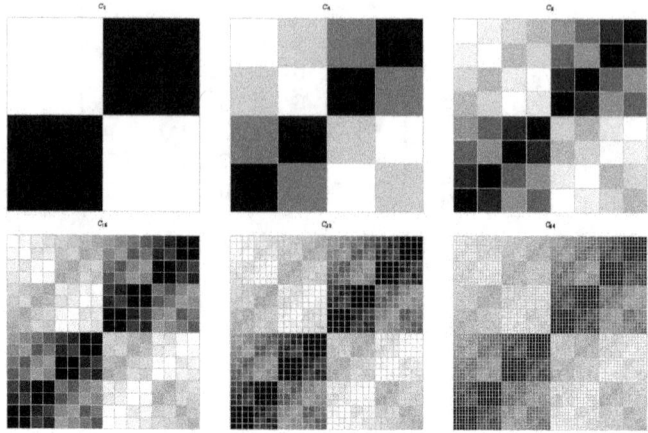

Figure 1. Group Multiplication Tables for Cyclic groups C_{2^i}, for $i = 1, \ldots, 6$.

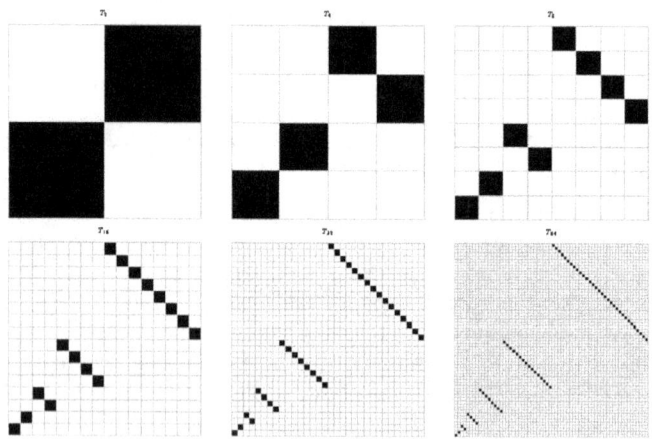

Figure 2. Matrix T_i, for $i = 1, \ldots, 6$.

We begin by defining subsets C_i of C indexed by integers. In particular, we take into account configurations concerning cells in positions $p \in [0, i]$:

$$C_i = \{(q, p, f) \mid q \in Q, p \in [0, i], f : [0, i] \to A'\}.$$

Let us note that $|C_i| = |Q| \, i \, 2^i$, and that by choosing $s = \log_2(i)$, and $q = \log_2 |Q|$, we have $|C_i| \leq 2^{N_i}$ where $N_i := s + q + i$; this also shows that $C = \bigcup_{i \in \mathbb{N}} C_i$ is denumerable. Let us consider a bijection $\phi : C \to G$. For any $c \in C$, we have one and only one $\phi(c) \in G$ and ϕ induces a correspondence between the configuration space C of μ and $\ell^2(G)$, such that for any configuration $c \in C$ one may define the corresponding element, denoted $\psi(c) \in \ell^2(G)$, in the following way:

$$\psi(c)(g) := \begin{cases} 1 & \text{if } \phi(c) = g, \\ 0 & \text{otherwise.} \end{cases}$$

We have that $\|\psi(c)\| = 1 < +\infty$; notice that $(\psi(c))_{c \in C}$ is automatically an orthonormal basis for a subspace of $\ell^2(G)$. Moreover, for any positive integer i, we consider the restriction $\phi_i : C_i \to G_{N_i}$ of ϕ defined as

$$\phi_i(c) := \phi(c).$$

From this restriction, we obtain a correspondence $\psi_i : C_i \to \ell^2(G_{N_i})$ as above. Let us observe in passing that the dimension of $\ell^2(G_{N_i})$ is 2^{N_i} which is $|C_i|$.

Let be (μ, c_0) a pair where μ is a Turing machine computing the function $f \in \text{DSPACE}(S(n))$, and c_0 is the initial configuration associated with input $x = (x_1, \ldots, x_n)$ such that $i = \|x\|$. Then, we consider the computation of μ starting from c_0 as the finite sequence

$$(c_0, c_1, \ldots c_k) \quad \text{where } k \leq 2^N$$

and we embed this sequence in the element

$$[[(\mu, c_0)]] := \sum_{j=0}^{k} \psi(c_j) \in \ell^2(G).$$

DEFINITION 7. For any TM μ and for any initial configuration c_0, we define the interpretation $[c_j]$ for any configuration c_j appearing in the computation

$$(c_0, c_1, \ldots c_k) \quad \text{where } k \leq 2^N \text{ as } [c_j] := \mu_N^j([[(\mu, c_0)]])$$

where μ_N is the operator corresponding to the action of the group generator of G_N.

By definition, it is clear that $[c_{j+1}] = \mu_N([c_j])$. So, the operator μ_N implements the TM μ in the sense that it makes the following diagram commute:

$$\begin{array}{ccc} C & \xrightarrow{\mu} & C \\ \downarrow{[\cdot]} & & \downarrow{[\cdot]} \\ \ell^2(G) & \xrightarrow{\mu_N} & \ell^2(G). \end{array}$$

This commutativity enable us to prove the following proposition:

PROPOSITION 8. *For any deterministic TM μ in $\mathrm{DSPACE}(S(n))$ and for any input x, $||x|| \leq n$ there is an interpretation $[.] : C \to \ell^2(G)$ such that there exists an operator $[\mu] \in \mathcal{R}$ such that $[\mu]([c_j]) = [\mu(c_j)]$.*

Proof. Let us consider the interpretation $[.] : C \to \ell^2(G)$ in Definition 7.

We define $[\mu] := J(\mu_{S(N)})$ of the operator μ_N associated with the generator of the group $G_{S(N)}$ by the embedding $J : \mathcal{N}(G) \to \mathcal{R}$ determined by Theorem 5. ∎

5 Space Bounded TMs

Let us denote by $\mathcal{E}$ the class of elementary functions, which was defined by Kalmár [7] as the least class of primitive recursive functions that contains the constant 0, all projections, successor, addition, cut-off subtraction and multiplication, and is closed under composition, bounded sum and bounded product.

Since $\mathcal{E}$ is closed under composition, for each m the m-times iterated exponential $2^{[m]}(x)$ is in $\mathcal{E}$, where $2^{[m+1]}(x) = 2^{2^{[m]}(x)}$ and $2^{[0]}(x) = x$. The elementary functions are exactly the functions computable in elementary time, i.e., the class of functions computable by a TM in a number of steps bounded by some elementary function. Two results are well known concerning the class $\mathcal{E}$:

PROPOSITION 9.

1. $\mathcal{E} = \mathrm{DTIME}(\mathcal{E}) = \mathrm{DSPACE}(\mathcal{E})$.

2. *If $f \in \mathcal{E}$, there is a number m such that for all $(x_1, \ldots, x_n)$,*

$$f(x_1, \ldots, x_n) \leq 2^{[m]}(||(x_1, \ldots, x_n)||)$$

where $||(x_1, \ldots, x_n)|| := \max_{1 \leq i \leq n} |x_i|$.

However, Proposition 9.2 tell us that $\mathcal{E}$ does not contain the iterated exponential $2^{[m]}(||x||)$ where the number of iterations m is a variable, since any function in $\mathcal{E}$ has an upper bound where m is fixed. This remark is useful to obtain the following result:

PROPOSITION 10. *Given a TM μ computing a Kalmar elementary function $f \in \mathcal{E}$, there exists an integer m such that for every input x, $|x| \leq n$, then the computation of the machine μ starting from the initial configuration associated with x is representable in $\mathcal{N}(G_{2^{[m^*]}(n)})$.*

Proof. By Proposition 9.1, we get that since μ computes $f \in \mathcal{E}$ there exists $g \in \mathcal{E}$ such that $f \in \mathrm{DSPACE}(g(n))$. Thus, for every input $(x_1, \ldots, x_n)$ $||x|| \leq n$, the machine μ has halt space $s \leq g(n)$, which by Proposition 9.2 implies that there exists m such that $s \leq 2^{[m]}(||x||)$.

For Proposition 8, by choosing $m^* = 2^{[m-1]}(n)$, we have that the computation of the machine μ starting from the initial configuration associated with x is representable in $\mathcal{N}(G_{2^{[m^*]}(n)})$. ∎

6 Conclusion

The ideas of this paper provide pointers towards the construction of a hyperfinite semantics of computation with an implicit complexity bound. In a recent paper, we have addressed the problem of finding the right perspective from which to conceive the interplay between computation and combinatorial aspects in operator algebras [15]. In fact, a similar construction of the interpretation of time bounded computation in the case of finite commutative von Neumann group algebras is reobtained in the noncommutative case inside the framework of the unique II_1 factor $\mathcal{R}$. To this end, it turns out to be crucial the notion of *binary shift*, that is a special family of unital $*$-endomorphism on $\mathcal{R}$ [14]. These mathematical objects make explicit the interaction between the *commutative* part and the *non commutative* one of $\mathcal{R}$. More methodologically, there is an aspect emerging from our approach that seems very promising: we have at disposal a range of combinatorial tools to try to come out with a sort of dynamical system whose dynamics is controlled by several combinatorial constraints.

On the other hand, discrete counterparts of operator algebras introduced to establish the computational setting, acquire a deeper mathematical status when they are embedded in the full-fledged framework of factor theory. A possible line of research is indicated by fundamental connections among conditional expectation, entropy and the Jones index, which are not tackled by the present work but are around the corner.

We may explain this novel approach as top-down: the sharp description of computational complexity in terms of a dynamical system implies in turn that we should focus on the understanding of the system in terms of *measure theory* in the hyperfinite II_1 factor $\mathcal{R}$. The next step is to detect those properties that are needed to catch *ergodic* aspects of the theory of computational complexity. The delineation of this theory will provide new insights into the relationships among logic, computation and physics.

BIBLIOGRAPHY

[1] B. Blackadar. *Operator algebras*, volume 122 of *Encyclopaedia of Mathematical Sciences*. Springer-Verlag, Berlin 2006.

[2] J.-Y. Girard. Geometry of interaction. I. Interpretation of system **F**. In *Logic Colloquium '88 (Padova, 1988)*, volume 127 of *Stud. Logic Found. Math.*, North-Holland, Amsterdam 1989, pages 221–260.

[3] J.-Y. Girard. Geometry of interaction. II. Deadlock-free algorithms. In *COLOG-88 (Tallinn, 1988)*, volume 417 of *Lecture Notes in Comput. Sci.*, Springer, Berlin 1990, pages 76–93.

[4] J.-Y. Girard. Geometry of interaction. III. Accommodating the additives. In *Advances in linear logic (Ithaca, NY, 1993)*, Cambridge Univ. Press, Cambridge 1995, pages 329–389.

[5] J.-Y. Girard. Geometry of interaction. IV. The feedback equation. In *Logic Colloquium '03*, volume 24 of *Lect. Notes Log.*, Assoc. Symbol. Logic, La Jolla, CA 2006, pages 76–117.

[6] J.-Y. Girard. Geometry of interaction V : logic in the hyperfinite factor. available at http://iml.univ-mrs.fr/~girard/GdI5.pdf, 2008.

[7] L. Kalmár. Ein einfaches Beispiel für ein unentscheidbares arithmetisches Problem. *Mat. Fiz. Lapok*, 50: 1–23, 1943.

[8] W. Lück. L^2-invariants: theory and applications to geometry and K-theory, volume 44. Springer, Berlin 2002.

[9] F.J. Murray and J. von Neumann. On rings of operators. *Ann. of Math. (2)*, 37 (1): 116–229, 1936.
[10] F.J. Murray and J. von Neumann. On rings of operators. II. *Trans. Amer. Math. Soc.*, 41 (2): 208–248, 1937.
[11] F.J. Murray and J. von Neumann. On rings of operators. IV. *Ann. of Math. (2)*, 44: 716–808, 1943.
[12] H. Nishimura and M. Ozawa. Computational complexity of uniform quantum circuit families and quantum Turing machines. *Theoret. Comput. Sci.*, 276 (1-2): 147–181, 2002.
[13] M. Ozawa and H. Nishimura. Local transition functions of quantum Turing machines. *Theor. Inform. Appl.*, 34 (5): 379–402, 2000.
[14] R.T. Powers. An index theory for semigroups of *-endomorphisms of $\mathcal{B}(\mathcal{H})$ and type II_1 factors. *Canad. J. Math.*, 40 (1): 86–114, 1988.
[15] M. Pedicini and M. Piazza. Computation and von Neumann algebras. Technical Report 138, CNR — Istituto per le Applicazioni del Calcolo Mauro Picone, January 2008.
[16] J. von Neumann. On rings of operators. Reduction theory. *Ann. of Math. (2)*, 50: 401–485, 1949.

Phenomenology and intuitionism: the pros and cons of a research program

MIRIAM FRANCHELLA

1 Introduction

In the last years mathematical intuitionism has been supplied with phenomenological methods. Richard Tieszen [16, 17, 18] was the initiator of this trend and Mark van Atten [19, 1] followed this path. So, it comes natural to ask how far they went on, how far it is possible to go on and which changes (or specifications), if any, are required from intuitionism to be supported by phenomenology, at least for what concerns its general structure. First of all, it is necessary to understand why there was the necessity of a phenomenological support for intuitionism.

2 An epistemology for intuitionism

Van Atten's *Brouwer meets Husserl* [2] is the best place where it is explained the ground for intuitionism to need a support from phenomenology, i.e. the fact that it lacks an explicit epistemology. We have to recall that Brouwer provided a mystical attitude for proposing his intuitionism: the man can obtain his happiness only by keeping himself closed inside the inner Self. Any attempt towards the exterior world is a source of pain: in particular the language and the sciences. Mathematics, in order to be morally acceptable (or, better morally harmless) should avoid both the use of language (and, hence, formalization) and applicative aims, and has to be produced inside the inner Self. The "material" that we have at our disposal for starting is the temporal intuition, that grants us the possibility of forming the two-ity (and, hence, all natural numbers), sets ("species", in intuitionistic terminology) and choice sequences (indefinitely proceeding sequences, including lawless sequences). As it was well established by W.P. van Stigt in his 1992 book *Brouwerian Intuitionism*, Brouwer could not explain in detail this viewpoint inside his 1907 Dissertation [4, pp. 11–111] because his supervisor "suggested" him to leave the issue out, in order to avoid criticisms by the mathematicians who had to judge the work. Now we have the rejected parts translated and published (in van Stigt's book [21, pp. 405–415]), and we can also read the texts of the lectures that Brouwer gave in 1928 "Mathematik, Wissenschaft und Sprache" [4, pp. 417–428] and in 1948 "Consciousness, Philosophy and Mathematics" [4, pp. 480–494], where he unfolded better his reflections about the inner Self, by explaining the phases consciousness has to pass through from its deepest home to the exterior world. He begins with consciousness that oscillates slowly between stillness and sensation; the status of sensation allows the move of time, by which consciousness retains a former sensation as a past sensation and, through the distinction past-present, becomes mind. As mind it takes the function of a subject experiencing past and present as objects (in this way the subject recedes from the object). Mind experiences causal attention that identifies iterative complexes of sensations, whose elements are not permutable in time, and such that if the first of them occurs, the second one is expected to occur too: they are "causal sequences". Iterative complexes of sensations, whose elements are permutable in time, that are extranged from the subject are called things. Among them we find the individuals, i.e. the human bodies, which are indissolubly connected with the whole of their egoic sensation, called soul. Causal attention allows cunning acts of the subject, by which he brings about the first element of a causal sequence in order to obtain the second. Groups of individuals can cooperate towards a common aim. Scientific thinking is an economical way to catalogue extensive groups of cooperative causal sequences. It often makes use of mathematics, because causal sequences can be manipulated more easily by extending their substratum "of-quality-divested" to a more comprehensive and surveyable mathematical system. At the end of this description, Brouwer stresses that beauty and wisdom cannot be reached this way: he recommends to refrain from exerting power both over the nature and over the fellow-creature, by specifying that "Eastern devotion has perhaps better expressed this wisdom than any western man could have done" [4, p. 486]. Hence, this description represented a specification of Brouwer's general mystical attitude. There is no further justification supporting such perspective. It is designed a mystical path towards wisdom and, if you become convinced of it or if you simply are willing to try, you can follow it. In particular, if you are interested in mathematics, you know (as this fact comes out as a consequence from such mystical premises) that you have to practice mathematics by developing it in your inner Self. Such mystical perspective was seen as a possible obstacle to the diffusion/acceptance of intuitionism, therefore Brouwer's pupil, Arend Heyting, tried[1] a different "marketing strategy", by stressing that intuitionism required no specific philosophy, but it was only a way for establishing how far mathematics could be developed by using only human means:

> Brouwer's program entails that we study mathematics as something simple, more immediate than metaphysics. In the study of mental mathematical constructions "to exist" must be synonymous with "to be constructed" [7, p. 2] ... In fact all mathematicians and even intuitionists are convinced that in some sense mathematics bear upon eternal truths, but when trying to define precisely this sense, one gets entangled in a maze of metaphysical difficulties. The only way to avoid them is to banish them from mathematics [7, p. 3] ... In order to construct mathematical theories no philosophical preliminaries are needed, but the value we attribute to this activity will depend upon our philosophical ideas [7, p. 9].

---

[1] We have to recall here, by considering the relationship intuitionism-phenomenology, that Heyting, during the first period of his intuitionistic production, when he was defining the logical constants, used a terminology borrowed from phenomenology. Still, he later abandoned it, probably because he considered any reference to philosophy as something that could only be an obstacle to peaceful collaboration. About Heyting and phenomenology see my [5].

Van Atten [2, p. 83] feels the need for a stronger philosophy (than the Brouwerian one) as a ground basis for intuitionism and explains that the weakness of Brouwerian philosophy lies in the uncapability of consciousness (as described by Brouwer) to be self-critical. I would like to add that there is the need for an alternative perspective (from which accepting intuitionism) as mysticism may be not so appealing. The difficulty is to find a philosophy that could be shared by many people.

3 A brief survey on phenomenology

Phenomenology seems to offer a key for solving, in a natural way, the above considered problems. As it is accurately explained in the Sixth of Husserl's *Logische Untersuchungen* [9, §§11–12], the I intentions the object. Categorial intuition is the source of what is called an object, and is based on given intuitions. The "object" is never given in its entirety but it is seen as the ideal end of a series of approximations, that are explained in terms of intentions and their fulfilment. We can consider at first place medium-sized objects of daily experience. They are only given in a perspectival manner: there can be indefinitely many percepts of the same objects, all differing in content. Some parts of the object are given and some are not, so this suggests the limiting case of an adequate perception in which the object is not given imperfectly. That is why the relation of fulfilment admits degrees in which epistemic value steadily increases. In case of fulfilment, a synthesis of identity takes place. Of course, it is also possible a disappointment of an intention: a "frustration", that however presupposes a partial fulfilment. Also the frustration is a synthesis, a synthesis of distinction. The same activity of knowledge allows us to get to ideal objects:

> The evidence of irreal objects, objects that are ideal in the broadest sense, is, in its effect, quite analogous to the evidence of ordinary so-called internal and external experience, which alone - on no other grounds than prejudice - is commonly thought capable of effecting an original Objectivation [8, p. 155].

As in their case we do not refer to perceptual stuff but to the data of categorical intuition, Husserl later realized that, in order to consider the possible different perspectives, we have to use a specific method: the free variation in imagination. What persists through this is some invariant, the essence common to all variants, the eidos:

> In this inquiry, the variation of the necessary initial example is the performance in which the "eidos" should emerge and by means of which the evidence of indissoluble eidetic correlation between constitution and constituted should also emerge. If it is to have these effects, it must be understood, not as an empirical variation, but as a variation carried on with the freedom of pure fantasy and with the consciousness of its purely optional character - the consciousness of the "pure" Any Whatever. ... But, precisely with this coinciding,what necessarily persists throughout this free and always-repeatable variation comes to the fore: the invariant, the indissolubly identical in the different and ever-again different, the essence common to all, the universal essence by which all "imaginable" variants of any such variant, are restricted [8, pp. 247–248].

Such variations are intentions that can be fulfilled or not, as in the case of medium-sized objects. The evidence is given in the (ideal) case of an adequate intuition of the object, and it should be distinguished from the feeling that can accompany it.

Absolute evidence is seen as a regulative idea for any object-constitution. As Husserl specified, the evidence that one has is the truth about that object, but it is a revisable truth [8, p. 164]:

> The possibility of deception is inherent in the evidence of experience and does not annul either its fundamental character or its effect; though becoming evidentially aware of [actual] deception "annuls" the deceptive experience or evidence itself. The evidence of a new experience is what makes the previously uncontested experience undergo the modification of believing called "annulment" or "cancellation"; and it alone can do so. [...] The conscious "dispelling" of a deception, with the originality of "now I see that it is an illusion", is itself a species of evidence, namely evidence of the nullity of something experienced [...] This too holds for every evidence, for every "experience" in the amplified sense. Even an ostensibly apodictic evidence can become disclosed as deception and, in that event, presupposes a similar evidence by which it is "shattered".

Absolute evidence is supposed only for the transcendental Ego. If the epoché is carried out, i.e. if we free ourselves from all the daily prejudices referring to the world of sciences and to psychical acts, then what remains is the transcendental Ego (the only original and apodictical evidence):

> By phenomenological epoché I reduce my natural human Ego and my psychic life — the realm of my psychological experience — to my transcendental-phenomenological Ego, the realm of transcendental-phenomenological self-experience [10, p. 65].

The transcendental Ego is not a piece of the objective world, its only crumb to be saved (as it was in Descartes' thought); on the contrary, it requires letting aside the objective world. Transcendental Ego is grasped as "intentioning", according constitutive types of thinkable objects.

It should be stressed that intersubjectivity is for Husserl necessary condition for objectivity, hence evidence is to be intended with respect to the transcendental Ego as "common Ego", shared monad. No discussion about the existence of objects is hypothesised by Husserl, because they are constituted by the transcendental Ego, i.e. by the human being deprived of its individuality, so they are constituted by what in each man is common to the others. The degrees of existence should be intended with reference to this common Ego, and not to what it appears to the single monad in its individuality. Husserl's truth, even if it is something revisable, is however the top which man can reach.

4 Positive aspects

Phenomenology can offer intuitionism a strong epistemology. Furthermore, it has surely two immediate positive effects: it explains the uniqueness of mathematics and allows to avoid intuitionism the charge of psycologism. As for the uniqueness of mathematics, we recall here that L.E.J. Brouwer had proposed his alternative foundations of mathematics for ethical reasons, that is to be consistent with his mysticism [4, p. 2]:

> Finally, you do know that very meaningful phrase "turn into yourself". This "turning into yourself" is accompanied by a feeling of effort ... If, however, you succeed in overcoming all inertia passions will be silenced, you will feel dead to the old world of perception, ... Your eyes, no longer blindfolded, will open to a joyful quiescence.

Consequently, both attempts to dominate the nature and the other men had to be condemned. Mathematics could be "saved", provided that it would not be developed for applicative purposes, by conceiving it as an inner experience of languageless self-unfolding of the primordial temporal intuition. Left aside the emotional aspect of mathematical experience (that seems to be linked to the individual subject), we can suppose that Brouwer had in mind a unique mathematics. Namely, in his 1905 "Leven, Kunst en Mystiek", the pamphlet where he openly expressed his mystical viewpoint and where he stressed his solipsism, he stated [4, p. 6]:

> Even in the most restricted sciences, logic and mathematics, no two different people will have the same conception of the fundamental notions of which these two sciences are constructed; and yet, they have a common will, and in both there is a small, unimportant part of the brain that forces the attention in a similar way.

So, even in such a paper, where the solipsistic component prevailed, he admitted that there was a "similarity" in the mathematical concepts from one person to another. In his other papers, he did not even stress the concept variation in different people. He only underlined the fact that mathematics was a languageless activity, that its exactness lay in the intellect (and not on paper). Furthermore, Brouwer, every time, when presenting the "first act of intuitionism" – consisting of unfolding the structure of the two-ity — did not say that it was only his own experience, a page of his diary that he (sinfully) wanted to publish. He spoke of the man in general, or, better, of the man as a man. Consequently, the problem arose immediately: how is it possible to talk about one mathematics if it consists of an interior experience of each single man? The transcendental Ego can be adequate to grant the uniqueness of mathematics. Namely, the notion of transcendental Ego allows to realize what is typical of a subject activity. Hence, as mathematics for intuitionists is the exploiting of human mental faculties, if it is ascribed to the activity of the transcendental Ego, its content is granted as something common to all de facto subjects.

As for the charge of psychologism, we recall that it came to intuitionism after Brouwer's introduction, in his *Cambridge Lectures*, of the term "creative subject" on the purpose of producing what will be later called "indefinitely proceeding sequences" (also called "free choice sequences" in the literature), i.e. (even lawless) sequences of mathematical objects. It was van Dantzig [20] that asked whether the creative subject is the author himself, an arbitrary human individual, a human individual possessing some (which?) qualification, an "infinite" sequence of such individuals, successively performing the activities, ascribed to the creating subject, or, finally, a more or less definite group of human individuals, for example, all mathematicians possessing some definite qualification. I.e., he put the question of the meaning of creating subject and considered as possible definitions only those in terms of psychological subjects. Furthermore, he stressed the weakness of creating subject, when used to construct real numbers, in the fact that a loss of unanimity (in the case of a group) or death (in the case of a single individual) could interrupt the constructing process. In this way the problem of Brouwerian psychologism entered the history, but Brouwer himself refused such a charge.

Namely, in a letter to van Dantzig (quoted by [1, pp. 75–76]), he wrote:

> I am glad to see that these developments make the essentially negative properties meaningful also to those who do not recognize the intuitionistic "creating subject", because with respect to mathematics they hold the psychologistic point of view, or in any case insist on a "plurality of mind".

That is, Brouwer attached van Dantzig a psychologist viewpoint, by considering it something that he could not share. The charge of psychologism can be solved by referring the production of mathematics to the transcendental Ego: it does not have the limits of the flesh-and-blood subject (for instance, it does not die), and, in general it avoids the charges of psychologism as it was thought of with such an aim.

5 A first problem

Phenomenology would seem to solve the main epistemological problem of intuitionism and also some collateral questions. Still there are some perplexities to consider. The principal one is due to the fact that Husserl always accepted classical mathematics and did not seem inclined to question it. Van Atten [2] stresses that Husserl's unwilling to sacrifice any part of classical mathematics is a fact but runs counter to his own general views on ontology. In order to show this, he proceeds along two steps. The first step is to see that Husserlian phenomenology allows (better requires) a form of revisionism in mathematics, i.e. the right to sanction or modify pure mathematical practice. Van Atten distinguishes [2, p. 53] between weak and strong revisionism: the former potentially sanctions a subset of mathematical practice, the latter potentially extends it. He argues [2, p. 55] that a weak revisionism comes out of Husserl's characterisation of the task of philosophy with regard to the sciences because the task is transforming them from techniques to insightful knowledge, by clarification of their concepts. Namely, the possibility of rejecting supposed objects is made manifest by Husserl in his introduction of *Formale und transzendentale Logik* as a consequence of such clarification. The second step by van Atten consists of providing an argument in favour of the statement that Husserl's weak revisionism implies a strong revisionism [2, p. 59]. The main core of such argument is that, for formal objects, transcendental possibility (= being conceptually possible and constituted with full evidence) is equivalent to existence. This fact allows to go beyond the weak-revisionist conclusion. This last would be the following: inside a class of objects that figure in actual practice of mathematics, some of them will be admissible, others may be not. The strong-revisionist conclusion will add: there can be admissible mathematical objects, that have not yet been entered in mathematical practice, whose existence is constituted by philosophical considerations.

6 A second problem

Still, even if we admit this defence of the possibility of revisionism in Husserl theorization, a further difficulty comes out of the fact that Brouwer's and Husserl's definitions of logic are completely different. In fact, in his *Formale und transzendentale Logik*, Husserl, at the end of a long and tortuous path of reflections and reassessments about the question, described transcendental logic as the a priori theory of science, that is as the inquiry about what is proper to the sciences, i.e. it fixes the general conditions for a science to be possible. Hence, in the first place it is analytic apophantic (predicative), i.e. morphology of judgements and logic of consequences, and, as such, it focuses the categorical objectualities in general [8, pp. 134]:

> What is judged in a judging is the judged — the judgingly meant or supposed — categorial objectivity. ... not until there is a judging on a second level does the proposition in logic's sense of the world — the proposition as a sense, the supposed categorial objectivity as supposed — become the object.

Formal logic is the objective aspect of logic, while transcendental logic is the "subjective" aspect of logic, i.e. where the focus is on a "theory of knowledge" around logic itself. The main aim of transcendental logic is pointing out the idealising presuppositions of formal logic, i.e. its surreptitious assumptions, and evaluating their ground, i.e. establishing whether there is some evidence in their favour. This inquiry takes place along the tripartition of logic into: pre-analytic (concerning the pure possibility of the judgements), consequence-logic (concerning the non-contradictoriness of the judgements) and truth-logic (concerning the truth of the judgements).

On his side, Brouwer considered logic as a description of the (linguistic) regularities present in mathematics:

> Man, inclined to take a mathematical view of everything, has also applied this bias to mathematical language, and in former centuries exclusively to the language of logical reasonings [=reasonings on relations of whole and part] : the science arising from this activity is theoretical logic. It is only in the last twenty years that people have started looking in the same way at mathematical language in general: this is the content of logistic, insofar it is studied without overrating its value. ... [In the classical syllogism] we have here one of the very symplest forms of mathematical reasoning ... However, looking at the *words* that accompany this primitive form of mathematics, we notice in them a surprising mechanism with a regularity which is not clear a priori. That is to say, it is possible to project on these words a new simple mathematical system; speaking about this system we explain the *theory of the syllogism* [4, pp. 74–75].

On the other side, for Husserl mathematics either is directly apophantic (it is the mathematics that treats propositional forms by computing with them like with numbers) or (this is the case of set-theory and cardinal numbers theory) it has as an object the "something in general", i.e. the object in general and, for this reason, it is defined "formal ontology" ("formal" because it leaves aside any concrete determination of objects). Hence, formal apophantic and formal ontology identify with each other through their referring to general objectualities, and so they fulfil the mathesis universalis designed by Leibniz. They do not ask about "truth", that would require to pass to a level of reflexion subjective-intersubjective-transcendental: the level of transcendental logic, asking about the evidence of logical principle. It is a viewpoint very different from the intuitionist: mathematics is seen as a formal discipline, that does not ask about truth, by limiting itself to non-contradictoriness and to the relationship "be consequence of". Mathematics remains out of the domain of evidence, that is the domain of truth. Apophantic-logic, objective, coincides with mathematics (without either of them absorbs the other), but transcendental logic lies at a higher level. Apart of the problem of the revolutionary vs. conservatory character of Husserlian phenomenology, it remains the fact that the relationship logic/mathematics is intrinsically different from the intuitionistic one.

7 A last doubt

A last analysis is deserved by the question of the law of excluded middle. Its Brouwerian reinterpretation and consequent criticism are fixed points of intuitionism. Many have stressed [13, 15] how Husserl was "conservative" in logic, and this fact can make difficult grounding intuitionism on phenomenology. An attempt of solving the question has been launched by Dieter Lohmar [11], as he pointed out some quotes from Husserl that would let suppose that the author wondered whether the law of excluded middle held and that he did not find reasons enough to support the validity of the law. As we have just seen above, the main aim of transcendental logic is pointing out the idealising presuppositions of analytic logic. The law of excluded middle is, according to Husserl, an idealization hidden in truth-logic. Namely, while the law of contradiction $\neg(A \wedge \neg A)$ if interpreted from a subjective viewpoint, i.e. by asking what evidence the subject has about it, does not give any problem (it only states that if a judgement can be brought to an adequation in a positive material evidence, then, a priori, its contradictory opposite not only is excluded as a judgement but also can not be brought to such an adequation; and vice versa), the law of excluded middle states that every judgement necessarily admits of being brought to an adequation. This is the crucial point [8, p. 201]:

> "Necessarily" being understood with an ideality for which, indeed, no responsible evidence has ever been sought. We all know very well how few judgements anyone can in fact legitimate intuitively, even with the best efforts; and yet it is supposed to be a matter of apriori insight that there can be no-evident judgements that do not "in themselves" admit of being made evident in either a positive or a negative evidence.

Lohmar stresses [11, p. 15] that Husserl, after posing this problem, did not supply us with a solution, that Husserl did not give justifications of such evidence. Moreover, according to Lohmar, accepting the validity of the law of excluded middle would have required, from a phenomenological perspective, the acceptation of the judgeability of any judgement, i.e. stating that we always possess evidences enough to establish whether an assertion is true or whether it is false. But, Lohmar remarks [11, p. 16], ethical questions testify against this acceptability, as they require a suspension of our judgement . Lohmar in any case clarifies [11, p. 16] that it is possible to find in Husserl some place for the law of excluded middle as a postulate introduced voluntary inside formal systems: it is not a law of thought but a possible postulate, among many others.

According to me, some specifications are required. If one looks through *Formale und transcendentale Logik*, Husserl explains that the laws of contradiction and of excluded middle are intended referred only to assertions that have a sense: therefore, assertions like "the sum of the angles of a triangle is red" are not a domain of application of logical laws. Only for senseful judgements [8, pp. 228-229] it is given a priori, by virtue of their genesis, that they relate to a unitary experiential basis. Precisely because of this, it is true of every such judgement, in relation to such a basis, either that it can be brought to an adequation and, with the carrying out of the adequation, either the judgement explicates and apprehends categorially what is given in armonious experience, or else that it leads to the negative of adequation ... But, for the broader realm of judgements, to which belong also the judgements that are senseless in respect of content, the disjunction no longer holds good. The "middle" is not excluded here; and consists in the fact that judgements with predicates having no senseful relation to the subject are, so to speak, exalted above truth and falsity in their senselessness.

Let us pay attention to the fact that Husserl recovers the validity of the law of excluded middle: by limiting it to assertions equipped with sense, he accepts such validity. And the ground is alleged in his expression: "it is given a priori, by virtue of their genesis, that they relate to a unitary experiential basis". On the contrary, by using a Husserlian terminology, we can say that for Brouwer the law of excluded middle requires/states an evidence of the fact that the subject has brought the assertion to an adequation and, hence, that he/she has grasped such experience as agreeing resp. disagreeing with the assertion. For Brouwer, it is not enough what Husserl considers enough, i.e. the a priori relationship of the assertions with experience. There must occur also an effective checking against the data of mental experience. The reason lies in the different concept of logic and mathematics that the two authors had. As we saw above, Brouwer considered logic as a description of the (linguistic) regularities present in mathematics. From this he derived the fact that, in order to accept a logical law, i.e. the affirmation of a regularity present inside mathematics, one should verify that what was expressed by the law really takes always place in mathematics. In the specific case of the excluded middle, in order to affirm it for any assertion, either we should have evidence enough to say that the assertion is true, or we should have evidence enough to say that its truth would lead to contradiction [4, p. 106]:

It claims that every supposition is either true or false; in mathematics this means that for every supposed imbedding of a system into another, satisfying certain given conditions, we can either accomplish such an imbedding by a construction, or we can arrive by a construction at the arrestment of the process.

Indeed, if each linguistic application of the principle of excluded third in a mathematical argument were to accompany some actual intuitionist mathematical construction, this would mean that each intuitionist mathematical assertion (i.e. each assignment of a property to an intuitionist mathematical entity) can be judged, i.e. can either be proved or be reduced to absurdity.

So, also in intuitionism the law of excluded middle means an evidence of the "positive or negative adequation" , but there is a passage from logic to mathematics as this latter is the world of mental experience, and the principle of excluded middle is not accepted.

8 A further problem: the notion of choice sequences

Brouwer arrived after many years and many reflections to the notion of indefinitely proceeding sequences, often called "choice sequences" as they include also the case of lawless sequences of mathematical objects. Brouwer realized that choice sequences, in their being unfinished but identifiable, were intuitionistically acceptable as a way of describing the points of the continuum. By reading Husserl's texts (after 1917), it seems that the omnitemporality of mathematical objects be a not negotiable condition [2, pp. 70–72], while choice sequences are surely temporal. So, it seems difficult to conciliate the intuitionistic acceptance of choice sequence with taking phenomenology as a support for intuitionism.

On the purpose of a phenomenological acceptance of choice sequences, we should also mention here Hermann Weyl, because he had contacts with both Brouwer and Husserl and developed a theory of the continuum that partially shared Brouwer's position. Namely, Weyl considered only lawlike sequences as mathematical objects and recognized the only status of concepts to lawless, due to the impossibility of representing them through a numerical code. Van Atten, van Dalen and Tieszen considered the question in their "Brouwer and Weyl: The Phenomenology and Mathematics of the Intuitive Continuum" [12], where they stressed the fact that Weyl was stimulated in this analysis by reading Husserl and that Weyl even informed Husserl about his own results. Still, after very articulated reflections, they [12, p. 221] concluded that

> Weyl does not conduct his investigation according to Husserl's methodology, for although there is in Weyl's work a phenomenological analysis of the continuum, there is no corresponding analysis of choice sequences as objects.

We can now come back to the question of the acceptability of choice sequence with respect to Husserl's viewpoint.

Van Atten [2] tried to solve the problem by two steps. Firstly, he showed that from a phenomenological viewpoint, choice sequences are objects [2, pp. 89–95]; secondly, he showed (always from a phenomenological viewpoint) that they are mathematical [2, pp. 95–101]. In order to show that they are objects, he stressed the analogy existing between melodies and choice sequences, as Husserl frequently referred to music in order to explain the constitution of objects. Melodies and choice sequences share the fact that they are experienced as an identity even though they have not yet been completed, they are distributed objects (i.e. they occupy an interval in time) and are constituted in a process of successive synthesis along the following steps: 1) keeping in retention the process till now; 2) re-presentation of the process (the process has to be thematised as extendable now); 3) choice of the next element; 4) retention of the sequence (see step 1), provided the apprehension of the identity of the process through its categorical form of ongoing process. In order to show how choice sequences can be recognised as mathematical objects, van Atten recalled [2, pp. 69–71], that Husserl passed from an initial conviction that mathematical objects are atemporal to the conviction that they are omnitemporal, because, by developing his genetic phenomenology, he realised that the constitution of any genuine object must be constitution in time and that the temporal flow is a condition for identity. But these reasons conclude only that mathematical objects must be temporal, not omnitemporal. Van Atten suggested [2, p. 97] that the original reason for Husserl to put omnitemporality as a necessary condition for mathematical objects was their monotonicity, i.e. the fact that what is proven for them once, is proven forever. But this is assured also by infinite temporality in the direction of future, and choice sequences allow it.

Temporality would seem to present a further problem with respect to the status of mathematical objects as, inside Husserl's perspective, mathematical objects should be formal, so they cannot depend on time, for the presence of time-sensations rules out complete formality. Still, van Atten stressed [2, p. 98], this viewpoint would belong only to the first Husserl.

Finally, a last objection to the acceptability of choice sequences as mathematical objects would be their subject-dependence: in fact, they are chosen from a subject, but 1) what the subject knows is shareable at any moment (hence, such dependence does not yield unshareable truths), 2) the freedom of the subject does not mean arbitrariness. Namely its choices will have to satisfy certain constraints, depending on their applications. For instance, in the case of analysis, there is the general constraint of determining nested intervals, i.e. a constraint motivated by the nature of the intuitive continuum.

A doubt remains to me. When we have to establish whether a notion is phenomenologically admissible, we have to wonder whether the result of its eidetic variations is evident to the transcendental Ego, who is the result of the eidetic variations on all the possible subjects. Now, choice sequences may be constructed (as Brouwer showed us) in relation to what the subject knows at a certain moment. Typical of Brouwer were, namely, examples where choice sequences were built as follows: Let α be a mathematical statement such that the "subject" does not know either if it is true or that it is absurd and $a_1, a_2, \ldots$ be a choice sequence whose terms are chosen as follows: if the knowledge of the subject about the statement remains the same, $a_n = 2^{-n}$; if between the step $s-1$ and the step s, the subject comes to know that the statement is true, $a_s = a_{s+n} = 2^{-s}$, if between the step $r-1$ and the step r, the subject comes to know that the statement is absurd, $a_r = a_{r+n} = -2^{-r}$.

We can specify that the "knowledge of the subject" is to be considered a common knowledge, i.e. the maximum that mankind knows about that issue. Still, either we consider the common knowledge as the future maximum of knowledge of mankind or we have to consider the actual maximum of knowledge of mankind. The first one is hard to be individuated (how many mathematical problems exactly could ever be solved by mankind?), while the second one introduces inside the transcendental Ego a temporal aspect, which is debatable. In order to perform such choice sequences, we have to admit that the knowledge of the subject, i.e., of the transcendental Ego, is in progress and we have also the problem to establish what is really the maximum of knowledge that the transcendental Ego has at any moment. Is this possible? I consider this an open question.

BIBLIOGRAPHY

[1] M. Van Atten. *On Brouwer*. Wadsworth, Southbank 2004.
[2] M. Van Atten. *Brouwer Meets Husserl. On the Phenomenology of Choice Sequences.* Springer, Dordrecht 2007.
[3] M. Van Atten, P. Boldini, M. Bourdeau and G. Heinzmann (eds.) *One Hundred Years of Intuitionism (1907–2007). The Cerisy Conference.* Birkhaeuser, Basel 2008.
[4] L.E.J. Brouwer. *Collected Works vol. I.* North-Holland, Amsterdam 1975.

[5] M. Franchella. Arend Heyting and Phenomenology: is the meeting feasible? *Bulletin d'analyse phenomenologique*, 3: 1–21, 2007.
[6] M. Franchella. Book review of M. Van Atten "Brouwer meets Husserl: On the phenomenology of choice sequences". *Philosophia Mathematica*, 16 (2): 276–281, 2008.
[7] A. Heyting. *Intuitionism: an Introduction*. North-Holland, Amsterdam 1956.
[8] E. Husserl. *Formal and Transcendental Logic*. Nijhoff, Den Haag 1969.
[9] E. Husserl. *Logical Investigations*. Humanities Press, New York 1970.
[10] E. Husserl. *Cartesian Meditations*. Kluwer, Dordrecht 1993.
[11] D. Lohmar. The transition of the principle of excluded middle from a principle of logic to an axiom: Husserl's hesitant revisionism in logic. *The new Yearbook for phenomenology and phenomenological philosophy*, 4: 1–16, 2002.
[12] R. Tieszen, M. Van Atten and D. Van Dalen. Brouwer and Weyl: the phenomenology and mathematics of the intuitive continuum. *Philosophia Mathematica*, 10 (2): 203–226, 2002.
[13] D.J. Mohanty. Review of R.S. Tragesser "Husserl and realism in logic and mathematics". *History and Philosophy of Logic*, 6: 230–234, 1985.
[14] T. Placek. *Mathematical Intuitionism and Intersubjectivity. A Critical Exposition of Arguments for Intuitionism*. Kluwer, Amsterdam 1999.
[15] R. Schmit. *Husserls Philosophie der Mathematik*. Bouvier, Bonn 1981.
[16] R. Tieszen. Mathematical intuition and Husserl's phenomenology. *Nous*, 18: 395–421, 1984.
[17] R. Tieszen. Phenomenology and mathematical knowledge. *Synthese*, 75: 373–403, 1988.
[18] R. Tieszen. Mathematics. In B. Smith and D. Smith (eds.), *Cambridge Companion to Philosophy: Husserl*, Cambridge University Press, Cambridge 1995, pages 438–462.
[19] M. van Atten. Brouwer as never read by Husserl. *Synthese*, 137: 3–19, 2003.
[20] D. van Dantzig. Comments on Brouwer's theorem on essentially negative properties. *Indagationes Mathematicae*, 11: 347–355, 1949.
[21] W.P. van Stigt. *Brouwer's Intuitionism*. North Holland, Amsterdam 1990.

A note on the circularity of set-theoretic semantics for set theory

Luca Bellotti

1. The fact that the ordinary semantics of formal languages, specifically of set theory, is itself essentially set-theoretic (though of course the relationship between theory and metatheory must respect Tarski's limitations, or at least some kind of hierarchical stratification), gives rise to the problem of singling out a semantics for set theory which is not, in a sense, *circular*. As we shall see, according to many authors, the only way out seems to be the admission that at the level of the last foundation of set theory one cannot do without a semantics which is, to a certain extent, *intuitive*. But what is an intuitive semantics for set theory? My main aim in this note is to show the apparently overwhelming philosophical difficulties arising when one tries to delimit this notion in a rigorous way (without renouncing either ordinary set theory or usual model theory). I shall argue that these difficulties stem from the peculiar nature (or even the irrelevance) that some problems, which naturally arise with other languages, take on in the case of mathematical languages.

2. First I shall hint at the iterative conception of sets as a possible intuitive semantics for set theory; this will lead us to the crucial problem of quantification over the universe of sets; then I shall consider a particular kind of mathematical realism, first expounded by Georg Kreisel ([14, 15]), as the possible source of a solution to our problem. I shall conclude that the prospected form of realism is *not* a way out, but that, fortunately, even the *ordinary* semantics of set theory does not necessarily bind us to this (metaphysical) realist philosophy of mathematics (with all its epistemological difficulties). As a consequence, giving up this view would have no serious impact on set-theoretic practice. It is important to point out that our problem has three different layers. The first one is purely mathematical, regarding the usual set-theoretic formulation of the model theory of set theory, and here the "circularity" is mathematically very interesting, but philosophically substantially unproblematic. The second one is the search for an ultimate, informal, intuitive semantics for set-theoretic discourse, and this touches on important foundational matters (iterative conception, quantification over the universe). The third one is the proper philosophical level, in which one tries to interpret and to justify (ontologically and epistemologically) the choice of the intuitive semantics of the previous level (this is usually done in terms of a realist view). Both "circularity" and "semantics" have slightly different meanings in each context, and one should never forget this, to properly evaluate the arguments below. Because of the space limitations of a short note like this, I shall only be able to give a quick sketch of the main arguments, and to merely state the main points of my view here (a thorough account can be found in [1]).

3. In order to avoid possible misunderstandings, I underscore at the outset that the object of this note is the usual *set-theoretic* semantics of ordinary Zermelo-Fraenkel set theory with the Axiom of Choice (ZFC), together with some consequences of its adoption. I take no position here in favour or against other semantical approaches to set theory: e.g., just to mention a few, the category-theoretic approach of Lawvere-Rosebrugh ([17]), the "operational" one of Feferman ([9]), the one based on "constructibility" of Chihara ([7]); or, finally, views based on absolutely unrestricted quantification, along the lines of McGee ([20]), or on Boolos' ([5]) plural quantification. I do not think these approaches are irrelevant to the problem of giving a non-circular semantics to set theory (in fact, they are extremely relevant), but I am interested here in the usual approach adopted in the ordinary model-theoretic metamathematics of set theory, as done by most mathematicians working on (and inside) the various models of the theory. As an excuse for my debatable choice, all the literature I deal with below is concerned with the same problem, arising when one accepts the *ordinary* (model-theoretic) semantics of *usual* (here, ZFC) set theory. Supporters of alternative theories can well take some of the arguments below as further reasons to abandon ordinary semantics (or even ordinary set theory), although this is not my purpose. I suspect (but this is only a conjecture) that some of the problems arising in this context are so fundamental for semantics in general, that they could resurface (though certainly with different features) in the other approaches.

4. Consider the following quotation, which I think neatly expresses a widely shared view among mathematicians and (some) philosophers of mathematics:

> There is a metaphysical element peculiar to semantics and, in particular, to the semantic conception of truth of mathematical statements, a metaphysical element represented by the postulation of a reality, whatever this might turn out to be, that mathematical statements are true of. ([8, p. 22].)

Well, this is perhaps one of the best possible formulations (in a nutshell) of the *opposite* position with respect to my own (but see also [22]). But I shall not attempt to refute it directly; rather, the content of this note should hint at a very different view, in which the "metaphysical element represented by the postulation of a reality" naturally appears to be utterly *irrelevant* to the semantics of mathematical theories.

5. If one considers the most basic facts about the relationship between semantics and set theory three immediate reasons to look for an *intuitive* semantics for set theory come out:

1. the need for a generalized notion of realization;

2. the need to apply the condition of the existence of a structure satisfying the axioms to the hierarchy of sets itself;

3. the fact that there is no truth definition for the universe.

The answer to this need of an intuitive semantics for set theory is usually given by the *iterative conception* of the universe of sets, suggested by the work of

Ernst Zermelo in the late 1920s (see [35]). Perhaps the most important attempts to justify the axioms of ZF on the basis of the iterative conception (after [10]) were made by Shoenfield ([31, pp. 238–240], and [32, pp. 322–327]), Boolos ([4]) and Wang ([33, 34]). On the other hand, Hallett ([11, Chapter 6]) noticed the *circularity* of the justification of the axioms on the basis of the iterative conception, and the ambiguity of the notion of *completability* underlying the conception itself (this notion is crucial because, in order to justify the axioms on the basis that they produce completable collections, one must explain in what sense the collections obtained are completed). Thus the iterative conception, considered as an intuitive semantics for set theory, has (in spite of its value per se) a basic defect which could raise doubts about its suitability to solve the problem of circularity in which we are interested. The problem is with the very concept of completability: either it is intended in a quasi-constructive way, contradicting the nature of the axioms of ZF which it was designed to justify; or it is intended in a broad sense, presupposing them.

6. The core of the problem of semantics for set theory is the question of *quantification over all sets*. This is, I claim, the central issue. There are two apparently incompatible features in the common description of the cumulative hierarchy of types. On the one hand, unlimited quantifiers over ordinals are intended to range over all ordinals less than some given ordinal α, and unlimited quantifiers in general are intended to range over sets of type less than α. Otherwise, as Kreisel observed ([13, p. 101]), the meaning of quantified expressions would be well defined only under the assumption that there is a collection consisting of all collections, which is false for the intended structure. On the other hand, the intention underlying unlimited quantification is clearly to deal with the whole universe of sets, whatever the latter might turn out to be. Of course, these features are not in contradiction at all: rather, their interaction is the driving force of the development of set theory and (inextricably) of its model theory. But there is still the need for a philosophical clarification of this striking example of "dialectics" between potentiality and actuality. Let us try to go more deeply into it.

7. When we use *classical* logic with sentences involving quantification over the universe of sets we seem to presuppose that it is perfectly definite what sets there are. This could be used as a sort of reductio for the use of classical logic in this context: since the multiplicity of all sets is irreducibly potential, we should not apply classical quantification to it. Two ways remain open: either it is simply a mistake to use classical logic in this case (this is the way chosen, with qualifications, by Lear; I shall presently discuss his arguments); or classical set theory is insufficient to cover all the possible ways of collecting manifolds into unities. In the latter case, classical set theory cannot distinguish the universe from a (large) set, which is built up precisely by all and only those ways of collecting which are embodied in the theory itself. *Reflection principles* are the rigorous formulation of precisely this idea. Parsons' formulation of the underlying problem cannot be improved:

> We seem to be confronted with an ambiguity in the notion of the intended interpretation of formalized set theory: the theory seems to be about a definite "universe" that we are tempted to conceive as the analogy of a set, and the formulae of the theory

can be given a sense so that this universe is a set, but to take it in this way falsifies the intent, and at the very least involves a use of set-theoretic language that would not be amenable to an interpretation with the same set as universe [23, p. 91].

This is precisely the central problem: semantics seems to require essentially the *sethood* of the universe; as soon as we try to catch the domain of quantification, it slips out of our hands, and we are left with a set, not with the universe as we wanted.

> From this point of view, if I take your quantifiers to range over "all" sets, this may only show (from a "higher" perspective) my lack of a more comprehensive conception of set than yours. But then it seems that a perspective is always possible according to which your classes are really sets [24, p. 219].

This is a good description of what is going on, and I fully endorse it. Nevertheless, I am reluctant to speak, with Parsons [24, p. 219], of a systematical *ambiguity* of the language of set theory. I think that this apparent "ambiguity" is only the symptom of a rather deep phenomenon, which typically occurs in set theory: the fact that no totality, no multiplicity of a set-like nature can cover comprehensively the unfolding of the higher and higher interpretations of the axiomatic theory. The word "ambiguity" is out of place, I claim, if it is used in opposition to the "uniqueness" of the universe of sets: we should question the idea, realistically understood, of a set-theoretic world whose "essence" will be, unfortunately, forever ineffable simply because of the weakness and context-dependence of our mathematical language. On the contrary, it is the mathematical nature itself of that universe which brings about the phenomena we are discussing, and the pretension to eliminate them, though tempting, is simply pre-scientific.

8. The basis of Lear's proposal (see [18]) is the idea that the extension of "set" is always capable of development (reflection principles can be seen as the technical realization of this idea). Lear proposes to use Kripke models of modal logic to take into account the development of the extension of "set". The semantics one obtains is obviously non-classical, in the sense that if p is false at t, we cannot infer that $\neg p$ is true at t; it is true at t if and only if p is false at all indexes accessible from t. Similarly, universally quantified sentences are true at t if and only if, however the extension of "set" may expand, all their instances are true in the expanded universe. But, as Lear points out, if we adopt this non-classical semantics we have to face a dilemma: either certain sets at certain times are collections that are not "grasped" (i.e., membership in them is not fixed once and for all); or the powerset axiom is false. Here the weakness of Lear's position becomes evident: the powerset axiom is false if we accept the second alternative of the dilemma (the first is arguably unacceptable); but it is false *on the semantics adopted by Lear* (see [18]): this casts a shadow of reductio on the dilemma, in the sense that it could simply be a reductio of Lear's proposal to use a nonclassical semantics.

9. It is now time to discuss a form of mathematical *realism* which is at first sight the best candidate for a perhaps definitive (dis-)solution of the philosophical problem of semantics for set theory. Yet, I shall argue that it has so many drawbacks that we should adopt an alternative view. Let us take as a provisional starting point Maddy's tentative definition of mathematical realism ([19, p. 14]): realism is the philosophical view according to which mathematics is a science, whose objects are mathematical entities which exist objectively (let us accept that this extremely vague characterization make sense), and whose sentences are true or false according to the properties of those entities, properties they possess independently of our language, our concepts, our theories, our knowledge in general. Realism in this sense seems a drastic but effective solution to the problem of semantics for set theory, since it seemingly allows one to jump out of the circularity of a set-theoretic semantics for set theory, postulating a well-determined reality set-theoretic statements must be true of: *the set-theoretic universe*, realistically understood. This gives a straightforward answer to the question which troubled us above: "What does 'all sets' mean?". "All sets" simply means: *all sets*, namely, all the members of the (realist's) universe of sets. Of course, I do not exclude here that other (non-realist) general philosophical interpretations of mathematics are capable of giving equally good (or even better) general solutions to our problem. But it seems that such solutions will hardly be suited for the usual formal systems of set theory, developed in the context of classical logic and so deeply compromised with the higher infinite.

10. Yet, it is highly questionable that the property we would attribute as an essential property to realism in this context, the property which makes it (seemingly) indispensable for giving a solution to our problem, namely, the fact that it would ensure *uniqueness* of the interpretation of the formal language (in other words: determinateness of the object of set-theoretic discourse) can be truthfully attributed to it. First, if the axioms are in fact presupposed in the construction of the cumulative hierarchy (see above), the hierarchy cannot infuse into them the determinateness they do not have. Secondly, it can be argued (but I shall not do this here) that the fact that the semantics of set theory is itself set-theoretic submits it to all the "pathologies" of formal languages, and uniqueness of interpretation is the first thing to be lost. Third, I assume here that the second-order alternative is not a way out. In sum, the need for uniqueness seems, at the same time, inescapable and not susceptible of fulfilment.

11. There is a perspective which is a radical alternative to realism, refusing the very setting of questions which follows from realist assumptions: I shall label it "the *transcendental* point of view". This does not imply a strict historical faithfulness to Kant's thought, nor to Marburg Neo-Kantianism (see, e.g., [6]), which nevertheless is the most natural candidate to be the starting point of a reflection about mathematics which could escape the well-established dichotomies of some contemporary philosophies of mathematics (e.g., the refusal to consider anything objective if it is not real, either physically/empirically or "Platonically"). I use the term "transcendental" because I would like to show a few consequences of a point of view in which ontology and epistemology are both set up by taking as a starting point the fact of mathematical knowledge, and going back to the conditions of its possibility, without imposing, from without, ontologies or epistemologies constituted in advance. For example, instead of asking (after [2]) how we can have epistemic access to mathematical entities, we could try to explain the reason why mathematics does not have any problem of access to the "entities" it deals with, and, on a deeper level, how (under what conditions) this activity can be possible: an activity which combines freedom of conceptual development and need for truth in a way such that there is nothing comparable in any other science. Another example: one could try to escape the impasse hinted at above, about the uniqueness of interpretation, by considering uniqueness as a sort of "regulative ideal" (in Kant's sense; see [12], henceforth: KrV, B 671ff.). In this sense, the universe of sets could be considered no more a mysterious substance, but rather something analogous to the "transcendental object" (in the sense of Kant's *Analytic of Principles*; see KrV B 295ff., A 253), or to an Idea of Reason (KrV B 378ff.). A fundamental aspect of the view which inspires my reflections is the following: axioms (past, present and future) are at the core of mathematics in that they are constitutive of mathematical reality. Here the task arises of understanding the synthetic nature of axiomatization, and the transcendental conditions of possibility of such synthesis. When considering the semantics of set theory, this means that we should identify the various levels of synthesis, in their forms and in their conditions, both with respect to axiomatization (strong hypotheses, etc.) and with respect to models.

12. On this view, the common set-theoretic nature of ordinary set theory and ordinary model theory should be the starting point of any reflection on the problem of semantics for set theory. Model theory is not a sort of naive inspection of the common "world" in which all different mathematical structures live. It is rather a mathematical theory in its own right, which needs its background set theory and, reciprocally, gives set theory much that it needs. This point is perhaps best clarified by observing the rather idiosyncratic character of the model theory of set theory. This is not a symptom of the fact that we have reached, at last, the "natural language" of mathematics (allegedly, set theory itself). Instead, it is a clear display of what happens (mathematically) when we can no longer keep separated a mathematical theory, its semantics, and the theory of sets upon which the latter is built. We can no longer naively think that these levels are "absolute" and have at the basis the very mathematical "reality". We can provisionally feign to believe this when we do, e.g., the model theory of algebraic theories, but with set theory we cannot avoid this apparent circularity. We are compelled to give up our previous (pseudo-)common sense and to deal with this circularity, at last, mathematically.

13. More generally, a consequence of the view I am adopting is that one should not look for a solution to the problem of the semantics of set theory accepting a general philosophical framework which, I think, prevents a priori any solution: a framework based on the dichotomy between "subject" and "object", between what is "internal" and what is "external", between the "mental" and the "real". I argue that this framework is philosophically wrong, and it is much more wrong in the case of mathematics, because mathematics is neither a product of the mind nor a description of an outside (perhaps hyperphysical) world. Instead, we should take seriously the idea that the mathematical world is a world of concepts, the world of the possibilities of thought in the most general sense. Here "thought" is not a mental activity, it is not the act of thinking, but it is simply what is thought, independently of any "subject". But neither is it an independent, "Platonic" reality, because, in a sense, it is no reality, it is only a logical (in the most general sense) condition of any possible, conceivable reality (here "possible" and "conceivable" are synonymous adjectives, neither "objectively" nor "subjectively" connoted). The reader might feel compelled, at this point, to ask whether these "logical conditions" are, in their turn, subjective or objective. This question is a symptom of an utter misunderstanding of my proposal: I just suggest trying to think of these problems without posing the question itself. There is a sense, I admit, in which the question is legitimate, but much less dramatic: the answer is that, of course, "logical conditions" are perfectly objective, and there is nothing subjective in them. This is the deep truth present in any form of realism, but it simply means that mathematics is not simply the subjective product of the mental activity of individuals (let alone of "communities"). It is important to notice that, in my perspective, the objective has a much wider extension than the real. Mathematics is the supreme, all-pervading form of objectivity, which logically precedes the constitution of any reality, and, a fortiori, any dichotomy between subject and object.

14. On the other hand, mathematical concepts are humanly understandable, because they are precisely what constitutes mathematics itself as a human activity: humans are capable of doing mathematics, inasmuch as mathematics is basically defined by whatever is done by them under this name. This does not mean, however, that one can call "mathematics" what one likes; it means only that the criteria by which we can distinguish mathematics are internal to mathematics itself. The point is that in this properly human activity, the concepts which arise are radically transcendent over the activity itself (in a different sense with respect to the concepts of physics). Once they are reached, these concepts have no connection whatsoever with human minds, not in the sense that they cannot be grasped (in fact, the possibility to grasp them, at least in principle, is their essence), but in the sense that nothing in their content presupposes anything having to do with cognition. I am aware that this phenomenon looks quite mysterious, but it is much more mysterious from those philosophical points of view which do not accept the fact of mathematical knowledge as a starting point. A Neo-Kantian "transcendentalist" perspective takes this decisive step. That this is not question-begging should emerge from the whole development of hard work that has yet to be done in philosophy and also in the history of mathematics. But I am persuaded that, at least, to highlight this "mystery" is philosophically more honest than to give a distorted image of mathematics in order to fit it into a pre-existing philosophical frame.

15. The last few sections are not intended as a rigorous argument for the conclusion I shall presently state, but should nevertheless give an idea of a general philosophical attitude about mathematics (of course, without any special claim of originality on my part) which, I claim, is the appropriate context in which these problems should be discussed. Coming back to our main topic, my conclusion is that we can keep doing set theory with its ordinary semantics, without committing ourselves to the realist philosophy allegedly imposed by that semantics, a philosophy which is unacceptable from the epistemological point of view.

Acknowledgements

I wish to thank the referees for helpful comments.

BIBLIOGRAPHY

[1] L. Bellotti. On the circularity of set-theoretic semantics for set theory. To appear.
[2] P. Benacerraf. Mathematical truth. *Journal of Philosophy*, 70: 661–679, 1973. Reprinted in [3, pp. 403–421].
[3] P. Benacerraf and H. Putnam (eds.), *Philosophy of Mathematics*, second edition, Cambridge 1983.
[4] G. Boolos. The iterative concept of set. *Journal of Philosophy*, 68: 215–232. Reprinted in [3, pp. 486–502].
[5] G. Boolos. Nominalist Platonism. *Philosophical Review*, 94: 327–344, 1985.
[6] E. Cassirer. *Substanzbegriff und Funktionsbegriff*. Berlin, 1910. English translation, Chicago 1923.
[7] C. Chihara. *A structural account of mathematics*. Oxford, 2004.
[8] H.G. Dales and G. Oliveri. Truth and the Foundations of Mathematics: an Introduction. In H.G. Dales and G. Oliveri (eds.), *Truth in Mathematics*, pages 1–37, Oxford, 1998.
[9] S. Feferman. *Notes on operational set theory I*. To appear.
[10] K. Gödel, K. What is Cantor's Continuum Problem? *American Mathematical Monthly*, 54: 515–525, 1947. Revised version (1964) reprinted in [3, pp. 470–485]. Both versions reprinted in K. Gödel, *Collected Works II*, Oxford 1990, pages 176–188 and pages 254–270.
[11] M. Hallett. *Cantorian Set Theory and Limitation of Size*. Oxford 1984.
[12] I. Kant. *Kritik der reinen Vernunft* (KrV). Akademie Ausgabe, Band III, Berlin 1911. English translation by N. Kemp Smith, London 1973.
[13] G. Kreisel. Mathematical Logic. In L. Saaty (ed.), *Lectures in Modern Mathematics*, pages 95–195, New York 1965.
[14] G. Kreisel. Informal Rigour and Completeness Proofs. In I. Lakatos (ed.), *Problems in the Philosophy of Mathematics*, pages 138–186, Amsterdam, 1967.
[15] G. Kreisel. Mathematical Logic: What Has It Done for the Philosophy of Mathematics? In R. Schoenman (ed.), *Bertrand Russell, Philosopher of the Century*, pages 201–272, London 1967.
[16] G. Kreisel and J.L. Krivine. *Elements de Logique Mathematique*. Paris, 1967.
[17] F.W. Lawvere and R. Rosebrugh. *Sets for Mathematics*. Cambridge 2003.
[18] J. Lear. Sets and Semantics. *Journal of Philosophy*, 74: 86–102, 1977.
[19] P. Maddy. *Realism in Mathematics*. Oxford 1990.
[20] V. McGee. Everything. In G. Sher and R. Tieszen (eds.), *Between Logic and Intuition*, pages 54–78, Cambridge 2000.
[21] Y.N. Moschovakis. *Descriptive Set Theory*. Amsterdam 1980.
[22] G. Oliveri. *A Realist Philosophy of Mathematics*. London 2007.
[23] C. Parsons. Informal Axiomatization, Formalization, and the Concept of Truth. *Synthese*, 27: 27–47, 1974. Reprinted in [27, pp. 71–91].
[24] C. Parsons. Sets and classes. *Noûs*, 8: 1–12. Reprinted in [27, pp. 209–220].
[25] C. Parsons. The Liar Paradox. *Journal of Philosophical Logic*, 3: 381–412, 1974. Reprinted in [27, pp. 221–267].
[26] C. Parsons. What is the Iterative Concept of Set?. In Butts and Hintikka (eds.), *Logic, Foundations of Mathematics and Computability Theory*, pages 335–367, Dordrecht 1977. Reprinted in [3, pp. 503–529] and [27, pp. 268–297].
[27] C. Parsons. *Mathematics in Philosophy*. Ithaca 1983.
[28] A. Paseau. Should the Logic of Set Theory Be Intuitionistic? *Proceedings of the Aristotelian Society*, 101: 369–378, 2001.
[29] A. Paseau. The Open-Endedness of the Set Concept and the Semantics of Set Theory. *Synthese*, 135: 379–399, 2003.
[30] D. Scott. Axiomatising Set Theory. In T. Jech (ed.), *Axiomatic Set Theory II*, pages 207–214, Providence, 1974.
[31] J.R. Shoenfield. *Mathematical Logic*. Reading 1967.
[32] J.R. Shoenfield. Axioms of Set Theory. In J. Barwise (ed.), *Handbook of Mathematical Logic*, pages 321–344, Amsterdam 1977.

[33] H. Wang. *From Mathematics to Philosophy*. London, 1974.
[34] H. Wang. Large Sets. In Butts and Hintikka (eds.), *Logic, Foundations of Mathematics and Computability Theory*, pages 309–333, Dordrecht 1977.
[35] E. Zermelo. Über Grenzzahlen und Mengenbereiche. *Fundamenta Mathematicae*, 16: 29–47, 1930.

The use of figures and diagrams in mathematics

VALERIA GIARDINO

1 Introduction: new approaches beyond logocentricity

1.1 "Logocentric views" of mathematics

The term "logocentric" was first introduced by Sheffer, in his review of the second edition of Russell and Whitehead's *Principia Mathematica*, where he explained how throughout the four volumes of *Principia*, the authors had no concern with the epistemology of mathematics, but were interested in its logical status — the question of the derivability or non-derivability of pure mathematics from formal logic alone [21, p. 226]. According to the "logocentric predicament", in order to give an account of logic, we must presuppose and employ logic. Sheffer is thus referring to logicism, and the idea that mathematics has to be founded on logic.

In a very different context, far from the foundationalist debate, Barwise and Etchemendy chose the same term to define what they considered the "dogma" of the standard view of mathematics, according to which the core issue of mathematics is to give proofs of mathematical statements, which are in turn syntactic objects consisting only of sentences arranged in a finite and inspectable way [3, p. 3]. By contrast, Barwise and Etchemendy took into account their students' performances and concluded that understanding semantic concepts can be of help for carrying out formal proofs in a deductive system. Indeed, according to them, reasoning is a *heterogeneous* activity: people use different representations of information while reasoning, and those representations are often in non-sentential forms, diagrams for instance. They began working on a project whose aim was to reconcile a seeming conflict between what goes on with our ordinary reasoning and what has been done in logic and mathematics, combining the merits of two possible poles and making sense of them. On the one hand, there is the practical power of multimodal reasoning, on the other modern logic and its formalization and rigour; to have both would bring to the unity of teaching and research [24]. Barwise and Etchemendy's motivation was to expand the territory of logic by freeing it from a mode of representation only.

Nevertheless, there is some sense in which their research can be considered "logocentric" as well. They explicitly express their intention of doing with visual/spatial reasoning something analogous to what Frege and his followers have done with the formal/linguistic one. Barwise and Etchemendy's work is indeed still motivated in great part by the proof-theoretic foundational tradition: their aim is to give an explicit syntax and semantics for diagrams, so as to provide a diagrammatic logic, that is a new system which makes use of both sentences and diagrams. Their work focuses on standard first-order analyses, and does not tackle the structural properties of diagrammatic systems in general. It is not clear how this strategy would deal with cases where representation is a more complex phenomenon than what we encounter in sentential systems; indeed, the logical approaches must be extended to cover also these cases.

It could be objected that their research on diagrams for first-order logic was only the first step in a most ambitious program. Nevertheless, as Mancosu explains, it looks like they had nothing to say about the criteria one can appeal to in order to distinguish linguistic systems from visual ones [17, p. 23].

1.2 Alternative views

It is easy to concede that logocentric views of mathematics reveal themselves to be inappropriate in order to describe mathematics as it actually is.

First of all, a logocentric conception of mathematics seems to *have moved away from the consideration of actual mathematical practice*, both in contemporary mathematics and in the history of mathematics before the foundational debate. This happened because of what Corfield defines as the "foundational filter" [10]. According to him, this idea of "filter" is precisely what is fundamental to all forms of neo-logicism. But it is an unhappy idea. Not only does the foundational filter fail to detect the pulse of contemporary mathematics, it also screens off the past to us as not-yet-achieved. The job of the philosopher is to dismantle this foundational filter, and go back to the consideration of real mathematical progresses.

A second possible criticism to the logocentric views is that they seem to *have moved away also from the consideration of actual reasoning processes when doing mathematics*. At the beginning of the 20th century, Wertheimer affirmed that both in mathematics and in logic, the standard tendency was to reduce logic to a game, governed by a sum of arbitrarily combined, piecemeal rules [28]. This could be a good method to study local mathematical problems such as the individuation of criteria for rigid, logical validity. Nevertheless, to avoid confusion, it should be pointed out that assessing questions of validity in an ideal, axiomatic system is very different from investigating what understanding mathematics and being productive mean [28, p. 137].

There some alternatives to these logocentric views of mathematics.

One of these alternatives is expressed by Brown [5]. In his view, the living philosophical issues for working mathematicians cluster around visualization and experimentation, since mathematics makes use of different formats to display information. Brown embraces a sort of "fallible" Platonism, according to which mathematics provides each time a new description of a Platonic world of objects outside time and space. Pictures are windows to Plato's heaven: it is possible to have a realist view of mathematics and its objects, without being at the same time committed to a realist view of pictures. As telescopes help the unaided eye, so some diagrams are instruments — rather than representations — which help the unaided mind's eye. Thus, mathematics can be compared to activities that are traditionally considered very far from it, art for instance, since both can have "figurative" and "symbolic" aspects, or poetry, because also in mathematics the choice of some particular expressions makes a difference. A mathematical diagram is like a painting, since it is simultaneously about something concrete and something abstract.

Nevertheless, it is questionable that ontological statements are less problematic than the difficulties they are intended to find solution to. The time-honoured debate about the ontological nature of mathematical objects shows that it is very difficult to settle ontological questions about mathematics once and for all. Furthermore, even if we accept that ontology has at least some methodological advantage for the mathematical practice, such an advantage cannot be settled in principle, since it heavily depends on the context.

Another alternative view is put forward by Giaquinto, who defends the epistemological value of visual thinking in mathematics, for example as a means of discovery. Actually, visual thinking is not just a heuristic aid: the content of visualizing can be an *operation-schema*, which schematically displays possibile arrangements, thus leading to general discoveries [13]. This kind of approach opens up to the consideration of not only the philosophical work on the nature of concepts, but also of empirical research on visual perception and mental imagery.

A point of view of this kind allows us to move away from the logocentricity which is typical of logicist and neo-logicist approaches, and to consider mathematical reasoning beyond first-order analyses, without at the same time committing ourselves to any ontological claim. Mathematics is the activity of solving and communicating problems of which the results are shared, and to this aim, it makes use of different tools and multi-modal representations. Following Corfield's and Wertheimer's suggestions, two possibile directions in the investigation are the following: (1) assuming a *historical* perspective, and (2) assuming a *cognitive* perspective.

The historical approach takes into consideration mathematical evolution, echoing Lakatos, and therefore case studies taken from the history of mathematics, now that the "foundational filter" has been removed.

The cognitive approach focuses on the cognitive processes that come into play when working with mathematics, and refers to other fields of research that study information displaying and thought processes, therefore taking into account empirical findings from disciplines such as cognitive science or computer science.

2 Cognitive approaches to diagrams

The differences between cognitive science on the one side and philosophy and logic on the other in their approaches to mathematics can be widely discussed, but for the purpose of this article it will be assumed that the use of figures and diagrams in mathematics represents an interesting case in which some psychological considerations can support more theoretical and general frameworks. Nevertheless, it is necessary to specify the extent to which this is possible.

According to the definition, a mathematical diagram is a figure, usually consisting of a line drawing, made to accompany and illustrate a geometrical theorem or a mathematical demonstration. My attempt will be to discuss the reasons why a mathematical diagram is effective in accompanying and illustrating a geometrical theorem or a mathematical demonstration.

I will discuss two simple cases and analyse the difference between using formal sentences and diagrams to display information.

2.1 The Venn diagrams case

Consider the simple case of Venn diagrams (Figure 1). Venn diagrams are

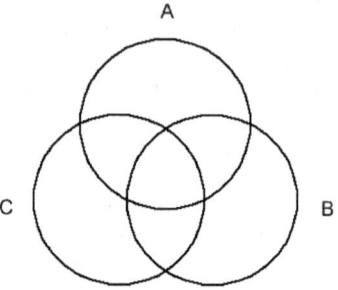

Figure 1. A Venn diagram.

massively used in teaching logic, since they seem to be helpful in communicating to the students what is at stake when they are learning under which circumstances a syllogism is valid or more in general under which circumstances some simple relation among sets holds. Different views can be advanced about what Venn diagrams are and how they work.

According to logocentric views, the diagram in Figure 1 is only a heuristic tool to prompt certain trains of inference, and is dispensable as a proof-theoretic device.

By contrast, Lakoff suggests that a formalist view of mathematics, according to which a set is any mathematical structure that satisfies the axioms of set theory, does not give an account of the way we commonly *speak* about sets [15]. Actually, we talk about them as "containing" their members: even the choice of the word "member" refers to this idea of "containment". Nonetheless, nothing in the axioms characterizes a set as a container, or defines what a container is. Lakoff adopts a cognitive perspective. According to him, mathematics starts from human experience, and metaphor plays a crucial role in it, as it involves the transfer of ideas between realms. Sets are commonly conceptualised in terms of Containment-schemas. A constraint follows automatically, which on the contrary does not follow from the axioms: sets cannot be members of themselves. To rule out this sort of possibilities, new axioms are introduced as new methaphors for thinking about sets, starting from our ordinary grounding metaphor which allows us to think about Classes as Containers. After enough layers of metaphors are given, the original conceptual grid is then forgotten. What Lakoff seems to suggest is that when we consider Venn diagrams, it is because of our Containment-schema that we can grasp their meaning.

Nevertheless, this is misleading. Consider for example the very simple relation depicted in Figure 1, which is the relation of having a non-empty intersection. How could we use Containment schema to grasp its meaning? Moreover, though Venn diagrams are similar to other kinds of logical diagrams such as for example Euler circles, they do not share with them the same meaning. More precisely, Venn diagrams are introduced to override the expressive limits of the Euler representation system. One of the main defects of Euler-Venn diagrams is their failure to visualize the difference between membership and inclusion relationships between sets. Therefore, Lakoff's point of view seems to oversimplify the way we use diagrams, focusing only on their perceptual constraints and underestimating the role played by other factors such as their interpretation.

Let us consider again what Barwise and Etchemendy suggest about how to deal with Venn diagrams. According to them, Venn diagrams provide us with formalism that consists of a standardized system of representations, together with rules for manipulating them. It is indeed possibile to work on the semantic analysis for a visual representation system of Venn diagrams [23]. This kind of approach is dedicated to the study of Venn diagrams as a primitive visual analog of the formal systems of deduction which are developed in logic.

As I have already discussed at the end of the previous section, this move, though appropriate to this case, does not as much adequately deal with other diagrammatic phenomena such as for example approximate representations, that can be well used in mathematical practice. Under these circumstances, psychological investigations are relevant in showing the constraints diagrams are subject to. According to Shin's reconstruction, Barwise and Etchemendy decided to work on their program in favour of heterogeneous reasoning precisely because they had observed the way students could easily learn first-order language using their new and experimental software *Tarski's world* [24]. Their research brought them to the creation of another software, *Hyperproof*, which not only used graphics to teach the syntax and semantics of first-order logic, but added to this also the teaching of inference. The new software incomporated heterogenous reasoning rules which moved information back and forth between graphical representations of blocks-worlds in the windowpane and the sentences of first-order logic below it.[1]

To consider mathematical practice it is now necessary to move beyond the visualization of first-order logic. The second example which is given in the following section takes into account this possibility.

2.2 Mathematical diagrams (I): grouping laws

Consider the formula for the solution of quadratic equations:

$$x^2 + ax = \left(x + \tfrac{a}{2}\right)^2 - \left(\tfrac{a}{2}\right)^2$$

---

[1] [25] provides a controlled comparison of the effects of teaching undergraduate classes using *Hyperproof* and a traditional syntactical method. The results show that individual differences in aptitude should be taken into account when choosing a teaching technique. According to the authors, the educational implications of these individual differences are far from clear: one possibility could be to devise and generally teach a domain-independent "graphics curriculum".

Let us represent the same information using diagrams. A diagram will be drawn to represent the elements which are present in the left side of the formula: if x and a are represented by two arbitrary segments, then their products will be represented by two figures, as indicated by the labels (Figure 2).

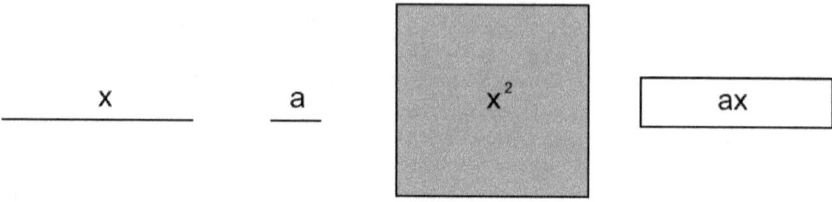

Figure 2.

Let us analyse what the two displays — the formula and the diagrams — have in common and what they have not. First of all, in both cases the spatial arrangement of the elements is relevant: it is only because both the formula and the diagrams are spatially organized in a particular way, that we are able to read off information from them. Nevertheless, the formula is a one-dimensional array, while the diagrams are two-dimensional objects displayed on a plane.[2] Moreover, in the first case, thanks to conventional rules, we are able to assign a meaning to each element of the formula in isolation and then to their composition. By contrast, in the case of the diagrams, we assign a meaning to the elements taking into account the global structure they are part of, such as a line or a rectangular figure. Furthermore, the use of labels which are inserted in the diagrams is important to consider the successful interaction among different formats.

The second step using the formula is to check that the left side is equivalent to the right side. To do that, some algebraic rules are applied. Let us consider instead how diagrams are used and arranged so as to obtain a second diagram that displays the left side of the formula (Figure 3).

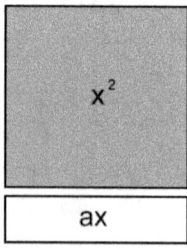

Figure 3.

---

[2] I am not claiming here that all formulas are linear. For example, formulas in tableaux calculi or in sequent calculi are not simply one-dimensional. Nevertheless, I would argue that not linear formulas of this kind could be considered as partly "diagrammatic".

This diagram is then changed into a new perceptual configuration, thus obtaining a new square (Figure 4).

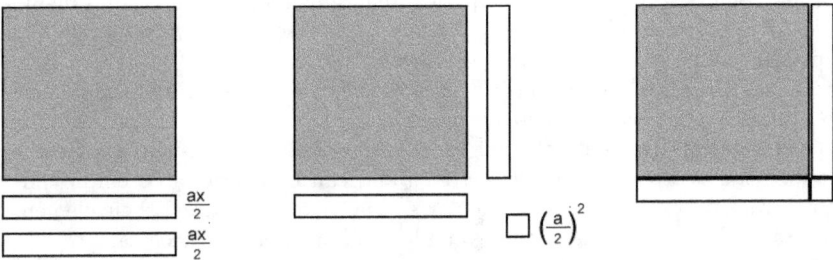

Figure 4.

The area of the square thus obtained is equivalent to $x^2 + 2\frac{ax}{2} + \left(\frac{a}{2}\right)^2$ which is equivalent in turn to $\left(x + \frac{a}{2}\right)^2$. The area of the first figure in the row, which is the one we are interested in, is thus showed to be equivalent to:

$$\left(x + \frac{a}{2}\right)^2 - \left(\frac{a}{2}\right)^2 \tag{1}$$

which is precisely the right side of the formula. The correctness of the formula is thus displayed by means of a new organization of the diagrammatic elements we had at the beginning. Moreover, such a visualization of the result gives a feeling of evidence that does not seem to be experienced by the simple application of rules.

Let us summarize how the diagrams were used. First, we had simple reception of visual data: some lines were sketched on a page or shown on a computer screen. Then, thanks to the spatial relations among the elements displayed, these local elements were grouped together, in order to detect some global configuration. From this point of view, perceiving a mathematical diagram is analogous to perceiving any other kind of visual percept. According to some empirical studies, there is evidential support for the idea that the perception of a diagram is subject to grouping laws [19], it can be influenced even by sensory motor information [2], and involves the diagram in its configurational globality, going beyond local information about its single elements [11]. This capacity to perceive configurations can be evaluated and empirically tested. Nevertheless, is it sufficient to give an account of the work with mathematical diagrams?

Consider again the manipulations performed. Given that a configuration was individuated, what role was played by its interpretation? Being able to configure diagram elements into groups is an *important precondition* for proper interpretation and is a key requirement of a well-designed diagram [9, p. 85]. Without considering the way a diagram is interpreted according to some background theory — Euclidean geometry, in this case — it would be very difficult to distinguish a well-designed diagram from a not-well-designed one. This does not mean that this interpretation is always explicitly given.

2.3 Mathematical diagrams (II): background assumptions

A cognitive approach to mathematics is inscribed in the context of the practice of mathematics, in which mathematical diagrams are introduced and manipulated to solve problems; more precisely, it is only in *specific* contexts that the solution of *specific* problems is needed.

First, diagrams are considered in their "literal" sense, simply looking at the spatial relations they display, as global configurations which are based on the spatial arrangements of their elements according to perceptual grouping laws. *This is already cognitive activity*, and can be investigated empirically: to extract information, the particular spatial relations among their elements are taken into account. In the example given above, it is necessary to "discern" different configurations in the diagram. These configurations are then interpreted (i) as corresponding to the elements in the formula; (ii) as subject to given rules of geometry such as how to obtain the area of a rectangle knowing the lenghts of its sides. The spatial representations are thus cognitively relevant as a medium to represent meaningful structures; once these relations are acknowledged, diagrams can then be interpreted, according to some theoretical assumptions which are based on background knowledge.

Therefore, we are able to use diagrams because diagrams are subject to two kinds of *constraints*. One one side, they are subject to *perceptual* constraints which depend on their spatial and structural relations; on the other, they are subject to *conceptual* constraints, which depend on interpretation. In some sense, this is similar to what happens with the formula, with the difference that its perceptual organization is in general simpler (concatenation, for istance), and its interpretation heavily depends on conventional rules. Moreover, the manipulation of the formula takes the form of a (more or less long) calculation. In the diagram case instead, it is possible to distinguish between the information that depends on the way the user is able to perceptually organize it, and information that depends on what the user knows about the interpretation of the representation system in use. In such a way, we may or may not be able to override the accidental features of the diagram and consider it as a structure of data. Diagrams appear to be highly specialised representations: the skills required for using diagrams effectively "must be *learned* and appear to be *highly domain-specific*" ([9, p. 86], my italics).

It is also possibile to go further. Consider one of the many possibile visualizations of the Pythagorean Theorem shown in Figure 5. History of mathematics has collected many visual proofs of this theorem. The reason is clear: to prove it by means of algebra only would bring into play complex numbers and the Euler formula, moving away from the clarity and the straightforwardness of the visual proofs.

A smaller square can be "seen in" a bigger square. The side of the smaller square is the hypothenuse of the triangle, while the side of the bigger square is the segment which is constituted by the legs of the same triangle. Thus, the bigger square contains the smaller square plus four triangles. Given this interepretation, it can be inferred that:

$$(a+b)^2 = c^2 + 2ab \qquad (2)$$

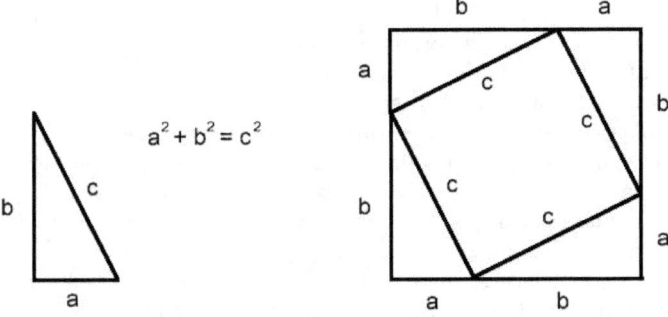

Figure 5.

Applying some simple algebraic rules, we obtain the Pythagorean Theorem.

Cellucci reports this case of visualization as an example of the application of the analytic method, for the very reason that to show that the square on the hypotenuse equals the sum of the squares on the two legs, we only need to show that the quadrilateral in the diagram is a square, which in turn contains the square on the hypotenuse and four triangles [7]. To do that, a new hypothesis is put forward, which states that the three interior angles of a triangle are equal to two right angles. This new hypothesis is a sufficient condition for the solution of the problem, and provides a possible reason why the square on the hypotenuse equals the sum of the squares on the two legs. Moreover, because of the format used in displaying the problem, the hypothesis is obtained from the information we had at the beginning, and possibly from other data, by some non deductive inference. The hypothesis must not only be a sufficient condition for the solution of the problem but must also be plausible, that is, compatible with the existing data. According to the analytic method, the solution of a problem is a potentially infinite process, since the plausibility of the hypotheses does not mean that these hypotheses are final: new facts may emerge which show that certain previous hypotheses are no longer plausible and as a consequence must be revised or even dropped. Diagrams can guide us in this process.

3 A pragmatic point of view: two strategies

In this third section, I will extend the cognitive approach to diagrams so as to include also a pragmatic evaluation of their role. The research hypothesis is that there is a cognitive advantage in using a diagram as opposed to the idea that diagrams are dismissible in the reasoning processes, but (i) this cognitive advantage is not determined by the fact that diagrams are simply "more visual" than linguistic sentences; (ii) this cognitive advantage can not be assessed in principle, but depending on the task in question. In fact, we go from cases in which diagrams represent only an auxiliary aid, to cases in which they structure the problem so as to direct the reasoning process of finding its solution. In mathematics in particular, it is necessary to consider the relationship that there is between what I defined *perceptual* and *conceptual* constraints.[3] According to Fischbein, there are reasons to consider beliefs and expectations, pictorial prompts, analogies and paradigms not as mere residuals of more primitive forms of reasoning, but as proper components of mathematical reasoning and, generally, of every kind of scientific reasoning. These properly mathematical features are genuinely productive. Nevertheless, practicality is not a criterion in accepting something as being mathematically valid: mathematical validity is based on definitions and consistency within some mathematical structure. Diagrams, which certainly contribute to the organization of information in synoptic representations, are nonetheless *post-conceptual* structures. For this very reason, they are tools for bringing together the practical expression of a certain reality and conceptual interpretation [12].

To assess my research hypothesis, I propose two strategies that reformulate the directions in the investigation presented in section 1.

The first strategy is to follow Mancosu's suggestion and focus on the notion of explanation in mathematics [18]. Mancosu points out that often pictures or informal arguments play an ideal "explanatory" role in mathematics, whereas a full proof will be no explanation at all in that context. Actually, a standard mathematical proof and an effective mathematical explanation often do not seem to be the same thing. This seems to be a peculiarity of mathematics, but as a matter of fact it is another consequence of the foundational filter. In fact, the question of mathematical explanation, as the one concerning mathematical proof, has up to now been studied only from a foundational perspective. On the contrary, it is correct to think of a new way of examining it. Also in this respect the short-sighted "logocentric" point of view reveals itself to be a detriment compared to the perspective of considering mathematics as the activity of "searching for reasons". The discovery of the hidden reasons for a theorem is the work of the mathematician.[4] Once such reasons are found, the choice of particular formal sentences expressing them is secondary. Different but exchangeable formal versions of the same reason can and will be given depending on circumstances. It can be investigated how these different formats are effective when the aim is to provide a successful explanation. Diagrams are more than useful heuristic tools and mathematicians "do not preach what they practice", since they are reluctant to acknowledge formally what they do in practice.

The second strategy is to focus on two sub-hypotheses that can be empirically tested.

The first sub-hypothesis is that diagrams confer cognitive advantage *over linguistic representations*. There is already a fair amount of empirical evidence in favour of this claim. According to Larking and Simon, for example, diagrams can be more effective than informationally equivalent sentential representations because of *locational indexing* [16]. Indeed, in the diagram, the information that is needed to solve the problem is usually found in adjacent locations. Diagrams and sentences do not differ in the information they contain,

---

[3] A similar distinction is made in [27] between *situated* elements, which make reference to concrete situations, and *formal* ones.

[4] Among others, see [20].

or in the kind of recognition or of inference they activate, but in the amount of search they demand to find the relevant information. In line with this idea, Stenning and colleagues distinguish between *computational* efficacy, i.e. low complexity of inference, and *expressive* efficacy, i.e. semantical properties such as consistency, or a restriction on the class of representable structures [26]. Diagrams are generally inexpressive, which is the reason why they lead to tractability of inference. The expressive restrictions on the diagrammatic representational systems arise from an interaction between topological and geometrical constraints on plane surfaces and the ways in which diagrams are interpreted. As a consequence, it is the nature of the interpretation of the medium, rather than the medium itself, which gives rise to the real differences between representation systems.

The second sub-hypothesis is that diagrams confer cognitive advantage *over mental computations*, for the very reason that they are extra-mental devices and thus reduce computational charge. There is a fair amount of empirical evidence also in favour of this claim. It has been surmised that certain cognitive and perceptual problems are solved more quickly, easily and reliably by performing actions in the world than by performing computational actions in the head alone. These actions and manipulations are performed directly on the diagram itself to promote new inferences. Kirsh and Maglio define these kind of actions as *epistemic* [14], and Shimojima and Fukaia introduce the notion of *hypothetical drawings* [22], which is the hypothetical trasformation of the graphic assumed by the agent to obtain a new conclusion.

4 Conclusions

In this article, I discussed the views that have been proposed to consider the use of diagrams and figures in mathematics, and I argued that "logocentric" views do not give an account of real mathematical activity. I introduced another approach to figures which focuses on the reasoning processes that are engaged in working with them.

Figures and diagrams are subjects to *perceptual* constraints, which are a necessary condition for them to convey information; nevertheless, these constraints are not sufficient, since interpretation — *conceptual* constraints — plays a very important role.

As a conclusion, to study diagrams it is more promising to stop asking what a diagram is, focusing too much on the uninformative dichotomy visual vs. non-visual, and to assume a pragmatic point of view, finding ways to assess the hypothesis according to which in some particular tasks there is a cognitive advantage in using a diagram. This brings to the consideration of case-studies taken from mathematics and to empirical investigations, which is matter of further research.

BIBLIOGRAPHY

[1] G. Allwein and J. Barwise (eds.). *Logical Reasoning with Diagrams*. Oxford 1996.
[2] R. Barone and P.C-H. Cheng. Interpreting lines in graphs: do graph users construe fictive motion? In [4], pages 333–336.
[3] J. Barwise and J.V. Etchemendy. Visual Information and valid reasoning. In [1], pages 3–25.

[4] A. F. Blackwell, K. Marriott and A. Shimojima (eds.). *Diagrams 2004, Diagrammatic Representation and Inference. Third International Conference. Lecture Notes in Computer Science*, 2980, Springer-Verlag, Berlin 2004.
[5] J.R. Brown. *Philosophy of Mathematics: an introduction to the world of proofs and pictures*. London 1999.
[6] K.R. Butcher and W. Kintsch. Learning with diagrams: effects on inferences and the integration of information. In [4], pages 337–340.
[7] C. Cellucci. The Nature of Mathematical Explanation. *Studies in History and Philosophy of Science*, 39: 202–210, 2004.
[8] B. Chandrasekaran, J. Glasgow and N. Hari Narayan (eds.). *Diagrammatic Reasoning: Cognitive and Computational Perspectives*. AAAI Press, the MIT Press, Cambridge, MA 1995.
[9] P.C-H. Cheng, R.K. Lowe and M. Scaife. Cognitive approaches to understanding diagrammatic representations. *Artificial Intelligence*, 15: 74–94, 2001.
[10] D. Corfield. *Towards a Philosophy of Real Mathematics*. Cambridge 2003.
[11] A. Desolneux, L. Moisan and J.M. Morel. Gestalt theory and computer vision. In A. Carsetti (ed.), *Seeing, Thinking, Knowing: Meaning and Self-Organization in Visual Cognition and Thought*, Kluwer, Dordrecht 2004, pages 71–101.
[12] E. Fischbein.*Intuition in Science and Mathematics*. Reidel, Dordrecht 1987.
[13] M. Giaquinto. Visualizing in arithmetic. *Philosophy and Phenomenological Research*, 53 (2): 385–396, 1993.
[14] D. Kirsh and P. Maglio. On distinguishing epistemic from pragmatic action. *Cognitive Science*, 18: 513–549, 1994.
[15] G. Lakoff and R. Nuñez. *Where Mathematics Comes From: How the Embodied Mind Brings Mathematics into Being*. New York 2001.
[16] J.H. Larkin and H.A. Simon. Why a diagram is (sometimes) worth ten thousand words. In [8], pages 69–109.
[17] P. Mancosu. Visualization in logic and mathematics. In P. Mancosu, K.F. Jørgensen and S.A. Pedersen (eds.), *Visualization, Explanation and Reasoning Styles in Mathematics*, Berlin 2005, pages 13–30.
[18] P. Mancosu. Mathematical explanation: problems and prospects.*Topoi*, 20: 97–117, 2001.
[19] P. Olivier. Diagrammatic reasoning: an artificial intelligence perspective. In *Artificial Intelligence Review*, 15: 63–78, 2001.
[20] G.C. Rota. The phenomenology of mathematical proof. In A. Kamamori (ed.), *Proof and Progress in Mathematics*, Kluwer, Dordrecht 1997, pages 183–196.
[21] H.M. Sheffer. Review of A. N. Whitehead and B. Russell, Principia Mathematica. *Isis*, 8: 226–231, 1926.
[22] A. Shimojima and T. Fukaya. Do we really reason about a picture as a referent? In *Proceedings of the Twenty-Third Annual Conference of the Cognitive Science Society*, 2003, pages 1076 – 1081.
[23] S.J. Shin. Situation-theoretic account of valid reasoning with Venn diagrams. In [1], pages 81–127.
[24] S.J. Shin. Heterogeneous Reasoning and Its Logic. *The Bulletin of Symbolic Logic*, 10 (1): 86–106, 2004.
[25] K. Stenning, J. Cox and J. Oberlander. Contrasting the cognitive effects of graphical and sentential logic teaching: reasoning, representation, and individual differences. *Language and Cognitive Processes*, 10: 333–354, 1995.
[26] K. Stenning, R. Inder and I. Neilson. Applying semantic concepts to analysing medias and modalities. In [8], pages 303–338.
[27] J. van der Pal and T. Eysink. Balancing situativity and informality: the importance of relating a formal language to interactive graphics in logic instruction. *Learning and Instruction*, 9: 327–341, 1999.
[28] M. Wertheimer. *Productive Thinking*. London 1961.

The role of epistemological models in Veronese's and Bettazzi's theory of magnitudes

PAOLA CANTÙ

1 Introduction

The philosophy of mathematics has been accused of paying insufficient attention to mathematical practice [15]: one way to cope with the problem, the one we will follow in this paper on extensive magnitudes, is to combine the "history of ideas" and the "philosophy of models" in a logical and epistemological perspective. The history of ideas allows the reconstruction of the theory of extensive magnitudes as a theory of ordered algebraic structures; the philosophy of models allows an investigation into the way epistemology might affect relevant mathematical notions.

The article takes two historical examples as a starting point for the investigation of the role of numerical models in the construction of a system of non-Archimedean magnitudes. A brief exposition of the theories developed by Giuseppe Veronese and by Rodolfo Bettazzi at the end of the 19th century will throw new light on the role played by magnitudes and numbers in the development of the concept of a non-Archimedean order. Different ways of introducing non-Archimedean models will be compared and the influence of epistemological models will be evaluated. Particular attention will be devoted to the comparison between the models that oriented Veronese's and Bettazzi's works and the mathematical theories they developed, but also to the analysis of the way epistemological beliefs affected the concepts of continuity and measurement.

2 Giuseppe Veronese

Giuseppe Veronese is well-known to mathematicians for his studies of projective geometry, but his epistemological contributions to the foundations of geometry have been mostly ignored by contemporary philosophers of science, although they were quite well-known at the beginning of the 20th century.[1] As I have shown in previous research [9], Veronese's epistemology is neither naive nor inconsistent: it justifies the acceptance of many non-Euclidean geometries, including elliptic, hyperbolic, non-Archimedean geometry and the theory of

---

[1] Felix Klein [25] and David Hilbert [23] mentioned Veronese's mathematical results, Paul Natorp [29] and Ernst Cassirer [13] discussed Veronese's non-Archimedean continuum, Bertrand Russell [32] praised Veronese's contribution to the history of the foundations of geometry.

hyperspaces. Moreover, Veronese's epistemological model, though apparently regressive for its recourse to synthetic tools and its refusal of analytical means, turned out to be fruitful from both a geometrical and an algebraic point of view.[2]

2.1 Veronese's epistemology

Veronese's mathematical theory of continuity and the geometry of hyperspaces contained in his main work — *Fondamenti di geometria* [36] — was influenced by his epistemology and especially by his conceptions of space and intuition, which are exposed in several of his writings, including various articles ([39], [35], [37]) and the prefaces to his geometry textbooks ([40], [38]). His epistemological model is compatible with the development of hyperspaces and non-Archimedean continuity, because it allows the representation of spaces with more than three dimensions and legitimates the intuition of infinitesimals, if not empirically, at least by means of an abstract intuitive capacity that one develops with time, experience and geometrical practice.

Unlike logic and mathematics, which were considered as formal sciences, and physics, which was considered as an experimental science, geometry was conceived by Veronese as a mixed science [36, p. vii], because its objects are partly abstracted from real objects and partly ideal and because its premises are partly empirical, partly semi-empirical and partly abstract [5]. Empirical premises are evident truths that one grasps by intuition when one observes certain physical objects: for example the property (usually attributed to rectilinear segments) of being determined by a couple of points derives from certain physical features of rigid rectilinear bodies. Semi-empirical premises have an empirical origin but they cannot be verified empirically because they assert something that goes beyond the observable domain: for example the geometrical properties of an unlimited line cannot be verified empirically, because the observable domain is finite, but they derive from an imaginary extension of the empirical properties of the object. Purely abstract premises concern ideal entities, such as infinitely great and infinitely small quantities, that are not related to any object of the observable domain.

Geometrical premises must satisfy both the requisite of mathematical possibility, i.e. logical soundness, and a specifically geometrical condition, that is, conformity to the intuition of space [36, pp. xi-xii]. According to Veronese, a mathematical theory that contradicts the elementary properties of spatial objects that one knows by intuition is not "geometrical", because geometry is the science of space. For example, Poincaré's theory of a one-sheeted hyperboloid or Hilbert's non-Arguesian geometry are perfectly sound mathematical theories, but they are not geometric theories, for they contain propositions that contradict our spatial intuition. A geometrical theory must have an empirical kernel and be compatible with our intuitions of space: the mathematician can

---

[2] The existence of a system of linear quantities containing infinitely small as well as infinitely great quantities was heavily criticised by Cantor [8], Vivanti [41], and Schoenflies [33], but it was praised by Stolz [34], and Bettazzi [4], and proved to be consistent by Levi-Civita ([27], [28]). The fruitfulness of Veronese's approach is clearly visible in the results of Hans Hahn [22], who built a complete non-Archimedean ordered system of linear quantities [17].

freely determine abstract hypotheses, provided that logical consistency and compatibility with empirical and semi-empirical hypotheses is maintained. The logical study of the independence of axioms is a main tool in order to define abstract hypotheses, for the introduction of new objects is accomplished by a change in the axioms. For example, non-Archimedean geometry arises from the negation of the axiom of Archimedes and the investigation of the properties of continuity that might be independent from it.

Veronese, who was strongly influenced by Moritz Pasch and Felix Klein, aimed at a common foundation of metric and projective geometry and was strongly involved in the project of establishing the theory of extensive magnitudes independently from numbers. His construction of a non-Archimedean geometry cannot be fully understood without considering his epistemological conceptions of space and intuition [11] and his familiarity with different models. Projective geometry led him to the introduction of ideal entities that might extend a system, while preserving its relevant properties. The empirical approach, strictly related to the interest for the origin, the history, and the teaching of geometry, together with the traditional insight that all geometrical properties should be somehow derived from our intuition of space, led him to the conception of "the rectilinear continuum" as an ordered system of segments rather than as an ordered system of points. The belief that geometry should be somehow distinguished from pure mathematics and therefore grounded independently from numbers also played a relevant role in Veronese's construction of a new geometrical theory. A further element that strongly influenced Veronese's construction of hyperspaces and infinitesimal quantities derives from Hermann Grassmann's epistemology: the belief that mathematical notions should be genetically connected to specific operations of thought.

Veronese's interest in an abstract and general introduction of the concepts of group (a general term for a multiplicity), equality, addition and order derives from the belief that the primitive mathematical concepts reflect relevant characteristics of the way we think. Veronese elaborated a general law that allows the construction of new ideal objects that extend any thinkable domain: "Given a determinate thing A, if we have not established that A is the group of all possible things that we might consider, then we can think of something else that is not included in A (that is outside A) and independent from A" [36, pp. 13–14]. This general law allows human thought to go beyond any given limit, because it makes it possible to assume the existence of a new entity outside the domain that was previously considered as the totality of the existing things.

2.2 Veronese's non-Archimedean continuum

Veronese's exposition of the non-Archimedean continuum is contained in the Introduction to his book *Fondamenti di Geometria*. The continuum is a system of segments endowed with an operation of addition and a relation of order. Unlike Dedekind's definition [16], Veronese's characterization of continuity is not Archimedean. According to the postulate of Archimedes — so called by Stolz [34] —, "given two quantities a and b, if a is bigger than b, there is a number $n \in N$ such that $nb > a$". On the contrary, if a system does not contain any multiple of b being bigger than a, this means that b is infinitely small with respect to a or, vice versa, that a is infinitely great with respect to b. In Veronese's theory the postulate of Archimedes does not hold, for the geometrical continuum contains infinitely small and infinitely great segments, which are introduced by two hypotheses: 1) if a segment AB is taken as a unit, there is a segment AA^1 that is infinite with respect to A, or rather a whole series of segments that are infinite with respect to A, but 2) this series, unlike Cantor's series of transfinite numbers, has no first element. To each segment of the new enlarged system Veronese associates a number, thus obtaining an enlarged numerical system II that preserves the operational properties of real numbers. A generalized version of Archimedes' postulate still holds if instead of $n \in N$ one considers a number $\eta \in II$: "if $a > b$, there is an η such that $\eta b > a$." Infinitely great and infinitely small numbers are introduced as symbols that can be assigned to infinitely great and infinitely small magnitudes (segments).

Veronese's continuity is a generalization of Dedekind's principle: if one does not assume Archimedes' postulate, there might not exist a segment being the limit of each partition of the straight line in two parts A and B so that each segment of A is to the left of each segment of B. This holds only if a further condition also holds: "There is a segment x in the part A and a segment y in the part B so that the difference between y and x becomes infinitely small". If a is infinitely small with respect to b, the difference between a and b does not become infinitely small and the continuum contains a gap, but if we restrict ourselves to finite segments, then Dedekind's condition holds and there are no gaps.

2.3 How the epistemological model affects continuity

The difference between the approaches of Dedekind and Veronese is relevant not only from a theoretical but also from an epistemological point of view. Veronese believed that the geometrical continuum should not be defined as a system of points but as a system of segments that should not and could not be reduced to a system of numbers. Refusing the idea of defining the continuity of space by means of the continuity of real numbers, Veronese did not assume the Archimedes' principle as a necessary element for the continuity of a geometrical system of magnitudes. If Veronese had assumed the real number system as the privileged model for the description of geometrical magnitudes, this would have hindered the discovery of an alternative description of spatial continuity.

Veronese's results, stemming from a combination of an empirical model for continuity, a thought model for order and equality, and a projective model for the foundation of the theory of extensive magnitudes, affected the meaning of the concepts of order, continuity, group, magnitude and number. Numbers were considered as essentially ordinal (a cardinal number being, as Cantor's power of a set, the result of a double abstraction from the nature and from the order of the elements of the set [8, p. 411]) and were introduced in two independent ways. Natural numbers were introduced as the result of an act of thought — the counting of the elements of an ordered set. Continuous numbers — real and non-Archimedean numbers — were introduced by association to a given system of geometrical magnitudes. The properties of numbers derive from the properties of magnitudes and not vice versa. According to Veronese, the continuity of numbers should be modelled on (since it is derived from) the continuity of the geometrical rectilinear line. Which numerical system should be associated to a given system of magnitudes depends on the properties of the magnitudes, that is to say, on the properties of the spatial continuum that one is not able to perceive but that one can represent to oneself by means of an abstract intuition.

3 Rodolfo Bettazzi

Rodolfo Bettazzi's mathematical works did not receive much attention from contemporaries and have been largely ignored both by historians of mathematics and by philosophers of science. Apart from some studies on Bettazzi's criticism of the axiom of choice [12] and from recent historical research on Peano's school, there is scarcely any literature on Bettazzi's writings.[3] Bettazzi's main work is a monograph on magnitudes that was awarded a prize by the Accademia Nazionale dei Lincei in 1888. Betti, Beltrami, Cremona, and Battaglini — members of the Prize Commitee — remarked that Bettazzi's *Teoria delle grandezze* was an original study in line with Grassmann's, Hankel's, Stolz's, and Cantor's writings [20].

The book [2] appeared just before Veronese's *Fondamenti di geometria*, but the two authors reached their results independently.[4] A comparison between the two works shows remarkable similarities in the results and in the epistemological background, but also a marked difference in the mathematical approach to the enlargement of the numerical domain and in the general aim.

3.1 The epistemological model

According to Bettazzi, the objects of a science are ideal and under-determined with respect to the properties of the objects of the real world, for only certain properties are defined and considered as relevant. Such objects are mere concepts and their properties might have similarities with the properties of real objects (for example in geometry) but might also be introduced independently according to certain specific aims. If the existence of an object is accompanied by the determination of the properties of the object, one has a definition of the object itself: so, if one says that there exists an object with certain characteristics, that is a definition of the object [2, p. 3 ff.]. Before Peano [31] commented on the topic, Bettazzi distinguished between a direct definition, that is to say a definition that aims at defining what an entity is in itself, from a relational definition that defines entities by their reciprocal

---

[3] The oblivion of Bettazzi's works might be partly due to the fact that he never attained an academic position nor published in international reviews. A recent historical study on non-Archimedean mathematics that dedicates a whole paragraph to Bettazzi and analyses the originality of his contribution to the topic is Ehrlich 2006 [18]. Since it is not focused on Bettazzi, it does not discuss Bettazzi's epistemology.

[4] Veronese remarked in a footnote that the work of Bettazzi came to his notice when his own book was already getting into print.

relations. Every definition is an existential definition asserting the possibility of the attribution of certain properties to a given concept: some properties are attributed to the introduced entity in itself or to its relations to other previously introduced entities; some properties express relations between entities that belong to the same category one wants to define.

Before introducing a precise definition of the concept of magnitude, Bettazzi makes some remarks on mathematical entities. All scientific entities need to be well defined, at least with respect to their relevant properties. Scientific entities are ideal because only some of their properties are taken as relevant. These properties might or might not be similar to the properties of certain objects of the real world, for certain entities derive from the observation of the external world while other entities are introduced according to special purposes. Like Veronese, Bettazzi makes a distinction between entities that are somehow connected to our experience and entities that are independent from it.

Scientific entities are pure concepts whose properties are expressed by contemporaneity of certain concepts with others: non-contradiction means "possible contemporaneity" of the concepts. Bettazzi's terminology here is similar to that of Grassmann, who used the expression "Vereinstimmung" to express the coherence of different acts of thought. The properties of the entities are called postulates and the existence of the entities is itself a postulate. Bettazzi is a conceptualist, because he considers scientific entities as ideal and believes that their properties might be arbitrarily chosen, provided that no contradiction arises. On the other hand Bettazzi, like Veronese, is very much concerned with experience and seems to believe that most mathematical concepts are derived from the observation of an external reality. Space and time cannot be a priori concepts but are rather relational concepts that have to be introduced by defining what it means that two spaces or two times are equal. Time cannot be defined in itself. Analogously all concepts should be introduced by defining relations of equality or inequality.

Refusing the idea of deriving the properties of magnitudes from the properties of the real numbers that are used to measure them, Bettazzi intends to build a rigourous system of magnitudes without presupposing the notion of number. He aims at deducing the properties of real numbers from the properties of magnitudes. In an article on the concept of number [1, p. 98 ff.] Bettazzi gives some reasons for introducing magnitudes independently from numbers. He recalls the distinction between two ways of introducing real numbers: a synthetic and an analytic way. According to the synthetic way, a number represents the ratio of a magnitude to a magnitude of the same species, the unity. According to this point of view, the number indicates the way a magnitude can be obtained from the unity of its category. Examples of magnitudes are aggregates of equal, separated objects, aggregates of their parts, segments, angles, surfaces, solids, times, weights, and so on. The notion and the properties of operations on numbers (such as commutativity or transitivity) must derive from the correspondent properties of magnitudes and have to be demonstrated as theorems rather than be assumed as definitions.

While in the synthetic approach numbers have a concrete meaning that derives from their being introduced as ratios of magnitudes, according to the analytic point of view, numbers are devoid of any concrete meaning. The properties of numbers depend on the formal properties of certain abstract operations, because numbers are first introduced as the elements of the given operations and can be generalized only if the properties of those operations are preserved and certain impossibilities eliminated. For example, natural numbers are generalized into integers so as to make subtraction possible, integer numbers are generalized into rational numbers so as to make division possible, rational numbers are extended by the introduction of certain real numbers so as to allow the operation of extracting the root of any positive number, and so on. A main difficulty of this approach consists, according to Bettazzi, in the fact that one does not know exactly where one should stop in this procedure of generalization or when one would have enough numbers to measure magnitudes.

Advantages and disadvantages of the synthetic and analytic approach are discussed in an article entitled "Sui sistemi di numerazione per i numeri reali" [3], where Bettazzi argues that the definition of real numbers as an extension of rational numbers is not convincing for two reasons: 1) it introduces a dishomogeneity, for it is not based on the closure of certain operations that should be made possible, but rather on a completely different concept: the limit; 2) it presupposes a property of extensive magnitudes, i.e. their undenumerable continuity. As a result, Bettazzi argues that those who intend to define the real numbers as successive enlargements of the natural numbers can never obtain a unitarian notion of number, but rather only give many different and separate constructions of rational, irrational, and negative numbers, so that including them all into a single concept of real numbers would be quite arbitrary. This criticism sheds doubts on the legitimacy of the arithmetization of analysis.

Similar remarks can be found also in Cesare Burali-Forti and Sebastiano Catania's works, which were, like Bettazzi's, influenced by Grassmann's writings. In his book on numbers and magnitudes [14, pp. vi-vii], Catania wanted to "deduce the whole class of absolute real numbers from magnitudes and the partial classes of integers and rational numbers therefrom. It is an inversion of the usual procedure, which first defines different entities in different ways and then identifies them afterwards to preserve the ordinary properties". Burali-Forti wrote similar remarks in his note on magnitudes [6] and in his book *Logica matematica* (especially in the 1919 edition) [7, pp. 323–4]. He argued that since defining real numbers from natural numbers is quite complicated and inconvenient, the simplest and clearest way to introduce numbers is to define them as corresponding to magnitudes.

3.2 Bettazzi's theory of magnitudes

In his book *Teoria delle grandezze* [2] Bettazzi defines magnitudes as the entities of a category that can be compared with respect to a relation of equality or inequality. A class of magnitudes is defined as a structure composed by a set and an additive operation that is associative, and commutative. In modern parlance, a class of magnitudes is an abelian additive semigroup. The introduction of an order relation allows a distinction between one-dimensional (linearly ordered abelian monoids), multi-dimensional (complex), and non-dimensional classes.

Bettazzi considers several properties of classes, such as that of being one-directional, limited, proper, isolated. One-directional classes correspond to what is now called positive or negative cone of a linearly ordered group. A class is limited if it has an inferior limit which is different from the neutral element. It is proper if the difference of two magnitudes belongs to the class whenever the minuend is greater than the subtrahend. A limited proper class can be ordered by a repetitive application of the additive operation to the limiting magnitude (it is right-solvable). A limited proper class is discrete: it contains the neutral element, a least element (the unit) and its multiples. A class is isolated if the module magnitude (the neutral element) is the only element which is smaller than any magnitude in the class and if the infinite magnitude (the absorbing element) is the only element which is bigger than any magnitude in the class (i.e. a class is isolated if the neutral element is the only least element and the absorbing element is the only greatest element). Should an isolated class be embedded into another class, any least element will be considered as equal to the neutral element 0 and any greatest element will be considered as equal to the absorbing element Ω:

(*) if $a^* < a$ for any a in G, then $a^* = 0$ and if $b^* > b$ for any b in G, then $b^* = \Omega$.

The procedure of isolating a class is very interesting, for it explains how the same class might be considered as containing or not containing infinitesimal magnitudes. For example, Veronese's non-Archimedean system would be Dedekind-continuous if the class containing the unit were considered as isolated. Bettazzi remarks that a new definition of equality is at stake when one considers a class as isolated: two magnitudes of a class H are equal to 0 when they diverge by a magnitude that is smaller than any magnitude of a subclass G and are equal to Ω if they contain a magnitude that is greater than any magnitude of the subclass G. If one does not want to modify the definition of equality, then one must assume the postulate (*) in order to consider a class as isolated. Bettazzi acutely observes that the postulate is implicitly assumed whenever one applies a specific name to the magnitudes of a certain category, because the exclusive name means that other things should not be considered as comparable to the given magnitudes. For example, if one defines segments as sets of consecutive points, one is thereby using an exclusive name that "isolates" the class of magnitudes that are called "segments": infinitely small or infinitely great entities are thus considered as not comparable to segments, that is, as equal to 0 or to Ω respectively.

Other features of dimensional classes are related to how an ordered class can be divided into subclasses. Connected classes contain only links or sections. Closed classes do not contain sections but contain the limit of every section of their subclasses. Continuous classes, being connected and closed, contain only links. Archimedean classes contain no gaps and are called classes of the 1st species, while non-Archimedean classes are called classes of the 2nd species.

Having introduced all these properties of classes of magnitudes independently of numbers, Bettazzi turns to the introduction of *numbers*: given a one-dimensional class of magnitudes, Bettazzi associates a number to each magnitude, then introduces a relation of equality and an operation of addition, and shows that these numbers form a class with the same properties of the correspondent class of magnitudes. Since numbers are associated to magnitudes, a relation of equality might be defined between numbers on the basis of the equality of the corresponding magnitudes. Bettazzi analogously defines other properties of numbers. In modern parlance, one could say that he introduces numbers by means of an isomorphism μ between a class of magnitudes G (which is an ordered abelian monoid) and a class of numbers K.[5] The system of numbers associated to a class of magnitudes is thus itself an ordered abelian monoid.

Bettazzi remarks that numbers are mathematically relevant because the same numerical system can be associated to different classes of magnitudes that have something in common — or, as Bettazzi expresses it, belong to the same category. Bettazzi defines two classes as belonging to the same category if they can be shown to have a correspondence that preserves the relation of order, the additive operation, the module magnitude and the infinite magnitude. In modern parlance, two ordered monoids belong to the same category, if they are isomorphic. The isomorphic function f that establishes the correspondence between the two classes of magnitudes is called a *metrical correspondence*; it allows a partition of classes of magnitudes in different categories: discrete, rational, continuous, and so on. The concept of metrical correspondence is an abstract algebraic notion that does not presuppose the notion of number: examples of metrical correspondence are both the mapping of a discrete (rational, continuous) class into any other discrete (rational, continuous) class and the mapping of a class of magnitudes into a numerical system.

Numerical systems are introduced as the systems that correspond to a class of a given category (and thus to all classes of the same category, for each class is homomorphic to each other) and can thus be used to represent distinct categories. The class of integer numbers is the class of numbers associated to discrete classes: it contains 0 (module magnitude), 1 (the unity), all multiples of the unity and ω (infinite magnitude):

$$I = 0, 1, (1+1), \ldots, (1+1+1++1), \ldots, \omega.$$

The class F of fractional numbers is the class of numbers associated to rational classes and it contains 0, the number associated to a rational magnitude a, its multiples and ω: $F = 0, a, (a+a), \ldots, (a+a+a++a), \ldots, \omega$. The class of fractional numbers contains the class of integer numbers as a subclass (for $a = 1$). A continuous class of numbers (real numbers) is associated to continuous classes of magnitudes.

Bettazzi finally introduces a representation theorem, which asserts that any continous class can be put into a metrical correspondence with the class of real numbers. Metrical correspondence is clearly distinguished, in Bettazzi's terminology, from measurement. The distinction is quite subtle but denotes

---

[5]Bettazzi does not explicitly say that the correspondence should be one-to-one.

a profound algebraic insight: a metrical correspondence is an homomorphism of a class of magnitudes (a specific set with a certain structure) into another class of magnitudes (a specific set with a certain structure), whereas measurement is the mapping of any class of magnitudes of a certain kind (a generic structure) into a class of numbers (a numerical structure). Measurement can thus be defined only after both metrical correspondence and numerical systems have been introduced: the representation theorem asserts that all continuous classes can be put into a metrical correspondence with the system of real numbers. In the last paragraphs of *Teoria delle grandezze* Bettazzi associates numbers to one-dimensional classes of 2nd species and generalizes the representation theorem to non-Archimedean classes of magnitudes.

3.3 How the epistemological model affects measurement and magnitude

An implication of the epistemological choice to introduce the properties of numbers synthetically is that they can be derived from the properties of magnitudes, which are assumed by definition. Bettazzi considers the synthetic method as more simple, intuitive, and comprehensible, but he acknowledges the risk of limiting the possible extensions of the notion of number, if the last is rooted to certain concrete classes of magnitudes. The risk might be avoided if one includes the study of classes of magnitudes that cannot be concretely imagined: this is exactly what Bettazzi does when he considers classes of more dimensions or one-dimensional classes of 2nd species. Bettazzi's synthetic approach is an abstract approach to the study of algebraic ordered structures. Although the epistemological background is similar to that of Veronese, Bettazzi's aim is quite different: a general investigation of magnitudes rather than a geometrical description of the intuitive continuum. Bettazzi is more influenced by the conceptualism of Grassmann than by the empiricism of Pasch.

Bettazzi extends the notion of measurement to non-Archimedean classes but assumes a continuous class of magnitudes to be Archimedean. The notion of measurement does not entail Dedekind's continuity nor monotonicity, but it cannot be defined in classes with n dimensions, because they lack an ordering. The definition of measurement presupposes the definition of a class of magnitudes as an abelian ordered monoid. That is a reason why Bettazzi's abstract approach marks a significant step towards the axiomatization of the theory of magnitudes, which is usually attributed to Otto Hölder [24].

4 The role of epistemological models

Both Bettazzi and Veronese adopt old epistemological models in an original and fruitful way. Bettazzi follows Grassmann's algebraic approach and develops a general theory of magnitudes independently of numbers, associating numerical systems to categories of magnitudes. Veronese follows Grassmann in the effort of developing geometry without numbers and associates a given system of numbers to a particular system of geometrical magnitudes. Both consider systems of numbers as something that has to be associated to previously defined classes of magnitudes.

Both Bettazzi and Veronese are concerned with the notion of ordinal number rather than with the notion of cardinal number. Veronese does not intend to derive cardinal numbers from ordinals: he explicitly introduces natural numbers as concepts deriving from the act of counting. Bettazzi tries to define real numbers without any reference to natural numbers, but he ends by presupposing their existence in several passages of his text, as Peano critically remarked [30].

In the writings of Veronese a new epistemological model begins to emerge: instead of the result of a successive enlargement of the domain of natural numbers, real numbers are considered as entities that can be defined in terms of richer systems of numbers: they are a subclass of the non-Archimedean numbers. This is due not only to the fact that attention is drawn to order but also to the fact that real numbers are considered as the final point of the enterprise rather than as its point of departure.

In the writings of Bettazzi the properties of continuous classes of magnitudes are similarly derived from abstract properties of general categories of classes of magnitudes. Nonetheless real numbers play a relevant role in Bettazzi's system, because Bettazzi, unlike Veronese, conceives of non-Archimedean numbers as hypercomplex numbers. Real numbers play a similar role in some works of Grassmann, especially in the second edition of the *Ausdehnungslehre* [21], where the philosophical approach is abandoned in favour of a widespread analytical notation [10].

Even if, from a strictly foundational perspective, neither Bettazzi nor Veronese develop a theory of magnitudes without numbers, what is radically new in their effort is the conception of numbers as a special case of an algebraic structure and the conception of the properties of real numbers as a special case of more general properties of ordered structures.

The abstract approach promoted by the synthetic models of Bettazzi and Veronese did not only contribute to a better understanding of the notion of magnitude but also induced an inversion of the defining techniques. The construction of the real numbers is obtained by a one-to-one correspondence with a previously given domain of magnitudes. The introduction of abstract categories of magnitudes allows the construction of new numerical systems that do not necessarily result from the analytical need to make certain operations possible, as in the usual procedures for enlarging the numerical domain.

The approach is top-down rather than bottom-up. Instead of enlarging smaller systems, one starts from larger systems and isolates subsystems by the introduction of new conditions. Following this approach real numbers can be identified as the largest Archimedean sub-field of an ordered field. This approach is radically different from the construction of hyperreal numbers by the enlargement of the system of real numbers: instead of assuming real numbers as a starting point and trying to insert new entities in the given domain, one starts from general properties of classes of magnitudes (ordered fields) and then isolates real numbers by means of the Archimedean property.

This approach has the advantage of avoiding ontological questions. Moreover, it is intrinsically devoted to the comparison of a plurality of models rather than to the search for "the" model of a categorical theory. Studies concerning the definition of real and hyperreal numbers as real closed fields are fruitful results of such an approach, which is interested not only in isomorphism but also in the study of common properties of non-isomorphic models (such as R and R*).

Acknowledgements

The author is grateful to Silvio Bozzi and to an anonymous referee for helpful comments.

BIBLIOGRAPHY

[1] R. Bettazzi. Sul concetto di numero. *Periodico di matematica per l'insegnamento secondario*, 2: 97–113, 129–143, 1887.
[2] R. Bettazzi. *Teoria delle grandezze*. Spoerri, Pisa 1890.
[3] R. Bettazzi. Sui sistemi di numerazione per i numeri reali. *Periodico di matematica per l'insegnamento secondario*, 6: 14–23, 1891.
[4] R. Bettazzi. Sull'infinitesimo attuale, *Rivista di matematica*, 2: 38–41, 1892.
[5] A. Brigaglia. Giuseppe Veronese e la geometria iperspaziale in Italia. In *Le scienze matematiche nel Veneto dell'Ottocento. Atti del Terzo Seminario di Storia delle Scienze e delle Tecniche nell'Ottocento Veneto*, Istituto Veneto, Venezia 1994, pages 231–261.
[6] C. Burali-Forti. Sulla teoria delle grandezze, *Rivista di Matematica*, 3: 76–101, 1893.
[7] C. Burali-Forti. *Logica matematica*. Hoepli, Milano 1919.
[8] G. Cantor. Mitteilungen zur Lehre der Transfiniten, 1887–1888. Repr. in G. Cantor, *Gesammelte Abhandlungen mathematischen und philosophischen Inhalts*. Olms, Hildesheim 1966, , pages 378–439.
[9] P. Cantù. *Giuseppe Veronese e i fondamenti della geometria*. (Italian) [Giuseppe Veronese and the Foundations of Geometry]. Unicopli, Milano 1999.
[10] P. Cantù. *La matematica da scienza delle grandezze a teoria delle forme. L'Ausdehnungslehre di H. Grassmann*. (Italian) [Mathematics from a Science of Magnitudes to a Theory of Forms. The *Ausdehnungslehre* of H. Grassmann]. Dissertation. Università degli Studi di Genova, Genova 2003.
[11] P. Cantù. Le concept de l'espace chez Veronese. Une comparaison avec la conception de Helmholtz et Poincaré. *Philosophia Scientiae*, 13 (2): 129–149, 2009.
[12] J. Cassinet. Rodolfo Bettazzi (1861–1941), précurseur oublié de l'axiome du choix. *Atti Accad. Sci. Torino, Cl. Sci. Fis. Mat. Natur.*, 116: 169–179, 1982.
[13] E. Cassirer. *Substanzbegriff und Funktionsbegriff*, 1910. Engl. Transl. *Substance and Function, and Einstein's Theory of Relativity*, Mineola, N.Y., Dover 2003, , pages 1–346.
[14] S. Catania. *Grandezze e numeri*. Giannotta, Catania 1915.
[15] D. Corfield. *Towards a Philosophy of Real Mathematics*. Cambridge University Press, Cambridge 2003.
[16] R. Dedekind. Stetigkeit und irrationale Zahlen, 1872. Engl. Transl. *Essays on the Theory of Numbers: I. Continuity and Irrational Numbers. II. The Nature and Meaning of Number*. Dover, New York 1963, pages 1–28.
[17] P. Ehrlich. Hahn's *Über die nichtarchimedischen Grössensysteme* and the Origins of the Modern Theory of Magnitudes and Numbers to Measure Them. In J. Hintikka (ed.), *From Dedekind to Gödel. Essays on the Development of the Foundations of Mathematics*, Kluwer, Dordrecht 1995, pages 165–213.
[18] P. Ehrlich. The Emergence of non-Archimedean Grössensysteme. The Rise of non-Archimedean Mathematics and the Roots of a Misconception. Part I: the Emergence of non-Archimedean Grössensysteme. *Archive for History of Exact Sciences*, 60: 1–121, 2006.
[19] G. Fisher. Veronese's non-Archimedean Linear Continuum. In P. Ehrlich (ed.), *Real Numbers, Generalisation of the Reals and Theories of Continua*. Kluwer, Dordrecht 1994, pages 107–145.
[20] L. Giacardi and C.S. Roero. La nascita della Mathesis (1895-1907). In L. Giacardi & C.S. Roero (eds.), *Dal compasso al computer*, Mathesis, Torino 1996, pages 7–49.
[21] H.G. Grassmann. *Die Ausdehnungslehre, vollständig und in strenger Form bearbeitet*, 1862. Engl. Transl. *Extension Theory*, American Mathematical Society/London Mathematical Society, Providence/London 2000.

[22] H. Hahn. Über die Nichtarchimedischen Grössensysteme, *Sitzungsberichte der kaiserlichen Akademie der Wissenschaften Wien*, math.-naturwiss. Klasse IIa, 116: 601–655, 1907.
[23] D. Hilbert. *Grundlagen der Geometrie* (1899). Engl. Transl. *Foundations of Geometry*, Open Court, Chicago 1987.
[24] O. Hölder. Die Axiome der Quantität und die Lehre vom Mass (1901). *Berichten der math.-phys. Classe der Königl. Sächs. Gesellschaft der Wissenschaften zu Leipzig*, 53: 1–64, 1901. Engl. Transl. in J. Michell and C. Ernst, The axioms of quantity and the theory of measurement, *Journal of Mathematical Psychology*, 40: 232–252, 1996 and 41: 345–356, 1997.
[25] F. Klein. On the Mathematical Character of Space-intuition and the Relation of Pure Mathematics to the Applied Sciences, 1893. In F. Klein, *Gesammelte mathematische Abhandlungen*, Springer, Bessel-Hagen/Berlin 1921, vol. 1, pages 225–231.
[26] T. Levi-Civita. *Opere Matematiche*, Zanichelli, Bologna 1954.
[27] T. Levi-Civita. Sugli infiniti ed infinitesimi attuali quali elementi analitici (1892–93). In [26], pages 1–39.
[28] T. Levi-Civita. Sui numeri transfiniti (1898). In [26], pages 315–329.
[29] P. Natorp. *Die logischen Grundlagen der exakten Wissenschaften*, 1910. Repr. Sändig, Wiesbaden, 1969.
[30] G. Peano. Sul concetto di numero. *Rivista di matematica*, 1: 256–257, 1891.
[31] G. Peano. Le definizioni in matematica (1921). Engl. Transl. in *Selected works of Giuseppe Peano*, Allen & Unwin, London 1973.
[32] B.A.W. Russell. *An Essay on the Foundations of Geometry* (1897). Repr. Dover, New York 1956.
[33] A. Schoenflies. Sur les nombres transfinis de Mr. Veronese. *Rendiconti Accademia Nazionale dei Lincei*, 6 (5): 362–368, 1894.
[34] O. Stolz. Zur Geometrie der Alten, insbesondere Über ein Axiom des Archimedes. *Berichte des Naturwissenschaftlich-Medizinischen Vereines in Innsbrück*, 12: 74–89, 1882.
[35] G. Veronese. Il continuo rettilineo e l'assioma V di Archimede. *Memorie della Reale Accademia dei Lincei, Atti della Classe di scienze naturali, fisiche e matematiche*, 6 (4) :603–624, 1889.
[36] G. Veronese. *Fondamenti di geometria a più dimensioni e a più specie di unità rettilinee esposti in forma elementare*. Padova: Tipografia del Seminario, 1891. German transl. *Grundüzge der Geometrie von mehreren Dimensionen und mehreren Arten gradliniger Einheiten in elementarer Form entwickelt*, Teubner, Leipzig 1894.
[37] G. Veronese. Osservazioni sui principii della geometria. *Atti della Reale Accademia di scienze, lettere ed arti di Padova*, 10: 195–216, 1893–94.
[38] G. Veronese. *Appendice agli Elementi di Geometria*. Drucker, Verona/Padova 1897.
[39] G. Veronese. *Il vero nella matematica.*. Forzani e C., Roma 1906.
[40] G. Veronese and P. Gazzaniga. *Elementi di Geometria, ad uso dei licei e degli istituti tecnici* (primo biennio). Drucker, Verona/Padova 1895–97.
[41] G. Vivanti. Sull'infinitesimo attuale. *Rivista di Matematica*, 1: 135–153, 248–255, 1891.

PART III

LIFE SCIENCES

Evolution by increasing complexity in the framework of Darwin's theory

PIETRO OMODEO

The theory of evolution, as elaborated by Ch. Darwin during 25 years of research marked by five main texts (1842, 1844, 1858, 1859, 1868), consists of a few propositions which will be examined in this article. This survey will not take into account the digressions made by the author to orient himself; moreover, I will also ignore various concessions of that he made in the hope of reconciling various opponents and doubters to his own views [20].

Some propositions that constitute the theory had been partly stated by precursors, but were recast in Darwin's own original blend. I will be briefly state them before proceeding to the main theme of this paper.

1 Propositions constituting Darwin's evolution theory

A detailed analysis of the logic underlying Darwin's theory of evolution theory has already been made by myself [17, 18] as well as — to cite a few — Sober [24], Brandon [1], Pievani [22]. Here I will summarize, with slight modifications, ideas on the subject expounded in earlier writings of mine.

1.1 First proposition, about the exponential growth of populations of living beings

Every population of living beings tends to grow exponentially till the environments capacity to provide support gets saturated; then a stationary state is entered in which said population necessarily looses a share of the offspring of each generation, who will either die precociously or fail to reproduce. Populations in nature are, almost always, in a stationary state; therefore this proposition is often stated as follows: due to subsistence depletion and to the ensuing fight for survival, every population is doomed to loose a share of the individuals of each generation.

The proposition cited here derives from ideas of the geographer and sociologist Th. R. Malthus [12] as well as from the principle of *struggle for survival* put forward by the botanist De Candolle [9, 10], and should be considered an axiom. Even so, within the framework of studies on population dynamics, it has been the subject of countless checks and experimental verifications and turned out to be true for all categories of living beings, ranging from bacteria to plants and higher animals.

It should be added that this postulate of Darwin's theory focuses on self-reproducing entities, excluding the evolution of entities which are exclusively driven by deterministic causes, such as chemical elements, stars, and rocks.

1.2 Second proposition, concerning mortality and prolificacy

In stationary-state populations, if environmental conditions stay fixed, mortality and prolificacy rates are not the same for all phenotypes, i.e. for all groups of individuals showing alternative characters, occurring *in it*.

This proposition too, despite having an axiomatic character, has been checked directly and indirectly in quite many ways, and got invariably confirmed. Outlines of it can be found in various authors before Darwin, but it was Darwin himself who attached a special emphasis on it, since 1839, in those pages of *Journal of Researches into the Natural History* [2] that treat an hereditary modification in cattle.

The above proposition gets often absorbed, according to many authors, into the following:

1.3 Third proposition: the selection principle

The population/environment interaction is such that every inheritance-based novelty, i.e. every mutation, either morphological, physiological, or behavioural, that shows up in one or more individuals of a population gets either accepted or rejected (i.e., it undergoes selection) in the course of one or several generations.

This proposition was stated by Ch. Darwin, who named it "selection principle", vigorously stressing from its very outset its hereditary basis.

Before Darwin, a proposition similar but unrelated to the genetic basis of physiological or morphological novelties had been put forward by E. Blyth and by E. Spencer. According to the ornithologist Blyth, interaction with the environment and thus causes the disappearance of plus- and minus-variants (in our terms), maintains species in the original integrity as established by the Creator. Spencer's position was instead closer to Darwin's one, and had been synthesized by the locution "survival of the fittest".

This locution, "survival of the fittest", was often ascribed, erroneously, to Darwin himself, whereas he cautiously adopted it only in the fifth edition of *The Origin of Species*. Moreover, as this preposition is often regarded, wrongly, as the most authentic and genuine statement of evolutionism, opponents declared it tautological and pointless, and hence declared the whole theory meaningless.

As a matter of fact, when stated in this fashion, the proposition can be judged as being true only *a posteriori*. Therefore, it is best not to state it in this form in order to avoid useless and annoying disputes.

Since the third proposition could be regarded as being devoid of predictive power, in order to make it clear that it may be also valid *a priori*, it is worthwhile to formulate it as follows: inheritance-based novelties showing up in populations of living beings in a given environmental context can either be accepted, provided they enhance reliability (measured according to the criteria of cybernetics), and/or the lower cost of the organism's functioning (measured according to the criteria of thermodynamics).

Assessing *a priori* the reliability and low-cost functioning of a mutant organism relative to a given environment is far too easy in certain cases, (e.g. albinism of an animal of the tropical steppe or blindness of a cave-dwelling animal), but it turns out to be problematic in most cases, and it becomes impossible when one is confronted with minor variants in an unstable environment.

1.4 Fourth proposition, about perpetual mutability

Every population of living beings has an inexhaustible source of hereditary variations, namely mutations.

This proposition ensues from an assumption to which Darwin devoted himself for long, well aware of the fact that if the selective processes relied simply on the variability existing at the outset of the evolutionary vicissitudes, they would soon come to an end. Indeed, once all features which are unfit for the original environment had been eliminated, the population would have become almost uniform and would be unable to cope with any new constraint arising from any further modification of the environment.

Only in 1867, around the end of his treatise on the variations among domestic animals and cultivated plants, did Darwin start investigating the sources of perpetual variability. He ascribed it to: 1) domestication; 2) unspecified sporadic events, or "sports"[1]; 3) the inheritance of acquired features.

The causes of hereditary variations proposed by Darwin, and particularly the last one, were soon subject to criticism, and they are still criticized today. Even harsher criticism was encountered by an explanatory theory which he called "pangenesis". For a long time only "sports" were accepted as the cause of mutation.

De Vries set up experiments to answer the question but met with little success, having chosen organisms unsuitable to the case. Not until 1927 did H. Muller find a physical cause of mutation in ultraviolet rays and in ionizing radiation (X rays, γ rays); this discovery earned him the Nobel prize in 1946. At the beginning of the second world war the geneticist Charlotte Auerbach identified yprite (*mustard gas*) the first known chemical compound endowed with a mutagenic action; later on, many more have been identified, and for many of them the intimate mechanism of reaction with DNA has been clarified. Around 1980 it was confirmed on experimental grounds that thermal shock increases genetic mutability and that shocks of other types may have the same effect, thus rendering Ch. Darwin's first hypothesis plausible.

New developments on the theme of mutations enabled identification of mutations affecting, not one or a few nucleotides of single genes, but entire genes, parts of chromosomes and even entire genomes. Moreover, molecular biology brought to light repair mechanisms which restore the condition altered by some mutagenic agents, and have also revealed the possibility that genes that have remained inactivated for a long time are irreversibly inactivated through methylation of a part of the constituent nucleotides.

To conclude, research on mutagenesis *sensu lato* has opened up various perspectives on evolution not all of which have been taken into due account by the scholars studying it, some of whom have entrenched themselves behind a heavy and limiting orthodoxy.

2 The theory as a whole

Darwinian theory based on these propositions, predicts that when a biological, physical or chemical parameter of the environment changes, the populations

---

[1] Sudden spontaneous deviations from type beyond the limits of individual variation (from the jargon of horticulturist and floriculturist).

that draw their subsistence from it respond by an *adaptive process*.

Therefore, individuals bearing characters alternative to those typical of the majority and more suitable for the new circumstances will be favoured as breeders, will obtain a reproductive premium. If these characters are inheritable, as is the rule, the cumulative hereditary pool (cumulative genome) gradually changes and the whole population becomes adapted. Selective processes would tend to impoverish the variability reserve of this patrimony which, however, is continuously supplied through the appearance, by mutation, of additional alternative novelties.

Darwin concluded that adaptive processes produce new races and, in the long run, new species. To illustrate this, from his earliest manuscript [3] on, he gave the example of the transformation of a population of "common dogs", i.e. generically adapted dogs, in a changing environment. This example, that happily summarizes the whole theoretical scheme, appears again and again in Darwin's works on evolution.

The population of common dogs lives by preying on rabbits, but when the latter, because of an adverse environmental change, slowly decline until they disappear, they are gradually substituted by hares which are much swifter and more difficult to catch. Only the quicker and more agile dogs will now succeed in appeasing their hunger and will become the more effective breeders (of the next generation). After an adequate numbers of generations the whole population of common dogs will be transformed into a population of greyhounds[2].

In this example it is tacit that the peculiarities conferring an advantage to the breeders are inheritable. It is also understood that the adaptation so acquired may lead to new species. On the origin of more dramatic transformations, which would render the descendants of the modified lineage such as to be classified as a new family, Darwin did not commit himself.

Darwinian theory applied to the evolution of living beings is as logically sound as the Newton's theory of gravitation. It has been experimentally confirmed and has decisively contributed to the development of taxonomy, genetics, biochemistry, molecular biology. Nevertheless it has been, and still is, the target of much more criticism than gravitation.

This is because determinism is more readily accepted in physics than in biology: if a plane takes off, or a rocket escapes from the earths gravitational field, nobody shouts that this is evidence for rebutting gravitation theory. Instead, when it was ascertained that in many, and perhaps in all, natural populations some biochemical traits show a stable genetic polymorphism, some specialists declared the selection principle to be false. The reason being that this principle, when bureaucratically applied, predicts that sooner or later only one among many alternatives characters, the "best one", must prevail and all others be eliminated.

---

[2] "Let hares increase very slowly from change of climate affecting peculiar plants and some other rabbit decrease in same proportion [...] a canine animal who formerly derived his chief sustenance by springing on rabbits or running them by scent must decrease too and might thus readily become exterminated. But if its form varied very slightly, the longlegged fleet ones, during a thousand years being selected, and the less fleet rigidly destroyed, must, if no law of nature be opposite to it, alter forms" [3, p. 64].

It is a fact, however, that for all populations, polymorphism is per se an effective defence against parasite micro-organisms and also against some predators. Thus, as the theory correctly interpreted predicts, selection has favoured precisely those devices which protect polymorphism.

3 New and old views on macroevolution and philogenesis

Let us hence lay aside such objections, which do not jeopardize the theory, and deal with a problem that is central to the field of evolutionary studies: Darwinian theory applies, obviously, to microevolution; yet, does it apply equally well to macroevolution?

Darwin did not answer this question explicitly, but pragmatically said: for the time being let us work in a field in which the theory holds true and can be verified, i.e. on the origin of species; later on we will explore other instances.

This attitude has been shared, until recently, by many researchers, and especially those with a background in genetics, some of whom have clearly stated that Darwinian theory is valid for the microevolution of "Mendelian" populations, i.e. for diplontic populations in which sexuality involves on an alternation of karyogamy and meiosis and that to venture into the evolution of classes, phyla and kingdoms is an unwise choice where one risks uttering senseless opinions.

Ernst Haeckel turned a deaf ear to o exhortations to prudence and, following Goethe and Lamarck, committed himself to unraveling the intricate problem of phylogeny. He was confident that the comparative method and his "ontogenetic law", i.e. the principle that embryonic development recapitulates the evolutionary history of a taxon, would provide a reliable guide. Among the evolutionary mechanisms called up by Haeckel was the inheritability of acquired characteristics [19].

Haeckel's ideas have been discredited, but phylogeny has been (and is still being) extensively explored by comparative methods, by considering either morphological and functional features, or nucleotide sequences in DNA. Often with conflicting results.

The question about the causes of macroevolutive process however has for long remained unanswered. A few decades ago certain well known biologists, including Albert Vandel and P.P. Grassé, maintained that the mutation/selection mechanism does not work in the case of transgeneric evolution, where other causes, other mechanisms must come into play.

In more recent times, researchers have suggested that macroevolution may depend on macromutations like the *Aristapedia* or *Bithorax* mutations described in *Drosophila* at the beginning of the past century. The "hopeful monsters" so produced could become the founders of new higher taxa. These lethal or semi-lethal mutations were eventually found to derive from genes controlling morphogenesis in all animals, from the lower invertebrates to the higher vertebrates with but little differences at a molecular level. As to the hopeful monsters, somebody claimed that, like all monsters, they are produced by *el sueño de la razón*. Recently Pievani [22] has written: "empirical evidence does not exist to justify a *discontinuity* between the mechanisms of microevolution and evolution on a broad scale". The following pages are devoted to this problem.

4 Progressive evolution, or evolution by complexification

4.1 The appearance of new genes and the origin of a new phyletic line

On various occasions [17, 18] I have emphasized the opportunity of introducing the two categories "evolution by adaptation" and "evolution by complexification"[3].

Evolution by adaptation was rationally supported by Darwinian theory and strengthened by convincing experiments. It has qualitative mutations of the genome as immediate causes and does not affect the complexity of the mutant phenotypes and their offspring. As it is well known and widely accepted, it is not necessary to discuss it further here.

Evolution by complexification — or progressive evolution, to employ a more widely used expression — is what is more frequently taken into consideration by botanists, zoologists, and even laymen. It concerns the origin of taxa higher than genera, and thus a range of problems that are hard to solve, like the origin of eukaryotes from prokaryotes, the origin of reptiles from fishes, the origin of vascular plants from algae, etc. It is a process that finds its decisive factor in an increase of genetic information, which promotes appreciable modifications of the frame, metabolism, biochemistry and genome of the phenotype, and thus of the way the genome works.

In principle, an increase in genetic information is necessary when a population is confronted with some dramatic modification of the environment which, so to speak, challenges it with contradictory demands. In such circumstances, no mutation, either of structural genes, or of genes controlling the expression of structural genes, would be adequate to meet the emerging requirements[4].

Contradictory demands occur, but they rarely appear as such. If the demand is to become heavier and quicker at the same time, to resist both the heat of the day and the cold of the night, to have good eyesight in water as well as outside, to have muscle fibres capable of contracting either briefly and quickly or over long tracts and slowly, or for the mammal foetus to obtain oxygen from its mother, in all such cases only innovative responses will be selectively advantageous and will indeed be acceptable. So the eye of *Anableps*, a freshwater fish, has two retinae, vertebrates, arthropods and some lower invertebrates have both striated and smooth muscle fibres, and the mammal

---

[3] "Progressive evolutions" or "evolution by complexification" are more precise and thus preferable to "macroevolution". Progressive evolution is largely in use and is relative to the quality of the process, while evolution by complexification is relative to the cause of the process.

[4] It is difficult to demonstrate a negative assertion, especially in biology. In fact a mutant enzyme for fur pigmentation is known that responds to contrasting demands: in the Himalayan rabbit, in some hamsters and in the Siamese cat, when hair grows on cold skin it is dark brown, so as to capture thermal radiation, whereas when it grows on warm epidermis it is white and does not interact with thermal radiations. Possibly other substances acting similarly will become known.

foetus is endowed with a haemoglobin having a stronger affinity for oxygen than maternal haemoglobin. Comparative physiology is rich in examples of this kind.

As for the appearance in a new protein, and thus of the corresponding genes, karyology and molecular genetics give convincing explanations. In many cases new genes originate from the malfunctioning of meiotic segregation and the duplicate gene thus produced may switch to a new function. For instance, a duplicate of the gene coding for insulin, the hormone controlling glucose uptake by cells, codes for relaxin, a hormone that loosens the pelvic ligaments in parturient mammals. A duplicate of the gene for lysozyme, an antibacterial protein, codes for lactalbumin.

The more complex the taxon, the more often it has to face contradictory demands from changing milieu. That explains the increase of both the speed and amplitude of the evolution of more advanced taxa.

4.2 Appearance of new genes in a phyletic line

Observations on new proteins stimulated investigations into the field of comparative biochemistry, a discipline that records the appearance of new proteins and thus of new genes in different phyletic lines.

Ohno in an excellent monograph [13] discussed the causes and evolutionary consequences of this phenomenon. Afterwards many voices joined in. For instance, Ohta [14, 15] proposed mathematical models to describe gene doubling and its subsequent specialization[5]. Many papers have been published concerning the phylogeny of protein families as well as the increase of genome size in many taxa.

Before venturing into the subject, however, I want to call attention to a remarkable phenomenon: in the prokaryote genome the repetition of nucleotide sequences longer than 12 or 15 nucleotides (corresponding to a sequence of four or five aminoacids) is forbidden because of physico-chemical constraints; thus, even more so, the repetition of a whole gene is forbidden (see [16, 21] for a comment). Only a few exceptions are recorded concerning ribosomal genes and genes involved in the photosyntethic apparatus in cyanobacteria. The prokaryotic genome, as a rule, has an upper limit of about 5.5 million nucleotide pairs. A few bacteria with more complex cells, however, break this ceiling: streptomycetes, which form long hyphes resembling those of fungi, have a genome roughly twice as large, and cyanobacteria, which form filaments composed of different sorts of cells, have a genome exceeding 6.4 million nucleotide pairs [25]; the maximum recorded is 13 million nucleotide pairs.

A similar relation was recorded in lower eukaryotes: red algae, which comprise the oldest known eukaryotes (*Bangiomorpha*) and the most primitive ones (*Cyanidioschyzon* and *Cyanophora*), have a genome of 14–47 millions nucleotide pairs (human genome encompasses over 3.1 billion nucleotide pairs).

To return to the main theme, diverse destinies have been recorded for duplicate genes in eukaryotes. The most common instances are: 1) the new copy is inactive (pseudogene); 2) the new copy codes for the very same protein;

---

[5]According to [14]: "Gene duplication could well have been the primary mechanism for the evolution of complexity in higher organism".

3) the new copy codes for a protein having a similar, but more specialized, activity; 4) the new copy codes for a protein endowed with quite a different function. The first instance is usually transitory and has no relevance in our context. The second instance may give origin to a multigene family of uniform proteins. The third instance gives origin to proteins that either operate simultaneously, as is the case of enzymes for melanine in man, or act in succession as in the well known case of vertebrate haemoglobins. The fourth instance concerns supergene families of proteins which have in common some physico-chemical properties but play quite a different physiological role as is the case of crystalline proteins forming the eye lens of invertebrates and vertebrates. Crystalline proteins derived from intracellular enzymes with highly hydrated and strongly refractive hyaline molecules (a typical case of evolution by *bricolage* to use the expression of Jacob [11]).

The multiplication of identical genes relates to the need for an organism to promptly get a large amount of RNA or protein molecules. This is the case of the histones that form the chromatin, and of the RNA and proteins that form the ribosomes. The most extreme case is that of the ribosomal genes of oocytes which multiply many thousands times before yolk synthesis; this phenomenon, called "gene amplification", is not due to mere chance, but is physiologically enhanced and controlled; it is also known as a pathogenic mechanism in carcinogenesis.

4.3 Other events that increase the genome size

There is no evidence that the appearance of extra chromosomes has played a role in progressive evolution. On the other hand, heteroploidy and polyploidy have certainly had a role in plants in withstanding harsh climatic conditions and in giving origin to new cultivars, races and species.

In animals, polyploidy is less common and it has probably been less important in adaptation. However there is evidence that the insurgence of tetraploidy may have had an important role in vertebrate evolution ([13] for a critical comment see [23]).

5 Novelties which have determined the appearance of higher taxa

When major evolutionary events are considered, it is frequently observed that they did not originate by the appearance of large scale novelties, but by the simultaneous appearance of a set of seemingly lesser modifications. The coincidence in time or even the rapid succession of modifications, is surprising.

On the whole, the process appears to depend not on some macromutation, of the type considered before (section 3), but on a *bricolage*, as suggested by Jacob.

5.1 Appearance of new substances in plant evolution

Vascular plants derived from fresh-water algae and colonized the subaereal milieu thanks to the evolution of new polymeric compounds.

The first compound to evolve was cutin which appeared in some bryophyte hepaticae as a protection for the epidermis exposed to the air: cutin is a polymer of lypids thet limits the loss of water and the ensuing desiccation. The second compound was lignin, an amorphous polymer of hydroxylated phenols. Long thin cellulose fibres are embedded in this compound thus forming a new material endowed with remarkable properties[6]. Through lignin higher plants acquired the support that an organism that leaves the aquatic environment requires. In fact, body weight is unimportant in water, but not out of it. Lignin is also of the utmost importance in the transport of sap.

A third important compound evolved in plants was cork, a polymer of organic acids which waterproofs and heat insulates the stem and roots of plants. To create these polymeric substances of paramount importance for plant evolution, no new raw material was necessary, but pre-existing catalysts adapted with some minor adjustments to carry out the new functions.

5.2 Motility and contractility in protist evolution

The utmost achievement of Protist Kingdom, which comprises mostly unicellular aquatic organisms, is the acquisition of motility. Thanks to flagellar motion, cells became capable of swimming in water, and thanks to amoeboid motion cell became capable of gliding on solid or incoherent substrates. Motility, necessarily assisted by adequate sensors and suitable programs, became a very important resource both for photosynthetic protists which must find an environment with optimal illumination, and for heterotrophic protists searching for their *pabulum*. Sensitivity and motility are also decisive to escape noxious substances.

In protists, motility depends on two different systems of proteinaceous molecules: the actin/myosin system confers contractility to the cytoplasm, while the tubulin/dynein system provides flagellar motion.

In both cases two polymeric filamentous proteins, actin and tubulin, pre-exist in the cell, where they have a role in mitosis and meiosis thanks to the fact that these macromolecules may elongate or shorten by adding monomers to the chain. Myosin and dynein are two polypeptides ubiquitous in eukaryotes, endowed with ATP-asic activity and tied to large proteinaceous supports whose origin is not known, but probably is not new.

Thus, in this case too, the raw materials required for cell motility pre-existed, and the novelty consisted in assembling these materials conveniently through a *bricolage* which produced composite structures endowed with new, selectively advantageous, properties.

5.3 The most important evolutionary leap in the history of life

The most important evolutionary leap was the one which determined the transition of the cell from the prokaryotic to the eukaryotic condition. It happened about 1,800 million years ago, when the oxygen concentration in the waters reached the level that allowed aerobic respiration. Respiration is a very advantageous metabolic process for heterotrophic cells which, thanks to it, may increase the energetic yield of ingested food by an order of magnitude.

---

[6]Such properties correspond to those intercellular substance of connective tissues of animals consisting of thin long fibres of collagen embedded in an amorphous mass of glycoproteins. They correspond also to fiberglass made by industry embedding glass fibres in synthetic resin and to similar materials.

Without going into technical details, it is possible to single out evolutionary steps which followed one another in a very short time, in geologic terms (see [16, 21]). The first step was the remodeling which took place in an archaean cell as a consequence of the association of histonic molecules with DNA. The genome, no longer made of "nude" DNA but of chromatin, acquired the possibility of breaking through the ceiling imposed by the physico-chemical properties of naked DNA, the possibility of accumulating redundant hereditary material and controlling its function.

The second evolutionary step took place when a cell with a genome made of chromatin engulfed and established symbiosis with a bacterial cell endowed with respiration and the enzymes that detoxify oxygen and especially the peroxides it produces.

The third evolutionary step took place when a cyanobacterium endowed with oxygenic photosynthesis became an endosymbiont of the chimeric proto-eukaryotic cell. Protist evolution went on mainly thanks to other endosymbiotic phenomena which have further enriched the set of structural and regulatory genes. Bricolage once more, but at a very large scale, with consequent conspicuous growth of the genome and of its self-control and consequent attainment of greater complexity.

6 Conclusions

Evolutionary transgeneric processes here reviewed differ from microevolutionary process, the ones most frequently considered in the literature, because they depend, not only on intragene punctiform mutations, but also on mutations that increase gene number. There is no evident relation between the apparent bigness of mutation and consequent evolutionary progress but in most cases, *a relation exists between the increase in genetic information and the evolution of a lineage.*

Thus if the spectrum of mutations considered by the evolutionist is extended to these not rare instances of gene and genomic mutation, and thus to major phenotypic novelties, even the mechanisms of progressive evolution fall within the principles of Darwins theory and in particular: every population of living beings has an inexhaustible source of hereditary variations.

Thus, *there is no discontinuity between mechanisms of microevolution and progressive evolution*, but a difference in the nature of the mutations involved.

BIBLIOGRAPHY

[1] R.N. Brandon. *Concepts and Methods in Evolutionary Biology.* Cambridge University Press, Cambridge 1996.
[2] C. Darwin. *Journal of Researches into the Nature History.* Murray, London 1839.
[3] C. Darwin. *A Sketch* 1842.
[4] C. Darwin. *Essay* 1844.
[5] C. Darwin. On the tendency of species to form varieties. *Journal of the Proceedings of the Linnean Society of London,* 3: 45-53, 1858.
[6] C. Darwin. *The Origin of Species.* Murray, London 1859.
[7] C. Darwin. *The Variation of Animals and Plants under Domestication.* Murray, London 1875.
[8] C. Darwin. *The Foundations of The Origin of Species. Two Essays Written in 1842 and 1844.* Cambridge University Press, Cambridge 1909.
[9] A.P. de Candolle. *Organographie.* Paris 1827.

[10] A.P. de Candolle. *Physiologie Végétale*. Paris 1832.
[11] F. Jacob. *La logique du vivant, une histoire de l'hèrèditè*. Gallimard, Paris 1970.
[12] T.R. Malthus. *An Essay on the Principle of Population*. J. Johnson, London, 1803 (2nd ed.).
[13] S. Ohno. *Evolution by Gene Duplication*. Springer Verlag, Berlin 1970.
[14] T. Ohta. Evolution by gene duplication and compensatory advantageous mutations. *Genetics*, 120: 841–884, 1988.
[15] T. Ohta. Role of gene duplication in evolution. *Genome*, 31: 304–310, 1989.
[16] P. Omodeo. Evolution of the genome considered in the light of information theory. *Boll Zool* 42:351–379, 1975.
[17] P. Omodeo. The theory of the living being and evolutionism. *Scientia*, 118: 51–64, 1983.
[18] P. Omodeo. Progressive evolution and increase of genetic information. In *International Symposium on Biological Evolution*, Adriatica Editrice, Bari 1987, pages 23–38.
[19] P. Omodeo. Phylogenetic concepts of the nineteenth century and the fundamental biogenetic law. *Boll. Zool*, 59: 17–21, 1992.
[20] P. Omodeo. Selezione, storia di una parola e di un concetto. *Nuncius* 19 (1): 143–170, 2004.
[21] P. Omodeo *Evolution of the Cell* (in preparation).
[22] T. Pievani *La teoria dell'evoluzione*. Il Mulino, Bologna 2006.
[23] M. Ridley *Evolution*. Blackwell, Oxford 2004.
[24] E. Sober *From a Biological Point of View: Essays in Evolutionary Philosophy and Biology*. Cambridge University Press, Cambridge 1994.
[25] N. Sugaya, H. Murakami, M. Satoh, S. Aburatani, and K. Horimoto. Gene-distribution patterns on cyanobacterial genomes. *Genome Informatics*, 14: 561–562, 2003.
[26] S.K. Swamynatha, M.A. Crawford, W.G.Robinson jr., K. Jyotshnabala, and J. Piatigorsky. Adaptive differences in the structure and macromolecular composition of the air and water corneas of the four-eyed fish Anableps anableps. *FASEB J.*, 17: 1996–2005, 2003.

Gene: an entity in search of concepts
STEFANO GIAIMO, GIUSEPPE TESTA

1 Introduction
Interrupted genes, alternative splicing, overlapping genes, nested genes, genome-wide transcription and regulatory RNA, regulatory sequences, somatic recombination in immunoglobulin genes and other molecular evidences represent serious challenges to the classic framing of the question "What is a gene?" In the first part of this paper, after showing that the DNA polymer is an entity without *bona fide* boundaries, we outline ways to extract gene functionalities from it by taking into account the most recent insights of molecular biology. In particular, we argue that a gene is not merely a continuous DNA sequence, but it is identifiable as a *functional* entity made of DNA which is part of a polypeptide/functional RNA coding process. Thus, a gene needs not be an isolated segment of DNA, but it can be productively understood as the sum of non spatially contiguous DNA sequences, since, even if it is scattered, it still constitutes a functional unit. Moreover, a single segment of DNA can be involved in many different polypeptide/functional RNA coding processes and can thus be a functional component of different genes. Conversely, distinct gene functionalities can be attributed, partially or totally, to the same DNA fragment. On the basis of this foundational analysis, we then develop in the last part of this paper an epistemological critique to expand the current popular distinction drawn by Lenny Moss [11] between Gene-P (preformistic gene) and Gene-D (developmental gene). We argue that some uses of the concept of 'gene' in experimental molecular biomedicine and evolutionary biology make Moss distinction not exhaustive, for example in the case of so-called "reporter" genes (i.e. luciferase gene) or genes predicted by many evolutionary models (i.e. Fisher's *runaway sexual selection*) that do not align neatly along the Gene-P/Gene-D dichotomy.

2 The problems of the classical molecular gene concept
"The classical molecular gene concept is a stretch of DNA that codes for a single polypeptide chain" [18, p. 132]. Recent empirical work shows that this is the definition of 'gene' contemporary biologists are more attached to [19]. Despite its widespread uptake, such definition faces several problems when confronted to the most recent advances in molecular biology [12]:

1. Functional RNA: some DNA sequences codify for RNA that becomes immediately functional and does not undergo translation into protein. This is the case of ribosomal RNA (rRNA), transfer RNA (tRNA) and small nuclear RNA (snRNA).

2. Interrupted gene and splicing: codifying (exons) and non-codifying (introns) sequences are alternated within the same gene [1, 2, 22]. Depending on the stage of development and in a tissue-specific manner different proteins are translated from distinct exon combinations obtained through the process mRNA splicing carried out by a sophisticated cellular machinery [14].

3. Regulatory sequences: some DNA sequences like promoters are never transcribed but they are active as regulatory mechanisms in transcription when bound by the appropriate factors. Such sequences, like enhancers, can be located far away upstream from the regulated nucleic acid chain or even within the controlled gene [16].

4. Somatic recombination: immunoglobulin genes are composed of distinct DNA fragments. In B-cells of the immune system those fragments are assembled by the somatic recombination process in order to be transcribed [6].

5. Overlapping genes: both in prokaryotes and eukaryotes the same DNA sequence can be transcribed with different reading frames thereby obtaining distinct polypeptide products [13, 23].

6. Nested genes: some genes can reside in the introns of other genes [5].

7. Junk DNA: extremely long DNA sequences do not codify for any protein and it is not clear what function (if any) they have.

8. Nucleosome positioning sequences: particular DNA sequences have a significant role in nucleosome positioning, and thus regulate the access of other proteins to DNA [15].

Apart from ignoring functional RNA, the main problem with the classical molecular gene concept is that it implies an active functional role of a unitary molecular structure, while evidences (2)–(8) all point to a non linear correspondence between function and structure of nucleic acids. If the gene function is specifying for a protein (or a nucleic acid, if we take into account a reasonable expansion of the classical molecular gene concept), then there is no identity between a gene and a unitary DNA sequence, because genes are composed by separated nucleic acid sequences and the same DNA stretch can host several different genes or none. There are then DNA sequences whose status is pretty questionable. Do, for instance, promoters or enhancers belong to the gene or not? They are indispensable in transcription but are never transcribed.

3 An ontological proposal

The weakness of the classical molecular gene concept has led some author to a sort of gene skepticism [21]. Kitcher, for example, argues that "in molecular biology research, talk of genes seems passé, a product merely of the accidents of history. There is no molecular biology of the gene. There is only molecular biology of the genetic material" [7, p. 357]. And in his opinion, if we would like to maintain the concept, then "a gene is [simply] whatever a competent biologist chooses to call a gene" [8, p. 131]. Without disregarding the importance of scientists' expertise, we believe that gene skepticism is not the necessary consequence of the most recent molecular discoveries we just reviewed. Difficulties in identifying the boundaries of an entity designed by a concept could be evidence of vagueness, but it is no reason to delete it from our inventory of the world or to expunge the corresponding concept from our theories. If we are able to solve the ontological problem regarding the lack of identity between linear DNA sequences and functional DNA sequence leading to proteins or functional RNA, then we can probably maintain genes as respectable biological entities. To overcome such problem, we appreciate both the idea proposed by Griffiths and Neumann-Held [4] that a gene is an entity linked to a process and Lewontin's bright insight that biological entities at levels lower than the organismal level can be grasped only in the dialectics between wholes and parts within an adopted epistemological framework [10]. Hence, we suggest the following solution articulated in five successive steps:

1. DNA molecule has a linear and directional chemical structure. It is a spatially extended entity that, within the chromosome, does not show spatial discontinuities on which we can establish *bona fide* boundaries.

2. Polypeptide or functional RNA coding processes unfold in space, involving DNA, RNA and proteins, and in time, from start of transcription to protein localization.

3. Genes are made of spatially extended nucleic acid segments that acquire functional meaning when they enter into polypeptide or functional RNA coding processes. Thus they can be regarded as functional entities that are part of these processes. If we appeal to an epistemological pluralism about the concept of "function", a certain degree of freedom is left here to the single scientist in order to better specify the considered process. This has positive consequences for our conception of genes. For instance, for those who are interested in the function of the DNA sequence in specifying by complementarity the primary transcript, introns will be included in the gene. Whereas those who focus on the rate of expression of a certain protein could assume that enhancers but not introns belong to the gene. Similarly, which exons should be considered as genuine parts of a gene then will depend on the cellular functional requirements in the developmental stage under study. None of such choices is forced or absolutely correct. But all are consistent with our proposal. They depend on the epistemological framework the scientist would adopt and on the phenomena under study.

4. Depending on the considered process, genes may or may not encompass contiguous segments of the DNA molecule. For instance, if the process taken into account is the one leading to a certain protein, then the corresponding gene will not include the introns in the DNA sequence that do not contribute to the translation for such protein. Therefore, genes can be scattered entities. This can sound counterintuitive, since

we are used to conceive entities as spatially continuous, like our body or a chair. And we typically concede the existence of scattered entities when they have at least some *fiat* boundaries, like a bikini, the US or an archipelago [17]. But functional entities are often scattered. Let us think about the immune system. It is scattered all throughout our body, but this does not prevent us to look at it functionally as a unit. It has to be noted that scattered entities are not a novelty in biology. One of the main biological concepts, the concept of "species", refers to a scattered entity, whether we agree with Ernst Mayr on sexual reproduction as its unifying force or with those authors who point to monophyly.

5. As functional entities, several genes partially or totally overlapped can reside on the same DNA stretch since the same spatial region can be functionally involved in different processes. The skeletal system, for instance, is the locus of production of new red globules, it is where calcium is stored and it serves as a physical and mechanical in sustaining the body. Analogously, what counts as a functional part in a polypeptide or functional RNA coding process does not necessarily count as a functional part in a different polypeptide or functional RNA coding process. Even if the two parts spatially coincide. This is due to the fact that the relation "being a functional part of" does not have the transitive property [20]: the handle of a home door is a functional part of the door, but it is not a functional part of the home to which it undoubtedly physically belongs.

4 Expanding Gene-P vs. Gene-D distinction

We hope to have shown that the ontological side of gene skepticism can be avoided and genes maintained as respectable biological entities. But another problem, a conceptual one, needs now to be faced. What concept can grasp such an entity? Recently, Lenny Moss [11] made an extremely influential proposal. He suggested to distinguish two concepts of "gene" with different historical roots and epistemological roles. Gene-P is the preformistic gene, or gene-for-something, as if it could determine a certain phenotypic trait. The epistemic value of Gene-P resides in its predictive power. In the case of BRCA1, namely the gene correlated to breast cancer, its mutated sequence is epidemiologically relevant, since it indicates a higher probability for its carrier to develop the disease, but although its sequence is known there is no complete molecular understanding of the mechanism leading from the sequence to cancer. On the contrary, Gene-D is essentially a molecular sequence that is used as a developmental resource by the cell for the RNA or protein production but it is indeterminate respect to the phenotype [11, p. 46]. Gene-D is one actor along others in the cellular dynamics and it is primarily distinguishable on the bases of the products for which it serves as a template. As an example of Gene-D, Moss puts forth the DNA sequence involved in the expression of N-CAM proteins, neural adhesion molecules that in different tissues and at a different stage of development give rise to a indeterminate array of phenotypic features. Moss' distinction rightly points to the division of explanatory and predictive roles between these two concepts of "gene". But in our opinion it is far from being exhaustive. There are at least two more concepts of "gene" that cannot be ignored:

- *Experimental gene*: a daily practice in experimental biomedicine consists in the introduction via vectors into model organism of DNA sequences of scientific interest. In order to check whether such sequences are actually transcribed by the model organism, the vector usually carries a reporter gene that codes for a chemiluminescent/ fluorescent protein (i.e. luciferase, *Green Fluorescent Protein etc.*) and is regulated by the same promoter as the sequence under study. By means of simple techniques, it is possible to elicit the characteristic light in those cell populations where the sequence is actively transcribed. There is no doubt that reporter genes are developmental resources (Gene-D) or phenotypic predictors (Gene-P) in the organisms where they were originally found, like GFP in Aequorea jelly fish. But once they are extracted and adopted by genetic engineering they are no longer developmental resources or phenotypic predictors, they become instead a tool along others for experimental practices in biomedical sciences.

- *Evolutionary gene*: in evolutionary biology it is customary to construct selection or drift models in order to explain character distributions in natural populations. Here the concept of "gene" is largely used neither as a template for protein or RNA synthesis (Gene-D), nor as predictor of individual phenotypes (Gene-P). Let us take the runaway sexual selection model sketched out by Ronald Fisher [3] to explain the presence of strongly marked secondary sexual attributes in most males, like long tails in birds, that are preferred by females. If there is covariance between the genes for long tail and the genes for female preference to long tails, then the character "long tail" will rapidly spread in the population and eventually it will reach fixation. Even if the character is not adaptively optimal from the ecological standpoint. Someone could argue that here Gene-P, the predictive gene, is involved, since we are speaking of "genes for long tail" or "genes for female preference". But, quite to the contrary, the distribution of the considered character in the population and the mathematical model that explains it are actually together predictive of genes, that are simply required to be faithful inheritance mechanisms that can recombine in meiosis (and possibly mutate in other evolutionary models). Of course, the model was originally developed in the 30's and it does not take into account forms of inheritance different from the genetic one. Now we know that there are other dimensions of inheritance involved in evolution [9] but still genes keep representing a good candidate for such role. Especially now that we possess a clearer understanding of their molecular details.

5 Conclusions

In this paper, we addressed both ontological and epistemological issues related to the concept of "gene". To make sense of most recent discoveries in molecular medicine, we advanced a notion of genes as entities composed of DNA fragments made cohesive by both the cellular functional requirements in the RNA and protein coding processes and the epistemological frameworks adopted by scientists. Then we developed a critique of the popular distinction between Gene-P and Gene-D to show that such distinction is not exhaustive, since biomedical experimental practices and evolutionary model construction ask for additional gene concepts.

Acknowledgements

We are grateful to Giovanni Boniolo for helpful discussion and to Giovanni d'Ario for his tecnical support.

BIBLIOGRAPHY

[1] S.M. Berget, A.J. Berk. T. Harrison and P.A. Sharp. Spliced segments at the 5' termini of adenovirus-2 late mRNA: a role for heterogeneous nuclear RNA in mammalian cells. *Cold Spring Harbor Symposia on Quantitative Biology*, 42: 523–529, 1978.
[2] T.R. Broker, L.T. Chow, A.R. Dunn, R.E. Gelinas, J.A. Hassel, D.F. Klessig, J.B. Lewis, R.J. Roberts and B.S Zain. Adenovirus-2 messengers–an example of baroque molecular architecture. *Cold Spring Harbor Symposia on Quantitative Biology*, 42: 531–553, 1978.
[3] R.A. Fisher. *The Genetical Theory of Natural Selection*, Clarendon Press, Oxford 1930.
[4] P.E. Griffiths and E. Neumann-Held. The many faces of the gene. *BioScience*, 49: 656–662, 1999.
[5] S. Henikoff, M.A. Keene, K. Fechtel and J.W. Fristrom. Gene within a gene: nested Drosophila genes encode unrelated proteins on opposite DNA strands. *Cell*, 44: 33–42, 1986.
[6] N. Hozumi and S. Tonegawa. Evidence for somatic rearrangement of immunoglobulin genes coding for variable and constant regions. *Proceedings of the National Academy of Sciences of the United States of America*, 73: 3628–3632, 1976.
[7] P. Kitcher. Genes. *British Journal for the Philosophy of Science*, 33: 337–359, 1982.
[8] P. Kitcher. Gene: current usages. In E. Keller and L. Lloyd, eds., *Keywords in Evolutionary Biology*, Harvard University Press, Cambridge, MA 1992.
[9] E. Jablonka and M. Lamb. *Evolution in Four Dimensions*, MIT Press, Cambridge, MA 2005.
[10] R. Lewontin. *The Triple Helix: Gene, Organism and Environment*, Harvard University Press, Cambridge, MA 2000.
[11] L. Moss. *What Genes Can't Do*. MIT Press, Cambridge 2003.
[12] P. Portin. The origin, development and present status of the concept of the gene: a short historical account of the discoveries. *Current Genomics*, 1: 29–40, 2000.
[13] F. Sanger, S. Nicklen and A.R. Coulson. DNA sequencing with chain-terminating inhibitors. *Proceedings of the National Academy of Sciences of the United States of America*, 74: 5463–5467, 1977.
[14] H. Sakano, J.H. Rogers, K. Huppi, C. Brack, A. Traunecker, R. Maki, R. Wall and S. Tonegawa. Domains and the hinge region of an immunoglobulin heavy chain are encoded in separate DNA segments. *Nature*, 277: 627–632, 1979.
[15] E. Segal, Y. Fondufe-Mittendorf, L. Chen, A. Thastrom, Y. Field, I.K. Moore, J.P. Wang, and J. Widom. A genomic code for nucleosome positioning. *Nature*, 442: 772–778, 2006.
[16] E. Serfling, M. Jasin, W. Schaffner. Enhancers and eukaryotic gene transcription. *Trends in Genetics*, 1: 224–230, 1985.
[17] B. Smith and A. Varzi. Fiat and bona fide boundaries. *Philosophy and Phenomenological Research*, 60: 401–420, 2000.
[18] K. Sterelny and P.E. Griffiths. *Sex and Death: An Introduction to the Philosophy of Biology*. University of Chicago Press, Chicago 1999.
[19] K. Stotz, P.E. Griffiths, P. E and R. Knight. How biologists conceptualize genes: an empirical study. *Studies in the History and Philosophy of Biological and Biomedical Sciences*, 35: 647–673, 2004.
[20] A. Varzi. A note on the transitivity of parthood. *Applied Ontology*, 1: 141–146, 2006.
[21] K. Waters. Molecular genetics. In E.N. Zalta, ed., *The Stanford Encyclopedia of Philosophy*, URL = http://plato.stanford.edu/archives/spr2007/entries/molecular-genetics/, Spring 2007.

[22] H. Westphal and S.P. Lai. Displacement loops in adenovirus DNA-RNA hybrids. *Cold Spring Harbor Symposia on Quantitative Biology*, 42: 555–558, 1978.
[23] T. Williams and M. Fried. A mouse locus at which transcription from both DNA strands produces mRNAs complementary at their 3' ends. *Nature*, 322: 275–279, 1986.

Categories, taxa, and chimeras
ELENA CASETTA

1 Biological entities

Biology is a large, heterogeneous field. If we look up "biology" in a standard encyclopedia, we can find several disciplines to which that label is attached (either disciplines that are sub-parts of biology proper or disciplines strictly connected to it): anatomy, biochemistry, developmental biology, cytology, ecology, ethology, genetics, and more. Each of these disciplines studies the living world at a different level of description, and focuses the attention on a different aspect of it. In doing so, each discipline countenances some entities or, better, some *types* of entity. Such types of entity figure in the formulation of scientific statements. Accordingly, the *truth* — or, at least, the meaning — of biological statements *depends* on the very *existence* of those entities. So we may ask: what are they? And what is their ontological status?

When we speak of *types* of entity, in biology, we are immediately concerned with *taxonomy*, that is, with the theory and practice of classifying living entities. The first attempt to put forward a taxonomy of the living world goes back to Aristotle. In the *Categories*, Aristotle introduces the difference between "species" and "genera", and in his biological writings (*History of Animals, Parts of Animals, Generation of Animals*) he works out the first classifying procedures and the first biological classification. But the father of taxonomy is unanimously considered the Swedish naturalist Carl Linnaeus, with his *Systema Naturae* (1735). The assumptions of his system were *creationism* and *Aristotelian essentialism*: species and other taxa are the result of divine intervention and, once a taxon is created, each of its members must have the essential properties of that taxon (compare Aristotle's theory of types). Briefly, Linnaeus' system — elaborated more than a hundred years before Darwin's *Origin of Species* — does not allow evolution, and its outdated theoretical assumptions undermine its ability to provide accurate classifications. Nonetheless, that system for naming, ranking, and classifying organisms is still in use today, even if with several refinements.[1]

The *categories* into which such taxonomies are organized are, however, to be distinguished from the relevant *taxa*. Broadly speaking, today categories are defined as "groups of taxa". Thus, for example, the *species* category is the group of all species taxa, whereas species taxa such as *Homo sapiens* or

---

[1] The changes do not concern the general system's structure, which remains valid, but just the number of ranks and, accordingly, the naming rules. For instance, contemporary taxonomists recognize six kingdoms organized into three domains, whereas Linnaeus recognized only three kingdoms; they have introduced the *trinomial name* for subspecies, whereas Linnaeus gave the binomial name for species, and so forth.

Drosophila melanogaster are groups of individual organisms (see [11, p. 3]). In what follows, my concern is with the ontological status and, relatedly, with the metaphysics of both categories and taxa: are they genuine denizens of reality? I'll take the case of species as *casus exemplar* mainly because species seem to enjoy a sort of ontologically privileged status; they are the "currency" of every biologist, and whereas the reality of higher classes is often denied, the reality of species is rarely in debate. Species are indeed the basic taxonomic units — for Linnaeus as for us — and they are generally considered to be the units of evolution (they would play a fundamental causal role in the evolutionary account, and this would be evidence for their existence). However, if what I say is valid for species, I believe it will in principle be valid for every taxonomic unit.

2   Aristotle vs. Kant

Let's start with categories. How are categories to be thought of? What is the relationship between our classifications and the classified reality?

We may identify two paradigmatic approaches to these questions that go through the history of philosophy: the first is Aristotelian realism (followed in our days by philosophers such as R. Grossmann, R.M. Chisolm, and E.J. Lowe [19, 3, 29]), according to which our classifications aim at mirroring the structure in which reality itself is articulated; the second is Kantian conceptualism (followed e.g. by P.F. Strawson [40] and other "descriptive" metaphysicians), according to which the content of experience is an unstructured whole shaped from the outside by our own conceptual schemes.

Aristotelian categories are meant to be real, "natural" articulations of the world out there. As Aristotle puts it in the *Categories*, a classification undertaken in this realist spirit would ideally list the highest genera of all entities (in the widest sense of the term). Thus, although Aristotle's departure point is language (predication), his concern is not — at least, not primarily — language, but rather *things* to which we refer by means of language. His categories are truly *ontological*: the ten genera of things that correspond to those linguistic expressions that can be predicated of something.

On the other hand, the twelve Kantian categories are thought of, not as intrinsic divisions in reality itself, but as the highest categories governing our conceptual schemes. Even if Kant declares that "our primary purpose is the same as his [Aristotle's], although widely diverging from it in manner of execution" (*CPR* A80/B105), it is clear that for him categories find their original source in the principles of human understanding, not in the mind-independent reality, and are discoverable by paying attention to the possible forms of human judgment, not by studying the world itself. (This is not surprising: Kant's categorial conceptualism is the obvious landing place of his Copernican revolution — and of the Cartesian assumption underlying it — according to which what is *inside* us is more certain than what is *outside*. We genuinely know only those "things" we made ourselves.)

Are biological categories (to be thought of as) Aristotelian categories or conceptual categories? An initially plausible answer is that biologists rely on categories of the Aristotelian sort. Indeed, one of the first tasks of biology is the classification of the natural world, which exists and which is what it is *mind-independently*. But the issue is not so straightforward. Because of the so-called "species-problem" (to which I'll come back in the next section), a Kantian approach to taxonomic categories is increasingly popular among biologists. By contrast, in the case of taxa, the realist approach doesn't seem to be in question: "The third option [according to which we should doubt the very existence of the category species] does not call into question the existence of *Homo sapiens* or *Canis familaris* or any other lineage that we call 'species'. The third option just calls into question the existence of the categorical rank of species" [12].

So here is my plan: (1) To show that taxonomic categories and taxa are strictly connected — more precisely, that taxa *depend* on categories for their individuation. It follows that taxa are derivatively affected from the species problem, which makes it difficult to embrace a realist approach. (2) To consider the difficulties and consequences of the alternative (Kantian) approach, and to reject it for taxa while accepting it for categories. (3) To defend a third view about taxa, which one might call "conventional realism". According to this view, biological entities — not only individual organisms but also entities at higher levels such as biological taxa — are among those entities that truly inhabit the world (contrary to Kantian conceptualism). However, their boundaries are not "natural" — as Aristotelian realism would have them — but conventional, i.e., produced by human conventions. (4) To support this view with some empirical evidence.

3 The species problem: a challenge to realism

The so-called "species problem" can be broadly articulated in three main issues:

(i) Do species *really* exist? In other words, are species the "natural joints" that the skilled Platonic butcher must *discover* and along which he must carve reality (realism), or are species just cognitive constructs, linguistic devices by means of which biologists — and the "street man" as well — dissect the external world (nominalism)?

(ii) If species really exist (as the realist holds), *what* are they? What is their *nature*? Two main accounts are on the field: *species as sets* and *species as individuals*. According to the first, the species *category* is the set whose members are all species taxa; and a species *taxon* is itself a set of organisms. By contrast, according to the species-as-individuals account, the species category is a set but species taxa are individuals, viz. mereological wholes made up of organisms causally connected in space-time [17, 23].

(iii) Should a realist stance go hand in hand with a monist view? Surely, as Hull put it, "it would seem a bit strange to argue that one and only one way exists to divide up the world, but that the groups of natural phenomena produced on this conceptualization are not "real" [23, p. 24]. But then, how can one account for the *fact* that a plethora of species definitions have been offered, definitions that produce different — often mutually incompatible — conceptualizations of the natural world?[2]

According to monism, *there is* one "right" species concept: maybe it is among the species concepts that biologists have already recognized, and they just need to identify which one it is; or maybe we need to wait for scientific progress. Briefly, the natural world has a structure in its own, and our taxonomy — sooner or later — will find that structure and will mirror it. However, several authors[3] are promoting a pluralistic approach to species definition: contrary to the monist hope, we cannot find *the* right species concept simply because there is *no* single correct species concept. Evolutionary mechanisms make it in principle impossible to pinpoint the essential property (or a cluster of such properties) that is needed in order to talk of *one* species category [13]; and natural fact-based remarks seem to support such a view [2]. Biology, it is argued, contains a number of legitimate species concepts. But such authors say more: species concepts are definitions of the species category, and not of species taxa. Hence, the multiplicity of species concepts that pluralism seems to require would not affect species taxa. Concerning species taxa, there is unanimous agreement to the effect that they are "lineages", that is, "either a single descendant-ancestor sequence of organisms or a group of such sequences that share a common origin" [10].

This sort of pluralism is attractive: it would forsake realism for categories while saving it for taxa. Unfortunately, I don't think it works. *Only* if we buy into a certain species concept — that is, a certain definition of the species category — can we individuate species taxa and, moreover, individuate those organisms that belong to a certain taxon. Consider the definition of lineage mentioned above. In order to say that species taxa are lineages, we must assume one of the following species concepts: the phylogenetic, the biological, or the ecological one. All of these are *historical* species concepts.[4] *Had we bought into a different sort of concept* (for instance a *structural* species concept, such as the phenetic one), *we would have individuated different taxa*. In other words, in order to countenance the species *category* we'd have to proceed roughly in this way: (a) partition organisms into species taxa, (b) identify all species taxa and seek if they — and only they — share a certain property that makes them members of a *unique* taxonomic category (the category of species). However, it is hard to see how one can determine *what* the species taxa are (and *what* the organisms belonging to them) without *already* having a certain species concept. It is hard to see how one can determine what taxa belong to the species rank — and not, say, to the genus rank, or to the subspecies rank — without relying on such category concepts. Taxa are *dependent* as for their individuation on categories. And this means that a monistic — and a realist — account is problematic for taxa, too.

---

[2] P. Kitcher [25] recognizes nine species concepts; R.L. Mayden [30] twenty-two definitions.

[3] See e.g. [25, 32, 5, 11].

[4] P. Kitcher [25] organizes the species concepts into two types: historical — which require that species are genealogical entities — and structural — which require that the organisms of a species (taxon) have important functional similarity.

4 Unaffordable costs of a Kantian solution

If I am right, then a Kantian approach would seem to be a solution. But at what costs? As we have seen, Kantian conceptualism claims that categories are "in our head". Not only are they produced by us; they are the way we actually shape the world. If categories belong to our understanding and not to the world, the problem of the multiplicity of species definitions can be bypassed: we can have as many concepts of species as we want — and, consequently, as many different taxa as we want — because categories do not allow us to access the world *itself*, but just the *phenomena*, that is, the world *as experienced* by us. In the end, categories allow us knowledge of ourselves, of our way of conceptualizing. (Of course, according to Kant, there are *transcendental schemata* that permit our categories to apply themselves to the content of experience, but — even without going into the details — the concept of schema is anything but uncontroversial.)[5]

Briefly, the cost of a Kantian approach is the embracing of a *descriptive* approach: we do not and cannot know the external world directly; we can just investigate the structure of our conception of the world. To be more precise, we can investigate reality only by means of an examination of its representation in our cognitive system, regardless of its correctness. Nothing wrong with this, if I am a descriptive metaphysician. But I think that a biologist, for instance, would look at me with suspicion if I said: "No problem, *Drosophila melanogaster* is just in my head". The biologist would rather begin to investigate how is it possible that some fruit flies are in my head and not on fruits. What I want to say is that, according to biology, categories and taxa are not just *conceptualizations*. Actually, there is a particular discipline in charge of investigating the structure of our thought about the natural world, but that is not biology; it is "folk-biology", a discipline that has more to do with anthropology and psychology than with biology. As a further confirmation of the fact that biological categories are not (thought of as) Kantian categories, we may note that species in biology and species in folk-biology rarely overlap. The core of any folk-taxonomy is the so-called "generic species"[6], a sort of "category mistake", would say a biologist. Indeed, folk-species often correspond to scientific *genera* (e.g., *oak*) or species (e.g., *dog*), sometimes to local fragments of biological families (e.g., *vulture*) or to order (e.g., *bat*), and so on.

5 The lesson from the platypus

In 1797, Captain John Hunter, the second governor of Australia, saw an Aborigine spear a strange animal in Yarramundi Lagoon near the Hawkesbury River, just north of Sidney. It was a very odd animal: it was covered with dense brown fur, it had a beaver tail, and it exhibited the perfect resemblance of the beak of a duck engrafted on the head of a quadruped. And his surprise

---

[5]To mention just one trouble: where do schemata come from? Either they are innate — in which case "she'd have to admit also the Ideas of grease, of dirty fingernails, and of scales" ([14] — or else schemata are built by abstraction from the objects of experience, hence the Conceptualist comes to a vicious circle.

[6]See [1, pp. 231–262].

would have been even bigger if he had known, for instance, that the animal spends part of its life under water and that the female lays eggs, as an amphibian, but its youngs are fed by the mother's milk even if the mother lacks teats (though possessing mammary glands).

So, our John was seeing *something*, but *what*, precisely? From a Kantian perspective, one can only say that John was not in posses of the concept of such an animal — because the platypus had not yet been "discovered" and classified. Accordingly, John was not able to see a platypus; he was merely looking at *that thing*. But, what does it mean to be seeing something if you don't have any idea of what you are seeing? If we embrace a Kantian approach, in order to see something we have to recognize it as a thing of a certain type, we need the type — the relevant concept — in the first place. From a Kantian perspective, then, the trouble is that, on the one hand, we cannot perceive a platypus without previously having the relevant concept and, on the other hand, we cannot *derive* concepts from experience (because they are *a priori*, that is, their source is the Understanding). Consequently, when we see something we've never seen before, we cannot see it as anything. Of course, this is an oversimplified picture of the issue [7], but I think it can give an idea of the difficulties concerned with "categorial conceptualism".

However, I think there is something enlightening in the platypus case, maybe not as much in our little thought experiment but, rather, in the "real story" of the discovery and of taxonomic arrangement of this weird animal (see [20]). Actually, this story is nothing else than the report of a lengthy *negotiation*. Let's see some stages of it.

At first, the British naturalist George Shaw tries to force the platypus into the Linnaeus' hierarchy, placing it among mammals, but without profound conviction (given the heterogeneous collection of the features the platypus shows). After him, it is Thomas Bewick who takes up the taxonomic challenge. He describes the platypus as a "three-fold nature" animal — a fish, a bird and a quadruped — and argues that there is no place for it inside the taxonomic system in force. The progress of science and the increase of empirical data are unhelpful: each new piece of information concerning the platypus seems just to confirm that there is no place for it in our taxonomy. Thus, in 1803, Etienne Geoffroy de Saint-Hilaire creates *ex novo* the category of Monotremes. But are Monotremes mammals or what? There seems to be no fact of the matter that can help one decide. Like mammals, the platypus has mammary glands, but it lays eggs, like amphibians or birds; its genital apparatus is similar to that of reptiles but — unlike reptiles — it is warm-blooded and it has a diaphragm like that of mammals. This is the beginning of a quarrel, on scientific gazettes, that will last more than eighty years. In 1886, the negotiation comes to an end: the Platypus belongs to the order of Monotremes, which belongs to the Mammalia class, and it is oviparous. A *convention* has been established.

6 Conventional boundaries

My claim is that species taxa are *conventional objects*.[7] They are conventional, yet they are *real*.

---

[7]My notion of a conventional object is in the spirit of [36] and [21]. On what a convention is, how it is established, and how it works, see [28].

Imagine having some dough and being baking cookies.[8] You can make cookies round, square, star-shaped... You can probably bake also just some huge unshaped cookies, or two big ball-shaped cookies. And it doesn't matter what portion of dough you choose for a certain cookie: cookies so obtained are perfectly real, you can eat them. At the same time, they are a product of your convention, and not only in the weak sense that conventions govern the meaning of the terms that refer to them (that we call "cookie" a certain thing made in such and such a way is established by language rules, that is, human conventions), but in the stronger sense that their *boundaries*[9] are conventional — they are the product of your deliberation.

Our cookies have *spatial, temporal,* and *modal* boundaries. Indeed, they have *surfaces*, they begin to exist (probably when they are "shaped") and they cease to exist (probably when we eat them, or when they go bad to a certain degree), and they may even have certain *essential* properties (could a star-shaped cookie have another shape and still be the same cookie?). But every boundary could have been different. Consider:

(i) *Spatial boundaries.* The cookies could have had a different size: if you need to bake them in ten minutes you won't make them thick, for instance;

(ii) *Temporal boundaries.* The cookies could have had a different shelf life, and in two ways: we decide *when* we can begin and stop to speak of a certain mass of dough as of a cookie ("weak" conventionalism); we decide which ingredients to put in the dough ("strong" conventionalism);

(iii) *Modal boundaries.* The cookies could have had different essential properties: if you are baking bar cookies, their shape will be among their essential properties (a bar cookie cannot be star shaped), whereas if you are baking brownies, it is *taste* that will be essential (brownies must taste like chocolate). In other words: we may *decide* that a certain property (such as a certain shape or taste) is *essential* to the cookies we are making, insofar as *those* cookies cannot loose that property without ceasing to be what they are, and they cannot lack that property and be what they are supposed to be.[10] As Mark Heller put it: "The supposedly essential properties are just the ones to which we have attached special significance" [21, p. 46].

(In what follows, I'll focus on *temporal* and *modal* boundaries, that is on the *persistence conditions* and the *essential properties* of an object. Since we do

---

[8] H. Putnam [33] is among the firsts to use the cookie cutter metaphor. (Actually, he criticizes the metaphor in the context of his "mereological argument" for conceptual relativism.) An analogue of the cookie-cutter metaphor can be seen in the scheme-content distinction [6]. For a discussion of these topics see, for instance, [34, 8].

[9] On boundaries, see [42, 43].

[10] This approach can be traced back to the medieval philosopher John Buridan. Buridan, on the basis of a distinction between "predicate essentialism" and "realist essentialism", was able to maintain a nominalist version of essentialism that is sufficient to provide a foundation for valid scientific generalizations without positing the existence of any common nature, or essence — only the existence of essential predicates. See [26]; for a conventionalist treatment of necessity, see also [36].

not usually think of spatial boundaries as being closely related to identity, in the present context we can leave them aside.)

My claim is that species taxa are not so different from our cookies. Here is why. The broadest definition of taxa tells us that they are "groups of organisms" (Section 1). According to which species concept we adopt, we'll have certain membership[11] condition for the relevant species taxa. To put it in other words, the resultant taxa will have certain essential properties. Some examples: If the species definition is the *phenetic* one [38], the relevant species taxa will be the sets of those organisms that have the property of being similar to a certain degree, and this will be the essential property of the taxon; if the definition is the *biological* one [31], the taxa will be the sets of those organisms[12] that have such properties as interbreeding and being reproductively isolated from other such sets; if the definition is the *ecological* one [41], the taxa will be those sets of organisms that share a certain ecological niche. Again, if we adopt the *filogenetic* concept of species, the relevant taxa will be the less inclusive monophyletic taxa, i.e., those lineages (sets of organisms tied by historical relations) consisting of an ancestor and all and only its descendants.

Generally, if we adopt a *typological* approach to the species category, we'll have — as an essential property of taxa — similarity (morphologic, genetic or whatever) among their members, while if we adopt an *historical* approach, we'll have as an essential property of taxa a certain relation of ancestry among their members. Which approach is the right one? It depends. Both similarity and historical relations are "real" features of the world. But choosing one or the other is just a matter of human interest, preferences, and beliefs. For instance, if I want to obtain a classification of organisms mirroring the evolutionary mechanisms, I'll opt for the historical relations: a certain set of organisms is a species taxon if and only if those organisms stand in a certain relation, that is — for instance — if some of them are the ancestors of the others and of no one else. If I prefer instead a theory-free classification, I'll go for "overall similarity", as in the phenetic approach. If I am a biologist working on insects, it will be very hard to use the biological concept of species — and consequently to define taxa as sets of interbreeding organisms — because several insects procreate by means of parthenogenesis: in such a case we'll say, for instance, that the several species taxa of aphids (a family) are composed of organisms that possess a certain cluster of essential properties — a certain shape, size, color, ecological needs, and so on.

Summing up: a species taxon *has* essential properties, i.e., modal boundaries that set the limits of what can and what cannot belong to the taxon, but such boundaries could have been different. They depend on a *choice* among several possibilities that are equally "grounded in reality". Of course, to say that the possibility of making a choice involves arbitrariness is not to say that anything goes: some choices yield taxa that are more "well-formed" than others. But more "well-formed" doesn't mean more real: it simply means better

---

[11] Remember we are treating groups as sets.

[12] Actually, Mayr's definition is: "Species are groups of natural populations that are reproductively isolated from other such groups" [31].

suited to a certain purpose. (Often, in the biological literature, a certain taxon is considered "more natural" than others, and "naturalness" is understood as a synonym for "realness". I don't know exactly what "natural" means, but I do not think that artifacts, for instance — which by definitions are not "created" by nature — are in any sense unreal.)

Now, what about a taxon's *temporal boundaries*? When does a species begin and when does it come to an end? Again, it depends. Already Darwin was aware that species gradually glide one into another, and that it is impossible to define an objective point of splitting: "Systematists will have only to *decide* (not that this will be easy) whether any form be sufficiently constant and distinct from other forms, to be capable of definition". (See [4], chapter XIV, italics added) As with modal boundaries, according to the species concept you choose to adopt, you will get different temporal boundaries. Speciation events are actually *gradual processes*, and to individuate the *beginning point* of a new species is just a matter of choice. Consider a population of tortoises living on a small peninsula.[13] Eventually, the ocean breaks through, causing the tortoises to live on an offshore island, geographically isolating from all other tortoises in the world. Over time the environment on the island comes to differ from that of the mainland and the mating habits of the island tortoises begin to vary from those of the mainland tortoises. According to the biological concept of species — the most popular concept — the island population forms a different taxon (a taxon of a different species) only if its organisms are reproductively isolated from the mainland tortoises. But *when* does that begin to occur? When the mechanisms that prevent interbreeding first arises in *one* islander tortoise? When the majority of tortoises on the island have that trait? When all of them have it? Even if we *settle on a* precise point of splitting — that is, even if we establish a convention answering these questions — other question arise, e.g., concerning *identity*. Consider a process of cladogenesis (Figure 1, left): how many species do we have after the splitting? Three — the original species + species a + species b? Or two — the original species that survives as species a (or as species b?) + a new species (species a or b)? Again, consider the other main process of speciation, anagenesis (Figure 1): *how* and *when* does species a become species b? As usual, the *how*-answer depends on which species concept one adopts: ceasing of interbreeding, ceasing of niche sharing, etc. And the *when*-answer depends on what we decide.

Now, if the modal and temporal boundaries of an object, that is, its persistence conditions and its essential properties, are conventional, then we are in the presence of a conventional object.

7 The rebelliousness of species boundaries

If taxa boundaries are a product of human conventions, we may expect that *non-humans* do not care about them. And actually it seems to be so. According to the most widespread conception of species taxa, they are sets of organisms interbreeding among them but *not* with other taxa. And this condition would define the boundaries of taxa. But then: how do we account for

---

[13] The example is from M. Ereshefsky [11]. This is an imaginary example, but actually geographical isolation is considered one of the main causes of speciation events.

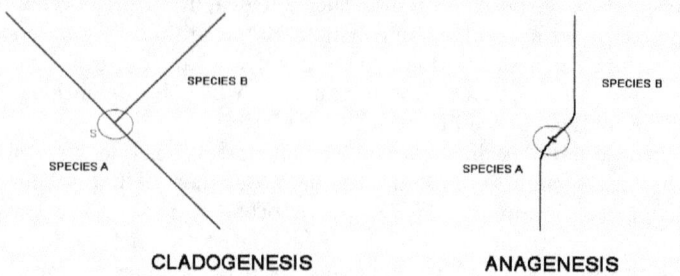

Figure 1. Cladogenesis and Anagenesis

hybrids and chimeras?[14] Actually, hybrids and chimeras show a manifest rebelliousness towards species boundaries. And they show more, in my opinion: they show that the claim that species boundaries are conventional is all but unsound. If boundaries are conventional, one can trespass them, exactly as one can trespass the frontier between two nations. Yet this does not imply that nations, or taxa, are unreal; it simply means that their reality depends on our intervention: as Frege put it, the objectivity of the North Sea "is not affected by the fact that it is a matter of our arbitrary choice which part of all the water on the earth's surface we mark off and elect to call the 'North Sea'" [15].

One could object that the case of hybrids and chimeras is not particularly significant, and this on the basis of three sorts of consideration.

First possible objection: hybrids and chimeras are produced in artificial environments (laboratories, cages of a zoo, and so on). Interbreeding among organisms belonging to (putatively) different species, if obtained in a lab, doesn't show that those organisms really interbreed. Actually, it shows that in a natural environment boundaries between species are fixed. *Reply*: I do not find this compelling. First, such organisms do interbreed — albeit in a lab. If this does not happen "in nature", it is because it is not needed. If a tiger and a lion were the only survivors to a great glaciation, it is likely that they would interbreed, in the same way they actually interbred in the Moscow zoo. (Actually, zookeepers have for years crossed tigers with lions, obtaining "tiglons" and "ligers" [37, 44].) Secondly, hybridization and chimerism are phenomena that frequently occur without human intervention.[15] Finally, what makes a certain environment "artificial"? Its manipulation by humans? Perhaps so. But we know that the human species is just one of the very many species inhabiting this planet, and that *every species* affects its environment in some way (organisms are "ecological engineers"). So, what is it for an environment to count as "natural"?

Second possible objection: hybrids and chimeras are rare cases, exceptions

---

[14] A hybrid is, by definition, an offspring of parents from different species or sub-species, while a chimera is an organism composed of two genetically distinct types of cells.

[15] See for instance [39, 41, 27].

in an otherwise stable and well-defined organism partition. *Reply*: first, even if this were true, are rare cases to be ignored? (Surely logicians cannot ignore the liar paradox because it does not arise in ordinary circumstances, just as a philosopher of mind cannot ignore the possibility of zombies or brain transplant scenarios on account of their being utterly unlikely.) Secondly, the objection relies on an assumption that is just not true. Among animals, think of such hybrids as mules and hinnies (the offsprings of a male donkey and a female horse and of a male horse and a female donkey, respectively), or think of cases of interbreeding between lions and tigers, or between goats and sheets. Or again, think of the *Lonicera* fly, a natural hybrid, the offspring of the blueberry maggot and the snowberry maggot [35]. Several other hybrids are recognized: hybrids among mammals, such as those between two different species (or sub-species) of bears [18], or among birds, e.g. between the Great Skua and the Artic Skua (which gives raise to the Pomarine Jaeger, a natural hybrid *now recognized as a distinct species*; see [16]) Not to mention the large variety of botanic cases: though relatively rare among animals, chimeras are quite frequent in the vegetal life, and hybridization is pretty widespread. (Just think of Rhododendrons: extensively hybridized in cultivation, natural hybrids also occur in areas where species ranges overlap. Or think of Orchids, the largest family of flowering plants: here hybridization is very common, and several hybrid species can interbreed with their parental species.)

Third possible objection: hybrids and chimeras are sterile, so they "interrupt" the gene transmission required for the continuance of species. *Reply*: even if so, why not just say that "hybrid species" are merely species with a shorter life than others? Moreover, the objection trades once again on an assumption that is not exactly true. Since 1527 there have been more than 60 documented cases of foals born to female mules around the world,[16] hybrids can frequently interbreed with their parental species, and they can sometimes interbreed among them, as with the *Lonicera* fly mentioned above. Finally, this objection misses the point: the question is not whether hybrids and chimeras can be partitioned into species; rather, it is just that the existence of hybrids and chimeras shows that *organisms often do not care about boundaries that we human beings recognize*.

8 Concluding remarks

I want to conclude by sketching some answers and solutions that conventional realism can offer to the species problem, showing how conventional realism does not face the troubles traditional realism must face and need not pay the costs conceptualism must pay.

Remember the species problem's first issue (Section 3): do species *really* exist? Conventional realism — like every realist approach — answers in the affirmative. Species are not just categories in our mind. As said, if we are realists, the issue we have to face is rather the following: *what* are species? Whereas the species category is a conceptual category, species taxa are *con-*

---

[16]This even made it in the news, see http://news.bbc.co.uk/1/hi/sci/tech/2290491.stm. It's worth noting that report such as these could be questioned. For an empirical and critical study about mules' fertility, see for instance [9].

ventional objects, i.e., real objects whose boundaries are functions of human conventions. On the one hand, a Kantian approach for the species category allows us to account for pluralism: if the species category is just a conceptual category, there can be different definitions or conceptions of it. Insofar as it just reflects our beliefs about the world, different people can have different beliefs. At this point one can object that I have shown that species taxa are dependent, as their individuation, on a certain species category and, consequently, that trouble for pluralism is not swept away. I don't think so: insofar as it stems from our beliefs, the species category enters somehow in the individuation of species taxa, but it is just one component. In order to "give rise" to species taxa, a belief has to be "supported" by conventions, otherwise it just stay what it is — a belief. On the other hand, in asserting that taxa are real conventional objects, our solution need not pay the cost of a Kantian approach. The fruit flies that worried our biologist are not in our head; they are in the world out there.

And, in this way, we can also account for the species problem's last issue, namely, how a *monist* approach can be reconciled with a *pluralist* perspective. We can be pluralists about the species category and realists (and monists) about species taxa. Of course, the metaphysical foundation of our monism is not the demand of carving nature at its joints. Rather, it is the demand of an accomplishment of our conventions. Establishing a convention is a process that takes time, and *only once* conventions have been well established will we have monism (of conventions and, consequently, of species taxa). To put it another way: is it true — as many monists argue — that we have to rely on science progress, but not in order to *find* that sole and true structure of the world. Rather, we have to rely on science progress in order to *achieve* good (that is, consistent, useful, practical, and unambiguous) conventions.

Acknowledgements

I am grateful to Maurizio Ferraris and Achille Varzi for helpful comments, suggestions and guidance. I would also to thank the participants to the 2007 SILFS Conference (Milan), in particular Giovanni Boniolo, Paul Griffiths, and Telmo Pievani, and the participants to the 2007 SEFA Conference (Barcelona), in particular Manuel García Carpintero, where earlier drafts of this work were presented.

BIBLIOGRAPHY

[1] S. Atran. The universal primacy of generic species concept in folkbiological taxonomy: implications for Human biological, cultural, and scientific evolution. In R.A. Wilson (ed.), *Species. New Interdisciplinary Essays*, pages 231–262. MIT Press, Cambridge, MA 1999.
[2] E. Casetta. Le tigri di Putnam. *Rivista di Filosofia*, 98 (1): 47–76, 2007.
[3] R.M. Chisholm. *A Realistic Theory of Categories*. Cambridge University Press, Cambridge 1996.
[4] C. Darwin. *The Origin of Species*. Penguin Books, London 1859/1985.
[5] J. Dupré. *The Disorder of Things: Metaphysical Foundations of the Disunity of Science*. Harvard University Press, Cambridge, MA 1993.
[6] D. Davidson. On the very idea of a conceptual scheme. *Proceedings and Addresses of the American Philosophical Association*, 47: 5–20, 1974.
[7] U. Eco. *Kant e l'ornitorinco*. Bompiani, Milano 1997; eng. trans. *Kant and the Platypus*, Harcourt Brace and Company, New York.

[8] M. Eklund. The picture of reality as an amorphous lump. In J. Hawtorne, T. Sider and D. Zimmerman (eds.), *Contemporary Debates in Metaphysics*, Blackwell, Oxford 2007, pages 382–396.
[9] F. Eldridge and Y. Suzuki. A mare mule — Dam of foster mother? *The Journal of Heredity*, 67: 353–360, 1976.
[10] M. Ereshefsky. Eliminative pluralism. *Philosophy of Science*, 59: 671–690, 1992.
[11] M. Ereshefsky. *The Poverty of the Linnaean Hierarchy: A Philosophical Study of Biological Taxonomy*. Cambridge University Press, Cambridge 2001.
[12] M. Ereshefsky. Species. In Edward N. Zalta (ed.), *Stanford Encyclopedia of Philosophy*, CSLI, Stanford 2006 (internet publication).
[13] M. Ereshefsky. Systematics and taxonomy. S. Sarkar and A. Plutynksi (eds.), *A Companion to the Philosophy of Biology*, Wiley-Blackwell, Oxford 2008.
In *The Blackwell Companion to Philosophy of Biology*. Blackwell, Oxford *forthcoming*.
[14] M. Ferraris. Di che materia sono fatti i brevetti. Sito Web Italiano per la Filosofia (SWIF), URL = http://www.swif.uniba.it/lei/rassegna//ferraris.htm, 2002.
[15] G. Frege. *Die Grundlagen der Arithmetik*. Breslau, Köbner 1884. Eng. transl. *The Foundations of Arithmetic*, Blackwell, Oxford 1950.
[16] R.W. Furness and K. Hamer. Skuas and Jaegers. In C. Perrins (ed.), *Firefly Encyclopedia of Birds*, Firefly Books, Buffalo, NY 2003, pages 270-273.
[17] M.T. Ghiselin. A Radical solution to the species problem. *Systematic Zoology*, 23: 436-544, 1974.
[18] A.P. Gray. *Mammalian Hybrids. A Check-list with Bibliography* (2nd edition). Commonwealth Agricultural Bureaux Farnham Royal, Slough, England 1972.
[19] R. Grossmann. *The Categorial Structure of the World*. Indiana University Press, Bloomington, IN 1992.
[20] B.K. Hall. The paradoxical Platypus. *BioScience*, 49 (3): 211–218, 1999.
[21] M. Heller. *The Ontology of Physical Objects: Four-Dimensional Hunks of Matter*. Cambridge University Press, Cambridge 1990.
[22] D.L. Hull. Are species really Individuals? *Systematic Zoology*, 25: 174–191, 1976.
[23] D.L. Hull. On the plurality of species: questioning the party Line. In R.A. Wilson (ed.), *Species. New Interdisciplinary Essays*, MIT Press, Cambridge, MA 1999, pages 23–48.
[24] I. Kant. *Kritik der reinen Vernunft*. Hartknoch, Riga 1781/1787. Eng. transl. *Critique of Pure Reason*. Macmillan, London 1958.
Critique of Pure Reason. Translated by N. Kemp Smith. Macmillan, London 1781/1958.
[25] P. Kitcher. Species. *Philosophy of Science*, 51: 308–333, 1984.
[26] G. Klima. The essentialist nominalism of John Buridan. *The Review of Metaphysics*, 58: 301–315, 2005.
[27] N.M. Le Douarin and A. McLaren. *Chimeras in Developmental Biology*. Academic Press, London 1984.
[28] D.K. Lewis. *Convention. A Philosophical Study*. Harvard University Press, Cambridge, MA 1969.
[29] E.J. Lowe. *The Four-Category Ontology: A Metaphysical Foundation for Natural Science*. Oxford University Press, Oxford 2006.
[30] R.L. Mayden. A hierarchy of species concepts: the senouement in the saga of the species problem. In M.F. Claridge et al. (eds.), *Species: the Units of Biodiversity*, Chapman and Hall, London 1997, pages 381–424.
[31] E. Mayr. *Populations, Species and Evolution*. Harvard University Press, Cambridge, MA 1970.
[32] B. Mishler and R. Brandon. Individuality, pluralism, and the phylogenetic species concept. *Philosophy & Biology*, 2: 397–414, 1987.
[33] H. Putnam. *The Many Faces of Realism*. Open Court, La Salle 1987.
[34] R. Rorty. Review: Putnam on truth. *Philosophy and Phenomenological Research*, 52 (2): 415–418, 1992.
[35] D. Schwarz. *Natural Hybridization and Speciation in* Rhagoletis *(Diptera: Tephiritidae)*. Ph.D. Thesis, Pennsylvania State University 2004.
[36] A. Sidelle. *Necessity, Essence, and Individuation: A Defense of Conventionalism*. Cornell University Press, Ithaca, NY 1989.
[37] M.H. Slater. *The Nature of Kinds*. Ph.D. Thesis, Columbia University 2006.
[38] R. Sokal and T. Crovello. The biological species concept: A critical evaluation. *American Naturalist*, 104: 127–153, 1970.
[39] G. Ledyard Stebbins. The role of hybridization in evolution. *Proceedings of the American Philosophical Society*, 103/2 (Commemoration of the Centennial of the Publication of "The Origin of Species" by Charles Darwin): 231–251, 1959.

[40] P.F. Strawson. *Individuals. An Essay in Descriptive Metaphysics*. Methuen, London 1950.
[41] L. Van Valen. Ecological species, multispecies, and oacks. *Taxon*, 25: 233–239, 1976.
[42] A.C. Varzi. Boundary. In Edward N. Zalta (ed.), *Stanford Encyclopedia of Philosophy*, CSLI, Stanford 2004 (internet publication).
[43] A.C. Varzi. Teoria e pratica dei confini. *Sistemi Intelligenti*, 17: 399–418, 2005.
[44] E.O. Wilson. *The Diversity of Life*. Penguin, London 1992.

Final, efficient and complex causes in biology

Flavio D'Abramo

1 Paley and Darwin: from final to efficient causes

During the XIX century the alliance between natural sciences and theology was great. I will consider theology as a particular kind of finalistic thought. Natural theology was a philosophy attempting to prove the existence of God appealing to natural phenomena such as adaptation. William Paley was the best example of a natural theologian. In 1803 he wrote:

> [...] it is in the construction of instruments, in the choice and adaptation of means, that a Creative Intelligence is seen. [...] The marks of design are too strong to be got over. Design must have a designer. That designer must have been a person. That person is God [16, p. 264].

Charles Darwin was greatly influenced by Paley's theory and what appealed to him most was his approach to functional analysis. However, between Darwin and Paley there is a fundamental difference: the Darwinian theory of evolution by natural selection is based on the analysis of functional causes involving final causes where the existence of God is not taken into account. With the evolutionary theory he put aside any scientific questions on the origin of life. Darwin hypothesizes that during evolution, the variation of organisms is caused by random mutations, so there is no place for teleological thought. The only feasible direction of the evolutionary process must be related to natural selection. The effect of natural selection is that adaptable organisms are selected. The randomness of mutations is one of the main concepts through which he criticizes the directionality of evolution.

In the same way, the concept of *random mutation* overthrows the *Scala Naturae* as seen by Aristotle, and directionality, as theorized by French biologist Lamarck where he maintains that organisms evolve progressively towards perfection. Similarly, the law of recapitulation proposed by Ernst Haeckel becomes devoid of sense, even if his studies are at the core of the connection between phylogeny and ontogeny [33, p. 13]. Haeckel asserted that during its embryonic development, the human being first progresses through all other animal forms; humans tend to the highest form of evolution. However, what was most significant in Darwin's day, was his evolutionism in contrast to creationism. In relation to natural selection D'Arcy Thompson wrote in his book *On Growth and Form*: "we have reached a teleology without a *telos* [...]: an 'adaptation' without 'design' a teleology in which the final cause becomes little more, if anything, than the mere expression or resultant of a sifting out of the good from the bad, or of the better from the worse, in short of a process of mechanism" [37, p. 4]. Natural selection is a process involving functional analysis, but considering only functional causes give rise to major problem. In fact, natural selection cannot explain homologous organs such as the fish fin, the arms of humans, or the anterior legs of quadrupeds. Nor can it explain those organs that have no function, where function changes over time, or the origin of organs. We cannot explain the emergence of novelties or innovations. Aware of the intrinsic limits springing by functional analysis, in the last chapter of *The Origin of Species* Darwin wrote: "Nothing can be more hopeless than to attempt to explain this similarity of pattern in members of the same class, by utility or by the doctrine of final causes" [8, p. 383]. Therefore, the origin of forms deprived of any functions, in other words the origin of structures, cannot be explain in functional terms. Maybe, Darwin avoided explaining the origin of biological forms and the origin of life in order not to generate conflict with theological beliefs.

2 D'Arcy Thompson and physical forces

Since the emergence of this stalemate — the inadequacy of functional analysis — some scientists such as D'Arcy Thompson, Karl Ernst von Baer, Conrad Waddington and Stephen J. Gould [see forward] tried to explain similarities and differences in organisms through a non-directed or a-finalistic analysis of the biological structures. Stephen Jay Gould in 1971, then Stephen Jay Gould and Richard Lewontin in 1979, and later Stephen Jay Gould with Elizazbeth Vrba in 1982 reconsidered the importance of structural analysis. They recovered the fundamental work of D'Arcy Thompson. The Scottish biologist in his *On Growth and Form* proposed an analysis of organisms' morphology using geometrical, physical and mathematical theoretical tools. In brief, he highlighted the analysis of efficient causes (in Aristotelian terms, the action of physical forces on matter) in his theory. I consider Thompson one of the most close successors of Darwin. He attached importance to Darwin's question about regularity, invariance and geometry, on natural patterns — such as homologous organs or ciliates' distribution in the Cilean sea, between Conception and the Archipelago of the Galapagos, where Darwin observed a symmetric discoloration of the sea from organic causes: "The colour of the water, as seen at some distance, was like that of a river which has flowed through a red clay district [...] where the red and blue water joined was distinctly defined" [7, p. 16]. That is a clear case of pattern formation, where the functional explanation couldn't be utilized, and where physical forces as 'currents of the air or sea' were taken into account [7, p. 17]. To answer such questions, D'Arcy Thompson considered organisms in the same way as inanimate objects. In both, inanimate and living worlds, physical forces shape structures before natural selection takes place. Physical forces are considered as structural constraints and these constraints shape and limit the ways that organisms can take. Comparing inanimate and living objects [Fig. 1] Thompson wrote: "We may use a hanging drop, which, while it sinks, remains suspended to the surface. Thus it cannot form a complete annulus, but only a partial vortex suspended by a thread or column [...] and the figure so produced, in either case, is closely analogous to that of a medusa or jellyfish,

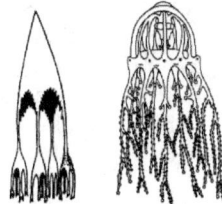

Figure 1. On the left a drop of amilic alcohol falls on paraffin. On the right Syncoryme jellyfish [37].

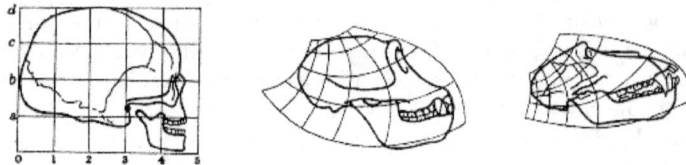

Figure 2. Human skull, chimpanzee skull and baboon skull [37].

with its bell or 'umbrella', and its clapper or 'manubrium'" [37, p. 345].

After the action of physical forces, such as gravity, surface tension, fluid friction, under balanced condictions of temperature, density and chemical composition, natural selection can occur; for instance, in order to keep up a circulation sufficient for the part and no more, to allow the circulation of the right quantity of oxygen, natural selection has not only varied the angle of branching of the blood-vessels to suit her purpose, she has regulated the dimensions of every capillary, through all the trials and errors of growth and evolution.

After much analysis Thompson proposed a theory of transformation in which he puts the different forms of homologous organs on a Cartesian grid deducing that variations usually occur in a modular manner and never alone [Fig. 2].

Using this kind of method, based on efficient and formal causes, D'Arcy ignored the historical side of evolutionary dynamics (the genetical side of biology) in order to consider only the temporal and visible aspects. This was a necessary step to formalize and describe the morphological variation. He considered physical and chemical forces as constraints of all the other biological dynamics. Today this analytical method is at the basis of complex studies where sophisticated biological phenomena are examined. The main philosophical revolution of D'Arcy Thompson was in the role he assigned to the physical forces. In his view physical forces are at the base of final causes. In other words, there are forces and dynamics such as gravity or viscosity, osmosis or surface tensions, solubility and diffusion of chemical compounds, that imposed their action on bodies. These forces are active or creative. What we can see as purposeful is the product of physical forces. To criticize phyletic gradualism D'Arcy Thompson applied a kind of typological thought in which biological shapes are seen as geometrical curves. As he pointed out, we cannot transform an helicoid into an ellipsoid; in a similar way we cannot transform an invertebrate into an invertebrate. He also showed the cases in which use, or not, this typological thinking:

> consciousness is not explained to my comprehension by all the nerve-paths and neurones of the physiologist; nor do I ask of physics how goodness shines in one man's face, and evil betrays itself in another. But of the construction and growth and working of the body, as of all else that is of the earth earthy, physical science is, in my humble opinion, our only teacher and guide [37, p. 8].

Therefore, he restricted explanational power of physical forces to construction, growth and working of the living body. These assumptions — the weight of the role assigned to physical and chemical forces and the discontinuity of evolution — are at the core of attempts to synthetize evolutionary and developmental biology (evo-devo) [5, 3].

Conrad Waddington was another Scottish biologist who took into account evolutionary phenomena, analyzing efficient causes in conjunction with formal methods. By changing the environmental factors such as temperature, he observed changes in the development of the organs of many generations of Drosophila. He also discovered that there were changes in the expression of the genes. Waddington connected the morphological analysis to the physiological one, with special emphasis on genetic and proteomic dynamics [35]. After his theoretical and experimental studies, mutations of the genes expression — also called epigenetic mutations — were considered to be an effect of environmental changes. Thus, environmental changes cause changes at the physiological level, which causes changes in the gene expression, which causes changes at the morphological level.

Conrad Waddington is one of the most important critics of teleological reasoning of embryonic development. Before Waddington, animal development was considered a linear progression from embryo to adult form. In this theory all organisms are *preformed* with all parts developed in the germ or in the egg, but in a reduced scale. This deterministic view was opposed by *epigenesis*, where all the morphologically homogeneous corpuscles forming the organisms, in time become structures with specific roles. With these two opposed views, biologist and scientists tried to explain the origin and organization of life. Naturalists such as Charles Bonnet or philosophers as René Descartes supported a preformistic view of life based on a mechanistic paradigm [1].

Since the second half of XVIII century some naturalists carried out experiments against the preformistic view. Observing the formation of organs in the embryo, Caspar Friedrich Wolff showed that preformation is in contrast to observed phenomena: each organ reaches its shape through different forms. Therefore immutability assumed in preformism was excluded in epigenesis. Through experimental studies inducing mutations in embryos, Geoffroy Saint-Hilaire pointed out that anomalies did not pre-exist in the fecundation. In this manner, teratogeny overthrows preformism [4]. Conrad Waddington, coupling genetics with epigenesis, put an end to the finalistic thought and preformation view concerning embryology and developmental biology. He considered environmental changes as the cause of the re-shaping of organisms; in other words, the adult form is not the only form possible [38]. Before the birth of molecular biology, one of the first studies against teleological and progressive development was by the Estonian naturalist Karl von Baer, who criticized Haeckel's recapitulation law and pointed out the autonomy of the developmental dynamics of each single species. Recovering von Baer's law for embryology, Darwin took into account the importance of deviation and resemblance in development. Darwin had a more devo-evo focus, concerned with using embryology to explain evolution [22]. As pointed out by Gould [13], in the last chapter of *The Origin*, Darwin used early embryonic stages as criterion in order to reconstruct phyletic-evolutionary lineage:

> In two groups of animal, however much they may at present differ from each other in structure and habits, if they pass through the same or similar embryonic stages, we may feel assured that they have both descended from the same or nearly similar parents, and are therefore in that degree closely related. Thus, community in embryonic structure reveals community of descent [6, p. 449].

3 Towards a new evolutionary developmental biology

Belyaev's experiments confirmed the theory of Waddington and reveal the great biases of the geno-centric view [19]. During the fifties Dmitry Belyaev began a time-consuming experiment: the taming of a population of foxes. After twenty generations he discovered that behavioural change also involves changes in the endocrynous values, such as longer mating time, altered sexual hormone value and variation of moulting time. He also discovered that the variation of the endocrynous systems' values is related to genetic expression and this in turn is related to morphological or phenotypical variations such as spotted coats, floppy ears, and curled tails. Thus Belyaev deduced that physiological shifts re-activate silent genes and these regulate morphological and phenotypical characteristics. According to Belyaev this is one of the possible interpretations of the genocentric gap, where silent genes are called *junk* DNA. In the XX century a huge dichotomy eclipsed the importance of experiments such as those of Belyaev or Waddington: the dichotomy of functional/evolutionary biology or proximate/remote causes. One of the most clear examples of this dichotomy is in Mayr's works. In his 1961 "Cause and effect in biology", published in *Science*, he wrote:

> The functional biologist deals with all aspects of the decoding of the programmed information contained in the DNA of the fertilized zygote. The evolutionary biologist, on the other hand, is interested in the history of these programs of information and in the laws that control the changes of these programs from generation to generation. In other words, he is interested in the causes of these changes. [...] The occurrence of a given mutation is in no way related to the evolutionary needs of the particular organism or of the population to which it belongs [25, p. 26].

The last sentence in the quotation highlights the central role of randomness. The lack of connection between mutations and evolutionary needs is, in other words, a random component of the theory. In some cases, like this one, the randomness obscures the analysis of the efficient causes that relate organism's mutations to evolutionary changes. We can interpret the lack of connection between organism's changes and evolutionary/ecological changes as the incapacity to comprehend the relation between genes and organisms — the concept of *randomness* sometimes can have the same meaning. In Mayr's article functional biologists "deals with all aspects of the decoding of the programmed information contained in the DNA of the fertilized zygote" [25], while the evolutionary biologists are depicted as interested in the history of the programs of information and in the laws and changes of these programs from generation to generation. Mayr was the first biologist who saw organisms as outcomes of DNA coding [25, 26]. Nevertheless in his 1961: "But let us not have an erroneous concept of these programs. It is characteristic of them that the programming is only in part rigid. Such phenomena as learning, memory, nongenetic structural modification, and regeneration show how open these programs are" [24, p. 1504].

If we look at the experiments made by Waddington or Belyaev, we can consider, at least, two theoretical levels, the individual and the environmental. In each one casual mutation or phenomena occur. Linking these two levels another one emerges. In this new level there is a structural coupling between organisms and the environment. The randomness of variations is such only if organisms and environment are separated. The organism/environment distinction is only an analytic splitting; organism and environment are mutually determined, each one exists thanks to the other [9, 21]. The coupling of so many levels causes problems concerning the prediction of biological phenomena. Predictions must involve dynamics of the environment, behaviour, genetic, epigenetic and morphological levels, contingent factors together with symbolic levels. In fact a symbolic cultural code such as speech, if applied with the same criteria, will bring about change on all other levels. In short, we need a multi-disciplinary approach. It is necessary to look at the formalization of biological phenomena to understand the need for a multi-disciplinary approach. The importance of treating biological phenomena in an a-temporal and a-historical way, emerges trough a theoretic *empasse*.

One of the most impressive problems relates to the formalization of dynamics interacting through many levels such as behavioural, endocrine, epigenetic, genetic, morphological — i.e. mutation in Belyaev foxes. In population biology, the tools used to analyze the causes of phenotypic variance are *segregation analysis* and *analysis of variance*. Lewontin [20] pointed out the limits of this linear analysis. These statistical tools separate the elements of a number of causes that interact to produce a single result — i.e. the interaction between environment and genotype in the determination of phenotype. In classical analysis it was supposed that the phenotype of an individual could be the result of either environment or genotype, whereas we understand the phenotype to be the result of both. The analysis of variation intended to consider environment and genetic effects on phenotype as homogeneous values. Considering these two values as homogeneous, we can hope to separate causes, but this is purely illusional. To do it, this two effects are substituted with the mean for a given environment averaged over all genotypes in the population and the mean for a given genotype averaged over all environments. The argument comes from a Cartesian world view that things can be broken down into parts without losing any essential information, that in any complex inter-

action of causes main effects will almost always explain most of what we see
and interactions tend to be of smaller importance. But this is a pure *a-priori*
prejudice.

> Analyzing causes, totally different objects has been substituted as the object of investigation, almost without noticing it. [...] The relations between phenotype and genotype and between phenotype and environment are many-many-relations, no single phenotype corresponding to a unique genotype and vice versa [...]. This is expressed in the *norm of reaction*, which is a table of correspondence between phenotype, on the one hand, and genotype-environment combinations on the other [20, pp. 403–404].

To formalize this complex phenomena expressed by *reaction norm* we need equations with square polynomials, in other words non-linear equations. We cannot solve this kind of equation in an analytical way, but only by using computers and graphic solutions [29].

The importance of non-linear equations has been known since Poincaré's "three-body problem" [3], while the solutions have been available for a few decades. Poincaré showed that a very tiny imprecision in the initial conditions would grow in time at an enormous rate. Thus two nearly-indistinguishable sets of initial conditions for the same system would result in two final predictions which differed vastly from each other. Poincaré mathematically proved that this transformation of tiny uncertainties in the initial conditions into enormous uncertainties in the final predictions remained even if the initial uncertainties were shrunk to smallest imaginable size. That is, for these systems, even if you could specify the initial measurements with great precision, the uncertainty in prediction would still remain huge. Even if large part of biological thought is constrained by this epistemological limit and biases — one of the most clear exemplars is the geno-centric point of view in which genotype is responsible for behavioural traits — nevertheless we have lots of counter example: Darwin hypothesized a complex causal net between species.

This causal complex connects living beings through mutual or competitive interaction. In one of the most popular examples, Darwin showed the way in which the number of cats and the number of red clovers are connected [8]. In a detailed description, he displayed that red clover is ferilised by humble-bee alone, and that the number of humble-bees is linked to the number of field mices, indeed mice destroy humble-bees' nests. Inasmuch as cats eat mices, the presence of a feline animal might determine, through the intervention first of mice and then of bees, the frequency of red clovers in that district. Therefore Darwin showed a causal web, or a causal complex, in which each element is related with all the others.

The necessity to move the analysis away from the historical-evolutionary aspect toward temporal-evolutionary one, regards not only the attempt to provide predictions but also to consider biological phenomena as complex. The more developed neo-darwinian address is far from taking into account the causal complex, and in many cases such a temporal-evolutionary analisys can be usefull in showing the inadequacy of the narrative — e.g. the application of Poincaré's "three-body problem" into the gene-centric theories. Such biological and complex dynamics are distributed on many levels, involving different disciplines. In this complex view such phenomena are effected by many parallel and serial causes. Poincaré's theory tells us that the longer the time the prevision becomes more inaccurate. Obviously in evolutionary biology, as in every science, the historical and narrative values are at the core of the devolpment of the theory. To be intelligible a model must have a narrative side where all the terms included are justified [23]. In formal models historical values are equal to the initial conditions. The most important meaning of genetics could be in taking into account the sedimentation of the most ancient historical conditions in the nucleotides. In Waddington's terms this value is expressed in *canalization*. Using dynamic systems to formalize biological phenomena we can see that starting conditions bring a substantial contributions to the future paths of the system. For instance, eukaryote genetics and epigenetic conditions — outcome of historical relations between ancestors and their environments — are the starting conditions of the (ontogenetic) system. This new ontogenetic system is again exposed to the (actual) environmental conditions. This means that evolutionary history acquired a crucial role to understand biological phenomena.

If the consideration of temporal and formal aspects of biological phenomena is crucial, recovering the historical and narrative sides will be equally important; in fact the formal model is based on historical and narrative aspects and sometimes it can bring to light details of natural history. We also notice that the evolution of genetic characteristics are slow, while the rate of social and cultural evolution, which determines biological dynamics is extraordinarily fast [19]. To limit the epistemological and ideological biases we need to put the significance of epigenetics into perspective. To rescale genetics and to prove it inseparable from the organism it is necessary to consider the relation between epigenetics and phenotypic expression such as morphogenesis.

The organismic systems approach (OSA) [3] rehabilitates D'Arcy Thompson approach and enriches it with Waddington's epigenesis and with the new tendencies of developmental, theoretical, evolutionary and ecological biology, pointed out by Scott F. Gilbert, Brian Goodwin, Stuart Kauffman, Richard Lewontin, Stephen J. Gould, Marion J. Lamb and Eva Jablonka. In OSA the causal flux running between genes expressions, proteins and phenotype is bi-directional. Between 1973 and 1977, S.J. Gould with others naturalists, wrote a series of articles, published in *Paleobiology*, in which there is a clear explanation of the mistake in invoking genetical determinism: many similar organisms can have different genotypes and many organisms with the same genotype can have different forms.

As pointed out by Werner Callebaut, Stuart A. Newman and Gerd B. Müller [3], the correlation of the organism's form with its genotype is a highly derived property, and during the course of evolution have been active few non-genetic causal determinants of biological morphogenesis. The phenotypic changes depend upon many environmental variables that determine epigenetic factors such as diet, pH, humidity, temperature, photoperiod, seasonality, population density or presence of predators [25], but also parental care [26]. In an indirect way, epigenetic variance derives from cultural and symbolic evolution [19]. According to OSA we can see phenotypic polymorphisms as sophisticated products of evolution [3]. This is a clear proximity to Thompson's theory. In this perspective homology is an organizational phenomena: "that initially arise from generic properties of cell masses, and later from conditionals interactions between cells and tissues, provide "morphogenetic templates" for an increasing biochemical sophistication of cell and tissue interactions" [3, p. 53]. Therefore beyond the comparative/historical methods, "our most powerful tool for investigating the evolution of form and function" [18] we can use also theoretical accounts to reach the organizational homology concept, where the role of developmental constraints and the active contributions of organising processes are highlighted. One of the most promising addresses of Thompson's approach is in morphometrics [2]. The new computational tools can represent the relationship between gene activation, cell behaviour, and morphogenesis [3] probes the mechanisms of epigenetic causation in morphological innovation. The theoretical and computational evo-devo programme: "has led to the development of computational tools for the three-dimensional reconstruction and quantification of gene expression in developing embryos, and the exploration of new mathematical methodologies for the analysis of such data" [26, p. 943]. This is a central point in the evo-devo agenda which can pragmatically demonstrate a communal research program. In fact, in evo-devo there are many centripetal and conflicting theoretical tendencies [1].

4 Conclusion

We have seen that functional analysis is the most ancient analysis in biology. Aristotle defined an organ as an animal's part that performs a function. Nevertheless during evolutionary time, the function can change, other structures can become organs and other organs can loose their function becoming mere structures. To analyse all these changes we need to observe the structure. We can interpret junk DNA as a DNA that has lost its functional role. To analyse the structures that emerge in the relation between organisms and environment during evolutionary time, we need to refer to physical and chemical laws and to ignore teleological thought. Of course, we also need historical explanation. In fact, the phenomena that trigger the physical and chemicals dynamics at the basis of mega-mutations are contingent, so we cannot anticipate these, but we can use these active and passive constraints to explain the evolutionary dynamics in a consistent manner. We can consider this kind of dynamics, such as epigenetics or morphometrics, the paradigm of evolutionary developmental biology.

BIBLIOGRAPHY

[1] R. Amundson. *The Changing Role of the Embryo in Evolutionary Thought. Roots of Evo-Devo*. Cambridge University Press, Cambridge 2005.
[2] F. L. Bookstein. *Morphometric Tools for Landmark Data*. Cambridge University Press, Cambridge 1991.
[3] W. Callebaut, G.B.Müller and S.A. Newman. The Organismic Systems Approach. In R. Sanson and R.N. Brandom (eds.), *Integrating Evolution and Development. From Theory to Practice*, MIT, Cambridge, MA 2007, pages 25–92.
[4] G. Canguilhem, G. Lapassade, J. Piquemal and J. Ulmann. *Du développement á l'évolution au XIXéme siècle*. PUF, Paris 1963.
[5] S.B.Carroll. *Endless Forms most Beautiful. The New Science of Evo-Devo*, Armoni, New York 2005.
[6] C. Darwin. *The Origins of Species*. First edition, (http://darwin-online.org.uk), 1859.
[7] C. Darwin. *Journal of Researches into the Natural History and Geology of the Countries Visited during the Voyage of H.M.S. Beagle round the World, under the Command of Capt. Fitz Roy R.N.* (http://darwin-online.org.uk), 1860.

[8] C. Darwin. *The Origins of Species*. Sixth edition, (http://darwin-online.org.uk), 1872.
[9] E. Favarelli, E. Cianci, E. Serrelli and D. Suman. *L'evoluzionismo dopo il secolo del gene*. Mimesis, Milano 2005.
[10] S.F. Gilbert. Evo-Devo, Devo-Evo, and Devgen-Popgen. *Biology and Philosophy*, 18: 347–352, 2003.
[11] B. Goodwin. *How the Leopard Changed its Spots. The Evolution of Complexity*. Princeton University Press, Princeton, NJ 1994.
[12] S.J. Gould. D'Arcy Thompson and the science of form. *New Literary History*, 2: 229–258, 1971.
[13] S.J. Gould. *Ontogeny and Phylogeny*. Harvard University Press, Cambridge, MA 1977.
[14] S.J. Gould, D.M. Raup, J.J. Sepkosky, T.J.M. Schopf and D.S. Simberloff. The shape of evolution: a comparison of real and random clades. *Paleobiology*, 3: 23–40, 1977.
[15] S.J. Gould and R.C. Lewontin. The Spandrels of San Marco and the panglossian paradigm: A critique of the adaptationist programme. *Proceedings of the Royal Society London B*, 205: 581–598, 1979.
[16] S.J. Gould. *The Structure of Evolutionary Theory*, Harvard University Press, Cambridge, MA 2002.
[17] S.J. Gould and E.S. Vrba. Exaptation — a missing term in the science of form. *Paleobiology*, 8: 4–15, 2008.
[18] P.E. Griffiths. Phenomena of homology. SILFS Conference 2007.
[19] E. Jablonka and M.J. Lamb. *Evolution in four dimensions. Genetic, Epigenetic, Behavioural and Symbolic Variation in the History of Life*. MIT Press, Cambridge, MA 2005.
[20] R.C. Lewontin. The analysis of variance and the analysis of causes. *American Journal of Human Genetics*, 26: 400–411, 1974.
[21] R.C. Lewontin. *Biology as Ideology. The Doctrine of DNA*. Anansi Press Limited, Concord, Ontario 1991.
[22] M.D. Laubichler and J. Maienschein. Embryos, Cells, Genes, and Organisms. In R. Sanson and R. N. Brandom (eds.), *Integrating Evolution and Development. From Theory to Practice*, MIT Press, Cambridge, MA 2007, pages 1–24.
[23] M.D. Laubichler and G.B. Müller (eds.) *Modelling Biology. Structures, Behaviors, Evolution*. MIT Press, Cambridge, MA 2007.
[24] S. Leonelli. What is in a model? Combining theoretical and material models to develop intelligible theories. In M.D. Laubichler and G.B. Müller (eds.), *Modelling Biology. Structures, Behaviors, Evolution*, MIT Press, Cambridge, MA 2007, pages 15–36.
[25] E. Mayr. Cause and effect in biology. *Science*, 134: 1501–1506, 1961.
[26] E. Mayr. *Toward a New Philosophy of Biology. Observation of an Evolutionist*. Harvard University Press, Harvard 1988.
[27] G.B. Müller. Evo-devo: extending the evolutionary synthesis. *Nature Reviews Genetics*, 8: 943–949, 2007.
[28] H.F. Nijhout. The control of growth. *Development*, 130: 5863–5867, 2003.
[29] H.F. Nijhout. Development and evolution of adaptive polyphenisms. *Evolution & Development*, 5: 9–18, 2003.
[30] P. Odifreddi. *La Matematica del Novecento. Dagli Insiemi alla Complessità*. Einaudi, Torino 2000.
[31] S. Oyama. Parlare della natura. In E. Favarelli, E. Cianci, E. Serrelli and D. Suman, *L'evoluzionismo dopo il secolo del gene*, Mimesis, Milano 2005, pages 107–127.
[32] J.-H. Poincaré. *Leçons de mécanique céleste professées à la Sorbonne*. Gauthier Villars, Paris 1905.
[33] A. Prochiantz. *Machine-esprit*. Odile Jacob, Paris 2001.
[34] D. Raup, S.J. Gould, T.J.M. Schopf and D.S. Simberloff. Stochastic models of phylogeny and the evolution of diversity. *The Journal of Geology*, 8: 525–42, 1973.
[35] R. Sanson and R.N. Brandom. *Integrating Evolution and Development. From Theory to Practice*. MIT Press, Cambridge, MA 2007.
[36] von L. Speybroeck. From epigenesis to epigenetics: the case of C. H. Waddington. *Annals of the New York Academy of Sciences*, 98: 61–81, 2002.
[37] W. Thompson D'Arcy. *On Growth and Form*. Cambridge University Press, Cambridge 1942.
[38] C.H. Waddington. *The Evolution of an Evolutionist*. Edinburgh University Press, Edimburgh 1975.

The concept of race and its justification in biology

LUDOVICA LORUSSO

1 Race and racial cluster

In biomedical research there is a wide discussion around the concept of racial cluster, where the prevailing notion of racial classification implies that phenotypic traits like skin colour and facial features can be used to categorize people into meaningful genetic subgroups.

Until now the most common problem about the concept of race has been the ontological problem — that is the problem concerning the existence of race — which for many people should be answered inside biology, since the race concept is a biological concept. For others, however, it has both a cultural and biological reality and the answer should involve philosophical considerations as well. The answer coming from biology is controversial. One of the main reasons for this disagreement among scientists is the *semantic vagueness* of the term. The term "race" is used with many different meanings and it is difficult to reach an agreement for the existence of something that means something different for different people. Therefore, there is another issue that should come before the issue about the existence of race: the issue about the definition of the term "race". Finally, there is the issue I am going to answer in this paper, which is not very well discussed in the scientific and philosophical literature. This is the epistemological issue, concerning the justification of the race concept in biological explanation. I have presented three kinds of issues concerning race: ontological, semantic, and epistemological. All these problems acquire a particular significance in relation to a fourth problem, that one concerning the use of racial categories in biology. In other words, all the three issues about race have become central in philosophy and science because the use of the race concept among humans is historically linked to ethical and social issues.

In philosophy almost any discussion about the use of race in biology is focused on the ontological issue. Roughly the most popular argument is: race do not exist, therefore the concept of race must not be used in biology. People who deny races can hold two different positions. In the first ontological position against races, races are considered as *social constructs*.[1] In the second one, instead, they represent the product of both social and biological properties.[2] While in the first position any biological value of race is denied, in the second one it is denied the possibility of a separation of social and biological

---

[1] About a discussion on race as social construct, see e.g. [6]
[2] About the concept of race as "bio-social" construct, see e.g. [5, 8].

properties in the definition of race. On the other hand, other philosophers try to reify races, by searching for viable conceptions of them.[3]

Unfortunately, the ontological argument does not work very well in criticizing or eliminating the use of race concept from biology. The reason for its failure should be found inside the vagueness in which this concept is used in science. In biology scientists hold an instrumentalist position and claim that races should be used because they are useful, no matter whether they really exist. Apparently they avoid the ontological problem, since they do not say explicitly that races exist. But I shall show that their neutrality about the existence of races is not genuine and it is due to the semantic and epistemological vagueness under which they operate. A clarification of the semantic issue and the epistemological role of races in biological explanation is needed for a justified refusal of a reification of the racial concept from the scientific reasoning.

Before starting my analysis, I need to make an important distinction, between race and racial cluster. I have just said that scientists like to claim that they use race without an explicit ontological commitment. When we find in scientific papers the word "race", this term has not a declared ontological meaning and it can just refer to the operational concept of racial cluster — that is, roughly, the concept of people's "self-declared ethnicity". The term "ethnicity" represents an alternative to the term race, suggested because it is thought to carry less of a strictly biologic connotation, implying that groups may differ by both biological and cultural heritage.

In my paper, however, I shall refer to the term "race" with a precise ontological meaning, which characterizes the genetic definition of race (see Section 2).

What is a racial cluster and in general a human cluster? A human cluster is simply defined as the set of humans having one or more properties. By using different kinds of properties one makes different clusters.[4] Note that a cluster in itself is not problematic, but the problem is in the meaning we give to it and how we use it.

Race is doubtless the most used human cluster in current human biology. Phenotypic and geographic properties are mostly used to make a racial cluster. While phenotypic properties are properties related to the colour of the skin, the length of bones, like for example "to be black", geographic properties refer to the origin from a specific continent or region, like for example "to be African". Racial clusters are constructed by using a combination of properties, usually correlated, like the colour of the skin and the geographical origin. The African-American population, for example, represents a group built on both phenotypic and geographical properties (see Table 1).

Even genetic properties can be involved in clustering humans, with the possibility of creating several genetic groups by choosing different genetic systems (e.g., nuclear DNA, mitochondrial DNA), genetic markers (e.g., SNPs, microsatellites), and sets of traits inside each genetic marker. In biomedical research like in epidemiology, pharmacogenetics, and forensic science, racial clusters are widely used with the justification that they have a heuristic pur-

---

[3]About possible reifications of the race concept see e.g. [1, 10].

[4]See about methodological issues in human cluster construction [9].

Population	Property
Mediterranean population	Geographical
Jewish population	Religious
African-American population	Phenotypic & Geographical

Table 1. Populations and properties

pose. In particular, in biomedical research, they are involved in the explanation for differences in diseases. Scientists who use racial categories, argue that the risk of social and cultural discrimination and that one of undervaluing the diversity among individuals within groups should be weighted against the fact that "in epidemiologic and clinical research racial and ethnic categories are useful for generating and exploring hypotheses about environmental and genetic risk factors, as well as interactions between risk factors, for important medical outcomes" [3, p. 1171]. What they claim is that human population is not homogeneous in terms of risk of disease, since this risk is uniquely defined by inherited constitution plus non-genetic or environmental characteristics acquired during life. They assume that this is true not only on an *individual* but also on a *population level*, which is for instance the race level. While the inherited constitution of an individual is given by its family background, the inherited constitution of a population is given by its ancestry, that is its peculiar evolutionary history, which can be represented by a clade (branch) in a phylogenetic tree. This means that they assume that there are genetic plus non-genetic differences among human races and that they are causally related to differences in risk of disease and drug response. Race-specific therapy is based on these two assumptions. However, these assumptions are not easily justifiable for many reasons. Firstly, in the light of the evidence coming from several studies on human genetic diversity, in which it has been demonstrated that human diversity is continuous and therefore not well represented by discrete groupings like clades or races: "Gradual variation and isolation by distance rather than major genetic discontinuities is typical of global human genetic diversity. Obviously, this does not imply that genetic discontinuities do not exist on a more local scale, for example, between people from different linguistic groups" [11, p. 1683]. This means that discontinuities are present only on a *local* level, but not on a *global* level, which would be between continents and racial clusters.

Besides the problem of representing human genetic diversity through a racial classification, there is that one concerning the knowledge of genetic causes of complex diseases.[5] This is an issue independent of the potentiality of race to represent the genetic variation among humans and it concerns the possibility of inferring a certain risk of disease from a given genotype of an individual. The fact is that, in biomedical research, we are currently ignorant

---

[5]By "complex" disease I mean a disease which is caused by the interaction of two or more genes and environmental factors.

about genes that substantively influence susceptibility to complex diseases, and so we cannot predict the risk of common diseases on the basis of genotype. In addition, with respect to complex diseases, there is a *very limited evidence* that specific susceptibility-gene variants are differently distributed among human populations, making meaningless any kind of criterion of human classification. This means that if you want to know whether someone has a particular genotype, you will have to do the test to find out! "Categorizing people on the basis of differences in allele frequencies is therefore not the same as apportioning the whole of human diversity into medically relevant categories. The more relevant outcome — that the sets of common functional polymorphisms[6] are distributed in discrete racial categories — has not been demonstrated" [4, p. 1166]. In summary, clusters based on the variation of genotypes within the whole humankind constitute the only ones really useful and projectible,[7] even if right now the knowledge of causal relations between genotypes and risks of diseases is very weak. The goal of medicine is to obtain a personalized medicine, that is the possibility of prediction of risk and the treatment of disease on the basis of a person's genetic profile. I am going to question the usefulness of racial classification in biology, and in particular the explanatory and projectible value of races in biology. Is racial classification needed in biology? Can it be projectible, in the sense of being useful for extrapolation, generalization, and prediction? I shall show that differences in diseases are consistent with different explanations, only one of which implies a genetic discontinuity among human clusters, which is here considered the necessary and sufficient condition for a racial cluster to be a race. Before facing the epistemological issue about the use and justification of race in biomedical research, I shall roughly present the problem concerning the definition of race and the problem concerning the existence of race, respectively the semantic and ontological problem.

2 The semantic problem

The first issue about race I am going to analyse is the semantic one, which is the problem of the meaning of the term "race". There are two main kinds of definitions:

- An instrumentalist definition that considers phenotypic properties as heuristically sufficient conditions for defining races.

- A realist genetic definition that uses genetic properties to define races. Genetic properties can be either differences in gene frequencies or genetic differences in the form of discontinuities.

In the case of a genetic discontinuity the same genotypes are rarely found in different populations. This means that from the genotype of a single individual it is possible to determine the genotype characteristic of the population to

---

[6]A polymorphism is a genetic trait that is present in human genomes with different forms.

[7]I use the term "projectible" in Nelson Goodman's sense of the property of an object to make successful projections [7].

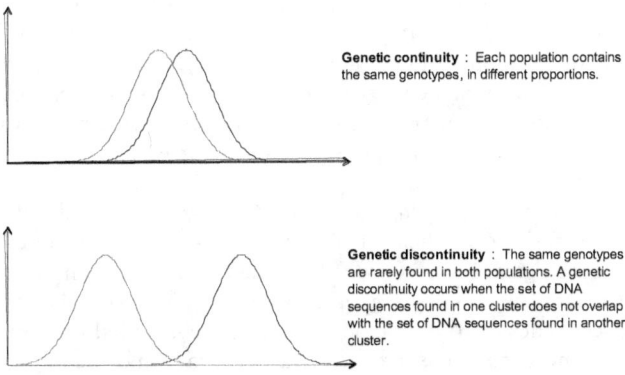

Figure 1. Continuity and discontinuity in genetic variation

which he/she belongs. And more than that, in the case of a discontinuity, "by attributing an individual's genotype to one genetic cluster, one would also obtain information on the individual's genome in general" [2, p. 15]. In the case of a genetic difference without discontinuity, only the frequency of different genotypes is different among human clusters. Therefore in this case from an individual's genotype it is not possible to determine the genotype of the population to which he/she belongs.

In Figure 1 it is shown an idealized representation of the concept of discontinuity, through a Gaussian distribution for different genotypes inside two human populations. A genotype can be characterized by a particular polymorphic trait or a combination of polymorphic traits inside specific DNA sequences. On the X axis there are genotypes and on the Y axis the number of people inside each sub-population.

The second definition involves shared genetic characteristics and it has a metaphysical implication. If one considers this kind of definition, then only two ways of clustering are possible and only one of these leads to a "clear cut" between individuals belonging to two different clusters. By "clear cut" I mean that only in the case of a discontinuity an individual with a certain genotype belongs unambiguously to a certain cluster.

3 The ontological problem

If we liked a realist and unambiguous concept of race, the ontological problem should concern the following question: Do genetic discontinuities exist? Can we, by attributing a specific genotype to one cluster, obtain information about the whole genome of any individual belonging to that cluster? The answer to this question comes from science and it is negative. For the biological evidence is that different genetic polymorphisms are differently distributed over the planet, but their distributions are not generally correlated. This is a consequence of the fact that gene flow, rather than isolation, is the main evolutionary force shaping human genetic diversity, which is therefore mostly dominated by a continuous change among populations. Genetic clusters exist but they cannot be constructed unambiguously since genetic variation is mostly continuous and discontinuous traits are not correlated in their distribution.

Since here I am not analysing the concept of race in general but the use of this concept in the specific field of biomedical research, it is worth introducing another ontological issue in order to make the picture complete: the issue regarding the fact that there are genetic differences among racial clusters that cause diseases. These two ontological issues are correlated only in the sense that if there is no genetic basis for a disease, there cannot be any genetic difference related to that disease among racial clusters, either as frequencies or discontinuities. Discontinuities among racial clusters are not correlated with the existence of a genetic basis for a disease and *a fortiori* with the existence of a genetic difference among races for that disease. For, discontinuities I would use to classify people are not necessarily related to the genetic patterns causally responsible for a specific disease. In other words, by admitting that human clusters based on some genetic discontinuities could be constructed, a discontinuity involving the whole genome is impossible at a race level. Therefore, how can the use of clusters built on some kind of genetic differences be justified in the explanation of differences in risk of a disease, given that this disease could be based on other genetic characteristics?

4 The epistemological problem: justification of races in biological explanation

I am going to analyse the possibility of a justification of races in explaining genetic complex diseases. The epistemological problem is important because the idea of developing different drugs for different human clusters, in spite of the instrumentalist position of scientists, actually depends on the assumption that races exist, that is that genetic variation is discontinuous between racial clusters. For what it would be the sense of developing drugs specific for the race of African-American, if only a percentage of these people had a genetic properties conferring to them a specific risk? I do not want here to consider all ethical problems correlated with the use of races in biomedical research, but I would like to show that the use of race cannot be justified and it does not represent a projectible concept.

Consider two clusters: Blacks & Whites. Consider then the case in which there are different genotypes producing different risks of heart disease.[8] Table 2 indicates that genotypes G1,G3 promote heart disease and genotypes G2,G4 help prevent heart disease. Consider four different hypotheses. In particular, I shall focus on the first two hypotheses, which I shall call the "strong

---

[8]It is worth noting that by saying that they produce I mean that they cause that disease, even if we are considering probabilistic causes, since causes of these kinds of diseases are not determined only by genotype, but genotype & environment.

G1,G3	Heart disease
G2,G4	No Heart disease

Table 2. Genotypes and risk of heart desease.

hypothesis" and the "weak hypothesis".

The strong hypothesis says that the genetic differences that cause the disease are discontinuous across racial boundaries, that is for example, when blacks are G1, G2, and whites are G3, G4. In this specific situation, given any individual, and his/her genotype for the disease, one can determine his/her race. The weak hypothesis, however, says that the genetic differences that cause disease are continuous across racial boundaries, that is for example when blacks are G1, G2 and whites G1, G2.

The strong hypothesis implies that races exist. The weak one, however, is neutral. Here, I do not want to argue that the strong hypothesis is false because races do not exist. I want to argue that the strong hypothesis is unjustified because the weak hypothesis is sufficient to explain the disease rates.

The third hypothesis says that there is no genetic difference between races that is causally responsible for racial differences in disease rates, even though there is a genetic basis for the disease within the whole human population. The correlation between race and disease can not be explained neither with differences in genotypes nor with differences in their frequencies, therefore the environment is needed in the explanation.

The fourth hypothesis is similar to the third one, except that it does not assume that there is any genetic basis for the disease, even in explaining an individual's susceptibility to disease. In the two latter hypotheses the correlation between races and disease is explained with a causal relation between environment and disease.

5 An example

I am going to introduce an idealized example, in which a difference in rate for a specific disease has been discovered inside a human population between the sub-population of blacks (B) and the sub-population of whites (W). We observe that 30 percent of B have heart disease and 20 percent of W have heart disease. This evidence — that is a correlation between heart disease and racial categories — can be explained with four different hypotheses, represented by the tables in Fig 2. The numbers in the right columns represent the number of people characterized by the properties indicated in the left column (i.e., "to be black" and "to have genotype G1"), inside the whole population of 200 people. In the first table on the left, for example, inside the population of 200 people, there are 30 blacks with genotype G1, 70 blacks with genotype G2,

B, G1	30
B, G2	70
W, G3	20
W, G4	80

The strong: genotypic discontinuities between human clusters.

B, G1	30
B, G2	70
W, G1	20
W, G2	80

The weak: differences in frequencies. Genetic causes are the same (G1,G2), only the frequency changes.

B, G1	20
B, G2	80
W, G1	20
W, G2	80

No genetic differences, therefore environmental differences must explain differences in disease rates.

B	100
W	100

No genetic causes, therefore environmental differences must explain differences in disease rates.

Figure 2. Correlation between heart desease and racial categories

20 whites with genotype G3 and 80 whites with genotype G4.

The strong hypothesis assumes that there are discontinuities in the genetic diversity between the two racial clusters. Here the difference in genotypes between blacks and whites is sufficient to explain the difference in disease rates. The weak hypothesis, instead, assumes only that there are differences in the frequency of genotypes in blacks and whites, but genotypes are the same inside the two sub-populations. Here these differences in frequency are sufficient to explain the difference in disease rates. The third hypothesis assumes that there are neither discontinuities in genotypes nor differences in their frequencies that can explain differences in disease rates. Genotypes can still be causes of that disease for any single individual inside the whole population, but in this case they are not involved in explaining the differences in disease rates between the two sub-populations. For this reason in the third hypothesis differences in the environment are needed in the explanation. In the fourth hypothesis there are no genetic causes involved in the explanation of the disease, but only environmental ones. This means that in the last case a genetic cause for the heart disease is denied also at an individual level. In the latter two cases, in which differences in disease rates among racial clusters are completely explained by environmental differences, no genetic difference is needed, and *a fortiori* no genetic discontinuity is needed. The third hypothesis can explain all those ones involving the so called "monogenic" diseases, in which specific genotypes are causally related to a specific risk for a disease or in other words diseases in which a mutation in a single gene is necessary and sufficient to cause disease. Any individual with a certain genotype for a disease can be affected by that disease, but the genotype distribution is the same among human clusters. The fourth hypothesis can only explain cases involving the so named "social" diseases, where individuals belonging to particular social classes and/or doing specific jobs are affected by particular diseases. In this case a clustering is possible, but on a social and not genetic basis. While examples from the third and fourth are well characterized and consequently it is relatively easy to identify the third and fourth hypotheses from the first two, to differentiate the weak and strong hypothesis is not so trivial. In order to clarify concretely the difference between these two hypotheses I am going to discuss a well known disease, the Mediterranean thalassemia.

In the United States, blacks are affected by this disease in a greater percentage than whites. How should the difference between black and white populations in this disease rate be explained correctly? Actually this example is simple, since we have enough information about this disease to discriminate among the four hypotheses and in particular between the strong and the weak one. For we know that there is a genetic basis for this disease and that the specific mutation related to the disease has a greater percentage in the black population than the white one. Therefore we know that there is a genetic basis with no racial discontinuity. But let us make a general reasoning and screening anyway. Given all this information, we can trivially exclude the fourth hypothesis, since it does not consider any genetic basis for the disease and the third one as well, since in the case of thalassemia a genetic difference is sufficient in the explanation of the difference in the disease rates between blacks and whites. Finally the decision would be between the weak and the strong hypothesis, but given that the mutation is present also in the white genotypes, the right hypothesis is clearly the weak one. We could still have the curiosity to know why the mutation for the disease is more frequent in blacks than whites. The reason of this ethnic specificity of the mutation it is that the trait for the disease offers some resistance to malaria, a very common disease in the regions in which some of the American black communities have their origin. In regions where malaria is present thalassemic trait has been positively selected and consequently the proportion of sick people coming from these places is greater. In this example there is a correlation between genetic trait distribution and colour of the skin and a clear causal relation between environment and genetic trait. As a consequence, it is easy to understand the fallaciuos reasoning that bring people to claim a causal relation between colour of the skin and genetic trait. Unfortunately, simple cases like this one are very rare in medicine, where for most of diseases a genetic cause is just assumed and just assumed is also a causal relation between genetic discontinuities and disease rates.

6 The *non causa pro causa* fallacy

In summary, clusters are defined on the basis of phenotypic, cultural, and geographical properties. Geneticists observe differences in disease rates among clusters, and clusters are often assumed to be based on genetic discontinuities related to that disease, leading to the following causal fallacy:

> Blacks have a greater risk of heart disease than whites ⟶ There are genetic discontinuities between blacks and whites that are causally related to this disease

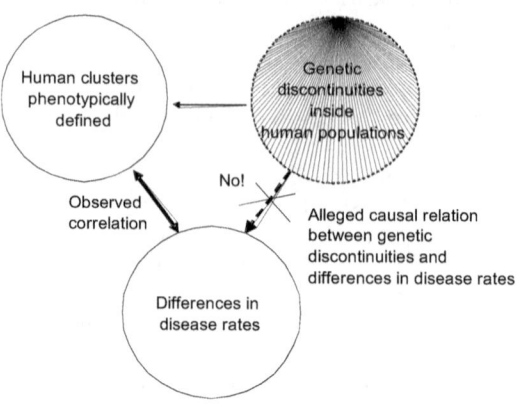

Figure 3. The non causa pro causa fallacy

This represents a special case of the fallacy of non causa pro causa (literally, *"non-cause for a cause"*). This fallacy represents a general, catch-all category for mistaking a false cause of an event for the real cause. In our specific case, as shown in Figure 3, the fallacy consists in assuming a causal relation between genetic discontinuities and racial differences in disease rates. In other words, one commits the fallacy when considers genetic discontinuity between racial clusters as needed in the explanation of racial differences in disease rates.

7 Conclusion: projectibility of race?

Biomedical research focus on statistically significant differences between ethnic groups with the aim of explaining those differences with a hypothesis of a different genetic inheritance among racial groups. However, I have shown that races are not easily justifiable as causes in biological explanation, since genetic discontinuities are sufficient to explain differences in diseases, but they are not needed in the explanation, plus there is a strong evidence against genetic discontinuities among human populations. While the ontological solution of the problem consists in claiming that races do not exist, the epistemological solution I have just offered here is to claim that, even if they existed, they would not be needed. At this point we could ask: Are races projectible? We have seen that races are used as proxies for undetected genetic patterns. This means that, besides the problem of existence of races, there is no evidence for the existence of genetic patterns causally related to diseases. If there is no evidence for genetic causes, *a fortiori* there is no evidence for genetic differences among racial clusters. The real projectibility is inside the variability of single genotypes causally related to specific diseases. For the aim of biomedical research should consist in searching inside genetic patterns related to diseases, and not in a blind searching for some kind of meaningless correlation between genotypes and racial clusters. I hope I have also made clear the point that where the ontological issue about race has become a dead end, the epistemological approach can finally reach the purpose of eliminating vague definitions and obscure uses of any kind of racial categorization.

BIBLIOGRAPHY

[1] R. Andreasen. A New perspective on the race debate. *British Journal for the Philosophy of Science*, 49: 199–225, 1998.
[2] G. Barbujani and E. Belle. Genomic boundaries between human populations. *Human Heredity*, 61: 15–21, 2006.
[3] E. Burchard, Z. Elad, N. Coyle, S. Gomez, H. Tang, A. Karter, J. Mountain, E. Pérez-Stable, D. Sheppard and N. Risch. The importance of race and ethnic background in biomedical research and clinical practice. *The New England Journal of Medicine*, 348: 1170–1175, 2003.
[4] R. Cooper, K. Jay and W. Ryk. Race and genomics. *The New England Journal of Medicine*, 348: 1166–1169, 2003.
[5] L. Gannett. The biological reification of race. *British Journal for the Philosophy of Science*, 55: 323–345, 2004.
[6] D. Goldberg. *Racist Culture*. Blackwell, Cambridge 1993.
[7] N. Goodman. *Facts, Fiction and Forecast*. Harvard University Press, Cambridge, MA 1955.
[8] I. Hacking. Genetics, biosocial groups & the future of identity. *Deadelus*, 51: 81–95, 2006.
[9] L. Lorusso and G. Boniolo. Clustering humans: on biological boundaries. *Studies in History and Philosophy of Biological and Biomedical Sciences*, 39: 163–170, 2008.
[10] M. Pigliucci and J. Kaplan. On the concept of biological race and Its applicability to humans. *Philosophy of Science*, 70: 1161–1172, 2003.
[11] D. Serre and S. Pääbo. Evidence for gradients of human genetic diversity within and among continents. *Genome Research*, 14: 1679–1685, 2004.

PART IV

ECONOMICS AND SOCIAL SCIENCES

Learning, generalization, and the perception of information: an experimental study

MARCO NOVARESE, ALESSANDRO LANTERI, CESARE TIBALDESCHI

1 Introduction

Sensorial perception, information processing, mental representation, and learning do not sound like the typical economics jargon. Microeconomic mainstream does not entertain with these concepts because its approach abstracts from real psychological properties and actual decision processes. According to the standard microeconomic approach, individual behaviour is rational — in a *substantive* sense — when it achieves the given goals of an agent within the exogenous limits of the choice environment. Individual preferences and meta-preferences, for instance egoism and altruism, are external to this approach and must be posited *a priori*. In order to realise given (egoistic or altruistic) goals, an agent must possess complete knowledge of the choice environment and be capable of perfectly computing all this information in an optimal fashion. Both conditions are hardly ever realised and the capacity of microeconomic models to explain individual behaviour are very scarce. Within the ranks of economics, however, on several occasions different scholars have called for an expansion of economic analysis to include more nuanced and plausible accounts of human agency.

For instance, Herbert Simon [19], suggested a concept of rationality which is bounded — i.e. with limited available information and limited capacity to process it — and which is based on *procedures* instead of substantive goals. His research, therefore, had a positive focus on the uncovering of actual decision processes, but inevitably took a normative lean in the definition of what are the best procedures available to real economic agents. Uncovering the way people think, decide, and learn affords a better understanding of the social world, but it also empowers the development of better choice aids and teaching methods.

This article falls within this approach, which may be called Cognitive Economics, and it experimentally explores the way in which the participants learn how to process and manage new information. Our experimental setting is abstract so that the participants cannot rely on any knowledge they already have and must instead learn everything from scratch. In such an abstract setting, our players should perform an everyday task: selecting information, making generalizations, distinguishing contexts. Can they learn how to consistently make the best choice in a new complex environment?

2 Literature Review

In standard economics, the pressure of competition [1] ensures that agents who do not make the best choices are forced out of the market in a fashion akin to natural selection [22]. Individual agents are therefore routinely modelled in such a way that they always make the best choices: this means that they are assumed to possess perfect information and unlimited computational skills, and to pursue their narrow material advantage. Although it is implausible that individuals are (or even can be) as microeconomic models represent them, it may be enough for economists to show that people become (or tend to become) such. Agents capable of improving their performance over time and of progressing towards ever more efficient decisions may uphold, and justify the recourse to, the assumption of perfect individual rationality. This requires the modelling of some individual capacity to learn.

Some examples of how this has been attempted are the Bayesian and the Least Square Learning (e.g. [13]). Both describe the optimal processing of available empirical data by individual agents. These data are then employed in subsequent decision making in a way that approximates the assumption of complete information. Though also the assumption of perfect processing of information is implausible, even psychological models which assume an imperfect processing of the information suggest that people can learn how to make the best choices. Reinforcement Learning models (e.g. [18]), for instance, suggest that agents repeat choices which allowed positive results in the past and consequently adjust their behaviour to empirical evidence in a way that makes it increasingly likely to observe a repetition of the same behaviour (although a, smaller and smaller, probability of making a different choice remains). In standard and stable contexts, reinforcement learning easily results in consistently optimal behaviour just like microeconomic models require.

Learning, however, should not be considered as a black-box mechanism that prompts automatic choices, but rather as a process of assigning specific meanings to different states of the world. Brian Arthur [2], for instance, observed learning cannot be reduced to the acquisition of new data, but it requires the construction of semantic categories that categorise the data. Moreover, individuals build mental models that organise large chunks of empirical evidence. Starting from observation, individuals generate hypotheses about causality and develop models that allow prediction and decision-making. These hypotheses and models are neither static nor unique. Choices are thus repeatedly tested against real world phenomena, associated with their observed outcome, and eventually reinforced or abandoned. The world presents traceable patterns and Arthur believes that the skill to detect these patterns is both a necessary and advantageous human cognitive skill.

Richard Nelson [14] suggests that the search for better ways of doing something is both oriented and constrained by what agents currently know. Current knowledge suggests some behaviour consistent with an agent's goal. The received feedback results either in a more efficient behaviour or in an improved understanding of the specific decision-making context. The agent "either needs to learn how to identify different contexts, as well as a set of context specific guides of action, or find a broad guide to action that works reasonably well in all or most contexts he will face" [14, p. 6]. Therefore, problem solving requires both trial-and-error learning and abstract theorizing.

The study of the capacity to manage information in a complex environment is also central to Ronald Heiner's [9, 10] model of behavioural entropy. According to Heiner, individuals more or less consciously make a choice between very few of the many different actions which are possible on each occasion. This subset consisting of "reliable" actions, or actions which typically afford satisfactory results, is a result of uncertainty — which can be defined as a lack of knowledge of (or lack of the skill to define) the link between contexts and optimal decisions. A reduction in the number of potential options may be a consequence of reacting only to some information, ignoring the rest, of disregarding the distinction among certain pieces of information, or of individual failures in the processing of information, resulting in somewhat generic rules of behaviour that disregard some context-specific variables.

In the presence of uncertainty, it can be expected that agents try out several alternative choices until they figure which ones are reliable. Therefore we observe high variability of behaviour and it is very hard to predict which option will an actor choose next. Over time, as agents learn to react to selected information, their behaviour should become less erratic and therefore more predictable. Heiner employs behavioural entropy as a measurement of the variability of behaviour. It can be computed as follows (see also the Appendix):

$$E^B = -\sum h_a log h_a \qquad (1)$$

where a is an element in the set of possible actions A, and $h_a = p(a)$ is the probability (relative frequency) of choosing a given action. The higher the number of different actions attempted in the same choice-context and the more uniform their frequency (for instance when an agent gives random answers), the higher an agent's behavioural entropy is and the harder it is to predict this agent's choices. Conversely, if an agent's behaviour is stable (because he always makes the same choice), entropy is zero.

Though he does not directly explores learning, from Heiner's reflections, and consistently with Arthur's and Nelson's above, learning may be considered a capacity to discover ever better reliable actions, which means to better use information and to better interpret decision contexts. This immediately translates in the abandonment of any concept of perfect rationality, which is instead replaced by a definition of bounded rationality (à la Simon, [20]) as the capacity to manage only some subsets of useful information. As people learn, they use larger and larger amounts of important information and they react in more specialised ways to subtle changes in environmental conditions. The overall variability of their behaviour therefore scales up, while its predictability is diminished. Within narrowly defined choice contexts, however, variability shrinks.

Since behaviour reflects individual cognitions, learning ultimately affects an agent's behaviour through a change in the type or in the amount of his processed information — e.g. concerning (un-)attainable or (un-)desirable outcomes; (un-)feasible, (in-)effective or (in-)efficient actions. In this sense, all learning modifies the knowledge agent possesses about the task he is facing. The two main vehicles of learning [5, 17, 23] are vicarious learning, which occurs via observation or imitation of the behaviour of others, and direct learning, which takes place when actors obtain information from the outcome of their own actions.

We now turn to the presentation of an experiment which analyses (direct) learning in a complex, but stable, choice environment with strong monetary incentives and where full feedback is immediately available.

3 The experiment

The experiment took place at the Centre for Cognitive Economics at the University of Eastern Piedmont in Alessandria (Italy) on 5 July 2000.

3.1 Participants and experimental design

The participants were twenty-three undergraduate students of Law, enrolled in the first-year optional Seminar of Economics. Each sat in a cubicle with a computer and was not allowed to take notes or to communicate with the others. After reading the written instructions (see the Appendix), the participants started the experiment, which lasted about one hour. The students were compensated with 40 ITL (€0,02) per point scored in the present experiment. They were told that the participation would have no impact on their academic career outside the Seminar.[1]

3.2 Task

The experiment was constructed around a fictional association, whose members fall within one of five age categories: *Children*, *Adolescents*, *Young*, *Adults*, and *Elderly*. The information about members is reported on a set of cards located on either of two shelves (*Right* and *Left*). Each card presents two features: one of four animals (*Cow*, *Horse*, *Goose*, and *Chicken*) and one of four shapes (*Square*, *Rectangle*, *Circle*, and *Oval*),[2] as in Figure 1. On each turn the participants were presented with a sequence of animal, shape, and shelf, and were asked to guess the corresponding membership category within ten seconds.[3] The logical relationship between the card features, the shelves, and the membership was based on a specified criterion (i.e. it was not random) and it remained constant throughout the 231 turns of the experiment, but it was not related to any real world fact and it explicitly did not require any academic knowledge (Figure 2). The connection could and

---

[1] The best performer thus earned €25.52, the worst performer earned €14.32. The average and median compensation were €18.62 and €18.08, respectively.

[2] The features needed be as neutral as possible. In a previous experiment [15], the employment of bright/dark colours and large/small sizes may explain why the subjects associated certain features with value judgements (i.e. insufficient to excellent). Here we also tested the features to ensure neutrality.

[3] The main results are similar to those in Novarese and Rizzello [15], where there was no time constraint.

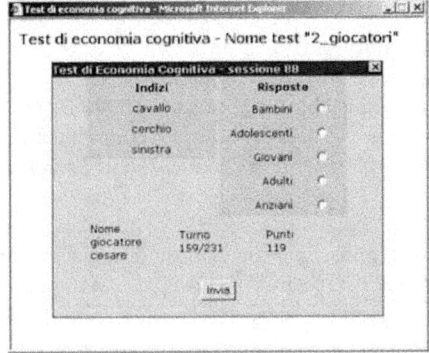

Figure 1. The computer screen of the game

should be learned during the experiment in order to fulfil the ultimate goal of scoring as many points as possible. At the end of each turn, subjects were given full feedback. The score was calculated with respect to the distance between the answer given an the correct answer: the distance is 0 when the answer is correct and in this case the score is 6, the distance is 1 when the answer given is one membership category above or below the correct answer (e.g. *Children/Adolescent, Elderly/Adults*) and in this case the score is 4, etc.

Animal	Shape	Shelf	Membership Group
Mammal (cow, horse)	With angles (square, rectangle)	Right	Children
Mammal (cow, horse)	With angles (square, rectangle)	Left	Adolescents
Mammal (cow, horse)	Without angles (circle, oval)	Right	
Mammal (cow, horse)	Without angles (circle, oval)	Left	Young
Bird (chicken, goose)	With angles (square, rectangle)	Right	
Bird (chicken, goose)	With angles (square, rectangle)	Left	
Bird (chicken, goose)	Without angles (circle, oval)	Right	Adults
Bird (chicken, goose)	Without angles (circle, oval)	Left	Elderly

Figure 2. Solution

4 Results

Earlier articles that investigated experiments such as the present one [15, 12] reveal a clear result: memorization does not explain individual performance (nor did we expect this to be the case on the basis of our background literature). There are so many sequences, which change with such frequency, that memorization is not cognitively speaking an option for the participants. It is both more natural and more efficient to develop actual theories about the experimental world that result in the repetition of choices consistent with these theories, both when they are correct and when they are wrong (because the revision of some theory is time consuming).

Participants progress from random choices at the beginning towards more stable (and therefore predictable) ones, based on a limited number of elements of the sequence — i.e. only Animal and Shelf, disregarding Shape [12] — and then on to a more complete and sophisticated representation of the experimental environment. The responses of each participant thus become ever more predictable, so that, given a sequence, we may forecast his responses with increasing accuracy. This is because, on the one hand, the number of correct answers increases, but so do the number of repeated mistakes. In the coming paragraphs we explain these trends.

4.1 The development of theories

When a participant gives several times the correct answer to a sequence, she must have understood the exact working of that portion of the experiment. If she often gives wrong answers, perhaps she has not yet uncovered the principle of that sequence. However, if the wrong answer is consistently the same, it is very likely that the participant has developed a mistaken theory.

In this section we study this phenomenon. Since it is possible that theories change over the experiment, as participants learn, we require that the repetitions occur at least for a period of time, and specifically for a third of the overall experiment, which gives us three phases: turns 1–77, 78–154, and 155–231. We only focus on the stable (which allows us to plausibly assume that it is principled, too) association of an answer with a sequence, therefore we only consider responses given 75% of the times. Any answer given to a sequence which only appears once would be given 100% of the times. This, however, does not seem enough to assume stability of behaviour. Instead, we require that a sequence has appeared at least four times during the experiment (but on average they appear ten times) and at least three times in each phase.

Participants indeed develop stable associations between sequences and responses, just like we expected. The number of theories that qualify for our analysis increases from 130 in the first phase to 235 in the third and they also become increasingly accurate going from a 56% rate of correct answers in the first phase, up to 67% in the third. Although participants get better and better, the number of stably mistaken theories is astonishing: 31% in the first phase and 42% in the third. Note that, though our condition was that answers were given 75% of the time, we have numerous observations with a 100% frequency. But only 57% of these were correct, while 43% were wrong.

How can this happen? Don't participants see their answers are wrong and change them accordingly? They do, but the reception, processing, and implementation of the feedback is imperfect.

4.2 The limited effect of feedback

Figure 3 reports the answers a typical participant gave to the sequence Chicken-Oval-Left, whose correct answer is Elderly.

On turn 12, the participant skips the answer, she observes feedback and on turn 26 he or she responds correctly. Also, she probably does so with some reason and not at random, provided that she repeats the correct answer

Turn	Answer given	Score
12	NA	0
26	Elderly*	6
56	Elderly*	6
60	Elderly*	6
109	Adolescents	-1
122	Adolescents	-1
135	Adults	4
144	Adults	4
171	Elderly*	6
181	Adults	4
199	Adults	4
209	Adults	4

* correct answer

Figure 3. Response to Chicken-Oval-Left by one Participant

on turns 56, and 60. One would then imagine that this participant grasped the criterion and is going to consistently give the correct answer from then on, but this is not what happens. On turn 109, the participant switched to a mistaken response, and then repeats it on turn 122. Her theory was probably undergoing some revisions. But the feedback warned her against that response. Indeed, she abandons the mistake and makes a different one! Although this mistake is less severe score-wise, she repeats it a few turns later and, after a single correct response on turn 171, from turn 181 until the end of the game she keeps repeating the mistake.

For the sequence under investigation, this participant incurred in a total of eight mistakes (including the missing answer on turn 12). Seven of these mistakes could be repeated (on turn 209 the mistake cannot be repetad because it's the last turn with this sequence). We consider repeated an error for which the same wrong answer is given in two subsequent appearances of the given sequence. This player repeats four times the same error. We can compute a mistake confirmation rate for this sequence (four out of seven, or 57%) and a mean overall value, for a given player during all the game. The mean value of this index for all player is 33%.[4] . The trend of this phenomenon, moreover, is counterintuitive: the number of confirmed mistakes increases, instead of decreasing: it is 27% on average in the first part of the experiment and 37% in the last part. With standard statistic tests, it is possible to demonstrate that this results can hardly be the result of random choices. Players are instead developing theories and representations of this world. These theories are often based on simplification and on reduced use of available information, as Figure 4 shows. Even mistaken answers are not given at random. Participants indeed employ (imperfect, incomplete, shifting) theories of the experimental world, so that even their mistakes become predictable.

Because of the score system, when the correct answer is Young, responding Adults and Adolescents is indifferent and the same happens for the answers Elderly and Children: both mistakes have the same distance from the correct answer and therefore results in the same score. There should thus be no specific reason to expect that, when the correct answer is Young, mistakes be not random. However, we observe that the mistakes are strongly clustered in

---

[4] For 20 out of 23 participants, we can reject the hypothesis that this happened by chance with a 90% confidence.

animal	shape	shelf	elderly	adults	young	adolescent	children
horse	circle	right	0%	4%	57%	13%	26%
horse	circle	left	4%	10%	54%	17%	14%
horse	oval	right	1%	9%	45%	24%	22%
horse	oval	left	4%	7%	59%	17%	12%
horse	square	right	2%	8%	20%	14%	55%
horse	square	left	0%	3%	43%	38%	16%
horse	rectangle	right	0%	3%	19%	14%	64%
horse	rectangle	left	0%	4%	48%	35%	13%
chicken	circle	right	17%	43%	26%	9%	4%
chicken	circle	left	57%	20%	18%	4%	0%
chicken	oval	right	18%	42%	28%	4%	7%
chicken	oval	left	70%	12%	17%	0%	1%
chicken	square	right	15%	33%	39%	7%	7%
chicken	square	left	30%	26%	26%	9%	9%
chicken	rectangle	right	17%	17%	48%	13%	4%
chicken	rectangle	left	32%	17%	33%	12%	6%
cow	circle	right	2%	7%	50%	23%	17%
cow	circle	left	0%	22%	48%	17%	13%
cow	oval	right	0%	13%	52%	17%	17%
cow	oval	left	8%	8%	52%	22%	10%
cow	square	right	0%	0%	17%	26%	57%
cow	square	left	0%	4%	17%	65%	13%
cow	rectangle	right	0%	4%	26%	22%	48%
cow	rectangle	left	1%	6%	30%	52%	10%
goose	circle	right	21%	48%	18%	12%	1%
goose	circle	left	51%	25%	19%	6%	0%
goose	oval	right	7%	55%	20%	14%	1%
goose	oval	left	50%	28%	15%	7%	0%
goose	square	right	17%	20%	48%	11%	4%
goose	square	left	26%	30%	35%	9%	0%
goose	rectangle	right	13%	39%	30%	9%	9%
goose	rectangle	left	23%	22%	43%	7%	4%

Figure 4. Distribution of Responses, Turns 154–231

an "almost-correct" direction: when the Animal is a Mammal, the mistakes group around the two youngest membership categories, while the vice versa is true for Birds.[5] This tendency may be explained by the fact that at least some of the participants disregard the second piece of information, i.e. Shape, for at least some of the sequences.

These observations serve as starting point for the analysis of the relationship between entropy and performance.

4.3 Performance and predictability

It is our goal in this article to analyse not performance itself, or best strategies, but learning.[6] In order to do so we divided the game in periods. Each period has a duration of 58 turns. The first period goes from turn 1 to turn 58, the second from turn 2 to 59, and so on. This way we obtain 175 periods, largely overlapping.

For each period, we compute:

1. the score of each player; and

---

[5] This is especially puzzling because, in reason of the score system, the most reliable option is always Young, which cannot be farther than two steps from any membership category and therefore always afford positive score.

[6] For deeper analysis of this experimental dataset, see [12].

2. the behavioural entropy for each animal-shape-shelf sequence which appears at least 3 times.

(1) is an indication of performance, which we could substitute e.g. with the number of correct answers, while (2) measures the stability of mistaken responses. We then calculate an average for each participant. This way, we obtain two figures per player for each period. In each period, using the data of all our subjects, we can measure how these two values are related. The simpler measure of linear relation is the correlation coefficient, which indicates whether there is a linear relation between two variables.[7] Figure 5 shows the evolution of this value. Except for a brief time at the beginning,

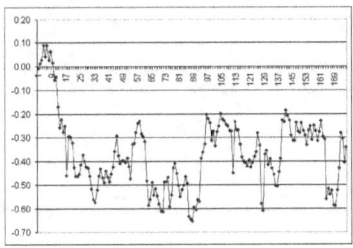

Figure 5. Evolution of the correlation between score and entropy

correlations are systematically negative (in 97 out of 175 cases the values are significantly different from 0), which means that the participants who perform best in that part of the game also have lower entropy. This result, on which we shall comment further below, is not trivial. The irregular trend depends in some measure on technical factors: the sequences appearing on each period differ. During some periods there are several sequences which were clearly understood by the participants. This, obviously, reduces the number of mistakes and makes the correlation lower and less significant.[8] Since the values remain negative despite this problem, on the other hand, the results are especially robust. This can also be confirmed by means of a different analysis. Consider now the participants who performed best in the last period and who, presumably, best understood the working of the experimental world. Call *Best* those who scored above the median and *Worst* the others. The *Best* have lower entropy in 159 out of 175 periods, of these 104 were statistically significant with the Kruskal Wallis test.

We can show what does this pattern represent in a very intuitive fashion, by means of Figure 6, which compares the behaviour of one of the *Worst* and one of the *Best* players. For each sequence in which the players made at least two

---

[7]The correlation coefficient can assume all values in the range between -1 and 1. It has negative values when the two variables move in different direction and therefore a large value of one of them implies a low value of the other. It has a value around zero when there is no linear relation among the two variables. Statistical tests are used to verify when there is evidence to support the idea of non-zero correlation.

[8]Like other indexes, ours is most meaningful when computed on a sufficiently varied sample. If all participants yield similar results, the index cannot unearth very meaningful relations.

mistakes in the period between turn 59 and 117, we calculate the frequency distribution of responses. It is quite evident — even without sophisticated indexes — that the worse player tends to have more heterogeneous mistakes, with many wrong answers only given once. The better player, on the other hand, tends to concentrate her mistakes on few sequences. Not only the mistakes are less numerous, but they are also more regular. The same results can be found throughout the experiment and for all players. In the very last turns, however, the *Best* players have so few mistakes that the comparison is meaningless.

			Worst						Best				
			Child.	Adol.	You.	Ad.	Eld.	TOT.	Child.	Adol.	You.	Ad.	Eld.	TOT.
Cow	Square	Right	-	-	-	-	-	-	1	1	-	-	2
	Rectangle	Right	-	1	-	1	1	3	-	-	-	-	-
		Left	-	-	1	-	1	2	-	-	-	-	-
	Circle	Right	-	-	-	-	-	-	1	2	-	-	3
		Left	1	1	-	-	-	2	-	2	-	-	2
	Oval	Left	-	1	-	2	1	4	-	2	-	-	2
Horse	Square	Right	-	-	2	-	-	2	-	2	-	-	2
	Circle	Left	-	1	-	3	-	4	-	5	-	-	5
	Oval	Right	-	2	-	-	-	2	-	1	-	3	-	4
		Left	-	1	-	1	-	2	-	2	-	-	2
Chicken	Circle	Left	-	1	1	-	-	2	-	-	-	-	-
	Oval	Right	-	2	-	-	-	2	-	-	-	-	-
		Left	-	-	1	-	1	2	-	-	-	-	-
Goose	Square	Right	-	1	-	-	1	2	-	-	-	-	-
	Rectangle	Left	-	1	-	-	1	2	-	-	-	-	-
	Circle	Right	1	-	-	-	2	3	-	-	3	-	-	3
		Left	1	-	-	1	-	2	-	-	2	-	-	2
	Oval	Right	-	-	1	-	1	2	-	-	3	-	-	3
		Left	-	-	1	1	-	2	-	-	-	-	-

Figure 6. Players' Comparison

Generally speaking, therefore, it seems that the capacity to give correct answers is associated with stable behaviour even with respect to mistakes, which is an indication of a tendency to apply rules even if these rules are wrong. On the other hand, the direction of causality is not clear, because both can in principle explain each other. Indeed, since this phenomenon can be observed very early in the game, but it is stronger in the central part of the experiment, and since it is larger for the *Best* group, suggests two interpretations.

* The participants who develop the most correct rules tend to apply rules even when they are not correct. In this case we imagine that people employ analogical reasoning and apply some "default" or "reliable" rule when they lack a context-specific rule.

* It is also plausible that the individual capacity or tendency to focus on some variables and the disregard of other variables (which produces steady behaviour and little entropy) facilitates the understanding of the solution and consequently results in a higher score.

It is not straightforward to understand whether the repetition of mistakes is caused by or is responsible for the high score. The observation that the participants who perform best at the end of (but not necessarily throughout) the experiment also have low entropy all along the game (and even at the beginning) suggests that it is the low entropy that favours a superior understanding of the experimental world. A deeper understanding of this issue is central to uncovering actual learning processes. Moreover, it may prove an important element towards defining better training and teaching techniques.

We may test this idea as follows. Consider the correlation between entropy in a period and score in an earlier period (e.g. 25 periods earlier). If low entropy is responsible for high score, which would mean that participants employ whatever rules they have learnt when lacking a better rule, this time-lagged correlation will be stronger than the normal correlation. Before they may apply a rule, participants ought to develop and work it out. Therefore a low observed entropy in a given turn should be a consequence of the correct answers given earlier in the experiment. In other words, under this hypothesis, if people do export to similar contexts the rules they have learnt in some decision contexts, there should be a slight delay in the correlation between high score and low entropy.

To study this phenomenon, we investigate to what extent does entropy in a given period depends on the score of the same period and how much does it depend on that of an earlier period. We confine the most technical parts of our analysis to the Appendix for the readers willing to dig deeper in the matter. Suffice it to say here that the correlation is highest between entropy and current score than with score of periods which started 12, 25, or 35 turns earlier. In fact, the effect of time-lagged score is opposite to what we expect: those who scored the most in previous turns have higher entropy. One plausible explanation for this pattern is that the participants who have found some simple strategies to respond, then try to elaborate on those by means of trial-and-error, therefore their behaviour is less stable (see also [12]).

5 Concluding remarks

The conclusions above, though perhaps not final, reinforce the two-headed interpretation. If the lower entropy of the best performing players depended on the application of past rules in the present, the effect of time-lagged score should be stronger. The fact that it is weaker suggests instead that best players tend to generalise rules, therefore reducing the information complexity of the environment. The tendency to generalize is common to all participants, but a decision context like that of our experiment certainly favours those for which the tendency is strongest.

The two procedures: "export rules beyond their context" (or analogical decision-making or learning spillover) and "reduce the amount of information employed" are not mutually exclusive. It is nonetheless better to keep them separated because they are conceptually distinct and both may prove either useful or dangerous.

Analogical problem-solving, on the other hand, amounts to a selective use of the information because it amounts to treating as identical or similar, two situations which differ in a number of respects. It is a strategy suggested by George Polya in *How to Solve It*, which can be described as perfectly rational. "Do you know a related problem?" is one of the first questions the Polya lists in his strategy to solve any problem, elaborating on a method derived from mathematical theory. The first step to devise a plan is to "ask these questions: 'Have you seen it before? Or have you seen the same problem in a slightly different form?' " [16, p. xvi].

On the other hand, the reduction of employed variables is part and parcel of theorisation: it is the very core of *ceteris paribus*. In order to investigate the effect of one variable, every other variable is excluded from the analysis by being held (or assumed) constant. Our best chances at understanding the effect of a variable is to investigate it in isolation.

More generally we may advance the following suggestion. Though we may not yet say for what specific reason, a high score is associated with low entropy for mistakes and therefore with a tendency to repeat mistakes over and over. Though the repetition of mistakes might be considered a failure to properly employ feedback or a bias, it may instead turn out as a viable and successful procedure.

BIBLIOGRAPHY

[1] A. Alchian. Uncertainty, evolution, and economic theory. *Journal of Political Economy*, 58: 211–221, 1950.
[2] W.B. Arthur. On learning and adaptation in the economy. *Santa Fe Institute Working Paper* 92-07-038, 1992.
[3] A. Ambrosino and A. Lanteri. Learning and the emergence of behavioural rules in a bargaining game. *Journal of Social Complexity*, 3 (1): 53–74, 2006.
[4] A. Ambrosino, A. Lanteri and M. Novarese. Preferences, choices, and satisfaction in a bargaining game. *ICFAI Journal of Industrial Economics* 3 (3): 7–21, 2006.
[5] A. Bandura. *Social Learning Theory*. Prentice-Hall, Englewood Cliffs, NJ 1977.
[6] J. Best. Knowledge acquisition and strategic action in "Mastermind" problems. *Memory & Cognition* 18 (1): 54–64, 1990.
[7] J. Bruner, J. Goodnow and G. Austin. *A Study of Thinking*. Norton, New York 1956.
[8] F. Hayek. *The Sensory Order: An Inquiry into the Foundations of Theoretical Psychology*. The University of Chicago Press, Chicago 1952.
[9] R.A. Heiner. The Origin of predictable behavior. *American Economic Review* 73 (4): 560–595, 1983.
[10] R.A. Heiner. Origin of predictable behavior: further modelling and applications. *American Economic Review* 75 (2): 391–396, 1985.
[11] G. Lakoff and M. Johnson. *Metaphors We Live By*. The University of Chicago Press, Chicago 1981.
[12] A. Lanteri and M. Novarese. Individual learning: theory formation and feedback in a complex task. *MPRA Paper* 3049, 2007.
[13] A. Marcet and T. Sargent. Convergence of least squares learning mechanisms in self referential linear stochastic models. *Journal of Economic Theory*, 48: 337–68, 1989.
[14] R. Nelson. Bounded rationality, cognitive maps, and trial and error learning. *Journal of Economic Behaviour and Organization*, 2007
[15] M. Novarese and S. Rizzello. A cognitive approach to individual learning: some experimental results. In R. Arena and A. Festrè (eds.), *Knowledge and Beliefs in Economics*, Edward Elgar, Aldershot 2006, . pages 203–219.
[16] G. Polya. *How to Solve It*. Princeton University Press, Princeton 1957.
[17] S. Rizzello and M. Turvani. Subjective diversity and social learning: a cognitive perspective for understanding institutional behaviour. *Constitutional Political Economy*, 13 (2): 197–210, 2002.
[18] A. Roth and I. Erev. Predicting how people play games: reinforcement learning in experimental games with unique, mixed strategy equilibria. *American Economic Review*, 88 (4): 848–881, 1998.
[19] H.A. Simon. From substantive to procedural rationality. In S. Latsis (ed.), *Method and Appraisal in Economics*. Cambridge University Press, Cambridge 1976, pages 129–148.
[20] H.A. Simon. *Reason in Human Affairs*. Stanford University Press, Stanford 1983.
[21] V. Smith. Experimental Methods in Economics. In J. Eatwell, M. Milgate and P. Newman (eds.), *The New Palgrave: Allocation, Information, and Markets*, MacMillan, 1987, pages 94–111.
[22] J.J. Vromen. *Economic Evolution: An Enquiry into the Foundations of New Institutional Economics*. Routledge, London 1995.

[23] U. Witt. Social cognitive learning and group selection. A game-theoretic version of Hayek's societal evolution. Presented at the *INEM-ASSA Session "Austrian Economics and Game Theory"*, 8 January, 2000, Boston, MA.

Appendix
A Instructions

In what follows we report the instructions of the game you will take part in. You will be compensated with real money (40 ITL per point).

A.1 The game

An association has different membership categories which pay different fees and have access to different services. The categories are: children; adolescents; young; adults; elderly.

Children and adolescents do not pay. Young pay a reduced fee. Adults and Elderly pay the full fee. Members' information are recorded on a set of cards, stored on different shelves.

Each card is characterized by: the drawing of an animal; the drawing of a shape; a shelf; adults; elderly (e.g. A card might have the drawing of a cow and a square, and be placed on the right shelf). You do not know the classification system and thus which cards corresponds to which category. The goal of the game is to understand this correspondence.

The game lasts 231 turns. In each turn you will be shown the information from one card, so you will see information about: animal; shape; shelf. Based on this information you shall indicate the correct membership category, keeping in mind that: there is a logical relationship between the information and membership categories; the relationship is constant throughout the game; the relationship is completely artificial (therefore it is neither necessary nor useful to have experience of actual filing systems or any other specialised knowledge). Obviously, the earliest answers will be given at random. Each turn, therefore, the game will take place in the following way: 1. You see the information; 2. You give your answer (note: you must choose within 10 seconds, after this time the system proceeds to the next turn); 3. You are told the correct answer and your score in the last turn; 4. You move on to the next turn and you start again. During the game you are not allowed to talk, nor to take notes.

A.2 The score

In order to calculate the score we define the distance between the answer you gave and the correct answer, as follows:

- if the answer is correct, the distance is 0 and you score +6;

- if the answer given is children and the correct answer is elderly (or vice versa), the distance is highest: 4 and you score -4;

- if the answer given is children and the correct answer is adult (or vice versa), or if the answer given is adolescents and the correct answer is elderly (or vice versa), the distance is 3 and you score -1;

- if the answer given is children and the correct answer is young (or vice versa), if the answer given is adolescent and the correct answer is adult (or vice versa), or if the answer given is young and the correct answer is elderly (or vice versa), the distance is 2 and you score +1;

- if the answer given is children and the correct answer is adolescents (or vice versa), if the answer given is adolescent and the correct answer is young (or vice versa), if the answer given is young and the correct answer is adults (or vice versa), or if the answer given is adults and the correct answer is elderly (or vice versa), the distance is 1 and you score +4;

- if you do not answer, you score 0.

A.3 Game dynamics

Each turn of the game can be divided into two parts. The first part requires that you choose one of the five alternatives offered, by means of selecting the corresponding button and then "Enter". It is important that you complete these operations within 10 seconds because, when such time has elapsed, the system moves on with the test and it records "No Answer" corresponding to zero points.

After you made your choice, or after 10 seconds, you move on to the second part of the turn. The screen will report the outcome of the present turn. It reminds you the choice you made, the correct answer, and your score for this turn. The window stays open for 6 seconds, after which the system starts the next turn.

B Time-lagged score and entropy

The correlation between entropy and time-lagged score has values closer to 0 and these values are less significant than the correlation between entropy and simultaneous score (79 vs. 91 values are significantly different from 0).

To assess more precisely the influence of simultaneous score compared with the time-lagged one, we shall employ a linear regression analysis, in which entropy is the dependent variable. The two scores are used as independent variables so to measure their joint effect and compare their relative strength. In most periods the correlation is not significant. It is sometimes significant, but neither independent variable is individually significant (this is also because of the correlation between time-lagged and simultaneous score). The sign of the variables SCORE and TIME-LAG SCORE are nonetheless noteworthy. Figure 8 does this and it also distinguishes the cases in which the two independent variables are significant at the 90% level. The SCORE variable is negative in 128 cases out of 150 and it is significant in 25 of these. The TIME-LAG SCORE variable is significantly different from 0 in just 9 cases, in 8 of which it has positive sign. Out of 150 repetitions, it is positive 71 times. We can thus make some inferences.

* The effect of simultaneous score is stronger and it is negative.

* The effect of time-lagged score is less strong. Perhaps during some periods it reinforces the other effect, but with an opposite sign: for a given score in

			TIME-LAG SCORE 25				
			not significant		significant		Tot
			-	+	-	+	
SCORE	not significant	-	56	47			103
		+	21		1		22
	significant	-	1	16		8	25
		+	0	0	0	0	0
	Tot		78	63	1	8	150

Figure 7. Score and Entropy (Time-lag 25)

a period, a higher the time-lagged score is associated with higher entropy for mistaken answers. During such periods, the players who previously performed better have higher entropy, possibly because they are confident with some portions of the solution and are more inclined to try and understand the remaining portions. The time-lag we use is obviously arbitrary, but we employ two more time-lags, 12 and 35 periods, to test the robustness of our inferences. In this former case, there are fewer significant cases, but the other conclusions hold. In the latter case, despite a smaller number of estimates (because we ought to skip the first 35 turns), the significant cases grow. The interpretation does not change and TIME-LAG SCORE is ever less significant and with a positive sign.

The larger the time-lag, therefore, the less significant is time-lagged score (and with a positive sign), but the more significant the simultaneous score (and with a negative sign) also because the correlation between the two is reduced and the estimates are more reliable.

Tacit knowledge and economics: recent findings and perspectives of research

ANDREA POZZALI

1 Introduction: tacit knowledge in social sciences

The concept of "tacit knowledge" has proven to be a very useful analytical tool in social sciences. Introduced into modern circulation by the scientist and philosopher of science Michael Polanyi [27] to refer to all those kinds of knowledge that can not be codified, it has been used afterwards in a wide range of disciplines. In addition to some applications in ethics, in studies on religious thoughts and in the debate on AI [34, 35], one of the most important lines of research on tacit knowledge has been developed in sociology of science, where this concept has been considered as a basic component of scientists' know-how. Among the many case studies that have worked in this direction, we can here mention the ones by Collins on TEA laser[5] or x-ray cristallography [6], and the reconstruction of the industry of nuclear weapons made by Mackenzie and Spinardi [20]. A separate field of research has been also developed by psychologists to analyse the characteristics of "practical intelligence" [37, 33] a concept that bears many resemblances with Polanyi's tacit knowledge.

For sure, economics has been one of the fields that has most relied on tacit knowledge as a heuristic tool, in particular for what concerns the study of processes of organizational change and of technological evolution. Nelson and Winter [22] have provided one of the most complete and well-known approach to this issue and their work can be considered as a matter of fact at the basis of the development of a whole new field in evolutionary economics. Other contributions, in some cases drawing on the work by Nelson and Winter, have then applied the concept of tacit knowledge to such things as knowledge management [24], the analysis of the role of technology in economic development [9] and the study of innovation models and technology transfer [15, 12].

This widespread usage has also given birth to some criticisms, pointing at the fact that in many cases tacit knowledge has been used as a kind of "black-box concept" that gives no real contribution to the advancement of knowledge and, indeed, "obscures more than clarifies" [7]. This objection seems perfectly shareable, while the subsequent argument that tacit knowledge that can not be codified "is not very interesting for the social sciences" [7] seems to be a little too strong, in light also of the fact that some recent findings in cognitive neurosciences clearly provide ground for an ever increased, and not diminishing, role for tacit knowledge in economics [23].

It is quite clear that if we want tacit knowledge to remain a useful tool in economical analysis we need to make a step further; in particular, a more in-depth clarification of the real meaning we want to assign to this concept seems to be needed. If we read the literature with a critical look, indeed, it seems that in many cases different authors have ended up by applying the same label of "tacit knowledge" to different things and this has surely contributed to increase the rate of terminological and conceptual confusion. In other words, it's not always the case that what a scholar means by "tacit knowledge" corresponds to what others do. In some previous works [28, 1] we have tried to give a first suggestion in order to address this kind of problems, developing a new classification of different typologies of tacit knowledge. In the next paragraph, I will briefly recall this classification, as it will be useful in developing the argument I would like to follow in this paper. In paragraph 3 I will then try to apply this framework to the specific field of cognitive economics, and in particular to the debate revolving around the dual-process account of reasoning. This line of thought, that is currently gaining ground among cognitive psychologists, analyses human reasoning by making reference to the presence of two different kinds of processes: deliberate reasoning and automatic "intuition". In my view, insights coming from the study of tacit knowledge can give some contributions in this field, that seems very promising for the development of a general framework in cognitive economics.

2 Disentangling the concept of tacit knowledge

As already mentioned, the literature concerning the role of tacit knowledge in economics has been somehow flawed by the lack of a precise definition of the meaning of this concept. As a tool that can aid both empirical and theoretical analysis, we have tried to develop a classification of tacit knowledge that considers the following three typologies [28, 1]:

- Tacit knowledge as a component of "skills"
- Tacit knowledge as "background knowledge"
- Tacit knowledge of a "cognitive type"

The first type of tacit knowledge embraces all those kind of physical abilities and skills that refer to the capacity of a subject to know how to perform certain activities or tasks. This knowledge can in turn assume two characters: it can be fully automatic and innate (think for example to unreflective activities such as breathing), or it can be the fruit of a conscious learning and training process that lead to the development of specific kinaesthetic abilities: in this second case, it can be considered as the main component of the so-called skilful performance.

For the sake of simplicity, we can put aside tacit knowledge of an innate type and focus on tacit knowledge as a component of skilful performance. This type of tacit knowledge is the one that has received the greatest attention in literature: first of all, it is mainly at this type of tacit knowledge that Nelson and Winter [22] refer to in their analysis of tacit skills. Secondly, tacit skills have been recognised to be a fundamental component of scientific knowledge in many high technology fields [6, 29]. In turn, as economic development is more and more dependent on technological advance, and as scientific and technology knowledge comes to play the role of an economic resource on its own, the economical relevance of this type of tacit knowledge will continue to rise [21]. Important as it may be, this type of tacit knowledge has somehow been neglected in epistemological analysis. In fact, as this kind of knowledge is largely (if not completely) non linguistic, it does not fit well into standard epistemological theories, that usually put at the centre of their analysis declarative knowledge, in the form of justified true belief, and tend to downsize the importance of other kind of knowledge, derived by competence or acquaintance [18].

In the second type of tacit knowledge (as a component of "background knowledge") we can include all those forms of interiorised regulations, of codes of conduct, of values and widespread knowledge that a subject knows by direct experience. This knowledge cannot be articulated or formalised because of its extremely dispersed nature. In economics, this type of knowledge can be considered as one of the main components of social capital and, more in general, of all those kinds of knowledge that are embedded in a given social and economic context [13]. It also bears some resemblances with the searlean concept of "background" [31].

The third type of tacit knowledge is the most problematic. As a matter of fact, for a long time the possibility that tacit knowledge could assume also a specifically cognitive character was ruled out. In the representational view of the mind that is sort of the dominant paradigm in cognitive sciences, it was difficult to admit the possibility of having a representation while at the same time not being able of articulating it. Nowadays, evidences coming from cognitive psychology and from neurosciences enable us to re-consider this issue. As a kind of tacit knowledge that can assume a cognitive character, we can consider for example linguistic knowledge, that does not represent a form of skill, but must be considered as an actual cognitive system, defined in terms of mental states and structures that cannot be articulated in words nor described in a complete formal system [4].

Other examples of these cognitive forms, not skill-like nor background-like, of tacit knowledge come from studies on implicit learning processes [30]. In the standard experiment, involving artificial grammar, subjects are given a list of alphanumeric strings, some of which generated from a hidden grammatical structure, others completely random. After completing this phase, the subjects are given other alphanumeric strings and they are then asked to distinguish between grammatical and non-grammatical ones. Subjects are usually able to perform this recognition task though they are unable to explain in an articulated form how they actually do it. These experiments have also been replicated by using different stimuli, for example by asking subjects to take managerial decisions in order to maximize firm profit [2].

Taken together, all these experiments seem to point to a general ability of human beings, that are able to use hidden structural characteristics that make up the essence of a given phenomenon, though they are not able to come to a complete explicit knowledge of these characteristics. The knowledge that subjects refer to, in these cases, can take the form of a sort of automatic reaction of the type: every time structural pattern x is presented then take decision y. This type of automatic reaction to a given stimulus, that is acquired with direct experience and without resort to deliberate reasoning, bears many resemblances with other well known phenomena that have been studied by cognitive sciences in the field of problem-solving [19], or in the case of games such as Tetris [14]. This type of experiments are often considered as one of the most important empirical proof for the presence of two different type of cognitive processes in the human mind [10]. As this dual-process account of reasoning seems to hold many promises for the development of a general framework in cognitive economics, it is worth discussing how the study of tacit knowledge can contribute to this debate.

3 Tacit knowledge, cognitive economics and the dual-process account of reasoning

The dual-process account of reasoning assumes that a significant part of neural activity involved in everyday activities can not be reduced to conscious processes of deliberate thinking, but is based on automatic and largely unconscious processes. The body of empirical evidence in support of this view is quite consistent and gathers together many different evidences involving phenomena of perception [38], blindsight, motor skills acquisition [25] and even more complex activities of problem-solving as the ones we have already seen. The point is how to interpret all this body of evidence in order to develop a general and coherent model of human cognition.

As a matter of fact, the literature on the dual-process account of reasoning is quite developed. Common among the authors is the belief in the existence of two systems of reasoning, even if the specific characteristic of these two systems may differ from author to author: while some authors prefer to use neutral terms such as System 1 and System 2 [32, 11] others differentiate between "automatic" and "controlled" processes: this last distinction is echoed also by Kahneman in his Nobel Lecture, where he speaks of intuition (automatic system) and of reasoning (controlled system) [17]. In recent developments, the distinction between controlled and automatic processes is considered as overlapping with the one between cognitive and affective ones, and this leads to the development of the two-dimensional categorization displayed in the following table [3]:

	Cognitive	Affective
Controlled	I	II
Automated	III	IV

Even if this refinement can add valuable dimensions to consider, the authors themselves recognize that their classification can be in some cases reduced to a simple dichotomy [3, p.19]:

> The four-quadrant model is just a way to remind readers that the cognitive-affective and controlled-automatic dimensions are not perfectly correlated, and to provide a broad view to guide exploratory research. For some purposes, reducing the two dimensions to one, or the four quadrants to two, will certainly be useful. Furthermore, noting all four cells is not a claim that all are equally important. It is just a suggestion

that leaving out one of the combinations would lead to a model which is incomplete for some purposes.

For the sake of simplicity, we can then limit ourselves to consider the more common distinction between automated and controlled processes, that is indeed at the basis of the dual-process account of reasoning. Among the many different issues that can arise at this regard, I would like to focus my attention on two specific problems: how can automatic processes be developed and how can we conceive of the relationship between the two different systems. As long as the first question is concerned, it seems to me that the mechanisms that can be held to be responsible of the development of automatic processes of reasoning are exactly the same that seem to be at work in the development of cognitive components of tacit knowledge, as the ones we have seen in the experimental results provided by Reber. In both cases we indeed can see, as Egidi puts it, that "elementary cognitive skills may be stored in memory as automatic routines or programs, that are triggered by a pattern matching mechanism and are executed without requiring mental effort or explicit thought" [10].

In other words, human mind seems to be endowed with the ability of drawing some kinds of implicit inferences from the mere exposition to stimuli that present a sort of hidden structure of regularity. Even if subjects are not able to explicitly grasp this structure, they can anyway learn to make their decisions in order to exploit these regularities for their own advantage. As the kind of knowledge that is developed in this way is completely implicit, every time subjects are presented with the same situation, they learn to react in an automatic way, without having to make reference to any kind of deliberate process of reasoning. Moreover, in these cases more deliberation may be even harmful, as long as it may not be needed in order to solve the problems and it can therefore have only the negative side effect of burdening the mind with a cognitive task that holds no benefit at all. In the experiments performed by Reber, the performances of subjects did not improve when they were given explicit hints about the hidden law that were responsible for the generation of the experimental material. In many other cases, both inside and outside of the lab, it has been proven that sometimes it's better to make some type of decisions using only our "intuition" and not trying to develop a deliberate reasoning: this has been labelled as the "deliberation-without-attention effect" [8].

In many cases this automatization of thought can represent a necessity, given the constraints of the cognitive system, and the urge of taking decisions in a short time. This automatization can also make the difference between experts and novices, as the processes of learning seems to involve exactly the ability of shifting from deliberate ways of reasoning to automatic ones. Many experimental evidences indeed point to the fact that, both in the acquisition of physical skills and in the development of decision-making routines "the brain seems to gradually shift processing toward brain regions and specialized systems that can solve problems automatically and efficiently with low effort" [3, p.24].

This last point leads us to the second question I have raised, that is the relationship between automatic and deliberate processes of reasoning. Since the first enunciations of the theory of probability and of modern logic, one of the great unresolved issues of the literature concerning human reasoning has been the existence of so many discrepancies between the normative and "rational" assumptions of the theory and the actual behaviour of subjects. Modern cognitive economics too has devoted a great effort in trying to find the source of decision-making biases. The presence of a dual-process account of reasoning seems to present at list two different opportunities to explain how biases may arise: the first is simply to attribute them to the fact that subjects rely almost esclusively on their "intuition" and do not fully employ their reason:

> The intuitive process, which automatically elicits prior knowledge, is therefore considered as a basic source of errors in reasoning: many experiments show that the cognitive self-monitoring, i.e. the control of the deliberate system over the automatic one, is quite light and allows automatic thought to emerge almost without control. According to Kahneman, errors in intuitive judgments involve failures in both systems; the automatic system, *that generates the error*, and the deliberate one, *which fails to detect and correct it* [10, p.10, emphasis added].

Even if it corresponds to a quite diffused view among economists and cognitive psychologists, this argument does not seem to be fully convincing, as long as, as we have already mentioned, many experimental proofs point to the fact that in many cases subjects can reach good performances by relying only on their tacit and "intuitive" faculties. The relationship between the automatic process and the deliberate one might then be more complex than a simple one-dimensional link in which "reason" has the duty to correct the faults of the "irrational side". Once again, experimental evidences coming from cognitive psychology may be helpful, as long as they have been able to show the phenomena of mechanisation of thought already cited.

In the dual-process account, automatic abilities acquired by experience are strongly domain-specific, and for this reason "performances will be strongly dependent upon the matching between the given problem and the capabilities acquired by previous experience" [10, p. 14]. In other words, decisional errors may then arise from the automatic and not pondered application of certain modules of thought (derived from the exposure to old problems) to new problems that should as a matter of fact require a different approach. By the way, this mismatching between the structure of a new problem and the application of old mechanisms of solution was also the way in which Tversky and Kahneman originally tried to explain the link between decisional heuristics and cognitive biases [16].

4 Conclusions

The literature concerning the role of tacit knowledge in economics has been centered on the analysis of tacit skills and know-how, while it has downsized the importance of other forms of tacit knowledge, that can not be considered as a kind of kinaesthetic abilities, but are of a fully cognitive type. The analysy of this kind of tacit knowledge is worth doing as long as it can contribute some useful insights to the debate on the dual-process account of reasoning, that is one of the most interesting developments in cognitive economics. The study of the relationship between tacit and explicit knoweledge is indeed strictly interwoven with the analysis of the relationship between automated and deliberate processes of reasoning.

At this regard, even if it is quite common to frame this relationship in terms of contrapositions (tacit vs. explicit, automated vs. deliberate), this may not be correct. It is much more probable that the relationship may be one of cooperation, where intuition and reasoning, tacit and implicit knowledge interact in complex ways to give birth to a sort of "cognitive duplicity of the mind" [36, p. 243]. In this frame, automated and tacit forms of knowledge are strongly domain specific, while explicit and deliberate forms of reasoning present a more general character.

Interpreting the relationship in a cooperative way is also more in line with the original work of Polanyi on tacit knowledge, where he always spoke of a sort of complementarity that based our explicit faculties on the presence of "tacit powers". This type of complementarity was to be found not only at the base of the development of our linguistic faculties and of scientific knowledge, but even in the social, political and economical life [26]. Polanyi used to address these complementarities with a model of focal/subsidiary awareness largely drawn from Gestalt psychology. Nowadays, we can start to analyse these same complementarities with the much more refined instruments provided by modern cognitive sciences, in order to develop a better understanding of the processes of human decision-making.

Acknowledgements

This paper was developed within the scope of a research project financed by the Italian Ministry of University and Research (FIRB project 2003 "A multidimensional approach to technology transfer for more efficient organizational models" - Prot. RBNE033K2R). Acknowledgements are made to the Ministry, to the general coordinator of the FIRB project and to all the partners.

BIBLIOGRAPHY

[1] M. Balconi, A. Pozzali and R. Viale. The "codification debate" revisited: a conceptual framework to analyze the role of tacit knowledge in economics. *Industrial and Corporate Change*, 16 (5): 823–849, 2007.
[2] D.E. Broadbent, P. Fitzgerald and M.H. Broadbent. Implicit and explicit knowledge in the control of complex systems. *British Journal of Psychology*, 77: 33–50, 1986.
[3] C. Camerer, G. Loewenstein and D. Prelec. Neuroeconomics: How neuroscience can inform economics. *Journal of Economic Literature*, XLIII: 9–64, 2004.
[4] N. Chomsky. *Reflections on Language*. Fontana, Glasgow 1976.
[5] H.M. Collins. *Changing Order*. University of Chicago Press, Chicago 1992.
[6] H.M. Collins. Tacit knowledge, trust, and the Q of sapphire. *Social Studies of Science*, 31 (1): 71–85, 2001.
[7] R Cowan, P. David and D. Foray. The explicit economics of codification and tacitness. *Industrial and Corporate Change*, 9 (2): 211–253, 2000.
[8] A. Dijksterhuis, M.A. Bos, L.F. Nordgren and R.B. Van Baaren. On making the right choice: The deliberation-without-attention effect. *Science*, 311: 1005–1007, 2006.
[9] G. Dosi, C. Freeman, R. Nelson, G. Silverberg and L. Soete. *Technological Change and Economic Theory*. Pinter, London 1988.
[10] M. Egidi. The dual process account of reasoning: historical roots, problems and perspectives. CEEL Working Paper 0706, 2007.
[11] J. Evans. In two minds: dual-process accounts of reasoning. *Trends in Cognitive Sciences*, 7 (1010): 454–459, 2003.

[12] W. Faulkner and J. Senker. *Knowledge Frontiers: Public Sector Research and Industrial Innovation in Biotechnology, Engineering Ceramics, and Parallel Computing*. Oxford University Press, Oxford 1995.
[13] M. Granovetter. Economic action and social structure: The problem of embeddedness. *American Journal of Sociology*, 3: 481–510, 1985.
[14] R.J. Haier, B.V. Siegel, A. MacLachlan, E. Soderling, S. Lottenberg and M.S. Buchsbaum. Regional glucose metabolic changes after learning a complex visuospatial/motor task: a PET study. *Brain Research*, 570: 134–143, 1992.
[15] J. Howells. Tacit knowledge. *Technology Analysis & Strategic Management*, 8 (2): 91–106, 1996.
[16] D. Kahneman, P. Slovic and A. Tversky. *Judgment Under Uncertainty: Heuristics and Biases*, Cambridge University Press, Cambridge 1982.
[17] D. Kahneman. Maps of bounded rationality: a perspective on intuitive judgment and choice. *The American Economic Review*, 93 (5): 1449–1475, 2003.
[18] K. Lehrer. *Theory of Knowledge*. Routledge, London 1990.
[19] A.S. Luchins and E.H. Luchins. New experimental attempts in preventing mechanization in problem-solving. *The Journal of General Psychology*, 42: 279–291, 1950.
[20] D. MacKenzie and G. Spinardi. Tacit knowledge, weapons design, and the uninvention of nuclear weapons. *American Journal of Sociology*, 101 (1): 44–99, 1995.
[21] J. Mokyr. *The Gifts of Athena. Historical Origins of the Knowledge Economy*. Princeton University Press, Princeton 2002.
[22] R.R. Nelson and S.G. Winter. *An Evolutionary Theory of Economic Change*. Harvard University Press, Cambridge, MA 1982.
[23] P. Nightingale. If Nelson and Winter are only half right about tacit knowledge, which half? A Searlean critique of " codification". *Industrial and Corporate Change*, 12: 149–183, 2003.
[24] I. Nonaka and H. Takeuchi. *The Knowledge-Creating Company: How Japanese Companies Create the Dynamics of Innovation*. Oxford University Press, Oxford 1995.
[25] R. Passingham. Functional organisation of the motor system. In R.S.J. Frackowiak, K.J. Friston, C.D. Frith, R.J. Dolan and J.C. Mazziotta (eds.), *Human Brain Function*, Academic Press, San Diego 1997.
[26] M. Polanyi. *The Logic of Liberty*. The University of Chicago Press, Chicago 1951.
[27] M. Polanyi. *The Tacit Dimension*. Routledge, London 1966.
[28] A. Pozzali and R. Viale. Cognition, types of "tacit knowledge" and technology transfer. In R. Topol, B. Walliser (eds.), *Cognitive Economics. New Trends*, Elsevier, Oxford 2006, pages 205–224.
[29] A. Pozzali. Can tacit knowledge fit into a computer model of scientific cognitive processes? The case of biotechnology. *Mind & Society*, 6: 211–224, 2007.
[30] A.S. Reber. *Implicit Learning and Tacit Knowledge. An Essay on the Cognitive Unconscious*. Oxford University Press, Oxford 1993.
[31] J. Searle. *The Rediscovery of the Mind*. MIT Press, Cambridge 1992.
[32] K.E. Stanovich and R.F. West. Individual differences in reasoning: implications for the rationality debate. *Behavioral and Brain Sciences*, 23: 645–726, 2000.
[33] R.J. Sternberg and J.A. Horvath (eds.). *Tacit Knowledge in Professional Practice: Researcher and Practitioner Perspectives*. Lawrence Erlbaum Associates, Mahwah, NJ 1999.
[34] S.P. Turner. Tacit knowledge and the problem of computer modelling cognitive processes in science". In Fuller S., De Mey M., Shinn T., Woolgar S. (eds.), *The cognitive turn. Sociological and psychological perspectives on science*. Kluwer, Dordrecht 1989, pages 83–94.
[35] S.P.Turner. Practice in real time. *Studies in History and Philosophy of Science*, 30: 149–156, 1999.
[36] R. Viale. Quale mente per l'economia cognitiva. In R. Viale (ed.), *Le nuove economie. Dall'economia evolutiva a quella cognitiva: oltre i fallimenti della teoria neoclassica*, Il Sole 24 Ore, Milano 2005, pages 233–252.
[37] R.K. Wagner and R.J. Sternberg. Practical intelligence in real-world pursuits: the role of tacit knowledge. *Journal of personality and social psychology*, 49 (2): 436–458, 1985.
[38] A. Zeman. Consciousness. *Brain*, 124: 1263–1289, 2001.

From individual well-being to economic welfare. Tibor Scitovsky explains why (consumers') dissatisfaction leads to a joyless economy

VIVIANA DI GIOVINAZZO

1 Introduction

Welfare economics is characterized by the analysis of the rationale of state intervention to increase welfare, whatever may be meant by the latter. Therefore, there will be no objection to begin a history of welfare economics from Jeremy Bentham, Adam Smith or even Aristotle and to identify "welfare economics" with a general theory of economic policy.

Early theorists (*e.g.* Bentham and Pigou) defined welfare as the sum of the satisfactions accruing to an individual through an economic system. Believing that it was possible to compare the well-being of two or more individuals, they argued that a poor person would derive more satisfaction from an increase in income than a rich person. Later writers (*e.g.* Robbins, Stigler) argued that making such comparisons with any precision was impossible, though they kept on endorsing the standard neoclassical assumptions to derive social welfare. A small but growing group of forerunners of the new welfare economics (*e.g.* Amartya Sen and Tibor Scitovsky) proposes a re-examination of the standard welfare economics principles by focussing on the *intrinsic value* of choice.

This paper is written with two goals in mind. The first is to point out the shortcomings of a welfare policy developed under neoclassical conditions. The second is to present the set of alternatives Tibor Scitovsky proposes in substitution of standard welfare economics.

Scitovsky's and Sen's approaches to the topic are similar: methodology is highly interdisciplinary in nature; the philosphical and conceptional reasoning are predominant over the modelling and formalizations. In Scitovsky's opinion, as well as for Sen, welfare has often been confused with "consumption" and, consequently, "growth". On the contrary, Scitovsky and Sen are resolute in claiming that human progress should be measured both qualitatively and quantitatively. Yet Scitovsky is less paternalistic in spirit than Sen. The Hungarian economist uses the psychologists' findings on motivation to demonstrate that focussing on the income distribution as the major issue for a welfare economics is a misleading strategy. In order to realize public happiness, policy-makers had better concentrate on the distribution of services, not for the actualization of certain expected potentialities (Sen's *desiderata*), but for a more libertarian self-determination of those potentialities the individual *has* chosen to develop. Therefore, Scitovsky suggests reconsidering the state intervention, which is quite different from Sen's "capability approach". Curiously enough, his claim for the role of the State matches well to Walras's claim: "Liberté de l'individu; autorité de l'Etat. Egalité des conditions; inégalité des positions" [43, p. 262].

The paper is organized as follows: Section 1 investigates the sources of (consumers) dissatisfaction. Section 2 analizes Scitovsky's grasp of "neural arousal system" as a feasible remedy to the puzzling situation of consumer's dissatisfaction in a welfare economy. Section 3 and section 4 give an overview of the consequences of economic welfare on social welfare. Section 5 explains Scitovsky's proposals to improve social welfare. The last section concludes the whole argument.

2 The desire to have desires

In *The World as Will and Representation*, Schopenhauer claims that human life "swings like a pendulum to and from between pain and boredom" [27]. Boredom sets in, Schopenhauer observes, when all our desires for determinate objects are satisfied and no new desire comes to agitate us. Thus, he describes boredom as an "empty longing", as a state in which the will, despite attaining some particular goal, continues to yearn, this time without any determinate intentional focus.

Boredom is a very common human disease; all of us have experienced it almost once in life. It is that subtle sensation occurring when we "lack objects of willing", not the *determinate objects* of particular desires, *but* rather the *objects to desire* even though nothing arouses our interest. By and large, we do not demand the complete satisfaction of desires that are, strictly speaking, unsatisfiable: we do want to get honors, fame, and fortune although we just want to possess them momentarily, in order to be able to desire them again. We want, in other words, never to be satisfied once and for all, but to be *moved* by desires that are perpetually rekindled.

Scitovsky indicates boredom as *the* disease of modern society and the major responsible for the dissatisfaction of the affluent consumer, who suffers from deprivation of substantial novelty.

Building on the findings of human brain psychophysiologists, he proves the philosophers' previous insights: novelty is as primary a need as hunger or thirst. Boredom arises when the massive production of modern economy begins to repress the novel part in product for the sake of more comfortable and easy marketable, but less innovative products; a severe lack of substantial novelty makes people feel bored and dissatisfied. Thus, in any case, choosing comfort as a goal, represents a joyless dead end. That is what Scitovsky holds: "could it not be that we seek or satisfaction in the wrong things, or in the wrong way, and are then dissatisfied with the outcome?" [36, p. 4].

3 The scientific explanation of novelty

Various theories have been proposed in economics and in psychology to explain how sensory events operate as motivators. By focussing on the concepts of (homeo)static equilibrium (the total supply equals the total demand) and the decreasing marginal utility of a self-sufficient and maximizing agent, the earliest twentieth century economists seemed to apply to economics the same rationale that led their colleagues way of arguing in psychology: the instinct theory, headed by Freud [9] and the drive theory, led by [49], in order to satisfy the needs later on hierachically ordered by [22].

To Freud the nervous system is "an apparatus which has the function of getting rid of the stimuli that reach it, or of reducing them to lowest possible level; or which, if it were feasible, would maintain itself in an altogheter unstimulated condition" [9]. From his side, Woodworth, accounts for a limited and well defined number of internal cyclical physiological conditions such as hunger and thirst and suggests that much conditions lead only to a generalized activation of behavior, rather than to a specific behavioral sequences or a preference for one type of object. Human behavior seemed under the control of external rather than internal stimuli.

In broad terms, the drive theory behavior implies that the organism is inert unless some disturbance or deprivation generates a drive leading to activity in order to eliminate disturbance. This conception is based on the assumption that the organism is quiescent except when bent on or engaged in satisfying some needs. Nerve cells and the central nervous system as a whole are also believed to be inert except when stimulated by some impulse. Both these theories stem from and support the Darwinian notion of evolution.

Now, this explanation of human behavior is generally aknowledged to be wrong [13, 41, 8]; nevertheless, present-day neoclassical economists keep on endorsing the outdated behavioral Stimulus-Response (S-R) psychology. Their behavior is inconsistent but comprehensible: by totally neglecting the active role of the emotional side of the mind in decision-making processes, S-R mechanism easily justified both the effectiveness of decreasing marginal utility law and the desirability of a certain level in whatever homeostatic equilibrium.

These shortcomings were likely to be justified in times when economics was under the leading forces of positivism in natural sciences and when the rational constructivism of concrete deductive logic [23] was the best aknowledged methodology. Today, the scientific development has seriously put into question the effectiveness of positivism approach. Such a steady persistence against the evidence has led many scholars to infer that the unwillingness to modify the assumptions of the economic theory might be due to the saving of their primateship. Other scholars go further and think it is aimed at "facilitat[ing] the development of a more general economic theory that is able to explain both when (in what sorts of conditions) and why anomalies occur and when and why behaviour is displayed that is consistent with standard economic theory's predictions" [44].

Scitovsky notices that many eminent economists in the history of economic thought had already clear in mind that the Stoic lifestyle, the absence of any sensation, was not the perfect bliss. He recalls a neoclassical economist like Alfred Marshall as one of the first to admit man's psychological need to engage in "activities ... pursued for their own sake" [21, III, II, p. 4], being that need as the main motivating force of all creative activity. Marshall mentions science, literature, the arts, athletic games, and travel as the principal futile but re-creative activities men pursue for their own sake.

John Maynard Keynes too, Scitovsky continues, was fully aware "of a spontaneous urge to action rather than inaction, and not as the outcome of a weighted average of quantitative benefits multiplied by quantitative probabilities",[1] but it was boredom he had in mind when he forecast general nervous breakdowns [18, p. 327]. And, roughly twenty-eight years later, Roy Harrod did not predict nervous breakdowns but rather a "return to the Middle Ages' warmongering, violence and bloodsports" [11, pp. 207–208].

Keynes' and Roy Harrod's predictions turned out to be correct, although the reasoning on which they were based was not proper. Labor productivity rose substantially but only increased output, while our forty-hour workweek has not been shortened since; although increased mechanization could well have created excess energy for workers to give vent to and seek for more stimulation. Indeed, recent evidence shows that workers' work time has even increased.

Such a puzzling situation is caused by the fact that "[t]he economist registers differences in what people consume and regards them as evidence of differences in what he calls "revealed preferences". [On the contrary, t]he psychologist is not content to accept these and stop there; he tries to penetrate beneath the surface to find the causes and explanation of the differences" [36, p. 28]. In view of the failure of neoclassical theory in predicting human behavior and the persistent consumer dissatisfaction in the affluent society, Scitovsky uses behavioral psychology to fill in gaps in the economists' understanding of consumer behavior. He suggests that in order "to develop a new and better theory of consumer behavior ... it will be helpful to look for guidance at the psychologist's work on welfare and motivation". Scitovsky notices that "the psychologist's approach to human welfare is very much like the economist's, in the sense that he, too, observes behavior, makes inferences from observed behavior, and builds up his theory from these inferences ... the most general such theory [physiological psychologist's theory of the motivation of behavior], and the one that seems most pertinent to economics, is the one that explains behavior in terms of *arousal*" [32, p. 9] (italics mine).

The brain activity, known as arousal, is manifested in electrical impulses which can be monitored by means of an electroencephalograph. Brain activity appears as waves in the elettroencephalogram (EEG waves): different brain waves correspond to different levels of agitation (Scitovsky synthesizes arousal system of the brain as the "level of excitement" [36, p. 28]). The activity of the nerve cells depends on the stimulation that the central system receives from outside through the senses, the muscles and internal organs (sensations), as well as within the brain itself (emotions), but it never sinks to zero as the

---

[1] "Most, probably, of our decisions to do something positive, the full consequences of which will be drawn out over many days to come, can only be taken as a result of animal spirit of a spontaneus urge to action rather than inaction, and not as the outcome of a weighted average of quantitative benefits multiplied by quantitative probabilities" ... "human decisions affecting the future, whether personal or political or economic, cannot depend on strict mathematical expectation, since the basis for making such calculations does not exist ... it is our innate urge to activity that makes the wheel go around ... " [17, pp. 161–162].

organism is alive.[2] So the zero point, the homeostatic equilibrium, must be avoided at any cost. Scitovsky too reminds us that "comfort is the absence of both pain and pleasure" [36, p. 137].

Significantly enough, in a report of a previous experiment on boredom (quoted in [36, pp. 22–23]), psychologist Woodburn Heron concluded: "[t]he pain of no stimulation was so great that most subjects, after the first four to eight hours, suffered headaches, nausea, confusion, fatigue, hallucinations, and a temporary impairment of various mental faculties".

The neuro-psychologists' findings on arousal system put into question early Freudian-like "unstimulated condition": the height of bliss, the perfect satisfaction of all needs and desires, actually was only a part of the story. The missing part became apparent when the psychologists discovered that just as too high an arousal and too little stimulation (i.e., too much comfort or the absence of any stimulus, say, boredom) are also unpleasant and motivate the seeking of stimulation that will raise arousal and bring it up to some optimal level of stimulation.

> In accepting that hypothesis [arousal mechanism], we must abandon the old-fashioned notion that pain and pleasure are the negative and positive segments of a one-dimensional scale, something like a hedonic gauge, calibrated from utter misery to supreme bliss, on which a person's hedonic state registers the higher the better off he is [36, p. 61].

Pleasure can be therefore activated into ways, either through a moderate increase in arousal (arousal-boost mechanism) or a decrease in arousal when this has reached an uncomfortably high level (arousal-reduction mechanism) [4, p. 282]; thus, a stable level of wealth gives the consumer comfort, but not pleasure. In short, change is necessary to feeling; pleasant sensations derive no more, as the supporters of hedonic happiness may claim, from the absence of stimulus; on the contrary, pleasant sensations originate from *contrast* and *discrepancies*. Perfect comfort and *lack of stimulation* are restful at first, but they soon become boring, then disturbing: "too much comfort may preclude pleasure" [36, p. 26]. In reaching historically high levels of comfort in their lives, modern consumers paradoxically have decreased their levels of pleasures, which derive from change in comfort levels.

> The theory that comfort depends upon the level of arousal and pleasure depends upon changed in that level fits well with conventional wisdom and our introspective knowledge. To begin with, it explains the fleeting nature of pleasure. It also explains the closely related belief that, in man's striving for his various goals and struggling to achieve them are more satisfying than is the actual attainment of the goals. The attainment of a goal seems, when the moment of triumph is over, almost like a letdown. Few sit back to enjoy it; in fact, most people seek a further goal to strive for,

---

[2] One group of facts, in particular, seems to stand in the way of this assertion, namely, the evidence that boredom, a condition resulting from a deficiency in the kinds of stimulation that elevate arousal, can be extremely distressing and that the termination of boredom by a renewed influx of such stimulation can be potently rewarding [...] Lack of stimulation or monotony is unpleasant, when it is, because it produces a rise rather than a fall in arousal". From this point of view, level of arousal is a *continuous variable*, fluctuating between the extremes of deep sleep and frenzied excitement, and constitutes an essential feature of the psychophysiological state of an organism at a particular moment" [2, pp. 307–308] (italics mine). Indeed, in medicine "death" has been now redefined as the stopped activity of the brain, not the heart.

preasumably because they prefer the process of striving toward a goal to the passive state of having achieved one [36, p. 62].

If things stand like that, Scitovsky argues, hedonic happiness is *not* happiness at all; on the contrary, it represents the first step to the suffering state of boredom: a desire to have desires (Schopenhauer's "empty longing"), which is frustrated by the satisfaction of all determinate desires (Scitovsky's comfortable goods).

The quotations here below highlight economist's and psychologists' deep understanding of the notion of *Vita Activa*:

> It is, again, the desire for the exercise and development of activities, spreading through every rank of society, which leads not only to the pursuit of science, literature and art for their own sake, but to the rapidly increasing demand for the work of those who pursue them as professions. Leisure is used less and less as an opportunity for mere stagnation; and there is a growing desire for those amusements, such as athletic games and travelling, which develop activities rather than indulge any sensuous craving ([21, III, II, p. 4], quoted in [36]).

> We systematically underestimate the human need of intellectual activity, in one form or another, when we overlook the intellectual component in art and in games. Similarly with riddles, puzzles, and the puzzle-like games of strategy such as bridge, chess, and go; the frequency with which man has devised such problems for his own solution is a most significant fact concerning human motivation ([14, pp. 246–247], quoted in [36]).

> Observation of animals and people ... indicates that much time and energy is taken up by brief, self-contained, often ripetitive acts which are their own reason, ... autonomously motivated, and not ... small contributions to some remote, critically important aim ([2, pp. 4–6], quoted in [36]).[3]

Scitovsky concludes that only when abandoning the emotion-averse and the utility-driven homeostatic Freudian explanation of human behavior, the economics of consumer will find more palatable solutions.

4 Scitovsky's threefold classification of needs

Virtually, all needs and desires, not only the biological, are arousing: they increase the arousal level of the nervous system, its alertness, tension and anxiety. This is a useful and functional reaction, because an *action is required* to satisfy a need or eliminate a discomfort; higher arousal usually increases the organism's speed and efficiency in responding to stimuli in deciding the requisite action, and carrying it out. At the same time, heightened arousal is also the immediate motivation to do what can satisfy the need, eliminate the cause of heightened arousal, and so reduce the arousal level again. In that case, excitement is the outcome of the increasing arousal of the nervous system that accompanies the process of being stimulated when the stimulus is fairly strong.

The ratio of the "best performance" has been detected early by the psychophysiologist Wilhelm Wundt [50]. Wundt's law says that the sensation is

---

[3] Berlyne led various researches on the intrinsic motivation of games, which may coincide with the work of art, strategic games, humor, curiosity, and exploration, to assess the inconsistency of a theory on motivation exclusively based on drive mechanism as the organism response to certain external stimuli.

the most pleasant for an intermediate degree of intensity: too much excitement is unpleasant and can become unbearable; too little excitement is boring and so unsatisfying. Thus, excitement is an essential part of unchanging human nature and it can be sought and obtained in many different ways (both with comfortable and stimulating activities).[4]

From the economic point of view, we have a consumer who tries to maintain optimal level of stimulation and is continuously looking for new things, *new stimuli*.

In putting this very general principle of human nature into an economic system, Scitovsky gives a plausible answer to the consumer dissatisfaction, because of his propensity to lavish in comfort-biased purchases. The economist notices that the findings of the neurophysiologists on arousal mechanism fit well into a threefold classification, which is a Maslow-averse[5] rendition of human needs and satisfactions *appetitive desires*, *social desires*, and the *need for stimuli*, "all of them being urgent and essential and with little scope for substituting one for the other" [37, p. 254]. He argues that the standard economic assumptions that consumers' tastes are given exogenously, do not change over time, and are well known to consumers, are reasonably apt only when applied to the bodily satisfactions on visceral needs (they are satiable needs, and consequently maximizable). Such assumptions, however, do not fit the other two equally urgent categories of human satisfactions, that is, social needs and the need for stimuli: gregariousness and curiosity, which are both of them unfillable, consequently, impossible to maximize. Furthermore, their sources of satisfaction are not only easily changeable but also include change itself (potential pleasure), owing to the fact that novelty is the crucial ingredient of all mental stimulation.

The shortcomings of the standard economic theory, as revealed by modern psychology, tap several dimensions of human action in the affluent societies. *The Joyless Economy* presents a comprehensive overview.

---

[4]Scitovsky cites the famous german psychologist Wilhelm Wundt; though, in developing his theory he builds on Berlyne's cognitive psychology of the arousal potential of the "collative variables" (i.e. novelty, variety, discrepancy and incongruity). See [3].

[5]The Maslow pyramid is associated with the hierarchy of need theory that Maslow originated in 1943. According to this theory the most basic need is related to physiological survival air to breathe, water to drink, food to eat and sex to procreate. Next in order of precedence comes a set of needs for such things as safety and security. Once an individual has taken care of his or her basic physiological needs and feels safe and secure some degree of need for love and belonging may well rise to the forefront of their concerns. If you don't have enough of something — i.e. you have a deficit — you feel the need. Maslow saw all these needs as essentially survival needs. Need for the respect of our fellow's, and for self-respect, are seen as being next in order of precedence. Maslow himself admits that his pyramid lies in uncharted waters "We must guard ourselves against the too easy tendency to separate these desires from the basic needs we have discussed above, i.e., to make a sharp dichotomy between 'cognitive' and 'conative' needs. The desire to know and to understand are themselves conative, i.e., have a striving character, and are as much personality needs as the 'basic needs' we have already discussed [...]. There are other, apparently innately creative people in whom the drive to creativeness seems to be more important than any other counter-determinant. Their creativeness might appear not as self-actualization released by basic satisfaction, but in spite of lack of basic satisfaction" [22, pp. 386-387].

5 The economic problem from the individual's point of view

Work is one of the best outlets for one's nervous energy; it can be really stimulating if testing our mental or physical ability. Scitovsky underlines that the problem of prolonged boredom arose when the nomadic tribes settled down to farming; this concerned for a strict minority of the leisure class; it was not the case of poor working classes whose hardships were too demanding to experience boredom. Having little to do during wintertime, farm workers had winter occupations to keep them busy; their parents used to teach them how to create, practice, and enjoy folk art, singing folk songs, painting and carving various objects (Scitovsky, *Unrelieved boredom*, undated typenote). Thus, before the advent of Industrial Era, man did not feel the need for stimulating and social activities: the hardships of work itself perfectly matched man's demand for mental and physical attention and commensality.

After the Industrial Revolutions farmers became factory workers, which meant no more leisure time for winter occupations. Since the eighteenth century, technical and economic progress have rendered most work far too mechanical and fragmented to be enjoyable. Consequently, in today's high specialized welfare economy, work has strong probabilities to become a primary cause of human dissatisfaction: increased specialization, mechanization, and automation have taken the fun and excitement out of much work whose difficulty and hazards were once a source of challenge and satisfaction. Boredom and alienation are the emotional consequences of frustrated creative powers and can be every bit as severe a privation as hunger. So, why not following Marx's advice: "each can become accomplished in any branch he wishes, [making] it possible for me to do one thing today and another tomorrow, to hunt in the morning, fish in the afternoon, rear cattle in the evening, criticize after dinner, just as I have in mind"? (Marx, *The German Ideology*, quoted in [36, pp. 90–91]). Because work is also the best token of one's status in society, and the individual will continue in performing it in spite of being dissatisfied, pursued by the anxiety to "keep up with the Joneses".

Since education qualifies people for more prestigious jobs, the ostensible requisite for the more important jobs is more education. So, the total demand for education is bound to *increase* and indeed it is for in developed countries the surge in the demand for education has been fully met by the surge in its supply. That is a good point, though we should bear in mind that by investing more on education, a society is likely to increase the number of qualified people looking for status, this does not imply an increase the *supply* of such jobs. This unmatched condition creates feelings of frustration once people realize that their labouriously skills acquired in such a laborious way are wasted; such an unmatched condition creates feelings of frustration.

Scitovsky is resolute in saying that *more* education does *not* necessary mean *better* education ("measured by number of man-years of schooling per head of population, we [Americans] are the world's most educated people. Are we also the best educated?" ([36, p. 224], see also [31, p. 63]). He specifies that the rise in the schooling rate is changed as a function of the increase in the demand for labour skills, that is, for productive and administrative skills, and neglecting the improving of leisure skills in response to the production's call.[6] As a result, to reply to the increasing demand for technical training, which are the typical skills required by the modern economy, the school curriculum is changed in the direction of improving the learning of the skills of making money, but whittling down the humanities part of the school curriculum, which is the part containing most of the training for leisure activity [36, p. 229].

The additive effects of: (1) the increasing productivity of modern economy, (2) the excessive supply in narrowly specialized people supply, together with (3) the shortening of weekly working hours (an achievement of a welfare State) create the awkward situation of having more leisure at disposal without knowing what making of it. As Scitovsky puts it:

> The changing aims of education are responsible for our increased productivity. They also explain the paradox that, as progress frees more and more of our time and energies from work, we are less and less well prepared to employ this free time and energy in the pursuit of an interesting and enjoyable life [31, p. 64].

6 The economic problem from the consumer's point of view

Since leisure skills are consumption skills [36, p. 229], whittling down leisure skills gives birth to an unskilled consumer. To better grasp the problem of the consumer dissatisfaction in the modern society, we first should look carefully at a representative list of the products modern economies put at consumer disposal. Ralph Hawtrey distinguished between defensive products and creative products [12, p. 189]. Defensive products (i.e. food, clothing, shelter, and all the thing that make one physically comfortable) aim at eliminating a discomfort or saving efforts; they are consumed in order to attain some comfort. On the contrary, creative products (i.e. sports, humor, entertainment, literature, art, and the creative ingredients of defensive products, such as the elegant and the artistic side in clothing, the decorative and applied arts, and the skilled preparation and selection of food whose satisfaction stems from their providing some form of stimulation), are consumed in order to attain pleasure. The difference between the two kinds of products lies in that defensive products yield merely the negative satisfaction of minimizing or eliminating pain, discomfort, and effort. As a consequence, our need for them is satiable and their ability to give satisfaction correspondingly limited.

The problem is that "we overindulge in comfort" [36, p. 289].[7] Scitovsky claims that modern technology creates more possibilities but it also drives to

---

[6]Scitovsky criticizes the proliferation of titles-without-substance and considers them as an undesired outcome of the changing in educational system. Scitovsky, Income and Happiness, unpublished paper, p. 7. All unpublished papers, undated typescripts and lectures, in Scitovsky papers, Rare Book, Manuscript, and Special Collections Library: Duke University: Durham, NC.

[7]The economist Ralph Hawtrey too concludes: "[i]t is possible for a rich man to incur heavy expenditure without any assignable purpose beyond securing the mini-mum of discomfort and the maximum of leisure. Those two aims between them will account for a stock of furniture and clothes, a staff of servants, a large house, and many other possessions. But the whole yields no positive good; it merely brings him to the zero point, at which he is suffering from no avoidable harm. He has weeded his garden, and still has to choose what he will plant in it, before he can't be said to have made anything at all of his life" [12, p. 191].

standardization and uniformity. Uniformity means lack of novelty which, as psychophysiologists now confirm, creates an excitement deficit. Thus, mass-production is responsible for the consumers' dissatisfaction because of the absence of any stimulus, and "not because mass-produced objects are expected to be inherently inferior to handmade ones but because they accumulate the inferiority of boring sameness only gradually, as more and more people acquire the same or similar items and so increase the frequency with which an individual possessor of an item encounters its identical twins" [12, p. 252]. If we remember the psychologists' indications that excitement (an enjoyable stimulation) is an important ingredient of human satisfaction, it is now clear that choosing comfort as a goal and neglecting creative activities make consumers dissatisfied.

A dissatisfied consumer has a lot of chance to be also a frustrated one. Producers cannot sell, of course, what consumers do not want to buy; nevertheless advertising provides the means to mould consumers' tastes to fit their production. The affluent consumers of the affluent society have to choose in dozens of markets from among hundreds of goods and services, whose ever-increasing complexity make informed choices very difficult (evidence in [16]): they do not know how to spend the high income they have earned and are willing to accept guidance from whoever offers it. Due to a production-oriented education, the lack of skills and knowledge required for enjoying particular forms of stimulation, renders people's choice from among the many alternatives fairly arbitrary and easily influenced by advertising.

The positive reinforcing effect generated by any pleasurable activity (arousal reducing) is accompanied by a secondary opposite effect. As Scitovsky puts it:

> by forming any kind of habit [by acquiring a taste], we acquire a distaste for breaking the habit [36, p. 131],

according to the fact that

> many of our wants are not innate and biologically determined, but acquired by learning. Once they are acquired, and once their ability to give satisfaction has been learned, they also become habitual and create drives to maintain or repeat the newly learned satisfactions [36, p. 67].

Scitovsky points that "addiction" or, the reduction in one's defensive consumption, is a much more general psychological phenomenon and not necessary confined to drug addiction. It would rapresent only an extreme example which is different just in the degree from addiction to many other sources of satisfaction (Scitovsky, *Income and Happiness*, unpublished typenote, p. 13).

Thus, *status consumption* can be as severe a drive as a physiological is, leading the consumer to a slavish dependence. The skilled producer has trapped the consumer into the so-called "salted peanut syndrome". The conspicuous consumption of the unskilled, but status-affected consumers will make him fall into the wretched plight of income-driven workaholic mania (which is the "treadmill effect", now well documented by [16]).

Rational or not, the careless spending of money not only hurts the person who wastes it, but also the whole consumer society as well. Though the business ethics promotes productivity and growth, it is consumers' choice that determines the direction taken by that growth; for the unskilled consumers purchases represent a biased signal for the seller, who will offer ever low-quality goods, since they are not stimulated by more demanding requirements.

The profit-oriented producer does not care about the "poor" affluent consumers who are dissatisfied with products despite being addicted to them.

7 The economic problem from the producer's point of view: growth

As far as producers are concerned, the economic problem for the producer arises when the demand for novelty amongst consumers has to be interpreted as the essential enabling force that allows innovating entrepreneurs to be successful [29]. In any case, the economic outcome is that the prices of those status symbol products raise. As those prices keep on rising, they begin to exert an inflationary pressure on all prices.

Scitovsky is not saying that affluence *is* the cause of *all* unemployment and inflation, but he argues that, as affluence continues to grow, these causes of unemployment and inflationary pressures are getting more important and make stagflation chronic and long lasting (Scitovsky, Ill. *State Lecture*, p. 12). Scitovsky quotes the works of arts as an example; if taken as an end (that is, positional goods), and not as a means to gain more knowledge, their limited supply renders the demand for them unfillable rather than unlimited and prevents the expenditure from creating output, employment and income (think, for example of the paintings by dead painters). The sale of these paintings creates a capital gain for the previous owners and turns some of their accumulated wealth into cash, the spending of which *would* increase output and employment *if* it were spent on personal comforts. Yet, that is not likely to happen, because the former owners of those paintings are usually pretty affluent themselves (and if they were impoverished and were forced to sell of their possessions, they would use the proceeds to compensate for debts. That, of course, will not generate current employment).

Scitovsky thinks that spending on positional goods is the main reason why we stuck with consumerism, even though our rising income virtually assure the whole population's comfortable survival. Competitive spending on positional goods is a zero-sum game that can never add to one's comfort as long as everybody else also invests on them. Moreover, spending on positional goods depresses the economy and reduces income, thereby diverting expenditure from the public domain yet further [34]. In short, additional demand for positional goods and services has almost the same economic impact as hoarding money, except that it also rises prices of positional goods and shifts wealth distribution in favour of the affluent former holders of those goods. Furthermore, it results that excess demand for some positional goods degrades their quality.

To prevent these shortcomings, Scitovsky proposes to render the positional goods public ("[b]y donating such valuable and rare possessions or collections of possessions for public use, the rich can further reassert their status" (Scitovsky, *Ill. State Lecture*, p. 9) and foster subsidies to the arts (both private and public), to avoid creative occasions of stimulation may turn into unaffordable status symbols and alllow them to become a *positive externality*, letting people gain more culture and, consequently, more knowledge.

8 The economic problem for the economic system: development

Novelty plays an essential role in explaining economic change. To Schumpeter, innovation is the key element for an economy to develop, though he aknowledges that the entrepreneur is only a vehicle of novelty. By putting the invention in the economic function the entrepreneur is the bearer of the mechanism of change [29], but he is not the inventor (here lies Schumpeter's paradox of the neoclassical exogenous explanation, that is invention, of an endogenous phenomenon, that is innovation). As Witt points out, theories of evolution have to satisfy three necessary conditions to explain self-transformation: "They must (1) be dynamical; (2) deal with nonconservative systems [...]; (3) cover the generation and the impact of novelty as the ultimate source of self-transformation" [47, p. 91]. Condition (1) and (2) are present in almost all modern economies, they are necessary but not sufficient to explain self-transformation. These two conditions cover the concept of economic growth, but not the one of development. Scitovsky explains why condition (3) does not appear in most economic systems.

Surprising enough, it is the very modern production that impedes development. Scale economies in production restrict the range of goods produced, those in distribution restrict even further the range of goods distributed. All these limit the economies' ability to cater to the *variety* (viz, a source of novelty) of consumers' tastes and discriminates against the minority consumers who are unable to satisfy some of their tastes. Moreover, economies of scale in distribution lower costs, but they also humper the distribution of products when there is no mass demand for. It is easy to get anything destined for nationwide mass consumption but hard to satisfy a need that is different. Is there any way out to the "tragic choice" between the comfort of standardized products and the stimulation of the original ones?

Stimulation is typically a non-exclusive or shared-source of satisfaction. By contrast, comfort usually lacks spillover effects of creative goods. Since many comforts come from the substitution of mechanical power to man's muscular power, they often have unpleasant side-effects, such as noise and air pollution. Hawtrey clarified that creative products and defensive products are not mutually excludible. Yet, Scitovsky ([36, p. 109]) points out that the needs (respectively, stimulation and comfort) aiming at satisfying people are all primary needs. So, when fashion, which is the touchstone for novelty, gives place to consumerism,[8] the new products are innovative only in their aesthetic issues, and changes tend to be more marginal than revolutionary in nature. It follows that a lack of stimuli arises again. Consequently, the

---

[8]Scitovsky defines "[c]onsumerism [a]s the usual term for a surfeit of comfort and insufficient stimulus, and our narrow specialization to the detriment of general knowledge and culture has very much to do with it" (Scitovsky, *What Went Wrong in Our Country?*, unpublished typenote, p. 2).

massive production of defensive but novelty-lacking products is self-defeating.

9 The social consequences of the lack of novelty: the mob rule

"The economist traditional picture of the economy resembles nothing so much as a Chinese restaurant with its long menu. Customer choose from what is on the menu and are assumed always to have chosen what most pleases them" [36, p. 149]. Economic data show what the public buys; they do not show whether it buys what it wants or only what it can get.

The flow of causality between the supply and the demand sides is an old question that has shifted economists from one side to the other along the history of economics. The structure of present economics legitimates Scitovsky's hesitation: "[d]o market prices bring the pattern of output into harmony with consumers' preferences as economic theory teaches us, or does advertising mold people's tastes to make them conform to the pattern of output?" [37, p. 253]. Present standard economics is forcing the direction by putting the demand side at the beginning of the economic development. By and large, as Schumpeter has explained very well, the producer is simply an intermediary between the consumer and his/her addiction to the status symbols of a society; he does not set fire to the addiction, but he blows the wind into his favour.

In Scitovsky's, consumer sovereignty — provided that there is one — in a standard free enterprise is a curious combination of plutocracy (the rule of the rich) and oclocracy (the rule of the crowd), where each consumer's influence on what is produced depends on how much he/she spends for a fully available massive product [36, p. 9]. Forced by the high cost of eccentricity, most consumers give up and conform instead.

On the economic level the loss of such a majority tyranny affects the whole community because the minority consumer is often the individual "person of genius"[9] who might lead the majority towards the new and the better. As Scitovsky puts it: "economies of scale in the modern economy impose the majority's tastes on the whole society, and when the majority chooses to sacrifices the stimulus of novelty for the sake of comfort, the creation of novelty and the minority' seeking new ways of attaining the good life are both impeded" [24, p. 289].

If it is so, democracy cannot be expected to counteract such manipulation since a democratic electorate consists of consumers who will tend to favor government policies because of their exposure to advertising; such policies should be instead designed to encourage the production of wealth so that

---

[9] "Persons of genius, it is true, are, and are always likely to be, a small minority: but in order-to have them, it is necessary to preserve the soil in which they grow. Genius can only breathe freely in an atmosphere of freedom. Persons of genius are, ex vi termini, more individual than any other people less capable, consequently, of fitting themselves, without hurtful compression, into any of the small number of moulds which society provides in order to save its members the trouble of forming their own character" [24]. Mill too was aware of the tyranny of the masses: "In sober truth, whatever homage may be professed, or even paid. To real or supposed mental superiority, the general tendency of things throughout the work is to render mediocrity the ascendant power among mankind" [24, p. 268].

consumers are enabled to purchase comforting consumer goods.

Since knowledge influences both consumers' lifestyle choices and producers' investments, the skilled consumer, who is searching after welfare, is urged to claim his/her sovereignty, whereas economics has to conceive consumers as involved in a continuous process of "learning to consume" in order to retrieve the necessary knowledge to make informed choices. If consumers were predominantly "lethargic dullards", growth would be limited not merely to a lack of interest in new things but also to a lack of new products in which to be interested [48].

10 The social consequences of the lack of novelty: violence

From sections 2 and 3 we learn that satisfaction depends not only on what money can buy but also on mental and bodily exercise ("just as starvation can make a person steal if he has no money to buy food, so boredom can lead to violence if a person finds no peaceful activity for enjoyment and keeping busy" [39, p. 32]. Thus, stimulating activities (the list of sources of stimulation including the arts, literature, sports, seeing the world, as well as work, artistic creation, and exploration) require skills that have to be learnt and matches sharing the satisfaction they yield. Many leisure activities are in fact less satisfying although economic progress has made them easier, less demanding and more accessible. Owing to more comfort and overproduction of defensive goods, jointly with a higher specialization man has been made not only more self-sufficient but also more isolated, which has consequently depressed his inborn need for social activities. The relative shortage of such pleasurable stimulating activities explains the human pursue to seek stimulation elsewhere. Scitovsky claims that violence is the natural outcome of the unexpressed energies by people who are unskilled in leisure activities and are unable to practice the unique activity they have been (and wanted to be) trained for, that is work.

> Violence is the result of too much leisure to people unskilled in its use. Thousands of jobless youngsters fresh out of school roam the streets and engage in violence and rowdyism. They are not starving, because they get reasonably generous unemployment benefits, and their behavior is explained partly also as an angry response to their rejection by society, but partly also as the normal reaction to boredom of energetic young people trained in the discipline and skills of work but left completely unprepared and untrained for leisure (Scitovsky, *Ill. State Lecture*, p. 18).

The scholar attributes the phenomenon of the great increase in crimes against the person (murder, forcible rape, vandalisms, aggravated assault) as sideeffects of people's inability to satisfy their increased need for excitement in more innocuous ways. Yet, he points out that serious crimes are not the only sources of stimulation and excitement. He places the vicarious violence inbetween them because viewers witnessing other people's violence in films and TV programs may learn violence and how to express it. Many people use television viewing as an escape. Children use it as an escape from loneliness due to the absence of their parents who are too tied up with their job to take care of them. Adults use it as an escape from the monotony of their daily working life. Indeed, despite the issue is dragging on for nearly 40 years between those who believe that television viewing is a significant cause of alienation, family disharmony, hostility, and tension and those who claim that viewing is a form of escape from these undesirable phenomena, the latest findings in this field[10] confirm Scitovsky's insights that television viewing functions like drugs and alcohol, driving man into isolation estranging him from both the environment and his fellow men, thus creating a lack of social comfort [36, p. 164].

11 The place of economic welfare in human welfare

Novelty is the key element for human welfare. Since economic development implies "steps between which there is no strictly continuous path" [28, p. 113] planning a novelty is impossible and a contradiction in terms, for it requires spontaneousness to emerge.

Because of revealed preference and full rationality, standard consumer theory lacks both the accuracy to explain and the flexibility to allow changes in (consumer) behavior. In advocating an abdication of a "bounded constructivism", Scitovsky proposes to replace Mill's concrete deductive logic with a sort of "rational a-logic" (Jasper's *vernünftige Alogik*). Only in this case Schumpeterian processes of self-transformation (creative destruction) may be engaged, obsolete routines (addictive habits) wiped out and new knowledge (development) built up.

The Hungarian economist warns that if the economics of welfare continues to focus on the production of comfortable but unchallenging artifacts rather than favoring risk-taking activities, as long as they are pleasurable, a relatively low growth rates should be expected due to low levels of entrepreneurial creativity. Moreover, modern economics is pushing man to develop an ever greater tolerance and desire for excitement and violence, because scientific and economic progress has dried up the source of stimulating activities by creating a stimulus deficit, which needs to be mitigated. The various sources of deliberately sought-for stimulation are leading people from dangerous sports and gambling to violence, crime and participation in collective hazardous actions (Harrod's prophecy).

So, what is the remedy?

The remedy is culture [36, p. 235].

> Anyone of innumerabile physical and mental activities can relieve boredom, provided it is sufficiently challenging to one's physical or mental aptitudes to make enjoyable. The challenge is to one's strenght, skill or knowledge, which means that almost all those activities only become enjoyable and relieve boredom if one has learnt their particular skills or acquired some of their relevant knowledge beforehand [40, p. 107].

Scitovsky defines culture as "that part of knowledge which provides the redundancy needed to render stimulation enjoyable", "as the training and skill necessary to enjoy those stimulus satisfactions whose enjoyment requires skill and training". "Culture, or consumption skills is the preliminary information we must have to enjoy the processing of further information" [36, pp. 226–227].

---

[10] For empirical evidence and a comprehensive list of references, see [6].

Culture is the key point for personal satisfaction and human development: being at the same time a means of satisfaction and a stimulating good she "satisfices"[11] our unfillable demand for curiosity, at the same time promoting social respect and fostering individual diversity.[12] Since choice is not so much a function of preferences, but a function of knowledge, of which references are a subset, in the economics of the consumer, culture provides the skills may expand the set of consumption possibilities [5]

In short, culture is a positive externality, being almost always the by-product of those goods and services which aim at providing entertainment, amusement, aesthetic pleasure which also are form of stimulations to others whereas comforts not only fail, typically, to carry external benefits, many of them generate external nuisances as well [36, p. 144].

How to diffuse culture? By education. Scitovsky "define[s] education in a broad sense, as that sort of schooling and training given to the young in preparation for the work of life, that development of powers and formation of character, as contrasted with the imparting of mere instructions or collecting information" ... "The educator is not infrequently expected to give guidance in manners and morals, to maintain order and prevent violence, to spur the indolent, to supervise the diet, to awaken ambition, to inculcate respect for law and decency in short to be at once priest, physician, policeman, parent, and more as well as teacher" [30, p. 1].

Social culture and "tacit knowledge" or the transmission of "learning how" are as important as "learning what" in teaching programs, because without the capacity to transfer patterns across fields, agents would be incapable of perceiving any kind of novel behavior.

In a recent paper, attribuing the effects to causes between democracy and education, the economist Edward Glaeser provided a model to illustrate the empirical link between education and democracy. Glaeser's paper suggests that education may increases the optimal size of effective uprisings [10, p. 30]. This seems to be supported by European and Latin American history. In less educated times and places, coups are generally small affairs including only small cadres of nobles or army officers [7]. As education grew, effective uprisings (like the American Revolution) became larger: broad swaths of society were included in attempts to overthrow a regime (even the Nazi takeover in Germany, which eventually led to a dictatorship, succeeded only after the Nazis had built a broad coalition, including students and earlier attempts at a narrow coup proved to be an embarrassing failure.

Seemingly, to Scitovsky the educational reform has to be done not in universities, but at the elementary and high-school levels, the task is not how much

---

[11]Here the term "satisficing" is used in the meaning Simon "to denote problem solving and decision making that sets an aspiration level, searches until an alternative is found that is satisfactory by the aspiration level criterion, and selects that alternative" [42, p. 168].

[12]Scitovsky warns political system from the lack of any discussion may derive from persons who, too proud for their too specialized technical knowledge, but too bold to admit any different proposal in their well planned programs. " [...] instead of debating the issues, defending their stand against their opponents' objections or suggesting modifications or objecting to particular parts of their opponents' positions, just call[ed] them liars or their statements all wrong and then proceed[ed] to restate verbatim their own position, repeatedly when necessary" [38, p. 147].

I have learned, but how many experiences I have done (see, Ryle's "knowing how" [26] and Polanyi's "tacit knowledge" [25]). Since many of those former skills are best learnt in early childhood, the best policy in that case seems to make children start out with as more and many skills as possible and cut down on their number later, when it becomes evident which activities children have a talent and liking for, and then keep up the learning process and practice of only the most promising ones. "Of course, the teachers themselves must be thought a new attitude to life before they can teach to the others [...] " because "it is not enough for the new generation to learn the leisure skills, they must also acquire a taste for them; and that again is very difficult to teach in a civilization whose work orientation also renders it excessively money-oriented" (Scitovsky, Ill. State Lecture, pp. 19–20). Both structured and informal education, plays an important role in personal and social development. Its most important function is teaching children to avoid hurting the others and other people's property, learn the skills of harmless but enjoyable stimulating activities. Consequently, the education system should not give a specific model, but rather provide the tools to build a personal method to acquire and process further information (i.e. learning how to learn), the cultivating of the self (i.e. "learning how", Scitovsky's rationale for "culture").

In that perspective Scitovsky makes clear that economic progress is not the cause of human development, as many today's scholars (see, for example, [1]) claim; on the contrary it is its natural outcome. Educating as opposite to teaching and formal training is the sole condition that allows the forming of "persons of genius" and, as a consequence, the following emergence of spontaneous novelty without knowing the meaning (*pax* according to Schumpeter's paradox). That is what Scitovsky proposes in order to make novelty emerge so that it can "satisfice" both consumers' unfillable needs and entrepreneurs' production *desiderata*.

On the individual level, education and culture teach how to search and find satisfaction in pleasant leisure activities. On the societal level, a primary aim of education is socialization, teaching students how to interact successfully with their fellows, to foster self-regulation of behavior and train them how to manage conflicts that may arise in sharing experiences. Such a successful interaction includes understanding and appreciating the others' point of view, as well as being able to effectively communicate one's own, through both writing and speech. Thanks to this interaction, convincing discussion in public (at work, in politics, in society) can be easily set up. That is how Mill's *desiderata* are realized through a non-Millian method, because it is not an *a priori* calculation, but only the far-sightedness of the "persons of genius" that can theorize and fulfill social, economic and educational policies that foster innovation without imposing it.

12 Conclusions

Economics is a means for human welfare and economic theories are instruments calling for direction. Mill's methodology of pure concrete logic may improve efficiency, but it never produces a new idea (Hume, 1739-1740).

Evolutionary economists argue that growth in economic systems is driven by the growth in knowledge [20]. Tibor Scitovsky reminds us that men are curious-born animals, hungry for novelty, variety, complexity and incongruity so that only the incessant and self-fomenting process of gaining more knowledge by means of culture can mitigate their "unfillable" desires to have desires and simultaneously *foster* development. When explaining the neurological basis of creative innovation, Scitovsky's *The Joyless Economy* apparently suggests only some devices for consumer satisfaction; actually it lays the foundations for an economic theory where novelty is the key element to assess the utility value and *knowledge* is the means to gain more novelty.

In 1926 the psychologist Graham Wallas suggested that creativity is realized through four stages: preparation, incubation, illumination and verification. Preparation is the acquisition of skills and knowledge allowing a person to create. [19] notes that many important discoveries are initiated by the observation of an anomaly. Although these discoveries are based on an anomaly, it is the "prepared mind" that enables creators to perceive the importance of the phenomenon they observed. Weisberg [46] suggests that creativity does not require great leaps (*e.g.* illumination) and the processes leading to many great discoveries is likely not to be subconscious incubation, but rather a series of conscious steps. Scitovsky points out that economics is the *last* step of a process that spontaneously emerges from an experience of knowledge. Thus, conventional economics keeps on appying deductive logic to endogenize (in the economic function) a naturally exogenous element (i.e. novelty) artificially, in order to be enabled to build up an *ad hoc* theory to justify economic planning for innovation. It is actually a self-defeating strategy, where the standard utility approach is a deceptive instrument since the possibility of any *substantial* (economic) development is prevented.

Scitovsky's scrutiny presents far-reaching implications, not only for the idea of rationality but also for the very concept of utility — by making it plural in nature — and the importance of freedom itself, including the freedom to change preferences). Considering that the primary channel of culture transmission is the family, Scitovsky's proposals involve group behavior as well as the instruments of the State: subsidizing leaves of absence for working parents and spreading liberal arts for the spontaneous self-determination of the individual:

> the solution of that [consumer dissatisfaction] problem awaits a reform of our educational philosophy and lifestyle [35, p. 269].

Only more parenting and a liberal education aimed at "stimulating individual curiosity about the world and mankind, and by encouraging self-reliance and independent thinking in the acquisition of knowledge" allow the emergence of "persons of genius" (the *generalist* character, as opposite to the *specialist* bureaucrat)[13] who, by fostering the claims of a minority, make a development possible.

That "person of genius" may be the innovative entrepreneur who fosters the economic production of stimulating *creative goods* (i.e. relational and re-

---

[13]Scitovsky defines generalist character (as opposite to the technical, but fragmented knowledge of the "specialist" character) a person with broad knowledge, long experience, good judgement, and wisdom [36, p. 247].

creative goods and services); he/she may be a skilled consumer who, through a careful choice, may guide the entrepreneur to improve his/her production. He/she may also be a far-sighted policy-maker who, by improving the diffusion of culture, may simultaneously promote personal *self-flourishing* and public happiness, thus preventing democracy from degenerating into disruptive oclocracy and society to turn into a violent, aimless and anonymous multitude.

Times are ready for further re-examinations of the principles of welfare economics. The improvement of social welfare requires to abandon the maximazing utility criterion. As Schumpeter stressed, the entrepreneur is only the bearer of an invention into the economic function. Scitovsky's inquiry proves that the reconstruction of social welfare starts from the individual contribution. To be born, authentic novelty requires more attention to the minority needs. Only when (if) economic and policy systems admit their limits of interfering in people's lives and understand their instrumental role of simply "setting the stage" for the spontaneous emergence of a novelty, novelty will really emerge and development will start.

BIBLIOGRAPHY

[1] R. Barro. Determinants of democracy. *Journal of Political Economy*, 107: 158–83, 1999.
[2] D.E. Berlyne. Exploratory and epistemic behaviour. In S. Koch (ed.), *A Study of a Science II*, McGraw-Hill, New York 1963 [1962].
[3] D.E. Berlyne. *Aesthetic and psychobiology*. Appleton-Century-Croft, New York 1971.
[4] M. Bianchi. Collecting as a paradigm of consumption. *Journal of Cultural Economics*, 21: 275–289, 1997.
[5] M. Bianchi. Novelty, preferences, and fashion: when goods are unsettling. *Journal of Economic Behavior & Organization*, 47: 1–18, 2002.
[6] L. Bruni and L. Stanca. Watching alone: relational goods, television and happiness. *Journal of Economic Behavior & Organization*, in press, 2007.
[7] F. Campante and Q.A. Do. *Inequality, Redistribution, and Population*. Mimeo: Harvard University, 2005.
[8] A. Damasio. *Descartes error: Emotion, Reason, and the Human Brain*. Putnam, New York 1994.
[9] S. Freud. Instincts and their vicissitudes. In *The Standard Edition of the Complete Psychological Works of Sigmund Freud*, vol. XIV. J. Strachey (trans. gen. ed.) in collaboration with A. Freud, assisted by A. Strachey and A. Tyson, Hogarth Press, London, 1957 [1915].
[10] E.L. Glaeser, G. Ponzetto and A. Schleifer. Why eoes eemocracy need education? *NBER*, working paper no. 12128, march 2006
[11] R.F. Harrod. The possibility of economic satiety. In *Problems of United States Economic Development*, vol. I, CED, New York 1958.
[12] R.G. Hawtrey. *The Economic Problem*. Longmans, London 1925.
[13] F.A. Hayek. *The Sensory Order: An Inquiry into the Foundations of Theoretical Psychology*. Routledge & Keagan Paul, London 1952.
[14] D.O. Hebb. Drives and the C.N.S. (Conceptual Nervous System). *Psychological Review* 62: 243–254, 1955.
[15] W. Heron. The pathology of boredom. *Scientific American*, 1957: 52–56.
[16] D. Kahneman, A.B. Krueger, D.A. Schkade, N. Schwarz and A.A. Stone. A survey method for characterizing daily life experience: the day reconstruction method. *Science* 306 (5702): 1776–1780, 2004.
[17] J.M. Keynes. The general theory of employment, interest and money. In D. Moggridge (ed.), *The Collected Works of J M. Keynes*, Macmillan, London 1973 [1936].
[18] J.M. Keynes. Economic possibilities for our grandchildren. In *Essays in Persuasion*, Norton, New York 1963 [1931].
[19] T.S. Kuhn. *The Structure of Scientific Revolutions*. UCP, Chicago 1962.

[20] B. Loasby. *Knowledge, Institutions, and Evolution in Economics*. Routlege, London 1999.
[21] A. Marshall. *Principles of Economics*. Macmillan, London 1947 [1890].
[22] A. Maslow. A Theory of Human Motivation. *Psychological Review* 50 (4): 370–396, 1943.
[23] J.S. Mill. *A System of Logic Ratiocinative and Inductive*. Routledge and Kegan Paul, London 1843. http://oll.libertyfund.org/?option=com_staticxt&staticfile=show.php%3Ftitle=246.
[24] J.S. Mill. On Liberty. In J.M. Robson (ed.), *The Collected Works of John Stuart Mill*, vol. XVIII, Toronto, 1988–1991 [1859].
[25] Polanyi, M. *The tacit dimension*. Routledge & Kegan Paul, London. 1967.
[26] Ryle, G. *The concept of Mind*. Hutchinsons University Library, London. 1949.
[27] A. Schopenhauer. *The World as Will and Representation*. Trans. by E. F. J. Payne, Dover Edition, New York, 1967 [1819].
[28] J.A. Schumpeter. Development. Festschrift offered to Emil Lederer in honour of his 50th birthday on 22 July. Tr. by M. C. Becker and T. Knudsen, or. tit. Entwicklung, arch. loc. SPE XMS Lederer, Box 1, 82.1. Lederer, Emil, Papers, German Intellectual Emigre Collection, M. E. Grenander Department of Special Collection and Archives, University Libraries: State University at Albany, State University of NY. 2005 [1932].
[29] J.A. Schumpeter. *The Theory of Economic Development*. Oxford University Press, New York, 1961 [1934].
[30] T. Scitovsky. Report of the San Francisco Curriculum Survey Committee, prepared for the Board of Education, San Francisco Unified School District, 1960.
[31] T. Scitovsky. What's wrong with the arts is what's wrong with society. *The American Economic Review* 62(2):62–69, 1972.
[32] T. Scitovsky. The place of economic welfare in human welfare. *Quarterly Review of Economics & Business* 117(2):7–19, 1974.
[33] T. Scitovsky. The desire for excitement in modern society. *Kyklos* 34:3–13, 1986.
[34] T. Scitovsky. Growth in the affluent society: Fred Hirsh memorial lecture. *Lloyds Bank Review* 1:1–14, 1987.
[35] T. Scitovsky. Hindsight economics. *BNL Quarterly Review* 178:251–270, 1991.
[36] T. Scitovsky. *The Joyless Economy: The Psychology of Human Satisfaction*. Revised edition, Oxford University Press, New York, 1992.
[37] T. Scitovsky. My Search for welfare. In M. Szenberg (ed.) *Eminent Economists: Their Life Philosophies*, Cambridge University Press, New York, 1992.
[38] T. Scitovsky. The need for stimulating action in rationality. In K. Dennis (ed.), *Economics*. KAP, Boston, 1998.
[39] T. Scitovsky. Boredom. An overlooked disease. *Challenge*, 42 (5): 5–15, 1999.
[40] T. Scitovsky. Memoirs. Undated typescript, in *Scitovsky Papers, Rare Book, Manuscript, and Special Collections Library*, Duke University, Durham, NC.
[41] H.A. Simon. From substantive to procedural rationality. In S.J. Latsis (ed.), *Method and Appraisal in Economics*, Cambridge University Press, Cambridge, pp. 129–148.
[42] H.A. Simon and A. Newell. *Human Problem Solving*. Englewood Cliffs, Prentice-Hall, NJ, 1972.
[43] L. Walras. Etudes d economie sociale. 1st edn. Pichon-Rouge, Paris-Lousanne, 1896.
[44] J. Vromen. Neuroeconomics as a natural extension of bioeconomics: how the proliferation of constrained maximization backfires, 2005. 2nd draft, July, presented at PHARE workshop march 2006.
[45] G. Wallas. *The art of thought*. Harcourt Brace, New York, 1926.
[46] R.W. Weisberg. *Creativity: Genius and other Myths*. Freeman & C., WH, 1986.
[47] U. Witt. Emergence and disseminaton of innovations: some principles of evolutionary economics. In R.H. Day and P. Chen (eds.), *Nonlinear Dynamics and Evolutionary Economics*, Oxford University Press, Oxford, 1993, pages 91–100.
[48] U. Witt. Learning to consume. A theory of wants and the growth of demand. *Journal of Evolutionary Economics* 11(1)23–36, 2001.
[49] R.S. Woodworth. *Experimental Psychology*. Holt & Co., New York, 1938.
[50] W. Wundt. *Grundzüge d. physiologischen Psychologie*. Engelmann, Leipzig 1874.

Explaining causal modelling. Or, what a causal model ought to explain

FEDERICA RUSSO

1 Introduction

One of the goals of the social sciences is to understand social phenomena, that is to exhibit the mechanism underlying and bringing them about. This task goes beyond description: to exhibit this mechanism requires identifying *causal* relations between variables of interest. In quantitative social research, causal models are used to provide such explanations of social phenomena. This paper investigates whether causal models can be seen as *models of explanation*, and argues that causal modelling, by modelling causal mechanisms, provides (or ought to provide) genuine causal explanations and should be considered as a model of explanation, notably a *hypothetico-deductive* model of explanation.

The paper is organised as follows. *Section two* presents and explains what a causal model is and what it is supposed to do. *Section three* analyses how the terms "explanation", "explanatory", or "explain" are used in causal models. *Section four* advances the view that causal modelling ought to be the modelling of causal mechanisms. *Section five* builds on the results of the previous sections and argues that causal models are a *model of explanation*, in particular, they are *hypothetico-deductive models*, where the H-D structure of the explanation is given by the H-D methodology of causal models. Finally, *section six* compares causal modelling with other models of explanation — notably, the deductive-nomological, statistical-relevance, the causal-mechanical, and the manipulationist model — and shows why those models of explanation are not fully satisfactory in the social sciences and particularly in quantitative social sciences. This enables us to highlight what causal modelling offers over and above traditional models of explanation.

2 Causal modelling

A causal model consists of a set of mathematical equations (also called *structural equations*) and/or of a graph laying down the hypothesised causal structure pictorially.[1] More technical and precise definitions of causal models are

---

[1] A number of causal models rely or employ structural equations in an essential manner — for instance, covariance structure models or multilevel models — but others do not — for instance, counterfactual models. However, although counterfactual models are rightly called "causal" because they seek to measure the average causal effect of a treatment or intervention, surely they substantially differ from structural equation models in that they do not aim at modelling the causal mechanism, which, as we shall see later, is an essential feature of structural equation-type models.

of course possible. For one account, and for detailed examples accessible to a non-specialised audience, see [18, ch. 3], [19, 13] and references therein.[2]

An important feature of causal models is that they rest on a number of assumptions, some of which are merely statistical and others have instead causal import. Among the statistical assumptions we find, for instance, linearity and normality, non-measurement error and non-correlation of error terms. Those are standard statistical assumptions also made in associational models. However, causal models are provided with a much richer apparatus that allows their causal interpretation. In this apparatus we find background knowledge of the causal context, the conceptual hypothesis, a number of extra-statistical assumptions and of causal assumptions. Among extra-statistical assumptions we can list the direction of time, causal asymmetry, causal priority, causal ordering, and the deterministic structure of the causal relation. Causal assumptions include: structure of the causal relation (separability), covariate sufficiency, no-confounding, non-causality of error terms, stability, and invariance. A large part of causal models used in social science, unlike associational models, use a hypothetico-deductive methodology, according to which causal hypotheses are confirmed or disconfirmed depending on the results of tests and on whether they are congruent with background knowledge.[3] For a detailed account of the features of causal models and for a comparison with associational models see [18, ch. 3]; some of these features will nonetheless be discussed later.

An important characteristic of causal models is that causal relations are *statistically modelled*. This aspect deserves attention because influential philosophers of causality such as Wesley Salmon believed that aleatory causality will give a better understanding than statistical causality even in the social domain [24]. On the one hand, aleatory causality bestows emphasis upon the *physical* mechanisms of causality, primarily uses concepts such us "process" and "interaction", and appeals to laws of nature such as the conservation of energy or momentum. In the Salmon-Dowe [25, 5] theory, causal processes are the key because they provide the link between the causes and the effects; causal processes intersect with one another in interactive forks, and in this interaction they are both modified and changes persist in those processes after the point of intersection. Causal processes and interactions are physical structures and their properties cannot be characterised in terms of probability values alone.

On the other hand, statistical causality puts emphasis upon constant con-

---

[2]Disagreement arises in causal modelling as to whether structural equations and directed acyclic graphs convey exactly the same information. Partisans of the former approach tend to give a negative answer, because in graphical models some assumptions are relaxed and relations between variables are not expressed with the mathematical precision of structural equations. On the other hand, supporters of graphical models, such as Bayesian nets, maintain that this formalism indeed provides a simplification with respect to structural equation modelling without loss of any relevant information.

[3]It is worth noting that causal models can also be used in an inductive way, e.g. data mining. This is, for instance, the approach of [27]. Inductivist approaches claim that causal relations can be bootstrapped from data without the burden of extra-statistical and causal assumptions made in their hypothetico-deductive counterparts. Unfortunately, it goes far beyond the goal of this paper to discuss the success of inductive causal models. Consequently the scope will be limited to causal models that employ a hypothetico-deductive methodology.

junction and statistical regularity and uses, above all, concepts such as statistical relevance, comparison of conditional probabilities, or screening-off relations. Those concepts can be defined solely in terms of statistical terms, without resorting to any physical notion. According to the received view, statistical regularities are the "symptoms" of causal relations. The conjunctive fork [16] gives the probabilistic structure of the causal relation and the screening-off relation alerts us about possible situations in which, given the correlation between two events A and B, a third event C may be responsible for their correlation.

But what do causal models do? Causal models *model* the properties of a social system. In particular, they model the relations between the properties or characteristics of the system, which are represented by variables. By "social system" I simply mean a given population, and "population" has to be understood here in the statistical sense, that is as a set of units, those units being individuals, households, firms, etc. In causal modelling, to model the properties of a social system means to give the scheme, or the skeleton, of how these properties relate to each other. In other words, the causal model *models* the causal mechanism governing the social system. However, this causal mechanism is not modelled in terms of spatio-temporal processes and interactions *à la* Salmon but is statistically modelled. This means that the concepts typical of statistical causality do help in identifying the types of relationships that hold among the variables of interest. In particular, causal models seek to uncover stable *variational* relations between the characteristics of the system. It is worth-noting that the received view of statistical causality, a heritage of Hume and represented, for instance, in [28, 7, 2], emphasises the role of statistical *regularities* for assessing causality.

In [17, 18] I challenge this view and argue that, instead, probabilistic theories of causality as well as causal modelling are governed by a rationale of variation, not of regularity. In a nutshell, the rationale of variation states that causal models measure and test joint *variations* between variables of interest, not regular sequences of events. To be sure, causal hypotheses are variational claims — that is they hypothesise how the effect would vary according to variations in the cause — and empirical testing aims at establishing whether variations are causal (rather than chancy), not whether regularities are causal. Of course, to ensure that variation can be interpreted causally, we have to impose further *constraints*, and the invariance condition is required in order to interpret variations causally. It is worth noting that the invariance condition is not a condition of regular occurrence of events, but of stability of the model's parameters across different environments. In other words, variations among variables of interest will be deemed causal if the parameters are sufficiently stable across different environments. In fact, the invariance condition ensures that accidental and spurious relations be ruled out.

3 Explanation in structural equations

Consider now a simple form of a structural equation:

$$Y = \beta X + \epsilon \tag{1}$$

where Y represents the putative effect, X represents the putative cause, β is a parameter quantifying the causal effect of X on Y, and ϵ represents errors or unmeasured factors. The scientific literature is not very homogeneous as to the vocabulary used. Y and X are called in a variety of ways depending on the specific discipline. Statistical textbooks will normally refer to X and Y as the dependent and independent variables, or as the explanatory and response variables, respectively; the econometric literature usually talks about exogenous and endogenous variables; the epidemiological literature spells them out in terms of exposure and disease, etc. Let us focus on the explanatory-response vocabulary, which perhaps constitutes the background of all disciplines that use causal models. In this case terminology is quite explicit: the Xs supposedly *explain* Y. But what do exactly the Xs explain? And how?

With much disappointment to the philosophers, in the scientific literature there is no explication of the terms "explanatory", "explanation", "explain". Intuitively, the Xs explain Y in the sense that they "account for" Y, namely the Xs are relevant causal factors that operate in the causal mechanism, which is formalized by the equations and the graph. Needless to say, this is a very unsophisticated explication of "explanation", yet intuitively clear. Let us leave aside, for the time being, the issue of what a good explanation is and of what causal modelling would offer over and above alternative models of explanation, and let us focus on what explanation in causal modelling consists of.

In causal modelling, the goal is to explain the *variability* in Y. Structural equations can be interpreted thus: variations in Xs explain variations in Y, or variations in Xs produce a variability in Y. Therefore, as long as we can control variations in Xs we can also predict how Y will accordingly vary. The βs quantify the causal impact, or the direct causal effect of each of the Xs on Y. So one can suggest that the more variability we can account for, the higher the explanatory power of the causal model. But how is this explanation "quantified"? The (statistical) answer lies in the coefficient of determination r^2. r^2 is the square of the correlation coefficient r and is a statistic used to determine how well a regression fits the data. It represents the fraction of variability in Y that can be explained by the variability in Xs; thus r^2 indicates how much of the total variation in Y can be accounted for by the regression function.

However, this statistical answer is insufficient. This is for three reasons. The *first* is that r^2 just measures the goodness of fit, not the validity of the model, and a fortiori it does not say how well the model *explains* a given phenomenon. So r^2 gives us an idea of whether the variability in the effect is accounted for, and to what extent, by the covariates we chose to include in the model. But, and here is the *second* reason, the coefficient of determination does not give any *theoretical* motive for that. *Third*, the coefficient of determination will give us an accurate quantification of the amount of variance of Y explained by the Xs only if the assumptions are correct. For instance, r^2 can be small either because X has only a small impact on Y (controlling for appropriate covariates) and/or because the relation between X and Y is not linear.

Instead, a more satisfactory (philosophical) answer lies in the specific features of causal models. Let us then know revert to the assumptions of causal model and examine their explanatory import. Among the features of causal models listed in the previous section, those having explanatory import are two causal assumptions — notably, covariate sufficiency and no-confounding — and background knowledge.

Covariate sufficiency assumes that the independent variables are direct causes of the dependent variable, and that these are all the variables needed to account for the variation of the dependent variable. No-confounding then plays a complementary role in assuming that all other factors liable to screen-off the causal variables are ruled out. Those two together convey the idea that the causal model includes all and only the factors that are necessary to explain the variability of Y. Those assumptions rely on the hypothesis of the closure of the system, namely causal modelling assumes, so to speak, that we can isolate a mechanism within the larger social system under consideration, and that this mechanism is not subject to external influences. Thus, we can account for Y — that is for its variability — just relying on the factors we decided to include. This is indeed a strong assumption but this is the only way to go in order to avoid an *ad infinitum* regression hunting for more and more ancestral causes, and in order to exclude that everything influences everything else in the system, thus making impossible to identify the causal relations to intervene upon. Covariate sufficiency and no-confounding also highly depend on which variables we choose to include in the causal model. This choice, in turn, depends on background knowledge. But what is background knowledge in the first place?

The notion of background knowledge belongs to most quoted and least explicated concepts in causal modelling. Anything could fit in it. Unfortunately, if anything can be background knowledge, we lack a sensible criterion to say when and why the covariates contribute toward the explanation of the response variable. So we'd better specify what it is in it. Background knowledge may include: (i) similar evidence about the same putative mechanism, (ii) general knowledge about the socio-political context, (iii) knowledge of the physical-biological-physiological mechanism, (iv) use of similar or different methodologies or of data. Different studies normally consider different populations. Differences can accordingly concern time, geographic location, basic demographic characteristics, etc. Background knowledge has to be used to justify the choice of the explanatory variables. This justification relies on the different aspects mentioned above. A detailed case study illustrating the explanatory role of covariate sufficiency, no-confounding, and background knowledge is offered in [18, ch. 4.3].

4 Modelling causal mechanisms

So far, I argued that causal models attempt to explain the *variability* of the effect variable by means of appropriate covariates. I also argued that the explanatory import is given by specific causal assumptions made in causal modelling — notably, covariate sufficiency and no-confounding — and by background knowledge. This philosophical answer complements the statistical answer according to which the coefficient of determination quantifies the explanatory power of a causal model. Let us now go back to what causal models do. Earlier, I briefly put forward the idea that causal models model the properties of a system and that we could conceive of them as the scheme or skeleton of the causal mechanism governing the causal system under investigation. It is now time to develop this idea further.

The notion of mechanism is often evoked both in the scientific and philosophical literature. No account seems to attract an unanimous consensus, yet various characterisations stick to the physical notions of process and interaction. That is to say, the most widespread conception sees mechanisms as made of physical processes, interactions, and of physical elements, somehow assembled together to behave like a gear. For instance, [11] opposes Humean causality, that sees causation as mere regularity, to a realist view, that sees causal mechanisms and causal powers as fundamental. According to the realist, says Little, "a mechanism is a sequence of events or conditions governed by lawlike regularities leading from the explanans to the explanandum". A partisan of this view is obviously Slamon [22], for he believes that causal processes, interactions and laws give the causal *mechanisms* by which the world works and that to understand why something happens we have to disclose the mechanism that brought it about, or Dupré and Cartwright [6] who, in the same vein, argue that discovering causal relations requires substantial knowledge of the capacities of things or events — i.e., their power to bring about effect. Likewise, Bunge [1] ultimately reduces mechanisms to physical process that interconnect with one another as is the case in most biosystems or physical systems. Also, in his account mechanisms are governed by causal laws.

Unfortunately, this view doesn't fit the case of the social sciences for two reasons. *First*, if the causal model only involves socio-economic-demographic variables, we cannot identify causal mechanisms in terms of *physical* processes and interactions (at least at that level of description). *Second*, if the causal model involves both social and biological variables, the causal mechanism will not be able to account for the "social" part. Let me explain these two reasons further. The problem is that, in social contexts, mechanisms are not always, or not necessarily, made of physical processes and interactions. Social scientists in [12] modelled a causal mechanisms involving the relations between regional mortality in Spain and the use of sanitary infrastructure. This model does not involve *physical* processes and interactions. The mechanism described by the authors rather explains the behaviour of a social system in terms of the relations between some of its properties. These properties, however, do no necessarily have "physical" reality as they might just be conceptual constructs, as for instance economic and social development. Consequently, the process leading, say, from economic development to mortality through the use of sanitary infrastructures does not correspond to a *physical* process, such as billiard balls colliding, but rather is our conceptualisation and schematisation of a much more complex reality.

Another difficulty is that if causal mechanisms are governed by causal laws, it is unclear where these laws come from, and if they come from causal modelling itself, then this leads to a vicious circle. So if we want to keep a physical notion of mechanism the price to pay is quite high — we would have to renounce to causal mechanisms in the social domain. We are not forced to this solution, though, if we are prepared to accept a wider concept of causal mechanism, in particular one that is based on causal modelling.

In a nutshell, causal modelling is, and ought to be, the *modelling of mechanisms*. A statistical characterization of mechanisms, along with a rationale of variation, is what mediates our epistemic access to causal relations. The net gain of this perspective is a non-physical characterization of causal mechanisms. In fact, mechanisms would then have observable components (corresponding to observable variables) and the only non-observed parts of causal mechanisms would be nodes representing latent variables. However, far from giving causal mechanisms a mysterious or epistemically inaccessible appearance, latent variables ought to be introduced to facilitate the interpretation of complex causal structures.[4] Most importantly, the modelling of mechanisms ought to rely on the rationale of variation rather than on the rationale of regularity: causal mechanisms are made of *variational* relations rather than regular relations. The components of the causal mechanisms are arranged depending on what variations hold. Agreed, those variational relations happen to be regular (or at least regular enough), but this depends on the fact that causal modelling analyses large (enough) data sets. Furthermore, regularity does not seem to be successful in constructing causal mechanisms, for the Humean view and the realist view eventually collapse in the same tenet. In fact, according to the realist, the sequence of events in causal mechanisms is, in the ultimate analysis, governed by a lawlike *regularity*.

Such characterisation of causal mechanism allows us to incorporate in the causal model both socio-demo-political variables and biological variables. It goes without saying that pathways in such a mixed mechanism have to be made explicit as there isn't homogeneity at the ontological level. Health variables do not cause changes in social variables (or vice-versa) *as such*. Socioeconomic status influences one's health through the possibility of accessing some sanitary infrastructures, but not directly. Arguably, the social sciences are interested in identifying causal mechanisms that involve different types of variables — this interdisciplinary stance is also a perspective undertaken in epidemiology (see for instance [29]).

To sum up, if causal models do not model the mechanism underlying the phenomenon being investigated, they lack explanatory power. To see why it is so, let us compare them with associational models. Associational models only investigate statistical associations between variables, but no causal interpretation is allowed for the parameters. This is due to several reasons. *First*, associational models don't have the rich apparatus of statistical, extra-statistical, and causal assumptions as causal models do — associational models are normally equipped just with statistical assumptions. *Second*, they do not employ a hypothetico-deductive methodology — there is no formulation of the causal hypothesis because it is not their goal to confirm or disconfirm hypotheses. Thus modelling mechanisms, that is identifying the causal inter-

---

[4] A similar view that emphasises the central explanatory role of mechanisms is advanced by R. Franck [8]. However, Franck's approach differs from mine in that it goes further in claiming that the modelling of a social mechanism ought to be completed with the modelling of the functions of the same mechanism.

relations between the variables of interest, becomes a necessary condition for the explanatory power of causal models. For a thorough discussion of this notion of causal mechanism in social science and for a detailed example, see [18, ch. 6.1]. The question then arises as to what kind of formal structure such explanation should have — this issue will be tackled in section 5.

5 Causal modelling as a model of explanation

The two previous sections argued that some specific features of causal models have explanatory import and that if causal models can be successful at all in the enterprise of explaining social phenomena, this is because they model causal mechanisms. I will now advance the view that causal models are models of explanation, in particular, they are *hypothetico-deductive* models of explanation.

The formal structure of the explanation is given by the hypothetico-deductive character of model-building and model-testing of causal modelling. Simply put, hypothetico-deductivism is the view according to which scientists first formulate hypotheses and then test them by seeing whether or not the consequences derived from the hypotheses obtain. K. Popper [14], who first developed the H-D methodology, was motivated by the need of providing a scientific theory in a non-inductive way. However, in causal modelling, hypothetico-deductivism takes a slightly different facet specifically concerning deduction, but does borrow from the Popperian account the primary role of the hypothesis-formulation stage. I shall get back to this point shortly.

According to the H-D methodology, model building and model testing essentially involve three stages: 1. formulate the causal hypothesis; 2. build the statistical model; 3. draw consequences to conclude to the empirical validity or invalidity of the causal hypothesis.

The hypothesis to put forward for empirical testing does not come from a tabula rasa, but emerges within a causal context, namely from background theories, from knowledge concerning the phenomenon at stake, and from preliminary analyses of data. This causal hypothesis, which is also called the "conceptual hypothesis", is not analysable *a priori*, however: its validity is not testable by a logico-linguistic analysis of concepts involved therein. On the contrary, to test the validity of the causal hypothesis requires building a statistical model, and then drawing consequences from the hypothesis. If the model is correctly estimated and fits the data, the hypothesized causal link is accepted — provided that it is congruent with background knowledge. The hypothetico-deductive structure of causal modelling is thus apparent: a causal hypothesis is first formulated and *then* put forward for empirical testing. That it to say, the causal hypothesis is *not directly inferred* from the data gathered, as is the case with inductive strategies, but accepted or rejected depending on the results of tests.

As anticipated above, hypothetico-deductivism in causal modelling does not involve deductions *strictu sensu*, but involves a weaker inferential step of "drawing consequences" from the hypothesis. That is to say, once the causal hypothesis is formulated out of the observation of meaningful co-variations between the putative cause and the putative effect and out of background knowledge, we do not require data to be *implied* by the hypothesis but just that data conform to it. Here, "conform" means that the selected indicators *adequately* represent the conceptual variables[5] appearing in the causal hypothesis. Thus, this way of validating the causal hypothesis is not, strictly speaking, a matter of deduction, but surely is, broadly speaking, a deductive procedure. More precisely, it is a *hypothetico*-deductive procedure insofar as it goes the opposite direction of inductive methodologies: not from rough data to theory, but from theories to data, so to speak.[6]

To sum up, a causal model attempts to *explain* a given social phenomenon — in particular, the variability of the effect variable Y — by means of number of explanatory variables X and the explanatory procedure is given exactly by the hypothetico-deductive methodology of causal models. How do we evaluate the goodness or the success of the explanation then? We have seen before that the coefficient of determination is insufficient to provide such an answer, which instead lies in the peculiar features of causal models. Statistical tests, notably invariance and stability tests, provide the accuracy of measurements but alone cannot guarantee the explanatory goodness of the causal hypothesis. In fact, non-sense correlations, such as the monotonic increase of both bread prices in England and sea-level in Venice, may well turn out to be stable or invariant and yet not causal nor explanatory at all. The goodness of an explanation cannot be assessed on statistical grounds *alone* — the story also has to be coherent with background knowledge and theories previously established, and has to be of practical utility for intervening on the phenomenon.

Thus, the problem of the goodness of explanation is mainly a problem of internal validity, with the caveat that, among various threats, coherence with the background plays a major role.[7] This, however, makes explanation highly context-relative simply because the causal model itself is highly context-relative. This could be seen as a virtue, as restricting the scope leads to more accurate explanations. But obviously this situation raises the problem of generalising results — that is the external validity of the causal model. It goes far beyond the scope of the present work to advance the criteria that allow the generalisation to a different population and/or different time.

Hypothetico-deductive explanations also exhibit a flexibility rarely found in other models. First, they allow a *va et vient* between established theories and establishing theories. Established scientific theories are (and ought to be) used to formulate the causal hypothesis and to evaluate the plausibility of results

---

[5] A conceptual variable is a variable that cannot be measured directly but from some "indicators". For instance, socio-economic status can be measured by taking into account income and years of schooling.

[6] For a discussion of the H-D method at work in the social sciences see [18, ch. 3.2], and also [3, ch. 2]. Cartwright, as many others both in the philosophical and scientific literature, calls the methodology of causal models hypothetico-deductive but she also warns us about the weaker form of deductivism hereby involved.

[7] Simply put, according to Cook and Campbell [4], *internal validity* establishes whether a relation is causal, or whether from the absence of a relationship between two variables we can infer absence of causality. *External validity*, instead, concerns the possibility of generalising a presumed causal relationship across different times, settings, populations.

on theoretical grounds. But causal models also participate in establishing new theories by generalising results of single studies. This reflects the idea that science is far from being monolithic, discovering immutable and eternal truths. If the model fits the data, the relations are sufficiently invariant and congruent with background knowledge, then we can say, to the best our knowledge, that we hit upon a causal mechanism that explains a given social phenomenon. But what if one of these conditions fails? A negative result may trigger further research by improving the modelling strategies, or by collecting new data, thus leading to new discoveries that, perhaps, discard background knowledge.

The hypothetico-deductive structure of explanations also allows us to control the goodness of explanation. We can exert (i) a statistical control by measuring, with the coefficient of determination, how much variability is accounted for. We can also exert (ii) an epistemic control, by asking whether results are coherent with background knowledge. (iii) A metaphysical control is also possible, as we have to make sure that there be ontological homogeneity between the variables acting in the mechanism. If such ontological homogeneity is lacking, this would trigger further research for indirect causal paths that would have been previously neglected. A detailed case study illustrating the hypothetico-deductive character of causal-model explanations is discussed in [18, ch. 6.1].

6 Causal modelling vs. other models of explanation

This last section aims at comparing causal modelling with other models of explanation and at showing why those models are not fully satisfactory in the social sciences and particularly in quantitative social sciences. This comparison will enable us to highlight what causal modelling offers over and above traditional models of explanation.

Contemporary philosophy has been debating explanation for about 60 years now. Salmon [23] has brilliantly summarised the first four decades, but much discussion followed since then and, in particular, a novel account — Woodward's manipulationist approach — has been proposed [30]. I direct the reader to detailed introductions to explanation [23, 15, 30] — here I only isolate four main contenders, namely the deductive-nomological model and more generally the covering-law model [9], the statistical-relevance model [21], the causal-mechanical model [22], and the manipulationist model [30]. The goal here is not to dismiss those accounts altogether. There is indeed much that can be learnt from them but there are some aspects peculiar to quantitative social science that they are not able, alone, to grasp or to account for.

According to the deductive-nomological model, an explanation is a deductively valid argument where the conclusion, or explanandum, states that the event or phenomenon to be explained occurred. A peculiar characteristic of the D-N model is that the premises of the argument, or explanans, have to contain at least a law. In a D-N explanation there are two types of conditions of adequacy. The first type is logical: (i) the explanation has to be a valid deductive argument, (ii) the explanans has to contain at least a law, and (iii) the explanans has to have empirical content. The second type is an empirical condition: statements in the explanans have to be true. Next to

D-N explanations, Hempel also recognised two other types: the deductive-statistical model, where the premises contain at least a statistical law, and the inductive-statistical model, where the explanans confers high probability on the explanandum event.

A first obvious (and well-known) problem concerns laws. First of all, it is a *vexata quaestio* of philosophy what constitutes a law and how we can discern between laws and accidental generalisations. However, this problem becomes even worse in the social sciences because, even if we are prepared to admit that there are laws in the natural sciences, arguably the social sciences do not have laws from which we can deduce the explanandum. A second difficulty in applying the D-N model in social contexts concerns prediction. In D-N explanations the explanandum is an occurred event or phenomenon. However, some causal models are used for forecasting, and the possibility to predict rests on the explanatory and causal power of the factors involved.

However, it might be objected that in spite of some similarity of structure, there is a fundamental difference between the deductive-nomological model and the hypothetico-deductive methodology of causal models. The difference, as Salmon [23, Introduction] himself points out, lies in their different goals: in order to explain phenomena, we use hypotheses, laws, or scientific theories that are highly confirmed, whereas in the hypothetico-deductive method the same inferential scheme is used to provide evidence for the hypothesis we want to establish. To this objection I would answer thus: in the hypothetico-deductive model of explanation, the explanatory enterprise becomes a dynamic process that involves hypothesising, deriving the consequences from the hypothesis, and testing the hypothesis against data; in this process, the interplay between establishing generalisations and using those generalisations as background knowledge is fundamental. Thus explanation is not reduced to an inference, but becomes the whole process by which we account for the variability of explanandum by means of the causal mechanism that brings it about.

In this respect, the statistical-relevance model later developed by Salmon [21] was a significant step toward approaching explanation and statistical modelling. The motivation behind the statistical-relevance model lies in two problems of the covering-law model in general and of the inductive-statistical model in particular. On the one hand, counterexamples exist showing that not all explanation are arguments, and, on the other, even if the explanans confers to the explanandum a high probability of occurrence, this is not ipso facto a guarantee of the goodness of explanation (see e.g. [20] and [10]). Salmon tried to develop an alternative model of statistical explanation where the principal concept was not that of high probability but that of *statistical relevance*. The main consequence of this shift was that a statistical explanation would now require two probability values and not only one. In this model of explanation, to explain a fact is to find the narrowest homogenous reference class the fact belongs to. However, the S-R model is too narrow in scope because it essentially applies to contingency tables but not to causal models broadly conceived.

The net advantage of causal modelling over the D-N and S-R explanations is that the generalisations involved need not to be laws. They can be empirical generalisations, weaker than laws but more suitable to the social sciences where we arguably don't have universal and necessary laws. The main flaw of the S-R model, however, as recognised by Salmon himself, is that statistical relevance is not a sufficient condition for causality nor for explanation, and in fact Slamon [22] developed the causal-mechanical model to solve this problem.

In the causal-mechanical model statistical relevance relations are only the basis upon which a *causal* explanation has to be built. A causal explanation has to appeal to notions such as causal propagation and causal interaction. Those are not explicable in mere statistical terms but require a characterisation in terms of physical notions. In a nutshell, a causal-mechanical explanation aims at tracing the spatio-temporal continuous process in which the cause and the effect occur. The basic concepts of the causal-mechanical model are those of causal process (vs pseudo-process) and of causal interaction. A causal process is a physical process that is able to transmit marks, namely modifications to the structure of the process that occur as a consequence of a causal interaction between two causal processes. For instance, two billiard balls colliding represent a causal interaction between two causal process, and the mark, i.e. the modification in one or the two trajectories of the balls, persists after the interaction takes place. Instead, the intersection between the shadows of two airplanes is not a causal interaction as no modification persists afterwards. This happens because the two shadows are not causal processes but pseudo-processes. One might wonder, however, whether the causal-mechanical model, that sees in physical processes and interactions the key to single out causal relations, is applicable in social scenarios too. This is questionable, as I argue in [18, ch. 1], because although complex socio-economic processes might well exist, it is not by means of concepts of aleatory causality that we model causal mechanisms in the social domain.

In many ways, the approach the best managed to account for the explanatory import of causal models is the manipulationist or interventionist account developed by Woodward [30]. According to Woodward, causes explain effects because they make effects happen. The bulk of his manipulationist account of explanation rests on the idea that causal and explanatory relationships are potentially exploitable for manipulation and control. More specifically, says Woodward [30, p. 191], explanation "is a matter of exhibiting systematic patterns of counterfactuals dependence". How and why the counterfactual element comes in will become clearer in a moment. There are two key notions in Woodward's account: causal generalisation and invariance. Causal generalizations are relations between variables and they have the characteristics of being change-relating or variation-relating. Of course, the problem of distinguishing causal from spurious generalisations immediately arises. We could hit upon a change-relating relation that is accidental: for instance, an increased number of storks might be statistically associated with an increased number of births, but arguably there is no causality going on there. Or the change-relating relation might be spurious: yellow fingers might be statistically associated with lung cancer but this is the case because they are effects of a common cause, that is cigarette smoking. So, change-relating relations have to show a certain invariability as prescribed by the invariance condition in structural models.

The role of generalisations is worth stressing. In Woodward's account, the role of generalisations goes beyond the role laws played in the D-N model. Here, generalisations not only (i) show that the explanandum was to be expected, but they also (ii) show how the explanandum would change if initial conditions had changed — this is where the counterfactual element comes in. Generalisations can be used to ask *counterfactual* questions about the conditions under which their explanandum would have been different. This sort of counterfactual information allows us to see that conditions in the explanations are in fact explanatory relevant. So the main advantage of counterfactual explanations over D-N explanations is that whilst the latter can only provide nomic grounds for explaining their explanandum, the former can answer *what-if-things-had-been-different* questions.

But not all counterfactuals will do. Relevant counterfactuals are those that describe the outcome of interventions. To causally explain a phenomenon is to provide information about the factors on which it depends and to exhibit how it depends on those factors. Dependence, and particularly counterfactual dependence, plays a crucial role in Woodward's account. Consider for instance the case of Mr Jones that takes birth control pills and does not get pregnant. In Woodward's account taking birth control pills has no explanatory import in the case of Mr Jones because there is no dependence between this factor and the explanandum (i.e., Mr Jones' not getting pregnant). No intervention on this factor will change whether or not Mr Jones becomes pregnant and therefore this factor lacks any explanatory power. The manipulationist account of explanation has the undisputed merit to tailor the concept of explanation to the actual scientific practice of causal modelling and, particularly, of emphasising the role of *variation*-relating generalisation in answering what-if-things-had-been-different questions. However, this exercise is not pushed far enough.

First, the manipulationist account overlooks the role of background knowledge and this opens the door to non-sense invariant generalisations to be explanatory. In fact, suppose that the relation between the increase of bread prices in England and sea level in Venice were found sufficiently stable, on what grounds could we possibly deny it explanatory power if not on background knowledge? Similarly, in the case of the missed pregnancy of Mr Jones, under Woodward's account we have to appeal to interventions to disclose the non-explanatory role of the factor; however, under the H-D account here developed background knowledge would be, as in the previous case, enough to deny birth control pills any explanatory power. *Second*, Woodward emphasises the explanatory role of counterfactuals that describe the outcome of interventions, but is this always appropriate in the social sciences? There are many factors, such as gender, on which we can't intervene, and yet they play an important causal and explanatory role. *Third*, Woodward seems to take for granted that we know what the causes are, and then focuses on what would happen if we intervened on them. However, a causal explanation is also meant to *provide* the causal factors, that is a causal explanation also is the search for causes, not only an answer to what-if-things-had-been-different-questions. Hypothetico-deductive explanations can account for this aspect because they seek to confirm causal hypotheses by including in the model explanatory covariates.

7 Conclusion

Causal modelling aims at explaining social phenomena by modelling causal mechanisms in which relations between variables exhibit a certain structural stability. Causal modelling also makes essential use of the explanatory vocabulary: for the variables, for their role, and for the interpretation of the coefficient of determination. However, in the scientific literature a detailed characterisation of the terms "explanation" or "explanatory" is missing. An unsophisticated meaning of "explanation" is that a phenomenon is explained by a causal model to the extent that we can account for the variation in the response variable by introducing relevant factors as explanatory variables. The coefficient of determination r^2 quantifies the amount of variation accounted for by the explanatory variables, but, I argued, this statistical answer is insufficient to understand why and how causal models have explanatory power.

I advanced the view that specific causal assumptions contribute to the explanatory import of causal models, notably covariate sufficiency and no-confounding, together with background knowledge. But causal models also participate in the explanation of social phenomena insofar as they model the causal mechanisms that bring them about. Thus causal models can be seen as *models* of explanation having a hypothetico-deductive structure. I emphasised the explanatory role of background knowledge in causal models and argued that alternative models of explanations more often that not, overlook it.

However, this is not tantamount to throwing out the baby with the bath water. The models of explanation here discussed pick out many aspects that contribute to the explanatory import of causal models: drawing consequences from generalisations, evaluating statistical relevance relations, identifying causal mechanisms, answering what-if-things-had-been-different questions. Conceiving of causal modelling as a model of explanation allows us to gather together all these features into a single account. However, this is not tantamount to saying that causal modelling is *the* model of explanation, but that, among various alternatives, this model fits well the case of quantitative social science.

Acknowledgments

I wish to thank Robert Franck, Phyllis McKay Illari, Jon Williamson, and Guillaume Wunsch for helpful and stimulating comments on an earlier draft of this paper. Financial support from the F.N.R.S. (Blegium) is also gratefully acknowledged.

BIBLIOGRAPHY

[1] M. Bunge. How does it work? The search for explanatory mechanisms. *Philosophy of the Social Sciences*, 34: 182–210, 2004.
[2] N. Cartwright. *Nature's Capacities and their Measurement*. Clarendon Press, Oxford 1989.
[3] N. Cartwright. *Hunting Causes and Using them: Approaches in Philosophy and Economics*. Cambridge University Press, Cambridge 2007.
[4] T. Cook and D. Campbell. *Quasi-Experimentation. Design and Analysis Issues for Field Settings*. Rand MacNally, Chicago 1979.
[5] P. Dowe. *Physical Causation*. Cambridge University Press, Cambridge 2000.
[6] J. Dupré and N. Cartwright. Probability and causality: why Hume and indeterminism don't mix. *Noûs*, 22: 521–36, 1988.

[7] E. Eells. *Probabilistic Causality*. Cambridge University Press, Cambridge 1991.
[8] R. Franck. Peut-on accroître le pouvoir explicatif des modèles? In A. Leroux, and P. Livet (eds.), *Leçons de philosophie èconomique*, Economica, Paris 2007.
[9] C. Hempel, P. Oppenheim. Studies in the logic of explanation. In C. Hempel, (ed.), *Aspects of Scientfic Explanation and Other Essays*, Free Press, New York 1965, pages 245–282.
[10] R. Jeffrey. Statistical explanation vs statistical inference. In N. Rescher (ed.), *Essays in Honor of Carl G. Hempel*, Reidel, Dordrecht 1969, pages 104–113.
[11] D. Little. Causal mechanisms. In M. Lewis-Beck, A. Bryman and T.F. Liao, (eds.), *Encyclopedia of Social Sciences*, Sage 2004, pages 100–101.
[12] O. López-Ríos, A. Mompart and G. Wunsch. Système de soins et mortalité régionale: une analyse causale. *European Journal of Population*, 8 (4): 363–379, 1992.
[13] M. Mouchart, F. Russo and G. Wunsch. Structural modelling, exogeneity and causality. In H. Engelhardt, H.-P Kohler and A. Prskwetz (eds.), *Causal Analysis in Population Studies: Concepts, Methods, Applications*, Springer, Dordrecht 2008, Chapter 4.
[14] K.R. Popper. *The Logic of Scientic Discovery*. Hutchinson, London 1959.
[15] S. Psillos. *Causation and Explanation*. Acumen Publishing, Chesham 2002.
[16] H. Reichenbach. *The Direction of Time*. University of California Press, 1956.
[17] F. Russo. The rationale of variation in methodological and evidential pluralism. *Philosophica*, 77: 97–124, 2006 (special Issue on causal pluralism).
[18] F. Russo. *Causality and Causal Modelling in the Social Sciences. Measuring variations*. Springer, Berlin 2009.
[19] F. Russo, M. Mouchart, G. Ghins and G. Wunsch. Causality, structural modelling and exogeneity. Discussion Paper 0601, Institut de Statistique, Universitè catholique de Louvain, Belgium 2006.
[20] W. Salmon. The status of prior probabilities in statistical explanation. *Philosophy of Science*, 32: 137–142, 1965.
[21] W. Salmon, R.C. Jeffrey and J.G. Greeno. *Statistical Explanation and Statistical Relevance*. Pittsburgh University Press, Pittsburgh 1971.
[22] W. Salmon. *Scientic Explanation and the Causal Structure of the World*. Princeton University Press, Princeton 1984.
[23] W. Salmon. *Four Decades of Scientific Explanation*. University of Minnesota Press, 1990.
[24] W. Salmon. Causal propensities: statistical causality vs. aleatory causality. *Topoi*, 9: 95–100, 1990.
[25] W. Salmon. *Causality and Explanation*. Oxford University Press, Oxford 1998.
[26] W. Salmon, R. Jeffrey and J. Greeno. *Statistical Explanation and Statistical Relevance*. Pittsburgh University Press, Pittsburgh 1971.
[27] P. Spirtes, C. Glymour and R. Scheines. *Causation, Prediction, and Search*. Springer-Verlag, New York 1993.
[28] P. Suppes. *A Probabilistic Theory of Causality*. North Holland Publishing Company, Amsterdam 1970.
[29] M. Susser and E. Susser. Choosing a future for epidemiology ii: from black box to chinese box and ecoepidemiology. *American Journal of Public Health*, 86: 674–677, 1996.
[30] J. Woodward. *Making Things Happen: A Teory of Causal Explanation*. Oxford University Press, Oxford 2003.

The epistemological statute of the rationality principle. Comparing Mises and Popper

ENZO DI NUOSCIO

1 The rationality principle as principle of order in human affairs

The individualistic methodology of Popper is based upon three fundamental "dogmas": the *ontological individualism*, the *unintentional consequences* and the *principle of rationality*. If the ontological individualism forces the researcher to take into account only the individuals and their actions, and if the notion of unintentional consequences permits to explain those social phenomena that are the result of the action but not of the human intention, the rationality principle is that vital epistemological tool that permits the construction of situational theories for the explanation of the human action. In the classic Popperian formulation, the rationality principle assumes that "individuals or agents always act *in manner appropriate* to the situation in which they find themselves" [37, p. 361]. It is supposed that the singles, given the circumstances in which they live, make their choices, according to the definition of Rescher, in the best possible way, in line with the "strongest reasons" [38, p. 14]. The *presumption of rationality* is, therefore, the presumption that the human action is ruled by some form of universal logic that can be decoded, that makes it understandable and explainable; it is the presumptuousness that human matters are not a kaleidoscopic irregularity lacking any sense accessible to an outside observer, but that their thick and endless plot is the result, sought or unsought, of regular human actions, regulated by a *principle* that gives it some kind of order; an order so limited and local from a social point of view, but also necessary and productive from an explicative point of view. It is thanks to this that we can try to cast a light upon those unique and unrepeatable events that are the human behaviours, but mainly upon actions that are more enigmatic, such as the adhesion to false or unfounded beliefs.

2 The epistemological statute of the principle of rationality: some critics against Popper

Conscious of the fact that the rationality principle is the basis of his *logic of the situation*, Karl Popper has been one of the methodological individualists that took an significant part in the attempt to define the epistemological nature of it. Nevertheless this attempt — as noticed by several critics — is plagued with contradictions and it does not lead to an unambiguous theoretical settlement of such principle.

Popper attempts above all to define *a contrario* the rationality principle: it is not an ontological theory of human rationality, nor an "empirical or psychological assertion" that men, always, or in the main or in most cases, acts rationally" [37, p. 359]. It should rather be considered as a "methodological postulate", that unlike other methods seems to allow explicative hypotheses — in other words conjectural/situational models — that can be best controlled [37, p. 362]. However, Popper appears to contradict himself when, in that same essay, suggests that the rationality principle is "a principle obviously *almost empty*" and this is anyhow "a good approximation to the truth" [37, p. 366]. It appears to be, in any case, "false", since, given "deep interpersonal differences also in terms of capacity", "some persons will act in an appropriate manner (to the situation), others will not" [37, p. 366]. We are in the presence of statements that are in clear contrast with the previous ones: if this principle is "almost empty" [37, p. 359] it is an hypothesis with an empirical content; if this principle is "false", it can't be a "methodological postulate", since the latter are *commands* of methods that do not put forward *descriptions* that are empirically controllable, but they impose prescriptions that are epistemological significant.

Popper tries to come out of such a contradiction proposing a solution that would cause more problems than it would solve: the rationality principle would be regarded as the "animating component of any social model", "integral part of every, or almost, social controllable theory"; and in the case of falsity of the situational conjecture "it is a sound methodological policy to decide not to make the rationality principle accountable but the rest of the theory; that is the theory" [37, p. 363]. It would cause, in this case, the falsification of the *logical connection* between the situational model and the rationality principle, that Popper would appear to consider almost as an *auxiliary hypothesis* necessary for the construction of the *situational metaconjecture*. This would be, though, an hypothesis always false, and consequently the Popperian decision of not to renounce to it would configure itself as a evident methodological mistake in comparison with the falsificationist and anti-instrumental Popperian rules. Assuming, as done by the Austrian philosopher, that such an hypothesis is *à la foi* false and necessary to explain the action, it would follow that every situational analysis — of which the principle is the *needed* "animating component" — would be untrue; which is an unacceptable conclusion, moreover paradoxal, for an individualist and a falsificationist such as Karl Popper.

3 Rationality: methodological principle or norm of action?

The contradictions which Popper comes across, in the attempt to provide to the rationality principle with an ambiguous epistemological place, are, in large part, due to the fact that he suggests an *objective* conception (from the perspective of the researcher) of such principle [30, pp. 446ff]. Popper implicitly puts aside the option of considering it *a parte* agent and he explicitly discards the possibility of considering it a principle a priori, as part of the same idea of action [37, p. 360]. Instead he assigns to the social scientist the task of construct situational models that identify actions linked to a base of information *deemed appropriate by the researcher* (Popper does not hesitate to speak of "complete rationality" and of "*knowledge* of all the information of the case"). And the recognition of the onset of actions based on a cognitive level, considered by the researcher inferior to the one established by the model, induces Popper to regard as "false" the rationality principle.

Having preferred the objective perspective in the definition of the rationality principle, Popper's analysis undergoes a sliding more or less imperceptible from a methodological dimension toward a valuating dimension: the notion of rationality tends to be not only an instrument used to investigate the situation, but also becomes a *criterion* to discriminate the rational actions from the irrational ones. According to Popper, in fact, human beings seldom act in a totally rational manner (they would if they could do the better possible use of the whole knowledge available to them for the achievement of their goals), but nevertheless they operate, more o less, in a rational way [34, p. 25]; and thus, when we speak about 'rational behaviour' or 'irrational behaviour', we refer to a kind of behaviour which is, or is not, in harmony with the logic of the situation [33, II, p. 230]. The researcher, for instance, could deem irrational — as argued by the Austrian philosopher — the behaviour of that car driver "desperately trying to park his car when there is no parking space to be found" [37, p. 361].

By suggesting a *normative* meaning of the situational models and the rationality principle, Popper is consequently induced to admit the existence of irrational actions: such irrationality is established by the observer through a *judgement of inadequacy*, or one of *non conformity*, of the action with respect to the situation, as the latter has been reconstructed by the observer. The interpreter hence regards that the single has not acted to the best of his possibilities, given that his action has not an adequate support of evidence, that the same interpreter, on the base of his analysis of that situation, regards as a minimal offer to the actor from the *background knowledge*, available at the moment the action has been carried out, of which the agent itself could take account of.

The *principle of rationality* becomes in such a way a *criterion of rationality*, on the base of which to establish the rationality or irrationality of the action by means of a *principle of objective optimisation*. And this normative interpretation of the rationality principle is confirmed with clarity by Popper, when, on *Models, Instruments, and Truth*, ads a third dimension to the situational analysis. He asserts that the above analysis describes not only 1) "the situation as it was in reality" and 2) "the situation as the agent actually pictured it, but it also has to establish 3) "the situation on the base of how the agent could have seen it (inside the objective situation), or perhaps on the base of how he should have". Moreover, "if there is a clash between 2) and 3), Popper specifies, then we can claim that the agent has not acted rationally" [37, p. 369].

It is evident that this slipping in a normative sense introduces an element of weakness in the Popperian situational logic, because it *clears* the way to the possibility of labelling as irrational certain actions, that will be ascribed to causes different from the reasons. By the light of this the fixing of standards of irrationality, which the agents has to follow in the several typical situations, is referred back to the discretionary power of the researcher.

4 The rationality of action as a principle *a priori*

Having referred the rationality principle to the situation *a parte* observant and the trivial ascertainment of the continuing emerging of actions not crowned with success, Popper has been induced to decidedly discard the hypothesis of it being a *principle a priori* as always true. Popper regards it a synthetic principle, endowed with an empirical content, to which, never mind its falsity, the Austrian philosopher tries to grant some sort of epistemological immunity, in the attempt to confer the highest empirical content, therefore the maximum explicative potential, to the situational models.

This *a priori* solution, that Popper deems unacceptable, considered by Ludwig von Mises as the real source of strength of the individualistic methodology. Mises hence follows an alternative path to the Popperian one; his *prasseology* is a doctrine completely *a priori* of the human action, with the intent of individualizing and singling out profound consequences for the social sciences, the *necessary attributes* of the action. In other words, it wants to trace those permanent characteristics of the action that are not derived from the experience but from the reason, them being prior to any factual instance; moreover it intends to track those attributes of the action often hidden and unknown, that are true *ex definitione*, thus necessarily present in every undergone action, without which, clearly, we would not be able to comprehend the subjective behaviour. By dealing not with the truth of the fact but with the truth of the reason, that is of terms a priori that, according to Mises, do not diverge from the ones characterising logic and mathematics, "prasseology is a theoretical and systematic, not an historical science. Its scope is the human action as such, irrespectively of all environmental, accidental and individual concrete acts" [28, p. 32]. The *postulates* of the praxeological science are, therefore, universal, and "without them we should not be able to see in the course of events anything than kaleoscopic chance and a chaotic confusion" [28, p. 32]. These principles can be as such deciphered: i) the human action is always the effort to remove an dissatisfaction (the condition of a permanent satisfaction would be an inhuman condition); ii) every individual gets rid of his dissatisfaction in the way he deems best; iii) to this end, every single individual, selects the means, to his judgement, more effective and convenient to the purpose.

5 Intentionality and rationality as necessary characteristics of the human action

By examining in details such postulates of prasseology, Mises identifies those characteristics of the action that are fundamental for the individualistic explanation of social phenomena. *If* the human action is always directed to the elimination of a dissatisfaction, *then* it is, by its nature, *intentional*, intentionally orientated to face the situation of uncertainty from which it originated. "The human action is purposeful behaviour", it "will put into operation and transformed into an agency"; and his *naturaliter* tendency to the pursuit of the aim "is the reacting to stimuli and the conditions of its environment, as is a person's conscious adjustment, to the state of the universe that determines his life" [28, p. 11]. Therefore: "the intentionality behind our acts is what turns them into actions" [29, p. 25].

The action, being always intentionally aiming to try and solve a problem of dissatisfaction, is also "by a definition always rational" [29, p. 35] from the perspective of the social actor. It is, in other words, the effort on behalf of the actor to come out of the state of discontent in which he is stuck, always trying to the best of his options. Let us assume that the action is the constant effort to go from the unpleasant to the more pleasant state, from a state "considered less important " to another "regarded more significant" as maintained by Mises [28, p. 41]. It then follows that it would be unthinkable that an individual acts intentionally and consciously in contrast to those that are his best reasons, hence choosing deliberately a solution inferior to those that he deems within his own possibilities in that particular circumstance.

It follows that such a Misesian solution allows to reveal the rationality principle as an *a priori* principle always true *a parte subjecti*. His tautological character, far from being sterile for the building of synthetic theories on the human action, as feared by Popper, reveals itself as an unavoidable methodological instrument for the *historical* sciences of the human action, to start with economics and sociology. It is, in fact, thanks to this *a priori* that is possible to explain even the rational actions *goal directed* with the evolutionist Popperian scheme, problems-theories-critics. If the action is intentionally and rationally, always meant to relieve or remove a state of discomfort, then it is essential to rebuild *a parte* actor such problematic situation, in order to clarify why the action was conceived by the single as the best solution.

The researcher will have to resolve the *rational calculus* from which any action originates; and, since every action is intrinsically rational, this calculus — that is pretty much always tacit — *will always be solvable given the availability of the essential situational information*. The *metaproblem* that the searcher has to set to himself, in order to understand the problem faced by the agent, hence the *metaconjecture* that he formulates upon the *situational conjecture* created by the latter to come out of the problematic situation in which he finds himself, are the instruments to solve this calculus, hence to trace the good reasons that induced the single to act the way he did; those reasons that, *ex definitione*, will be voluntarily and rationally *problem-oriented*. In applying, even for the account of the action, the evolutionistic scheme of *trial and error-elimination*, the methodological individualist, that cares about the construction of situational models that do not fall into the psychology and that depict an acceptable level of empirical content, has to transform, as illustrated by Popper, the elements of such calculus (objectives, preferences, information and relevant knowledge, etc.) in situational, empirically controllable, "objective" data [26, pp. 15–20].

6 The rationality of the actions which fail to achieve their goals

The character *a priori* of the rationality principle unravels to the root the problem of the supposed inexplicability of the actions that fail, in the whole or in part, to achieve their objective. R. Collingwood has put forward the objection against an individualistic method free of empathy with the following reasoning: if every action corresponds to the effort of cracking a problem, a social scientist can only explicate the action by reconstructing the problematic situation. "But the fact that we are able to single out his (of the single) problem proves that he has solved it; in fact we get to know what a certain problem was only by tracing back its origin from its solution" [10, p. 50]. It follows that an action that does not achieve its aim would represent, according to Collingwood, a mistaken solution that puts the researcher out of track, causing him to search a problem different from the one actually faced by the actor.

These conclusions reached by Collingwood, that are affected by the firm empathic method adopted by the idealist English philosopher, would bring to the elimination, from the horizon of the individualistic explanation, those same problematic situations that should be, instead, the first concern of the social scientist. Seen as an permanent attribute, hence incorporated in the same concept of action, the rationality principle allows to easily overcome these collingwoodian objections. Being the action always rational, even the actions *objectively* (for the researcher) inadequate to the situations will be rational. The problem will then be one of evaluating why the agent has had good reasons to believe the false, and this could be explained, as supported by Popper himself, by highlighting the divergence of the situation as perceived by the actor and the situation as observed by the researcher, by pointing out, in this way, in the decisional strategy of the single some cognitive deficiencies that caused him to fault. In other terms, *the individual behaves in a way appropriate to his improper perception of the situation in which he finds himself.* The latter, as known, is the approach utilised by authors as Spencer, Weber, all the way to Boudon, in order to offer a enlightening individualistic-rational account of phenomena, such as the one concerning magic and more generally embracing false beliefs, traditionally interpreted in collectivistic-irrational way.[1]

Moreover such Misesian solution permits, in this context, to underline all the limiting factors of the Paretian distinction between l *ogical actions* and *non-logical ones*. The logical (or rational) actions are, according to Pareto, those "operations that sensibly link the actions to the goal, not only with regard to the subject that executes them, but also concerning those that present a more comprehensive knowledge" [32, p. 150]. These are actions shaped by a calculus of means/ends *objectively and unanimously shareable*, and that consequently present a low level of problematic factors, as is the case, exemplifies Pareto, of the behaviour of those Greek sailors that used to row to make their ship move: the act of rowing is considered by Pareto as a logical action because it is directly linked to its objective (to sail the ship forward). On the other hand, those actions based upon a calculus means/ends only *objectively valid* (as, for instance, the offerings that those same Greek sailors used to make to Poseidon, God of the sea, before setting for sailing) are considered not logical (or irrational), hence not explainable through a logical-

---

[1] About the individualistic explanations of magic the reader should refer to Boudon ([4, pp. 34ff])

experimental approach as far as Pareto is concerned. They can be rationalised *ex post* by the use of *derivations* hence justified with misleading motivations that, as acknowledged, Pareto defines "logical paint" [32, p. 150].

Amongst the methodological individualists, the positivist Pareto is one of the most orthodox: he only and systematically applies a principle of rationality *a parte objecti*, making rationality coincide with the *optima reali* and pushing the *optima apparenti* back in the limbo of irrationality. In addition, he does not conceive such principle as a regulative ideal, as Popper does when he talks about "zero coordinate", but he puts forward a realist version, with a large discriminatory power, that has turned out to be methodologically weak, having hampered the application of the rationality principle *a parte subjecti* to that vast and very interesting territory (characterised by actions not met by success, false or unfounded beliefs, feelings, moral values), that escapes the use of the rationality principle *a parte objecti*. The renunciation to the employment of the principle of rationality, in such domains, is tantamount *ipso facto* to the renouncement of a good part of the explanatory power of the situational analysis, hence of the individualistic methodology.

7 Logical context of the action and the sociological context of the reasons.

If the second postulate of praxeology (every individual removes his own dissatisfaction in a way he deems best) leads to regarding the human action as being always rational, the third (every individual utilizes more effective means to eradicate his own discomfort) implies that every human action is also, by definition, always *latu sensu* economic. This is because the choice of the more suitable methods entails *ipso facto* selecting the cheapest ones. "The whole of the human action, Mises writes in *Human Action*, consists in the economising of the available means for the realisation of the chosen goals. The basic law of the action is represented by the economic principle. Every action is within its domain" [28, p. 86]. It follows that: the spheres of the rational action and of the economical action are coexisting [28, p. 153]. Concerning the human actions, which are necessarily rational and economical, Mises regards as "purely economical" those actions that are based upon a "calculus" expressed "in monetary terms" [28, p. 153]. It this particular class of actions (we could think in terms of the actions characterising the buyer, the entrepreneur, the saver) that, for Mises, must be subject of study of the economic science.

The third postulate of the Misesian theory of action presents quite a few problems. In fact, it seems to go against that tendency spread amongst several contemporary individualistic methodologists — starting from Boudon — who has decided to separate the individualistic method from the far too restricted notion of economical rationality, in order to enhance the explicative capacity, mainly in the ethical field and in the more wide-ranging one concerning beliefs. As it is known, Weber himself has conducted a real and deep-rooted division between *Zweckrationalität* and *Wertrationalität*, emphasizing as these two dissimilar dimensions of the human rationality could be reconciled with one another.

The *querelle*, if the methodological individualist must keep himself anchored to the economical dimension of rationality, or if, instead, it must take into consideration other forms of human rationality (Boudon has talked about *axiological rationality, cognitive rationality, traditional rationality*) [4, pp. 13ff], has given rise to an intense debate over the past decade. Moreover the querelle on the Rational Choice Model has been centred on the degree of extension of the subjective rationality, that authors such as J. Coleman [9, pp. 40ff] and G. Becker [3, pp. 35ff] consider as pretty much equivalent to the economical one, while others, for instance Boudon, attempted to demonstrate the explanatory restrictions of an orthodox conception of the their decisions.

Nevertheless, even in this case, what could appear to be a radical contradiction, born at the heart of the logic of the individualistic explanation itself, can be, if not completely broken up, drastically shrunk due to the distinction between the *logical context of the action and the sociological context of the reasons*. When Boudon differentiates several forms *of rationality*, he clearly *refers to the framework of the reasons*: such form corresponds to a typology of good reasons that, for the French sociologist, is as useful on the explanatory level as it is more articulated. According to Boudon, rationality is a "language's game" [5, p. 351] by means of which the researcher ascertains *good reasons* that are causally feasible to illustrate the action; of such subjective rationality, though, we can only give a "semantic definition" [4, p. 464]: "X has acted in a rational way, he had good reason to do Y, because". Given the existence of different kinds of good reasons that, thanks to the situational analysis, will permit to complete this definition, we will have various types of rationality (economical, axiologicall cognitive, etc.). But the construction of a typology of reasons, that will undoubtedly enhance the situational analysis, is made possible thanks to the fact that there is rationality inherent to the action. We can have various degree of extension of the rationality of the reasons (more restricted with Coleman, but wider with Boudon) because the postulates of praxeology hold, stating that the perspective of who carries out the action is always rational.

8 Beyond the contrast between values and interests

The distinction between *framework of the action* and *framework of the reasons* allows to settle the controversy on the nature of rationality and to understand the harsh critics directed by Mises to the classical economists, responsible to unduly have made the individual to coincide with "the phantom of the homo oeconomicus" [28, p. 44], by using "a fictitious image of a man only led by economic' motives", while instead "real men are influenced by several other 'non-economic' motives" [28, p. 63]. By considering the economic reasons as characteristics inherent to the action, classical economists had fallen, according to Mises, in the "more common misunderstanding" that consists in recognising, in the economic principle, a declaration on the matter and the content of the action [29, p. 53]; whereas the permanent attributes of the behaviour, to start with his economic nature, are to be seen in their "formal meaning" and are divested of any "material content" [28, p. 64]. Mostly the orthodox utilitarists, as J. Bentham, have confused the logical framework of the action and the sociological one of the reasons, as if they had discovered, in the strictly economic motivation (utilitaristic calculus), an immanent dimension on human behaviour that allowed to explain and plan every social context; whereas, instead, in its "formality", utility is, for Mises, a necessary attribute of the action. Considering a "human being exclusively determined by economic' motives", hence by the desire of "becoming as rich as possible", leads the Austrian economist to think about a "phantom" that "does not have and never did have a counterpart in reality" [28, p. 60]. "Everything that we say state about action — Mises admits [29, p. 34] — is independent from what causes it and from the goals towards which it strives in the individual case; it makes not difference whether actions spring from altruistic or from egoistic motives, from a noble or from a base disposition; whether it is directed towards the attainment of materialistic or idealistic ends" [28, p. 63].

By the light of such considerations it is possible to tackle with success another old problem: the contrast between interests and values. If every action is rational, and thus if any kind of holding of a belief is rational, then choices of value will also be rational, whose explanation is seen as the overcome testing ground that cannot be overcome by the individualistic methodology. Due to the praxeology it will hence be possible to avoid such confusion at the basis of this antiindividualistic controversy, between the *rational foundation* and the *rational content of values*. Values are not based on logic, but this does not imply they are irrational: they are *logically unutterable*, but they are not r *ationally unutterable*. Ethics is without truth, but not without reason. Values can be explaining through the individualistic viewpoint only to the extent that have a rationality content that can not be eliminated, rooting them back to the reasons behind individuals' choices. Also in this case the postulates of praxeology appear decisive, conferring a inalienable logical base for those attempts, starting with Boudon's one, that have subjected to the individualistic explanation also the field of axiological believes. On one side, they have also fought against the rationalism of those keen on explaining values by ascribing them to causes different from the reasons[2] and, on the other, against those regarding values as mere irrational intuitions that reason is unable to exert any surveillance on.[3]

9 The rational choice of reason

Equally to the other axiological options, also the choice of reason is not based upon any *fundamentum inconcussum* nor it is a purely irrational decision. Being a *choice*, sticking to the reason displays all the praxeological characteristics of the action: it is intentional, rational and "economical". The persona adopting such decisions has reasonable motives to do so, according to the expected consequence of such means.

---

[2] The "economic structure" according to Marx, the "sense of guilt" for Freud, the "feelings" of Pareto, the "biological order" for J.Q. Wilson.

[3] It is the case, obviously for Nietzsche, but also for several of the analytical English philosophers, that, as G.E. Moore, were intuitionists in ethics. It has to be noticed that, often, the non-rationalistic outcome in ethics and the relativistic one in epistemology is a perverse effect of foundationalism, that is the hopeless search for an Archimedean holding stance. W.W. Bartley III, has not hesitatated to support that "rationalism and irrationalism have in common the justificationism" [2, p. 249].

Therefore Popper is right in stating that reason does not self-justify itself nor self- assert, thus its choice has to necessarily be of a *critical* nature, linked to the consciousness that the acceptance of the rationality tools — mainly "logical argumentation" and "experience" [33, II, pp. 241ff] — is a decision logically groundless, that has got nothing of compelling to it, and that can also not be implemented. But given Mises praxeology, the conclusion Popper gets to, in talking about "irrational *faith* in *reason*", becomes unsustainable. Shifting, *de plano*, from the idea that the choice of reason can not be justified in a logical manner to the conclusion that it stems from an irrational choice, Popper appear not take into account this crucial distinction between rational *foundation* and rational content of a value option. Equally to other tools, reason is also chosen by individuals that have good reasons to do so, in the attempt to solve some problems. It is therefore feasible to agree with Rescher, in stating that "nothing forces' us to be rational but our rationality itself" [38, p. 49]. A argued by the American philosopher, it is certainly possible to decide not to make this choice, but the point is that "if we abandon reason, there is not a better place where we can (rationally) go" [38, p. 49].

10 The universal logic of the action

The *a priori* perspective of Mises allows, moreover, to provide an answer to what has been a decisive question for several social scientists: what do the social actor and the research have to have in common in order for the latter to be able to explain actions that could be also profoundly different culturally, geographically and temporally, amongst themselves? What are the features common to all individuals, without which it would be impossible to identify an order, a meaning, in the human behaviour? If, in other terms, account for an action, as supported by Robert Nozick, implies to favour a "reasoning for analogy" [31, p. 699], in some way, what do the observer and the person observed have to have of "analogous"?

The answers provided to this questions have often referred to a rather binding notions, quite frequently of tricky definition. Wilhelm Dilthey has supposed the existence of a "common trait existing amongst individuals, "arising in the identity of reason, in the sympathy of the emotional life, in the mutual obligation of duty and right, accompanied by the conscience of what needs to be" " [12, pp. 227–228]. Gadamer has illustrated the same concept talking about "uniquity of human genus" [16, p. 463]; while according to M. Hollis there is "an epistemological unity of the human race" [20, p. 320]. P.K. Feyerabend supports the existence of a "human nature similar for all individuals" [15, p. 10]. R. Boudon supports the thesis of "cognitive universal mechanisms" [7, p. 230] and Popper himself refers to a "rational unity of mankind" [33, II, p. 238].

The employment of these, rather engaging, notions by several influential individualists, is driven by a clear worrying motive: to demonstrate that *individuum non est ineffabile*. They have in fact, tried to reveal the existence of cognitive universals or cultural variants to save the individualistic explanations from a collectivistic relativism, that denies the possibility for an intercontextual (individualistic) account of social phenomena. Eventually it is possible to get to the extreme cases of those that have hypothesised the existence of an actual "poliogism", that is a logic of action and thought that is a function of the cultural environment, that is, as a matter of fact, one of the main critical target of Mises [28, pp. 72–88]. Even a philosopher not exactly adhering to the methodological individualism, as Peter Winch, apparently main support of that collectivistic interpretation of the "second Wittgenstein" from which the relativism of some post-Popperian epistemologists is not that unrelated, has stated that the way to avoid the collectivistic relativisms of "separate and incommensurable worlds" is to detect an universalistic ("constant factor") balance, that he deems as attainable from the solutions worked put by every community to face the universal problems inherent human behaviour, those related to birth, death, feeding, sex, etc.

What makes these anti-relativistic responses less challenging, Winch's proposal less convincing and, in general, all attempts to find universal elements linked to human nature, is the ambiguity of such notion. It is, in fact, a notion that is elusive and inevitably "cultural" hence evolving, given that fact that it has, historically, progressively changed content.

The weakness of these stances is the endeavour to try and find an antidote to the cultural ineffability at the cultural level, while it would be much more effective trying to define it at the ontological level to start with. It is foremost necessary to clear out the field from any collectivistic forms of ontology, stating, as done by Mises, that "only individuals exist" and consequently that for "a social community there is no existence or realities beyond the actions of the single members" [28, p. 41]. If social entities ("life forms", "cultural contexts", "economic structures", "cultural norms", etc.) ontologically autonomous from the individual behaviours existed, they would obviously follow operation logics independent from the actions of the single members, and the individualistic explanation would simple be not feasible.

But this philosophical assumptions pushes only as far as the singling out, in the human action, the object of study of social sciences; to leap from the *ontological individualism* to the *methodological individualism* it is essential to identify some features of the action that makes it accountable. To this regard the aprioristic prospective of Mises becomes vital, more than useful, providing the social scientist with crucial information on some permanent characteristics of the action as such. The intentionality, the rationality, the economical nature and the theological-causal character of the action are thus not *factual truths*, but rather *truths of reason*, that are true *ex definitione*: "what we know about the fundamental categories of acting", clearly outlines Mises, "does not depend on experience, but on reason" [28, p. 31]. "The aprioristic reasoning", typical of praxeology, is "purely conceptual and deductive. It can not produce anything apart from tautologies and analytical judgments. All of its implications logically arise from the initial assertions and were already contained in it" [28, pp. 36–37]. The outcomes of praxeology are " *a priori* logical and mathematical truth without referring to any experience", they have, therefore, the statute of "apodictic certainties" [28, p. 41]; it follows, as Mises concludes, that "the term 'rational action' is pleonastic and as such it has to be rejected" [28, p. 18].

However the establishment of tautologies does not imply, as mistakenly assumed, that praxeology does not let our knowledge to advance, also because "all geometrical theorems are implicit in their axioms. The concept of the rectangular triangle already implies Pythagoras's theorem [..]. Yet none would object that geometry in general, and Pythagoras's theorem in particular, do not broaden our knowledge" [28, p. 37].

Making clear those permanent peculiarities of the human action that would have been otherwise unknown to the social scientist, Mises' aprioristic reasoning identifies the logic form of individual behaviour, providing social sciences with very precious information, that have the task of capturing the *unintentional empirical content* that progressively fills this form of *universal logic*.[4]

Therefore praxeology allows to answer the anti-relativistic qualms of individualists, highlighting how human action is characterised by a unitary logical form, independently from the cultural context, without which "we would not be able to discern in events nothing but a kaleidoscopic change and chaotic confusion" [28, p. 31]. The *homo sapiens* is by his own nature a *homo agens*, that, according to the different cases, becomes *homo oeconomicus, homo sociologicus, homo ethicus, homo religiosus*, etc. By singling out some unchanging peculiarities of the object of study of social sciences (human actions), praxeology allows to fight not only, as illustrated, the ethical irrationalism, but also the epistemological relativism, that is sometimes the consequence of collectivism of a cultural kind.

Goodman and Quine have addressed philosophers of mathematics with a crucial query: "how much mathematics is possible to construct by only using an ontological apparatus of individuals" [18, p. 276]. The same question can be passed onto social scientists: "how much social science is feasible won the basis of the ontological individualism?". It is possible to argue that due to praxeology a rather significant outcome can be achieved: to produce as much social science as possible by reducing to the minimum the ontological engagements.

BIBLIOGRAPHY

[1] J. Agassi. Methodological Individualism. *British Journal of Sociology*, 11: 224–70, 1960.
[2] W.W. Bartley III. *The Retreat to Commitment*. Open Court, la Salle 1984.
[3] G. Becker. *Accounting for Testes*. Harvard University Press, Cambridge and London 1996.
[4] R. Boudon. *L'art de se persuader des idées douteuses, fragiles ou fausses*. Fayard, Paris 1990.
[5] R. Boudon. *Le juste et le vrai*. Fayard, Paris 1995.
[6] A. Bouvier. *Philosophie des sciences sociales*. PUF, Paris 1999.
[7] R. Boudon. Razionalità e religione. *Biblioteca della Libertà*, 158: 3–33, 2001.
[8] A. Boyer. *L'explication en histoire*. Presses Universitaires de Lille, Lille 1992.
[9] J. Coleman. *The Foundations of Social Theory*. Harvard University Press, Cambridge and London 1990.
[10] R.G. Colligwood. *An Autobiography*. Routledge, London 1944.

---

[4]F. Machlup, that has been a student and then the assistant of Mises, has rightly stated that "his [Mises] basic theories a priori were linked to a crucial crierion, the way of applying them requires an empirical judgement. Some critics of apriorism, Machlup underlines, have been rather unfair when the have failed to see how, although having constructed a model a priori, an empirical judgement was nevertheless needed to guide us throughout its application" [27, p. 9].

[11] P. Demeulenaere. *Normativité et rationalité dans l'analyse sociologique de l'action*. In R. Boudon, P. Demeulenaere and R. Viale (eds.), *L'explication des croyances*, PUF, Paris 2001.
[12] E. Dilthey. *Der Aufbau der Geschichtlichen Welt in der Geisteswissenschaften*, Parte III: Allgmeine Sätze Über den Zusammenhang der Geisteswissenschaften, Abhandlungen der königlich Preussischen Akademie der Wissenschaften , 1910. In *Gesammelte Schriften*, Leipzig und Berlin, Teubner, vol. VII, 1927, pp. 49–123.
[13] E. Di Nuoscio. *Le ragioni degli individui*. Rubbettino, Messina 1996.
[14] A. Faludi. Why in planning the myth of the framework Is anything but that. *Philosophy of Social Sciences* 28: 381–399, 1998.
[15] P.K. Feyerabend. Contro l'ineffabilità culturale. *Il Mondo* 3 (2): 6–11, 1994.
[16] H.-G. Gadamer. *Whrheit und Methode*. Mohr, Tübingen 1960.
[17] L.A. Gerard-Varet and J.L. Passeron (eds.). *Le modèle et l'enquete. Les usage du principe de rationalité dans les sciences sociales*. Editions de l'Ecole des Hautes Etudes en Sciences Sociales, Paris 1995.
[18] N. Goodman and W.V.O. Quine. Steps toward a constructive nominalism. *Journal of Symbolic Logic*, 12, 1947.
[19] P. Hedström, R. Swedberg and L. Udéhn L. Popper's situational analysis and contemporary sociology. *Philosophy of the Social Sciences*, 28: 339–364, 1998.
[20] M. Hollis. *The Epistemological Unity of Manking*. In S.C. Brown (ed.), *Philosophical Disputes in the Social Sciences*, Harvester Press, Brighton 1979.
[21] I.C. Jarvie. *Concepts and Society*. Routledge and Kegan Paul, London 1972.
[22] I.C. Jarvie. Situational logic and its receptions. *Philosophy of the Social Sciences* 28: 365–380, 1998.
[23] N. Koertge. Popper's methaphisical program for the human sciences. *Inquiry*, 18: 437–462, 1975.
[24] N. Koertge. The methodological status of Popper's rationality principle. *Theory and Decision*, 10: 251, 1979.
[25] L. Langeux. Popper and the rationality principle. *Philosophy of the Social Sciences*, 23: 468–480, 1993.
[26] M.T. Leeson and P.J. Was Mises Right? *Review of Social Economy*, 64 (2) 2, 2006.
[27] F. Machlup. *Tribute to Mises 1881–1973. The session of the Mont Pelerin Society at Brussels on 13th September 1974*. Quadrangle Publications, Kent 1974.
[28] L. von Mises. *Human Action: a Treatise on Economics*. Yale University Press, New Haven 1949.
[29] L. von Mises. *Epistemological Problem of Economics*. New York University Press, New York and London 1976.
[30] R. Nadeau. Confuting Popper on the rationality principle. *Philosophy of the Social Sciences* 23:446–467, 1993.
[31] R. Nozick. *The Nature of Rationality*. Princeton University Press, Princeton 1993.
[32] V. Pareto. *Trattato di sociologia generale*. Edizioni di Comunità, Milano 1916.
[33] K.R. Popper. *The Open Society and Its Enemies*. Routledge, London 1945.
[34] K.R. Popper. *The Poverty of Historicism*. Routledge, London 1957.
[35] K.R. Popper. *Objective Knowledge. An Evolutionary Approach*. The Clarendon Press, Oxford 1972.
[36] K.R. Popper. *Intellectual Autobiography*. In P.A. Schillp (ed.), *The Philosophy of Karl Popper*, Open Court, La Salle 1974.
[37] K.R. Popper. *Selections of Writings*. Princeton University Press, New York 1985.
[38] N. Rescher. *Rationality: a Philosophical Inquiry into the Nature and the Rationale of Reason*. Oxford University Press, New York 1988.
[39] J. Wellersten. Styles of Rationality. *Philosophy of the Social Sciences*, 25: 69–98, 1995.

Evolution, cooperation and rationality: some remarks

ALBERTINA OLIVERIO

This article reviews a series of issues that have been dealt with in recent years and that, taken together, provide a consistent picture of the possible interactions between biological and cultural evolutionism, cooperation and rationality in the social sciences. It focuses in particular on what can be seen as one of the central themes of social inquiry: explaining the emergence of social cooperation. Social theory offers three main strategies for explaining the emergence of cooperation: rational choice theory, based on the so-called *homo economicus* model; theories featuring a collectivist methodology, based on *homo sociologicus*, which we disregard here; and evolutionary theory, based on biological and cultural evolution.

As we know, the theory of rational choice is the dominant framework in the social sciences for explaining the rise of cooperation and the observance of cooperative social norms. To simplify, we can say that conventional rational choice theory (the expected-utility variant, [68]) explains the existence of cooperation postulating the main characteristics of "economic man" — instrumental or normative or economic rationality (which can be reduced to a cost/benefit calculation with perfect information on the alternative choices and their consequences, including their probability of occurrence in the case of risk or uncertainty) and the ranking of preferences according to personal self-interest (optimization, or the maximization of expected utility).

From a theoretical standpoint, the hypotheses of cost/benefit calculus and utility maximization imply an explanation of the genesis of social cooperation via three main mechanisms: coordination through conventions or norms, sanctions, and repeated games. Going by the first of these mechanisms, the emergence and persistence of social cooperation is explained in the context of coordination games (see among others: [44, 67, 18, 48, 40, 4, 5, 70, 71]). The basic hypothesis is that so-called norm-guided behavior is rational in a "folk-economic" sense, in that the individual choice to comply with a norm or convention or not depends on a cost/benefit calculus driven by preference and self-interest in abiding by that norm. Norms, that is, act on the preferences (utility) of individuals. In this sense it is maintained that the social function of norms is solving problems of coordination in collective action.

The classical example is choosing which side of the road to drive on, a choice that can result in a stable outcome either by explicit pact or by spontaneous convergence on a "focal point" [56], i.e. something that is "common knowledge," a norm or convention that is diffused throughout the society, from which no one has an interest in deviating and in whose existence everyone has an interest [44].

Applying the sanctions mechanism, cooperation is explained by punishment for non-cooperative behavioral choices [39, 18, 40]. The assumption is that there is an endogenous sanctioning machinery that punishes cheating — free-riding — or, alternatively, that rewards those who do cooperate. Thus the individual who is threatened or affected by the sanction is prompted to modify his behavior, acquiring an interest in avoiding the sanction or its repetition. It is worth noting that the sanction/incentive mechanism may remain at a totally "virtual" level and act as a force of persuasion or deterrence via individual psychological dynamics working on the desire to be or to appear honest, to respect others, to show trust and loyalty, to avoid guilt feelings, shame or social disapprobation [30, 62]. This also allows us to explain the fact of compliance with norms even in the absence of external controls or sanctions (such behavior as not littering even when no one can see you, [22, 8].

In the repeated game framework, the emergence and continued existence of cooperation is explained by positing that the possibility of repeating a game shapes individual choices. When a number of repetitions are possible, ones choices are no longer limited to one of just two strategies (cooperation or cheating) but now extends to other, longer-term strategic alternatives. Specifically, it is hypothesized that the repetition of a game creates a set of implicit norms that stimulate players to follow long-term strategies that tend to be cooperative, because this produces an advantage for all the players.

The classical example here is the repeated prisoner dilemma. In this game, cooperation is one possible equilibrium solution, sustained by the "tit-for-tat" strategy [1, 2], i.e. reciprocity. The simple tit-for-tat schema, briefly, consists of two precepts: cooperate in the first round, and afterwards do what the adversary does. Historically, a number of real-world episodes of cooperative conduct in the trenches during World War I followed this pattern, such as that of not firing at mealtimes. Since the players know that in the long run cooperation is more advantageous for all, they have an interest in reciprocal non-betrayal. But it is precisely when a cooperative climate sets in that one of the two may imagine that cheating can be a winning strategy. At this point, the other responds with an equally non-cooperative attitude, until a cooperative climate is restored.

When rational choice theory resorts to this mechanism to explain the genesis of a cooperative social order, it is referring to evolutionary game theory [47, 4, 5, 7, 59, 3, 69, 71]. In an evolutionary approach, the argument goes, the most successful behavior tends to be dominant (that is, it is a strategy that is useful to the survival of the gene or the species). It is important to recall that if a behavior becomes dominant it is not solely because it has been selected over unsuccessful behaviors but also because social actors imitate and, more importantly, learn successful behaviors.

The key features of *homo economicus* and thus of rational choice theory — instrumental rationality and self-interest — have been widely criticized. There is very substantial experimental evidence that most individuals do not conform to instrumental rationality or to the theorys routine assumptions concerning preferences (see, among many, [43]), so much so that it has been called into question even within the discipline of economics itself [63, 49]. There have also been a good many critiques of the hypothesis of self-interest as the sole motive for action, instead highlighting such motives as love of others or sense of duty, or instincts — like hunger — that may conflict with self-interest and hence with the folk-economic sense of rationality. An instance might be the taste for sweet, fatty foods, which evolved in parallel with modern man during the Pleistocene Era (from 1.8 million to 11,000 years ago), when hunger was a threat to survival and eating the most nourishing food available was therefore essential. Today, though, we know that if consumed in large amounts these foods can be unhealthy, so the instinct to eat them often conflicts dangerously with self-interest, i.e. with the physical well-being of the individual.

To bring out the limitations of rational choice theory, let us consider situations in which individuals act cooperatively under "strong reciprocity," engaging with others both to punish violations of the norms (altruistic punishment) and to reward compliance (altruistic reward), even when this has a cost but brings no direct benefit and thus contrasts with personal self-interest in the economic sense [10, 29, 34, 27, 25, 26, 31, 12, 9, 35]. Some experiments in this area have shown that when people are treated with generosity or friendliness by strangers, they tend to respond in like fashion, even at a cost [26]. Further, comparable experiments have also shown that people generally respond in kind to an ungenerous or hostile act, even when it brings no advantage [27]. The results of these experiments show how cooperative habits can be maintained through reciprocity.

All these types of behavior are based on reciprocity, a notion that differs from that of cooperation or cheating in the framework of repeated interactions. The latter is motivated exclusively by future gain or advantage, whereas in the case of reciprocity individuals respond in kind to friendly/generous or to hostile/selfish behavior when they can expect no gain or benefit, and indeed even when they must sustain a cost. The framework is one of one-shot games (non-repeated interactions).

The example that best highlights the dynamics underlying reciprocal behavior is the "ultimatum" game. Two players have to divide a sum of money, say ten dollars. The Proponent must lay down an ultimatum in the form of a proposal to the Counterparty, who can accept or refuse. If he accepts, the money is divided as proposed; if he refuses, neither player gets anything. Under rational choice, we would expect an equilibrium corresponding to a division of 9.99 to the Proponent and 0.01 to the Counterparty; or, if the amount is in one-dollar bills, nine dollars and one dollar. Actually, however, the empirical evidence on this game, from experiments in a number of different countries, is that the most common outcome is a division of five and five or six and four. Further, when the Counterparty is offered less than 30 percent of the total to be divided, the probability that the proposal will be rejected is extremely high; and even quite substantial increases in the amount of money at stake have at most only a marginal effect on these experimental outcomes [53, 13, 52, 60, 14, 41, 42, 35].

Most people, then, play the ultimatum game according to norms of reciprocity, honesty and fairness both in making proposals and in punishing offers that they consider unfair. Such conduct, which does not tend to maximize expected monetary gain, violate the theory of rational choice. This kind of empirical evidence, that is, implies a model in which people are not motivated exclusively by their own self-interest but also by norms of reciprocity. Recent studies in the neurobiology of cooperation and sanctions corroborate such model [50, 55]. Magnetic resonance imaging of the brain activity of Counterparties in the ultimatum game has shown that the most unfair offers stimulate particular areas of the brain, for instance the frontal insula, connected with such negative emotions as discomfort and disgust. So the more these areas are stimulated — the greater the activity of their neurons, the more an innate sense of justice will prevail over a strictly utilitarian calculus, and the likelier it is that the unjust proposal will be rejected.

There is a lack of consensus among experimental researchers on the main sources of reciprocal and cooperative behavior. Referring to evolutionary theory, we can try to trace their origins. It is plausible that they are the fruit of a sort of "evolutionary rationality" whose epistemology differs from that of instrumental rationality. As recent research has shown, reciprocity seems to have been fundamental to the evolution of human cooperation; it can be hypothesized that reciprocity and cooperation are the fruit of biological and cultural co-evolution [12, 23, 9, 28]. The evolutionary interpretation, then, means trying to understand how the general psychological capacities for adopting and complying with some forms of reciprocity or specific norms of cooperation came to evolve, and how they have been maintained through imitation and learning.

At this point one may reformulate the initial question on the emergence of cooperation to ask how cooperation is possible between individuals who are not genetically related — cooperation which in human society depends mainly on social norms. Human society, in fact, can be considered as a "company of strangers" [57]. In city life, for instance, most of the individuals who are essential to our survival (providers of food, clothing, health care, defence, etc.)are not relatives but actually total strangers. This large-scale dependency on others is based on cooperation and the sharing of tasks; interaction of this kind between members of a species without genetic correlation is an exclusively human phenomenon. For while human beings share a large number of genes, this does not mean there are no differences depending also on socalled polymorphism, i.e. the different forms in which the same gene may occur. The genomes of human beings and chimpanzees, recently sequenced [64], are 98 percent similar. But this resemblance does not necessarily carry the implications that are emphasized in arguing that the two species answer to a common genetic program and so share most of their phenotypes [45], including cerebral ones. The fact is that even extreme similarity between coding genes means little, in that not all the genes are expressed, those that are expressed are not always expressed, and they are not expressed simultaneously in the same quantities [21].

As a consequence, there are quite substantial differences within species. Species, in fact, are characterized by great variability, and as Darwin argued this is what the mechanism of natural selection depends on. Individual variability is the reservoir that evolution taps to foster more or less sudden changes: the brains of different groups and species of animals are marked by notable individual diversification of structure and behavior. This individual diversification and variability in the structure of the nervous system — true cerebral polymorphism, i.e. differing extension and qualitative characteristics of some nuclei or nerve structures, differences in levels of nerve mediators, etc. — indicates that variation has played an important role in the evolution of behavior, both as regards such rigid processes as instinct and as regards more elastic processes such as memory, learning, and cultural transmission.

At this point, allow me a brief digression. Forms of cooperation are also found in other animal species, although only where there is a very close genetic relationship (everyone knows about the rigid cooperative behavior of such social insects as bees and ants, which have a very high percentage of genes in common). According to the theory of kin selection [46, 37], cooperation among genetically related individuals is favoured by natural selection.

The theory of "reciprocal altruism" [66] highlights the fact that there are also animal species featuring forms of cooperation between individuals that are not genetically related, in highly specific tasks essential to the survival of the species, even though this may seem to be contradictory, insofar as according to the logic of evolution altruistic individuals should succumb. In nature, instances of this sort are numerous; sometimes female lions, for example, even when they are not genetically related, pool their cubs and split the time devoted to hunting and that devoted to defence and nutrition of the cubs. This feline capacity is sometimes observed in pet cats as well. The same occurs with mares and colts (ethologists sometimes speak of "aunts").

There are even forms of reciprocal cooperation between different species (symbiosis). The most famous instance is the symbiosis between the Egyptian plover and the crocodiles of the Nile. The crocodile even keeps its jaws wide open for the bird to carry out its work of ridding its teeth of residues of food and parasites. For the plover, the relationship is not just a guaranteed source of food but also a shield against any imprudent would-be predators.

In all these cases, however, cooperative behavior occurs in very small groups; and qualitatively, it is radically different from human cooperation, which is based on the ability to establish and maintain social norms and takes highly complex forms. It is likely, therefore, that original instincts such as reciprocity, which played a marginal role in the evolution of forms of cooperation among animals, have instead been central to human cooperation [23, 38, 61, 24].

To see why the theory of evolution applies, we must consider the characteristics of the environment in which human beings evolved. The relevant period is the Pleistocene, when the selective pressure for some types of behavior (such as reciprocity) and for some preferences apparently favoured reproductive success in that environment. Modern humans have probably kept some of those preferences and can adopt some of those behaviors.

In any case, the period since the dawn of agriculture (some 10,000 years ago) is too brief to have caused significant genetic, evolutionary behavioral changes. That is, the genetic behavior of *Homo sapiens sapiens* cannot have adapted to the new social environment [15, 16, 17]. So while some behaviors did evolve genetically in the Pleistocene Era, others were transmitted or evolved culturally as a result of observed advantages. In this sense culture, hence cooperative norms, can be seen as a highly efficient adaptive mechanism.

Now, natural selection favoured genetic mutations that encouraged helping and reciprocal behavior with respect to genetically related individuals. But cross-cultural empirical evidence indicates that reciprocal behavior vis-à-vis strangers is also common within human nature. It is plausible to assume, then, that the forebears of *Homo sapiens sapiens* had developed trust and engaged in cooperative behavior with known non-relatives before moving on the strangers by imitation and learning. If you expect to meet a person again, you have more of an incentive to abide by agreements, so as to ensure cooperation in the future, as long as it is not too far off in time [57].

The evolution of cooperative behavior, in other words, can be regarded as a sort of "cultural explosion", a convergence of social interactions, diversifications, division of roles, cooperation, which caused a progressive cultural acceleration that far outpaced biological evolution. The constant progress of the cultural capacities of the genus *Homo* has been explained by some scholars as correlated with the progressive increase in the size of the brain [65]. The correlation between the two phenomena — brain growth and cultural progress — is certainly suggestive, but this thesis probably reflects a view of evolution in which the tendency is towards constant improvement. Such an interpretation takes no account of the fact that three million years ago the brain already had quite nearly human characteristics and that the brain of *Homo erectus*, to say nothing of Neanderthal Man, resembled that of *Homo sapiens* even more closely. This means that one can only explain the extraordinary cultural development of the last 10,000 years as the result of the attainment of a critical social mass, not of changes in the brain; the timeline of evolution is measured in not in thousands but in hundreds of thousands of years.

The close relationship that has come to be created between biological evolution and cultural evolution is such that culture is now an increasingly decisive factor in biology: medicine, environmental intervention, resource exploitation, the use of medicines, the production of polluting substances are all factors which, for good or ill, modify the environments selective pressures and introduce new, not necessarily desirable ones. It is quite clear, in short, that in the last few thousand years men have adapted the environment to their own genetic traits, not the other way around. For excepting a small number of genes subjected to unusually strong selective pressure in the last 10,000 years — such as the gene for tolerance of lactose (the ability to digest milk) in adults; northern Europeans have undergone a series of genetic changes because it was advantageous for evolution to be able to assimilate dairy products — the time frame is just too short, as we have said, for cultural changes to have evolved genetically.

In order to understand the evolution of behavior, therefore, we must consider the sphere of evolutionary adaptation, and here an important contribution comes from evolutionary biology, in that reciprocity and cooperation between strangers have helped to give rise to a physiological and a psychological structure that have evolved to deal with a set of challenges: problems faced at first by small bands of hunter-gatherers who devised forms of cooperation for hunting and defence. These challenges prompted changes whose cultural potential was enormous (though minimal genetically), changes that made human beings capable of abstract and symbolic thought to fabricate tools to serve specific bodily and communication needs. The earliest evidence of these changes are the cave paintings in Lescaux or Altamira, or burial objects and man-made implements of Cro-Magnon Man, 60,000 to 70,000 years ago. These cultural mutations, which at first enabled small bands to organize their hunting and defence activities, appear to have then made possible both the movement towards agriculture and settlement and the erection of social norms and habits of cooperation to curb violent instincts, thus paving the way to a larger, regulated society and the accumulation of knowledge.

Reciprocity and cooperation may well have been advantageous from the evolutionary point of view by enhancing Darwinian "fitness". They gave rise to trends that are favoured by natural selection, chiefly specialization, risk-sharing and the accumulation of knowledge [57]. It is reasonable to imagine that specialization began to spread among hunter-gatherers when cooperation between unrelated individuals first began. For while specialization means an increase in risk (less adaptability when environmental conditions change, as for pandas, which can survive only if bamboo shoots are available), the risk can be reduced thanks to the increased security and risk sharing that are possible in larger, richer communities.

Among the forebears of *Homo sapiens sapiens*, then, the larger groups had better chances of obtaining greater advantages in hunting and defence, which explains the evolution of cooperation among unrelated individuals. The evolution of cooperative norms can thus be explained by the plausible hypothesis that the groups or societies in which such norms are present are more likely to prevail in expansion, conquest or migration than those that lack them [71]. And this stimulates the groups or societies that do not have cooperative rules to emulate, to learn the practices of the more successful societies founded upon cooperation [51, 11, 24].

Cooperation, that is, can be seen as the product of biological and cultural evolution, i.e. the capacity for rational calculation of its relative costs and benefits, and of a tendency to reciprocity even when calculation would advise against it [42, 57, 6, 35]. It is likely that natural selection favoured the evolution of an equilibrium between the two inclinations for the development of social life, such that it was reasonable to treat strangers as family relatives.

Let us go back to the sanctioning of other peoples transgressions (which, via introjection, is also a curb on the actions of those who impose the sanction). Rational choice theory sees this in terms of the individual self-interest in avoiding the costs of the sanctions for failure to comply with the rules for cooperative behavior. Punitive behavior is not a specifically human prerogative but is common in other species, and not just primates. From the evolutionary standpoint, we can read these behaviors as designed to protect the interests of the species (the "gene pool"). Among humans, punishment has the purpose of fostering and preserving cooperative behavior, which depends on a series of moral norms, essentially bound up with culture. This explains why there exist forms of punishment that are "altruistic" in the sense that they entail a personal cost to those who execute them but protect the cooperative interests of the community. Cooperative behavior is useful in making and enforcing decisions that have an impact on the community. Nevertheless, we know that together with cooperation there also arise violations of the rules, which bring advantage to the individual but damage the community. A common deterrent is punishing non-cooperative behavior that benefits only the perpetrator. In the long run this pays, if the person meting out the punishment has to interact again with the one being punished, but it has a cost if this does not occur. Altruistic punishment is widespread in many cultures and often involves the witnesses to the violation of a norm, who feel authorized to punish the violator. In all these cases the ultimate aim is to foster cooperation, even if for purposes of evolution these behaviors are not evident to the person who practices or is subjected to them [10, 12, 9].

From the neurological point of view, punitive behavior can be considered in a variety of ways, ranging from the neurobiological bases of motivation to the study of such behavior patterns as learning and the selection of actions that affect other people. When a person notes improper behavior or perceives, to his own mind, a situation in which others behave improperly, this activates a nervous area located deep in the cortex, between the parietal and the temporal lobes, the insula. This region of the cortex is a functional part of the limbic system (it is connected to the amygdala, and in the course of emotional responses it induces the activation of the vegetative nervous system and of the "somatic markers" that make a person aware of emotional dynamics [20]. Subsequently the orbitofrontal cortex is activated, strictly associated with the amygdala (which makes motivational assessments and thus judges whether an act is advantageous and satisfying). Magnetic resonance imaging has shown that in addition to these structures that are involved in the emotional and motivational component of a judgment, the striatum, which is involved in the representation of punitive actions, is also activated [58].

From the evolutionary viewpoint, then, we can offer a rational explanation (under evolutionary rationality) for individual behaviors based on cooperation and reciprocity that conflict with ones personal self-interest and that violate the core of the notion of instrumental rationality. To understand why so many human behavior patterns violate instrumental rationality, the background in which our behavioral, evaluative and decision-making capacities evolved is crucial. Some studies on evaluation and choice take account of highly evolved human abilities and look to heuristics as adaptive strategies in complex environments that evolved simultaneously with fundamental psychological mechanisms and that assisted survival and reproduction [19, 32, 33, 36, 54]. An example is the recognition heuristic, a process based on the hypothesis that something known must be classed higher than something unknown. This would make it plausible and reasonable, for instance, that a known food is very likely to be edible; that strategy must have been very widely employed in the course of hundreds of thousands of years of human evolution.

The models based on rational choice theory are not models that individuals actually use in making decisions, which explains why they are so frequently disconfirmed by experimental evidence. More "frugal" mechanisms may be considered as near optimality in the environment in which they are employed.

Essentially, our emotional and cooperative responses to the events of life are the fruit of a biological evolution that has endowed our minds with a series of adaptive strategies, each appropriate to particular types of objective and areas of adaptation, and of a cultural evolution that has made these strategies more powerful, triggering a virtuous circle.

BIBLIOGRAPHY

[1] P. Axelrod and P. J.Hamilton. The evolution of cooperation. *Science*, 211: 1390–1396, 1981.
[2] R. Axelrod. *The evolution of cooperation*. Basic Books, New York 1984.
[3] J. Bendor and P. Swistak. The evolution of norms. *The American Journal of Sociology*, 106: 1493–1545, 2001.
[4] C. Bicchieri.. *Rationality and Coordination*. Cambridge University Press, New York 1993.
[5] C. Bicchieri. *The Grammar of Society: The Nature and Dynamics of Social Norms*. Cambridge University Press, New York 2006.
[6] K. Binmore. *Natural Justice*. Oxford University Press, Oxford 2005.
[7] K. Binmore and L. Samuelson. An economists perspective on the evolution of norms. *Journal of Institutional and Theoretical Economics*, 150: 45–63, 1994.
[8] R. Boudon. *Le juste et le vrai: études sur lobjectivité des valeurs et de la connaissance*. Fayard, Paris 1995.
[9] S. Bowles and H. Gintis. The evolution of strong reciprocity: cooperation in heterogeneous populations. *Theoretical Population Biology*, 65: 17–28, 2004.
[10] R. Boyd and P. J. Richerson. Punishment allows the evolution of cooperation (or anything else) in sizable groups. *Ethology and Sociobiology*, 13: 171–195, 1992.
[11] Boyd R. and P. J. Richerson. Group beneficial norms can spread rapidly in a structured population. *Journal of Theoretical Biology*, 215: 287–296, 2002.
[12] R. Boyd, H. Gintis, S. Bowles and P. J. Richerson. The evolution of altruistic punishment. *Proceedings of the National Academy of Sciences (USA)*, 100: 3531–3535, 2003.
[13] C. Camerer and R. Thaler. Anomalies: Ultimatums, dictators and manners. *Journal of Economic Perspectives*, 9: 209–19, 1995.
[14] L.A. Cameron. Raising the stakes in the ultimatum game: Experimental evidence from Indonesia. *Economic Inquiry*, 37: 47–59, 1999.
[15] L.L. Cavalli-Sforza and M. Feldman. *Cultural Transmission and Evolution*. Princeton University Press, Princeton 1981.
[16] L.L. Cavalli-Sforza. *Geni, Popoli e Lingue*, Adelphi, Milano, 1996. En. Tr.: *Genes, Peoples and Languages*, North Point Press, New York 2000.
[17] L.L. Cavalli-Sforza. *Levoluzione della cultura*. Codice Edizioni, Torino 2004.
[18] J.S. Coleman. *Foundations of social theory*. The Belknap Press of Harvard University Press, Cambridge, MA 1990.
[19] L. Cosmides and J. Tooby. Cognitive Adaptations for Social Exchange. In J. Barkow, L. Cosmides and J. Tooby (eds.), *The Adapted Mind*, Oxford University Press, New York 1992.
[20] A.R. Damasio. *Decartes error. Emotion, reason and the human brain*. G.P. Putnam, New York 1994.
[21] E. Di Mauro. Radical reductionism. *Journal of Anthropological Sciences*, 83: 133–135, 2005.
[22] J. Elster. Social norms and economic theory. *Journal of Economic Perspectives*, 3: 99–117, 1989.
[23] E. Fehr and J. Henrich. Is Strong Reciprocity a Maladaptation? On the Evolutionary Foundations of Human Altruism. In P. Hammerstein (Ed.), *The Genetic and Cultural Evolution of Cooperation*. MIT Press, Cambridge, MA 2003.
[24] E. Fehr and U. Fischbacher. Social norms and human cooperation. *Trends in Cognitive Sciences*, 8: 185–190, 2004.
[25] E. Fehr, U. Fischbacher and S. Gächter. Strong reciprocity, human cooperation, and the enforcement of social norms. *Human Nature*, 13: 1–25, 2002.
[26] E. Fehr and S. Gächter. Fairness and retaliation: the economics of reciprocity. *The Journal of Economic Perspectives*, 14: 159–181, 2000.
[27] E. Fehr and S. Gächter. Cooperation and punishment in public goods experiments. *American Economic Review*, 90: 980–94, 2000.

[28] E. Fehr and S. Gächter. Altruistic punishment in humans. *Nature*, 415: 137–140, 2002.
[29] E. Fehr and K. M. Schmidt. A Theory of fairness, competition, and cooperation. *Quarterly Journal of Economics*, 144: 817–868, 1999..
[30] C. Fershtman and Y. Weiss. Why do we care what others think about us? In A. Ben-Ner and L. Putterman (eds.), *Economics, Values and Organization*, Cambridge University Press, New York 1988.
[31] E. Fehr, U. Fischbacher and S. Gächter. Strong reciprocity, human cooperation, and the enforcement of social norms. *Human Nature*, 13: 1–25, 2002.
[32] G. Gigerenzer. Ecological intelligence: an adaptation for frequencies. In D. Dallarosa Cummings and C. Allen (eds.), *The evolution of mind*, Oxford University Press, New York 1998.
[33] G. Gigerenzer. and P.M. Todd. Fast and frugal heuristics: the adaptive toolbox. In G. Gigerenzer, P.M. Todd and the ABC Research Group, *Simple heuristics that make us smart*, Oxford University Press, New York 1999.
[34] H. Gintis. Strong reciprocity and human sociality. *Journal of Theoretical Biology*, 206: 169–179, 2000.
[35] H. Gintis, S. Bowles, R.T. Boyd and E. Fehr (eds.). *Moral Sentiments and Material Interests: The Foundations of Cooperation in Economic Life*. MIT Press, Cambridge, MA 2005.
[36] D.G. Goldstein and G. Gigerenzer. Models of ecological rationality: the recognition heuristic. *Psychological Review*, 109: 75–90, 2002.
[37] W.D. Hamilton. The genetical evolution of social behaviour II. *Journal of Theoretical Biology*, 7: 17–52, 1964.
[38] P. Hammerstein. Why is reciprocity so rare in social animals? A protestant appeal. In P. Hammerstein (ed.), *The Genetic and Cultural Evolution of Cooperation*, MIT Press, Cambridge, MA 2003.
[39] M. Hechter. *Principles of Group Solidarity*. University of California Press, Berkeley e Los Angeles 1987.
[40] M. Hechter and K.-D. Opp (eds.). *Social Norms*. Russell Sage Foundation, New York 2001.
[41] J. Henrich, R. Boyd, S. Bowles, C. Camerer, E. Fehr, H. Gintis and R. McElreath. In search of homo economicus — Behavioral experiments in 15 small scale societies. *American Economic Review*, 91: 73–78, 2001.
[42] J. Henrich, R. Boyd, S. Bowles, C. Camerer, E. Fehr and H. Gintis (eds.). *Foundations of Human Sociality: Economic Experiments and Ethnographic Evidence from Fifteen Small-Scale Societies*. Oxford University Press, New York 2004.
[43] D. Kahneman, P. Slovic and A. Tversky (eds.). *Judgment under uncertainty: heuristics and biases*. Cambridge University Press, New York 1982.
[44] D. Lewis. *Convention: A Philosophical Study*. Harvard University Press, Cambridge, MA 1969.
[45] J. Marks. *What it means to be 98% chimpanzee*. University of California Press, Berkeley 2002.
[46] J. Maynard Smith. Group selection and kin selection. *Nature*, 201: 1145–1147, 1964.
[47] K.-D. Opp. The evolutionary emergence of norms. *British Journal of Social Psychology*, 21: 139–49, 1982.
[48] E. Posner. *Law and Social Norms*. Harvard University Press, Cambridge, MA 2000.
[49] M. Rabin. Psychology and economics. *Journal of Economic Literature*, 36: 11–46, 1998.
[50] J.K. Rilling, D.A. Gutman, T.T. Zeh, G. Pagnoni, G.S. Berns and C. D. Kilts. A neural basis for social cooperation. *Neuron*, 35: 395–405, 2002.
[51] A. Robson and F. Vega-Redondo. Efficient equilibrium selection in evolutionary games with random matching. *Journal of Economic Theory*, 70: 65–92, 1996.
[52] A. Roth. Bargaining Experiments. In J.H. Kagel and A. E. Roth (eds.), *Handbook of Experimental Economics*. Princeton University Press, Princeton 1995.
[53] A. Roth, V. Prasnikar, M. Okuno-Fujiwara and S. Zamir. Bargaining and market behaviour in Jerusalem, Ljubljana, Pittsburgh and Tokyo: an experimental study." *American Economic Review*, 81: 1068–1095, 1991.
[54] P.H. Rubin. *Darwinian Politics. The Evolutionary Origin of Freedom*. Rutgers University Press, New Brunswick 2002.
[55] A.G. Sanfey, J.K. Rilling, J.A. Aronson, L.E. Nystrom and J. D. Cohen. The neural basis of economic decision-making in the ultimatum game. *Science*, 13: 1755–1758, 2003.
[56] T. Schelling. *The Strategy of Conflict*. Harvard University Press, Cambridge, MA 1960.
[57] P. Seabright. *The Company of Strangers: A Natural History of Economic Life*. Princeton University Press, Princeton 2004.

[58] B. Seymour, T. Singer and R. Dolan. The neurobiology of punishment. *Nature Reviews Neuroscience*, 8: 300–311, 2007.
[59] B. Skyrms. *The Evolution of Social Contract*. Cambridge University Press, Cambridge 1996.
[60] R. Slonim and A. Roth. Learning in high stakes ultimatum games: an experiment in the Slovak republic. *Econometrica*, 66: 569–596, 1998.
[61] J. Stevens and M. D. Hauser. Why be nice? Psychological constraints on the evolution of cooperation. *Trends in Cognitive Sciences*, 8: 60–65, 2004..
[62] R. Sugden. "Normative Expectations". In A. Ben-Ner and L. Putterman (eds.), *Economics, Values and Organization*, Cambridge University Press, New York 1998.
[63] R.H. Thaler. *The Winners Curse: Paradoxes and Anomalies of Economic Life*. Free Press, New York 1992.
[64] The Chimpanzee Sequencing and Analysis Consortium. Initial sequence of the chimpanzee genome and comparison with the human genome. *Nature*, 437: 69–87, 2005.
[65] P.V. Tobias. Numerous apparently synapomorphic features in Australopithecus robustus, Australopithecus boisei, and Homo habilis: Support for the Skelton-McHenry-Drawhorn hypothesis. In F.E. Grine (ed.), *Evolutionary History of the "Robust" Australopithecine*, Aldine de Gruyter, New York 1988.
[66] R. Trivers. The evolution of reciprocal altruism. *Quarterly Review of Biology*, 46: 35–57, 1971.
[67] E. Ullmann-Margalit. *The Emergence of Norms*. Oxford University Press, Oxford 1977.
[68] J. Von Neumann and O. Morgenstern. *Theory of games and economic behavior*. Princeton University Press, Princeton, NJ 1944.
[69] H.P. Young. The evolution of conventions. *Econometrica*, 61: 57–84, 1993.
[70] H.P. Young. *Individual Strategy and Social Structure*. Princeton University Press, Princeton NJ, 1998.
[71] H.P. Young. Social Norms. In S.N. Durlauf and L. E. Blume (eds.), *The New Palgrave Dictionary of Economics*. Macmillan, London 2008.

Self-organization of the mind and methodological individualism in Hayek's thought

FRANCESCO DI IORIO

1 Introduction

As widely known, Hayek has been one of the main theorists of methodological individualism. In contrast with the holist approach and in line with authors such as Max Weber he has upheld the necessity to explain social phenomena starting from an interpretation of the "sense" that generates and orientates the actions of individuals [25, pp. 1–48]. Hayek rules out the possibility for the action to be determined by factors exogenous to the individual and, in particular, to the social context.

By analysing Hayek's contribution to the theory of methodological individualism, scholars have underlined the crucial relevance of the work *Scientism and the Study of Society*, completed in 1944. Apart from rare exceptions, however, they have probably underestimated the value of another interesting essay that Hayek had partly devoted to the epistemology of social sciences: *The Sensory Order*, published in 1952. This work has been largely devoted to issues concerning the sphere of cognitive psychology. However, it has mainly been considered as an effort to support, on neurophysiologic bases, the interpretative method (*Verstehen*) of methodological individualism.[1]

The content of this work needs to be carefully taken into account first of all for the scientific relevance of Hayek's theory of the mind. As highlighted by neurobiologists and cognitive scientists as Gerald Edelman, Joaquin Fuster or Barry Smith, *The Sensory Order* represents a fundamental contribution to the psychology of the 20th century (see [15, 16, 17, 34]). According to them, it is a greatly innovating work and its relevance in terms of studies on perception has been erroneously, and for too long, underrated. In particular they agree on the fact that Hayek needs to be regarded, together with Donald Hebb, as the father of one of the nowadays most quoted approaches on the studies of the mind: the "connectionist" paradigm.[2] Such paradigm is based upon the idea of the self-organisation of complex systems, central concept within the framework of the overall epistemological and scientific reflection of Hayek[3]

It has to be added that Hayek has worked out the core of the innovative theses in *The Sensory Order* well before the circulation of this work. They

---

[1]This has been underlined by [8, pp. 270–277]. See also [7, pp. 20–22].

[2]For an introduction to connectionism please refer to: [14, 3, 30, 36].

[3]For an analysis of the role covered by the concept of self-organisation within Hayek's epistemology please see [28].

had in fact already been outlined in a short essay of the 1920s, as outcomes of the intuitions of a young Hayek, crossed between the passion for economics and psychology. Titled "Contributions to a Theory of How Consciousness Develops", this essay has never been published due to scientific and personal issues.[4] In The Sensory Order the theories contained will be taken up again and enriched, also in line with the contributions from cybernetics and the systems' theory [14]

Why does Hayek state that his connectionism implies arguments in favour of methodological individualism? The crucial point is that according to the theory of self-organization of the mind, cognitive processes are non deterministic. If cognitive processes are non deterministic, it is impossible to consider the action as a mechanical effect of the social context as envisaged by methodological holism. According to Hayek, connectionism legitimizes the idea, supported by Weber and the other methodological individualists, that the explanation of the action goes through the reconstruction of its *sense*. In other words, he argues that if the mind is a self-organized order, the cause of action can only lay inside the individual; it cannot be found outside him.

The present article is structured in two main sections. In the first one (paragraphs 1 and 2) Hayek's connectionist theory is schematically delineated. In the second part (paragraph 3) the methodological consequences of such theory are analysed. In particular, the way he resorts to it, to justify the necessity for a *Verstehen* method in social sciences, is investigated.

2 The Hayekian connectionism

Hayek works out a theory of the mind that well matches with the connectionism as intended by nowadays authors like Petitot and Varela who link their approach to the Husserlian Phenomenology [3, 30, 36]. According to those scholars, the mind cannot be reduced to a logical machine like a computer. Its conscious and logical skills are not the only ones available. Those skills presuppose tacit or meta-conscious abilities. As we will see, the latter can be described as effects emerging by self-organization from neuronal networks. For those connectionist scientists, the ability of logical deduction is based on tacit processes of interpretation and meaning construction.

The starting point in *The Sensory Order* is precisely the idea that the mind operates as an apparatus of *interpretation*. According to Hayek there is no correspondence between the sensorial world and the external world: "Every sensation [...] must be regarded as an interpretation of an event in the light of the past experience of the individual or the species" [18, p. 166]. Perceptions are thus interpretations that depend on memory: both on the biological memory (the way natural selection has shaped the nervous system and the receptive organs of stimuli) and the personal memory (what the individual has learned over his/her life). It follows that sensory qualities as, for instance, the blue of the sky can not be regarded as an objective property of reality, but as a mental construction.

---

[4]Many thanks to Professor William N. Butos from the Foundation for Economic Education of New York for having provided me with a copy of this essay.

Hayek states that sensory constructions coincide with "acts of classification" [18, p. 78] of external stimuli. In other words, such interpretations consist in processes that tie typical sets of stimuli to typical meanings. Hayek expressed this concept in terms of "Primacy of Abstract" : the possibility to acknowledge that a given object is, for instance, a "car" depends on tracing a certain typical set of stimuli back to an abstract class "cars" (Hayek also talks about *pattern recognition* or *rules of perception*). It follows that he totally opposes the idea, that is at the basis of inductivism, that knowledge is acquired by a mind that is, at the beginning, a *blank sheet of paper*. The perception of a tangible detail always presupposes interpretative schemes, a sort of theoretical knowledge selecting and interpreting the external reality, linked to past experience: "In the mind the abstract can exist without the concrete, but not the concrete without the abstract" [22, p. 35]. As his fellow-countryman and friend Popper, the author of *The Sensory Order* thus regards that: "*all* we know about the world is of the nature of theories and all 'experience' can do is to change these theories" [18, p. 143]. Regarding analogies between Popper's criticism towards *observationism* and the connectionist approach please refer to [4, pp. 75–79].

Hence Hayek holds that abstraction, the tendency to order the phenomena in typical classes, is not a purely rational and conscious ability. It is first and foremost a tacit or meta-conscious capability; a property of categories through which the mind operates: "the richness of the sensory world in which we live [...] is not — Hayek states — the starting point from which the mind derives abstractions", but "the product" [22, p. 44] of meta-conscious abstractions.

To fully grasp the above point it is necessary to specify that the classification processes generating perceptions are not procedures of simple classification: they are, instead, modes of "multiple classification" [18, p. 50]. This implies that, in spite of what assumed by behaviourism, they never concern a single stimulus, but always groups of stimuli or events. Within the framework of these processes: "at any moment a given event may be treated as a member of more than one class, each of these classes containing also different other events; and a given event may also on different occasions be assigned to different classes according to the accompanying events with which it occurs" [18, p. 50]. Moreover, the minds also carries out a "a third type of multiple classification: namely one in which successive acts of classification follow upon each other in relays, or on different 'levels'; in this type the distinct responses which effect the grouping at a first level become in turn subject to a further classification (which also may be multiple in both the former senses)" [18, p. 51].

The mind is thus based upon a logic of multiple classification. The detailed analysis of its functioning does not constitute part of the scope of the current work. It is sufficient to specify that this particular classifying activity constantly rectifies the interpretations, to which it leads according to the continuous flow of new experiences. Moreover, it allows to experience a large abundance of sensory qualities. This is in accordance with the fact that it allows the overlapping of several abstract schemes of meanings. It is precisely due to the logic of multiple classification, on which the mind is based that, by observing a certain object, we can simultaneously gather a large number of particular characteristics. We can realise, for instance, that this item is a house, that it is a yellow house, that it is a house in an *art nouveau* style, that it is a house with three floors, that it is a house with a garden and so on. In the child or in the animal this capacity in terms of overlapping of abstract schemes is less developed. The sensory world of the child or the animal is simpler "because of the much thinner net of ordering relations which they posses — because the much smaller number of abstract classes under which they can subsume their impressions makes the qualities which their supposedly elementary sensations posses much less rich" [22, p. 44].

According to Hayek, the sensory order is thus created from classifications and consists in the order of analogy and difference relations amongst the sensory qualities. It is the way though which these qualities, and not the objective facts, differ from each another (in terms of dimension, colour, weight, etc.) and it is also the whole set of meanings of the phenomenal reality. The sensory order is a simplification of the physical one. It reproduces the relations objectively in existence in an imperfect and partial way. The latter are drawn from natural sciences, nevertheless it does not imply that the latter analyse a world that is more real, because also the physical world is an "abstract" construction [18, p. 143]. It is the outcome of an *alternative* classification compared to the mental one that produces new meanings and operates according to experimental theories [18, pp. 145–146].

Concerning the most complex sensory phenomena, Hayek maintains that his theory of the mind "leads indeed to conclusions very similar to those of the gestalt school" [18, p. 77]. However, while the exponents of the *Gestalt* school hold that the unconscious organisation, leading to the recognition of analogous forms also in the case of objects not having any identity in terms of physical structure, concerns elementary sensory qualities that are directly communicated to the mind from basic nervous impulses (that is a sort of basic "qualitative" information), Hayek holds a different point of view. He states that the impulses do not incorporate any sensory quality: in other words, the sensory qualities are not linked to attributes of the impulses. He proposes a "connectionist" approach.

For the Austrian author the mental interpretations depend solely upon the way impulses are channelled through the neural networks, the connections amongst neurons that the impulses are able to activate. For every single *kind* of perception there is a peculiar *kind* of canalization of the impulses. Both the most and less "elementary" qualities are created via these processes. Impulses do not incorporate any basic "qualitative" information: according to Hayek, the theory of an original pure core of sensation is wrong. Here is his quote: "my theory maintains that the sensory (or other mental) qualities are not in some way originally attached to, or an original attribute of, the individual physiological impulses, but that all of these qualities are determined by the system of connexion by which the impulses can be transmitted from neuron to neuron; that it is thus the position of the individual impulse or group of impulses in the whole system of such connexions which gives it its distinctive quality" [18, p. 53]. Mental classifications thus depend on the activation of several neurons' chains. They are the product of a system effect. This theory is confirmed by the fact that a precise location of perceptions in the brain does not exist [18, p. 148].

According to this connectionist paradigm the way neurons work is not controlled by a central unit, but is simply based upon certain "rules" of activation. Such rules define, in an abstract mode, the modalities and the conditions for the activation of the neurons. So whether or not a neuron becomes part of a chain of connections, carrying nervous impulses, depend solely on these rules. It follows that perceptions emerge according to a logic of self-organization.[5] Due to the above, Barry Smith has underlined that a strong analogy does exist between Hayek's market theory and his theory of the mind because they are both based upon the model of a *spontaneous* or *self-organised order* [34].

Through the self-organisation of the mind's neural connections a "*map* of the relationship between various kind of events in the external world" is produced, that "will not only be a very imperfect map, but also a map which is subject to continuous although very gradual change" [18, p. 110]. It will be partly modified by new experiences. This is due to the fact that perception and memory act according to a circular causality logic (see also [17, p. 87]): perception, created by the memory, affects the latter and changes in the memory loop back on perception. It follows that a learning process is incessantly running, developing under a *trial and error* fashion and consisting in a substitution of classification modalities, inbred or acquired, with new classification modalities; this substitution is based upon a partial restructuring of the neural connections system (see also [4, pp. 75-79]). It is thus relatively easy to understand how an individual, after a time lag, can perceive the same fact or object differently.

By adopting a similar perspective, it is necessary to exclude the existence "of elementary and constant sensations as ultimate constituents of the world" [18, p. 176] and assumes "the inconstancy of sensory quality" [18, p. 173]. For a connectionist as Hayek, learning is a *creative* act: "To acquire the capacity for the new sensory discrimination is not merely to learn to do better what we have done before; it means doing altogether new. It means not merely to discriminate better or more efficiently between two stimuli or groups of stimuli: it means discriminating between stimuli which before was not discriminated at all". For instance, it makes no sense to state "that, if a chemist learns to distinguish between two smells which nobody has ever distinguished before, he has learnt to distinguish between given qualities: these qualities just did not exist before he learnt to distinguish between them" [18, p. 156].

Mental *maps* are thus partly modified by experience. Given that and taking into account that the biological evolutionary logic makes a perfect correspondence of the individuals' anatomic structure rather impracticable, a complete identity of human minds is in turn impossible. Human minds will in fact be sufficiently similar to allow the mutual comprehension and interaction amongst individuals, but "they will not be identical" [18, p. 110].

---

[5] See [36, pp. 60-61]. Hayek does not illustrate in depth the activation modalities of neurons. He goes as far as to agree with Hebb, who has been the first to analyse in details this issue. Please see: [18, note 1 p. 64 and note 1 p. 114]. See also [23, p. 20].

Having analysed the modalities behind the elaboration of perceptions, Hayek can emphasize the relationship between perception and action. To his judgement, perceptions rules have to be regarded as closely related to action rules: the classification of *typical meanings* has to be conceived as functional to an adaptation effort consisting in the implementation of *typical and appropriate answers*. A trivial example is represented by the consequences of a driver's perceptions of red traffic lights. Such acknowledgment of a typical meaning allows the driver to develop an adaptive answer that is in turn typical: that is to stop when the traffic lights are red. The abstract rules of perception allow to single out typical problematic situations and incorporate, in the meanings they create, information about the rules of action that are useful to face them. These action rules can also be defined as "abstract" [22, pp. 35 ff]. (see also [28, pp. 51–53]) because they can also be applied to abstract classes of events.

Hayek's analysis leads to a rather important conclusion: *know-how* is the pillar of consciousness and of the rational thought. There is a tacit or a not-articulated dimension of knowledge (see also [12]). Such dimensions will allow us to use certain skills without being able to verbally explain what exactly entails the capability to do them. It embraces not only the meta-conscious competence to apply rules for the elaboration of the sensory world, but also the capacity to fully master some *practical* skills (as swimming, riding a bike and painting). These *practical* skills, not dissimilarly from those behind perception, can not be verbally described (it is impossible, for instance, to illustrate in a manual the guidelines to be followed in order to acquire capabilities to ride a bike without loosing balance) and they also partly depend on instinct (the biological memory) and partly on learning (the personal memory). Therefore Hayek's connectionism lets us comprehend that the mind is not limited to the sole logic and rational capabilities. Consciousness is the *tip* of an iceberg: "what we consciously experience is [...] the result [...] of processes of which we cannot be conscious" [22, p. 45]. The *tacit* dimension of knowledge produces the necessary assumptions of the intentional choice. It defines alternatives on the basis of which the individual decision is arrived at.

3 The impossibility of an all-inclusive explanation for the sensory order

Hayek specifies that the theory on the configuration of conscience that he proposes, related to the self-organisation idea, represents a mere "explanation of principle" [18, p. 182], thus only taking into account a general logic of an extremely complex phenomenon. Hayek excludes the possibility for the mind to eventually arrive at a comprehensive self explanation. This is due to the fact the he denies the possibility for an *explanation of detail* of its running that would permit "to substitute a description in physical terms for a description in terms of mental qualities" [18, p. 189] (see also [27, pp. 15-20]). To his judgment this possibility needs to be discarded for three reasons.

First of all he considers that, given that the mind emerges from the activity of 100 billion neurons able to interconnect with one another according to a virtually unlimited number of combinations, it belongs to the "complex phenomena" [20, p. 55] category. It is impossible to identify all the interdependent variables contributing to determining such a complex order as the mind is. It is not feasible to master the whole of its physical causes. Moreover, the chance to only resort to an explanation of principle presents severe drawbacks in terms of forecasting. In fact, only for the "simple phenomena", caused by a limited number of variables, it is possible "to predict particular events" [18, p. 185].

The second reason is, instead, strictly linked to the adaptive logic of a self-organised order. This kind of order is not characterised by a fully knowable and foreseeable behaviour also because, employing a concept widely used by Maturana and Varela, it is endowed with "autonomy" [26, pp. 77-87]. As increasingly clarified by the connectionist cognitive scientists, the behaviour of a self-organised order is neither determined by a programme introduced from the outside and followed mechanically (behaviour of a self-organised order is not comparable to the one of a machine that has been previously programmed, as a computer), nor by the effects applied by the surrounding environment. Basically an order of this kind does not passively undergo such effects (its behaviour has nothing to do, for instance, with the one of a pool's ball that is, on the opposite, entirely determined by the external forces acting on it) [18, pp. 122–127]. The cause behind the behaviour of a self-organised order is not to be sought outside it. An order as the mind actively employs random novelties that appear in the environment in a continuous and unpredictable manner, to constantly self-reprogramme itself. Its logic is to safeguard its autonomy from the outer environment and thus its capability to adapt. An order of this kind is not predetermined: because it is not possible to foresee the random novelties affecting its behaviour, nor the way they will impact on the outcomes of its processes of self-organisation and upholding of autonomy. These processes are acts of pure creation [2, pp. 157 ff.]. By virtue of its working modalities, a self-organised order is the "cause of itself" (see also [13, pp. 109–124]).

The third and last reason taken into account by Hayek is of a logical kind (Hayek explicitly speaks of a goedelian limit). As we have seen, Hayek regards the mind as an apparatus for classification. To his judgement, we cannot totally account for our interpretative categories also because, being them the source of meanings, they can not find a place in the order of meanings: "There is [...] on every level, or in every universe of discourse, a part of our knowledge which, although it is the result of experience, cannot be controlled by experience, because it constitutes the ordering principle of that universe by which we distinguish the different kinds of objects of which it consists and to which our statements refer" [18, pp. 169–170] (see also [20, pp. 60–63]). To get around this problem we should place ourselves outside our own mind. A classifying apparatus of a higher complexity compared to the one of the human mind would in fact be required — an apparatus that, moreover, would in turn be, for the same logical reasons, unable to exhaustively explain its functioning [18, pp. 184–190] (see also [28, pp. 60–61]). Also due to this motive, we are faced, according to Hayek, with the impossibility to replace an explanation based on mental skills with one characterised by mere physical terms.

4 The indeterminism of human action

Hayek is, in a way, a monist: "in some ultimate sense mental phenomena are 'nothing but' physical processes" [18, p. 191]. According to his view: "this, however, does not alter the fact that in discussing mental process we will never be able to dispense with the use of mental terms, and that we shall have permanently to be content with a practical dualism". Being the mind an emergent, complex, self-organized system and being impossible to detail its working, it is not feasible to seek the causes of thought in the physical characteristics of the external environment and in the physical processes induced by the former in the nervous system. It stems "that we shall never achieve a complete 'unification' of all sciences in the sense that all phenomena of which it treats can be describes in physical terms"[18, p. 191]. Given that an *explanation of principle* of the sensory order is the only feasible route, it follows highly impracticable to reduce social sciences to physics.

Bearing the mind an emergent complex system based on *autonomy*, as intended by Maturana and Varela, and being impossible to fully comprehend the way it functions, the idea stating that cognitive processes can be conceived in deterministic terms needs also to be ruled out. The prospect according to which action is nothing less than a mechanical product of the context, shared by the behaviourism of as much as by the methodological holism, is not compatible with the Hayekian connectionism. By reason of the mind's complexity and the fact that such order is the "cause of itself", the "data" for the explanation of the action can not be external to the individual. Both methodological holists and Hayek maintain that consciousness is not the only important aspect of human nature. But, while holists assume the action as determined at an unconscious level (consider, for instance, Marxs theory that collective beliefs are determined by the economic structure), Hayek states the contrary. For him, the tacit and meta-conscious processes that are at the basis of knowledge legitimate methodological individualism. Their indeterminism rules out the possibility to use the holist approach and to consider the individual ideas as determined by causes which are external to individuals like the social or the economic system.

The conclusion to which his theory leads, Hayek stresses, "is [...] of the greatest importance for all the disciplines which aim at an understanding and interpretation of human action" [18, p. 193]. The adoption of the collectivistic paradigm and the search for the action's causes outside the individual is, according to Hayek, the result of a *Hybris*. In other words it is a scientism-type and over simplifying conceit neglecting logical and epistemological limits, that is necessary to come to terms with in accounting for the outcomes of mental operations.

The Austrian author regards the explanation for the action as necessarily resulting in a reconstruction of the ideas motivating the individuals: he supports that the "data" of social sciences are internal to the actors. To support the thesis, as holists do, considering the "sense", the action has for individuals, as irrelevant is illegitimate and misleading, according to Hayek: "Unless we can understand what the acting people mean by their actions any attempt to explain them [...] is bound to fail" [19, p. 53]. The possibility to apply such hermeneutical procedure calls for existence of a *quid* common to the researcher and social actors: it needs, therefore, an invariant element in spite of the variability of beliefs and knowledge. Hayek states that such element does exist by virtue of a genetic predisposition, constituted by the knowledge a priori of the mind's logical structure and of its perceiving basic categories [19, pp. 99–103].[6]

Holism has originally developed in close correlation with the naïve realism, one of the mistakes of the positivist vision of science. This helps understanding the reason why such paradigm has come to consider the mind as a deterministic mechanism instead of as apparatus for interpretation, taking the action's causes as objective or external rather than internal to individual minds. The naïve empiricism denies the dichotomy between sensory and physical order. Hayek demonstrates, instead, that precisely due to this dichotomy and the non reducibility of the mental to the physical, it is necessary to rule out the determinism of action and support a "subjectivist method". By the light of its "practical dualism", it become clear that it is not possible to set aside an explanation of the action in terms of universal a priori and effects of perceptive interpretations: "In the study of human action [...] our starting point will always have to be our direct knowledge of the different kinds of mental events, which to us must remain irreducible entities" [19, p. 191].

The approach supported by Hayek thus concerns a "*verstehende* psychology" [19, p. 192]. According to this approach the powers of human sciences are bounded. The scientist has no other possibility but to "use our direct ('introspective') knowledge of mental events in order to 'understand' [...] the results to which mental processes will lead in certain conditions". However "such a *verstehende* psychology, which starts from our given knowledge of mental process, will [...] never be able to explain — Hayek writes — why we must think thus and not otherwise, why we arrive at particular conclusions". Moreover "assertion that we can explain our own knowledge involves also the belief that we can at any one moment of time [...] act on some knowledge" due to the fact that it is associated to the idea that it is possible to acquire "some additional knowledge about how the former is conditioned and determined" [19, p. 192]. This is a further reason behind the close association between methodological holism and constructivist rationalism.

According to Hayek the error of conceiving in deterministic terms cognitive processes is plainly illustrated in Karl Mannheim's approach: "In particular, it would appear that the whole aim of the discipline known under the name of 'sociology of knowledge' which aims at explaining why people as a result of particular material circumstances hold particular views at particular moments, is fundamentally misconceived". In opposition to Manneheim, Hayek reaffirms that, given the sensory order's characteristics, the causes of actions can not be traced back to the influences coming from the external contest: "To us — he writes — human decision must always appear as the result of the whole of a human personality — that means the whole of a person's mind — which, as we have seen, we can not reduce to something else" [19, pp. 192–193].

---

[6] As widely known Hayek has been influenced by Mises. See [37, pp. 1-102].

Therefore *The Sensory Order* represents the attempt to transform methodological individualism from a mere epistemological principle to an approach centred around a scientific theory of the mind, able to adequately account for the logic behind the creation of sensory perceptions; a theory that legitimizes the indeterminism of human action and the *interpretative* method and that, thus, well combine with the weberian tradition. While methodological holists consider the action as determined at an unconscious level, Hayek considers the *tacit* and *meta-conscious* processes, that are at the basis of knowledge, as legitimating methodological individualism.

It is worth noting how, by developing a connectionist approach similar to the Hayekian one, cognitive scientists as Dreyfus, Petitot and Varela have recently confirmed the unfeasibility in explaining the action without taking account of the sense that the action carries for the individual. These authors define the connectionist approach as a "phenomenological" approach and question another paradigm of cognitive sciences: the logical-symbolic paradigm. It compares the mind to a computer running a programme and reducing it to only logical skills. It does not deem the mind as an interpretation apparatus, based on the self-organisation logic, and does not take into account the issue relative to the sense's explanation nor the role played by tacit capabilities in perceptions and action. The experimental researches connectionists have implemented over the past few years have reasserted Hayek's intuitions on cognitive limitations connected to the study of the mind and, indirectly, also the epistemological consequences derived by Hayek [12, 14, 29, 36].

5 Conclusive remarks

It is necessary to briefly analyse some additional issues before coming to a conclusion. First and foremost it has to be underlined that Hayek, despite some ambiguity, does not appear to deny the nomological nature of the action's explanation. In the *Sensory Order* he writes that the logic behind knowledge acquisition is universal and that science, similarly to the mind, works through a classification procedure: it explains phenomena grouping them in typical classes, thus in a nomological fashion.[7] As clarified by Weber himself, the *interpretative* method is not in contrast with the nomological-deductive paradigm (for details see [38]). Contrarily to what supported by the anti-nomological individualists, such paradigm does not inevitably require the use of deterministic laws. It is also compatible with probabilistic laws and hence with the idea that action is non-deterministic. As clearly illustrated by Popper and Hempel, the social scientist can not refrain from using this kind of laws in reconstructing the logic of the situation, in other words the *reasons* of the actor.[8]

Hayekian theory of the sensory order, in addition to legitimize the *interpre*-

---

[7]In *The Counter-revolution of Science* Hayek explicitly states that trying to explain individual actions implies "to subsume them under rules which connect similar situations with similar actions" [19, p. 53]).

[8]See [24, pp. 359 ff.], [32], and [33]. Also the "explanation of principle", the only type of explanation available given complex phenomena as the mind, is, as illustrated by Hempel, compatible with the nomological-deductive method. See [24, pp. 359 ff.]. See also [11, p. 234].

tative method, is relevant also for another reason. It establishes the subjectivism of values from a neurophysiologic viewpoint. As already mentioned, it excludes the possibility for a complete identity of the different maps occurring in the individual minds. It follows that it also excludes a perfect matching of the personal assumptions of the subjective evaluations (for more details see [9]). The relation between the action and the environmental context is structured according to more complex terms compared to what maintained by those advocating the *objectivistic* and *collectivistic* approach in social sciences.

A further issue to be stressed revolves around the careful analysis of Hayek's cognitive psychology, that will clearly highlight the groundlessness of the criticism coming from authors as Victor Vamberg, claiming that Hayek, by heavily relying on the role of cultural tradition and on action governed by acquired rules, has ended up in theorising a cultural determinism that is quite inconsistent with the pillars of methodological individualism (see [35, p. 48 ff.]).

These kinds of criticisms do not take into account the fact that Hayek, as already illustrated, considers the mind in a rather different way compared to a large part of durkheimian sociology. He does not regard it as a machine that mechanically implements a programme previously acquired via the process of socialization. From the perspective of his psychology, an individual follows certain ethical rules because he *accepts* them by an evaluation which is based on non-deterministic tacit processes. He follows them and keeps using them until he believes that they are an appropriate way to act or can be useful to solve problems [10, pp. 177–178, fn. 36]. For him, the cause of action is always represented by the sense the latter has for the actor, never by the rules he follows. Moreover, according to Hayek, also applying a certain response rule in specific situations will never be characterised by a mechanical nature because, as already seen, it relies on non deterministic interpretative processes (interpretative process that might concern the same meaning of cultural rules, that could sometimes be ambiguous and contradictory)[21, pp. 72–73]. Moreover it has to be reminded that Hayek considers the mind, being based on the logic of self-organisation, as endowed with the capacity to modify, al least partly, the effects of prior learning (and thus also of the socialisation process). The connectionist approach regards the mind as having the capacity to develop, according to new experiences, new perceptive and adaptive rules in line with the safeguarding of its autonomy (see also [6, pp. 527–534]).

The implications of Hayek's cognitive psychology are of great relevance not only for the "static" analysis of the social process, but also for the study of the causes behind social changes and the cultural evolution. His theory of the mind implies a criticism of holism also on these grounds. In particular, Hayekian connectionism rules out the chance of identifying laws governing social changes; that are laws establishing a deterministic relation between a certain types of variations in the environmental conditions and the upcoming of a distinct type of social structure (consider for instance Parsons' theory on the tendency towards the nuclearization of families in industrial societies).[9] Hayek's approach helps us in understanding why the teories of this kind have always been falsified by opposite historical cases.

---

[9] Concerning this point, Hayek's opinion is similar to Boudon's [5].

Acknowledgements

I would like to thank Dario Antiseri, Jean-Michel Besnier, Raymond Boudon, Raimondo Cubeddu, Sylvain Charron, Pierre Demeulenaeire, Jean-Pierre Dupuy, Robert Nadeau, Philippe Nemo and Jean Petitot who discussed with me the subject of this article, for their valuable remarks and suggestions.

BIBLIOGRAPHY

[1] D. Antiseri. *Popper's Vienna. The Davies Group*, Aurora, Colorado 2006.
[2] H. Atlan. *Entre le cristal et la fumée. Essai sur l'organisation du vivant*. Seuil, Paris 1979.
[3] J.-P. Barthelemy, M.D. Glas, J.-P. Descles and J. Petitot. Logique et dynamique de la cognition. *Intellecta*, 2–23, 1996.
[4] J.-M. Besnier. *Les théories de la connaissance*. Puf, Paris 2005.
[5] R. Boudon. *Theories of Social Change*. Polity Press. Cambridge, 1991.
[6] R. Boudon and F. ois Burricaud. *A Critical Dictionary of Sociology*. Univer- sity of Chicago Press, Chicago 1990.
[7] W.N. Butos and R.G. and Koppl. Does the sensory order have a useful economic future? In E. Krecke and K. Krecke (eds.),*Advances in Austrian Economics*, volume 8, JAI Press, Oxford 2006.
[8] B. Caldwell. *Hayeks Chal lenge. An Intel lectual Biography of F.A. Hayek*. The University of Chicago Press, Chicago 2004.
[9] R. Cubeddu. *The Philosophy of Austrian School*. Routledge, London 1993.
[10] E. Di Nuoscio. *Epistemologia dell'azione e ordine spontaneo. Evoluzionismo e individualismo metodologico in Herbert Spencer*. Rubettino, Soveria Mannelli 2000.
[11] E. Di Nuoscio. *Tucidide come Einstein. La spiegazione scientifica in stori- ografia*. Rubettino, Soveria Mannelli 2004.
[12] H.L. Dreyfus and S.E.. and Dreyfus. *Mind Over Machine. The Power of Human Intuition and Expertise in the Era of Computer*. Paper Back, New York 2000.
[13] J.-P. Dupuy. *Ordres et désordres. Enquete sur un nouveau paradigme*. Seuil, Paris 1990.
[14] J.-P. Dupuy. *The Mechanization of Mind. On the Origins of Cognitive Sci- ence*. Princeton University Press, Princeton and Oxford 2000.
[15] G.M. Edelmann. Through a computer darkly: Group selection and higher brain function. *Bulletin — The American Academy of Arts and Sciences*, 36 (1): 20–49, 1982.
[16] G.M. Edelmann. *Neural Darwinism: The Theory of Neuronal Group Selec- tion*. Basic Books, New York 1987.
[17] J. Fuster. *Memory in the Cerebral Cortex: An Empirical Approach to Networks in the Human and Nonhuman Primate*. The MIT Press, Cambridge, MA 1995.
[18] F.A. Hayek. *The Sensory Order. An Inquiry into the Foundations of Theo- retical Psychology*. Routledge & Kegan Paul Ltd, London 1952.
[19] F.A. Hayek. *The Counter-Revolution of Science Studies on the Abuse of Reason*. Liberty Press, Indianapolis 1953.
[20] F.A. Hayek. *Studies in Philosophy, Politics and Economics*. The University of Chicago Press, Chicago 1967.
[21] F.A. Hayek. *Rules and order. In Law, Legislation and Liberty*, volume 1. University of Chicago Press, Chicago 1973.
[22] F.A. Hayek. *New Studies in Philosophy, Politics, Economics and the History of Ideas*. Routledge & Kegan Paul, London 1978.
[23] F.A. Hayek. Préface. In *Lordre sensoriel. Une enquete sur les fondements de la psychologie théorique*. CNRS Editions, Paris 2001.
[24] C.G. Hempel. *Aspects of Scientific Explanation and Other Essays in the Philosophy of Science*. The Free Press, New York 1966.
[25] L. Lachmann. *The Legacy of Max Weber*. The Ludwig von Mises Institute, Auburn, Alabama 2007.
[26] H.R. Maturana and F.J. Varela. *Autopoiesis and Cognition. The Realization of the Living*. D. Reidel Publishing Company, Dordrecht 1980.
[27] R. Nadeau. Friedrich hayek et la théorie de lesprit. In J.-P Cometti and K. Mulligan (eds.), *La philosophie autrichienne de Bolzano à Musil*. Vrin, Paris 2001.
[28] P. Nemo. *La société de droit selon Hayek*. Puf, Paris, 1988.
[29] J. Petitot. *Physique du sens. De la théorie des singularités aux structures séemionarratives*. Editions du CNRS, Paris 1992.

[30] J. Petitot, F.J. Varela, B. Pachoud and J.-M. Roy. Beyond the gap : an introduction to naturalizing phenomenology. In J. Petitot, F.J. Varela, B. Pachoud, and J.-M. Roy, *Naturalizing Phenomenology. Issues in Contemporary Phe- nomenology and Cognitive Science*. Stanford University Press, Stanford, California 1999.
[31] K.R. Popper. *Objective Knowledge: An Evolutionary Approach*. Oxford University Press, Oxford 1979.
[32] K.R. Popper. *The Logic of Scientic Discovery*. Routledge, London 2000.
[33] K.R. Popper. *All life is Problem Solving*. Routledge, London 2001.
[34] B. Smith. The connectionist mind: A study of Hayekian psychology. In S.F. Frowen (ed.), *Hayek Economist and Social Philosopher: A Critical Retrospect*. McMillan, London 1997.
[35] V.J. Vamberg. *Rules and Choice in Economics*. Routledge, London 1994.
[36] F.J. Varela. *Connaitre les sciences cognitives*. Tendances et Perspectives. Seuil, Paris 1989.
[37] L. von Mises. *Human Action. A Treatise on Economics*. Ludwig von Mises Institute, Auburn, Alabama 1998.
[38] M. Weber. *Rocher and Knies: The Logical Problems of Historical Economics*. Free Press, New York 1975.

Can ethics be naturalized?
SIMONA MORINI

1 The problem

According to Hilary Putnam "Naturalism [...] is often driven by fear" and "to the extent that the appeal of 'naturalism' is based on fear, the fear in question seems to be a horror of the normative. In the case of logical positivism, there was a not-dissimilar horror, the horror that the slightest trace of realism about scientific objects was tantamount to the acceptance of 'metaphysics'. We got over that horror when we realized that talk of unobserved entities did not need either metaphysical interpretation or positivist reinterpretation. We need to learn that the same is true of normative language" [12, p. 70].

So why be afraid of the normative? Because — says Putnam — naturalism assumes that if we cannot eliminate the normative, or reduce it to the non-normative, then we shall allow the "occult" and the "supernatural" to enter our discourse. Actually, this is the original *philosophical* meaning of naturalism that dates back to the seventeenth century and that is defined by the Oxford English Dictionary as "a view of the world and of mans relation to it, in which only the operation of natural (as opposed to supernatural or spiritual) laws and forces is admitted or assumed". This kind of definition can be either about ontology (it is the view that there are no entities or laws outside those of the natural world) or about methodology (if all facts are natural facts the method by which we investigate all facts must be the scientific method).[1] Methaphysical naturalism entails a kind of "explanatory" naturalism (if everything that exists is composed of natural stuff and constrained by natural law, then everything that is not described in the language of a natural science must ultimately be describable in such terms) even if, as noted by Jesse Prinz, "the explanatory naturalist can be an antireductionist".[2] In any case naturalism appears to exclude assessments from scientific discussion that are not descriptions or definitions of reality (or natural facts). But if the appeal of naturalism, today, amounted to this, it would be hard to understand why we speak of a "naturalistic turn".[3] We are confronted with an

---

[1] [11], pages 2–3 further refines the analysis of naturalism distinguishing four kinds of naturalism: metaphysical, explanatory, methodological and transformation (quinean) naturalism.

[2] [11], further refines the analysis of naturalism in his "Preamble. Naturalism and Hume's Law", pages 1–10.

[3] The classical expression of naturalism is to be found in Willard van Orman Quine's *Word and Object*: "The philosopher's task differs from the others', then, in detail; but in no such drastic way as those suppose who imagine for the philosopher a vantage point outside the conceptual scheme that he takes in charge. There is no such cosmic exile. He cannot study and revise the fundamental conceptual scheme of science and common sense

honoured philosophical tradition, certainly not with a "turn". If we consider naturalism as a turn, it is probably because many of the characteristics of human nature that seemed impossible to describe are today at the centre of scientific research, particularly of the so-called "cognitive sciences".

Some important discoveries in neurophysiology and in brain studies are about aspects of human mind and of subjectivity that seemed to be intractable, "irreducible" to scientific analysis. Jealousy, avarice, greed, altruism... every day we learn something new about our most intimate emotional states and even the normative — what we consider as beautiful, ugly, pleasant, unpleasant, good or bad — is the object of scientific analysis and experimentation. The "naturalistic turn", then, seems to consist in the scientific conquest of the as yet unexplored territories of emotions, conscience, the mind and the self.[4]

We now face a new, different fear, the fear of unveiling something secret, deep and very intimate: our identity, our self. Confronted with these extraordinary and revolutionary discoveries, philosophers, scientists, men of faith and laypeople often react in a fearful and conservative way. The same happened when someone dared to unveil the secrets of the heavens, or to compare men to apes or to decode the human genome. Fear of science strengthens religion, which strengthens in scientists the fear of irrationality and the dogmatic denial of any criticism. In the end many fears are conjured up. One would say that we are in need of an epistemological psychoanalysis. But I am not interested in psychoanalysis, neither normal nor epistemological. There is no reason to be frightened at all.

2 The status of normative reasoning

The question "Can ethics or culture be naturalised?" appears to be a question on the philosophical implications and on the possible limits of the application of cognitive sciences and of some of its results. I would like to focus now on the "terror of the normative", on the normative nature of intentionality and on the naturalisation of ethics (and of rationality, as I consider ethics as a branch of the general theory of rational behaviour). A good starting point for the discussion is the "open question" argument raised by George Edward Moore in Section 13 of his *Principia Ethica* (1913) [10]:

> In fact, if it is not the case that "good" denotes something simple and indefinable, only two alternatives are possible: either it is a complex, a given whole, about the correct analysis of which there could be disagreement; or else it means nothing at all, and there is no such subject as Ethics. [...] Whoever will attentively consider with himself what is actually before his mind when he asks the question "Is pleasure (or whatever it may be) after all good?" can easily satisfy himself that he is not merely wondering whether pleasure is pleasant. And if he will try this experiment with each suggested definition in succession, he may become expert enough to recognise that in every case he has before his mind a unique object, with regard to the connection of which with any other object, a distinct question may be asked. Every one does in fact understand the question "Is this good?" When he thinks of it, his state of

---

without having some conceptual scheme, whether the same or another no less in need of philosophical scrutiny, in which to work."

[4]Daniel Dennett, for one, in his book *Sweet Dreams* [4] goes on the offensive against the "new mysterians", those who argue that the problem of consciousness is fundamentally unsolvable or requires an explanatory framework outside that used by observational science.

mind is different from what it would be, were he asked "Is this pleasant, or desired, or approved?" It has a distinct meaning for him, even though he may not recognise in what respect it is distinct. Whenever he thinks of "intrinsic value", or "intrinsic worth", or says that a thing "ought to exist", he has before his mind the unique object — the unique property of things — that I mean by "good".

Let us reformulate it in the following way: "I'm attending a conference. Outside there is fine weather. I feel a certain disposition to go out for a walk. But *should I do it?*" My reply "No, I shouldn't do it" expresses a mental condition, an *intention*, which is different from my *disposition* (desire) to go out for a walk. It is something that is "added" to my disposition to act in a certain way and that, moreover, inhibits it, forbids me to satisfy it. From this kind of reasoning, Moore reached the conclusion that evaluative sentences such as "no, I shouldn't go out for a walk" are assertions about a "non-natural" world or ascribe "non-natural" properties to the objects of the natural world (to desires, or impulses, for example, judging them as inappropriate). To use Wittgentein's words — and following Saul Kripke's interpretation of it — the important fact is that "my present mental state does not determine what I *ought* to do in the future". So in Wittgenstein's (and especially Kripke's [9]) interpretation, intentional states (beliefs, intentions, etc) are internal "oughts" (or "ought-nots"); or, better, "commitments" to act in a certain way (in the terminology of decision theory one would say that they are preferences subject to "rational constraints").

As Akeel Bilgrami has remarked in his convincing interpretation of Saul Kripke:

> to desire something, to believe something, is to think that one ought to do or think various things, those things that are entailed by those desires and beliefs in the light of certain normative principles (whether those codifying deductive or inductive or decision theoretic rationality). It is not to be disposed to do or think those things, it is to think one ought to do and think them [1, p. 128].

3 Dispositions and intentions reformulated as individual/social preferences

In our example, the fact of going out for a walk instead of attending the conference would be inconsistent with my desire to belong to a scientific community, or to maintain good relations with the organizers of the conference or with the speakers. I am therefore "committed" or "constrained" by my general system of preferences to stay at the conference. Moore's "open question" argument targeted properties such as social aggregated utility in its Benthamite naturalist versions that interpreted utility as "happiness" of the greatest number and therefore as a natural sentiment, as a disposition. Economists redefined individual utility as a *coherent* ordering of preferences (thus imposing a set of axioms defining the rationality of our system of preferences). More recent utilitarian models use cardinal utilities, that is utilities that measure the *intensity* of our preferences, and explicitly abandon the edonistic interpretation of utility as a "mental state". As Harsanyi explicitly recognised, humans are not merely interested in their mental states, in their personal feelings of pleasure or pain; they also have "transcendent" preferences, that is desires concerning the mental states of others or the external world in general. Utility is not necessarily identified with happiness, as it can be independent of the achievement of our goals. In some cases it can depend on the mere act of having these goals, in others we don't even know if our actions will have the desired effects (for example when we worry about the welfare of future generations, even if we have no way to control the beneficial effects of our action). This conceptual shift in the notion of utility, from the "naturalistic" idea of the "maximum happiness" of society to the definition of a "social welfare function" as the arithmetic mean of the utility of the individuals that compose society, is one of the most interesting contribution of the "unorthodox utilitarianism" of John Harsanyi.[5] In his model *dispositions* would correspond to personal preferences (conceived as ordered sets of cardinal preferences), while *intentions* would correspond to our social preferences, that are to be "constructed" in a very "unnatural" way, that is, by putting oneself in an impersonal and impartial position[6] and deciding from this uncomfortable and unnatural "original position" what is the act (if we are act utilitarians) or the rule (if we are rule utilitarians) that maximises everybody's utility. In other words, as a rule utilitarian — for example — I do not go out for a walk because if it were to become a social rule that people leave conferences to go out for a walk when there is fine weather outside, nobody would attend conferences any more (supposing that walks with fine weather are more agreeable — i.e. preferable — than philosophy conferences!). Moreover, according to the Wittgenstein-Kripke position, my preference for the walk would be incompatible[7] with my desire not to offend my collegues, to be invited to other conferences, and so on. (Of course, if I thought that conferences were completely useless, I could decide to go out. In that case I would consciously violate a convention and I should be ready to suffer the consequences of my choice).

I believe Donald Davidson was right to stress the *holistic* nature of this idea of intentionality (or rather of the rationality of preferences) and to maintain that intentional states are mental dispositions that are "constrained" or "governed" by normative principles of rationality and therefore irreducible to physical states [2]. However, the notion of normative rationality principles "constraining" or "governing" our actions needs a more precise definition.

4 Moral intuitions versus moral judgements

In general, one of the basic ideas underlying normative models of social or ethical decision (in their neoutilitaristic versions, at least) is that moral judgements are not the expression of our "moral intuitions" or value judgements on a particular behaviour (as when we think that "lying is bad") — that is they

---

[5] For example, in [7].

[6] This impersonal and impartial position consists, for Harsanyi, in choosing as if one had a probability $\frac{1}{n}$ of being in one of the n possible positions of the n individuals composing the society, or who are affected by the decision. While in the "original position" of John Rawls one has to choose as if one was in the position of the more disadvantaged person in the society. We shall adopt here Harsanyi's position, considering the well known problems of Rawls' maximin rule , as discussed in [6].

[7] According to the model's normative rationality axioms that thefine the coherence of a set of preferences.

do not express "natural" sentiments — but express a *judgement*, for example the acceptance of some fallible and revisable norms, that we follow in order to decide if certain moral "sentiments" or "intuitions" are justified or not. (This is the position taken, for example, by Allan Gibbard [5], following John Harsanyi). Norms are chosen on the basis of rational criteria that can be different — and therefore lead to different choices (as in the case of Harsanyi's neoutilitarianism and John Rawl's neocontractualism). This presupposes the acceptance of rationality criteria as the basis for choice (it would be difficult to discuss ideas or norms with someone using the Coran or the Bible as a basis for choice, that is, with people belonging to traditions basing their moral "judgements" on a sacred text or on a religious authority).

Therefore, we must ask if there are *universally shared* reasons for accepting the rational discussion of norms and principles: a discussion where every person must take into account the interests and sentiments of the others in order to choose actions that are acceptable to everybody. This kind of ethics defines critical and carefully considered reactions to moral disagreement. It does not seek a definition of what is "naturally" good, right or just. In this sense it is normative and not descriptive. According to utilitarianism, the reasons for accepting these rational constraints are based on two characteristics of human behaviour: the fact that it aims to achieve goals (or fulfill desires, whether they are personal or "transcendent") and a basic "empathy" or "benevolence" of human nature (in principle if all human beings were sadists, utilitarianism would not be able to exclude the choice of a society following sadistic rules!). Moreover, ethical models of this kind affirm a basic assumption about human nature, that is, that people are *rational* (or try to be rational). Since this assumption is a normative one, this seems to exclude a naturalistic interpretation of the theories of moral choice. Is this a serious problem? Does it mean that we are introducing something "occult" into our discourse? I do not believe so, and neither do I believe that scientificity is in any sense menaced.

5 Conclusion

I agree with Davidson when he maintains that from the fact that the normative, holistic and externalist elements in psychological concepts cannot be eliminated without radically changing the subject, it does not automatically follow that there are no scientific models of rationality or of moral choice, but that "rationality and morals cannot be *reduced* to physics, nor to any other of the natural sciences" [2, p. 157]. This doesn't mean either that scientific research cannot influence our moral thinking in a deep way. Not only psychology and the brain sciences could very well explain the mechanisms of formation of our dispositions or of our preferences, or of our sentiments, or of our risk aversion or probability assignments. Neuroscientific research also has the potential to shed light on important methaetical problems, concerning the very nature of moral judgement. As we all experience, moral judgements are accomplished in an intuitive, effortless way. The "obviousness" of the fact that we shouldn't kill, torture or cause the suffering of other human (or living) beings makes us believe that some actions are *really* right or wrong, "vicious or virtuous" independent of what any particular person or group thinks about it. But, as David Hume already remarked:

> Take any action allow'd to be vicious: Wilful murder, for instance. Examine it in all lights, and see if you can find that matter of fact, or real existence, which you call vice. In whichever way you take it, you find only certain passions, motives, volitions and thoughts. There is no other matter of fact in the case. The vice entirely escapes you, as long as you consider the object. You never can find it, till you turn your reflection into your own breast, and find a sentiment of disapprobation, which arises in you, towards this action. Here is a matter of fact; but 'tis the object of feeling, not of reason. It lies in yourself, not in the object. So that when you pronounce any action or character to be vicious, you mean nothing, but that from the constitution of your nature you have a feeling or sentiment of blame from the contemplation of it. Vice and virtue, therefore, may be compar'd to sounds, colours, heat and cold, which, according to modern philosophy, are not qualities in objects, but perceptions in the mind [...] Nothing can be more real, or concern us more, than our own sentiments of pleasure and uneasiness; and if these be favourable to virtue, and unfavourable to vice, no more can be requisite to the regulation of our conduct and behaviour [8, III.i.i.].

Evidence from neuroscience seems to confirm Hume's idea that moral judgements that motivate our *dispositions* are not "matters of fact" but lie "in the eye of the beholder", like colours, heat or cold.

The mistake consists in thinking that what cannot be reduced to a natural science does not deserve to be called science or, even worse, in thinking that the psychological description of people's behaviour in everyday life could represent a refutation of normative models of rationality or of ethical choice. The serious risk, in my opinion, consists in using naturalistic or "scientific" descriptions of our moral intuitions in order to confer scientific value on them, instead of subjecting them to rational criticism — as suggested by a normative approach to ethics, which at least has the advantage of avoiding the difficulties of defining moral concepts. In my opinion, a critical attitude — and not science as such — can free us from fundamentalisms. One should fight against dogmatism (including scientific dogmatism), not against people who interpret the world in a different way from us.

BIBLIOGRAPHY

[1] A. Bilgrami. Intentionality and norms. In [3], pages 125–151.
[2] D. Davidson. Could there be a science of rationality? In [3], pages 152–172.
[3] M. De Caro and D. Macarthur. *Naturalism in Question*. Harvard University Press, Cambridge, MA 2004.
[4] D.C. Dennett *Sweet Dreams: Philosophical Obstacles to a Science of Consciousness*. MIT Press, Cambridge, MA 2005.
[5] A. Gibbard. *Wise Choices, Apt Feelings: A Theory of Normative Judgement*. Harvard University Press, Cambridge, MA and Oxford University Press, Oxford 1990.
[6] J.C. Harsanyi. Can the maximin principle serve as a basis for morality? A critique of John Rawls' theory. *American Political Science Review* 59: 594–606.
[7] J.C. Harsanyi. *Rational Behavior and Bargaining Equilibrium in Games and Social Situations*. Cambridge University Press, Cambridge 1977.
[8] D. Hume. *Treatise on Human Nature*, edited by P.H. Nidditch, Oxford University Press, Oxford 1978.
[9] S. Kripke. *Wittgenstein on Rules and Private Language*. Harvard University Press, Cambridge, MA 1982.
[10] . G.E. Moore. *Principia Ethica*, Cambridge University Press, Cambridge 1903.
[11] J.J. Prinz. *The Emotional Construction of Morals*, Oxford University Press, Oxford 2007.
[12] H. Putnam. The content and appeal of "naturalism". In [3], pages 59–70.
[13] W.V.O. Quine. *Word and Object*. MIT Press, Cambridge, MA 1970.

Searle's collective intentionality and the "invisible hand" explanations

STEFANO VASELLI

1 Introduction

The purpose of this paper can be summed up into the following points, which form, each of them, the structure of this argument:

1. A very brief exposition of John Searle's theory of "Collective Intentionality" (or "We Intentionality"), as it is stated in *The Construction of Social Reality* and in other writings[28, 29], and a second brief exposition, conveniently in context, of a socio-historical explanation of the *Invisible Hand* type — or explanation of the "IH" type, based on methodological individualism' ontology (page 4).[1]

2. Why IH' explanations of social behaviour is so much important for the same theoretical context of Collective Intentionality in social sciences. What, in Searle's theory, can be useful to "IH" type of explanations, beyond the explicit positions of Searle and his commentators on what refers to a redefinition of "IH" in terms of *analytical social ontology*.

3. Why, though, Searle's theory cannot be compatible with the "IH" model of explanation and why, then, if one is valid the other cannot have validity and vice versa, (*tertium non datur*). The validity of the example of the Condorcet's "election paradox"[4] as the extreme extension of Arrow's theorem of (im)possibility.

4. In the *Conclusion* we shall elaborate a final argument which is a request for an "exit strategy" from the theoretical impasse analyzed in precedent points (1–3). The pivotal issue is that we need for alternative social ontology model, beyond the IH and the Collective Intentionality claims. We'll attempt to focus this idea exploiting a "medium-concept" between the classical methodological-individualistic thesis of unintentionality of the best social and institutional order of F. A. Von Hayek, and the Searlean idea of Collective Intentionality: the praeter-intentionality concept.

2 Collective intentionality. From physical to social facts

According to John Searle, *facts* (whatever these may be, from an ontological point of view: obtaining states of affairs that *à la* Wittgenstein, Fregean

---
[1] An excellent criticism against any ontological-radical interpretation of individualism, as a social methaphysics is very well expressed in [26, Ch. 1–4].

propositions etc.) can be, roughly, divided into two groups: *brute* or physical facts and *mental* facts.[28, pp. 138–144]. Because of their biological complexity, mental facts derive from physical facts, but they have a greater power in causing events in the social world (apart from natural catastrophes: 67 millions of years ago a meteorite destroyed the whole biological ecosystem of our planet, causing the extinction of many animal "societies" and, on a wider scale, the extinction of the different species of dinosaurs).

Brute facts have a strictly *unintentional* origin, they are essentially based on (or they are caused by) physical events (i.e.: material, energetic events, which are situated in space and time): they can be ontological entities such as *electric shocks* (natural or atmospheric electromagnetism), the *forming of clouds, light, warmth*, the *flowing of rivers*, the *force of gravity*, but also *heartbeats*, the *activation of synaptic buttons* and their related brain's circuits and consequently with *movements* and/or *cognitive processes*. In fact, according to Searle, who refuses any sort of dualism, mind is, too, with its intentionality, an essentially biological and, for this reason, a natural process, which is then caused by a synergy of brute facts (see Fig. 2) [28, p. 139].

Mental facts originate from brute facts by means of the unique and extremely particular capacities of superior organisms like *homo sapiens*. Then, they can be divided into different evolutionary levels, indicated by Searle through an apposite hierarchical scheme (see Fig. 2).

In this essay we will not deal with all the problems deriving from the already debatable definition of "fact" given by Searle, who sometimes uses this concept as a synonymous of "causal relevant event" (ignoring or without taking into consideration the metaphysical dualism fact/event) sometimes as a precise synonymous of "proposition" (an identification which is not accepted by all the theorists of facts, especially by Wittgensteinian [28, pp. 227–231.] and Russellian critics, for whom *fact is an obtaining state of affairs*, or a positive state of affairs) [36, para 1.21]. Socio-institutional or conventional facts are the product of a creation of human intentionality, and so they have a mental origin (i.e., *à la Brentano*, they are provided with an intentional reference), even if they transcend this origin.

The relevant intentionality which causes objects or events, *by stipulation* attributes to certain properties of one or more objects, one or more events (one or more relations among objects, among events, or among objects and events) the capacity of working as a *function of status*, in a given *institutional* context. This way, the first fundamental function of Searle's collective intentionality finds its definition: the one of *function of status*, according to which, X, in a social and institutional way, counts as Y, under the aspect C (or in the aspectual context C, which has a social and institutional value, too).[28, pp. 92–129].

The most clear example is paper money: the rectangular piece of paper X (which is, by definition, an object artfully created, with brute materials such as cellulose and inks) is transformed, by means of a stipulation $X : C$, that is by laws passed by the various competent International Organisations (International Financial Institutions/World Bank, IMF, CEB etc., National Governments with their different Departments, such as National Banks, Treasury

Departments etc.) into and object which possess the value Y: the possibility if being immediately paid to the person who possess it, in a given context or *aspectual shape* [27]; as, in this case, the context of monetary interchange "C".

Symbolically, we could formalize Searle's definition as follows:

$$X : C \to Y$$

That is: the function of status considered above *maps* the dominion of objects and brute facts X into the co-dominion Y, given the by-stipulation context C.

This way, collective intentionality can be defined as that type of intentionality that, by means of the conventional stipulation of at least two individuals' intentionality, *defines and recognises the existence, the nature and the possibilities of action of given functions of status*, through an intentional act which has, stipulation for stipulation: for the same references, the same function of status; for the same institutional or conventional contexts, the same validity.

3   "Invisible hand" explanation model

There are at least three models and at least three theoretical traditions of the IH type of socio-economical explanation and prediction. The "descriptive-allegorical" model, by Adam Smith — whom we owe the terminological baptism of the IH model. This model is very important, rather fundamental and always theoretically stimulant [31, pp. 248–249], [32, p. 39], [18, Ch. I]; see also: [25, Ch. I]; the model proposed by the Austrian economical school of Carl Menger and Ludwig Von Mises (this model is very important, too, and it derives someway from the first one). We'll not analyze it. The model of the post-war monetarist and libertarian economical *mainstream* of London and Chicago represented by Von Hayek, Stigler (who has modified this same expression and definition), Friedman and, in analytic epistemology, by Robert Nozick.[2] In this essay, we will consider as analytical target the third *model of IH explanation*. It is an explanatory model that explicitly grounds on the following ontological and epistemological assumptions (A–C), by interpreting them in a *non-allegorical* way.

A) The social mechanism that is probably the most able to generate a state of balance and development from an economical aggregation (like a market of goods and services moved by agents/competitors/consumers/savers) is the process that emerges from a spontaneous series of attempts and errors. This process has as a pivotal claim, from a socio-economical view, the total freedom of economical operators to discriminate their own function of well-being, their own heuristics of problem solving. On the whole, this process has its own field of action or *behavioral setting* in a state of free competition among ideas aiming at magnifying profits and minimizing losses.[3]

---

[2][11]; Regarding the distinction between freedom as process and freedom as opportunities-rights: [30, Ch. 1–3].

[3]According to Hayek only rules can unify an extended order. Neither all pursued purposes nor all used tools or devices are known, or must be known by everyone, to be taken

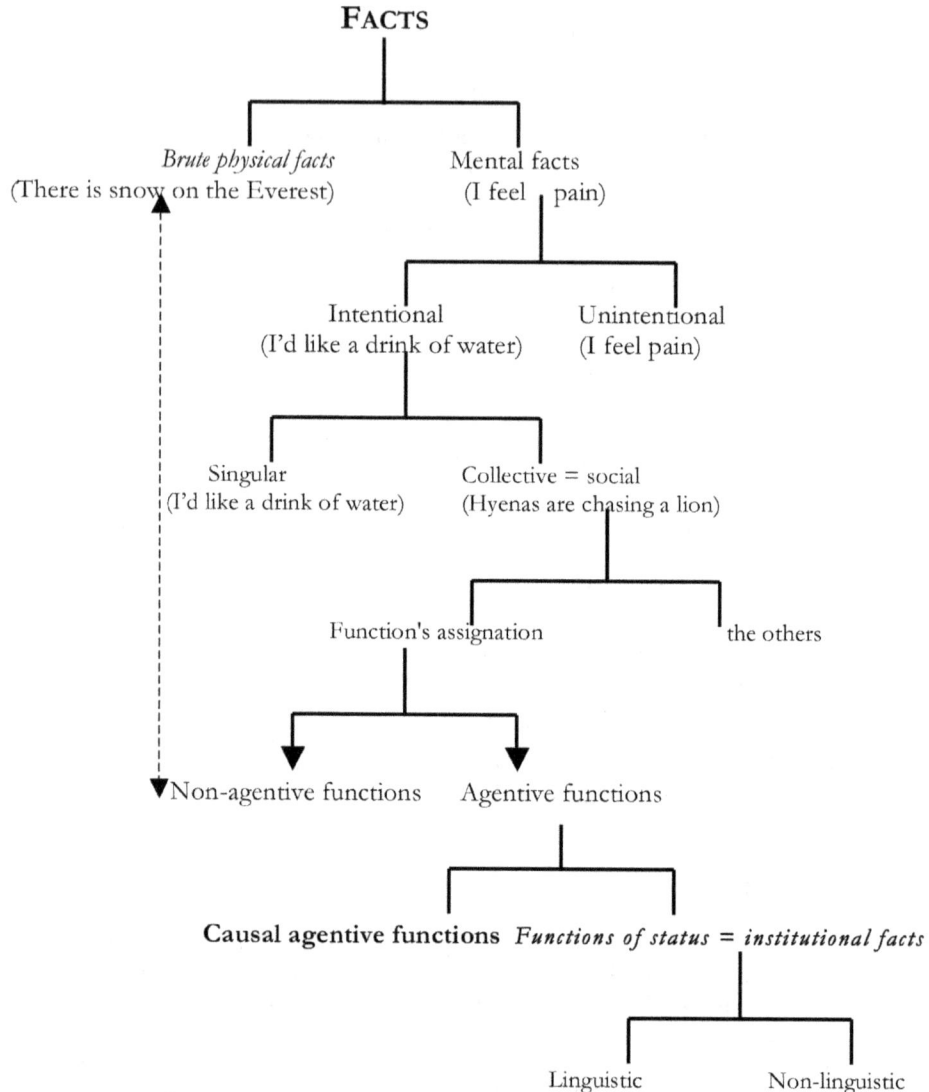

Figure 1. Hierarchical taxonomy of facts, from brute facts to non-linguistic institutional facts (e.g. money)

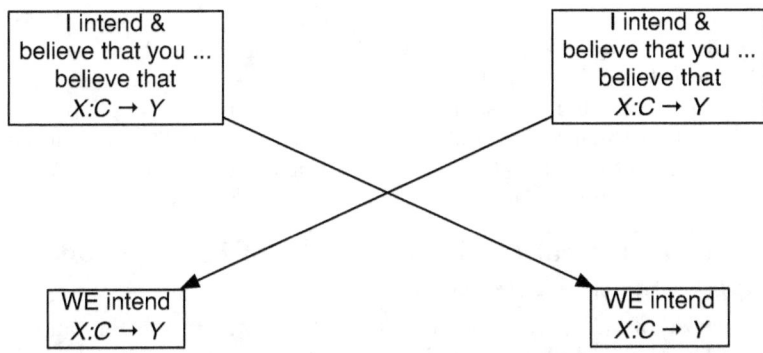

Figure 2. From the reciprocal attribution of second level intentionality (I believe that he believes that $X : C \to Y$)to the institutional creation of collective intentionality

B) Societies, social groups, social classes, their representatives, State, political parties, institutional associations, nation, *strictu senso* do not *exist* except from being "flati vocis" or, if we prefer, fictional conceptual constructions, deprived of objective ontological references.[4]

C) The invisible hand is (at the same level as what we have seen in B) among the super individual institutional organism, an instrumental predictive function useful to calculate the whole process of emergence seen in A, *ergo* it does not belong to the "extra mental equipment of the objective world".[5]

(C) raises some serious problems, because it is difficult to understand how a mechanism of the IH type can be physically responsible for such predictable

---

in to consideration into a spontaneous order. An order of this sort grows up by itself. If we had to apply the unmodified and not repressed rules (rules to take care of, in order to obtain a visible social "wealth") of the little band or small group, or our families rules to the extended order of cooperation by means of *markets*, as *our instincts and emotional desires, often, would drive us to*, we could destroy this order. In a very same way, if we would apply the (competitive) rules of the extended order to our most personal groups, we could break these ones. See also [11, pp. 50–53]. The same process cannot, by no means, be considered as a system of deliberate, intentional and fully conscious construction and consequently proper of a model of "constructivistic" and "planned" model of rationality. See also [12, pp. 24–26, 210].

[4]K.R. Popper: "Talking about 'society' is extremely misunderstanding. Of course, it is possible to use concepts as 'society' or 'social order" but we can't forget that they're just fictional concepts. What really exists are men, good and bad men indeed — let's hope that these latter are not too many. What really exists, anyway, are human beings, who are, in part, dogmatic, in part lazy, or diligent, or anything else. This is what is really existent. But, for these reasons, what does not exist is society. People, instead, believe in society's existence and, consequently, people believe that the cause of every evil in the world should be imputed to society or to social order". Quoted in [1, p. 51]. According to Popper and Hayek, the only kind of acceptable metaphysics is the metaphysics of individuals; the only kind of social ontology of individuals is the ontology of individuals belonging to our species. See also [13].

[5]This is true for the explanations of the "catallaxy"-type, too, introduced by Hayek; see also two Hayek's fundamental works such as [14, 15].

processes on the basis of its algorithms of succession and balance if not for its belonging, someway, to the physical world (Principle of causal enclosure of the physical world) [22]. Briefly: how can it be that a mechanism of real explanation does not correspond to something that is more than a simple prediction and, then, something which is *objective, real, extra-mental*? If it is a purely conventional process, then (A–C) risks to remain a mere instrumentalistic model, *à la Bellarmino-Osiander*. Popper, maybe, would not agree....

4 The strategical (and tacit) value of IH explanations for Collective Intentionality

The purpose of this paper's section is further motivate the relevance of a concept like that of Invisible Hand in a research program of social ontology. The first point to focus a good investigation strategy is to observe that any form of Collective Intentionality in the methodology of social sciences *must to satisfy*, as a necessary (but not sufficient) explanatory condition the fundamental criterion of "catallaxy explanation" as provided by Hayek's classical formulation of IH explanation model [14]. The logical force of this purpose is established by the same definition of Invisible Hand model furnished by "literalistic" or "radical" supporters of this conception of social world, the London School of Economics and Chicago University' radical individualist authors, like the same Hayek, and remarkably, into his philosophical and sociological legacy, in the well known contemporary theory of Douglass C. North and Paul David about *path-dependence* structure of social and historical facts view as *not-ergodic* or not algorithmically deterministic social phenomena [21, 5], [3, p. 5].

Some little remarks on the logical structure of IH' hayekian interpretation and in the Collective Intentionality's one, is indispensable. According to Hayek's view, the unintentional process which brings a set of possible problem-solving euristic goals to wave a real, diachronic *path* of behavioural (i.e. *economical, institutional, historical*) settings, forms the so-called *catallaxy process*. This is a completely unintentional *auto-equilibration* mechanism utilized by a set of individual agent trends to structure a praeter-individual collective context of interactivity from individual well known preferences and desires to social (unknown or partially known but optimized) goals and solutions. Catallaxy is, in few words, the best spontaneous order from the most natural unintentional "social entropy" of free individual activities of many and many "exempla" of *homo economicus* at work.

The best, classical, example is the free-market way to solve the usual problems of customers, producers, and distributors to reach an equilibrium point of common satisfaction. The question is: it is possible to conceive all this process like a sort of *growing up institutional whole*? According to hayekian methodological individualism the answer is absolutely positive, because the "catallaxy process" is *the most primitive form of proto-institutional dynamical mechanism existing among individuals, considered as rational agency-subject*. The analogy and the similarity between IH'model in the Hayekian perspective and the "primitive form of intentionality" invoked by Searle is very strong, because for Searle's model too *Collective Intentionality represents the most primitive ability to construe social references in a collective habitat* by means of *status functions*; we can, of course, postulate a sort of *explanatory isomorphism* between the two form of explanation models, stating that all these models need, in a very analogous way, to be founded by mean of some primitive, or basic, status function. The status function for free market equilibrium process by "invisible hand"'s catallaxy would be defined like "X counts as Unintentional Optimezer of economical desires and needs in the Context M (where M is a market)" ; therefore C.I. is expressed by "we intend to optimize in the most free and not-planned way the best economical problem solving in a given freechange context". But, as it is very simple to observe, this definition is *entailed* or presupposed, in some way, in Collective Intentionality' way to work, in its most basical meaning. In fact, in order to satisfy the most important property of C.I. — to be the most universal way to define institutional and social fact's status functions — C.I. must necessarily to be a not-constructivistic and a not-explicitly (not-unintentionally) planned status function, in order to avoid a very vicious form of circularity! But, not-constructivism and untintentionality, token and brought together, are the most basical properties of catallaxy processes [12, pp. 24–26,210], i.e. the minimal metaphysical properties of Invisible Hand models "token and brought in action" — according to Hayekian theory.

Anyway, it's very important — even if so much trivial, after all — to underline that *unintentional* in the case of IH's Hayekian models is not a synonymous of *not-intentional* in the brentanian-husserlian-searlean meaning of word. One thing is an intentional *will* (the full intentional range of human free or determined will) another is intentionality — the property to have, subjectively, an objective reference for every psychological state of mind; i.e.: the brentanian and post-brentanian mark of mental events and states. In this sense Collective Intentionality can be a property of a lot of social, but *unintentional* (not-voluntary) behaviour, acts and individual events, like unconscious preferences, desires, trends, tendencies; the only, necessary, condition for all these kind of mental state to be a sort of C.I. is always the same: to belong to a wide range of collective designation, as given in the typical intentional genre of "we accept/intend/desire/believe that X count as Z in the context Y".

5 Why the methodological radical individualism implicit in the model would find confirmation from the success of Searle's theory?

Even though Searle does not make any kind of explicit and direct hint at the IH model of socio-economical explanations, his purpose to "bring to light" what he defines as *"an immense invisible ontology"* [28, Ch. I–II] which informs, structures and gives birth to all facts implying social intentionality, appears to be absolutely involved in the theme of the existence and of the social efficacy of IH [33]. According to Searle, only an ontology founded on collective intentionality, in fact, can be accounted for the existence, *de facto* and not only *de dicto*, of a complex of social and institutional mechanisms perfectly invisible but, nevertheless, perfectly efficient and able from a causal point of view, as in an economy of exchange.

In this case, a valid example can be the one of the Parisian pub [28, p. 9], and Barry Smith interpretations, which consider, critically, Searle's theory as a possibility to interpretate and to explain the phenomenon of the creation and global diffusion of the financial capital as a "virtual reality", physically not dependant on the "brute" occurrences of the availability of liquid money or of gold and oil [28, pp. 8–12]. The clearest example is conemporary capitalism [6].

Caveat: if this attempt should fail, then the problem raised above about the causal efficiency of the IH would not only remain unresolved, but would risk to get more difficult, because we would lose a very rich solution. A fortiori, the ontology of the IH models would risk to tremble.

Moreover, with the ontological and epistemological model of the IH explanations, Searle agrees about an extremely important premise: the *ontology of the radical methodological individualism*. According to this premise, and to Searle, too, as we have already seen, *the only effectively existent entities in the inventory of the human social world are only the individuals belonging to the homo sapiens-sapiens species*, as they only can live an existence of the same physical type as "brute facts". All the rest appears to be "observer dependent", i.e. "dependent on the conditions of the observer's possibility of experience" [28, pp. 213–219].

Also, inside this "rest" there are facts that, as we have seen, are further more complex, as non-linguistic institutional facts, that represent the conceptual peak of the articulations of the collective intentionality.

In fact, according to Searle, the social ontology of collective intentionality not only is far from contradict the fundamental axiom of radical methodological individualism, but helps to understand how and why it is possible to pass from a world in which, nominalistically (metaphysically speaking) we can only talk about individuals and for this reason (physically speaking) only of particles, waves and force fields, to a world in which, as there also exist aggregations of particles and particular force fields, as human beings are, there may also exist interesting and relevant institutional facts, as those that are instituted and ruled by collective intentionality [28, Ch. II–III].

Even though the objects of such ontology are not formed by "brute" or "rude" physical circumstances (material, energetical, informational), they are surely as objective as the individually existent objects, objects like those that are allowed to be the only *obiecta realia* of individualistic ontology. The example Searle refers to is taken from Frege: the fact that the Northern Sea is only a region of the planet individuated and defined *ad arbitrium* on the map of the boreal hemisphere does not make the Northern Sea or the North Pole less objective. They remain extra-mental entities and facts, anyway[9, Section 26].

6 Why, then, Searle's theory and the IH model of explanation result in being reciprocally incompatible (and what are the consequences of this incompatibility)?

Let's suppose that Ψ is an electoral community of a small town in South Dakota's, where only three persons live: α, β, γ, These three electors can only choose among three candidates to the United States Congress: A, B, and C — respectively a republican, a democratic, and a pro-life independent candidate.

According to Searle's theory of Collective Intentionality, even if Ψ is an extremely small group of individuals it can express, *hypothetically*, a Collective Intentionality which, in turn, is able, by definition, to express *collective preferences*. An intuitive way to develop a scale of these preferences could be the assumption that one choice is preferred to another one *if the majority of the community prefers the first to the second one*, and that it would prefer the first to the second if the two choices were the only ones available.

Let's assume that A, B, C are the three alternatives and that α, β, γ, are the three electors. *Ex hypothesis*, it happens that α prefers A to B and B to C (and that, consequently, α prefers A to C, by Transitivity Law), that β prefers B to C, and C to A (and that, consequently, B to A), and that γ prefers C to A, and A to B (and, consequently, C to B). It follows that a majority prefers A to B and that another majority prefers B to C. If we follow Searle's theory of the Collective Intentionality, we should conclude that the collectivity of individuals Ψ (that is the only way in which the nominalistic ontology of methodological individualism allows us to talk about this small community of South Dakota's mountains, *sine multiplicare praeter necessitatem* — prefers A to B and B to C.

If the Invisible Hand of the Collective Intentionality, apart from being a sort of "guardian angel" of that community, could have a power of making the intentional choices of α, β, γ, *capable of reaching, involuntarily, the maximum of rationality* (as it should be *ex hypothesis*), then it would be reasonable to conclude that the involuntary but also the best consequence of the rational choices of α, β, γ, is that the majority of Ψ does prefer A to C (according to the Transitivity Law when it is applied at *the level of the whole majority*). However, what really happens is that the majority of Ψ prefers C to A, as it is possible to verify through a complete check of the (intentional) votes (C has been preferred to A by only two votes).

This is the well knows *paradox of the election*, discovered by Condorcet but formalized during the 19th century and that, through the explicit acknowledgment of Kenneth Arrow, it is the foundation of the intuition from which the Yale economist derived its famous demonstration of 1951 of the theorem of the impossibility for systems of preferences with a transitive function of well-being.[6]

Moreover, according to Arrow this paradox is only a particular case, predicted by the demonstration of his theorem. This way, *the simplest method*

---

[6][20, pp. 197–240]. Quoted in [2, p. 4–5]. This theorem was successively extended to quasi-transitive ordering functions by the 1998 Nobel Prize A. Sen.

of calculation which can be used to pass from the counting of the individual preferences to the form of itemization that the Collective Intentionality should recompose in the most rational preferences of the collective intentionality falls into the following dilemma:

(a) Either it does not maintain, at a super-individual level, a minimal condition of rationality as the Not-Contradiction. (b) Or it attributes to the choices of the collective intentionality beliefs of preferences which are completely false from the point of view of the single persons (and, for this reason, it is completely incompatible *contra Searle* with (radical) methodological individualism)[7]. If instead of electors we had simple consumers, and instead of electoral lists we had products (such as mobile phones brands or operative systems for PC), the final situation would remain the same. The Invisible Hand, considered as Collective Intentionality *à la Searle*, is not able to preserve the individual rationality of economical subjects (the sole expression of rationality that, *ex hypothesis*, IH would maintain in its ontology). This should not amaze us. The main reason for this impossibility is that, being the collective intentionality *a form of intentionality*, no "spontaneous order" could be at the same time (a) rational effect resulting from intentional forces (b) non-intentional rational effect, maintaining the principle of non-contradiction.

7 Conclusions: beyond the Invisible Hand and the Collective Intentionality ontological models: *praeter* (beyond)-intentionality

Even if John Searle has never been directly interested with all the problems linked to a kind of explanation like the IH model's, the remarks excerpted in precedent pages permit us to derive some relevant conclusions and implications, with special regard to compatibility of Collective Intentionality (CI) model — that Searle never defines as a "olistic" or a "anti-individualistic" explanation model in social sciences — with IH explanation model.

A first conclusion — only *prima facie* not relevant for our analysis of CI — that may appear just hardly evident from our previous analysis, can be summarized as follows: the two models of the IH explanation and of the *radical* methodological individualism, which apparently serve one another as theoretical and metaphysical pivotal claims, are two explanation-prediction and metaphysical ontological models too strong or too much radical, and, if considered with attention, are hardly compatible because of their "ideological force". It is always possible to see in CI an attempt to unify them logically using a notion as the one proposed by Searle; but for all our observations and objections this attempt does not work, because even though Searle's CI can multiply among them the values of a collective matrix of individual intentionalities, a product of collective intentionality cannot ever be, by definition, the isomorphic correlate of a not planned and unaware un-intentionality, *without changing*, in fieri, *the meaning of the words*. In facts, if something like an Invisible Hand is only a useful fiction (or a useful social allegory, as, actually, Smith believed), a fiction that does not hide any real, extra-mental,

---

[7] *Infra*.

objective mechanism, then it's rather difficult to understand how some real social dynamics could incorporate it into their proper models of realization of a *socio*-economical equilibrium. However, on the other hand, if the Invisible Hand is a process that affects some real social wholes (enterprises, societary assets, etc.) *then it's impossible to understand how the "austere" nominalistic ontology invoked by authors like Popper and Hayek, could have reason to have a future.* We should also, at least, accept in our social ontology the economical operators' mental states (intentional or not), the logical-mathematical structures implied in their description and the relationships among individuals and among individuals and social aggregations, and so on. In this way, metodological individualism would disappear, giving space to a quasi "baroque" social realistic ontology.

The fact that methodological individualism could present some relevant weakness is, for example, the theoretical basic issue of classical methaphysical criticism of David-Hillel Ruben. According to Ruben "The chances of methodological individualism being true see to be very slim indeed", (where Ruben for "metaphysical individualism" roughly means the denial that there are social entities and substantive social properties). In the first of the *The Methaphysical of Social World's* four chapters [26, Ch. 1)] he contends that there are social entities which cannot be reduced without circularity. His example is France, a social substance — according to methodological individualism — which is not identical with the group of French. Why not, asks Ruben? French persons are not all and only those inhabiting a particular geographical area or those descended from a specified individual. To group them as "French" is merely to relate them to France. To group them by their beliefs and attitudes is to miss the point that the beliefs and attitudes nee to be about *France*. Furthermore, according Ruben (Chapt. II), the relation of members to groups is not always that of parts to wholes. A part of a part of a table is a part of a table, whereas France's membership of the United Nations does not make French persons members of UN — or, strictly speaking in an husserlian-mereological language,[8] it doesn't make French persons *not-indipendent* members of UN (differently from the case of European citizenship, by which a French citizen of course is a member of UE!).

Following our opinions, the real problem lies on the "fundamentalistic" stress put on the importance (at this point, maybe, "unimportant"), of the level of the "ontological radicalism" of the individualistic assumptions, and of the concept of intentionality implied into the evaluation of social processes. The consequences of an economical investment on the long term, in facts, should not be considered either as something unintentional or as something strictly intentional (because the intentional references of an investment's gain can surely change during the time, or just be not valid anymore). The most apt concept is of an intermediate type and it is different either from the concept of *unintentionality* or the concept of *intentionality* and it is the concept

---

[8]In fact, as a hole in the pattern of my shirt in the suitcase is not a hole in the suitcase, (the hole is an Husserlian independent "missing part" of the suitcase, but it is not a not-independent part of the bag) in the same way, an individual citizen could form an independent part of a social group without circularity [17]: [34, pp. 15–109].

of "*prater-intentionality*" (*beyond-intentionality*). The *causal* consequences (and so, not necessarily reducible to a only rationale, following the IH model) of investments, bets, social risks, financial savings, and of other forms of economical acts, should be defined as prater-intentional events, that is as something that does not cancel the ontological co-participation of the individual intentionalities of the single agents, but, at the very same time, as something that recomposes them structurally in a real *social vectorial structure* more-than-simply-intentional, that is, that goes further than the primal intentionality. From a biological perspective, the best model to be candidated to the position of the praeter-intentional operator of the primal or direct intentionality's transformation is the model of *ex-aptation*[9], that is of a re-usability of a primal adaptative function of an organism into a secondary function, inside a given behavioural setting of a niche or, even of a given ecological habitat. This point forces us to redefine our ontology of social wholes in a further more complex way than Searle's, by adapting it to "ecological" and neodarwinian models of comprehension.[10]

This is not a overtaking of the individualism *tout-court*; on the contrary, *two subjective intentionalities do not result*, for us, *into a collective* (hypostatised) *intentionality*, but always into two individuals' subjective intentionality. We can only accept the moderate individualistic *mantra* of John Stuart Mill, according to which "When men put them together they do not transform themselves into a substance of a different kind",[11] modifying some of its presupposition: two or more individuals that form an association do not form *ex-novo* a social "individual" (a sort of "collective Leviathan" or an "arithmetic reification" of Rousseau's General Will) but do activate new *praeter-individual properties* which derive from basic individual properties (praeter-intentionality and intentionality). This way, if we take a *mereological* look at the socio-institutional composition of the so-formed new whole, we will discover in it a mere effect of *diachronic recomposition*, whose consequences, as regards the *original, ancestral* intentions and references, will be evaluated only from the point of view of the prater-intentional history. New interventions of human intentionality will have new prater-intentional consequences, which *a priori* will not be completely predictable, and so on, *ad libitum*, putting the statistical knowledge of effects and the degree of their incalculability in a relation of inverse proportionality. Thus, societies, groups, small associations (such as the so called "intermediate bodies", like families and little associations) or bigger ones (such as multinational enterprises, nations, governments, churches etc.) cannot "spontaneously" avoid (as Arrow himself did predict [2, Chap. III]), dictatorial systems of control of the exercise of individual pref-

---

[9]According to the evolutionistic biology, DNA's mutations, i.e. variations that, in their turn do provoke changes on the genomic product, are usually not favourable, while neutral mutations can represent what today we can define "exaptations", i.e. evolution's characteristics which originally had other purposes (or no purpose at all) and which, later, have been co-opted for their actual role, as explained in [10, pp. 8, 4–15].

[10]"More complex" does not mean *ipso facto* "absolutely incompatible": between the definition of "institutional fact by means of a *function of status*" and "social *fiat* object", as it can be seen infra, there are some interesting analogies, anyway.

[11]Quoted in [19, VII, I].

erences, either an issue of the type "we intend/we believe/we accept" or of the type "public virtues from private vices". Nevertheless, such structure will continue, prater-intentionally, to emerge, granting a re-collocation tending to an optimum of the behavioural resources, which have been object of investment.[12]

BIBLIOGRAPHY

[1] D. Antiseri. *Relativismo, individualismo e nichilismo*. Rubbettino, Soveria Mannelli 2005.
[2] K.J. Arrow. *Social Choices and Individual Values*. Wiley, New York, 1951-1963. Italian translation: *Scelta sociale e valori individuali*, ETAS, Milano 2003.
[3] W.B. Arthur. Self-reinforcing mechanisms in economics. In *The Economy as an Evolving Complex System*, Santa Fe Institute Studies in the Science of Complexity, Addison-Wesley, Redwood City, CA 1988.
[4] N. de Caritat (Condorcet). *Essai sur l'application de l'analyse á la probabilité des décisions rendues á la pluralité des voix*. Paris 1785.
[5] P.A. David. Path-dependence, its critics and the quest for "historical economics". In P. Garrouste and S. Ioannides (eds.), *Evolution and Path Dependence in Economic Ideas: Past and Present*, Edward Elgar, Cheltenham 2000.
[6] H. De Soto. *The Mistery of Capital. Why Capitalism Triumphs in the West and Fails Everywhere Else*. Basic Books, New York, 2000; Italian translation: *Il Mistero del Capitale. Perchè il capitalismo ha trionfato in Occidente e ha fallito nel resto del mondo*, Garzanti, Milano 2001.
[7] J. Dewey. *Experience and Nature*. Open Court, La Salle 1925.
[8] P. Di Lucia (ed.). *Ontologia sociale; potere deontico e regole costitutive*. Quodlibet, Macerata 2003.
[9] G. Frege. *Die Grundlagen der Arithmetik*. Breslau, Köbner 1884.
[10] S. J. Gould and E.S. Vrba. Exaptation. A missing term in the science of form. *Paleobiology*, 8: 4–15, 1982.
[11] F.A. von Hayek. *The Fatal Conceit*. Chicago University Press, Chicago 1988.
[12] F.A. von Hayek. Scientism and the study of Science. In *The Counter-Revolution in Science*, The Free Press, Glencoe, ILL 1952.
[13] F.A. von Hayek. *Individualism and Economic Order*. Routledge and Chicago University Press, London and Chicago 1948.
[14] F.A. von Hayek. *The Constitution of Liberty*. Routledge and Chicago University Press, London and Chicago, 1960.
[15] F.A. von Hayek. *Studies in Philosophy, Politics, and Economics*. Routledge and Chicago University Press, London and Chicago 1967.
[16] F.A. von Hayek. *New Studies in Philosophy, Politics, Economics and the History of Ideas*. Routledge and Chicago University Press, London and Chicago 1978.
[17] E. Husserl. *Logische Üntersuchungen*. Martinus Nijoff, Den Haag; III Ünters, 1900–1901.
[18] B. Mandeville. *The Fable of Bees or Private Vices, Public Benefits*. Capricorn Books, New York 1926.
[19] J.S. Mill. *A System of Logic, Ratiocinative and Inductive*. Harper and Brothers, New York 1874.
[20] E.J. Nanson. *Transactions and Proceedings of the Royal Society of Victoria*. Vol. 19, 1882.
[21] D.C. North. *Understanding the Process of Economic Change*. Princeton 2005.
[22] D. Papineau. The rise of physicalism. In M.W.F. Stone and J. and Wolff (eds.), *The Proper Ambition of Science*, Routledge, London 2000.
[23] A. Quinton. Social objects. *Proceedings of the Aristotelian Society*, 75: 1–27, 1975.
[24] A. Reinach. Die Apriorischen Grundlagen des Bürgerlichen Rechtes. *Jahrbuch für Philosophie und Phänomenologische Forschung*, 1: 695–847, 1913.
[25] A. Roncaglia. *Il Mito della Mano Invisibile*. Laterza, Roma and Bari 2005.
[26] D.-H. Ruben. *The Methaphysics of Social World*. Routledge and Kegan Paul, London 1985.
[27] J.R. Searle. *The Rediscovery of Mind*. The Mit Press, Cambridge, MA 1992.
[28] J.R. Searle. *The Construction of Social Reality*. Free Press, New York 1995.

---

[12][18], quoted in [16].

[29] J.R. Searle. Ontologia sociale e potere politico. In [8].
[30] A. Sen. *Rationality and Freedom*. Harvard 1999.
[31] A. Smith. *The Theory of Moral Sentiments*. A. Millar, London, 1759. D.D. Raphael and A.L. Macfie (eds.), Oxford University Press, Oxford 1976; Italian translation: *Teoria dei Sentimenti Morali*, Istituto dell'Enciclopedia Italiana, Roma 1991.
[32] A. Smith. *An Inquiry on Nature and Causes of the Wealth of Nations*. W. Strahan and T. Cadell, London, 1776. R. H. Campbell and A.S. Skinner (eds.) Oxford University Press, Oxford, 1976; Italian translation: *La Ricchezza delle Nazioni*, Newton Compton, Roma 1995,
[33] B. Smith. Un'aporia nella costruzione della realtà sociale. In [8].
[34] B. Smith and K. Mulligan. Pieces of a theory. In B. Smith (ed.), *Parts and Moments. Logic and Formal Ontology*, Philosophia Verlag, Munchen 1982.
[35] B. Williams. Formal and substantial Individualism. *Proceedings of Aristotelian Society*, 85: 1984 -1985.
[36] L. Wittgenstein. Logische-Philosophische Abhandlung. *Annalen der Naturphilosophie*, 1921. Engl. tr. by F.P. Ramsey and C.K. Ogden, *Tractatus Logico-Philosophicus*, Kegan Paul, London 1922.
[37] C. Znamierowski. O przedmiocie i fakcie spolecznym [Social objects and social facts]. *Przeglad Filozoficny*, 24: 1–33, 1921.

The naturalness of religion and the action representation system

SERGIO LEVI

1 Introduction

During the last two decades attempts to outline an evolutionary framework for the study of religious universals (and other cross-cultural traits) have focused on two aspects and followed two main strategies. One strategy is to consider the degree to which a given religious concept approaches a cognitive optimum which is judged crucial to the concept's chance to stabilize and spread. A different strategy puts more weight on the role played by different sorts of human memory systems in the transmission of religious forms. As we shall see the former corresponds to the naturalness-of-religion thesis; the latter to the modes-of-religiosity model.

Both the naturalness-of-religion thesis and the modes-of-religiosity model are meant to explain the origins of religious universals by an appeal to cognitive structures in the human mind. Both approaches regard our cognitive structures as a key to explaining the persistence of religious beliefs (or why they have such a strong hold on the human minds) and both try to show that religious traditions can count for their persistence on the same cognitive functions on which non-religious cultural forms also depend. However, the naturalness of religion involves a much stronger claim on the role played by cognitive optimality in constraining (and shaping) our capacity to understand religious doctrines and narratives and rituals.

It is important to note at the outset that claiming that religion may be a matter of natural cognition is not meant to defeat the possibility that its objects may involve a supernatural dimension. The old-fashioned natural history of religion usually conceived of supernatural entities as not so different from human persons or other human-like legendary characters. Instead the naturalness of religion is not a thesis on the nature (or existence) of religious entities, but on the status of religious concepts — the concepts we actually use (as opposed to those we believe we use) in reasoning about religious entities.

At the heart of this approach lays the assumption that human individuals come equipped with an evolved Action Representation System (ARS) — a complex machinery whereby humans generate detailed representations of observed, imagined or planned actions — and this paper is also an attempt to stimulate discussion on the explanatory adequacy of the ARS as a unified cognitive system. Old-fashioned speculations over the origin or function of religious behaviour have much benefited from the introduction of this method for the analysis of action in general, and of ritual action in particular. And the investigation of how the ARS is supposed to work has greatly boosted the maturation of interest in the cognitive science of religion.

2 The "naturalness-of-religion" thesis

Over the last two decades an increasing number of religious scholars and psychologists have been endorsing the view that religious cognition might be not so different from non-religious, ordinary (social) cognition. The basic idea behind the "naturalness-of-religion" thesis [2, 7, 8, 14] is the observation that although theological concepts may involve radical violations of fairly natural, ontological assumptions, the concepts used in reasoning about supernatural entities by religious and non-religious people tend to be much simpler and more similar to ordinary concepts of human-like agents.

Thus in a famous empirical survey Barrett and Keil [3] found that

> when adults in India and the United States reflected on their theological ideas about supreme beings, they generated abstract, theologically correct, descriptions of gods that have no physical or spatial properties, are able to know and attend to everything at once, and have no need to rely on sensory inputs to acquire information. However, when comprehending narratives about the same deities, the same adults mistakenly remembered the god of the narratives as having a single location in space, as being unable to attend to multiple events at once, and as needing to see and hear in order to complete otherwise fallible knowledge [2, p. 89].

Dozens of developmental studies have shown that human beings (from an early age) entertain tacit, intuitive assumptions about the distinctive properties of different categories of living, moving, and thinking things. Such assumptions appear to be violated both by complex theological concepts as well as by more ordinary religious concepts. But while theological concepts are designed to violate various basic-level assumptions (thus becoming hardly graspable), religious concepts tend to violate only a few feature-level assumptions (thus being highly memorable). Some violations may amount to discounting some feature-level property: ghosts, although thinking, human-like creatures, are allowed to penetrate solid walls. Other violations may imply a cross-domain transfer of feature-level properties: a statue, although inanimate, may be able to hear people's prayers and shed tears [9].

A related claim involved in the "naturalness-of-religion" thesis is that only by approaching a cognitive optimum (that is, only by being easy to grasp and recall) can the concept of a supernatural entity hope to stabilise and spread over time. Pascal Boyer, who firstly speculated on the contrast between religious and theological concepts, has dubbed the former — that is, the concepts involved in religious reasoning — "minimally counterintuitive concepts" [7, 8].

The most represented category of things is that of intentional agents (most of them deities) and for good reasons. As Justin Barrett pointed out, the importance of being capable of detecting any source of agency in the environment is the best explanation for why our cognitive mechanism for detecting agency is still so sensitive as to make us impute agency when no real agent is around. As he famously put it [2] the human mind evolved an "hyperactive agent detection device" (HADD).[1]

---

[1] To support this claim one needs to have a prior theory about what it is to have the

What is the relation between religious concepts (in this fairly technical sense) and religious rituals? According to Barrett "people spread religious concepts in the context of shared religious actions"; on the other hand, religious actions are executed "in response to those concepts" [2, p. 92]. One would expect that religious ritual actions should correlate with theological, not religious concepts, because there is a clear parallel between ritual exploration and theological exploration. Both attempt to transcend the limits that our tacit assumptions impose on much of our ordinary cognition — including those religious concepts we use when we need to remember narratives, or when forming expectations on their basis. Barrett, on the contrary, goes on to say that "cognitive scientists of religion argue that [...] ordinary cognition both structures religious practices and underlies the representation (and thus the execution) of religious actions in participants' and observers' minds" [2, p. 92].

When it comes to explaining the function of religious behaviour, the proponents of the naturalness-of-religion thesis generally concede that rituals are a special type of actions: "Someone does something to someone or something in order to bring about some non-natural consequence. That is, rituals are actions that are performed to accomplish something that would not normally follow from this specific action" [4, p. 216]. On the other hand, the relevant participants are supposed to be playing the same sorts of roles (Agent, Instrument, Patient) and to bear the same relations to one another as they do in ordinary social action: "Structurally, religious rituals mirror social actions: someone performs some kind of action in order to motivate another's action or change in disposition. It just so happens that the person being motivated to act is a god or other non-natural agent" [4, p. 217]. It should be noted, at any rate, that the alleged similarity is more a matter of action representation than a matter of action execution; despite their unusual qualities, conclude Barrett and Lawson, ritual actions "are cognitively represented as actions [...] using ordinary resources" [4, p. 216].

In order to prove the existence of that similarity Barrett and Lawson interviewed a group of "ritually naive individuals" (128 North American Protestant college students) who were asked to select which one from among a dozen ritual variations they considered more likely to achieve the intended non-natural effects. The results seem to confirm two hypotheses about what is judged most important for the ritual to succeed:

(1) the possibility to represent superhuman agency *somewhere* in the ritual structure;

(2) the presence of an *appropriate agent*, that is, of an agent which is "capable of the right intentions" [2, p. 94].

A host of scholars had assumed that more than the agent's intention the crucial ingredient of ritual action was the agent's overall power — as this can be expressed by how well the action is executed.[2] Yet the notion that an

---

concept of a natural (human-like) agent. Studies about the cognitive mechanisms allowing us to detect self-propelled movement are also relevant to complete a picture of the human capacity to deal with agency-related facts in the environment [5].

[2]Perfect execution is considered critical — or the very purpose of one's actions — in many

essential feature of ritual is perfect execution is dubbed "the folk notion" of ritual [4, p. 218] or "the popular conception of magic" [2, p. 94]. These radical (if empirically testable) claims on what is essential to ritual (but perhaps we should say, to ritual representation) are of a piece with the above distinction between theological and religious concepts, as I will try to argue.

3 Ritual competence and ritual form

The hypotheses discussed and tested by Barrett and Lawson [4] — and by many others since then — were inspired from the theory of ritual competence proposed by Thomas Lawson and Robert McCauley in their 1990 groundbreaking work, *Rethinking Religion* [14]. There, Lawson and McCauley, in the context of lying the foundations of a cognitive science of religion, were endeavoring to figure out how an idealized religious participant might tacitly represent the structure of the ritual acts she is participating in.

The theory of ritual competence they proposed describes the workings of the Action Representation System, which is the cognitive system every ritual participant is supposed to utilize to understand (and potentially execute) the relevant actions. The system yields structural descriptions of seen or executed actions; their general form is always comprised of at least the following *roles*: Agent, Action (perhaps by means of some Instrument), Patient — to which a few categories for goals, times, locations or other relevant conditions can be added.[3]

Consider the example of a parishioner blessing himself. And suppose also that the way he blesses himself is by making the sign of the cross after dipping his fingers into a font of holy water. Much simplified the structure can be represented by the following diagram:

Agent		*Action Complex*		*Patient*
Parishioner	$\Longrightarrow$	Bless with water	$\Longrightarrow$	Parishioner

The above diagram is understood as providing a structural description of the action representation produced by an idealized participant — the representation, that is, of the parishioner blessing himself (as I have supposed) by making the sign of the cross after dipping his fingers into holy water.

Now the first thing to note is that in order to tell how the action may ever achieve a non-natural goal we need to specify that the action complex includes an action condition (an Instrument). Moreover, a full description of the ritual action will need to report that the water was made into a ritually

---

non-religious social contexts. And there must be some reason if religious interpretations the world over tend to place so great a emphasis on perfect execution. As Daniel Dennett puts it, "There is artifice in the design and execution of religious practices, as anyone knows who has ever suffered through an ineptly conducted religious ceremony. A stammering and prosaic minister and boring liturgy, shaky singing from the choir, people forgetting when to stand and what to say and do — such a flawed performance can drive away even the best-intentioned congregants. More artfully celebrated occasions can raise the congregation to sublime ecstasy" [11, pp. 153–154].

[3]The origin of the categories of agent and patient dates back to Aristotle; they were used throughout the middle ages by grammarians, and during the twentieth century have been applied to the analysis of the logical form of action sentences [18].

efficient Instrument by means of some preceding action (an enabling ritual) — for example, by being purified by some priest. Again the diagram for this enabling ritual action will be the following:

Agent		Action Complex		Patient
Priest	$\Longrightarrow$	Purify	$\Longrightarrow$	Water

In this case the special quality responsible for the ritual success is possessed by the priest, and it is here that the connection with superhuman agency becomes essential. To offer another example, consider baptism. As Barrett puts it: "A baptism is only a man wetting an infant except that the man is understood to be acting in the place of a superhuman agent" [2, p. 94]. The formal system provided by Lawson and McCauley [14] includes also rules that generate complex descriptions (in the form of tree diagrams) of the likely representations operating in the minds of religious idealized participants. Such representations are believed to reliably capture tacit intuitions about those features of ritual actions that are most critical for their chance to succeed [20]. Taken together, these action representations express the general ritual competence of an idealized ritual participant, together with his biases and predictable expectations.

The theory of ritual competence first developed by Lawson and McCauley [14] was later improved in their *Bringing Ritual to Mind* [16]. As the authors make clear, the ritual form model there deployed is not much different (in both method and assumptions) from the theory of ritual competence. The new model maintains that ritual actions are capable "of triggering tacit intuitive judgments about the appropriate forms that [religious] rituals should take" [22, p. 33]. McCauley and Lawson are ever more convinced that theorizing about ritual forms from a cognitive viewpoint involves modeling cognitive processes and showing their influence on behavior, without paying attention to the alleged meanings of ritual actions. They are confident that "While the meanings associated with rituals may vary, such variability typically has no effect on the stability of the ritual actions' underlying forms" [16, p. 9].

According to this model, rituals bring about their non-natural goals by virtue of the fact that they involve transactions with culturally postulated superhuman agents (CPS-agents) *independent of* the participants' states of mind. This would seem to clash with Barrett and Lawson's claim that the agent's special powers depend on his actual intention [2]. To deal with this problem McCauley and Lawson say that in order to decide which quality is essential we need "to specify, when necessary, what makes the agent eligible to perform the action, what properties a particular act must possess, as well as the qualities of the patients that make them eligible to serve in that role" [16, p. 17].[4]

---

[4] In many ritual systems the eligibility of participants (whether they serve as agents or patients) is an issue to be settled through ritual action, not a conclusion based on ministers assessing credentials. Recognizing genuine believers from imposters is among the very functions of ritual itself [10, 21]. Every ritual system has a number of clues to assess the authenticity of credentials advertised by self-appointed ministers. For "If [the priests] are imposters, ritual failure looms" [16, p. 17].

By and large the ritual form model fails (no less than the theory of ritual competence) to recognize the importance of memory and motivational dynamics to issues of cultural transmission and evolution; and it also tends to simply identify the problem of action execution with that of action representation. Justin Barrett (as we saw) went so far as to make the following statement: "cognitive scientists of religion argue that [...] ordinary cognition both structures religious practices and underlies the representation (and thus the execution) of religious actions in participants' and observers' minds" [2, p. 92].

Now there may be some truth to the claim that theological stipulations tend to be disregarded (or even "violated") when religious people need to process information about gods and supernatural beings acting in space-time. Problems arise when claims (or even findings) on how supernatural concepts are represented by means of ordinary cognitive resources are uncritically assumed to support the further claims that this very same ordinary cognition (1) structures religious practices and (2) underlies the representation and thus (3) the execution of religious actions. The third step (from action representation to action execution) will require more argument if it is to appear less questionable. But on closer inspection, the three claims can all be questioned in the light of the fact that our emotional and motivational systems tend to be more efficient than pure cognition in guiding the construction and transmission of ritual behaviour. It would be no objection to observe that this has more to do with religious transmission than ritual form because (as we shall see presently) what a given ritual form is like is not independent of how the ritual action is memorised and recalled and actively passed on to the next generation.

A major problem for both the theory of ritual competence and the ritual form model has to do with the underlying assumption that cognitive optimality is all there is to religious transmission. For one thing, defining religion exclusively in terms of traits that cluster around the cognitive optimum position has the consequence of rendering very hard to exclude phenomena such as beliefs in fairies and compulsive behaviors from the province of genuinely religious phenomena. In fact beliefs and narratives that cluster around the cognitive optimum position are often considered by their own entertainers as more about fictional characters than really existent deities. In a previous paper [15] I suggest that this paradigm may leave open many issues also as an explanation of the quasi-stability of human traditions.

On the other hand, the notion of cognitive optimality is incapable of accounting for the panoply of cults and religious systems over the world in which the gods and supernatural entities are not at all easy to grasp. Very often supernatural agents are such that a great deal of cognitive investment is required to grasp and transmit them. Understanding these concepts may presuppose highly elaborated bodies of cultural knowledge and require costly support in terms of both memory and motivation. And very often the latter concepts are quite difficult to grasp, remember and transmit. They presuppose complex bodies of exegetical knowledge and the mastery of technical details, and "require special mnemonic support in the form of routinized narrative rehearsal"

[22, p. 55]. Another flaw of the naturalness-of-religion thesis is that it conceives of memory as a general, all-purpose cognitive device, whereas there are important differences between implicit and explicit memory, and (with respect to explicit, long-term memory) between semantic and episodic memory.

4 Memory systems and modes of religiosity

Bearing on the different types of memory systems, Harvey Whitehouse has elaborated a theory of religious and ritual transmission that revolves around two divergent "modes of religiosity": he has called them the doctrinal and the imagistic mode of religiosity [22].

A *doctrinal* mode of religiosity is one in which ritual actions and doctrinal teachings tend to be highly routinized. Frequent repetition leads to the activation of two types of memory systems: implicit memory for ritual actions (that get rehearsed like behavioral scripts) and long-term semantic memory for religious doctrine (which allows the transmission of complicated bodies of knowledge). In order to prevent the rise of a "tedium effect" and the consequent fall of motivation the faithful are induced to believe in otherworldly sanctions and rewards by the use of rhetoric and logically integrated theology. This in turn favors the emergence of religious leaders and of methods for safeguarding the orthodoxy and reducing the volume of personal exegetical reflection (central authority).

An *imagistic* mode of religiosity is characterized, on the other hand, by highly arousing and infrequently rehearsed rituals; for example, initiation rites. Rarely performed rituals tend to trigger vivid and enduring episodic memories (called "flashbulb memories"). If the high levels of arousal provide a motivating force for the transmission of the ritual within the community, infrequent repetition hinders the establishment of doctrinal knowledge, and fosters spontaneous exegetical reflection (SER) which is often experienced as personal revelation. The ensuing lack of orthodoxy and centralization and leadership makes room for the establishment of emotional bonds that render the ritual group more and more exclusive and its traditions less capable of spreading outside the group (without substantial mutation).

Modes of religiosity should be thought of as "attractor positions". Whitehouse's idea is that there is a tendency for religious systems (or more likely, for single religious rituals) "to gravitate toward" this or that divergent position [22, p. 76]. Elements belonging with the two modalities are capable of interacting with one another within the same religious system. For example, members of a religious community considered doctrinal (e.g. Turkish Islam) may also participate in rituals considered imagistic (e.g. Muslim circumcision) [19]. Being more concerned with the various ways in which religious forms can be actively passed on to the next generation, the modes-of-religiosity theory can offer a more concrete picture (than that provided by the naturalness of religion) of the strategies adopted by religious leaders to preserve the traditional forms from disruption. And the theory's empirical adequacy enhances its explanatory power. Instead of positing cognitive structures to explain how what we observe is generated over and over again, Whitehouse is in a position to reconcile the cognitive science of religion with the multifarious dimensions of really existent religious systems to be found all over the world [12]. The wealth of forms is thus rescued both from the scientific need to homogenize and from the anti-scientific decision to show deferential respect to any fragmentary non-conceptualized datum. Thus the claim that ritual participants tend to consider the intentional states of the agent to be the crucial requisite for the success of the ritual [4] is falsified by observing the actual behavior of participants in ritual actions over the world: they behave (and know they behave) as if what matters is the *correct performance* of the action, as the following argument seems to suggest.

> What I think is really happening is that [...] rituals are being processed implicitly as automated procedures in which actions are all that matters, and intentional states are quite irrelevant. This, indeed, is probably a typical feature of all circumstances of ritual routinization. By contrast, exegetical knowledge for these kinds of rituals is organized and transmitted in a more or less exclusively explicit way, through the endless reiteration of the meanings and significance of rites... This explicit level of understanding is profoundly disconnected from the level of procedural competence: people's knowledge of why they perform their rituals is cognized very differently from their knowledge of how to perform them. [22, pp. 56–57].

The explanatory and heuristic power of this model also springs from the light it sheds on the evolution of religious systems at large. Instead of the all-too-obvious idea of a *transition* from illiterate groups of free believers to the more complex congregations monopolized by literate guilds basing their truth claims on texts [8] White-house depicts a different story. He hypothesizes that "the presence of doctrinal orthodoxies favors the subsequent development of writing systems, rather than being caused by them" [22, p. 80].

> Through elevated arousal, cognitive shocks, and the creation of consequential events, the rituals of the imagistic mode set off trains of exegetical thinking that are enduring and (over time) capable of generating highly elaborate semantic knowledge [...]. These are the general conditions, I would suggest, in which the great philosophers of Aboriginal Australia, Amazonia, Africa, and Melanesia [...] come into existence [...]. But another kind of philosopher was also born with the advent of the doctrinal mode of religiosity. What made this new breed of religious experts different was that their knowledge could be transmitted verbally, via highly repetitive regimes of teaching and reminding [22, p. 81].

5 Concluding remarks

When it comes to theorizing about cultural evolution (whether one is concerned with human or animal traditions) an important issue to consider is what drives (or guides) the mutation. As Avital and Jablonka [1] made clear in their analysis of animal traditions, if your topic is cultural evolution you cannot stick to the Darwinian assumption that all variation is random, and simply concentrate on what operates the selection. What one has better be able to do is showing how, for example, religious norms shape or constrain the evolution of religious (and non-religious) forms. In the case of religious traditions, this implies that ease of learning and remembering could not be the only (or main) factor contributing to determine variations. Religious individuals are not like passive bystanders presented with advertisements about this or that deity. They need to be motivated to engage in activities which are more demanding than picking up a product from the shelf in the supermarket. So in order to spread, a ritual must exhibit more than a fashionable pattern of moves. In other words a ritual form must be more than merely memorable — it must be repeatable and worth repeating.[5]

In this sense, cultural variation is never completely blind, random, directionless. There are strategies of cultural transmission that in the long run prove to be better than others at preserving concord and continuity within the group — and this will help those strategies and mechanisms of transmission to be passed on to the next generation [1]. But when considered as cultural gadgets those strategies are reflected on more and more and are made the subject of increasingly conscious modification — as we pass from unconscious to methodic selection [13].

The theory of the naturalness of religion fails to account for the panoply of complex religious beliefs that can be found everywhere and the bodies of explicit knowledge they presuppose. Anthropological findings outline a more complex religious landscape. The prediction that any supernatural concepts that appear too strange will not be capable of spreading is disconfirmed by a number of god concepts over the world. So as an explanation of the origins and meanings of religious rituals both the theory of ritual competence and the theory of ritual form will leave much to be desired, especially for cultural anthropologists. But the notion of Action Representation System on which these models hinge offers the material for a challenging research program which could be further extended to include the exploration of moral domain and the analysis of action representations associated with moral judgment.

BIBLIOGRAPHY

[1] E. Avital, E. Jablonka. *Animal Traditions: Behavioural Inheritance in Evolution*. Cambridge University Press, Cambridge 2000.
[2] J.L. Barrett. Exploring the natural foundations of religion. *Trends in Cognitive Science*, 4: 29–34, 2000. Reprinted in D.J. Slone (ed.), *Religion and Cognition*, Equinox, London 2006, pages 86–98.
[3] J.L. Barrett and F.C. Keil. Conceptualizing a nonnatural entity: anthropomorphism in God concepts. Cognitive Psychology, 31(3): 219–247, 1996.
[4] J.L. Barrett, E.T. Lawson. Ritual intuitions: cognitive contributions to judgments of ritual efficacy. *Journal of Cognition and Culture*, 1: 183–201, 2001. Reprinted in D.J. Slone (ed.), *Religion and Cognition*, Equinox, London 2006, pages 215–230.
[5] S.J. Blakemore, P. Boyer, M. Pachot-Clouard, A. Meltzoff, C. Segebarth and J. Decety. The detection of contingency and animacy from simple animations in the human brain. *Cerebral Cortex* 13: 837–844, 2003.
[6] M. Bloch. Ritual and deference. In H. Whitehouse, J. Laidlaw (eds.), *Ritual and Memory: Toward a Comparative Anthropology of Religion*, AltaMira Press, Walnut Creek 2004, pages 65–78.
[7] P. Boyer. *The Naturalness of Religious Ideas: A Cognitive Theory of Religion*. University of California Press, Berkeley 1994.
[8] P. Boyer. *Religion Explained: The Human Instincts that Fashion Gods, Spirits and Ancestors*. Random House, London 2001.
[9] P. Boyer and H.C. Barrett. Domain specificity and intuitive ontology. In D. Buss (ed.), *Handbook of Evolutionary Psychology*, John Wiley & Sons, Hoboken, Nj 2005, pages 96–118.
[10] J. Bulbulia. Religious costs as adaptations that signal altruistic intention. *Evolution and Cognition*, 19: 19–42, 2004.
[11] D.C. Dennett. *Breaking the Spell: Religion as a Natural Phenomenon*. Viking, New York 2006.

---

[5]Not necessarily easy-to-repeat, because in-group competition requires individuals to be able to compete on increasingly difficult tasks. And their difficulty is also enhanced by the need to distinguish cheaters from sincere believers [10, 21].

[12] J. Goody. Is image to doctrine as speech to writing? Modes of communication and the origins of religion. In H. Whitehouse, J. Laidlaw (eds.), *Ritual and Memory: Toward a Comparative Anthropology of Religion*, AltaMira Press, Walnut Creek 2004, pages 49–64.
[13] C. Knight. Sex and language as pretend-play. In R. Dunbar, C. Knight, C. Power (eds.), *The Evolution of Culture*, Edinburgh University Press, Edinburgh 1999, pages 228–247.
[14] E.T. Lawson and R.N. McCawley. *Rethinking Religion: Connecting Cognition and Culture*. Cambridge University Press, Cambridge 1990.
[15] S. Levi. Religious forces behind cultural traditions. In H. Høgh-Olesen, P. Bertelsen, J. Tønnesvang (eds.), *Human Characteristics: Evolutionary Perspectives on Human Mind and Kind*, Cambridge Scholars Publishing, Newcastle 2009, pages 297–313.
[16] R.N. McCawley and E.T. Lawson, *Bringing Ritual to Mind: Psychological Foundations of cultural Forms*. Cambridge University Press, Cambridge 2002.
[17] A. Norenzayan. Evolution and transmitted culture. *Psychological Inquiry*, 17: 123–128, 2006.
[18] G. Preyer and G. Peter. *Logical Form and Language*. Oxford University Press, Oxford 2002.
[19] R.A. Rappaport. *Ritual and Religion in the Making of Humanity*. Cambridge University Press, Cambridge 1999.
[20] J. Sørensen, P. Lienard and C. Feeny. Agent and instrument in judgments of ritual efficacy. *Journal of Cognition and Culture*, 6: 463–482, 2006.
[21] R. Sosis. The adaptive value of religious ritual. *American Scientist*, 92: 166–172, 2004.
[22] H. Whitehouse. *Modes of Religiosity: A Cognitive Theory of Religious Transmission*. AltaMira Press, Walnut Creek 2004.

On value judgement and the ethical nature of economic optimality

ANDREA ZHOK

In the following pages we are going to address the issue of value judgment and value measurability with an essential concern for its ethical significance. This issue, as we see it, represents the main explicit joint between economics and ethics and underlies the idea of economic efficiency as something per se valuable. We must stress in advance that the criticisms we are going to raise may make sense only insofar as our concern is with a general intuitive notion of value, while they may fade away if we should circumscribe our understanding of "value" to some narrow technical acceptation.

1 On the ethical nature of economic optimality: a tacit model

Let us start with an outline of the model reasoning that, sometimes explicitly and more often implicitly, supports the idea that economic efficiency is also axiologically good, whenever such claim makes its appearance. This reasoning rests on five props, which we are going first to briefly expose and then to knit together.

1) The *Consumer's Surplus Principle* argues that, if an exchange is voluntarily accepted by both transactors, each one must be better off after the exchange. The argument can be explained with reference to two qualifying points: *a)* should one of the transactors not think that he is better off by making the transaction, he would never go for it; *b)* since each transactor is the only authority in judging what is good for herself, no external evaluation of the profitability of the exchange is relevant. Therefore, we apparently cannot but conclude that after each voluntary exchange there is *ceteris paribus* an increase in overall positive value (which is tantamount to subjective appreciation).

2) The individual's authority we have just mentioned rests on *Axiological Subjectivism*, which is the approach to value that economics inherits from classical Utilitarianism. This argument can be also split into two components, which are, though, independent from each other. We can name them respectively as "non-paternalism" and "emotivism". *a) Non-paternalism* assumes that each private mind is a black box, the value judgments of which nobody else is entitled to discuss. This argument is in itself pre-eminently a political or moral stance towards individuals, and does not provide autonomous power to the thesis. *b) Emotivism* says that each private mind is the last authority upon the value judgments that concern that mind itself, because the last substrate of value judgments is just "feeling" of a sort (propensities, inclinations,

etc.). In its classical version emotivism[1] conceives of the essential substrate of value judgments as units of pain and pleasure (or, more generally, of aversive and favourable feelings): we are supposed to weigh on the scale of our immediate sentiments the amounts of agreeable and disagreeable feelings that we expect from a certain decision, and if the net amount is positive, we go for it. This way of producing value judgments is apt to provide us with *orderings of preference* among entities.

3) According to a widely accepted definition of *measurement* in the social sciences, being able to univocally determine an ordering is tantamount to producing a measurement: "Measurement is the assignment of numbers to objects or events according to a rule" [18, p. 679]. That is, provided that *value* is what is determined by subjective value judgments, and provided that we can have a rule that allows us to *assign ordinal numbers* to value judgments, then we will have a *method to measure value*.

4) Granted that value judgments would take place inside private minds and that their constitutive matter be units of feeling (pain and pleasure), then it should always be possible to compare and order two different valuable items. The *homogeneity* of internal feeling, to which any object of judgment is reduced, allows any comparison to take sensibly place. This is what is implied by the *Completeness Axiom* in Consumer Behavior Theory: given any pair of items from a set, inclusive of anything axiologically significant for a subject, they can be mutually ranked by this subject as better, worse or equally valuable. This argument supports the idea of general *value commensurability*.

5) If we are able to rank any two items according to its subjectively appreciated value, then it seems that we can produce any sequence of inequalities we like and that, by stringing them together, we can extend our evaluations into a unitary homogenous scale: if A is better than B, and B is better than C, then A is better than C. This is what is stated by the so-called *Transitivity Axiom*. The transitivity of value judgments allows us to build complex orderings of valuable items, and therefore, given the above definition of measurement, it allows us to *ordinally* measure value.

From these five points a Paretian notion of economic optimality can be derived, and precisely a notion such that it entails axiological optimality. Let's call "society" any finite set of subjects. Let's suppose that there is in society an original distribution of valuable items (goods, resources). Let's now ask ourselves how we could produce, given such conditions, the best distribution of such valuable items in society.

α) Because of *Consumer's Surplus*, we should promote all possible voluntary exchanges, or, if the transaction costs of free exchange are too high, we should move resources according to the following principle (consistent with Consumer's Surplus): whenever it is possible to make somebody better off, without making anybody worse off, the relevant reallocation is justified. After

---

[1] Our understanding of emotivism in the present context is only partially coincident with the customary acceptation of the term, which is not strictly connected to Utilitarianism. By mentioning emotivism we are focussing on the philosophical tradition on the nature of rational deliberation that stems from Humean ethics and is embraced by classical Utilitarians.

each exchange or transfer we would have at least one member of the society better off and nobody worse off (according to their own judgment). Since value as such is the outcome of subjective evaluation, after each exchange or transfer we will have a straightforward increase of value in society. Had we reached a distribution where it were impossible to move valuable items from one member to another without making somebody worse off, we would have reached a non-improvable situation, that is, what is known as Pareto optimality. Under these conditions the distribution of resources in society would be economically efficient, and, insofar as we accept a Utilitarian framework for value, also *axiologically optimal*.

β) In this framework, the *Completeness Axiom* warrants that there are no "irrational" situations in value comparison, that is, no deadlock, nor a plurality of overrunning (absolute) values.

γ) The *Transitivity Axiom* warrants that we can progressively reach the best distribution through successions of distinct comparisons (evaluations and exchanges).

δ) As to *Axiological Subjectivism*, it warrants that no higher principle of evaluation than private choice can be called for, and actually forbids overrunning any private evaluation in the name of alleged higher goods. This is the basis for all the approaches to Utility Theory in terms of "revealed preferences", rather than in strictly hedonic terms: we do not need to take position about the specific nature of pain and pleasure as it was discussed by Bentham [1], Mill [11], Sidgwick [16], etc. but we can be content by granting that the last criterion of goodness is immediate subjective evaluation.

The main consequence of this position is the notorious limit that Paretian optimality sets on any forced transfer: regardless of how beneficial a transfer appears and for how many people, if it has to be paid by making worse off even a single individual in society, it seems to be unjustifiable. Since value is purely subjective, no external interpersonal comparison of value judgments is possible, since we can never be certain that the experienced loss of an individual will be greater or smaller than the overall gain for others.

Although Utility Theory, as it is used in economics, has made all possible efforts to refrain from any commitment to specific Utilitarian theses, this does not neutralize the fundamental Utilitarian bias of Neoclassical economics. The nexus between economic optimality and ethical optimality is *neither theoretically contingent, nor marginal as to its consequences for political economy.*

It is not theoretically contingent insofar as the core elements of Utilitarian ethics that we mentioned are regarded as true or even just as approximately plausible. The whole famous debate on higher and lower pleasures that contrasts Bentham's quantitative view of pleasures and pains with Mill's qualitative analysis is perfectly irrelevant to the translation of economical conceptuality into economics. Indeed, even if in the black box of our mind we would always and regardless of quantitative determinations prefer poetry to pushpin, this is just a matter of our actually "felt" dispositions thus far: it does not change anything with regard to the subjective immediate quasi-sensuous nature of value. The introduction of qualitative thresholds in our ranking of pleasures can be significant for our opinion of ourselves, but it does not change the points that are qualifying for the translation from value judgment to economical quantification. It does not change the fact that the preferences that we reveal in our actual choices are unappealable by external judgments; and it does not change the fact that both poetry and pushpin are ranked on a homogenous scale: different qualities of pleasure are anyway more or less agreeable. [17, p. 24–5], [15].

No less irrelevant for the economical use of Utilitarian ethics is the other intensely debated point about the nature of interpersonal comparisons of value: regardless of how much we believe that people are able or legitimated to read the others' evaluative mind, the ultimate sovereignty of first-person judgments is not touched. It must be noticed that even if we embrace the maxim of classical Utilitarians, according to which "*it is the greatest happiness of the greatest number that is the measure of right and wrong*", [2, p. 393] it remains up to us (i.e. to our subjective "feelings") to determine whether to attribute value to the others' preferences and, in case, how much. After all, in Bentham and Mill the basis on which the claims of others' happiness rests is sympathy, which is itself just a private feeling ruled by the laws of greater or lesser pleasure. The others have no claim on us beyond our privately felt dispositions. Once the emotivist roots of Utilitarian ethics are granted, the way to equating value with the preferences revealed by actual transactions is paved, and consequently what is optimal in terms of voluntary transactions must count as optimal also in general axiological terms. The conceptual framework provided by Paretian optimization raises private evaluations to the public dimension of value proper: this happens just in the form of the ideal addition of all voluntary transactions, which are assumed by definition to be manifestations of adequate private evaluations.

Although such a thesis is rarely, if ever, openly stated in the previous terms, it is nevertheless pervasively operant. In the history of economic thought, an understanding of action and value that closely resembles the just outlined conceptual framework can be found in Ludwig von Mises' *praxeology*. According to Mises, each action is "an attempt to substitute a more satisfactory state of affairs for a less satisfactory one" [12, p. 97], therefore all meaningful human action constitutively has the character of an exchange driven by private preferences, even before interpersonal transactions are implemented. Thus, the passage from the general axiological dimension of goal-directed action to proper economic transactions is smooth and spontaneous, and prices ideally become an appropriate measurement of social value. This lesson of Mises', even when it is not followed in detail, remains recognizable across the whole multifarious legacy of the Austrian School, from Hayek's notion of catallaxy as spontaneous social order [8, Ch. 10], to Homans' theory of social exchange [9], down to the theorists of the Law & Economics School. We claim that, although an explicit argumentative connection of the above described passages (1 to 5) is not to be easily found in the available literature, this reasoning represents an internal and obvious line of thought underlying a very significant part of contemporary orthodoxy in political economy. It concerns all the theoretical inclinations to regard the order of economic transactions as ideally self-sustaining, self-justifying and *never legitimately constrained by further moral instances* (because such instances are not properly recognised) [5, p. 21].[2] The idiosyncratic level of private evaluation by "feeling" is, somewhat paradoxically, mirrored in social transactions by the working of money: as feeling makes all possible evaluations homogenous and comparable, as well money (or, better, monetary practice) makes all goods commensurable.

Now, we think that some version of the reasoning we have just outlined is at the core of any explicit or implicit justification of the ethical goodness of economic efficiency. In the rest of the paper we are going to show that even the less demanding versions of that reasoning are untenable.

2 Violating transitivity without irrationality: an example

Let us start from the hotly debated transitivity issue.[3] Transitivity may seem a light requirement for value judgments, or, at least, a requirement equivalent to a general demand for rationality: it is a basic logical requirement that, given a quality p, if A is more p than B and B is more p than C, then A must be more p than C, regardless of what the quality p is.

However, we should keep in mind that transitivity is a requirement derived from the *timeless* realm of logical relations, and it is not self-evident that you can safely apply it to judgments that take place over time. Let us formulate the following example:

> Suppose that I am invited for an official fish dinner by a close friend of mine and I want to buy him a good bottle of wine. I enter a wine shop, where I find wine A, an excellent white wine, which I personally appreciate. Then I remember that my friend once told me that he put the red wine B above all other wines, and I look for it, since I want to please my friend. But when I find B, I see, next to B, the red wine C, which I think is akin to B but even better, and I want to buy it, because I want my friend to make a new oenological discovery. However, a moment before taking C, I remember that it is a fish dinner, and I am afraid that, if I bring a red wine, the other guests might think that I am so uncouth not to know that white wines are more appropriate with fish; so, fearing a poor figure, I go back to wine A, and, since it is getting late, I just stop browsing and buy A.

Now, this is, I think, a realistic description of a deliberative process, where an open violation of transitivity takes place, without any irrationality being involved. How is that possible? Well, we propose that it is possible on the following grounds:

---

[2]For a comprehensive analysis of this issue, we refer the Italian-speaking reader to [20].

[3]Since we could not hope to do justice here to the vast literature on the issue of transitivity, incommensurability and incomparability of values, we have chosen in the following to discuss just one position and to be silent on many bright others. The reader who needs to be introduced to the issue will find a good introduction in [6]. Our choice is mainly due to the need to provide in the limited extension of the present text the whole "gestalt" of a reasoning, which can presently aspire to be just a rough outline, but wants to be a blueprint for further research.

1) Because the "currency" in which real value comparisons take place is not some formless homogenous "feeling", but is constituted by "*reasons for action*";

2) Because reasons for action do not just choose among *present* items, but among *present items in a temporal story*, which is open-ended in the direction of future;

3) Because reasons for action are *required* to turn into actual actions, since they have an *inbuilt time-constraint*: the sheer need and timing of decision is itself an overrunning reason for action.

3 The aspectual nature of reasons for action

We said that no irrationality is involved in the wine deliberation; however standard objections to intransitivity assert that this is impossible, because, if intransitivity were permitted, we would end up browsing around in the wine shop forever. To escape from such a possibility an ingenious solution is the one proposed by J. Broome in *Weighing Goods*: he would say that the wine A I envisaged at the beginning was not properly the same thing as wine A at the end of my browsing. Broome elaborates on the following example:

> Maurice prefers visiting Rome (R) to mountaineering (M), since mountaineering frightens him; he prefers staying home (H) to visiting Rome, because sightseeing bores him; but he prefers to go mountaineering than staying home, because he does not want to look coward. This is apparently a case of intransitivity not unlike the one we have exemplified above: according to Maurice (R) > (M), (H) > (R), and, against transitivity, (M) > (H). For Broome the solution, in order to maintain transitivity, is to say that in reality the alternatives are not three, but four: staying home is to be split in H(1), staying home *without having turned down* a mountaineering trip, and H(2), staying home *instead* of mountaineering [4, p. 100–101]. For Broome the correct move is to consider *only* H(1) and not H(2), since it is H(1) that happens to be the original object of comparison with the option of visiting Rome.

But is it so? Is H(1), as it were, the *pure essence* of staying home for Maurice, regardless of "interfering" comparisons? Well, it seems that we should at least say that H(1) is just ⟨staying-home-by-turning-down-sightseeing⟩. In order to support Broome's solution, we should believe that the items among which we operate our choices are something like substances that exist independently of the context of life and choice in which they occur. If it were so, we could imagine carving out of our comparisons the pure essence of an item, regardless of the alternative (actual and potential) options. But this is hardly tenable. Alternative options are essential parts of the context of choice, together with plans, desires, people's opinions, etc. You can never abstract the content of the objects of choice from the context in which they occur: this is the main lesson to be drawn from the notorious water-diamond paradox: to see if water is more or less valuable than diamonds we have to evaluate, for instance, if we are in a desert without water or in a jewellery to leisurely buy presents.

The point here is that what we compare are not metaphysical substances, but items occurring in contexts, and what we call "reasons for action" are the way we articulate how items occur in impending contexts.

Let's go back to the wine shop: were we just muddled in our deliberative reasoning, by raising each time a new reason for choosing the relevant wine? Shouldn't we force ourselves to judge each wine according to the same qualities and from the same point of view? Indeed, if I had reduced my judgment to how wine tastes in my mouth, then I would have likely found a homogenous dimension where transitivity and rationality would have gone together. All possible reasons for choice would have levelled down to subjective taste and actual flavour in a quasi-atemporal judgment. This is certainly an option, but to force all deliberations into this cast would mean conquering formal rationality at the price of stiffening into unrealistic and hopeless obtuseness. In fact all choices are concerned with three dimensions: the qualities of the compared items (that are always a multiplicity), the possible points of view from which to evaluate those qualities, and the personal tastes with regard to those qualities. If we confront precisely the same items, with the same qualities, as in the wine example, the objective side of the choice is kept, by definition, steady. If we suppose that the relevant choices take place over a short time (as in the wine-shop browsing) we can also sensibly assume that the subjective taste remains constant. However, regardless of the stability of subject and object, the points of view from which we evaluate the items can *rationally* change even over a very short time.

The point is that, whenever we enter a process of deliberation, we elaborate small possible contexts, in the form of *narrative action units, small stories*, where the items we evaluate should occur, and then we try to see if some resulting scenarios are more convincing than others. Since the items among which we choose are bundles of qualities (aspects), and since the possible scenarios give prominence from time to time to different aspects of the items, then it may well happen that for one scenario an item is more convincing from a certain point of view (because of a certain quality), while for another scenario the same item is more convincing from a different point of view (because of another quality). These narrative action units are just *possible developments of our lives*, and they are never completely predetermined by our past and present. At the same time, the qualities we appreciate in an item owe their value for us to the specific way they fit certain scenarios (projects, commitments).

When I considered wine A the first and the second time, I just saw the same item, with the same qualities, under different points of view. I have not *weighed* A(1) first, and then A(2), that is A(1) plus a further quality, which tipped the balance in its favour. I just looked at A's quality of being white from a different point of view, that is, *on the background of a different possible story*, a story *that is up to me to make real*. The same review could take place with wine B, while regarding it under a further perspective, and similarly with wine C, and so on. So, transitivity can never be reinstituted by splitting and hypostatizing reasons for action, as if they were different substances,[4]

---

[4] A wider analysis of the nature of value in relationship with action can be found in [19].

unless we metaphysically assume that our prospective lives are always devoid of alternatives. If we enjoy some degree of freedom to act (or just if we enjoy the subjective belief in such freedom), then our meeting with the very same object of choice is not bound to generate the same outcome.

4 The open-endedness of narrative action units

The last reasoning leads us to the commensurability issue. The Completeness Axiom states that we must be always able to rank two items as greater, smaller or equal. The standard objection to commensurability is made by resorting to the opinion that some entities should be counted as infinitely valuable. Thus, such entities could not be properly compared with each other, nor could we ever offset the gap between a finite and an infinite value by adding more finite values together. For instance, we might say that, regardless of how many ice-creams we pile up, their value will never equal a single human life. However, such reasoning is faulty, if it is applied to real valuable items and not to abstract substance-like values. Perhaps we may say that the value of Life is incommensurable with the value of Ice-Cream, and this may well mean something profound and unquestionable, however, if it is supposed to mean that under any circumstance we are going to prefer an instantiation of human life to an instantiation of ice-cream, well, this is certainly untrue. I am confident that I can make up a plausible story where we have to choose between the life of an evil human being, obnoxious and harmful, and a supply of ice-cream, which can help a large group of children to grow up well-fed and not stunted for lack of nourishment; under similar circumstances I would bet on the ice-cream and against the relevant human life [cf. 7, p. 82f.]. Once again, we are looking at the problem from the wrong point of view: in our actual deliberations we are never comparing two eternal substances (Life and Ice-Cream), but two specific items in specific temporal contexts.

Nevertheless, there is a brand of *infinity* that is relevant to incommensurability. When I chose wine A in the previous example, I did it because of the way it was fitting in a narrative action unit of mine: I thought, for instance, that I was going to meet a stiff and formal gathering of guests, while not being able to explain the reasons of my gift, so that I could have cut a poor figure. However, I could have consistently thought otherwise: for example, that I might be able to see first my friend alone, before meeting the other guests, having so the opportunity to explain why I wanted him to make the oenological discovery of wine C. According to this story I would have rather gone for wine C. Regardless of the strict plausibility of such psychological motivations, the point here is that the stories, in which the valuable items we choose are bound to occur, are constitutively *open-ended stories*: the future is yet to be decided, and, what is more, it is partly *up to me* to decide which story will host in the future the valuable item I am presently choosing. This means that, *while choosing among items, we simultaneously commit ourselves to a specific future in which the relevant item is supposed to fit*.[5]

---

[5] In the present context we cannot appropriately discuss the interpretation of Narrativity, action and personal identity that is presupposed by the present reasoning. In the main, we as-sume an understanding of the role of Narrativity, which is close to the one expounded

Therefore, if we are able to rank items with completeness, this is only because we provide a completion for the reasons of choice, completion without which reasons for action are generally underdetermined. This is something that systematically escapes the eyes of rational choice theorists: a choice, in order to be rational, has to be made rational by the agent's commitment to a certain future. Without that step, if we just look at the inherent qualities of valuable items, we might well stare at them forever, waiting for them to tell us what they are worth. By slavishly following the intrinsic features of our objects of choice, we are bound to end up with a stalemate. The Completeness Axiom is at best a Completeness *Imperative*.

5 Time-constraints and random commitments

Now we are also in a position to clarify why it is important to consider time-constraints as an essential part of reasons for action. If we were to live forever, without any time-constraint, we would never have the necessity to choose a course of action: we might simply refrain from acting, since, if we did nothing, just nothing would happen to us, or we might even act at random, since, if we did so, probabilistically in an infinite time everything would happen. But under ordinary deliberative conditions, as to the past, no past conditions *dictate* future courses of action; and as to the future, we have to limit the range of possible futures we have to confront. Even in the most stale and unadventurous lives one is steadily called to choose among alternative courses of action, trivial and not. From a rational point of view, if we disregard time-constraints, there is never objectively enough argumental matter to settle our decisions. In the absence of any reason to reach a definite deliberation we could go on and on speculating about the scenarios we like to consider and always formulating new ones that bypass the conclusions of previous possible stories. As it has been often depicted in literature, a "dreamer" character can weave virtually infinite courses of implications from the most trivial act, with the result of never reaching the point of decision (which, after its Latin etymology, is indeed a cut in the weaving of consciousness). In the absence of time-constraints our decisions are always motivationally underdetermined. In order to reach an actual decision we have to cut off the process of speculation by *"betting" on a certain future of ours*. This bet cannot be completely settled either by my past and present experiences, or by my propensities for the future. However, in real life all our decisions have an (implicit or explicit) expiry date: if I want to buy a bottle for dinner, I have to do it before dinner; more generally, whatever I want to do, I have to do it within my life expectancy. But, in a sense, a pure time-constraint does not provide us with any specific reason for acting one way rather than another, since it does not determine in which narrative action unit our present choices are going to be embedded. However the very existence of a general time-constraint forces us to cut short with our elaboration of scenarios, by plunging into action whenever the options we have in mind look prima facie good enough. This means that, whenever we find ourselves before a potential deadlock, due

---

by Paul Ricouer [13,14] and akin to the one proposed in *After Virtue* by Alasdair McIntyre [10].

to the ubiquitous underdeterminacy of conditions, we can always bootstrap ourselves into a solution: we will solve our Buridan's stalemate by a *random commitment* [cf. 3, p. 11–17]. We just have to take a decision, since no decision is anyway a bad option, and in order for the decision to be rational, we *have to be able to commit ourselves to it*, since only our effort to realise it will justify our present choice. It is important to notice that such random commitments *are perfectly rational, while being at the same time impossible to anchor in the actual objects of choice*. They will become rational through my future action.

6 Being measured without being

At this point we can also easily see why Stevens' definition of ordinal measurement is inadequate. Measurements are ways in which we extract some characters from an entity, by a rule that generates numbers while being applied to that entity. Stevens' only concern was to exclude random assignments of numbers, which in his eyes meant non-rules. But this condition is absolutely not enough: we have also to ascertain in advance *that the object to be measured does exist*, and that *it exists in a fashion that allows it to be measured by the relevant rule*. In the case of ordinal ranking of value judgments, utility theorists are presuming the existence of values, which are conceived as qualities, akin to the classical pain-pleasure model. But since it turned out that the object of our evaluations are not such qualities, but rather reasons for action, then our numbering, even if there never occurred instances of intransitivity or incompleteness, would never provide any representation of its alleged object. The object of such supposed measurement is actually a moving target, which changes together with the way in which the agent's life unfolds. Even if the consumer's taste (his utility curve) was always constant, the agent's past and future, which give sense to any specific choice, are continuously changing. It is worth noting that, in order for such supposed measurements to have any plausibility at all, all consumers should commit themselves not just to constancy of tastes but to a completely standardized conception of their lives: dullness (or should we call it "reliability"?) is indeed a most highly appreciated economic virtue.

7 The inconsistency of the pain-pleasure model

Let us come to the last point of our discussion. Something akin to the hedonic principle, characteristic of classical Utilitarianism, is epistemologically required by any reasoning that wants economic efficiency to spell axiological goodness. As we saw, without the idea of a homogenous "substance" on which value judgments rest there is no common ground for "weighing" evaluations against each other. Although classical hedonism is no longer fashionable, and although very few rational choice theorists would openly defend the pain-pleasure model of Utilitarianism, that model is still far from being dispensed with. The central point of that model does not concern the specific nature of pains and pleasures, but implies that pain and pleasure represent positive and negative feeling in a symmetrical way: they are supposed to be the very same experiential substrate, just with inverted sign. Furthermore, both positive and negative feeling must be interpreted as something whose content can be appreciated independently of context and attributions of meaning. However, if we had the time to provide a phenomenological analysis of how actual pains and pleasures are experienced, we would find an interesting asymmetry between them: while pains can be experienced in a thoroughly passive way, pleasures must be actively enjoyed, must be accepted in order to carry their positive meaning to us.[6] We can impose pain to others, so much so that we can torture them, but we can at most lure the other into pleasure. This is the case precisely because enjoyment involves active participation, acceptance, whereas pain is experienced as something foreign, spatially locatable in my body, something that can impose itself on me: I can say that *my foot aches*, but it is *me*, as a whole person, the one who *enjoys* the chocolate. This means that only pains (by which we mean physical pains) can be conceived as bearers of (subjective) disvalue independently of the interplay of personal reasons and contextual judgment. These considerations should make us aware of the peculiar character of the Utilitarian and post-Utilitarian conception of value as quasi-sensuous appreciation (or aversion). It is not true of the whole dimension of evaluative feeling that it can "transmit its content" regardless of context and self-understanding: this is in fact the case only for physical pains, whereas the whole area of positive appreciation, as well as the area of spiritual distress, are highly sensitive to rational self-positioning, to the interpretation of oneself in a temporal context. The point here is that, in order for "feeling" of a kind to be the "currency" of value judgments, it should always behave more or less after the model of pain: *only something that can carry its meaning regardless of context and active participation can be the embodiment of the "value units" that Utility Theory needs*. This observation has an interesting entailment: the model of value judgments, which utility theory and microeconomics in general make use of, does work relatively well only for pain avoidance, but not at all for the purposive positive side of value.

8 Conclusions

Now we may find ourselves wondering how it is possible that such a blatantly wrong approach to value can have had any chance to be accepted as a sensible way to conceptualize human actions and choices, which is a primary matter of concern for economics. Although the fact that this underlying reasoning is rarely made explicit can be regarded as a mitigating circumstance, this is not quite enough to justify the easiness with which such an approach can spread; it is plausibly embraced at least as an approximation, very much like *homo oeconomicus* is regarded as an approximation to human kind. Even if few theorists would claim today that ethics can be resolved into economics, or that market economy is or can be a perfect embodiment of people's values, most would still say that these are anyway good approximations, leaving out probably just the sphere of allegedly non-rational behaviour (religion, customs, etc.). We believe that the apparent rationality of this line of thought is sustained by two extrinsic props, which can be mentioned as a conclusive suggestion.

---

[6]We have provided a relevant phenomenological analysis of pain and pleasure in [19, p. 152–165].

1) On the one hand the picture of value implicit in Utility Theory can be accepted as a sensible approximation because of various *ad hoc* adjustments that have been introduced. The best known and most important one is the so-called Law of Diminishing Marginal Utility. This so-called Law is a way to partially introduce a regard for the temporal and agent-bound nature of value into Utility Theory. Out of all the countless ways in which a valuable item can occur in the framework of human action, the fact of being less and less wanted the more of it is available is a (very abstract and partial) way to consider the item's position in the agent's life, while preserving the ideally quantitative nature of value units. The Law of Diminishing Marginal Utility gives a flavour of sensitivity to context to the choices of economic agents, such that the general abstractness of value judgment remains concealed.

2) But more important than all *ad hoc* adjustments is another factor. Economical evaluations of axiological import do not strike us as pure nonsense because of their *normative* power. As we have seen, in order to evaluate an entity, we have to consider how it fits in an open-ended narrative unit of action. To provide determination to the constitutively underdetermined situation in which we are called to make our choices, we have to *commit ourselves* to a certain future, to a certain scenario. Now, economic theory, and especially the correlative economic habits and material conditions, allure or push agents to shape their lives in ways that are coherent with most economic dictates. We are steadily called to commit ourselves to futures, which are consistent with commensurability, transitivity, axiological subjectivism and much more. We may not be prone to accept to quantitatively compare costs and benefits of any alternative, but drawing up a health insurance or paying a monetary compensation for an insult is just general commensurability *in rebus*. We may not spontaneously believe that our lives are naturally shaped as a progress towards better and better achievements, but we are expected to browse market options while respecting value transitivity, on pain of pauperization. The very institution of monetary practice forces each economic agent to provide cardinal estimates of value, to cardinally compare every good with any other, and to prefer items that are more monetarily valuable to less valuable ones. These are all conditions, which we are called to respect not in the realm of pure axiology, but in the earthly kingdom of monetary transactions, where the price to be paid for non-compliance is not the rebuke of economists or philosophers, but social failure and utter disempowerment. So, although the basic principles of economics are descriptively a failure, they get closer to actual behaviour by forcing behaviour to get closer to themselves.

BIBLIOGRAPHY

[1] J. Bentham. An Introduction to the Principles of Morals and Legislation. in *The Collected Works of Jeremy Bentham*, edited by J.H. Burns and H.L.A. Hart, Oxford University Press, Oxford 1970.
[2] J. Bentham. *A Comment on the Commentaries and A Fragment on Government*. The Athlone Press, London 1977.
[3] M. Bratman. *Intention, Plans, and Practical Reason*. CSLI Publications, Stanford 1999.
[4] J. Broome. *Weighing Goods*. Basic Blackwell, Oxford 1991.
[5] J.M. Buchanan. *Liberty, Market and State*. Harvester Press, Brighton 1986.

[6] R. Chang (ed.). *Incommensurability, Incomparability, and Practical Reason.* Harvard University Press Cambridge, MA 1997.
[7] J. Griffin. *Well-being.* Clarendon Press, Oxford 1986.
[8] F.A. Hayek. *Law, Legislation and Liberty.* Routledge and Kegan Paul, London 1973.
[9] G.C. Homans. *Social Behavior. Its Elementary Forms.* Routledge and Kegan Paul, London 1961.
[10] A. MacIntyre. *After Virtue: A Study in Moral Theory.* University of Notre Dame Press, 1984.
[11] J.S. Mill. Utilitarianism. In *On Liberty and Other Essays*, Oxford University Press, Oxford 1991.
[12] L. von Mises. *Human Action. A Treatise in Economics.* New Haven, Yale University Press 1949.
[13] P. Ricoeur. *Time and Narrative.* 3 vols; trans. by K. McLaughlin and D. Pellauer. University of Chicago Press, Chicago 1984, 1985, 1988.
[14] P. Ricoeur. *Oneself as Another.* Trans. by K. Blamey, University of Chicago Press, Chicago 1992.
[15] J. Ryberg. Higher and Lower Pleasures — Doubts on Justification. *Ethical Theory and Moral Practice*, 5: 415–429, 2002.
[16] H. Sidgwick. *The Methods of Ethics.* Hackett Pub Co Inc. 1981.
[17] J.J.C. Smart. An outline of a system of utilitarian ethics. In J. J. C. Smart and B. Williams, *Utilitarianism: For and Against*, Cambridge University Press, Cambridge 1977, pages 3–74.
[18] S.S. Stevens. On the theory of scales of measurement. *Science*, 103: 677–680, 1946.
[19] A. Zhok. *Il concetto di valore: dall'etica all'economia.* Mimesis, Milano 2001.
[20] A. Zhok. *Lo spirito del denaro e la liquidazione del mondo.* Jaca Book, Milano 2006.

Carl Menger, Ludwig von Mises, Friedrich A. von Hayek and Karl Popper: Four Viennese in defense of methodological individualism

DARIO ANTISERI

1 Karl R. Popper, theoretician of methodological individualism

The social scientist — the sociologist, economist or historian, for example — must continually deal with what is called *collective concepts* (*Kollketivbegriffe*) such as "society", "party", "class", "state", "revolution", "people", "nation", and so on. There are two main rival currents of thought on the interpretation of these concepts: *methodological individualism* and *methodological collectivism*. And the debate involves three problems: an *ontological* problem, a *methodological* problem, and a *political* problem.

1. *The ontological problem*: What do these collective concepts really correspond to? The individualists (including B. Mandeville, D. Hume, A. Ferguson, A. Smith, C. Menger, L. von Mises, and F.A. von Hayek) reply that they correspond only to individuals. Only individuals exist and only individuals reason and act. The collectivists (Saint-Simon, Comte, Hegel, Marx, the neo-Marxists, structuralists, etc.), on the contrary, think that collective concepts refer to substantial realities — autonomous bodies that are independent of individuals and, like the "church" or the "army" or the "native land", take action to shape individuals and make them conform.

2. *The methodological problem*: Where do social studies — inquiries intended to explain social events and institutions — begin? Since the individualists believe that only individuals exist, they maintain that research on the origins and changes in social events and institutions necessarily must start from the actions of individuals (in order to explore, in detail, their unintentional consequences). The collectivists, on the other hand, due to their belief in the reality of *collectives*, try to discover the laws (of progress and decline, dialectics, and so on) that would govern the origins and development of such collective bodies.

3. *The political problem*: Is the objective a collective body like the party or the nation, or is it the individual with more freedom and responsibility? If the reality actually is a collective body like the State, it is obvious that individuals are at the service of this body and are, as the collectivists claim, agents for collective objectives. On the contrary, the individualists maintain that the end is the individual, not the State, the class, or the party; and, they claim, if an individualistic concept of society is eliminated there is no justification for democracy.

In the first volume of *The Open Society and Its Enemies*, Karl Popper draws the reader's attention to the following prospect [17, vol. I, p. 100]:

a) Individualism is opposed to a) Collectivism
b) Egoism is opposed to b) Altruism

He comments: "Collectivism is not opposed to egoism nor it is identical with altruism or unselfishness. Collective or group egoism, for instance class egoism, is a very common thing [...] and this shows clearly enough that collectivism as such is not opposed to selfishness. On the other hand, an anticollectivist, i.e. an individualist, can, at the same time, be an altruist" [17, vol. I, p. 100].[1] Though very brief, this counteracts the "ethical" objection of those who are hostile to individualism because it would be the same as an egotistical position. Individualism is, on the contrary, a philosophical concept in contrast to collectivism but not to altruism. Individualism maintains that only individuals exist, and they may be either egoists or altruists. What does not exist is *society* if it is intended as an independent body that exists prior to and is unrestricted by individuals.

In 1990, in a conversation with Guido Ferrari, Karl Popper explained this point in more details:

> Moreover, I'd like to add, if I may, that it is extremely misleading to talk about *society*. Naturally one can use a concept like society or the *social order*, but we shouldn't forget that they are only auxiliary concepts. What really exists are people, good ones and bad ones — let's hope there aren't too many of the latter, in any case human beings, some of whom are dogmatic, critical, lazy, diligent or something else. This is what really exists [18, pp. 24–25].

There are men who have ideas and act on them, and their actions have intentional and unintentional consequences. Men exist, "what doesn't exist is *society*. However, people believe it exists and so blame everything on society or the social order" [18, p. 25]. So, Popper says, "one of the worst mistakes is to think that something abstract is concrete. This is the worst kind of ideology" [18, p. 25].[2]

The error the collectivists have always made is to replace a theoretical abstract construction with concrete things. Therefore, for Popper, society does not exist. Not even the police exists: "[...] the laws regarding the police can be changed but the police as such can not. The police as such does not exist. Laws can change because they are written down in codes and therefore exist." There are police officers but not the police. The army does not exist either. In *The Poverty of Historicism*, he says:

> Most of the objects of social science, if not all of them, are abstract objects; they are *theoretical* constructions. (Even 'the war' or 'the army' are abstract concepts, strange as this may sound to some. What is concrete is the many who are killed, or the men and women in uniform, etc.) These objects, these theoretical constructions used to interpret our experience, are the result of constructing certain *models* (especially of institutions), in order to explain certain experiences [..]. Very often we are unaware of the fact that we are operating with hypotheses or theories, and we therefore mistake our theoretical models for concrete things. This is a kind of mistake which is only too common [19, pp. 135–136].

---

[1] For a clear view of the differences between individualistic and collectivist concepts in social philosophy in the period that goes from the Middle Ages to Adam Smith, see [25, pp. 1–18]. On [25], see [6, pp. 90–92]. See al so [7, pp. 66–114].

[2] For similar considerations, see [26, pp. 57–61] and [4, p. 656].

Actually, Popper adds, "the task of social theory is to construct and to analyse our sociological models carefully in descriptive or nominalist terms, that is to say, in *terms of individuals*, of their attitudes, expectations, relations, etc. — a postulate which may be called 'methodological individualism' "[19, p. 136].[3]

There is nothing corresponding in facts to the word society that is different from individuals with certain specific ideas who act according to these ideas. There are really only men — "good ones and bad ones"; what really exists are human beings [18, pp. 24–25]; only they can think and act. From this it follows that "institutions (and traditions) must be analysed in individualistic terms — that is to say, in terms of the relations of individuals acting in certain situations, and of the unintended consequences of their actions" [17, vol. II, p. 324]. Individuals exist, and it is they who act: "institutions don't act, only individuals act in or for the institutions" [20, p. 12], so that what is needed is "to construct a theory of the intentional and unintentional institutional consequences of purposeful actions. This might lead to a theory of the origin and development of institutions" [20, p. 12].[4]

In Popper's opinion, once these premises have been established, social events can be explained and understood by situational analysis, a research strategy in which a human action — interrelating with other actions of other individuals — is considered an attempt to solve some problem.

2 Three methodological individualists: Carl Menger, Georg Simmel and Max Weber

"If in the end I have become a sociologist [...], it is because I wanted to rule out those exercises based on collective concepts. Sociology itself cannot proceed from anything but actions of single individuals. For this reason, sociology must adopt strictly 'individualist' methods' "[29]. Max Weber wrote this in 1920. This individualistic concept is clear; so is his rejection of the claims of "so-called 'organic sociology', which attempts 'to explain social action by starting from the 'totality' " — for example, from the totality of the "economy of peoples" — and then, within it, to interpret the individual and his behavior, similar to how physiology considers the position of a bodily "organ" in the functioning of the organism from the point of view of 'preserving' it" [30, ch. 1]. Obviously, Weber points out, the consideration of the functional aspects of the "parts" of a "totality" may play a useful role "in practical explanation and temporary orientation", but "it also may happen that its cognitive value is overestimated because very detrimental false conceptual realism has been accepted" [30]. All this reasserts that collective concepts should be interpreted individualistically, that is, the social sciences must proceed in an individualistic way, from the actions of "a single individuals, of a few individuals, or of many individuals". Therefore, "even a socialist economy should be *included* in sociology in virtue of a procedure interpreted "individualistically" — that

---

[3] On the logic of the situation and methodological individualism in Popper, see the critical remarks by A.M. Petroni, [15, pp. 64–76]. In the essay L'individualismo metodologico in [16, p. 144], Petroni maintains that "some of Popper's weakest pages are devoted to the principle of rationality".

[4] On the differences between the positions of Popper and Hayek see [28, pp. 215–220] and [1, pp. 39–41].

is, *based in the actions of individuals*" [30]. So much for Max Weber. In 1897 Georg Simmel was already critical of the pernicious illusion whose view of the evolution of the law and other social institutions would lead to thinking of society as an "impersonal being," independent of individuals [27]. According to Simmel, society cannot be considered something "conceived for itself". "It is certain", Simmel writes, "that only individuals exist, that human products have a reality apart from men only if they are of natural materials, and that since the creations we are referring to are spiritual, they live only in personal brains [...]" [27].

In Vienna, Carl Menger, the Austrian father of the marginalist theory, had become an advocate of *methodological individualism* even before Weber and Simmel. Resources are scarce and individuals seek the greatest satisfaction of their needs and desires. Menger believes it is precisely these *actions* and these *individuals* that economic theory should take into consideration and adopt as a point of departure. There is nothing else economic theory can do, and since — Menger says — "*the collectivity as such* is *not like a great subject*, who has his needs, who works, competes; what we call "social economy" is not the economic activity of a society, strictly speaking: it is not analogous to individual economies and does not exist as opposite to individual economies. In its phenomenic form it is a particular multiplicity of particular economies [8, pp. 86–87]. Therefore, there are no specific realities independent of individuals that correspond to collective concepts used in economic theory — and even more widely in the social sciences. Menger is convinced that the facts of social economics are not the immediate manifestations of the life of a people as such, immediate direct products of its economic activity, but rather the result of the countless efforts of collectivity. Therefore, anyone who wants to arrive at a theoretical understanding of the more complex human phenomena — for example, the phenomena that we call "social economics" — must appeal to the real elements of the individual economies of the collectivity and attempt to understand how social economy derives from individual economies [8, p. 87]. Anyone who insists on following the opposite route, does not understand — Menger claims — the nature of social economy and proceeds through a fictitious method. And, as we shall see later on, Menger shows how the individualistic method works to explain the origins of social rules and institutions, which, in his opinion, can arise in two ways: either in a *pragmatic* way (as the result of the will of individuals directed toward an objective) or as *unintentional results* of individual actions that had other objectives (this is an invisible hand mechanism which manages to account for the origins of changes in many social events, even very important ones) [8, pp. 84–85].

3 Ludwig von Mises: "It is only the individual who thinks. It is only the individual who reasons. It is only the individual who acts!"

Ludwig von Mises was one of those who followed and developed Menger's teachings, and, in turn, exerted a strong influence on Friedrich von Hayek. Popper knew of Mises in Vienna in 1935, and several months later in London met Hayek [21, p.10]. Perhaps no one managed to express the fundamental principle of methodological individualism better than Mises: "It is only the individual who thinks. It is only the individual who reasons. It is only the individual who acts" [11]. Consequently, the course of history is determined by individual actions and by the effects of these actions [12]. But from the prospect of research methodology, all this means that in science we must always move from individual actions. The idea of a society which works independently from the individual actions is absurd. The fact that one is a member of a capitalistic society, or is a citizen, or is a member of a particular association cannot be shown but through his individual actions [13, p.43]. This is how methodological individualism applies social studies to *observable experience*. Experience does not show us things like the "society" or the "nation". Experience always introduces us to individuals, it makes us listen to their words, read what they write and *see their actions*. It is merely an illusion to think it is possible to visualize collective unities. The knowledge of such unities is the result of the understanding of the individual actions and the significance that the individuals give to their actions [12]. One must continually fight against the temptation to hypostatize and the tendency to attribute substance and real existence to our mental constructs [14, p. 78]. The most striking case of this fallacy is the use of the term society in pseudoscientific schools. Society is not itself a substance, a power, an agent. Society does not exist independently from the thoughts and actions of individuals. As all the other collective unities, it has no proper interests or aims. The tendency to hypostatize collective concepts is the most dangerous enemy for scientific knowledge and helps those political projects that ascribe more dignity to collectivities rather than to individuals, or that conceive individuals as mere abstractions. Instead, everything we can know about society can be achieved only studying the actions of the individuals. Methodological collectivism is a *mythological* conception based on the idea that human actions are led by mysterious forces [14, p. 78–82].

4 Friedrich A. von Hayek: "It is a serious error to treat as facts what are at best vague popular theories"

Friedrich von Hayek, like von Mises, believed that the typical element of methodological collectivism is that "it treats social phenomena not as something of which the human mind is a part and the principles of whose organization we can reconstruct from the familiar parts, but as if they were objects directly perceived by us as wholes" [5, p. 94]. This deep-rooted philosophical position springs from the fact that "the existence, in popular usage of terms like *society* or *economy* is naively taken as evidence that there must be definite 'objects' corresponding to them. The fact that people all talk about the nation or capitalism leads to the belief that the first step in the study of these phenomena must be to go and see what they are like, just as we should if we heard about a particular stone or a particular animal" [5, pp. 94–95]. That is why, Hayek rightly affirms, "the error involved in this collectivist approach is that it mistakes for facts what are no more than provisional theories, models constructed by the popular mind to explain the connection between some of the individual phenomena which we observe" [5, p. 95]. It is a serious mistake "that of treating as facts what are no more than vague popular theories"[5, p. 95]. Unfortunately, this mistake — naive realism — "is so deeply embedded in current thought about social phenomena that it requires a deliberate effort of will to free ourselves from it" [5, p. 96]. This is why, it is necessary, once the error of naive realism has been exposed, to resolutely and clearly confirm that "the social sciences do not deal with 'given' wholes but their task is to constitute these wholes by constructing models from the familiar elements" [5, p. 98]. "The mistake of treating as definite objects wholes that are no more than constructions, and that can have no properties except those which follow from the way in which we have constructed them from the elements, has probably appeared most frequently in the form of the various theories about a "social" or "collective" mind and has in this connection raised all sorts of pseudo-problems" [5, pp. 100–101]. And, at the same time, the most ruthless social atrocities.

An escape from this trap of naive realism — which *reifies* collective concepts and makes them become things — is to distinguish between *motivating or constitutive opinions* on the one hand and *speculative or explicative concepts* on the other.

Is it the ideas which the popular mind has formed about such collectives as society or the economic system, capitalism or imperialism, and other such collective entities, which the social scientist must regard as no more than provisional theories, popular abstractions, and which he must not mistake for facts? That he consistently refrains from treating these pseudo-entities as facts, and that he systematically starts from the concepts which guide individuals in their actions and not from the results of their theorizing about their actions, is the characteristic feature of that methodological individualism which is closely connected with the subjectivism of the social sciences [5, p. 64].

But these concepts that lead men to action are precisely *constitutive ideas*: for example, we recognize that the cause of changing opinions toward certain goods is in price changes of those goods; or beliefs or opinions that "lead a number of people regularly to repeat certain acts, for ex ample, to produce, sell, or buy certain quantities of commodities, are entirely different from the ideas they may have formed about the whole of the 'society', or the 'economic system', to which they belong and which the aggregate of all their actions constitutes" [5, p. 63].

Like the natural sciences, social studies requires contact with experience. *Motivating or constitutive ideas* are, for Hayek, the *empirical base* of social studies. "It is the so-called wholes, the groups of elements which are structurally connected, which we learn to single out from the totality of observed phenomena only as a result to our systematic fitting together of the elements with familiar properties, and which we build up or reconstruct from the known properties of the elements" [5, p. 67]. It should be emphasized that "in all this the various types of individual beliefs or attitudes are not themselves the object of our explanation, but merely the elements from which we build up the structure of possible relationships between individuals" [5, p. 68]. All this is the same as saying that individual beliefs and behavior are the "data" of the social sciences, so it is not the job of the social sciences" [5, p. 68] to explain conscious action: "this, if it can be done at all, is a different task, the task of psychology" [5, p. 68]. Therefore, Hayek says, the method of the social sciences becomes a "compositive" or synthetic method [5, p. 67].[5] What are given to the social scientist are "elements from which those more complex phenomena are composed that he cannot observe as a whole" [5, p. 66]. To put it briefly, "for the social sciences the types of conscious action are data and all they have to do with regard to these data is to arrange them in such orderly fashion that they can be effectively used for their task" [5, p. 68]. And for Hayek this task — which is the task of the social sciences — "is done, as we shall soon see, by analyzing the unintentional consequences of intentional human actions" [5, p. 68].

5   Karl R. Popper: the spontaneous growth of social institutions invalidates both psychologism and the conspiracy theory of society

Only individuals reason and take action. This is the cardinal principle of methodological individualism. It is precisely from the actions of individuals, says Popper, that the social scientist must start in order to explain the origins of and changes in institutions and social events. The crucial point here is the awareness that intentional human actions constantly have unintentional consequences.

In section 21 of *The Poverty of Historicism*, Popper says, "the piecemeal technologist or engineer recognizes that *only a minority of social institutions are consciously designed while the vast majority have just 'grown', as the undersigned results of human actions*" [19, p. 65].[6] This theory, which was only mentioned in *The Poverty of Historicism*, later became increasingly important in Popper's thoughts on society and its functioning, especially on the role of the theoretical social sciences. Thus, in the first volume of *The Open Society and Its Enemies*, Popper — concerned with discovering the fate or real role of institutions in the development of history, in the sense that they are seen as willed by God or destined by Fate or as obedient to important historical trends, etc. — counterpoises to the historicist the gradualist social technologist, who will not forget that the institutions grow similarly to natural organisms. In Chapter XIV of the second volume of *The Open Society and Its Enemies*, Popper elaborates on this subject and launches an extensive attack against *psychologism*, arguing in favor of the autonomy of sociology.

---

[5]In note 4 on p. 67, Hayek says "I have taken the term 'compositive' from a manuscript note of Carl Menger, who, in his personal annotated copy of Schmoller's review of his Methoden der Sozialwissenschaften wrote it above the word deductive used by Schmoller. Since writing this I have noticed that Ernst Cassirer in his *Philosophie der Aufklärung* (1932) uses the term 'compositive' in order to point out rightly that the procedure of the natural sciences presupposes the successive use of the 'resolutive' and the 'compositive' technique. This is useful and links up with the point that, since the elements are directly known to us in the social sciences, we can start here with the compositive procedure". However, R. Cubeddu disagrees with Hayek. He says [2, p. 325] that the correlation between the use of the term "compositive" in Menger and the "resolutory ", "compositive" method that Cassirer speaks of would be misleading because it "would imply that Menger's 'compositive method' is related to the method of the modern natural sciences. However, in Menger's text there is an obvious influence of Aristotelian philosophy".

[6]According to Popper, the two theories that respectively assert that social institutions either are 'designed' or grow spontaneously are typical, on the one hand, of the theoreticians of the Social Contract and, on the other, of their critics, such as David Hume.

Certainly, Popper says, "it must be admitted that the structure of our social environment is man-made in a certain sense; that its institutions and traditions are neither the work of God nor of nature, but the results of human actions and decisions" [17, vol. 2, p. 93, 50]. Nevertheless, "this does not mean that they are all consciously designed, and explicable in terms of needs, hopes, or motives. On the contrary, even those which arise as the result of conscious and intentional human actions are, as a rule, *the indirect, the unintended and often the unwanted by-products of such actions*" [17, vol. 2, p. 93]. The easiest way to understand the meaning of the idea of *unintentional consequences* of an action is to use an example: "If a man wishes urgently to buy a house in a certain district, we can safely assume that he does not wish to raise the market price of houses in that district. But the very fact that he appears on the market as a buyer will tend to raise market prices. And analogous remarks hold for the seller" [17, vol. 2, p. 96].[7] Therefore, only a small number of social institutions are or have been intentionally planned; the majority of them simply grew, or sprung up, as the unintentional result of intentional actions [17, vol. 2, p. 91]. And this is not all, because, Popper states:

> [...] we can add that even most of the few institutions which were consciously and successfully designed (say, a newly founded University , or a Trade Union) do not turn out according to a plan — again because of the unintended social repercussions resulting from their intentional creation. For their creation affects not only many other social institutions but also "human nature" — hopes, fears, and ambitions, first of those more immediately involved, and later often of all members of the society [17, vol. 2, pp. 93–94].

If all this is true — if it is true that most social institutions are not planned and if it is true that even institutions that are planned do not realize the plan as it was conceived — there are two inevitable consequences, according to Popper: the first is that *psychologism* fails and the second is that the *conspiracy theory* of society is untenable.

According to the doctrine of *psychologism*, the study of society should be limited to psychology, in the sense that the origins of and changes in all social events and institutions would be explained by the intentional actions and projects of individuals [17, vol. 2, p. 90]. But this interpretation of psychologism does not hold up because "it fails to understand the main task of the explanatory social sciences" [17, vol. 2, p. 94], which is to explain the unintentional, perhaps even unexpected and undesired, consequences of intentional human actions [17, vol. 2, p. 93]. There are innumerable vitally important social institutions and effects of institutions that are not in the least due to intentions, hopes, fears, and conscious projects; and psychologism does not know what to make of them.[8] However, the *conspiracy theory of society* does not hold up either. It maintains that a social phenomenon is

---

[7] Menger mentions this example in passing, but Popper returns to it several times. For example, see his essay [22, p.452].

[8] The inability of *psychologism* to offer explanations was well understood by Marx. Says Popper, "To have questioned psychologism is perhaps the greatest achievement of Marx as a sociologist. By doing so he opened the way to the more penetrating conception of a specific realm of sociological laws, and of a sociology which was at least partly autonomous" [17, vol. 2, p. 88] ; see also [22, p. 452]. In any case, Popper maintains that "my arguments against psychologism should not be misunderstood. They are not, of course, intended to show

explained only if it manages to discover the men or social groups that have planned or conspired to promote it [17, vol. 2, p. 95]. In other words, according to the *conspiracy theory, every* social event, particular those like war, unemployment, poverty and famine (events that people generally abhor) is *always* the result of successful direct intervention of powerful individuals or groups, of real conspirators. In Popper's opinion, this theory is "a typical result of the secularization of a religious superstition. The belief in the Homeric gods whose conspiracies explain the history of the Trojan War is gone. The gods are abandoned. But their place is filled by powerful men or groups — sinister pressure groups whose wickedness is responsible for all the evils we suffer from — such as the Learned Elders of Zion, or the monopolists, or the capitalists, or the imperialists [17, vol. 2, pp. 97–98].[9]

Of course there are also conspiracies, but this does not mean that all social events and institutions are the result of conspiracies. Even if we admit there are conspiracies, we must also acknowledge that "few of these conspiracies are ultimately successful. *Conspirators rarely consummate their conspiracy*" [17, vol. 2, p. 95]. That the results achieved differ greatly from those sought is so "because this is usually the case in sociallife, conspiracy or no conspiracy. Social life is not only a trial of strength between opposing groups: it is action within a more or less resilient or brittle framework of institutions and traditions, and it creates — apart from any conscious counteractions — many unforeseen reactions in this framework, some of them perhaps even unforeseeable" [17, vol. 2, p. 95]. Consequently, it is obvious that "to try to analyse these reactions and to foresee them as far as possible is, I believe, the ma n task of the social sciences. It is the task of analyzing the unintended social repercussions of intentional human actions — those repercussions whose significance is neglected both by the conspiracy theory and by psychologism" [17, vol. 2, p. 95]. We can understand that this task is essential for the social sciences if we think that "an action which proceeds precisely according to intention does not create a problem for social science (except that there may be a need to explain why in this particular case no unintended repercussions occurred) [17, vol. 2, p. 96].

6 Karl R. Popper and the task of the social sciences

In the 1948 essay *Prediction and Prophecy in the Social Sciences*, Popper, referring primarily to Marxism, criticizes "the doctrine that it is the task of the social sciences to propound historical prophecies, and that historical prophecies are needed if we wish to conduct politics in a rational way" [22, p. 452]. This doctrine, which Popper calls historicism — and of which Marxism is

---

that psychological studies and discoveries are of little importance for the social scientist. They mean, rather, that psychology — the psychology of the individual — is one of the social sciences, even though it is not the basis of all social science. Nobody would deny the importance for political science of psychological facts such as the craving for power, and the various neurotic phenomena connected with it. But 'craving for power' is undoubtedly a social notion as well as a psychological one: we must not forget that, if we study, for example, the first appearance in childhood of this craving, then we study it in the setting of a certain social institution, for example, that of our modern family. (The Eskimo family may give rise to rather different phenomena.)" [17, vol. 2, pp. 97–98]

[9]On the same subject see [22, pp. 460–461].

a case — is, in Popper's view, untenable. One of the reasons for this is that the *historicist* — and therefore also the Marxist — is incapable of distinguishing between a *scientific prediction* and *unconditional historical prophecy* [22, p. 454]. The predictions of science are conditional. "They assert that certain changes (say, of the temperature of water in a kettle) will be accompanied by other changes (say, the boiling of the water)" [22, p. 456]. However, the historicist cannot make these conditional predictions because society is not a well-isolated, stationary and recurrent system like the solar system. "These systems are very rare in nature; and modern society is surely not one of them" [22, p. 457]. The solar system and cyclic biological systems, where conditional predictions can be made, are exceptional cases. But "society is changing, developing. This development, in the main, is not repetitive" [22, p. 457][10] so it is impossible to make predictions, especially long-range ones. Actually, "the most striking aspects of historical development are non-repetitive. Conditions are changing and situations arise (for example, in consequence of new scientific discoveries) which are very different from anything that ever happened before. The fact that we can predict eclipses does not, therefore, provide a valid reason for expecting that we can predict revolutions" [22, pp. 457–458]. Political prophecy is not a scientific prediction.

Furthermore, it is not the task of the social sciences to study "the behaviour of social wholes, such as groups, nations, classes, societies, civilizations, etc." [22, p. 459]. Many sociologists consider these social wholes as empirical objects to study in the same way a biologist studies animals and plants. This view, Popper warns, "must be rejected as naive. It completely overlooks the fact that these so called social wholes are very largely postulates of popular social theories rather than empirical objects" [22, p. 459].[11]

Popper adds, as we have already seen, that another widespread erroneous view is that social events are the results of *conspiracies* and that the social scientist's task should be to discover and explain them. This theory doesn't work because "not all consequences of our actions are intended consequences". This means that "the conspiracy theory of society cannot be true because it amounts to the assertion that all events, even those which at first sight do not seem to be intended by anybody, are the intended results of the actions of people who are interested in these results" [22, p. 460]. And it is remarkable that many Marxists have completely accepted the conspiracy theory, which "objectively" sees the fatal capitalist conspirator behind every event, especially if it is negative. This is remarkable because "Marx himself was one of the first to emphasize the importance, for the social sciences, of these unintended consequences. In his more mature utterances, he says that we are all

---

[10]Of course, Popper points out that in so far as the development of society is repetitive "we may perhaps make certain prophecies. For example, there is undoubtedly some repetitiveness in the manner in which new religions arise, or new tyrannies; and a student of history may find that he can foresee such developments to a limited degree by comparing them with earlier instances, i. e. by studying the conditions under which they arise. But this application of the method of conditional prediction does not take us very far" [22, p. 457] because in the history of society situations and conditions change. Since the changing of conditions and their combinations are not predictable, neither are their effects

[11]Actually, Popper adds, "while there are, admittedly, such empirical objects as the crowd of people here assembled, it is quite untrue that names like 'the middle-class' stand for any such empirical groups".

caught in the net of the social system. The capitalist is not a demoniac conspirator, but a man who is forced by circumstances to act as he does; he is no more responsible for the state of affairs than is the proletarian" [22, p. 460].[12] However, "perhaps for propagandist reasons, perhaps because people did not understand it" [22, pp. 460–461]. This view of Marx's has been abandoned and an infinite number of people have taken up "a Vulgar Marxist Conspiracy theory". It is, Popper points out, "a come-down-the come-down from Marx to Goebbels. But it is clear that the adoption of the conspiracy theory can hardly be avoided by those who believe that they know how to create heaven on earth. The only explanation for their failure to produce this heaven is the malevolence of the devil who has a vested interest in hell" [22, p. 461].

What then is the task of the theoretical social sciences if it is not to make historical predictions, if the theory of sociological collectivism is not valid, and if the conspiracy theory does not work? Popper answers this question in familiar language: "the main task of the theoretical social sciences [...] is to trace the unintended social repercussions of intentional human actions" [22, p. 460].[13]

It is important to note that understanding the unintentional, and possibly undesirable, repercussions of social actions brings the social sciences close to the natural ones. Natural laws can be defined as warnings of a technical nature since they can be formulated as practical technological rules stating what *we cannot do* [22, p. 461]. The same is true for the social sciences. Examples of these rules in the social sciences are: "You cannot, without increasing productivity, raise the real income of the working population" and "You cannot equalize real incomes and, at the same time, raise productivity" [22, p. 461]. This shows that, although the historicist doctrine is untenable, science and reason can benefit us in practical life by "helping us choose our actions more wisely" [22, p. 461].

In works later than those examined here, Popper substantiates his criticism of the conspiracy theory and the proposition that the basic task of the social sciences is to recognize the *unintentional, especially the unwanted*, repercussions of certain actions. For example, in the essay *Towards a Rational Theory of Tradition*, once again he stresses that "we wish to foresee not only the direct consequences but also these unwanted indirect consequences" of human actions [22, p. 167]; "either because of our scientific curiosity, or because we want to be prepared for them; we may wish, if possible, to meet them and prevent them from becoming too important. (This means, again, action, and with it the creation of further unwanted consequences.)" [22, p. 167].

Later on, Popper developed the theory of World 3. This seems to be no more than a generalization for the entire world of culture — especially for the logical province of World 3, which is made up of problems, theories and scientific argumentation — of results such as the autonomy of sociology

---

[12] Popper says in *The Open Society and Its Enemies* that for the idea that Marx conceived of social theory as the study of the unintentional repercussions of almost all our actions, he is indebted to K. Polanyi who emphasized this aspect of Marxism in private discussions (1924) [17, p. 323].

[13] Note that once again Popper clarifies this idea of the unintentional repercussions of intentional human actions with the example of the man who wants to buy a house and so, by appearing on the market, involuntarily raises the price of the house.

and the unwanted consequences of intentional actions. According to Popper, there is a World 3 of books in themselves, theories in themselves, problems in themselves, arguments in themselves. A great part of the objective World 3 of theories, books, and arguments is an unintentional product: they are a sub-product of human language. Language itself — Popper argues — is an unintentional sub-product of actions that were aimed at other goals [23].[14]

7 Carl Menger: "Not all social events are the consequence of explicit contracts or legislation"

While Popper maintained that analysis of the unintentional consequences of intentional human actions is the *principle task* of the social sciences, Hayek believed that the task of the social scientist is *completely* fulfilled in that analysis. The problems that the social sciences "try to answer", Hayek says, "arise only insofar as the conscious action of many men produce undersigned results, insofar as regularities are observed which are not the result of anybody's design" [5, pp. 68–69]. In other words, the task of the social sciences is to explain the unintentional effects of intentional actions. This explains the reason for the autonomy of sociology. In fact,

> [...] if social phenomena showed no order except insofar as they were consciously designed, there would indeed be no room for theoretical sciences of society and there would be, as is often argued, only problems of psychology. It is only insofar as some sort of order arises as a result of individual action but without being designed by any individual that a problem is raised which demands a theoretical explanation [5, p. 69].

Well before Hayek and Popper, in Vienna Carl Menger had stated very clearly that the solution of the most important theoretical social sciences and theoretical economics in particular is strictly connected with the comprehension of social institutions, grown not intentionally but spontaneously.

Of course, no one can deny that there are social institutions and events that arise from conscious explicit agreements, pacts, or conventions between people. There are social and economic events and institutions that come into being because certain groups of humans want to achieve specific objectives: for example, insurance companies, urban development plans, press agencies, price controls, etc. The fact that there are some social institutions and events resulting from intentional agreements among people has given credence to the belief in a general theory that all social and economic events can be explained as consequences of deliberate collective activity. Menger calls this interpretation, "*the pragmatic theory* of the origin of social phenomena; it is the explanation of the nature and the origin of phenomena based on the aims, opinions and means of human associations and their representatives" [5, pp. 164–165].

---

[14] The idea of unintentional consequences of intentional human actions was then extended to the products of World 3: these also have unexpected consequences, and this is obviously true also for scientific theories. In *Unended Quest* Popper writes: "There is an *infinity of unforeseeable nontrivial statements belonging to the informative content of any theory*, and an exactly corresponding infinity of statements belonging to its logical content. We can therefore never know or understand all the implications of any theory, or its full significance" [24, p. 26]. In a certain sense, we never know what we are talking about. A similar situation is presented in the hermeneutics of Gadamer at the beginning of the history of effects (*Wirkungsgeschichte*); see [3].

So there do exist social phenomena that originate in agreements among people and we give a pragmatist interpretation to them: we analyse the aims that have guided the people who have created certain institutions or we analyse the means that were available to them. These phenomena — Menger adds — are subjected to pragmatistic-historical examination in so far as in each concrete case we evaluate the real ends of the associations or their representatives in the light of the needs of the associations, while we judge the use of auxiliary means of social activity in the light of the fulfillment of the social needs.

Therefore, all this applies to those social phenomena of a pragmatistic origin. But a pragmatistic interpretation of social phenomena cannot explain "a long series" of social phenomena as language, religion, law, State, market and currency. Such institutions are not the result of any agreement or legislation, and this presents a strange, interesting problem, "a curious problem" — perhaps the most curious of all the problems in the social sciences [5, p. 163]. The problem is: "How is it possible that institutions which are so important for collective well-being grow without a collective will or stipulation¿'

To dispel any doubt about the significance of what he wanted to demonstrate, Menger cites one of the many possible examples, the example of *social prices of goods*. Normally, prices develop without any regulative influence of the State or collective agreement, as a spontaneous result of social evolution [5, pp. 172–178].

It is clear, therefore, that there is a whole series of very important social phenomena that arise spontaneously or non-intentionally. These phenomena belie the universalizing claims of pragmatist theory in the sense that *not all social* phenomena arise from explicit agreements or positive legislation. If this is the case, it is obvious, Menger believes, that the solution of the most important problems of theoretical social sciences and, in particular, theoretical economics, is narrowly linked to the comprehension of the growth of the institutions which developed in an "organic" way, i.e. in an unintentional and spontaneous way.

8 Carl Menger: examples of Institutions arising "organically"

Actually, the idea of explaining the nature and development of social institutions as the result of an agreement between individuals or legislation was, says Menger, the first hypothesis for the understanding of social institution. This theory or pragmatist explanation was not realistic, but had the advantage of reducing the comprehension of every institution to the same principle of interpretation.

However, hypotheses are not true merely because (certainly to be desired) they are simple and standardized. This is also applies to *pragmatistic theory* on the origin of social phenomena. For example, even though history shows us that *new localities* have been formed because a certain number of people of different abilities and professions joined together with the purpose of founding a city, usually new cities arise without a precise general intention. A common intentional will develops only at advanced stages of the collective life: it does not cause the birth of a social institution, but only its perfecting.

What was said for new localities also applies to the *state*. If the theory attributing the origin of the state *exclusively* to an organic formation is unilateral, it is much more mistaken the idea that States were born through an agreement of powerful people.

Actually, Menger points out, that institution that we call State is in, its original form, the undersigned result of actions led by individual interests.

In the same way, in Menger's view, it can be demonstrated that other social institutions as language, law, economic institutions were born without an explicit agreement, without legislation, without concern for collective interest, but only by the impulse of individual interests. Other examples are the separation of professions, the division of work, commercial customs, etc.

Certainly in the course of social evolution public powers have intervened to create new institutions and change or develop those which arose unintentionally, but at the beginning of society, the creation of institutions was unplanned. Consequently, an explicit willful act for a specific purpose by institutions whose origin was not contemplated — says Menger — means that those institutions are the result of a combined action of both unintended and teleological forces, i.e. of both *organic and positive factors* [9, p. 250].

9 Carl Menger: the origin of money

Chapter IX of Menger's *Principles of Economics* is dedicated to the *Theory of Money* and a good part of the chapter deals with the origin of money. Actually — argues Menger — the fact that one gives to another person a part of his properties and receives back something else is clearly comprehensible. But the fact that in modern societies, thousands of people exchange useful properties with pieces of paper is certainly enormously curious.

Menger's inquiry concerns the nature of those pieces of paper or metal which represent a huge power on markets and the lives of people. Are they desired intentional creations attributable to conventions or legislative acts of men for explicit objectives, or is this also a case of an institution which "grew up" unintentionally as a result of other intentional actions? This is a problem that science must solve [8, p. 172]. The idea that goods become money as a result of an explicit convention or legislative act is not a "basically false" opinion since history offers examples of certain good which have been declared to be money by legislation. However, Menger points out that the legislative act did not give to a particular good the status of money, but only the official acknowledgement, while the good was already used as money. In any case, the "pragmatist explanation" of the origin of the social institution of money does not work in all those important cases where money is not clearly the result of legislative acts but was born spontaneously (or organically). In other words, science must explain how, with the development of economic civilization certain goods are raised to the status of money without explicit agreement or legislative act.

Menger explains the phenomenon using the following line of reasoning. As long as a simple commerce of exchange (barter) is dominant among a population, every individual tries to exchange excess goods for those needed immediately and refuses those goods not needed or which are in abundant supply. Now, however, in order for a person who brings surplus goods to the market to be able to exchange these superfluous goods with those he wants, he must not only find someone who asks for his goods, but also who can offer him the goods he pursues. But this is precisely what makes a pure barter regime difficult and limited.

How can such a drawback to exchange be overcome? What effective remedy could be applied? Everyone could observe that some goods were requested more than others. So, in a nomadic people everyone knows that cattle could be exchanged more easily than other goods, and this made easier to find in return the desired goods, among the many individuals who wanted to acquire cattle. Those who had to offer goods which were less in demand, tended to acquire goods that perhaps they did not really need, but that were more exchangeable. In this way, they did not reach their immediate goal (the good that they needed), but gradually approached to get them. This happened without any agreement or legislative force or concern for collective interest. People tended to prefer goods more comfortable and useful for exchange. The goods which were more safe, lasting and more easily transferable began to be accepted by everyone and were called "money".

In conclusion, money did not originate by a legislative act: it was rather the unintended product of the individual actions of a collectivity. This explanation is a typical example of how, with time, it is possible to classify as controllable phenomena which at other times were attributed to religious powers or metaphysical entities.[15]

BIBLIOGRAPHY

[1] N.P. Barry. *Hayek's Social and Economic Philosophy*. Macmillan, London 1979.
[2] R. Cubeddu. Sul tema dell'individualismo metodologico. *Il Politico*. LIV, 2, 1989.
[3] H.G. Gadamer. *Wahrheit und Methode*. Mohr, Tübingen 1960. English trans. by G. Barden and J. Cumming, *Truth and Method*, Sheed, London 1975.
[4] A.W.Green. The reified villain. *Social Research*, 35, 1968.
[5] F. A. von Hayek. *The Counter-Revolution of Science. Studies on the Abuse of Reason*. Liberty Press, Indianapolis, PLACE YEAR.
[6] W.M. Johnston. *Oesterreichische Kultur und Geistesgeschichte im Donauraum 1848 bis 1938*. H. Böhlaus, Wien-Köln-Grare 1974.
[7] R. Kerschagl. *Einführung in die Methodenlehre der national Ökonomie*. Holder-Pichler-Tempsky, Wien-Leipzig 1925.
[8] C. Menger. *Untersuchungen über die Methode der Sozialwissenschaften und der politischen Ökonomie insbesondere*. Duncker & Humbold, Leipzig 1883.
[9] C. Menger. *Grundsätze der Volkswirtschaftslehre*. BraumuHer, Wien 1871.
[10] R.K. Merton. The unanticipated consequences of purposive social action. *American sociological Review*, 1936.
[11] L. von Mises. *Socialism*. Liberty Classics, Indianapolis 1981.
[12] L. von Mises. *Human Action. A Treatise on Economics*. Hodge, London 1949.
[13] L. von Mises. The task and scope of the science of human action. In *Epistemological Problems of Economics*. New York University Press, New York-London 1981.
[14] L. von Mises. *The Ultimate Foundation of Economic Science*. Sheed Andrews and McMeel, Inc., Kansas City 1978.
[15] A.M. Petroni. Introduzione to K.R. Popper, *Il pensiero politico*. Le Monnier, Firenze 1981.
[16] A.M. Petroni. L'individualismo metodologico. In A. Panebianco (ed.), *L'analisi della politica*, il Mulino, Bologna 1989.
[17] K.R. Popper. *The Open Society and Its Enemies*. Routledge, London 1999.

---

[15]See [10, pp. 894–904]. It is interesting that in this important essay, which is almost completely overlooked by sociologists and methodologists of the social scientists, Merton also cites Vico and Bossuet

[18] K.R. Popper. *La scienza e la storia sul filo dei ricordi.* Interview by Giulio Ferrari, Jaca Book Edizioni Casagrande, Bellinzona 1990.
[19] K.R. Popper. *The Poverty of Historicism.* Routledge, London 1989.
[20] K.R. Popper. *Die Logik der Sozialwissenschaften.* In T. Adorno, H. Albert, R. Dahrendorf, J. Habermas, H. Pilot and K.R. Popper, *Der Positivismusstreit in der deutschen Soziologie*, Luchterhand, Darmstadt 1969, pages 103–123.
[21] K.R. Popper. The communist road to self-enslavement. *Cato Policy Report*, May–June 1992, vol. XIV, n.3.
[22] K.R. Popper. *Conjectures and Refutations.* Routledge, London 2006.
[23] K.R. Popper. *Objective Knowledge: an Evolutionary Approach.* Clarendon Press, Oxford 1972.
[24] K.R. Popper. *Unended Quest.* Routledge, London YEAR.
[25] K. Pribram. *Die Entstehung der individualistischen Sozialphilosophie.* C. Hirschfeld, Leipzig 1912.
[26] M.N. Rothbard. *Individualism and the Philosophy of the Social Sciences.* Cato Institute, San Francisco 1979.
[27] G. Simmel. *Comment les formes sociales se maintiennent. L'annèe sociologique*, 1987
[28] K.J. Scott. Methodological and individualism. In J. O'Neill (ed.), *Modes of Individualism and Collectivism*, Heinemann, London 1973.
[29] M. Weber. Letter to to R. Liefmann (1920). In R. Boudon. *La place du desordre.* PUF, Paris 1991.
[30] M. Weber. *Wirtschaft und Gesellschaft.* Mohr, Tübingen 1922.

PART V

NEUROSCIENCE AND PHILOSOPHY OF MIND

Jaegwon Kim and the threat of epiphenomenalism of mental states

FABIO BACCHINI

Jaegwon Kim [8] has given new strength to the thesis of mental epiphenomenalism. He has furnished powerful reasons to think that, if you truly accept physicalism and if you truly admit that mental properties are distinct from physical properties, you cannot concede that mental properties have any authentic causal power. He has offered two main arguments to support this implication. The first he baptized the "Supervenience Argument": this is a dilemma argument showing that — whether you accept mind-body supervenience or not — you must conclude that mental causation is unintelligible. The second asserts that the causal inefficacy of mental properties descends from their being functional properties; we may call it the "Dormitivity Argument". Although I recognize both of them to be effective and convincing, I will try to show that they are also questionable, and that mental epiphenomenalism is not as inevitable as Kim wants us to decree. In particular, I will show that the Supervenience Argument meets with difficulties when ruling out the possibility of some of the effects of mental properties being systematically overdetermined; and, on the other hand, that the Dormitivity Argument rests on an analogy between mental properties and typical functional second-order properties that may perfectly well be discussed and rejected.

1 The Supervenience Argument: first appeal to the NOP

I mean to show that the Supervenience Argument makes three crucial appeals to a No Overdetermination Presumption (NOP): the presumption that systematic overdetermination cannot be considered as a part of an acceptable explication of the relation between the mental and the physical. I will sustain that these appeals are either unargumented or supported by weak arguments, so that the whole Supervenience Argument is undermined.

Let us see how the Supervenience Argument goes in Kim's original formulation:

1. Either mind-body supervenience holds or it fails;

2. If mind-body supervenience fails, there is no visible way of understanding the possibility of mental causation;

3. If mind-body supervenience holds: suppose that an instance of mental property M causes another mental property M* to be instantiated;

4. M^* has a physical supervenience base P^*;

5. M^* is instantiated on this occasion: (a) because, ex hypothesis, M causes M^* to be instantiated; (b) because P^* is instantiated on this occasion;

6. M caused M^* by causing P^*;

7. M itself has a physical supervenience base P;

8. P caused P^*, and M supervenes on P and M^* supervenes on P^*;

9. The M-to-M^* and M-to-P^* causal relations are only apparent, arising out of a genuine causal process from P to P^* (mental epiphenomenalism);

10. If mind-body supervenience fails, mental causation is unintelligible; if it holds, mental causation is again unintelligible. Hence mental causation is unintelligible [8, pp. 38–47].

The fact that the Supervenience Argument is a dilemma argument is clearly revealed by (10). We first encounter an indirect appeal to the NOP in (2), that alone constitutes the first horn of the dilemma. Why indeed should we accept that "if mind-body supervenience fails, there is no visible way of understanding the possibility of mental causation"? Kim's answer is that "the simplest and most obvious reason for the physicalist to accept (2) lies, I think, in her commitment to the causal *closure of the physical domain*" [8, p. 40]. We rapidly discover that a "principle of causal closure" can be enunciated: "If you pick any physical event and trace out its causal ancestry or posterity, that will never take you outside the physical domain. That is, no causal chain will ever cross the boundary between the physical and the nonphysical" [8, p. 40]. Such a principle cannot be rejected, under pain of abandoning physicalism itself: "If you reject this principle, you are *ipso facto* rejecting the in-principle completability of physics — that is, the possibility of a complete and comprehensive physical theory of all physical phenomena. For you would be saying that any complete explanatory theory of the physical domain must invoke nonphysical causal agents" [8, p. 40]. As a final step in the reasoning, it is not difficult to see that "if mind-body supervenience fails — that is, if the mental domain floats freely, unanchored in the physical domain, causation from the mental to the physical would obviously breach the physical causal closure" [8, p. 40].

According to Eric Marcus [11], there is a leak in Kim's line of reasoning. Kim is guilty of confusing a Completeness Principle of physics with a Closure Principle. Completeness is the idea that all physical events have complete physical causal histories, and that we simply never *need* to causally explain physical phenomena by nonphysical ones: "There is some true physical theory capable of fully explaining why physical processes unfold in precisely the way they do. At each stage in a physical causal chain, the causal connection between it and earlier and later stages can be completely accounted for by a true physical theory" [11, p. 28]. On the other hand, when we invoke Closure, we are referring to the idea that all causal histories of all physical events must contain just physical events or physical properties, and that we therefore *cannot* causally explain physical phenomena by nonphysical ones: "According to Closure, physical events cannot interact causally with nonphysical events, or with physical events by virtue of their nonphysical properties. Nothing nonphysical can affect the physical" [11, pp. 28–29].

Which principle is Kim referring to? The simple fact that he is calling it "principle of causal closure" is not revealing, since his description of it is ambiguous. He first says that "no causal chain will ever cross the boundary between the physical and the nonphysical", and this is what Marcus calls Closure. But he also says that "if you reject this principle, you are *ipso facto* rejecting the in-principle completability of physics", and this correctly amounts to what Marcus calls Completeness. So Marcus seems right in claiming that Kim conflates two distinct principles.

Now, Completeness is weaker than Closure; still, it is strong enough to guarantee physicalism. After all, a physicalist should be satisfied by the thesis that we do not *need* to depart from physical explanations to explain physical events. But Completeness does not require the unintelligibility of mental causation. Completeness is consistent with mental causation. So, if what Kim is doing is appealing to a principle such that, if you reject it, you are *ipso facto* rejecting the in-principle completability of physics, he is *not ipso facto* referring to a principle such that, if you are postulating that mental properties do exist and float freely and are causally efficacious, you are rejecting it. Mental causation without mind-body supervenience just involves rejecting Closure, not Completeness. As Marcus says, Completeness is plausible, and innocuous to mental causation. Closure is dangerous; but it clearly awaits adequate justification, since it is not required by physicalism, and still entails that all psychological explanations are false.

Maybe Kim would try to defend (2) by claiming that the truth of Completeness implies the truth of Closure. But, as Marcus shows, the route from Completeness to Closure is not so immediate. You need to make an additional hypothesis to find this route: you need to assume that systematic overdetermination is unacceptable — that is, you need to appeal to the NOP. In fact, if you are accepting Completeness and rejecting Closure, you think that every physical event that has a mental cause is overdetermined. It seems that, if you take Completeness and add the NOP, you obtain Closure. Kim is not manifestly appealing to the NOP in justifying (2): but he should. That is why I am detecting in the justification of (2) Kim's first appeal to the NOP. Not only is this appeal concealed; it is completely unargumented.

2 Second appeal to the NOP

We are now under the hypothesis that mind-body supervenience holds. After asserting (5), Kim writes:

> I hope that you are like me in seeing a real tension between these two answers: Under the assumption of mind-body supervenience, M^* occurs because its supervenience base P^* occurs, and as long as P^* occurs, M^* must occur no matter what other event preceded this instance of M^*. [...] This puts the claim of M to be a cause of M^* in jeopardy: P^* alone seems fully responsible for, and capable of accounting for, the occurrence of M^* [8, p. 42].

It seems to me that the status of M as a cause of M^* is in jeopardy only if we are appealing to the NOP. Without the NOP, we could consider the possibility of M^* to be overdetermined: its presence is guaranteed by P^*, but is also caused by M. In other words, it is true (after the very definition of the supervenience relation) that P^* — as M^*'s supervenience base — absolutely guarantees the presence of M^*. But it does not follow — unless appealing to the NOP — that "given this, the only way anything can have a role in the causation of P^* would have to be via its relationship to M^*'s supervenience base P^*" [8, p. 42]. Kim enunciates a "plausible general principle", which in his opinion "is by itself sufficient to justify (6) even if you do not see any tension in (5), and it is this: *To cause a supervenient property to be instantiated, you must cause its base property (or one of its base properties) to be instantiated*" [8, p. 42]. My claim is that both for seeing a tension in (5) and for accepting this latter principle as "plausible", you need to have accepted the NOP. On the contrary, if you are prepared to accept systematic overdetermination, you will not see any tension in (5) and you will not feel forced to accept Kim's "plausible general principle".

I would like to add that appealing to the NOP for claiming that there is a tension in (5), and for justifying (6), is particularly preposterous. In fact, the alleged tension in (5) would be due to systematic overdetermination, but not to systematic *causal* overdetermination. And systematic *non-causal* overdetermination is not so much a problem.

We could start by recalling that Kim himself introduced supervenience as a determination-free relation: not necessarily the instantiation of the base property *determines* the instantiation of the supervenient property. Here you have Kim's definition of mind-body supervenience:

> Mental properties supervene on physical properties, in that necessarily, for any mental property M, if anything has M at time t, there exists a physical base (or subvenient) property P such that it has P at t, and necessarily anything that has P at a time has M at that time [8, p. 9].

And here you have the passage where Kim admits that supervenience — and mind-body supervenience in particular — is not necessarily a determination relation:

> The relation of dependence, or determination, is asymmetric: if x depends on, or is determined by y, it cannot be that y in turn depends on or is determined by x. [...] But mind-body supervenience isn't asymmetric; in general, the supervenience of A on B does not exclude the supervenience of B on A. The notion of supervenience we introduced simply states a pattern of covariation between two families of properties, and such covariation can occur in the absence of a metaphysical dependence or determination relation [8, p. 11].

Then we should not see any tension between: (i) M^* is instantiated on this occasion because M caused M^* to be instantiated, and (ii) M^* is instantiated on this occasion because P^*, the physical supervenience base of M^*, is instantiated on this occasion. We would be wrong in interpreting the second "because", the one connecting the occurrence of P^* and the occurrence of M^*, as a "because" involving a determination relation. Not only M^* could be *not* causally overdetermined; it could also be *not* overdetermined *tout court*.

We surely have to register that, having just said that supervenience is not necessarily a determination relation, Kim rapidly assumes in his book — indeed without adequately justifying this important step — mind-body supervenience to be a dependence/determination relation:

> We will simply follow the customary usage and understand supervenience to incorporate a dependence/determination component as well. In fact common expressions like "supervenience base" and "base property" all but explicitly suggest asymmetric dependence [8, p. 11].

What we have, then, is that supervenience is just covariation plus dependence/determination: "[supervenience] merely states a pattern of property covariation between the mental and the physical and points to the existence of a dependency relation between the two" [8, p. 14].

Our question is: provided that M^* is caused by M and supervenes on P^*, is there any tension on M^*? Well, M^* is surely *overdetermined*, since both causation and supervenience are forms of determination. But it seems to me that the only kind of systematic overdetermination that we happen to find hard to accept is systematic *causal* overdetermination. And the kind of overdetermination weighing on M^* is not causal overdetermination. In fact, Kim himself denies that supervenience can be thought of as a causal relation:

> The relation from P to M is not happily thought of as a causal relation; in general, the relation between base properties and supervenient properties is not happily construed as causal [8, p. 44].[1]

What we have is that M^* is not causally overdetermined. It is just a supervenient property that is not uncaused. Should we see a problem in M^*'s being (non-causally) overdetermined, we should admit that every supervenient property must have no cause, and that causal powers can only be correctly attributed to base properties which are not in turn supervenient properties of some lower base properties on deeper supervenience relationships. A consequence of this vision would be that causal powers would drain away, unless there is a fundamental ontological level of reality which can only contain causal effects — and postulating such a bottom ontological level of reality might seem to us harder than accepting systematic non-causal overdetermination. In any case, whether a bottom ontological level exists or not, causal powers would drain away down the supervenience relationships, and no supervenient level would show authentic causal effects.

Kim's answer is that causal powers do not drain away down the levels of reality, because supervenience does not track the micro-macro hierarchy: mental, physiological, chemical, molecular, atomical and subatomical properties are all macroproperties, i.e. properties of the same thing, i.e. properties of the same level of reality. But as Ned Block [3] remarks, although Kim is free to use the notion of 'level' he prefers, he cannot deny that another

---

[1] Kim motivates his claim by saying that "for one thing, the instantiations of the related properties are wholly simultaneous, whereas causes are standardly thought to precede their effects; second, it is difficult, perhaps incoherent, to imagine a causal chain, with intermediate links, between the subvenient and the supervenient properties [8, p. 44]. Karen Bennett writes that "no one but Searle thinks that causation is the relation between the mental and physical [1, p. 479].

legitimate notion of 'level' exists, keyed to relations among properties, according to which mental properties are at one level, physiological properties at a lower level, and atomical properties at a lower level. Kim can now object that the causal powers of the levels under the mental do not drain away by virtue of the fact that they — differently from the mental level — are made up by micro-based properties: and micro-based properties are identical to mereological configurations of lower level properties, whose causal powers they inherit. But Block [3] is right in claiming that Kim needs to assume that no micro-based property can be micro-based in alternative ways: while there are reasons to suspect that some causally efficacious micro-based properties admit multiple decompositions, as is the case for rigidity, heat, temperature, water, and many others. In similar cases, Kim's strategy is fragmenting the macro properties: but once again this might seem to us more unpalatable than accepting systematic non-causal overdetermination as unproblematic.

Kim [9] admits that Block's worry is serious, and simply puts its significance into perspective. First, he says, reductionism blocks causal drainage. If the atomical level is reducible to the subatomical level, it inherits its causal powers, and it is not true that causal powers drain away. In the same way, if mental properties are reducible to physical properties, they do have authentic causal powers, i.e. the causal powers of the lower physical level. After all — Kim underlines — the Supervenience Argument has a fundamental premise: mental properties are not identical, nor reducible, to physical properties. Since the whole Argument is a *reductio*, what it aims at proving is that if you want to stay within physicalism, you have just two options: mental epiphenomenalism, or reductionism. But many philosophers would consider reductionism not as a vindication of mental causation, but rather as another way of proposing mental epiphenomenalism. After all, mental properties would have causal powers, but these causal powers would be the causal powers of subvenient physical properties. No mental property would cause anything by virtue of its being a mental property. Reductionism could be seen as blocking the causal drainage; but, also, it could be seen as confirming it, since it would be true that level L_n can only have causal powers if it collapses into the lower level L_{n-1} where causal powers only can be found, and so on down the hierarchy of levels. In any case, all irreducible supervenient levels of properties would be deprived of causal powers, and all occurrences of irreducible supervenient properties would stay uncaused.

Kim [9] is right in specifying that a causal collapse to the level below would occur only if the lower level is causally closed,[2] as his Supervenience Argument requires. This means that "there is no step-by-step devolution of causal relations from level to level" [9, p. 174], since the biological level, the chemical level and the macrophysical level are probably not causally close. But as the fundamental level of microphysics is causally close, causal relations at every irreducible supervenient level would give way to causal relations at the microphysical level. As for reducible supervenient levels, causal powers would only remain at the price of systematically fragmenting the macro properties

---

[2] Kim's point here does not suffer from the effects of his conflation of Closure and Completeness.

in order to confront multiple decomposition. Is not accepting systematic non-causal overdetermination a better option?

We can check how easily we normally accept systematic non-causal overdetermination with the help of an example in which supervenient, irreducible, non-causally overdetermined properties are not mental properties. Let us take taxonomical/phenotypical properties like *being a dog, being a cat, being a giraffe*. According to all of Kim's definition of supervenience,[3] such taxonomical/phenotypical properties are supervenient on DNA chemical properties. I also think it reasonable to assume that taxonomical/phenotypical properties are neither identical nor reducible to DNA chemical properties. If an occurrence of the property of *being a dog* supervenes on an occurrence of the DNA chemical property C, then it must also be accepted that an occurrence of the property of *being a dog living in my apartment* supervenes on an occurrence of the property of *C-in-my-apartment*. Now, let us suppose I found an injured dog along the road on my way home and that I decided to bring it to my apartment. I claim that an occurrence of the property of *being a dog living in my apartment* has been *caused* by an occurrence of the property of *being a dog found injured by me along the road*. But it is also true that that very same occurrence of *being a dog living in my apartment* supervenes on an occurrence of *C-in-my-apartment*. And this latter supervenience relation does not compete with the former causal relation. The occurrence of *being a dog living in my apartment* is non-causally overdetermined: but this is not a problem. There is no "tension" on the property of being a dog living in my apartment coming from its being caused by one property and its being necessarily correlated with, and dependent on, another different property. My suggestion is that we should see no tension on M^* either.

Let me remark that my example depicts a *systematic* non-causal overdetermination, since every occurrence of a property of the kind *being a member of species Y in situation Z* would be non-causally overdetermined: and unless we sustain that occurrences of such properties are systematically uncaused, it seems to me that we must concede that what we have is a systematic unproblematic overdetermination. But the overdetermination of M^* that Kim asks us to ban in order to accept (6) is exactly the same: systematic non-causal — and therefore unproblematic — overdetermination.

Kim acknowledges that considering systematic non-causal overdetermination as problematic is "the fundamental idea that drives the Supervenience Argument" [9, p. 153]. He states this idea in the form of a dictum named after the 18th Century theologian Jonathan Edwards:

> Edwards' dictum: There is a tension between vertical determination and horizontal causation. In fact, vertical determination excludes horizontal causation [9, p. 153].

Kim trusts that the reader cannot but see the tension provoked by non-causal overdetermination, and admits not believing "that the invocation of any gen-

---

[3]Take, for example, Kim's definition of strong supervenience: "A strongly supervenes on B just in case, necessarily, for each x and each property F in A, if x has F, then there is a property G in B such that x has G, and necessarily if any y has G, it has F" [7, p. 65]. According to this definition, taxonomical/phenotypical properties like *being a dog* and *being a cat* strongly supervene on DNA chemical properties.

eral principle will help persuade anyone who is not with me here" [9, p. 156]. I think that it is perfectly possible to see no tension at all weighing on M^*, and that Edwards' dictum has to be rejected.

Kim goes further saying that the alleged tension on M^* can only be relaxed by accepting:

6. M caused M^* by causing P^*. That is how this instance of M caused M^* to be instantiated on this occasion [8, p. 42].

But as we do not detect any tension on M^*, we do not see any need to decree that M can only cause M^* by causing P^* either. Of course, we can accept that, by its causing M^*, M also caused P^*. But Kim is arguing something different. In his opinion, M can only cause M^* *more indirectly* than its causing P^*: that is, M can only cause M^* through its causing P^*. This passage seems unjustified. We are not forced to agree with him. We can very well sustain that M caused both M^* and P^*; or, that M caused M^* without causing P^*. Kim has furnished no decisive argument to reject these possibilities.

3 Third appeal to the NOP

Kim's most crucial appeal to the NOP is the third. He first introduces an obvious consequence of mind-body supervenience:

7. M itself has a physical supervenience base P [8, p. 43].

He then shows that P qualifies as a cause of P^*. If we take causation as grounded in nomological sufficiency, we have that P is sufficient for M (as P is M's supervenience base) and M is sufficient for P^* (as (6), that we have questioned, states), so P is sufficient for P^*. If we take causation as grounded in counterfactual evaluation, we have that "if P had not occurred M would not have occurred (we may assume, without prejudice, that no alternative physical base of M would have been available on this occasion), and given that if M had not occurred P^* would not have occurred, we may reasonably conclude that if P had not occurred, P^* would not have either" [8, p. 43].

Now, Kim deplores, we have "an overabundance of causes: both M and P seem severally eligible as a sufficient cause of P^*. And it is not possible to escape the threat of overdetermination" [8, pp. 43–44]. This is not just overdetermination: this is *causal* overdetermination. Why should we think of causal overdetermination as a threat? Here's how Kim justifies this crucial passage:

> And, finally, it is not possible to take this simply as a case of causal overdetermination — that the instance of P^* is causally overdetermined by two sufficient causes, P and M. Apart from the implausible consequence that it makes every case of mental causation a case of overdetermination, this approach encounters two difficulties: first, in making a physical cause available to substitue for every mental cause, it appears to make mental causes dispensable in any case; second, the approach may come into conflict with the physical causal closure. For consider a world in which the physical cause does not occur and which in other respects is as much like our world as possible. The overdetermination approach says that in such a world, the mental cause causes a physical event — namely that the principle of causal closure of the physical domain no longer holds. I do not think we can accept this consequence [8, pp. 44–45].

We have three distinct arguments in favour of the NOP. One argument is that mental causes are *dispensable*. But this cannot be considered as a good argument against causal overdetermination. Causal overdetermination involving the subsistence of two (or more than two) different sufficient causes, it is by definition describable as a situation in which each single cause is dispensable — in the sense that the other cause would cause the effect as well if it occurred without the first. Perhaps Kim is assuming some form of Ockham's razor here. But he should argue for it. Kim surely does not deny that *some* causal overdetermination does exist — for example, death caused both by an heart attack and by a rifle shot. In such a case, the heart attack is clearly *dispensable* as a cause of death, since in its absence the rifle shot would do the causal work as well. But this does not amount saying that the heart attack is not causally efficacious here, nor that there is no causal overdetermination here. So the *dispensability* argument must be rejected.

Another argument is that causal overdetermination would conflict with physical causal closure. This is true; but Kim once again confuses completeness and closure here. Causal overdetermination is perfectly consistent with physical completeness; and as closure is what you obtain if you take completeness and add the refusal of causal overdetermination, you should not argue in favour of the NOP using an unargumented appeal to physical closure. Actually Kim argues in favour of his appeal to physical closure asking us to imagine "a world in which the physical cause does not occur and which in other respects is as much like our world as possible". If we really take this no-P world as a world which "in other respects is as much like our world as possible", we should demand that mind-body supervenience continue to hold; and, if mind-body supervenience holds, the disappearance of P must have one of two possible consequences: (i) M does not occur because of a missing physical supervenience base; (ii) M does occur because of the substitution of P with another supervenience base P^{**}. If (i) is the case, M does not cause all alone a physical event in the imagined world, and completeness is safe. If (ii) is the case, we cannot say whether P^* occurs or not. But there are just two possibilities, and both are consistent with completeness: (iia) P^* does not occur; (iib) P^* occurs, and it is "regularly" causally overdetermined by M and P^{**}. So there is no violation of completeness at all.[4]

Kim's third argument is an appeal to common sense: systematic causal overdetermination is *implausible*. It would not be, obviously, if it were also systematically designed. But as we can assume that the alleged systematic causal overdetermination of M^* would not be a designed one, we cannot escape this way from the implausibility accusation. We can nevertheless wonder whether it is true that *any* non-designed systematic causal overdetermination is implausible to our eyes.

The particular kind of non-designed systematic causal overdetermination exemplified by deaths caused both by a heart attack and by a rifle shot surely appears implausible to us. But why is this so? I think the key is our idea of a *coincidental* event. E.J. Lowe [10] actually defines the notion of "occurs by coincidence" in terms of the notion of multiple independent causes:

---

[4] A similar point has been made by [4].

I take an event to be one which "occurs by coincidence" if its immediate causes are the ultimate effects of independent causal chains [10, p. 579].

I think we can go further, and claim that an event is one which "occurs by coincidence" if *and only if* its immediate causes are the ultimate effects of independent causal chains. My claim is that systematic occurring by coincidence is what we perceive as implausible whenever we happen to perceive systematic causal overdetermination as implausible, and that occurring by coincidence means having as immediate causes the ultimate effects of independent causal chains. You have systematic coincidence (and consequent implausibility) *without* causal overdetermination if, and only if, you have an event whose immediate, singularly non-sufficient causes are systematically the ultimate effects of independent causal chains; and you have systematic coincidence (and consequent implausibility) *plus* causal overdetermination — that is, *implausible* systematic causal overdetermination — if, and only if, you have an event whose immediate, singularly sufficient causes are systematically the ultimate effects of independent causal chains. But, in the latter case, implausibility is provoked by systematic coincidence, not by causal overdetermination — just as, in the former case, implausibility is provoked by systematic coincidence, not by causation.

I assume that the reason why we usually judge systematic causal overdetermination to be implausible is this: we (correctly) take systematic occurrence by coincidence to be implausible, and we (incorrectly) conclude that *any kind* of systematic causal overdetermination is implausible. The correct conclusion is simply that the type of systematic causal overdetermination is implausible where the overdetermining causes are *independent*. If the overdetermining causes are not independent, in fact, we cease to see any coincidence in systematic causal overdetermination. Consider, for example, the taking of an aspirin systematically provoking the end of a headache. Let us suppose that the taking of an aspirin *also* systematically causes (provided that the subject believes she is taking an aspirin) a placebo effect. What we have is systematic causal overdetermination of the provoked end of a headache, which is normally caused both by the taking of an aspirin and by the belief that the aspirin has been taken. This is systematic causal overdetermination. But this is *not* implausible — the reason being that the overdetermining causes are not independent. And that is the case with M and P causing P^*. M and P are dependent causes, as M supervenes on P, and supervenience is (among other things) a *dependence* relation. No systematic coincidence, no implausibility: Kim's third argument must be rejected.

We can then conclude that Kim's Supervenience Argument makes some crucial appeals to the NOP, and that Kim does not support this presumption with any adequate arguments. His Supervenience Argument as a whole reveals itself therefore to be as weak as the NOP it heavily depends upon.

It is worth specifying that we are not identifying causal overdetermination with an effect's having more than one sufficient cause. As Goldman [6] and Yablo [12] have remarked, it may be that every cause in a causal chain is a sufficient cause of its effect: although this means that any effect of the causal chain has (by transitivity) many different sufficient causes, this does not mean that any such effect is causally overdetermined. An effect's having more than one sufficient cause is a necessary condition of its being causally overdetermined, but it is not a sufficient condition. Nor can we say that a necessary and sufficient condition for an event to be causally overdetermined is for it to have at least two different sufficient simultaneous causes, or for it to have at least two different sufficient causes such that neither is causally sufficient for the other: these stipulations exclude the effects of causal chains, but they also exclude some causally overdetermined effects, like for example the effect e at t_3 having one sufficient cause c_1 at t_1 and another sufficient cause c_2 at t_2, provided that c_1 is also a sufficient cause of c_2 [1, pp. 478–479].

The fact that an effect of a causal chain is not causally overdetermined in spite of its having many sufficient causes has led some philosophers to conjecture that the subsistence of some close relations between the causes can *defuse* causal overdetermination, producing events that do have more than one sufficient cause but are still not causally overdetermined. Karen Bennett [1] has tried to rescue mental causation from the accusation of involving systematic causal overdetermination by saying that, although the effects of mental causes systematically have more than one sufficient cause, they are not systematically causally overdetermined. I do not think her results are convincing, and I prefer to say that — whenever the sufficient causes of one systematically causally overdetermined effect are in relations of dependence — it is no longer implausible to have systematic causal overdetermination, and it is no more required by reason to appeal to the NOP. Although I think we cannot accept Goldman's suggestion that mental states and neurophysiological states are simultaneous nomic equivalents,[5] I appreciate his line of argument — if mental states and neurophysiological states are simultaneous nomic equivalents, then the fact that a neurophysiological state is a sufficient cause for a bit of behavior does not require that some mental state is not a necessary or even sufficient cause of that very same bit of behavior.

I agree with Bennett [1] that what is important is to break the analogy between the mental/physical case and standard textbook examples of causal overdetermination involving "houses that are struck by lightning at the same moment that someone tosses a lit cigarette into the draperies", but I disagree that it "is just a terminological issue" to say that the effects of mental causes are not overdetermined, or, to say that they "*are* always overdetermined, just not in the bad way — the overdetermination is perfectly acceptable, unsurprising, and unproblematic" [1, p. 474]. This would amount to saying that, every time we have a case of unproblematic causal overdetermination, we may also truly say that it *is not* a case of causal overdetermination. But take the example of an effect of the kind e systematically having one sufficient cause c_1 and another sufficient cause c_2, such that c_1 is also a systematic sufficient cause of c_2: this may be *unproblematic* systematic causal overdetermination, but it is still not possible for us to say that this *is not* systematic causal overdetermination.

---

[5] See [6, p. 473]: "Suppose there is a (contingent) law saying that for any object of kind H and any time t, the object has property P at t if and only if it has property Q at t. Then if a particular object a has properties P and Q at a particular time t_1, I shall say that a's having P at t_1 is a 'simultaneous nomic equivalent' of a's having Q at t_1".

Bennett never specifies what's wrong in bad causal overdetermination. She rejects the thesis that what makes the systematic causal overdetermination of the effects of mental causes acceptable is its not being a coincidence; but if, as I sustain, being coincidental is what is unacceptable in "bad causal overdetermination", then showing why the systematic causal overdetermination of the effects of mental causes is not coincidental may point to a relevant difference.

Bennett proposes a test for (bad) causal overdetermination: "e is overdetermined by c_1 and c_2 only if: (O1) if c_1 had happened without c_2, e would still have happened, and (O2) if c_2 had happened without c_1, e would still have happened" [1, p. 476]. She aims at proving that, in the mental/physical case, either O1 or O2 is false or vacuously true, i.e. its antecedent cannot be true. But her arguments to support the claim that the vacuous truth of just one counterfactual is sufficient to rule out causal overdetermination are weak. The test seems constitutively inappropriate to deal with *dependent* overdetermining causes, and in particular with overdetermining causes such that one supervenes on the other, since it requires evaluation of possible worlds where each cause occurs *without the other*. Bennett encounters embarrassing difficulties when evaluating such possible worlds, in particular when arbitrarily establishing which possible worlds are closer to us than others; her thesis that a counterfactual relation between c_1 and c_2 is not relevant to the truth of O1 and O2 is unconvincing; her bizarre concern that the truth-value of O1 and O2 "should not be held hostage to facts about the essence of p" [1, p. 495], nor to facts about the essence of m, opens the door to some vague and *ad hoc* concepts such as "anything relevantly p-like" and "anything relevantly m-like". This is the first reason why I do not accept Bennett's conclusion, that the effects of mental causes are not causally overdetermined since they do not pass her test.

The second reason is that her claim, that the effects of mental causes do not pass her test, is very problematic. All she has is that, (i) if we consider p = a physical property of the brain, O2 ("if p had happened without m, e would still have happened") is false because p without m would not cause e; and, (ii) if we consider P = the complication of p "conjoining in the laws of nature, and other facts about the physical world" [1, p. 486], O2 is vacuously true because P necessitates m. I think that (i) is insufficiently argued for; and as it literally denies that p is causally sufficient to e, it begs the entire question. As p's alleged insufficiency to cause e is due to its requiring further background physical conditions, then it is P that we have to look at as the physical property which is causally sufficient to e — and speaking of p we have wasted our time. If all we have is (ii), we do not have so much: as I said, I am not convinced that the vacuous truth of just one counterfactual is sufficient to rule out causal overdetermination. Moreover, I am not sure that (ii) is true. Does P really necessitate m? Are not the putative good reasons we could have to subscribe (ii) also good reasons to think that $P = m$? The truth of (ii) is doubtful, but if it were not it would represent an argument against the distinctness between the mental and the physical, and therefore against the thesis that mental properties have causal powers by virtue of their being mental properties.

4 The dormitivity argument

You can see the Supervenience Argument as a particular version of the Causal Exclusion Argument, which displays the general idea that causal powers of physical properties preempt the causal powers of mental properties. The Supervenience Argument is the version of the Causal Exclusion Argument that exploits the supervenience relation. What I will call the 'Dormitivity Argument' is the version of the Causal Exclusion Argument that assumes that mental properties are functional properties: that is, second-order properties defined in terms of causal/nomic relations between first-order physical properties. On this account, which Kim baptizes "physicalist functionalism" or "physical realizationism", a mental property M is nothing above the property of having some physical property P (which we call "the realizer of M") "specified by causal roles, that is, in terms of causal relations holding for first-order physical properties" [8, p. 21]. Mental properties would be similar to *dormitivity*: "a substance has this property just in case it has a chemical property that causes people to sleep. Both Valium and Seconal have dormitivity but in virtue of different first-order (chemical) realizers — diazepam and secobarbital, respectively" [8, pp. 20–21].

If mental properties are functional second-order properties like as dormitivity, then their causal powers are in danger, and mental epiphenomenalism is an obvious conclusion. As Ned Block puts it:

> If a dormitive pill is slipped into your food without your noticing, the property of the pill that is causally relevant to your falling asleep is a (presumably first-order) chemical property, not, it would seem, the dormitivity itself. Different dormitive potions will act via different chemical properties [2, p. 45].

There is no further causal work left for dormitivity once diazepam or secobarbital have caused your falling asleep. There is no further causal work left for the provocativeness of the cape once its color has caused the bull's anger. And there is no further causal work left for mental properties once their physical realizers have been causally efficacious. Also if we concede that systematic causal overdetermination is not implausible, there is no room for causal overdetermination here. Kim says that "there is a real problem, the exclusion problem, in recognizing second-order properties as causally efficacious in addition to their realizers" [8, p. 53], and enunciates the Causal Inheritance Principle:

> If a second-order property F is realized on a given occasion by a first-order property H (that is, if F is instantiated on a given occasion in virtue of the fact that one of its realizers, H, is instantiated on that occasion), then the causal powers of this particular instance of F are identical with (or are a subset of) the causal powers of H (or of this instance of H) [8, p. 54].

If mental properties are second-order properties, it follows that their possible causal powers are just the causal powers of their realizers. We simply have no space to argue that an M-instance has a causal efficacy in addition to that of the P-instance that realizes it. M turns out to be completely epiphenomic.

In order to react to this threat of mental epiphenomenalism, we could try to argue against the thesis that mental properties are just functional second-order properties which are subject to the Causal Inheritance Principle and, in particular, against the analogy between mental properties and typical functional second-order properties subject to the Causal Inheritance Principle like provocativeness and dormitivity.

We can start by remarking that, while provocativeness and dormitivity are entirely definable in terms of the causal roles of their realizer, many mental properties are not. In fact, many mental properties have a phenomenal component, and Kim himself states that "the functionalization of qualia won't work". So, it is not so correct, after all, to develop an analogy between mental properties on one side and provocativeness and dormitivity on the other.

A similar point is that whether an entity has dormitivity or not is a *logical consequence* of whether its first-order properties cause people's falling asleep or not, while on the contrary whether an entity has a mental property or not *is not a logical* consequence of whether its first-order properties have certain causal roles or not. It is maybe a consequence, but not a *logical* consequence.

Let us now consider this excerpt from Daniel Dennett:

> The power of the intentional strategy can be seen even more sharply with the aid of an objection first raised by Robert Nozick some years ago. Suppose, he suggested, some beings of vastly superior intelligence — from Mars, let us say — were to descend upon us, and suppose that we were to them as simple thermostats are to clever engineers. Suppose, that is, that they did not need the intentional stance — or even the design stance — to predict our behaviour in all its detail. They can be supposed to be Laplacean super-physicists, capable of comprehending the activity on Wall Street, for instance, at the microphysical level. [...] Our imagined Martians might be able to predict the future of the human race by Laplacean methods, but if they did not also see us as intentional systems, they would be missing something perfectly objective: the patterns in human behavior that are describable from the intentional stance, and only from that stance, and which support generalizations and predictions. Take a particular instance in which the Martians observe a stockbroker deciding to place an order for 500 shares of *General Motors*. They predict the exact motions of his fingers as he dials the phone, and the exact vibrations of his vocal cords as he intones his order. But if the Martians do not see that indefinitely many different patterns of finger motions and vocal cord vibrations — even the motions of indefinitely many different individuals — could have been substituted for the actual particulars without perturbing the subsequent operation of the market, then they have failed to see a real pattern in the world they are observing. Just as there are indefinitely many ways of being a spark plug — and one has not understood what an internal combustion engine is unless one realizes that a variety of different devices can be screwed into these sockets without affecting the performance of the engine — so there are indefinitely many ways of ordering 500 shares of *General Motors*, and there are societal sockets in which one of these ways will produce just about the same effect as any other. There are also societal pivot points, as it were, where which way people go depends on whether they believe that p, or desire A, and does not depend on any of the other infinitely many ways they may be alike or different [5, pp. 81–82].

Here we have an important difference between dormitivity and mental properties: for considering causal chains, we can find out that there are some "pivot points" where, it seems to us, a particular physical property occurs *(also) by virtue of* its being a subvenient physical property of a particular mental property. Of course, the occurrence of that physical property is caused by the previous occurrence of another physical property. But it is often probable that that particular node in the physical causal net be occupied by the occurrence of a physical property that has the property of subvening to the same particular mental property also in the case we physically alter the whole frame. Let us suppose that we can make laplacean predictions about the particular physical property that will occur in that cell at t_2. Let us now suppose that we physically perturb the physical causal chain in a time t_1 preceding the time t_2 at which we are expecting the predicted physical property to occur. It is often probable that whatever different physical property occurs in the place of the predicted one, it will have a property in common with it: they will both be subvenient physical properties of the same mental property.

Nothing similar happens to dormitivity: there is no physical causal chain such that we can perturb it at t_1 and expect that the *new* physical property that occurs at t_2 will have in common with the physical property it is "substituting" the property of subvening to dormitivity. If we randomly change things at a physical level, dormitivity will consequently randomly remain or disappear. But if we randomly change things at a physical level, mental properties will often *not randomly* remain, "absorbing physical perturbations". This difference is an argument against the analogy between mental properties and dormitivity, and against the thesis that a mental property is, like dormitivity, a typical functional second-order property subject to the Causal Inheritance Principle.

Actually Block's thesis is not that second-order properties like dormitivity are never causally relevant. First of all, he just questions their causal relevance to the effects in terms of which they are defined. Maybe — he concedes — dormitivity is causally relevant to some other effect, for example to getting cancer. Kim strongly denies such a possibility: in Kim's view, whatever dormitivity can seem causally relevant to has to be caused by dormitivity's first-order realizers, and there is no causal work left for dormitivity.

However, Block also weakens his thesis in another direction. He does not claim that "second-order properties are *never* causally relevant to the effects in terms of which they are defined", and rather says that those "second-order properties are not *always* causally relevant to the effects in terms of which they are defined" [2, p. 46]. Why? Because he admits that *sometimes* dormitivity can be causally relevant to sleep, independently from its first-order realizers being causally relevant too (and we get causal overdetermination here) or not. This is when we know that a pill is dormitive; and it is usually called *placebo effect*:

> If a dormitive pill is so labeled, thereby causing knowledge of its dormitivity, this knowledge can cause sleep (though the truth and justification of the knowledge are of course causally irrelevant). So dormitivity can be causally relevant to sleep [2, p. 45].

Dormitivity can be causally relevant to sleep in the case the pill contains one of dormitivity's first-order realizers causing sleep *too*. And dormitivity can be causally relevant to sleep also in the case the pill does not contain any of dormitivity's first-order realizers. This is a pill whose dormitivity just "requires its own recognition" [2, p. 45]. Block declares that:

> The only cases that I can think of in which second-order properties seem to be causally efficacious are those where an intelligent being recognizes them [2, p. 46].

If mental properties are to be considered as second-order properties, then mental properties their owners are aware of are second-order properties "which an intelligent being recognizes". Why could they not be as causally efficacious as recognized dormitivity? In this perspective, mental causality could be seen as a special kind of placebo effect involving the mind making itself causally efficacious.

BIBLIOGRAPHY

[1] K. Bennett. Why the exclusion problem seems intractable, and how, just maybe, to tract it. *Noûs*, 37 (3): 471–497, 2003.
[2] N. Block. Can the mind change the world? In G. Boolos (ed.), *Meaning and Method: Essays in Honor of Hilary Putnam*, Cambridge University Press, Cambridge MA, 1990; now also in C. Macdonald and G. Macdonald (eds.), *Philosophy of Psychology*, Blackwell, Oxford 1995.
[3] N. Block. Do causal powers drain away? *Philosophy and Phenomenological Research*, 67 (1): 133–150, 2003.
[4] Th. M. Crisp and T. A. Warfield. Kim's master argument. *Noûs*, 35: 304–316, 2002.
[5] D. Dennett. True believers. In A.F. Heath (ed.), *Scientific Explanation*, Oxford University Press, Oxford 1981; now in W. G. Lycan (ed.), *Mind and Cognition*, Blackwell, Oxford, 2nd ed. 1999.
[6] A.I. Goldman. The compatibility of mechanism and purpose. *The Philosophical Review*, 78 (4): 468–482, 1969.
[7] J. Kim. Concepts of supervenience. *Philosophy and Phenomenological Research*, 45: 153–176, 1984; now also in *Supervenience and Mind*. Cambridge University Press, Cambridge, MA 1993.
[8] J. Kim. *Mind in a Physical World*, MIT Press, Cambridge, MA 1998.
[9] J. Kim. Blocking causal drainage and other maintenance chores with mental causation. *Philosophy and Phenomenological Research*, 67 (1): 151–176, 2003.
[10] E.J. Lowe. Causal closure principles and emergentism. *Philosophy*, 75 (4): 571–585, 2000.
[11] E. Marcus. Mental causation in a physical world. *Phylosophical Studies*, 122: 27–50, 2005.
[12] S. Yablo. Mental causation. *The Philosophical Review*, 101: 245–280, 1992.

Philosophy of mind between reduction, elimination and enrichment
WOLFGANG HUEMER

1 Introduction

Philosophers of mind and neuroscientists arguably describe the same range of phenomena: they formulate theories of how human beings gather information about relevant aspects of their environment, how they form beliefs and desires, and what kind of processes are going on within them that allow them to act on the world. They do so, however, in very different ways: while neuroscientists focus on the causal processes that take place in our nervous system at a sub-personal level, philosophers focus on mental episodes; they describe human beings as *persons* who have propositional attitudes that stand in rational relations (of justifying and being justified) to one another.

These differences in approaches can be explained by the diverging interests of philosophers and neuroscientists, respectively, which can be best illustrated with Wilfrid Sellars' distinction between the *manifest* and the *scientific* image of man. The manifest image, according to Sellars, is a "sophistication and refinement of the image in terms of which man first came to be aware of himself as man-in-the-world" [9, p. 18]. Human beings came to be aware of themselves as having mental episodes, which they experience from a first-person point of view. In the attempt to explain this phenomenon, they soon started to develop a theory that was step by step refined over the centuries — a theory that is (nowadays) often referred to as "folk psychology". This theory describes (a considerable part of) mental episodes as propositional attitudes that justify or are justified by other mental episodes. In short, the manifest image leads to a theory that describes human beings in a conceptual framework of *persons* who understand themselves as occupying a certain position in this world and who have perceptions of, hold beliefs about, and act on objects in their environment, and who can imagine scenarios they have so far not (yet) encountered.

The scientific image, on the other hand, is characterized by the development of scientific theories that postulate "imperceptible objects and events for the purpose of explaining correlations among perceptibles" [9, p. 19]. When we formulate a theory of the mind within this scientific image, we describe causal processes that take place in our nervous system; processes, that is, that are not perceptible (at least not directly, i.e. without the means of sophisticated medical imaging techniques) and that provide a basis for explaining the observable behaviour of human beings at a sub-personal level.

A good part of work in the philosophy of mind, especially in the twentieth century, focused on the question of how these two levels of description are related. Scientific theories are often considered to be more exact than the ones formulated within the manifest image for they are based on empirical data and experiments and are formulated in the language of mathematics. This raises the question of whether philosophers should aim to incorporate the results of the scientific image into their theories. Should the manifest image be reduced to or replaced by the scientific image? Can the two images coexist and complement each other, thus providing a more comprehensive understanding of the mind? Or do we rather have to accept the fact that the two frameworks provide two descriptions of different aspects of the phenomena in question; two descriptions that are so different that we cannot merge them into one unified account?

In this paper I will argue that there are specifically philosophical desiderata that a theory of mind should satisfy; desiderata that scientific theories cannot account for — at least not in their actual shape. The question, I will suggest, is not how to reduce philosophical theories of the mind to the neurosciences, nor how to eliminate them from our scientific vocabulary, but rather whether and how scientific theories can be enriched to account for those aspects that are relevant in the philosophical debate. In the first section of this paper I will spell out the philosophical desiderata for a theory of the mind. I will then discuss two philosophical approaches that emphasize the importance of the results of scientific theories for a philosophical understanding of the mind, reductionism and eliminativism, with the aim to show that a scientifically oriented approach to the philosophy of mind should bet on eliminativism rather than on reductionism. In the concluding section I will argue, however, that radical eliminativist strategies must fail for they cannot account for the very fact that we formulate theories that satisfy the standards of rationality. I will then discuss the question of whether, at the current state of research, we should focus our efforts on the question of how the scientific and the manifest image are related and whether or how they can — at some point in the future — be merged into a synoptic vision that provides a more comprehensive theory of the mind.

2 Philosophical desiderata for a theory of the mind

Philosophers conceive of human beings as *rational agents* who interact with their (social and physical) environment: they collect information which they receive, at least in good part, through their sense organs, and perform actions on objects around them. By forming beliefs about their environment as well as about themselves and generalizing from empirical observations they come to represent the world in a (more or less) coherent way. Moreover, they are able to communicate their picture of the world to other persons and so confront it with their ways of seeing the world, a process which can result in their revising, adapting, enriching, or sophisticating their own views.

In order to explain these facts, philosophers typically describe persons as having a large number of propositional attitudes that stand in rational relations to one another. Rationality, however, contains an intrinsically normative element, which manifests itself in the possibility of error and corrections: we can *mis*perceive the world around us, make *illegitimate* inferences, and hold *wrong* beliefs. Moreover, propositional attitudes have a content that can be expressed linguistically and can so be communicated to other persons. In case of disagreement, other members of our linguistic community can criticize our beliefs. They can point out our errors or inconsistencies in our system of beliefs and justify this critique with arguments, which we are free to accept or not to accept — after all, we are free to insist in our false beliefs.

Scientific theories — at least the ones we know today — on the other hand, are characterized by the fact that they abstract from all normative concerns: science is (supposed to be) value-free. Scientists are interested in the causal workings of the machinery, as it were; they describe causal relations that take place in our nervous system, but they do not have the means to state that these causal relations take place *wrongly* or that they ought to be *corrected*. Thus, the attempt to explain the normative aspect of the mental in terms of the causal relations studied by the sciences constitutes, as Wilfrid Sellars has pointed out, "a radical mistake — a mistake of a piece with the so-called 'naturalistic fallacy' in ethics" [10, p. 19, §5].

This does, of course, not show that the results of neurophysiological research is irrelevant to the philosophical understanding of the mind; after all, our cognitive processes take place in the brain and a better understanding of our cerebral processes can illuminate our understanding of the enabling conditions for our having mental phenomena. Moreover, the fact that we can describe mental episodes in two very different ways does not entail a form of ontological dualism. I do want to emphasize, however, that philosophers and neuroscientists study very different aspects of the mind. Knowing that a certain stimulation of the retina can cause a certain cerebral process does not explain why a certain perceptual experience justifies a certain belief. This suggests that a scientific approach cannot be sufficient for formulating a comprehensive theory of the mind; it will (in the best case) need to be complemented by a theory that can account for the normative and the social aspects of the mental. Philosophers who argue that neuroscientific theories are complete will have to argue that a more elaborate and sophisticated neurological theory of the mind, one that will be formulated in the future, will be able to account for these features, or else provide an argument that shows that they can be neglected and thus eliminated from our scientific study of the mind.

3 Reductionism versus eliminativism

The philosophy of mind of the twentieth century was strongly characterized by the debate of whether and how philosophical theories of the mind can be substituted by or at least reconciled with scientific theories. The urge to do so was arguably the result of the widespread scientism at the beginning of the century as well as of a post-Vienna Circle conception of the unity of the sciences that was shared by most analytic philosophers until very recently[1],

---

[1]A conception, that is, that was often attributed to the Vienna Circle, but was in fact elaborated by Paul Oppenheim and Hilary Putnam in [7] as well as Ernest Nagel in [5] and many others. The canonical conception of the unity of the sciences of the Vienna Circle

according to which all special sciences, including biology, psychology, sociology, and economics, etc., could be reduced to more basic sciences or else should be eliminated from our scientific discourse.

In the early twentieth century, most (analytic) philosophers of mind suggested that psychological descriptions of mental episodes — including propositional attitudes, the existence of which was not questioned — can in principle be reduced to scientific descriptions of events that take place in our nervous systems. The first theories, the (nowadays) so-called type-type identity theories, argued that specific types of mental episodes (e.g., pain, or beliefs like the belief that the earth moves around the sun) were identical with specific types of neurological processes (the firing of C-fibers or a specific activation pattern in the brain, respectively). This gave raise to the hope that a philosophical or psychological theory of the mind could in principle be reduced to neurology. The proponents of this view were ready to admit that we would have to wait for further progress of the neurosciences to perform this inter-theoretical reduction, but suggested that, once the reduction is completed, we will be able to account for all philosophical desiderata for a theory of the mind in terms of neurology.

It is worth noting, however, that according to this conception the fact that a theory can be reduced to another, more basic theory does not show that the reduced theory is obsolete or false. It rather shows that there are systematic relations between the reduced theory and the theory to which it is reduced — and that the latter has (at least) the same explanatory power as the former. This conception of intertheoretical reduction does allow for the possibility, however, that some aspects of the reduced theory might be slightly transformed — and thus rendered more precise — in this process; just as the notion of temperature was slightly altered when it was reduced to that of kinetic energy of molecules. In consequence, reductionism in the philosophy of mind does not entail that it is wrong to attribute propositional attitudes to human beings, nor that folk psychology is a theory that should be substituted by a more basic theory. Reductionists, thus, can admit that our having propositional attitudes that stand in rational relations to one another is a crucial fact that needs to be explained by a theory of the mind — but they do not spell out how a neurological theory could address these aspects; they merely express their hope that future progress of neurology will make this reduction possible. Various proponents of this view do suggest, however, that there are systematic relations between folk psychology and neurology and that a better understanding of these relations will also lead us to reformulate — and thus develop a more precise version of — folk psychology.

These early identity theories were soon criticized by functionalists and anomalous monists (among others), who pointed out that they were too narrow, ascribing, as they do, mental episodes only to organisms that have a

---

did not insist in the reduction of theories, but put its emphasis on the unity of language that allows for an encyclopedia, a patchwork that combines results of various scientific disciplines; Otto Neurath, for example, explicitly discarded the importance of a reduction of theories in [6, p. 362]. In the last two decades of the last century, an increasing number of analytic philosophers began to take a critical stance towards the idea that the unity of the sciences is to be achieved by means of intertheoretic reduction.

nervous system (more or less) equivalent to ours. If type-type identity theory was right, we could not hold that, for example, intelligent extraterrestrial life forms or non-human animals like octopuses, whose nervous system is substantially different from ours, could experience pain or hold the belief that the earth moves around the sun.[2] It was argued that mental episodes could be realized in various different ways and that they are nothing but functional states of the brain (or some other system). While these theories avoid the pitfalls of ontological dualism, they also raise a serious challenge to the hope that a neurological theory will be able to address the philosophical desiderata for a theory of the mind.

A more radical attempt to establish a "neurophilosophy", i.e., to replace the philosophy of mind with a neuroscientific theory of the nervous system, was proposed by Paul Churchland, who criticizes reductionist theories for making too many concessions to what I have called, using Wilfrid Sellars' terminology, the *manifest image*. It is, Churchland argues, a mistake to hold on to propositional attitudes and try to reconcile them with a scientific description of the nervous system — as reductionists do; they should rather be eliminated from our scientific vocabulary and replaced by more adequate and precise descriptions of our actual cerebral processes at a sub-personal level.

Churchland argues that in our brain we do not find a linear series of activations of single (groups of) neurons — which could be identified with propositional attitudes or parts thereof — but rather networks of neurons that can be represented by connectionist networks that are composed of units at different levels, where all units at one level are connected with all units at the subsequent level. According to connectionism, these units have a (numerical) activation value; their connections have a certain weight, which, again, is represented numerically. The whole network, thus, consists of a pattern of numerical values that can be represented by a matrix. We can feed the system with a (numerical) input, which, when being propagated through the system, is transformed by the units' activation values and the weights of their connections and thus produces a specific (numerical) output. By adjusting the activation values of the units and the weights of their connections, a system can be trained to react to a certain kind of input by producing a certain kind of output. In this way, the system can, as connectionists argue, *learn* to perform certain tasks.

The surprising success of early connectionist systems has soon raised the plausibility of this approach. Churchland describes a connectionist network with 13 units at the input level, seven units at the hidden level, and two units at the output level which, after a learning process, could be trained to decide whether a (numerically encoded) echo sonar reading was reflected by a rock or by a mine.[3] The learning process involves a great number of signals of which it was known by the scientists who performed the training whether they were reflected by rocks or mines. These signals were run through

---

[2] In a classic passage, Hilary Putnam states: "Thus if we can find even one psychological predicate which can clearly be applied to both a mammal and an octopus (say 'hungry'), but whose psycho-chemical 'correlate' is different in the two cases, the brain-state theory [*i.e.*, type-type identity theory] has collapsed" [8, p. 288].

[3] Cf. [2, pp. 262ff].

the system several thousand times. By continuously adjusting the values of the units and the weights of their connections (by means of a mathematical formula, the so-called *delta-rule*), the system could be trained to produce the desired output: after the training process, the system, when confronted with a hitherto unknown echo sonar reading, is (almost always) able to correctly decide whether it was reflected by a rock or a metallic object (i.e., a mine). This result is relevant because, as Churchland explains, experienced soldiers are able to *hear* whether the sonar reading was reflected from a rock or a mine — and that, even though there are no specific characteristica that would allow us to distinguish the two: even the soldiers were not able to explain how they come to their conclusion; they "just hear" the difference. This suggests that the system works in a way similar to (a relevant part of) the human brain. Churchland even goes so far as to say that the system "knows" whether there is a mine or a rock, using quotation marks to highlight that this is not a form of propositional knowledge.

It is central to Churchland's argument, however, that the system does not contain symbolically represented information on rocks or mines (or rather: rocks and metallic objects; the system cannot distinguish between sonar readings reflected by mines and those reflected by other metallic objects), nor does it possess the concepts ROCK or METALLIC OBJECT. Moreover, it is impossible to say in which part of the system this "knowledge" is located. All we get is a matrix of (numerical) activation values and weights of connections. Thus, Churchland concludes, the "paradigm of symbolic representation" can be given up: known propositions are not located in a determinate part of the system, but rather "embodied" in the system as a whole. In consequence, Churchland argues, we can overcome our inclination to hold on to propositional attitudes and rational relations that hold between them and consequently can eliminate folk psychology in favor of the neurosciences.[4]

The connectionist system described by Paul Churchland was developed some twenty years ago. I do not, of course, want to suggest that no relevant progress has been made since then. In particular, it was pointed out that in order to understand the workings of the brain it does not suffice to focus on one subsystem in isolation, we rather need "to study many levels of organization, from molecules, to synapses, neurons, micro networks, macro networks and systems" [3, p. 187], which suggests that a promising neurophysiological account will have to be far more complex than the connectionist system described. I did present this example in some detail for it illustrates very well that a neurophilosophical perspective invites a shift in paradigms that results in an elimination of the idea that propositional attitudes and the rational relations that hold between them are core elements in a description of the mind

---

[4] It might be noteworthy that Churchland applies this picture also to scientific theories, suggesting that we should conceive of theories as connectionist networks that do not consist of a set of propositions that stand in rational relations (of justifying and being justified) to one another — this would mean to fall back into the "paradigm of symbolic representation" — but rather as system that produces a certain output when being fed with a certain input. This makes it impossible, however, to criticize and revise single propositions of a given theory; one could only adopt or give up theories as a whole; a move Churchland explicitly endorses. Cf. [2].

— a tendency that is also characteristic for more recent developments in the field of "neurophilosophy".

Like reductionists, eliminativists are happy to acknowledge that at the current state of the art neurological theories are not yet able to substitute philosophical theories of the mind. They are convinced, however, that once we are able to formulate a mature neurological theory we will understand that folk psychology is nothing but an immature and inadequate attempt to describe the human mind that is based on a series of highly questionable presuppositions. Unlike reductionists, however, eliminativists take note of the most recent developments of neurophysiology. They do not hold on to a philosophical picture of the mind nor do they limit themselves to express their hopes that at some point in the future we will be able to reformulate this picture in terms of neurology, but take the idea serious that a neurological theory of the mind — a theory, that is, that takes into account the actual architecture of the nervous system — might propose a different picture; a picture that might substantially revise our conception of the mind. If one admits that the results of neurology are pertinent to our philosophical understanding of the mind, this seems to suggest that eliminativists are in a better position, since they do take into account the most recent results of neurology.

4 On the prospects of a unified account

But why, we might ask, should we expect that a faithful description of the processes that actually take place in our nervous system is able to illuminate our philosophical understanding of the mind? I do not want to deny that they are crucial for our understanding of the workings of our brain, nor do I want to diminish the importance of this project — I do not mean to take a critical stance towards the neurosciences. I do want to raise the question, however, what kind of *philosophical* illuminations we can get from these theories? It seems obvious to me that Churchland's position — but also the strong identity theories — are committing a fallacy that is analogous to the psychologistic fallacy that was criticized by Frege and Husserl more than one hundred years ago. A detailed study of our brains can at best show how our brain actually proceeds certain kinds of stimulus. It can show what causal processes take place in our nervous system, but cannot say anything about the nature of our mental episodes, their propositional contents and the rational relations that hold between them.

To do so it would have to be able to address the normative level of the rational relations that hold between our mental episodes: beliefs, for example, can be true or false, they have a truth-*value*, and justify or are justified by other beliefs — by inferences (in a large sense) which we *can*, but do not *have to* draw. Our mental episodes, to come back to Sellars' terminology, are positions in a game of giving and asking for reasons. As rational beings we strive for truth (however you want to define "truth"); and it is part of the very concept of "belief" that we *should* hold true and revise wrong beliefs — and similarly for the concept of inference. Moreover, a person who holds a belief takes *responsibilities*: the responsibility to justify or revise her beliefs when asked to do so. All these aspects belong to the core of the concept of person that plays a crucial role in the manifest image.

In addition, eliminativism cannot account for the very fact that we formulate scientific theories like neurology. After all, also theories contain a normative element: also theories strive for truth, aiming, as they do, to describe (a relevant part of) the world *correctly*. Moreover, a scientific theory has to fulfill certain standards: in order to qualify as a scientific theory it has to obtain data in the ways required by the methodological standards that hold in this discipline and justify its hypotheses by arguments that are sensitive to the rules of rationality. This shows that even to account for the very fact that human beings are able to formulate scientific theories, we need to account for the intrinsically normative aspect of rationality.

The neurosciences (at least in their current shape), however, cannot address these normative issues. For methodological reasons they have to limit themselves to describe the causal relations that take place in our nervous system with scientific necessity and to describe the nervous apparatus at a sub-personal level. Thus, they can at best explain how a certain stimulation of the retina *causes* a certain neuronal activity, but not *why* a perceptual experience can *justify* an empirical belief. Similarly, they might be able to explain how we come to hold a certain theory of the mind, but cannot explain why this theory is preferable to another less adequate, less coherent, or less simple theory.

Churchland, I should note, explicitly reacts to the argument concerning the intrinsically normative aspect of the mental. The arguments he proposes, however, aim into two opposite directions. On the one hand he seems to downplay the strength of the argument by denying that the normative aspect is essential to the realm of the mental. He argues that

> the fact that the regularities ascribed by the intentional core of FP [*Folk Psychology*] are predicated on certain logical relations among propositions is not by itself grounds for claiming anything essentially normative about FP. To draw a relevant parallel, the fact that the regularities ascribed by the classical gas law are predicated on arithmetical relations between numbers does not imply anything essentially normative about the classical gas laws. [1, p. 82]

This argument is based on a confusion of two different levels, though, namely that of the description of relations and that of the relations described. There is no doubt that there are rational relations between various descriptions of scientific facts; scientific theories, as we have seen above, contain a normative element. The relations described by these scientific theories, however, are causal, and not rational. The relations described by folk psychology, on the other hand, are rational. In the case of the latter, the normative aspect belongs not only to the theory, but also to the entities described by the theory — not only our theories about beliefs, but also the beliefs described by the theory have a truth-value. Churchland misses this point when he suggests that a "normative dimension enters only because we happen to *value* most of the patterns ascribed by FP" [1, p. 83].

On the other hand, Churchland does seem to acknowledge the pertinence of the argument concerning the intrinsic normativity of the mental, but suggests that scientific theories are, or better: will be, at some point in the future, able to account for this aspect better than folk psychology, which, as he suggests, is not without flaws. He notes, for example, that "the laws of FP ascribe to us only a very minimal and truncated rationality, not an ideal rationality as some have suggested" [1, p. 83]. Churchland's point shows at best, however, that folk psychology describes beings who are not perfectly rational; moreover, it still leaves open the possibility that some time in the future, a matured folk psychology will be in a position to provide an even more adequate description of the mind. Churchland does not take this possibility into account, he rather bets on the future development of scientific theories suggesting, as he does, that these normative issues "will have to be reconstituted at a more revealing level of understanding, the level that a matured neuroscience will provide" [1, p. 84]).[5]

Churchland, thus, suggests that in the future scientists will be able to develop a neurological theory that is able to address these issues. With this, he adopts a line similar to that of his teacher Wilfrid Sellars, who argued that in order to get a comprehensive theory of the mind, or, as he formulates it, a synoptic vision, aspects of the manifest image need to be *added to* the scientific image of man:

> Thus the conceptual framework of persons is not something that needs to be reconciled with the scientific image, but rather something to be joined to it. Thus, to complete the scientific image we need to enrich it not with more ways of saying what is the case, but with the language of community and individual intentions [9, p. 40].

But Sellars also admits that he is not able to provide a clear idea of how this task is to be achieved. He concludes his essay with the following statement:

> We can, of course, as matters now stand, realize this direct incorporation of the scientific image into our way of life only in imagination. But to do so is, if only in imagination, to transcend the dualism of the manifest and scientific images of man-of-the-world [9, p. 40].

So the question remains: how can we enrich the neurosciences to put them into a position to address the normative issues of the mental? For the moment being, I want to suggest, we should not get stuck with this question and accept, for methodological reasons, that we are currently facing a pluralism of descriptions: philosophers and neuroscientists do describe the very same set of phenomena, but they do so in very different ways. Philosophers, it seems to me, will have to go a long way to clear conceptual issues related to the philosophy of mind; and also neurologists will have to work hard to formulate more precise theories of the causal workings of our nervous system. Since we do not yet have a clear idea how and at what level of analysis the philosophical description of persons as rational agents and the description of their nervous apparatus at a sub-personal level can be reconciled and, in consequence, which parts of the respective theories will have to be refined or altered, it seems to me that both disciplines will optimize their results if they focus on their own projects.

---

[5] For a more detailed discussion of Churchland's arguments concerning the normative aspect of the mental cf. [4, pp. 41ff].

The question will turn out to be pressing, however, once we are able to formulate more advanced neurological and more precise philosophical theories of the mind. At this point, I think, we will need to pursue two strategies at the same time: we will need to show how propositional attitudes can emerge from neural networks in the brain an thus give an exact description of enabling conditions for our having mental episodes. A bottom-up strategy alone will not do, however: to get a synoptic vision, we will also need to show how the normative dimension can influence the sub-personal level. After all, this dimension might be crucial for our training up neuronal networks to perform certain tasks. This normative aspect, however, is rooted in our capacity to follow rules, i.e. to act in conformity or contrary to standards that are essentially social: a person can be said to follow a rule — which, moreover, holds within a social community — only if her actions can be corrected by the members of this community. If this observation is correct, it shows that the neurosciences will be able to formulate a comprehensive theory of the mind only if they succeed in capturing this social dimension, i.e. in understanding the nervous apparatus of an individual in its interaction with that of other individuals. Recent research on mirror neurons hints in this direction, but scientists will be able to reach this goal only if they give up the methodological individualism that is still inherent in their research programs; rather than studying the nervous apparatus of one person they might have to turn to describe those of a community of persons in their interaction.

Even when pursuing this strategy, however, we might find ourselves in a position that we are not able to formulate a scientific theory that explains how the normative level emerges from the causal order of the world; we might have to accept it as a brute fact that the level of norms is *always already* there. In other words, we might not be able to explain where the normative aspect comes from — at least at a phylogenetic level; we will be able to do so on an ontogenetic level — or how a certain set of values can be justified scientifically or whether or why it is preferable over another one and might have to conclude that "it is there — like our life" [11, §559]. This does not seem to be a major problem, however: explanations have to come to an end somewhere. By adding the normative and the social aspects to our scientific theory of the mind, we might not be able to solve all problems. In trying to do so we will, however, gain a more comprehensive understanding of propositional attitudes, consciousness and the mind. In short, we will get a more profound understanding of a central philosophical problem: what is the nature of the mind; and what does it mean to be a person.

BIBLIOGRAPHY

[1] P.M. Churchland. Eliminative materialism and the propositional attitudes. *The Journal of Philosophy*, 78: 67–90, 1981.
[2] P.M. Churchland. On the nature of theories: A neurocomputational perspective. In J. Haugeland (ed.), *Mind Design II*, MIT Press, Cambridge, MA 1997, pages 251–292.
[3] P.S. Churchland. Neurophilosophy: the early years and new directions. *Functional Neurology* 22: 185–195, 2007.
[4] W. Huemer. *The Constitution of Consciousness. A Study in Analytic Phenomenology.* Routledge, New York 2005.

[5] E. Nagel. *The Structure of Science. Problems in the Logic of Scientific Explanation.* Hartcourt, Brace & World, New York 1961.
[6] O. Neurath. Empirical sociology. The scientific content of history and political economy. In M. Neurath and R. Cohen (eds), *Empirism and Sociology*, pages 319–421. Reidel, Dordrecht 1973.
[7] P. Oppenheim and H. Putnam. Unity of science as a working hypothesis. In H. Feigl, M. Scriven and G. Maxwell (eds), *Minnesota Studies in the Philosophy of Science*, Vol. 2, University of Minnesota Press, Minneapolis 1958, pages 3–36.
[8] H. Putnam. The nature of mental states. In Ned Block (ed.), *Readings in Philosophy of Psychology*, vol. 1, Harvard University Press, Cambridge, MA 1980, pages 223–231.
[9] W. Sellars. Philosophy and the scientific image of man. In *Science, Perception and Reality*, Ridgeview, Atascadero 1963, pages 1–40.
[10] W. Sellars. *Empiricism and the Philosophy of Mind.* Harvard University Press, Cambridge, MA 1997.
[11] L. Wittgenstein. (*OC*): *On Certainty.* G.E.M. Anscombe and G.H. von Wright (eds), trans. by Denis Paul and G.E.M. Anscombe. Blackwell, Oxford 1969.

Discourse and action: analyzing the possibility of a structural similarity

LAURA SPARACI

1 Introduction

Lately major attention has been given to the study of forms of non-verbal communication such as gestures. Gestures are a very peculiar type of non-verbal communication, their nature is to linger between spoken words and performed actions. As a two faced Janus, gestures in conveying meaning look on one side to the grammatical construction of verbal communication and on the other to an organization of thought in a sort of space of performance, where the gesture is enacted. Studying gestures may enhance our performance in certain critical situations (for example gestures are inserted in most military manuals as an essential part of training procedures), may shed light on our understanding of narrative construction [8] and may even help in analyzing specific deficits [7]. Notwithstanding their applications, most studies on gestures lack to underscore that they rely on an implicit understanding of an existing relation between speech and action and to clarify this relation's internal dynamics. Therefore the overall attempt of this paper will be to gain a new perspective on the relation between speech and action that may inform gesture studies, while it might also prove to be of use in other fields of research.

The presence of a relation between verbal communication and action may be new to the field of gestures studies, but it is not new to the field of philosophy. Specifically I shall call to aid Speech Act Theory and a more recent account given by Ricoeur to explain certain dynamics. But since gestures are often studied within the near area of cognitive psychology I shall also refer to authors from this field.

The work will be organized as follows. In the first paragraph I shall offer an overview of Speech Act Theory within philosophy. A short history of the theory will be given and its scope will be to highlight the problems that this theoretical perspective tried to solve and the method used. This paragraph will also highlight the concept of language as essentially dynamic, with reference to the work of Vygotsky. Finally a first question will be raised on possible applications of this view outside the field of philosophy of language.

The second paragraph will try to answer this question by analyzing the application of the philosophical method previously described within developmental psychology, specifically in Bruner's work. This will bring to a second question as to the true relation existing between discourse and action.

In the third paragraph I shall propose an answer to this question referring largely to Ricoeur's work.

Finally the fourth paragraph should sum up the conclusions that may be obtained from this work and consider its relevance for future research as well as for the elaboration of further theoretical perspectives on gestures.

2 Language: form vs. use

Probably the first author to indirectly point out the dynamics of the relation between discourse and action was Ludwig Wittgenstein. In his *Philosophical Investigations* Wittgenstein clearly states that the definition of an utterance is in its use [24]. Notwithstanding Wittgenstein's indication studies on language have focused rather on form than use.

2.1 Saussure and the link between signifier and signified

One of the reasons for this lack of attention to the relevance of use may be ascribed to certain aspects of the Structuralist approach initiated by Ferdinand de Saussurre, which influenced most of the later studies on language. Saussure distinguished within the system of language between *langue* (i.e. the system of symbols that constitute a code within a language) and *parole* (i.e. the linguistic act of a speaker of that same language) and within the sign between *signifier* (i.e. the sound that is pronounced in uttering a word) and *signified* (i.e. the concept that the word conveys) [18].

The relation between signifier and signified is conceived as an arbitrary one, i.e. as the variety of languages demonstrates there is no natural law that establishes the necessity of a link between a given signifier and a certain signified. Still signifier and signified are the inseparable faces of the same sign-coin, a coin that is forged by social context. In Saussure's conception *langue* is a socially built convention and it is this convention that acts as glue for the arbitrariness that is at the heart of signs. In this perspective the relation between a word and its meaning is essentially associative and meaning is seen as static. Once the coin is forged its use becomes of less relevance than its form [18]. Therefore due to its conceiving meaning as essentially static Saussure's influential nomenclature contributed to underestimate the importance of language use, while giving relevance to form.

2.2 Speech acts theory within philosophy of language

Wittgenstein's indication was later picked up by a group of analytic philosophers that focused their attention on the pragmatics of language mainly Grice (1989), Austin (1962), Searle (1969) and Strawson (1964). Their studies led to the formulation of what has since been called Speech Acts Theory.

Overall Speech Act Theory may be subdivided into two parts. A definitional part which outlines a new way of looking into communication and an analytic part which contains the specific taxonomy that should be applied to the analysis of communicative acts. It is the presence of this second aspect of the theory that importantly distinguishes it from Wittgenstein's initial warning. While Speech Act Theory preserves the core instance that language should be analyzed in its use, since it is in use that meaning manifests itself, it also provides a specific taxonomy for the analysis of how meaning is conveyed.

Such taxonomy is purposefully absent in Wittenstein's work, which elucidates the emergence of meaning in use rather through the conceptual framework of language games than through a full-blown categorization [24].

As to the definitional part, speech acts are acts of communication, whereas to communicate is to express a certain attitude, and the type of speech act being performed corresponds to the type of attitude being expressed. For example, a statement expresses a belief, a request expresses a desire, and an apology expresses a regret. As an act of communication, a speech act succeeds if the audience identifies, in accordance with the speaker's intention, the attitude being expressed [2, p. 1].

In its explanatory aspects this new scheme supported the idea that language should be seen as possessing a broader structure, which goes beyond form and projects itself into a social world. But unlike Saussurre here the role of the socio-cultural environment is not to limit the arbitrariness of signs. In fact the structure of speech acts and therefore their meaning is not considered to be arbitrary at all, therefore there is no need for a 'glue' that may keep together signifier and signified and the social context is free to play a very different role which will be made clear as we proceed. The non-arbitrariness of speech act structure is demonstrated through an analysis of the taxonomy of speech acts.

Austin describes three levels of communication within a given speech act:

1. The act *of* saying something (i.e. the locutionary act);

2. What one wishes to communicate *in* saying it (i.e. the illocutionary act);

3. What causal consequences one wishes to bring about *by* saying it (i.e. the perlocutionary act) [1].

Kent Bach provides a very British example to elucidate this nomenclature. The example goes something like this: lets suppose that a bartender says "The bar will be closed in five minutes". He is saying that the bar is closing in five minutes (i.e. performing a locutionary act), he is informing his clients that the bar will be closing and possibly urging them to order their last drinks (i.e. performing an illocutionary act) and finally he wishes to perform the further effect of making his clients believe that the bar is closing in five minutes and getting them to want to order their last drinks (i.e. performing a perlocutionary act) [2]. All these acts are contained at different levels within the bartender's words. It should be noticed at this point how this taxonomy projects the spoken words into the external environment determining a specific structure of abstract acts within a social context.

More important than the performance of a speech act is the analysis of its success. The success of a speech act may be evaluated at different levels, of which the most important ones for the present discussion are the illocutionary and the perlocutionary ones. A speech act will be generally considered to succeed at the illocutionary level if the audience recognizes the intention with which it was performed. But, and here lies the catch, this *recognition* is not a mere act of decoding and it is at this point that the different authors seem to have different perspectives on how recognition takes place.

At the perlocutionary level a speech act will succeed if the addressee will believe what is being stated or perhaps do what is being requested. So unlike the other levels the perlocutionary one is projected towards the future. Speech acts may be further divided into types and in ways in which they are performed. But for the purpose of the present discussion it is sufficient to know that these distinctions exist and relate to the main distinction among levels described above. In describing Speech Act Theory I have stated that this approach ascribes a different role to the social environment than the one encountered in Saussurre. It is now time to elucidate what is meant here, but in order to do so it is best to refer briefly to Lev Semënovič Vygotsky's work and his dynamic conception of language.

2.3 Vygotsky's dynamic view of language

As stated above Saussure conceived the association between signifier and signified as static and established through socio-cultural convention. This perspective was challenged by Vygotsky's view. In his 1934 book *Thought and Language* Vygotsky clearly states that word meaning has a dynamic rather than a static nature. Not only word meanings change throughout child development, they also change "with the various ways in which thought functions" [23, p. 217]. Vygotsky's main objective seems to be lifting the curtain and showing what lies beyond speech and he finds that as meanings change and develop the relation between thought and word changes too. Therefore the image of the two-faced coin seems outdated and what we come to look upon is truly a process.

> Behind words, there is the independent grammar of thought, the syntax of word meanings. The simplest utterance, far from reflecting a constant, rigid correspondence between sound and meaning, is really a process [23, p. 222].

Some may ask what is the relevance of having a dynamic approach to meaning within our analysis of discourse and action. The truth is that the dynamical perspective is essential as it not only lifts the curtain to show the wizard behind thought, but also projects itself into the outer, social world, as Vygotsky indirectly demonstrates in one of his examples.

In the prologue to his play *Duke Ernst von Schaben*, Uhland says, "Grim scenes will pass before you". Psychologically, "will pass", is the subject. The spectator knows he will see events unfold; the additional idea, the predicate, is "grim scenes." Uhland meant: "what will pass before your eyes is a tragedy".

> Analysis shows that any part of a sentence may become a psychological predicate, the carrier of topical emphasis. The grammatical category, according to Hermann Paul, is a petrified form of the psychological one. To revive it, one makes a logical emphasis that reveals its semantic meaning. Paul shows that entirely different meanings may lie hidden behind one and the same grammatical structure. Accord between syntactical organization and psychological organization is not as prevalent as we tend to assume — rather, it is a requirement that is seldom met [23, p. 221].

This long quote is essential. Here in Vygotsky's play-example we may trace a distinction that we have seen as essential within Speech Act Theory between the act *of* saying something and what is meant to be communicated *in* saying it. Furthermore the Russian psychologist is able to throw a bridge between thought as a dynamical psychological form, grammatical categories as momentary and ever fleeting petrification of thought and finally meaning, as being born through the entire process.

At this point the question that stays open is: what of social conventions? Isn't meaning to be somehow related to the socio-cultural environment? Vygotsky's answer is yes, but society is not conceived as a generator of conventions any more, it is rather an environment or better a context in which language is embedded. Rather than acting as glue between signifier and signified context provides the stage setting were the play of thought and language takes place. To use a term that finds a vast use in Vygotsky we may say that context is the social scaffold where speech acts dynamically take place. This same way of conceiving the role of society may be found in Speech Act Theory.

Now that a method of analyzing the relation between discourse and action has been clearly stated and that the problem area has somehow been set out we may proceed to consider the following question: can Speech Act Theory find an application outside the field of philosophy of language?

3 Looking into developmental psychology

The answer to this question is yes and an explicit attempt to apply Speech Act Theory to the field of developmental psychology has been made by Gerome Bruner in a 1974 paper. As I have stated above Vygotsky underlined the dynamical nature of speech by stating, among other things, that word meaning changes throughout child development. Even if Bruner never cites Vygotsky in this work, he attempts to study the development of meaning through the observation of the ontogenesis of speech acts. His main scope is to demonstrate that the development of meaning relies heavily on the child's possession of certain patterns of action [4].

Bruner refers explicitly to Speech Act Theory in philosophy of language, stating that there are two ways of conceiving meaning in language: analyzing the actual utterance and considering its effectiveness. This proposal, translated into the language of psychology, becomes for Bruner equal to stating that the formal structure of language is not totally arbitrary, but it rather reflects the psychological events and processes, which it encodes. This passage is somehow analogous to the one we have seen performed by Vygotsky, but Bruner goes one step further, since he tries to bring proof of this theoretical hypothesis by referring to specific mechanisms present in cognitive psychology.

Specifically Bruner describes two mechanisms that may be taken as candidates of a possible application of Speech Act Theory outside the realm of philosophy of language, i.e. the isomorphism present between predication and attention processing and the relation between case structure and action organization. This second example is particularly interesting in order to answer the question I have posed at the end of the previous paragraph.

Bruner states that the primitive categories of grammar may be brought to correspond to certain structures of action, in particular the structure of action establishing joint-attention. Therefore he looks into very early forms of joint attention in children. His hypothesis is that

> The facts of language acquisition could not be as they are unless fundamental concepts about action and attention are available to children at the beginning of learning [4, p. 6].

This hypothesis, which interestingly he traces back to David McNeill, raises the following question

> How precisely does the child's knowledge of action and how does his way of attending lead him to grasp concepts embodied in language [4, p. 8]?

To answer this question he conducts a close analysis of mother-infant interaction during play activity, resembling Trevarthen's studies on joint attention mechanisms [22]. Bruner's conclusion is that the child's knowledge of action and attention provides bench-marks that will later be used in interpreting order-rules in grammar. Various processes contribute to building these bench-marks. It may be useful to spell these out for the purpose of clarity:

1. The child learns segments of joint action with the mother such as agent-action-object and how to manipulate these segments;

2. The child gradually learns routines that bring to joint attention such as using eye-to-eye contact or common foci of attention which are gradually replaced by simple forms of semanticity through early vocalizations;

3. The child learns phonological patters that act as place-holders through imitation such as interrogative, indicative or vocative contours that accompany action and that often require a rise in intonation indicating early forms of prosody.

All these action patterns and their links with early verbalization bring Bruner to state that

> the child is grasping initially the requirements of joint attention at a pre-linguistic level, learning to differentiate these into components, learning to recognize the function of utterances placed into these serially ordered structures, until finally he comes to substitute elements of a standard lexicon in place of the non-standard ones [4, p. 17].

Bruner's work therefore not only answers the question posed at the end of the previous paragraph as to the possibility of applying Speech Act Theory outside the field of philosophy of language, it also establishes an ontogenetical link between discourse and action. In this analysis action seems to be the cradle of language, a necessary instrument that is entwined with language development and that supports its first steps. Therefore action seems to be closely linked with speech and its use in early phases of development.

Now, a further question could be how this relation develops. If not only can action contribute to the development of meaningful speech, but traces of the structures of speech, outlined within Speech Act Theory, may also be found within action. To answer this last question I will refer to the work by Paul Ricoeur.

4 Ricoeur's insight on discourse

In his essay *The Model of the Text: Meaningful Action Considered as a Text* Paul Ricoeur wishes to demonstrate how meaningful action may be considered as a proper object of science through an objectification that delineates its inner patters in a way that is analogous to discourse. His attempt to propose action as an object for social sciences seems particularly interesting for the purpose of answering the question posed at the end of the previous paragraph, since, in describing the essential constituents of action, Ricouer draws on his conception of discourse and language use.

Before moving on to describe Ricoeur's analysis it is worthwhile to point out that this philosopher's work has the further merit of providing a definition of the term "discourse". Up to now I have used interchangeably the words language, speech and discourse. I have done so on purpose since the authors described until now do not provide a clear-cut nomenclature for these terms. Not so Ricoeur, who pointing out to the main characteristics of discourse seems to come up with the appropriate definition for our object of investigation. Without going to far off on this topic it will be sufficient to say that discourse according to Ricoeur has the fallowing traits:

1. It is realized in time, while language as a system is virtual and outside of time;

2. It refers back to its speaker through personal pronouns, whereas language lacks a subject;

3. It is always about something, while signs in language only refer to other signs;

4. It is in discourse that messages are exchanged, while language is only a condition for communication.

After this brief parenthesis we may return to Ricoeur's further theories keeping in mind that I will from now on be referring to discourse since this is *the* language-event, language in use.

According to Dauenhauer's analysis of Ricouer's work, action is considered as analogous to discourse "because, to make full sense of any action one has to recognize that its meaning is distinguishable from its occurrence as a spatiotemporal event. Nevertheless, every genuine action is meaningful only because it is some specific person's doing at some particular moment" [9]. In this position held by Ricoeur we may trace both a dynamical conception of meaning and the relevance of context in language use that has been highlighted above.

Ricoeur carries the analogy between the structure of discourse and that of action even further when he makes explicit reference to Speech Act Theory and proceeds to demonstrate that the same taxonomy that this theory traces within speech may be ascribed to action itself. It is this very taxonomy that enables him to propose the objectification of action that will make it into an appropriate object for science. As Ricoeur clearly states

This objectification is made possible by some inner traits of the action that are similar to the structure of the speech act and that make doing a kind of utterance [17, p. 151].

Action in this perspective is found to have:

1. The structure of a locutionary act, since it has a "propositional content" that may be identified and reidentified;

2. Illocutionary characteristics, since each action has constitutive "rules" that make it into a specific type of action;

3. A perlocutionary nature since as a social phenomenon it has effects that may or may not be anticipated.

Finally actions follow speech acts also in their being subject to interpretation.

According to what has been just described Ricoeur's work is a good candidate in offering an answer to the question formulated at the end of the previous paragraph. His work seems to provide evidence that the relation between action and discourse is a two way thing and that if Bruner's work has helped us in tracing the ontogenesis of speech acts within action, Ricoeur analysis provides the appropriate framework to consider action as possessing some of the essential characteristics of discourse.

5 Informing gesture studies

The overall purpose of this work was trying to shed some light on a question triggered by David McNeill, that is: what is the relation between discourse and action? And can they truly be thought as having a similar structure? McNeill's question derives from his interest in a very special type of non verbal communication: gestures. Traditional views on gestures alternatively consider the relation between gestures and speech as:

1. Separate communication systems, gestures simply support speech when it is disrupted or unavailable [3];

2. Reciprocally linked but only at the phonological encoding stage of speech production (i.e. gestures are used to retrive words from lexical memory) [14].

But recent studies, especially those analyzing gesture-speech timing, have demonstrated how these classical interpretations are unable to capture in full this relation [16]. Therefore McNeill and others have offered the following alternative hypothesis:

1. Gesture and speech are so tightly connected that they constitute a single system of communication based on a underlying common thought process [12, 15, 16].

This new perspective alongside with a reconsideration of the gesture-speech relation also implies a different understanding of the nature of speech. In fact, it is necessary to conceive language use and meaning as dynamically built even with the support of a different medium, such as gestures. Philosophy of language and and psychology seem now to offer some possible support to this view. For example considering the various aspects of the relation between discourse and action highlighted above:

1. The possibility of considering different levels of communication within discourse (such as the locutionary, illocutionary and perlocutionary acts);

2. The role of the social environment as scaffolding the dynamical constitution of speech acts;

3. The existence of an ontogenetical link between discourse and action;

4. The possibility of a two-way relation between discourse and action.

All this may shed new light on McNeill's conception of language as a dynamical process in which speech and thought impact each other and meaning is not fixed but continually developing, explicitly following Vygotsky [16, pp. 80–86]. But most importantly this allows to support his further step, i.e. stating that the construction of meaning passes through the coordination of language and gesture. A coordination that to him is so tight that it can be seen as the two sides of the same process (retrieving in some way Saussurre's coin image). In fact underscoring the importance of language use, different ways in which spoken words project themselves in the external environment have been described, now it may be stated that during this process meaning is built not only through words, but also through visible actions in the form of gestures. Aside from allowing a clarification of the relation between speech and action, often implicit in gesture studies, the description of this relation given above may also inform these studies on specific aspects of gestures.

It has been stated above that discourse and action are closely related and that this is due not only to the nature of discourse, but to the nature of action as well. Now gestures are one specific aspect of discourse, i.e. according to Kendon an utterance is essential unit of communication conceived as any ensemble of action that may count for an interpreter as an attempt by the actor to provide information of some sort and a gesture is the visible body action that plays a role in utterances [13]. So one important question may be: can gestures be considered as possessing at least some of the aspects of discourse described above?

Speech acts have been described above as possessing a three-level-of-communication structure, can the same be said of gestures? Can we distinguish a locutionary, illocutionary and perlocutionary aspect in gestures as well? Maybe our British pub-keeper can give us a hand. We can imagine him looking at his clients and accompanying his words (or even substituting them) by raising his left fore arm in the air and shaking his left hand sideways while he shapes it as if he were holding a rope from an imaginary bell. This may happen because his wife has taken the real bell he usually uses to call the final rounds to give it a fine polishing and hasn't returned it yet. That gesture would be saying that he is ringing an imaginary bell as if he were ringing a real one (locutionary act), that he is calling the final rounds so you better get your order in if you want another beer (illocutionary act) and making his clients believe that the final rounds are being called (perlocutionary act).

This holds to say that gestures accompanying or substituting speech may be analyzed using the same levels proposed for Speech Acts, which may be useful in the in the analysis on complex gestures. Let's consider the second aspect listed above, i.e. can the social environment serve as a scaffold to gesture behavior?

On this point Bruner, follows Vygotsky's footsteps, and helps clarify the role of context for gestures. According to Bruner the adult-interpreter, operates not as a corrector or as a reinforcer, but rather as a provider, expander and idealizer of early communicative forms. The adult's "as if" interpretation, enables the child to grasp the basic elements of joint action first at a pre-verbal level, distinguishing different components and learning to recognize their functions. Only later the child will finally substitute these pre-verbal forms of communication with elements of standard lexicon [4]. This mechanism is made possible by the fact that children have very few situational contexts during early communication, which constitute a series of fixed formats, within which the child interacts with the mother. These formats are therefore a sort of framed microstructure a "constrained and segregated transaction between child and adult with a goal, a mode of initiation, and a means-end structure that undergoes elaboration" [5, p. 162]. Within this view gestures have the important role of triggering and supporting early forms of contextualized communication. For example pointing behavior is considered to trigger and support the first forms of phonological utterances loosely modeled on adult speech [5].

Early pre-verbal communicative forms gradually build up and become more complex until a gesture will eventually be accompanied by a single word utterance [10]. Gradually a deictic gesture, may be substituted by a deictic word, i.e. a word that picks out or points to specific objects in relation to the participants in a given speech situation, indicating a continuity from gestures to speech [6].

This brief description also highlights the ontogenesis of gesture behavior and its relation to action patterns. In conclusion it seems that at least some aspects of discourse uncovered by the former analysis may characterize gestures as well. This not only strengthens the link between speech and gestures, but may also provide some new insight for gesture analysis, highlighting parameters that should be taken into account.

6 Conclusions

Summing up, I have provided a description of the main aspects of Speech Act Theory as compared to the traditional linguistic approach derived from Saussure and I have also tried to enrich this problem area with the perspective elaborated by Vygotsky. Two different questions have been raised as related to this method: if it could be used outside the field of philosophy of language and if the relation that it establishes between discourse and action may be considered as a two way one. I have attempted to answer these questions relating respectively to the work of Jerome Bruner and Paul Ricoeur.

The overall purpose was that of trying to shed some light on a question triggered by recent studies on gestures done by David McNeill, that is: what is the relation between discourse and action? And can they truly be thought as having a similar structure? Philosophy of language and psychology seem to offer some possible replies to this interrogative. This theoretical landscape may be used to enrich current attempts to study gestures as acts that stand on the boarder between discourse and action. Mainly this work will have been useful if it has helped in clearing out some possible doubts on the possibility of using a similar taxonomy for discourse and action and on the possibility of objectifying meaningful action within a speech act perspective.

BIBLIOGRAPHY

[1] J.L. Austin. *How to Do Things with Words*. Harvard University Press, Cambridge, MA 1962.
[2] K. Bach. Speech Acts. In *Routledge Encyclopedia of Philosophy*, http://online.sfsu.edu/~kbach/spchacts.html, 1998.
[3] B. Butteworth and U. Hadar. Gesture, speech, and computational stages: a reply to McNeill. *Psychological Review*, 96 (1): 168–174, 1989.
[4] J.S. Bruner. The ontogenesis of speech acts. *Journal of Child Language*, 2: 1–19, 1974.
[5] J.S. Bruner. The social context of language acquisition. *Language and Communication*, 1 (2/3): 155–178, 1981.
[6] E.V. Clark. From gesture to word: On the natural history of deixis in language acquisition. In J.S. Bruner and A. Garton (eds.), *Human Growth and Development*, Clarendon Press, Oxford 1978, pages 85–120.
[7] E.S. Colgan, E. Lanter, C. McComish, L.R. Watson, E.R. Crais and G.T. Baranek. Analysis of social interaction gestures in infants with autism. *Child Neuropsychology*, 12: 307–319, 2006.
[8] K. Dautenhahn. The narrative intelligence hypothesis in search of the transactional format of narratives in humans and other social animals. In M. Beynon, C. L. Nehaniv and K. Dautenhahn (eds.), *Proceedings of the Fourth International Cognitive Technology Conference, CT2001: Instruments of Mind*, Springer Verlag, Berlin 2001, pages 248–266.
[9] B. Dauenhauer. Paul Ricoeur. In E.N. Zalta (ed.), *The Stanford Encyclopedia of Philosophy (Winter 2005 Edition)*, http://plato.stanford.edu/archives/win2005/entries/ricoeur/, 2005.
[10] H. Gray. Learning to take an object from the mother. In A. Lock (ed.), *Action, Gesture and Symbol: The Emergence of Language*, Academic Press, London 1978, pages 159–182.
[11] H.P. Grice. *Studies in the Way of Words*. Harvard University Press, Cambridge, MA 1989.
[12] J.M. Iverson and E. Thelen. Hand, mouth and brain. The dynamic emergence of speech and gesture. *Journal of Consciousness Studies*, 6 (11-12): 19–40. 1999.
[13] A. Kendon. *Gesture: Visible Action as Utterance*. Cambridge University Press, Cambridge 2005.
[14] R.M. Krauss. Why do we gesture when we speak? *Current Directions in Psychological Science*, 7: 54–60, 1998.
[15] D. McNeill. *Hand and Mind: What Gestures Reveal about Thought*. University of Chicago Press, Chicago 1992.
[16] D. McNeill. *Gesture and Thought*. University of Chicago Press, Chicago 2005.
[17] P. Ricoeur. *From Text to Action: Essays in Hermeneutics II*. English translation by K. Blamey and J.B. Thompson. Northwestern University Press, Evanston 1991.
[18] F. de Saussure. *Course in General Linguistics*. English translation by R. Harris, Open Cout Publishing Company, La Salle, ILL 1983.
[19] P.F. Strawson. Intention and convention in speech acts. *Philosophical Review* 73: 439–460. 1964.
[20] J. Searle. *Speech Acts: An Essay in the Philosophy of Language*. Cambridge University Press, Cambridge 1969.
[21] B. Smith. Towards a history of speech act theory. In *Speech Acts, Meanings and Intentions. Critical Approaches to the Philosophy of John R. Searle*, de Gruyter, Berlin and New York 1990, pages 29–61.

[22] C. Trevarthen and P. Hubley. Secondary intersubjectivity: confidence, confiding and acts of meaning in the first year. In A. Lock (ed.), *Action, gesture and symbol: The emergence of language*, Academic Press, London 1978, pages 183–229.
[23] L.S. Vygotsky. *Thought and Language*. Revised and edited by A. Kozulin. MIT Press, Cambridge, MA 1986.
[24] L. Wittgenstein. *Philosophical Investigations*. Blackwell Publishing, Oxford 1953.

Visuomotor representations: Jacob and Jeannerod between enaction and the two visual systems hypothesis

ALESSANDRO DELL'ANNA

1 Introduction

On the track of Milner and Goodale's hypothesis of the two visual systems [14], Jacob and Jeannerod (JJ) wrote an important book for philosophy of mind, from both a methodological and a substantial point of view [9]. From a methodological point of view, they show how far the cooperation between a philosopher and a neuroscientist can reach in the field of (in this case visual) cognition. Their robust and punctual use of available empirical results is guided by always rigorous and subtle arguments, as rarely happens in scientific books. On the other hand, their theoretical elaboration is continuously supported by the force of the evidence gathered in neuroscience in the last twenty years. From the point of view of contents, JJ's contribution is remarkable. Their framework is the two visual systems hypothesis, according to which the ventral pathway (that projects from V1 to the inferior-temporal lobes) is devoted to what in traditional vision science (and in common sense) is called "visual perception", whereas the dorsal pathway (that projects from V1 to posterior-parietal lobes) is devoted to "vision for action". By means of the former homo sapiens (like his primate relatives) would see shapes, colours, faces and complex scenes, by means of the latter he would see what is necessary to aptly interact with objects in real time, that is, shapes still, but

1. in egocentric coordinates, that is, in relation to the body of the agent, rather than to the object itself [9, pp. 103–104], and

2. in their absolute size, location and slant, deprived, then, from typical context effects [9, pp. 112–113], like chromatic and motion induction, perspective etc..

The two visual systems hypothesis is supported by a huge amount of experimental results in the domains of physiology, neuropsychology and psychophysics, that JJ review in some details. What they reject of Milner and Goodale's interpretation is their too rigid dichotomy between the two pathways. What they try to supplement to it is an analysis of the kind of representations characterizing the dorsal system, which they call visuomotor representations (VMRs), contrasting them with percepts, characterizing the ventral system. I will focus my paper on this subject, even if the book provides a lot of other issues to discussion.

2 Visuomotor representations

JJ's is a representational theory of the (visual) mind [4], i.e. a theory that takes the activation of a given neural area to be the representation of something internal or external to the organism. For example, the firing of a given group of neurons in V1 would represent the edge between an object and its background in the external world (in line with something like Marr's [12] model). Moreover JJ's is a teleo-semantic theory, because it claims that the property represented by a given neural network (its content) is such by virtue of an evolutionary process which makes it advantageous, for the animal, to be endowed with the capacity to detect such property (edge, slant, texture, shape). In this sense, it would seem that, according to JJ, the visual system has a limited plasticity, evolution determining most of the contents to be detected[1].

Now let's consider visuomotor representations. It is almost unanimously accepted in neuroscience that the ventral pathway processes or detects the properties of objects like location, slant, size and shape. Milner and Goodale's hypothesis (MG) concerning the anatomical duality between the dorsal and ventral systems differs from previous interpretations (e.g. [21]) in that it considers the properties detected by the former to be the same as those detected by the latter, but exploited differently. Whereas the ventral pathway processes slant, size and shape in order to elicit planning and conscious reasoning (digitalizing the information implicit in percepts, and so becoming epistemic perception , in Dretske's words) or selects potential objects for action, according to Clark's assumption of the experience-based selection of the action to be accomplished [2], the dorsal pathway drives the animal's action toward those very properties, avoiding, reaching, grasping or manipulating objects. *Percepts* are the products of the ventral system, *visuomotor representations* are the products of the dorsal system. But, whereas MG claim that the dorsal system does not participate in conscious processes, JJ remark that at least the inferior-parietal lobes play some role in conscious acting. What I am interested in, in this paper, is if and how exhaustive JJ's characterization of visuomotor representations is.

First JJ compare VMRs to Gibson's [7] *affordances*, properties on which Gestalt psychology threw light at the beginning of last century[2]. Properties of this kind are the walkability of a solid plain of rock or the graspability of a stick of wood, i.e. the possibilities for action offered by a given object or context. We are dealing, then, neither with purely *descriptive* representations, like percepts or beliefs, nor with purely *directive* representations, like intentions or desires. JJ draw, indeed, on Millikan's [13]idea of a *pushmi-pullyu representation* (PPR), a kind of representation that she thinks must logically precede both the previous kinds. In VMRs, as in PPRs, there is an implicit tie between seeing something and acting on it, like in seeing the shape of a stick and gearing the motion of the arm in its direction and the grip of the

---

[1]This is, actually, what Dretske thinks too [4, chap. 2].

[2]Being neither primary qualities (like shape, size, motion) nor secondary qualities (like colours or sounds), according to the classical lockian-galileian distinction, Gestalt psychologists used to called them also third properties [11, chap. 8]. Koffka, by the way, was one of Gibson's teachers).

hand to its potential handle. In this sense, VMRs are intermediary kinds of reaction between a reflex and a representation that is totally independent from action, e.g. between closing your eyes and moving your head when a ball quickly runs towards you, on the one hand, and contemplating the shadows of a drawing, on the other hand. In Millikan's own words:

> Pushmi-pullyu representations are more primitive than either purely directive or purely descriptive representations. Representations that tell only what the case is have no ultimate utility unless they combine with representations of goals and, of course, representations that tell what to do have no utility unless they can combine with representations of facts. It follows that a capacity to make mediate inferences, at least practical mediate inferences, must already be in place if an animal is to use purely descriptive or purely directive representations. The ability to store away information for which one has no immediate use, and to represent goals one does not yet know how to act on, is surely more advanced than the ability to use simple kinds of pushmi-pullyu representations [13].

From a naturalistic point of view, Millikan's reasoning sounds quite obvious. Knowledge has no goal in itself, except for contributing to the survival of the organism. We also know that, at a sub-cortical level, the perception-action tie is the rule, rather than the exception. I have just given an example of reflex. Another one is the by now classic mechanism of bug-detection in the frog that obliges it to launch its tongue whenever something like a bug crosses its visual field. Even if it is a sub-cortical mechanism, the bug-detector is a VMR.[3] .

One of the most impressive example of VMRs JJ offer, is drawn from the study of D.F. visual agnosia, on which MG based much of their two visual systems hypothesis. Visual agnosia was the consequence of lesions to her ventral pathway due to inhalation of carbon monoxide. Though she was no more able to see the shape, size and slant in depth of, say, a pencil, D.F. turned out to be able to grasp it with a precision grip, i.e. without using the whole hand or injuring herself. JJ claim that D.F., thanks to her still intact dorsal pathway could form a VMR of the parameters she could not consciously see, so that she could guide her arm and aptly gear her fingers toward the pencil. As I said, VMRs detect the same parameters as the ventral system would, but, rather than producing a belief about the scene (the pencil), they elicit a motor scheme.

As JJ admit [9, pp. 178, 212], this kind of hypothesis cannot appeal to phenomenology, to what we ordinarily say we are perceiving, like shadows, optical illusions or colour contrasts. Rather, we have to rely on the behaviour of the experimental subjects and on the correlational brain imaging data. But, then, how can JJ defend their claim that dorsal and ventral parameters are the same, given that we identify the latter exactly on the basis of phenomenology?

3 Varieties of enaction

Here a contrast emerges between positions like JJ's and MG's on the one hand, and positions like Gibson's, Varela's, Thompson's, Noe's, Rizzolatti's

---

[3]Its representational nature has been discussed at length, and it has been identified in its capacity to misrepresent, e.g. making the frog launching the tongue when it sees a black moving dot, rather than a bug [5].

and Gallese's, on the other hand. According to the former, action remains *instrumental* to perception, whereas, according to the latter, action is *constitutive* of perception (see [8], for such distinction). The latter position seems incompatible with the dichotomy between dorsal and ventral systems, because its proponents tend to reduce every aspect of perception (and cognition) to action. But I shall show that this outcome is not necessary at all.

If we take into account Gibson's ecological optics [7], we find that affordances (VMRs, in JJ' words) are detected in an *optic flow* produced by the animal's exploration of its ecological niche. This means that the properties represented by a given organism should not be pre-specified by the scientist. Pigeons, for example, can see different colour hues from humans, being tetrachromatic, i.e. endowed with one more kind of photoreceptor than we are [20, 19]. In the same way, pigeons will probably detect different properties from we do, being able to produce different optic flows from ours, as they are endowed with lateral eyes, different points of retinal resolution, continuous motion of the head, and so on. On the other hand, we can be certain, on the basis of behavioural studies, that dorsal VMRs are suitable for the useful motor actions of an animal in its ecological niche (D.F could not see, but she could aptly move in the environment, thanks to her dorsal VMRs). That is one of the reasons why Gibson tended to underestimate cases of perceptual errors (inferring, wrongly, that talking of mental representations should be banished). A careful exploration of a scene or an object should, most of the time, give a veridical perceptual outcome. The perspective illusion of the Ames'room, for example, is inescapable *if and only if* it is seen monocularly, without moving the head and from a narrow aperture. But moving the head or opening the other eye is enough to reveal the trick. Similarly, the grasping of the central disk of the Titchener-Ebbinghaus illusion (discussed at length by JJ), even if it looks bigger than that surrounded by bigger circles, implies a movement of the arm and hand that is an already explorative action, turning out to be less influenced by the context than (ventral) vision for perception[4].

The exponents of the enactive approach (supposing they agree sufficiently with one another[5]) accepted Gibson's suggestion, naming sensorimotor contingencies the regularities that emerge from the optic flow of a given organism. Starting from saccades until the motion of the whole animal, regular couplings occur between motor action and perceptual outcome (and *vice versa*). The motion parallax is one of the most important and instructive among such couplings: whenever homo sapiens (and animals with a similar perceptual system) walks looking at one side, the scene flows more quickly in the lower part than in the upper part (revealing different distances from the observer). This is a wonderful example of what a dorsal VMR could actually detect.

There is, by now, plenty of evidence showing the plasticity of such coupling. Visual adaptation, for example, shows that after some weeks of training, persons wearing inverting goggles become able to accomplish actions typical of a

---

[4]I won't enter the debate about if and by what VMRs are influenced, being themselves subject to mistakes [9, p. 199].

[5]In the following pages I will review some major point of disagreement among Rizzolatti, Gallese, O'Regan, Noe.

normal subject. People that recover their sight after surgery, do not initially see but confused dots that acquire meaning only after some months of active exploration, like in sensory substitution devices [16].

In all these cases, like in the VMRs illustrated by JJ, active movement and exploration, are constitutive, not only instrumental elements for perception. If this is true, then we should infer that JJ's choice of shape, slant, size as the parameters represented by the dorsal system is somewhat arbitrary, because they lack a constitutive tie to action. Such a tie allows also to valorise JJ's remark that those parameters would be coded in *egocentric* coordinates by the dorsal system, whereas the ventral system would do it in *allocentric* coordinates. At the same time, it makes it possible to reject the idea that the former detects the *absolute* properties of objects, the latter the *relative* ones.

What JJ mean with the egocentric-allocentric distinction [9, pp. 103-104] is that, whenever I look at something in view of some interaction with it, the object has to be codified in relation to the subject, to his possibilities for action (which was exactly the concept of *affordance*). On the contrary, if I look at it simply in order to know what it is, or to contemplate it, or to report some of its properties to someone, the object has to be codified in terms as detached as possible from the subject. If I had to describe the shape of a spoon to someone on the phone, I would consider its subtle and lengthy shape rather than the graspability of its handle from my point of view now. In any case, I think it is worth stressing the link that still exists between egocentric and allocentric frames of reference. As suggested by Millikan's idea of PPR, both can be taken as poles of a continuum that starts with VMRs and, after the evolution of the ventral system, characterising our primate relatives, gradually frees itself from the pressure of real-time actions (see again[2]).

If this is true, saying that, contrary to the ventral, the dorsal system detects the absolute properties of the objects [9, pp. 112–113] no more makes sense. JJ claim that percepts are *relative* to the context of observation, so becoming non veridical in cases like optical illusions, mirages, colour contrasts, etc. VMRs, on the contrary, need to compute the *absolute* properties of the objects, in order for the action to be effective, avoiding possible illusions. But, if we accept the idea of an egocentric frame of reference, we should admit that a visually guided action makes those very properties *relative* to the body of the agent. By the way, don't JJ seem to say something like that: " Coding the position and/or distance of a visually presented object in an egocentric frame of reference is representing the object position's and/or distance relative to the observer's body" [9, p. 113]? It doesn't exist an absolute size, an absolute position, an absolute slant, no more than an absolute colour, but only properties relative to some frame of reference, being it egocentric or allocentric.

This does not imply that I agree with enactive theories, when they insist that every aspect of perception (and cognition) can in principle be tied up again to action. Rather, I think we should draw the right consequences from Millikan's suggestion that purely descriptive (like purely directive) representations cannot but derive from PPRs. To this end, we could assume that, in the animal kingdom, at the beginning there was the reflex, i.e. an automatic reaction to a possibly noxious stimulus, that could be very slightly modified with experience[6]. VMRs followed, and made the animal free from an almost fixed response to the stimulus, allowing it to modulate its action. Finally, with the development of the ventral system, purely visual (descriptive) representations set in, tied to no kind of action at all (only in this sense allocentric). But, saying that they are not tied to, does not mean that they do not have their roots in VMRs, that is in PPRs. In this sense, we could go on and hypothesize that the properties detected by the ventral system, identified at a phenomenological, first-person level, are nothing but "frozen" properties of an optic flow. Gibson used that term to reject the assumption of the visual properties traditionally studied in vision science, but here we can see how they can be fully accepted in the light of the two visual systems theory. In the course of evolution, the ventral system could, indeed, have acquired the function to abstract (and extract) properties like size, slant, shape, figure against a background, from the flow of sensorimotor contingencies that, on the contrary, are still exploited by the dorsal system, by means of VMRs.

Before facing the next point, anyway, I should mention some major point of disagreement among enactivists. First, sensori-motor contingencies described by Noe and O'Regan ([16], see also [15]) involve not only sensori-motor representations of the dorsal stream. Indeed, Noe and O'Regan's favorite example is color perception and the color processing, as generally known, takes place in the ventral stream. In fact, Noe and O'Regan's criticism to the two visual systems hypothesis refers to the view that in order to understand perception one has to look beyond cortical mechanism and take into account that the body is able to dominate sensori-motor contingencies on which depend the possibility itself to perceive something as a colored object. Color and brightness of the light reflected from an object change in lawful ways as the object or the light source or the observer move around, or as the characteristics of the ambient light change [16, p. 942]. Percievers must have mastery (i.e. practical knowledge, *knowing-that*) of such sensori-motor contingencies, integrating them with reasoning and action-guidance.

Rizzolatti and Gallese [17], on the other hand, suggest a rather different approach, that aims to question the significance and the nature of the notion of sensori-motor representation itself. To this regard, it should be noted that both mirror neurons and canonical neurons are located in cortical areas which belong to the dorsal stream that, according to the two visual systems hypothesis, might not play any role in perceiving, helping in driving action only (see next section). However, the function of canonical and mirror neurons shows how the so-called visuomotor representations could not be restricted to the role of a guide for the action, but involve perception itself. For example, space is not represented in some cortical area, but it depends on the activity of neural circuits whose primary function is to organize interactions with the environment. The close/far dichotomy itself would depend on the bodily activity of the observer [18, chap. 3]. Maybe the problem with JJ's view of

---

[6] Here I draw somewhat on Dennett's [3, chap. 13] hierarchy of cognitive abilities, starting from what he brightly calls skinnerian creatures, passing from popperian creatures and arriving to gregorian creatures.

visuomotor representation is that JJ didn't analyze the role such representation plays in forming the percept, missing the fact actions shape perceptions even where actions are not actually executed, as I'm going to show.

4 Visuomotor mirror systems

I think that the interpretation of the dichotomy between dorsal and ventral system outlined above is coherent with another fundamental discovery made in the past fifteen years of neuroscience, i.e. mirror neurons, discovered by Rizzolatti's team. JJ deal with this subject in the beautiful part IV of their book. Mirror neurons were first discovered in F5 monkey's motor cortex, then in inferior parietal lobes, pre-motor cortex and Broca's area in humans. They are *visuomotor* neurons, i.e. they fire either during the subject's action or during the sight of the execution of the same action by a conspecific. The congruence between the accomplished and the observed action can be very high, so that, for example, a given mirror area fires only if the animal sees someone grasping an object with its index and thumb, or tearing it with its mouth, or reaching it with two arms.

This evidence shows that the sensorimotor domain goes far beyond the dorsal pathway, extending to areas that some years ago were considered to be merely executive areas, like the (frontal) motor cortex. I will focus neither on mirror neurons nor on the perception of social actions (what JJ call the social perception system), but rather on the premises of that discovery, i.e. on the circuits formed by the anterior intra-parietal lobes (AIP) and F5. Neurons in those areas are called "visuomotor" in a slightly different sense from the dorsal neurons discussed above. The former neurons fire, indeed, not only when someone (a man or a monkey) somehow interacts with an object, but also when he/it simply looks at it. Rizzolatti and Gallese's [17] hypothesis is that even in this static situation the affordance of the object, processed by AIP, elicits a visuomotor response in F5, that, anyway, can be inhibited (probably by prefrontal control areas[7]). Continuous feedback between motor actions and perceptual outcomes, taking place since birth, build up a sort of *vocabulary of actions* that can be seen as a *semantic*, not simply a *pragmatic* system, contrary to what Jeannerod himself [10] proposed for the dorsal system.

Exactly like mirror neurons, visuomotor neurons in F5 can be very specialized, codifying actions such as grasping with two fingers, rather than with the hand, or tearing with the mouth, rather than with the fingers, etc. The actions someone can accomplish on a given object can be sharpened in various chains of action, leading to professionals such as craftsmen, musicians or dancers. Could we deny the attribute of "semantics" to these repertories[8]? But, naming a system like this "praxic", JJ [9, p. 215] (see also [10]) denies exactly this possibility. Their conclusion follows from another of their major tenet: the domain of concepts, i.e. the semantic domain, is apart from that of perception, but a little farther from VMRs than it is from percepts. In-

---

[7]Lesions to these areas causes ecopraxia. People affected by ecopraxia, indeed, can't avoid imitating whatever action they happen to see around them.

[8]Here, of course, a revision of the dicotomy between know-how and know-that is required, but it goes far beyond the scope and limits (look at JJ's subtitle) of the present article.

deed, according to Dretske's model, shared by JJ [9, chap. 5], percepts can be digitalized, producing either concepts or epistemic seeing. On the contrary, VMRs serve as inputs to motor intentions and to causal indexicals [9, pp. 202–208], thus remaining informationally encapsulated from concepts and semantics.

Nevertheless, Rizzolatti and Gallese's [17] (see also [18]) vocabulary of actions shares some important features with language, the semantic system *par excellence*:

1. it is made up of simple elements (reaching, grasping, modulating grips and so on);

2. different elements can be combined in order to produce more complex motor schemes (playing tennis or saxophones);

3. the virtual combination of different elements can generate something like an inference, i.e. the production of new motor schemes, starting from old ones. This is what happens during a fight, whenever you have to anticipate and react to the opponent's moves;

4. finally, the possibility of sharing with conspecifics (but not only) the common repertory of actions needed to cooperate (or to better fight each other).

Apart from the rising of new hypotheses about the sensorimotor origins of language, fostered by the discovery of the mirror system[9], here I insist on the phenomenological burden of JJ's theory (and a lot of neuroscience). Indeed, ordinary experience suggests that vehicles of meaning are words, images, icons. Those seem like the primitives first coded by the ventral system (textures, 2 D sketches, shapes) and then conceptualized by higher cognitive processes. The presence of visuomotor neurons in the circuits formed by AIP-F5, instead, points to the possibility of a *pre-linguistic pragmatic semantics*, firmly anchored to action. Far from denying the functional specificity of the ventral system, I simply wished to stress the pervasiveness of action even in domains once considered to be very far from it, like semantics. JJ's crucial contribution to the clarification of the idea of VMRs will have to take into greater account this pervasiveness in the future.

BIBLIOGRAPHY

[1] M.A. Arbib. From monkey-like action recognition to human language: an evolutionary framework for neurolinguistics. *Behavioral and Brain Sciences*, 28(2): 105–124, 2005.
[2] A. Clark. Visual experience and motor action: Are the bonds too tight? *Philosophical Review*, 110, 2001.
[3] D.C. Dennett. *Darwin's Dangerous Idea: Evolution and the Meanings of Life*. Simon and Schuster, New York 1995.
[4] F. Dretske. *Naturalising the Mind*. MIT Press, Cambridge, MA 1995.
[5] J.A. Fodor. *Psychosemantics*, MIT Press, Cambridge, MA 1987.
[6] M. Gentilucci. Grasp observation influences speech production. *European Journal of Neuroscience*, 17: 179–184, 2003.

---

[9]See, for example, [6] or [1].

[7] J.J. Gibson *The Ecological Approach to Visual Perception*. Houghton Mifflin, Boston 1979.
[8] S. Hurley. Perception and action: alternative views. *Synthese*, 129: 3–40, 2001.
[9] P. Jacob and M. Jeannerod. *Ways of Seeing: The Scope and Limits of Visual Cognition*. Oxford University Press, Oxford 2003.
[10] M. Jeannerod. The representing brain. Neural correlates of motor intention and imagery in *Behavioral and Brain Sciences*, 17: 187–245, 1994.
[11] K. Koffka. *Principles of Gestalt Psychology*. Harcourt Brace, New York 1935.
[12] D. Marr. *Vision: A computational investigation into the human representation and processing of visual information*. Freeman, New York 1982.
[13] R. Millikan. Pushmi-pullyu representation. *Philosophical Perspectives 9: AI. Connectionism and Philosophical Psychology*, ed. J. Tomberlin, 1995.
[14] D.A. Milner and M.A. Goodale. *The Visual Brain in Action*. Oxford University Press, Oxford 1995.
[15] A. Noe. *Action in Perception*. The MIT Press, Cambridge, MA 2004.
[16] J.K. O'Regan and A. Noe. A Sensorimotor Account of Vision and Visual Consciousness. *Behavioral and Brain Science*, 24: 939-10, 2001.
[17] G. Rizzolatti and V. Gallese. From action to meaning. In J.-L. Petit (ed.), *Les Neurosciences et la Philosophie de l'Action*, J. Vrin, Paris 1997.
[18] G. Rizzolatti and C. Sinigaglia. *Mirrors in the Brain: How Our Minds Share Actions, Emotions*. Oxford University Press, Oxford 2008.
[19] E. Thompson. *Colour Vision*. Routledge Press, London 1995.
[20] E. Thompson, A. Palacios, and F. J. Varela. Ways of coloring: Comparative color vision as a case study for cognitive science. *Behavioral and Brain Sciences*, 15 (1): 1–74, 1992.
[21] L.G. Ungerleider and M. Mishkin. Two visual systems. In D.J. Ingle, M.A. Goodale and R.J.W. Mansfield (eds.), *Analysis of Visual Behavior*. MIT Press, Cambridge, MA 1982.

Mirror neurons and the "radical view" on simulation

DANIELA TAGLIAFICO

The recent discovery of the so-called "mirror neurons" and other mirroring phenomena (for a review see [6, 7, 9, 27]) has represented a fundamental step in the evolution of the debate about the nature of our folk psychology, i.e. the capacity to understand and predict the behaviour of other agents, as well as our own, by ascribing mental states such as beliefs, desires, emotions, intentions, and the like.

The identification of several different brain areas which are activated not only when we accomplish a certain movement or when we feel a certain emotion, but also when we observe another person accomplishing the very same movement or undergoing the same emotion, has been considered as an important proof of the fact that our understanding of other minds would be possible through their "mental simulation" [11].

In this paper I will take into consideration the interpretation that Robert Gordon [17] — the main proponent of the so-called "radical view" of simulation — has recently offered of these phenomena. Gordon understands mirroring phenomena as constituting a part of a specific capacity, the capacity to implicitly recognize other human beings as *intentional agents like myself* [23, 24], i.e. to interpret the behaviour of other people as if it were our own, under the same "intentional scheme" of reasons, purposes and object-directedness that we apply to ourselves.

As I will try to show, however, the interpretation proposed by Gordon for mirroring phenomena is affected by several problems and can be seen, in the end, as a sort of *ad hoc* solution, designed to solve a more general problem, which is characteristic of the kind of simulation proposed by Gordon — the so-called "ascent routine" — i.e. the problem of explaining how we can attribute mental states without appealing to our repertoire of mental concepts.

As a more general conclusion, I will claim that, if mirroring phenomena certainly provide evidence in favour of the existence of some processes of "mental imitation", this still does not mean that they constitute a clear proof in favour of the simulation theory. On the contrary, a convincing integration of this data within a simulationist paradigm like Gordon's one is still a long way off.

In what follows, I will first sketch the theoretical background of Gordon's simulation theory (section 1) — i.e. the hoary debate about the nature of our folk psychology — as well as the main interpretation that has been given about mirroring phenomena (section 2). Then I will take into consideration the specific interpretation that Gordon has proposed for them (section 3), and finally (section 4) I will try to show both the problems it gives rise to and why I consider this proposal as a kind of *ad hoc* solution, designed to solve a more general problem, which is characteristic of his kind of simulation theory.

1 The nature of folk psychology: theory-theory vs. simulation theory

As humans, we all share a capacity known as "folk" or "naive psychology", i.e. the capacity to understand and predict the behaviour of other agents, as well as our own, by ascribing mental states such as beliefs, desires, emotions, intentions, and the like. For example, if I say that "Hugo has taken the umbrella, because he *believes* that it will rain today", or if I say that "I will go to the cinema tonight, because I do not *want* to miss the last film by David Lynch", what I am doing is simply to explain or to predict a certain behaviour by ascribing to the subject some mental states, such as beliefs and desires or, in other terms, I establish some causal links between certain states of the world, certain mental states of the subject, and her actions into the outer world.

This capacity has long been explained as a kind of *theory* [22, 1, 12], i.e. as a body of knowledge concerning the psychological domain, on which I can operate with my inferential mechanisms, in order to produce some causal explanation of a certain behaviour. For example, I could possess in my mind a certain piece of information such as "If a subject *wants* that p, and he *knows* that, in order to obtain p, it is necessary to do x, she will do x"; relying on this knowledge, with the help of my inference mechanisms, I could reason about singular cases, coming to an explanation of a certain behaviour or to its prediction from an observed situation. For example, if I know that Giulia desires to eat a sweet and she knows that there is one in my bag, I can easily predict that she will rummage in my bag, in order to get the sweet.

Although the great individual differences between the singular theories, this is, in sum, the idea shared by all the so-called "theory-theorists", i.e. by all the people who believe that our naïve psychology has a theoretical nature.

Since the second half of the '80s, however, a new paradigm has emerged. In contrast with the theory-theorists, the so-called "simulation theorists" [15, 20, 12] have claimed that the theory-theory approach cannot satisfactorily account for the rapidity and spontaneity of our mental states attributions. In other words, if every time I wanted to understand or predict another's behaviour, I should appeal to some theory of the mind and make use of my inferential mechanisms, this task would take me a relatively long time and would not be so easily carried out. But this contrasts with the immediateness and the spontaneity we all experience in observing the other' behaviour: as it is often said, when we look at somebody, we have the impression to perceive her mental states — the beliefs and desires which underlie her action — in the same manner we perceive the colour of her eyes, her actions, or the expression of her face.

Simulation theorists of mindreading thus have claimed that our folk psychology should be understood, rather than as a theory, as an *heuristics*, i.e. as a practical ability to "put oneself in ones' shoes", to *simulate* the thoughts and feelings one would undergo if she were in a certain situation. A good example is given by Gordon himself. When I play chess, he says, and I have to predict my opponent's next move, what I do is simply to pretend to be him: I put myself, imaginatively, in his situation and I decide which move is the best to be accomplished in that situation; then, I simply have to ascribe my decision to him. A similar procedure can also be employed for an explanatory or "retrodictive" task: for example, if my opponent makes an unexpected move, I can infer that she is applying a different strategy and so, I have to imagine which antecedent states could have led me to adopt such a strategy — if, of course, I were in her situation — and if that strategy could have really led me to make such a move.

The typical simulative heuristics can be thus summed up in the following threesteps:[1]

1. First of all we have to assume — physically or, whenever this is not possible, in our own imagination — the perspective of a certain subject; so, we have to imagine the situation she is in and what she could think, desire, and feel in that situation;

2. Then we can feed our own cognitive mechanisms with these (pretend) mental states and let them run *off-line*, i.e. let them work on these pretend inputs *as if* they would work on our own mental states;

3. Finally, we can ascribe the output of this process — i.e. the result of this computation performed on pretend inputs — to the subject which is the target of our simulation process.

The general idea of simulation is thus that, since all human beings are provided with the same cognitive mechanisms, once I have assumed the perspective of a certain subject, I can make use of my own cognitive mechanisms to understand or anticipate what is going on in the other's mind. In this sense, the idea that we can put ourselves in the other's shoes is more than a metaphor: we can literally put ourselves — at the cognitive and, presumably, neural level — in the other's mind. Now, as hinted before, mirroring phenomena have been considered precisely as the first empirical evidence in favour of this thesis. More precisely, an increasing amount of data, coming mainly from the field of neurophysiology and neuropsychology, has shown the existence, in our brain, of a wide spectrum of mirroring phenomena, i.e. phenomena of "inner" imitation — imitation at the neural level — of the others' mental life, in consequence of the observation of their behaviour. To better understand what this means, I will consider in more detail some of the evidence which scientists refer to as "mirroring phenomena". My aim, of course, will not be to provide a complete review of mirroring phenomena — these could be hardly done, considering the huge amount of data that is already available — but only to give an idea, both of the kind of evidence which simulationists appeal

---

[1] It is important to stress that, on the simulationist account, this heuristics is not limited to third-person mindreading but can be equally applied to the case of first-person ascriptions, i.e. to the case we wanted to understand or anticipate our behaviour in a future or possible situation.

to, and of the general interpretation of these phenomena that is actually at the centre of the debate.

2 Mirroring phenomena

The first evidence for the existence of mirroring phenomena was obtained in the early '90s by a group of researchers at the University of Parma [25, 10]. During the study of the motor properties of neurons located in a ventral premotor area (area F5) of the macaque monkey's brain — an area associated with hand and mouth movements — they found a particular class of neurons — later called "mirror neurons" — which showed a very surprising characteristic: although these cells were located in a motor area, in fact, they seemed to possess not only motor, but also visual properties, i.e. they were activated not only when the monkey performed a certain type of action (e.g. grasping, placing, manipulating), but also when she was completely immobilized and could only observe the same action performed by an experimenter, as if she were mirroring — at a motor level — the action performed by another individual.

Subsequent experiments made clear that the "mirror" activation of these neurons was induced by the observation of actions, and not merely of movements without a goal, and ruled out, at the same time, the possibility that it could be caused by other factors, such as an expectation for food or a motor preparation to execute the same action. Moreover, it was showed the existence of a high degree of congruency between the neurons which were activated for a certain type of action — i.e. a grasping movement — and those which were activated by the observation of the very same action. For example, several neurons which were activated for an action of grasping, and more precisely a precision grip, also discharged when the monkey saw a precision grip performed by the experimenter [11].

The existence of analogous mirroring phenomena in humans has been proved by several studies, using different techniques, such as the registration of motor potentials evoked by the observation of the others' actions [6], the registration of electroencephalographic and magneto-electroencephalographic activity [3, 9], and the positron emission tomography (briefly PET: [26, 18]).

Moreover, mirroring phenomena have been identified not only for actions but also for sensory and emotional states like pain, disgust, fear and anger. For example, an fMRI (functional Magnetic Resonance Imaging) study (cf. [28]) has shown that the same neural structures — in particular, the anterior insula — that are activated when we experience disgust, are also activated when we observe the same emotion facially expressed by another individual. The proof is — as shown by another important study [2] — that when a subject is selectively impaired in feeling disgust, she is also selectively impaired in recognizing expressions of disgust in other individuals.

Analogously, it has been shown that the fact of being aware that another individual is receiving a painful stimulus induces in the observer a response which is analogous to the one she has when she experiences the same painful stimulus (for a review cf. [4]).

At a functional level, the evidence that all (or, at least, most of) the times we perceive another human being performing a certain action or expressing a certain emotion, we respond by mirroring what is happening in her brain, has led the neuroscientists to formulate the hypothesis that these phenomena would be part of an *inner representation* of the other's behaviour (cf. [21]). In other words, our neural activation would be an important component of the representation either of the motor plan that has guided a certain action or of the feeling that underlies a certain expressive behaviour, and this representation would be necessary in order to *understand* the meaning of that behaviour [10]. In the case of motor neurons, for example, since they respond only to goal-directed actions, a possible interpretation would be that they are able to detect the intention that underlies a certain action (which type of intention is still a matter of debate); and analogously, in the case of sensations or emotions, mirroring would be necessary in order to recognize them and to have an "engaged representation" of the other's behaviour [28].

This makes clear, of course, why simulation theorists have interpreted these phenomena as an experimental evidence in favour of their position: what these phenomena seem to demonstrate, in fact, is that we understand others' behaviour because we are able to reproduce in ourselves, at the neural level, what is happening in another's mind, thus performing some kind of *embodied simulation* [6, 7].

3 Gordon's interpretation of mirroring phenomena

In a recent paper, also Robert Gordon [17], one of the main proponents of the simulation theory of mindreading, has offered his own interpretation of mirroring phenomena. Gordon does not think that mirroring phenomena could underpin a full-blown folk psychology, but rather believes that they could be the basis of a protopsychology; more precisely, he says, they could constitute the neural basis of a capacity, possessed by all humans, to implicitly recognize the others as *intentional agents like myself* [17, p. 95].

The existence of this capacity, Gordon observes, has been recognized or somehow implied in the work of several scientists, such as Meltzoff, Tomasello, and Gallese; nevertheless, it still has not been adequately understood. In particular, Gordon wants to challenge the explanation given by Meltzoff [23], who interprets this capacity in terms of analogical inference, by proposing an alternative account, which appeals, instead, to a direct experience of what is happening in the other's mind.

Although mirroring phenomena can be generally characterized as "responses brought about by b"s perception of a, in which b comes to have property p because a has property p", Gordon says [17, p. 95], there is a distinction to be drawn between two different kinds of *mirroring*: an imitative mirroring — about which Meltzoff is mainly concerned — and a *constitutive* mirroring — which is instead at stake in Gallese's works. While the former is a neural matching which arises in consequence of my imitation, i.e. in consequence of my attempt to match the other's behaviour, the latter is already in place when I perceive her movements — at least those movements for which I possess the corresponding mirror neurons — and is part of the *representation* of her behaviour [17, p. 96]. In other words, while imitative mirroring supposes that the observer already has some representation of the other's movement, constitutive mirroring is part of this representation and, more precisely, it is a representation of the other's behaviour at the motor and visceral level, i.e. it is an internal replica of the motor plans and visceral responses that lie behind her behaviour [17, p. 96]).

Motor plans and visceral responses arising from mirroring, however, are not *endogenous*, i.e. they are not produced by my own cognitive mechanisms — such as the decision-making system or the emotion-formation system — but rather, they are *exogenously* induced by the observation of another's behaviour. Nevertheless, as Gordon says, I do not understand them as they would be my own mental states, but I map them onto another human or humanlike body [17, p. 96]. For example, I see your expression of disgust and I project this emotion on your face, thus understanding that you — and not me — are disgusted. But how can this be accomplished?

What Gordon wants to contend with Meltzoff is precisely the idea that such a process of recognition could be accomplished in virtue of an analogical inference from my experience to yours. According to Meltzoff's account, in fact, the infant would be able to make — although implicitly — some reasoning from analogy, from the observed behaviour to her own behaviour and its underlying experiences, such as: "When I produce behavior of type x, I feel a certain way f; therefore, when a similar body does x, the behavior was probably produced by another subject — another "I" — that feels the same way f" [17, p. 98].

As Gordon points out, Meltzoff's explanation thus requires that we ascribe to the baby at least two capacities [17, p. 98]:

> i. being able to identify one's own behavior in a way that allows comparison with the observed behaviour of another body and ii. being able to identify one's own feeling or experience *as such* (i.e., interpret it as something that is going "within me", in the appropriate sense; that is, subjectively, as opposed to "out there in the world" or in someone else).

None of the two, however, seems to be reasonably ascribable to a newborn, or even to an infant of a few months. The first ability, in fact, would require that the newborn had a visual representation of her own behaviour (e.g. of her own facial expression when she feels in a certain way) — but this seems to be quite implausible, since newborns have no access to their visual aspect.

The second, moreover, would imply that the newborn was perfectly able to distinguish her own feelings and sensations — feelings and sensations which have arisen spontaneously (i.e. endogenously) — from those which were elicited by the observation of the other individual, but also this capacity of differentiation, Gordon argues, seems to be too far from the actual possibilities of a newborn. Furthermore, as Gordon rightly remarks [17, p. 99], if such a capacity of distinction between what happens in me and what is happening in another subject should be ascribed to a newborn, there would be no more reason to understand a phenomenon such as emotional contagion — i.e. the baby's crying in response to another baby's crying — as an indistinct affect sharing, without any awareness of the other's distress as the source of my own distress (cf. [4]). In other words, if newborns were able to appeal to an argument from analogy, why still postulate the existence of an undifferentiated imitative behaviour such as emotional contagion?

A reasoning from analogy, Gordon remarks, seems to be more appropriate for an imitative mirroring than for a constitutive one: if I imitate a certain behaviour, I can become aware of certain feelings, intentions and desires one could have had while accomplishing that behaviour. For example, if I imitate a subject who is manipulating a box, I can have, at a certain moment, the intention to open the box, and then I can ascribe such an intention to the subject who is manipulating the same box, reasoning that something like what I feel at this moment may have inspired also his action [17, p. 99]. But in the case of constitutive mirroring, a reasoning by analogy seems to be highly implausible.

One should search for an alternative explanation to the argument from analogy, in Gordon's view [17, p. 100], in the kind of embodied simulation suggested by Gallese [6, 7, 8]. More precisely, mirroring mechanisms should be understood, in his opinion, as mechanisms which force us to interpret others' behaviour as if it were our own, meaning, "under the same scheme that makes my own behavior, along with the intentions, motor plans, and visceral feelings that underlie it, intelligible to me; namely, the intentional scheme of reasons, purposes, and object-directedness that we apply to our own behaviour" [17, p. 100].

Gordon first tries to explain what this could mean for endogenous states like, for example, the emotions which arise in me when I gaze at the Grand Canyon. While watching the Grand Canyon, Gordon says, a great part of its emotional coloration comes from the feelings it elicits in me. This happens, on Gordon's account, because the brain is able to pick up these feelings and to ascribe them to the object that elicited them, probably, he says, by *consulting* the cognitive system — in this case, the emotion-formation system — that produced them in the first place [17, p. 100].

The same happens in the case of endogenous intentions and motor plans. When dealing with these states, the brain applies a strategy to make sense of them, by fitting them into a scheme of reasons and goals: the result, at a personal level, is an intentional explanation of our behaviour [17, p. 100]. For example, if it is raining and I am running, I can explain my behaviour by saying that: "I am running because it is raining, and doing so in order to avoid getting drenched": although this explanation does not employ any mental concept, it is — at least implicitly — an understanding of my action in intentional terms, in the sense that it necessarily implies the (auto-)ascription of some mental states such as a belief (the belief that it is raining) and a desire (the desire to avoid getting drenched) [17, p. 105]. At the subpersonal level this understanding would be realized, again, thanks to a *consulting mechanism*, which would be capable of querying that cognitive mechanism — in this case, most probably, the decision-making system — which was responsible for the decision to run in the first place.

Gordon also specifies that this alleged consulting mechanism should not be conceived as some kind of introspective mechanism (something like the mechanism proposed by Goldman [13]), but rather as "a hypothetical neural capacity to do a 'trace' of the pathways and processes that led to a particular outcome" or, in alternative, as "a hypothetical capacity of decision-making and emotion-formation systems to conduct 'what if' experiments on themselves" [17, pp. 100-101]. This second hypothesis, in particular, appeals to the central idea of simulation theory, i.e. the idea that we can use our own cognitive mechanisms off-line, by providing them with some pretend states, in order to simulate our thoughts, desires, and feelings in certain possible situations. What is new, from this point of view, is only that, in addiction to this off-line simulation, there should be something like a consulting mechanism, which would be able to query our cognitive systems, i.e. which could keep track of the entire computational process accomplished by a certain cognitive mechanism, providing us, as its result, with an implicitly intentional understanding of our behaviour.

The process we have described so far also applies, on Gordon's account, to the case of exogenous states, i.e. states arisen from mirroring phenomena: "The brain treats the exogenous replicas of another's motor plans and visceral responses in the same way it treats their like-coded endogenous counterparts. It seeks to make them unsurprising, to make sense of them, by fitting them to the "intentional" scheme of reasons, purposes, and object-directedness" under which it understands our endogenous states [17, p. 101].

In the case of the exogenous states, however, the problem is that they arise in an "unmotivated" way, since they are not the outputs of our own cognitive mechanisms: for example, a mirrored motor plan does not result from a process of decision-making, but arises, instead, whenever I observe somebody performing a certain action, and so it is not immediately understandable for the brain. To deal with these cases, the brain must create, first of all, an endogenous replica of the exogenous state: this latter, arising from a "typical" causal chain, i.e. having been caused by some of our cognitive mechanisms, can be understood by the consulting mechanism and consequently interpreted under the same intentional scheme that applies to our endogenous states [17, p. 101]. Gordon proposes the following example [17, p. 102]. I see my colleague reaching out and picking up the phone while it's ringing. What she is doing is clearly answering the phone. Now, let us suppose that the action observed activates my premotor cortex, giving rise to a mirroring phenomenon: my visual perception thus produces the activation of a certain motor plan to reach out and pick up something like a phone. Contrary to the motor plans which are produced in an endogenous way, however, this one is totally unmotivated — it has no reason or purpose — and so it is not understandable to me. Still, if my brain can produce an endogenous replica of this state, the consulting mechanism can query the system that produced the replica (the decision-making system), thus providing me with an explanation of this state. For example, if I hear the phone ringing, it is probable that I feel an impulse to answer it and — although I am able to inhibit my behaviour — it is probable that I activate, in an endogenous way, a motor plan which matches the one that has arisen in an exogenous way. At this point, since the consulting mechanism can query my own decision-making system, I can get an intentional explanation: for example, it is obvious that I would not have been inclined to pick up the phone if it were not ringing, and that my purpose was to answer the phone (and not, e.g., to start a new conversation). So, in this way, I can have a ready answer if somebody asks me why my colleague reached for and picked up the phone.

Something analogous can happen in the case of emotions [17, pp. 102–103]. I can see a face showing disgust and this perception can elicit a certain typical visceral response in me. This response, however, is an exogenous one and cannot be directly understood. So, first I map this response onto the other's face, thereby isolating it from my own visceral responses. Then, my brain can start looking out, into the world, in search of something that could produce the same kind of inner response. For example, I will look around until I find something which disgusts me, i.e. something which produces a visceral response which matches with the exogenous one. Since in this case the consulting mechanism has access to my emotion-formation system, I can understand this state, giving reasons for it, and then I can attribute this intentional interpretation to other people.

In conclusion, Gordon's proposal is to understand mirroring phenomena as a sort of "switch mechanism", one by which a copy of a certain mental state — be it a motor plan or some kind of visceral response — is reproduced in ourselves in consequence of the observation of someone else's behaviour. Since this exogenous replica arises in a surprising way, however, it cannot be immediately understood by our brain. This compels us to try to create an endogenous copy of it, whose production could be analysed by our consulting mechanism. This latter, by querying the systems which have generated the endogenous replica, can finally understand this state, thus providing us with an implicit intentional interpretation of the other's behaviour.

4 Gordon's interpretation: an *ad hoc* solution

In this last section I would like to clarify the reasons why I consider Gordon's proposal an implausible one. There are, at least, four points which are particularly problematic, I think, in his account.

The first problem is surely that Gordon's theory seems to be a very expensive solution — in terms of cognitive costs — since it states that, in order to understand an exogenous state, we must first produce an endogenous replica of this state, to which we can then apply our consulting mechanism. Moreover, this replica is not something like a direct by-product of the exogenous state, but it is probably the result, as he specifies, of the capacity possessed by our cognitive mechanisms to conduct "what if" experiments on themselves [17, p. 101].

Now, if, on the one side, we can concede that the brain does not always work in an economical way, on the other side, if things really go as Gordon tells us they do, then it is very diffcult to explain the immediateness and automaticity which often characterize our understanding of other minds. In other words, if, in order to understand some exogenous state, we should first produce an endogenous replica of it, and moreover, if, in doing so, we should proceed in a temptative way, our understanding of other minds would be arguably slower and perhaps harder than it generally is.

Secondly, the way in which the brain could produce an endogenous replica of the exogenous state is — at least in some cases — not clear. For example, if in a certain visual scene there is nothing that induces a sensation of disgust in me, how can I individuate the target of the other's emotion? Gordon suggests that in this case I have to assume the other's perspective — so I have to imagine what could disgust me if I were the other person — but this is not, again, a simple task, one that we can accomplish nearly instantly. Nevertheless, also in those cases in which we do not understand why somebody is disgusted, we are perfectly able to recognize her disgust and to have some intentional understanding of her state.

Thirdly, also the process of mapping envisaged by Gordon seems to be quite mysterious. Gordon says that, as soon as the exogenous state arises in us, we map it onto an external object, in order to avoid confusing it with our endogenous states [17, p. 103]. But how can we map the exogenous state onto the external world if we still have not understood it, i.e. if the consulting mechanism still has not recognized e.g. the perception of a certain object as the plausible cause of a certain inner state?

Finally, although Gordon himself recognizes that his hypothesis is a speculative one [17, p. 101], there is no empirical evidence that something like a consulting mechanism really does exist. Moreover, it is not clear at all how this mechanism should be described on a computational level: which are, exactly, its characteristic inputs and outputs? Should we conceive it as a mechanism that takes in input the operations executed by a certain mechanism (e.g. the decision-making system) and gives in output some kind of intentional representation?

At this point, one might wonder why Gordon has offered such an audacious and also complicated interpretation of mirroring phenomena, rather than keeping with more prudent positions. My hypothesis is that, by this interpretation, in fact, Gordon has tried to solve a central problem of his "radical" theory of simulation, i.e. the problem of the identification of the type of mental states that have been simulated.

As I have already pointed out, Gordon's model is radical in the sense that it excludes from the process of simulation — at any stage of this process, from the moment of perspective taking to the attribution of the simulated mental state — the use of mental concepts like "belief", "desire", "intention", and the like. In Gordon's view, in fact, any appeal to these concepts would necessarily imply to leave a door open for the intervention of a theory into the process of simulation — and this is precisely what Gordon rejects: for him the exercise of a folk psychology is simply a matter of simulating, not of possessing a theory. How mindreading could be accomplished avoiding any appeal to our mental concepts repertoire is explained by Gordon with his famous *ascent routine* [15, 16].

Let us first consider the case of belief attribution. In Gordon's view, when we have to answer a question about our own beliefs, for example when we have to answer the question: "Do you believe that Italians like pizza?", we do not interrogate ourselves about our actual beliefs, but rather go down one level and answer a simpler question about a certain state of affairs as: "Do Italians like pizza?". Once we have given an answer to this question, all we have to do is put it into a more complicated syntactical structure of the type: "Yes(No), I do(not) believe that __ (Italians like pizza) __". So, in Gordon's opinion, the autoascription of beliefs does not necessarily require that we possess and apply some mental concept like "belief", but only that we can deal with certain syntactical structures.

Things happen in exactly the same way for third-person attributions. For example, when we have to answer a question about the mental state of another person, such as: "Does Mary believe that tomorrow will be a sunny day?", what we really do, in fact, is not to reason about her mental state — about her beliefs — but, again, about a state of affairs. The only difference is that, in this case, we have to change our perspective, i.e. we have to assume Mary's perspective, to pretend to be in her situation. Then, as in the case of the first-person attribution, we first go down one level and ask ourselves: "Will tomorrow be a sunny day?" and then, once we have found an answer, we fit it into a syntactical scheme as: "Yes(No), Mary does(not) believe that __".

The problem with this strategy, however, is that it works well only in the case of beliefs, but not for other types of mental states. For example, if I have to answer the question: "Does Mary want tomorrow to be a sunny day?", answering the lower-level question "Will tomorrow be a sunny day?" will not be of any help to give an answer to the first question. And this is also true for other states such as emotions, intentions, motor plans, and so on. In other words, in cases different from beliefs, it seems absolutely necessary for us to reason about the subject's mental states — and consequently, to make use of some mental concepts — if we want to answer the question.

Now, it seems to me that the consulting mechanism proposed by Gordon [17] offers a solution precisely to this problem, since, as we have seen, it is a mechanism which is sensitive to the computations executed by our cognitive systems, and so it must somehow keep track of the inputs from which a certain state originated or the outputs it gave rise to. In a word, it must keep track of the cognitive causal chain in which a certain state was embedded. But how should this causal chain be interpreted, if not as the functional role of a certain state, which is defined, precisely, by the causal relations it entertains with other cognitive states and mechanisms? So, in this sense, from my point of view, the consulting mechanism should be seen as a mechanism which identifies — although in an implicit way — the type of mental state that has been simulated and this is why, I think, it is right to say that the overall interpretation of mirroring phenomena given by Gordon is, in the end, an *ad hoc* solution, one by which he tries to give a solution to the old problem which affected his radical theory of simulation.

In conclusion, although the existence of mirroring phenomena is certainly a proof in favour of the existence of some kind of "mental imitation" of the others' minds, I think that a convincing explanation of the meaning of these phenomena is still not available. In particular, as I have tried to show, their integration within the simulation theory — at least, within the radical model proposed by Gordon — is still very problematic.

Acknowledgments

I wish to thank professors Diego Marconi and Cristina Meini for their useful comments on an earlier version of this paper.

BIBLIOGRAPHY

[1] S. Baron-Cohen. *Mindblindness: An Essay on Autism and Theory of Mind*. The MIT Press, Cambridge, MA 1995.

[2] A.J. Calder, J. Keane, J. Cole, R. Campbell and A.W. Young. Facial expression recognition by people with mobius syndrome. *Cognitive Neuropsychology*, 17 (1-3): 73–87, 2000.

[3] S. Cochin, C. Barthelemy, B. Lejeune, S. Roux and J. Martineau. Perception of motion and qeeg activity in human adults. *Electroencephalography and Clinical Neurophysiology*, 107: 87–295, 1998.

[4] F. de Vignemont and T. Singer. The empathic brain: how, when and why. *Trends in Cognitive Sciences*, 10: 435–441, 2006.

[5] L. Fadiga, L. Fogassi, G. Pavesi and G. Rizzolatti. Motor facilitation during action observation: a magnetic stimulation study. *Journal of Neurophysiology*, 73: 2608–2611, 1995.

[6] V. Gallese. The "shared manifold" hypothesis: from mirror neurons to empathy. *Journal of Consciousness Studies*, 8 (5-7): 33–50, 2001.

[7] V. Gallese. The manifold nature of interpersonal relations: the quest for a common mechanism. *Philosophical Transactions of the Royal Society*, 358: 517–528, 2003.

[8] V. Gallese. Being like me: self-other identity, mirror neurons, and empathy. In S. Hurley and N. Chater (eds.), *Perspectives on Imitation. From Neuroscience to Social Science*, volume 1, the MIT Press, Cambridge, MA and London 2005, pages 101–118.

[9] V. Gallese. Before and below "theory of mind": embodied simulation and the neural correlates of social cognition. *Philosophical Transactions of the Royal Society*, 362 (1-3): 659–669, 2007.

[10] V. Gallese, L. Fadiga, L. Fogassi and G. Rizzolatti. Action recognition in the premotor cortex. *Brain*, 119: 593–609, 1996.

[11] V. Gallese and A.I. Goldman. Mirror neurons and the simulation theory of mind-reading. *Trends in Cognitive Sciences*, 12: 493–501, 1998.

[12] A.I. Goldman. Interpretation psychologized. *Mind and Language*, 1989; reprinted in M. Davies and T. Stone (eds.), *Folk Psychology: the Theory of Mind Debate*, Blackwell, Oxford 1995, pages 74–99.

[13] A.I. Goldman. The psychology of folk psychology. *Behavioral and Brain Sciences*, 16: 15–28, 1993.

[14] A. Gopnik and A.N. Meltzoff. *Words, thoughts, and theories*. The MIT Press, Cambridge, MA 1997.

[15] R. Gordon. Folk psychology as simulation. *Mind and Language*, 1: 158–171, 1986; reprinted in M. Davies and T. Stone (eds.), *Folk Psychology: the Theory of Mind Debate*, Blackwell, Oxford 1995, pages 60–73.

[16] R. Gordon. Simulation without introspection or inference from me to you. In M. Davies and T. Stone (eds.), *Mental Simulation*, Blackwell, Oxford 1995, pages 53–67.

[17] R. Gordon. Intentional agents like myself. In S. Hurley and N. Chater (eds.), *Perspectives on Imitation. From Neuroscience to Social Science*, the MIT Press, Cambridge, MA and London 2005, volume 2, pages 95–106.

[18] S.T. Grafton, M.A. Arbib, L. Fadiga and G. Rizzolatti. Localization of grasp representations in humans by pet: 2. observation compared with imagination. *Experimental Brain Research*, 112: 103–111, 1996.

[19] R. Hari, N. Forss, S. Avikainen, S. Kirveskari, S. Salenius and G. Rizzolatti. Activation of human primary motor cortex during action observation: a neuromagnetic study. *Proceedings of the National Academy of Sciences of USA*, 95: 15061–15065, 1998.

[20] J. Heal. Replication and functionalism. In J. Butterfield (eds.), *Language, Mind, and Logic*, Cambridge University Press, Cambridge 1986; reprinted in M. Davies and T. Stone (eds.), *Folk Psychology: the Theory of Mind Debate*, Blackwell, Oxford 1995, pages 45–59.

[21] M. Jeannerod. The representing brain: neural correlates of motor intention and imagery. *Behavioural and Brain Science*, 17: 187–245, 1994.

[22] A.M. Leslie. Pretense and representation: The origins of "theory of mind". *Psychological Review*, 94 (4): 412–426, 1987.

[23] A.N. Meltzoff. Imitation and other minds: the "like me" hypothesis. In S. Hurley and N. Chater (eds.), *Perspectives on Imitation. From Neuroscience to Social Science*, the MIT Press, Cambridge, MA and London 2005, volume 2, pages 55–77.

[24] A.N. Meltzoff. The "like me" framework for recognizing and becoming an intentional agent. *Acta Psychologica*, 124: 26–43, 2007.

[25] G. Rizzolatti, L. Fadiga, V. Gallese and L. Fogassi. Premotor cortex and the recognition of motor actions. *Cognitive Brain Research*, 3: 131–141, 1996.
[26] G. Rizzolatti, L. Fadiga, M. Matelli, V. Bettinardi, E. Paulesu, D. Perani and G. Fazio. Localization of grasp representations in humans by pet: 1. Observation versus execution. *Experimental Brain Research*, 111: 246–252, 1996.
[27] G. Rizzolatti, L. Fogassi and V. Gallese. Mirrors in the mind. *Scientific American*, 295 (5): 54–61, 2006.
[28] B. Wicker, C. Keysers, J. Plailly, J.-P. Royet, V. Gallese and G. Rizzolatti. Both of us disgusted in my insula: the common neural basis of seeing and feeling disgust. *Neuron*, 40: 655–664, 2003.

Multiple realizations of the mental states: hunting for plausible chimeras

VINCENZO G. FIORE

1 The framework of multiple realizability

The key elements characterising the functionalist approach to mind studies are commonly identified (e.g. see [5]) with claims concerning:

1. The cognitive creatures' essential feature (they are all computational systems);

2. The object of the research in the fields of cognitive psychology and artificial intelligence (abstract functional states and novel physical realizations for these states respectively);

3. The irreducibility and consequently the autonomy of special sciences;

4. The inefficacy of the empirical research on the neural structure, because of the merely contingent relation established between the neural structure and the functional states it realizes.

The objective of this paper is to support a reductionist perspective in mind studies, disputing the soundness of the claims 3 and 4 in particular. Therefore, since it is easy to concede that the theory of multiple realizability of mental states plays the role of the hub, binding all the four claims one another, this paper aims at showing the weaknesses of the grounds on which the theory has been built.

The Multiple Realizability Theory (MRT) has been first formalized in the late sixties by Hilary Putnam in a famous series of papers (for a collection see [12]). In the article commonly recognised as the most representative of that period [11], it is assumed that every animal, independently of the species it belongs to, is capable of feeling pain: the mental state of pain is not species-specific. Therefore the identification of the mental state with a certain C-Fiber activation (or any other neural correlate) leads to the conclusion that all species should be found sharing the same neural structure and the same neural activation at the right moment. Even if we consider that parallel evolution might lead to the same physical structure, once the argument is extended to other psychological predicates (such as, for instance, hunger or sexual attraction), it becomes *overwhelmingly plausible* (Putnam's words) that these multiple realizations across species simply cannot be explained in terms of a theory grounded on the identity between mental and physical states. After all, even if parallel evolution could be proved in all known creatures, the conceivability of artificial silicon based systems capable of feeling pain, would definitively discard any attempt to establish an identity. Putnam's famous proposal is then to conceive a different approach to the mind, grounding it on the concept of a virtual machine analogous to the Turing Machine, but characterised by a few strategic differences.

It is useful to remind briefly what these devices are: a Turing machine (TM) is a computational — seria l — device that is instructed by a program (set of instructions) to process a symbolic input in order to give a symbolic output as a result. These processes may have the following schematic representation:

$$\{x_1, x_2, x_3, ...x_n\} \rightarrow A \rightarrow B \rightarrow C \rightarrow D \rightarrow ... \rightarrow [\textit{final status}]$$

The input assigns a value to each of the n variables $\{x_1, x_2, x_3, ...x_n\}$, then the virtual machine computes these values as it is described by its set of instructions, reaching its first state (A). The new state gives life to a new series of processes that allows the machine to change again state in favour of the second one (B): the operation is replicated until the virtual machine reaches the final state described by the instructions in relation to the values assigned to the variables.

This mechanism implies that a TM is characterised by an assignment of probabilities 1 or 0 to every transition. On the contrary, if the instructions allow the machine to change its status from the original one to a series of target ones, with probabilities assigned to each of them, (e.g. starting from the functional state A the machine may change in favour of B with 30% of chances or C with 70%) then the machine is called Probabilistic Automaton. Finally, there are devices capable of processing sets of inputs in order to generate new sets of instructions: this ability allows simulating any possible TM generating a so-called Universal Turing Machine (UTM). In other words, the UTM is directly programmed by the input, which instructs the machine about the processes to apply thenceforth. The MRT assumes that the combination between a probabilistic automaton and a UTM gives in return a virtual device whose processes are consistent with the living beings' ones.

All these devices (TM, UTM and probabilistic automaton) are known as virtual machines because of their nature which makes them completely independent of any specific physical structure: it doesn't matter if the computation required by the set of instructions is performed by a neural system, a CPU or a series of cogs wheels. The focus is on the functional organization realized by the device (i.e. the instructions concerning its state transitions) and the functional state it can consequently reach, once the device has received a specific symbolic input. Furthermore, since the states are also independent, it is not even necessary for two systems to be functionally isomorphic (i.e. it is not necessary that they realize the same set of instructions) to reach the same state: different programs may lead to the same functional state.

In conclusion, the MRT entails that two generic neural structures A and B may realize a mental state M, but they can never be identified with the mental state itself: the relation between the physical system and its mental realizations is always contingent and there can be infinite physically different systems realizing the same mental state. The focus changes from the reductionist study of the neural correlate to the functionalist study of the realized functions.[1]

Putnam's early argument has been originally applied to different neural structures belonging to different species, but few years later Jerry Fodor [7, 8] generalised the value of the MRT, presenting his assumption as the necessary consequence of Putnam's conclusions. The generalised version of the MRT has started appealing to the 70s studies on brain mapping and to the notions of neural degeneracy and plasticity: the key argument coming from these studies is that the nervous system of higher organisms is able to accomplish a single psychological task in a wide variety of ways by means of several neurological parts of the whole structure. As a consequence, the relation between physical and mental states proves to be contingent even when it is applied to the same species or a single neural system:[2] time becomes a legitimate variable to take into account when considering the contingency of the causal relation between the physical system (the implementer) and the functional state (the implemented).

2 The computability issue and the overestimation of the UTM

The superimposition of the processes performed by a virtual machine on the ones realized by cognitive organisms has been attractive since the very beginning: even those who have tried to discard the functionalist approach have rarely questioned the argument of the multiple realizations of mental states and have preferred to focus their attention on the implications the theory has on reductionism [5, 9, 10, 4]. A few exceptions are represented by those [17, 15, 1] who have challenged the likelihood of the argument by means of theoretical reasoning or stressing the failures of the predictions implied the generalised MRT. Nonetheless, I think a computational approach to this matter has been surprisingly ignored: the theory relies on the identification of the mind with the TM; should this identification be computationally inadequate, the MRT would be proved ill-grounded. As a matter of fact, there are three reasons that lead to this conclusion.

The first reason is the limited range of Turing-computable algorithms. To put it simple, the computational capacities of a TM are widely overestimated and they are usually erroneously attributed to Turing himself. There is a huge list of philosophical misconceptions about Turing's virtual machine [6] and they are all grounded on the erroneous assumption that in his articles Turing may have mathematically demonstrated how a UTM can compute any algorithm (i.e. the mathematical function that formally describes the set of instructions or program of the virtual machine) performed by any other machine with any architecture, given enough time and memory.

---

[1]Subsequent articles (e.g. see [2, 12]) have also dealt with the problem of the realization of more than a single functional state (or psychological predicate) at the same time. The solution proposed assumes complex living beings are able of realizing the processes of several virtual machines at the same time (i.e. in parallel).

[2]E.g. a single human being realizes the same mental state of pain during childhood and adulthood, despite the differences characterising the same neural structure in the two periods.

What Turing did demonstrate is that a UTM can realize any algorithm characterised by the following requirements (which define the "mechanical method"):

1. finite number of exact instructions (each instruction expressed with a finite number of symbols) to make the machine change from one functional state to the following one.

2. Finite number of state transitions to produce the expected result.

3. In principle, a human being can carry it out only aided by paper and pencil.

4. It does not require insight or ingenuity to be carried out.[3]

For the purpose of this article, it is sufficient to point out that the set of hypothetic algorithms realized by any TM is countable, that is to say, it is characterized by the same order of infinite of the integers. On the contrary, the number of all the hypothetic computable algorithms is uncountable (i.e. of a higher order of infinite): hence, there is an infinite number of algorithms which have a mathematical description and cannot be realized by a UTM, even if they are realized by differently structured systems.

If the algorithms implemented by neural systems are not found to meet at least one of the four requirements for Turing-computability, it must be concluded that a UTM may not simulate or even describe information processes in living beings. Consequently, it is necessary to study the way biological neural systems process their data, before formulating any hypothesis about the possibility to realize such processes by means of a virtual machine. Under these circumstances, the hypothesis of multiple realizations of mental processes may be empirically falsified: MRT cannot be established a priori.

It may be argued that even if we could find out that neural systems do not realize Turing-computable algorithms, this finding by itself would not be enough to discard multiple realizability. A new hypothetical and more powerful virtual machine might be conceived: different from the known Turing machines, it might widen the range of realizable algorithms, overcoming some of, if not all, the weak points of the classic machines.

Nonetheless, it seems that such a powerful virtual machine is unlikely to come and it is usually considered mathematically implausible.[4] Even if it were plausible, this objection would not lead far from the prospected path: these new hypothetic systems would not be asked to simulate a generic new set of

---

[3] These notions have a formal and rigourous equivalent[16, 3]: for the purpose of this paper it is sufficient to refer to their informal version.

[4] The existence and the features of devices that may result to be able to implement such Turing-incomputable algorithms have been debated at least for five decades. An essential bibliography and a brief account of this debate can be found in section two of the cited Copeland's article [6]. As a matter of fact, the probabilistic automaton already represents a virtual machine which is able to realize a wider set of algorithms, if compared to a TM. I mainly refer to the TM for the convenience of the reasoning, but the criticism is valid for the probabilistic automaton as well: the set of algorithms realized is still countable and the algorithms themselves are characterized by similar features.

algorithms but those specific of the parallel distributed — neural — systems. Once again, in order to be sure that the proper set of algorithms is part of the domain of these new machines (proving the soundness of MRT), it would be necessary to know beforehand what sort of algorithms are implemented by neural systems.

This conclusion leads to the second reasoning against the plausibility of the MRT. There is a particular causal relation between the physical structure of a neural system and the algorithm it implements: a neural network realizes a sheaf of sets of mathematical functions[5] defined by its architecture and by the computation performed by each single node of the network. The values assigned to the other variables, such as the weights of the synapses (i.e. the electrochemical conductibility of the synapses), fix the constants for any specific set of algorithms within this sheaf. Every modification in the architecture of the network or in the processes of the single nodes leads to a system that can or cannot solve a specific given task.[6]

If we use simple connectionist models, the sheaf of algorithms implemented can be mathematically described with ease: in these conditions, the analysis of the relation between the neural structure and the implemented algorithm makes us conclude that the former has a causal influence on the latter. Nonetheless, even if the systems show a higher order of complexity (such as those proper of biological networks), it is possible to have an idea of the sheaf of algorithms determined by the architecture, especially considering that, though extremely complex, single neurons compute their electrochemical signals in a way that can be described by adequate mathematical functions. In a few words, different neural systems realize different algorithms, require different amount of energy and time to perform the same task and — due to differences in vector conversion — differ in the way the information is encoded or stored, in the categories developed and in their resistance to physical damages. Thus, mathematical analysis of neural systems is telling us a different story from the one told by the MRT: in order to be able to process information — precisely — in the same way, two neural systems must be physically identical (i.e. two biological neural systems can hardly ever be functionally isomorphic due to the known structural differences across species and within the same one).

It is still possible to claim that whether or not two neural systems may perfectly match their processes implementing the same algorithm, this would not affect the hypothesis that a serial device may be conceived realizing neural processes. Once a probabilistic automaton were shown simulating the information processes of a neural system, the possibility to separate single states in the virtual machine would make it irrelevant for the MRT the whole second reasoning. Yet, the problem with this criticism is that it does not consider

---

[5]E.g. the equation $(ax+by=k)$ describes a sheaf of straight lines. If we fix the constants (in this case: a, b, k) attributing them a value, the result is the equation of a single straight line (e.g. $2x+3y=1$). A set of straight lines describes the equations combined in single or multiple systems.

[6]The logical operator XOR is often cited in literature: it is known that there is no way to realize this computation with a single layer neural network (e.g. see [14, chap. 19, sect. 3]).

both the arguments so far described at the same time:

A. Whether or not a virtual machine may realize the set of instructions implemented by a neural system can only be established a posteriori.

B. The physical structure in neural systems is directly responsible for the processes implemented.

The two premises A and B lead inevitably to:

C. In order to support an anti-reductionist path (MRT), it is necessary to use a reductionist strategy, seeking the knowledge concerning the processes realized by a neural system.

When everything is taken into consideration, the proof in favour of the multiple realizability of the mental states would be reached after it had become irrelevant.

The third reason against the plausibility of the MRT is grounded on the computational inadequacy of serial systems in simulating the unique features of biological neural systems. Biological systems deal with continuous and infinite inputs, processes and outputs, processing information in a flow; on the contrary, a virtual machine necessarily works with discrete and finite data and state transitions, following a step-by-step procedure. External data can reprogram a UTM to make it change its processes (once the input has changed the set of instructions, the device can also apply its rules to previously incomputable data), but the neural systems are able to change their processes both depending on and independently of the input. For instance, biological systems based on neural structures require a specific amount of energies in order to activate their systems: a lack of energy modifies the computational processes by means of a change in the computation performed in the single neurons of the network. This change takes place independently of both the awareness and the perception of such a lack in the organism. This feature is not limited to the energy requirements: any physical alteration[7] directly modifies the way the information is processed by the system, but cannot be considered as part of the input.

A simulation with a Universal Turing machine can hardly give an account of these phenomena, despite the fact that they are very frequent in all living beings based on neural systems. Interestingly, Fodor [7] has used the argument of plasticity and degeneracy to propose his generalised version of the theory, but I think that this argument can be of use also against the virtual machine hypothesis, at least until these systems will be able to realize algorithms which can only be reprogrammed by input information.

Lastly, such differences make the parallel neural systems more robust in respect of time and energy requirements: if the processes are suddenly interrupted due to a lack of time, these systems are still able to give an output, even if it will probably differ from the one the system would have reached

---

[7]E.g. structural damages or any other alteration of the neural architecture, chemical or electrical interference in electrochemical synapses, modification of the metabolic state of the neurons, etc.

having sufficient amount of time. On the contrary, the mechanical method implies that a serial system needs to follow all the given instructions in order to perform its transition among states: the lack of the time required to accomplish it would cause a failure in giving an output.

3 Making it through the MRT

It may be argued that it is here discussed the multiple realization of a whole set of instructions, but the object of the MRT is a single, independent and isolated functional state, which has its equivalent in the mental state/psychological predicate of a living being. Nonetheless, the supposed isolation of single psychological predicates such as pain, hunger, etc. is acceptable within the context of the known virtual machines, such as the UTM and the probabilistic automaton: these machines are characterised by serial processes and therefore allow the existence of autonomous functional states. Once the identification of the mind with virtual machines is disputed, the existence of states of this sort in the mind is challenged too: our self-beliefs about them may be misleading.

Let us push this line of thought a little farther. This article has outlined the following proportion:
Set of instruction: Turing machine = algorithm: system whose processes are mathematically describable
It may be argued that this proportion implies the following:
Functional state: Turing machine = assignation of values to all variables in the algorithm: system whose processes are mathematically describable
In the set of parallel neural systems (which is a subset of the mathematically describable systems), this proportion would imply that a particular kind of activation pattern would take the place of the third term in the second proportion. Though different from the "C-nerve activation" correctly defined as *philosopher's fiction*[1], this would be anyway a completely theoretical object: a sort of photography of the entire structure, taking into account the whole network, the activation and metabolic status of all neurons and the disposition of every synapse to propagate its signals. Consequently, any change in any of the variables involved, would generate a different assignation to the variables as well as a different mental state, a conclusion that may seem to lead to an unusable theoretical object.

The problem is that biological neural networks are dynamical information processing systems, and consequently this perspective brings forth the concept of a theoretical object (the photography of the whole structure) characterised by an unavoidable incoherence. If the new definitions imply a concept of mental state which is both unusable and incoherent, then it seems it would be a good idea to discard the whole thesis, on the basis of its implications.

I think this is not a good reasoning: an analogy with the field of analysis in mathematics should help in this case. A sheaf of straight lines can be studied both independently of the assignations of values to its constants and after the partial or complete assignation of the same values; the variables also contribute to locate specific parts or single points on the line analysed. As a consequence, it is perfectly plausible to imagine general rules that can be applied to parallel neural systems (e.g. the computation performed by a single neuron is almost the same in every organism showing a central or distributed neural system: this is the assignation of value to a constant), other rules that are species specific (the macro structure of the neural network shows its similarities) and finally those rules which are single-structure specific and vary within a single organism depending on its natural development, experience and accidents. The use of the fine and coarse grain of analysis [1], should make it possible to relate the new born theoretical mental states — indeed a dynamic concept, far from the static serial equivalent, but still usable — to the variances here described across species or within the single organism.

This use of the mathematical descriptions does not lead to a hyper local reductionism: the single events in the flow of continuous processes of the system are still comparable within the same species with an acceptable fine grain of analysis and the tool that allows such a comparison relies again in the mathematical description of the algorithms realised by the neural processes. Furthermore, there are many advantages in pursuing the use of this tool to understand mind processes. The algorithms describe the way every possible signal is computed by a system: they are not influenced by the presence of a specific stimulus or a combination of stimuli, neither they rely on the analysis of visible behaviours or other forms of output. As it was originally conceived by Putnam concerning the set of instructions of a probabilistic automaton, the specific study of the algorithms implemented by neural system would allow to describe every possible process these system perform in each of their layers, reaching important results in the understanding of the observable and hidden phenomena.[8]

4 Conclusions

This paper states a methodological problem. There is no computational device able to realize all the uncountable possible algorithms: as a consequence, if the object of mind studies are the psychological predicates, it is necessary to study the specific processes that generate them. Whether or not these will result to be multiply realized, the computational study of neural structures is the necessary first step of a realistic approach to the mind. Furthermore, contrary to what expected by the MRT, the more science gives us tools to investigate neural systems, the more it seems that the processes they implement are supervened by the physical matter and are characterised by a series of unique features.

Whenever the processes realized by a particular system are inaccessible, the only way to attempt an analysis consists in assuming that another system, whose processes are accessible, is realizing some of the processes of the first inaccessible system. This procedure creates a useful analogy allowing an analysis narrowed to a part of the whole set of processes of the accessible system: as a consequence, the new aimed description is partial and indirect, because it refers to the supposed analogous system rather than to the original one.

---

[8] Along this path, the main obstacle is represented by the epistemic indeterminacy due to the order of complexity of the biological neural systems, but I assume that grounding the models on the findings in neuroscience, a better explanatory value will be granted.

My claim is that when multiple realizability is applied to neural systems, it is useful to conceive it as a tool giving access to incomplete descriptions of the psychological predicates: a similar constraint does not entail to discard the procedure as a whole, because there are still cases in which there is no or little access to complete descriptions. Nevertheless, if a complete description is accessible or if a better analogy is established (due to an accessible system which is closer to the unaccessible one), then the new description must be preferred to the partial one formerly achieved. In the field of mind studies, in the past few years, the mental processes are becoming more and more accessible and consequently new descriptions will be formalized thanks to this change: on this new ground, new explanatory theories will be built, showing substantial divergence if compared with the ones formerly inferred on the ground of the MRT.

In the attempt to save the MRT from Shapiro's remarks [15], Rosenberg has stated that this theory has been proposed to explain *the absence of discoverable psychophysical laws in a way compatible with physicalism*[13]. It seems today that we are moving towards the finding of these laws: should this happen by means of the mathematical description of the processes realised by the neural systems, the prediction here supported is that the multiple realizability tool will see the fields it has been applied so far restrained, in favour of the new tools.

BIBLIOGRAPHY

[1] W. Bechtel and J. Mundale. Multiple realizability revisited: Linking cognitive and neural states. *Philosophy of Science*, 66: 175–207, 1999.
[2] N. Block and J. Fodor. What psychological states are not. *Philosophical Review*, 81, 1972.
[3] A. Church. An unsolvable problem of elementary number theory. *American Journal of Mathematics*, 58: 345–363, 1936.
[4] P.M. Churchland. Eliminative materialism and the propositional attitudes. *The Journal of Philosophy*, 78: 67–90, 1981.
[5] P.M. Churchland. Functionalism at Forty: a critical retrospective. *Journal of Philosophy*, 1: 33–50, 2005.
[6] J.B. Copeland. The Church-Turing thesis. *Stanford Encyclopaedia of Philosophy* [available online: http://plato.stanford.edu/entries/church-turing, (last modified: August 2002, last consulted: July 2007).
[7] J. Fodor. Special sciences (or: on the disunity of science as a working hypothesis). *Synthese*, 28: 97–115, 1974.
[8] Jerry Fodor. Special sciences: still autonomous after all these years. *Noûs*, 31: 149–163, 1997.
[9] J. Kim. The myth of nonreductive materialism. *Proceedings and Addresses of the American Philosophical Association*, 63: 31–47, 1989.
[10] J. Kim. Multiple realization and the metaphysics of reduction. *Philosophy and Phenomenological Research*, 52: 1–26, 1992.
[11] H. Putnam, Psychological predicates. In *Art, Mind and Religion*, University of Pittsburgh Press, Pittsburgh 1967, pages 37–48.
[12] H. Putnam. *Mind, Language and Reality. Philosophical Papers*, vol. 2. Cambridge University Press, Cambridge 1975.
[13] A. Rosenberg. On multiple realization and the special sciences. *The Journal of Philosophy*, 98: 365–373, 2001.
[14] S. Russell and P, Norvig. *Artificial Intelligence: A Modern Approach*. Prentice-Hall, 1995.
[15] L. Shapiro. Multiple realization. *The Journal of Philosophy*, 97: 635–654, 2000.

[16] A.M. Turing. On computable numbers, with an application to the Entscheidungsproblem. *Proceedings of the London Mathematical Society*, series 2, 42: 230–265, 1936-37.

[17] N. Zangwill. Variable realization: Not proved. *The Philosophical Quarterly*, 42: 214–219, 1992.

The embodied meaning and the "unfolding" of the mind's eyes

ARTURO CARSETTI

With the assistance of the new methodologies introduced by Synergetics and by the more inclusive theory of dissipative systems, we are now witnessing the effective realisation of a number of long-standing theoretical Gestalt assumptions concerning, in particular, the spontaneous formation of order at the perceptual level. This realisation endorses lines of research Kanizsa has been conducting along the entire course of his life [13]. However, a careful analysis of Kanizsa's experiments, particularly those dedicated to the problem of amodal completion, presents us with incontrovertible evidence: Gestalt phenomena such as those relative to amodal completion can with difficulty find a global model of explanation by recourse only to the methodologies offered by order-formation theories such as, for example, those represented by non-equilibrium thermodynamics and the theory of non-linear dynamical systems.

Indeed, it is well known that particular neural networks, inhabiting the border-area between the solid regime and chaos, can intelligently classify and construct internal models of the worlds in which they are immersed. In situations of the kind, the transition from order to chaos appears, on an objective level, as an attractor for the evolutionary dynamics of networks which exhibit adaptive properties, and which appear able to develop specific forms of coherent learning. However, natural-order models of the kind, while appearing more adequate than Koehler's field-theoretical model, are still unable to provide a satisfactory answer to the complexity and variety of problems regarding the spontaneous order formation at the perceptual level. A number of questions immediately arise. What cerebral process, for example, constitutes and "represents" perceptual activity as a whole? How can we define the relationship between the brain and the mind? How can we explain the direct or primary nature of the perceptual process when we know that at the level of underlying non-linear system dynamics there exists a multiplicity of concurrent mechanisms? How can we speak in terms of stimulus information if the measure of information we normally use in psychological sciences is substantially a propositional or monadic one (from a Boolean point of view) like that introduced by Shannon? A percept is something that lives and becomes, it possesses a biological complexity which is not to be explained simply in terms of the computations by a neural network classifying on the basis of very simple mechanisms (the analogue of which is to be found, for example, in some specific models studied at the level of statistical mechanics, such as spin-glass models).

In a self-organising biological system, characterised by the existence of cognitive activities, what is self-organising is, as Atlan states [1], the function itself with its meaning. The origin of meaning at the level of system-organisation is an emergent property, and as such is strictly connected to very specific linguistic and logical operations, to specific procedures of observation and self-observation, and to a continuous activity of internal re-organisation. In this context, the experimental findings offered, for example, by Kanizsa remain an essential point of reference, still constituting one of our touchstones. The route to self-organisation, which Kanizsa also considers in his last articles the route of primary explanation, is ever more universally accepted. Yet questions remain: *via* what informational means and logical boundaries is self-organisation expressed? What mathematical and modelistic instruments can we use to delineate self-organisation as it presents itself at the perceptual level? What selection and elaboration of information takes place at, for example, the level of amodal completion processes? What is the role, in particular, of meaning in visual cognition (and from a more general point of view, in knowledge construction)?

Problems of the kind have for some years been analysed by several scholars working in the field of the theory of natural order, of the theory of the self-organisation of non-linear systems, and of the theory of the emergence of meaning at the level of biological structures. They have recently received particular attention (albeit partial), from a number of scientists investigating connectionist models of perception and cognition. The connectionist models, as developed in the eighties, may be divided into two main classes: firstly, the PDP models first posited by Hinton (1985) and Rumelhart (1986), based essentially on a feed-forward connectivity, and on the algorithm of back-propagation for error-correction. These models require a "teacher": a set of correct answers to be introduced by the system's operator. A second class, posited by, in particular, Amari (1983), Grossberg (1987), and Kohonen (1984), replaces the back-propagation and error-correction used by PDP models with dense local feedback. No teacher is here necessary: the network organises itself from within to achieve its own aims. Perception is here no longer viewed as a sort of matching process: on the contrary, the input destabilises the system, which responds by an internal activity generated via dense local feedback.

Freeman's model of olfactory perception, for instance, belongs to this second class [8]. It contains a number of innovative elements that are of particular interest to the present analysis, in that for Freeman perception is an interactive process of destabilisation and re-stabilisation by means of a self-organising dynamics. Each change of state requires a parametric change within the system, not merely a change in its input. It is the brain, essentially, which initiates, from within, the activity patterns that determine which receptor input will be processed by the brain. The input, in its turn, destabilises, for instance, the olfactory bulb to the extent that the articulated internal activity is released or allowed to develop. Perception thus emerges above all as a form of interaction with the environment, originating from within the organism. As Merleau-Ponty maintained, it is the organism that selects which stimuli from the physical world it will respond to: here we find a basic divergence with respect to the theory of perceptual organisation as posited by Synergetics. Freeman's system no longer postulates an analogy-equivalence between pattern formation and pattern recognition. While in other self-organising physical systems there exists the emergence of more ordered states from less-ordered initial conditions, with precise reference to the action of specific control and order parameters, at the brain level, according to Freeman, a specific selective activity takes place with respect to the environment, an activity which lays the foundation for genuinely adaptive behaviour. What happens inside the brain can therefore be explained, within the system-model proposed by Freeman, without recourse to forms of inner representation. At the perceptual level we have the creation of a self-organised internal state which destabilises the system so as to enable it to respond to a particular class of stimulus input in a given sensorial modality. Perception is thus expressed in the co-operative action of masses of neurones producing consistent and distributed patterns which can be reliably correlated with particular stimuli.

It should be emphasised here, however, that if we follow the route indicated by Freeman, the problem of the veridical nature of perception immediately takes on a specific relevance. As we have just said, we know quite well that, for example, Boolean neural networks actually classify. Networks of the kind possess an internal dynamics whose attractors represent the asymptotic alternative states of the network. Given a fixed environment, from which the network receives inputs, the alternative attractors can be considered as alternative classifications of this very environment. The hypothesis underlying this connectionist conception is that similar states of the world-surroundings are classified as the same. Yet this property is nearly absent in the networks characterised by simple chaotic behaviour. At this level the attractors as such are unable to constitute paradigmatic cases of a class of similar objects: hence the need to delineate a theory of evolutive entities which can optimise their means of knowing the surrounding world via adaptation through natural selection on the edge of chaos. Hence the birth also of functional models of cognition characterised in evolutionary terms, capable of relating the chaotic behaviour to the continuous metamorphosis proper to the environment.

This line of research, while seeming a totally natural direction, is not without its difficulties. There is the question, for example, of the individuation of the level of complexity within existing neural networks capable of articulating themselves on the edge of chaos, at which the attractors are able to constitute themselves as adequate paradigms to cope with the multiple aspects of external information. How can we specify particular attractors (considered as forms of classification), able to grasp the interactive emergence proper to real information as it presents itself at the level of, say, the processes of amodal completion? How can the neural network classification-processes manage to assimilate the information according to the depth at which the information gradually collocates itself? And what explanation can be given for the relationship between the assimilation of emergent "qualities" on the one hand, and adaptive processes on the other? How to reconcile a process having different stages of assimilation with perception's direct and primary nature as described by Gibson and Neisser? What about the necessary interaction between the continuous sudden emergence of meaning and the step by step development of classification processes? And finally, what about the necessary link between visual cognition and veridical perception or, in other terms, between cognitive activity, belief and truth?

To attempt even a partial answer to all these questions, it should first be underlined that the surrounding information of which Gibson speaks is immense, and only partly assimilable. Moreover, it exists at a multiplicity of levels and dimensions. Then, between the individual and the environment precise forms of co-evolution gradually take place, so that to grasp information we need to locate and disclose it within time: we have progressively to perceive, disentangle, extract, read, and evaluate it. The information is singularly compressed, which also explains why the stimulus is a system stimulus. Actually, information relative to the system stimulus is not a simple amount of neutral sense-data to be ordered, it is linked to the "unfolding" of the selective action proper to the optical sieve, it articulates through the imposition of a whole web of constraints, possibly determining alternative channels at, for example, the level of internal trajectories. The intrinsic characteristics of an object in a given scene are compressed and "frozen", and not merely confused in the intensity of the image input. If we are unable to disentangle it, we are unable to see; hence the need to replace one form of compression for another. A compression realised in accordance with the selective action proper to the "optical sieve", producing a particular intensity of image input, has to be replaced by that particular compression (costruction+selection) our mind constructs from the information obtained, and which allows us to re-read the information and retrieve it along lines which, however, belong to our visual activity of recovery-reading. What emerges, then, is a process of decodification and recodification, and not merely analogy-equivalence between pattern formation on the one hand, and pattern recognition on the other. This process necessarily articulates according to successive depth levels. Moreover, to perceive, select, disentangle, evaluate, etc. the mind has to be able autonomously to organise itself and utilise particular linguistic instruments, interpretative functions, reading-schemes, and, in general, specific modules of generation and recognition which have to be articulated in discrete but interconnected phases. These are modules of exploration and, at the same time, of assimilation of external information; they constitute the support-axes, which actually allow epigenetic growth at the level of neural cortex.

From a general point of view, depth information grafts itself on (and is triggered by) recurrent cycles of a self-organising activity characterised by the formation and the continuous *compositio* of multi-level attractors. The possibility of the development of new systems of pattern recognition, of new modules of reading will depend on the extent to which new successful "garlands" of the functional patterns presented by the optical sieve are established at the neural level in an adequate way. The afore-mentioned self-organising activity thus constitutes the real support for the effective emergence of an autonomous cognitive system and its consciousness. Insofar as an "I" manages to close the "garland" successfully, and imprison the thread of meaning, thereby harmonising with the ongoing "multiplication" of mental processes at the visual level, it can posits itself not only as an observer but also as an adequate grid-instrument for the "reading-reflection" on behalf of the Source of itself (but in accordance with the metamorphosis in action), for its self-generating and "reflecting" as *Natura naturata*, a Nature which the very units (monads) of multiplication (the final result of this specific metamorphosis) will actually be able to read and see through the eyes of mind.

When we take into consideration visual cognition we can easily realise that vision is the end result of a construction realised in the conditions of experience. It is "direct" and organic in nature because the product of neither simple mental associations nor reversible reasoning, but, primarily, the "harmonic" and targeted articulation of specific attractors at different embedded levels. The resulting texture is experienced at the conscious level by means of self-reflection; we actually sense that it cannot be reduced to anything else, but is primary and self-constituting. We see visual objects; they have no independent existence in themselves but cannot be broken down into elementary data. Grasping the information at the visual level means managing to hear, as it were, inner speech. It means first of all capturing and "playing" each time, in an inner generative language, through progressive assimilation, selection and real metamorphosis (albeit partially and roughly) and according to "genealogical" modules, the articulation of the complex semantic apparatus which works at the deep level and moulds and subtends, in a mediate way, the presentation of the functional patterns at the level of the optical sieve.

What must be ensured, then, is that meaning can be extended like a thread within the file, constructing a "garland"; only on the strength of this construction can an "I" posit itself together with a sieve: a sieve in particular related to the world which is becoming visible. In this sense, the world, which then comes to "dance" before my eyes, is impregnated with meaning. The "I" which perceives it realizes itself as the fixed point of the garland with respect to the "capturing" of the thread inside the file and the genealogically-modulated articulation of the file which manages to express its invariance and become "vision" (visual thinking which is also able to inspect itself), anchoring its generativity at a deep semantic dimension. The model can shape itself as such and succeed in opening the eyes of the mind in proportion to its ability to permit the categorial to anchor itself to (and be filled by) intuition (which is not, however, static, but emerges as linked to a continuous process of metamorphosis). And it is exactly in relation to the adequate constitution of the channel that a sieve can effectively articulate itself and cogently realize its selective work at the informational level. This can only happen if the two selection processes (i.e. the selection linked to the full expression of the original incompressibility and the selection performed within an ambient meaning) meet, and a telos shape itself autonomously so as to offer itself as guide and support for the task of both capturing and "ring-threading". It is the (anchoring) rhythm-scanning of the labyrinth by the thread of meaning which allows for the opening of the eyes, and it is the truth, then, which determines and possesses them [5]. Ariadne is a lesson in how to "think by forms": how to order and unify (according to a semantic representation process) generative thoughts and functional patterns in order to see. Hence the progressive construction of an "I" as a fixed point: the "I" of those eyes (an "I" which perceives and which exists in proportion to its ability to perceive). What they see is a generativity in action, its surfacing rhythm being dictated intuitively. What this also produces, however, is a file that is incarnated in a body that posits itself as "my" body, or more precisely, as the body of "my" mind: hence the progressive outlining of a meaning, "my" meaning which is gradually pervaded by life.

The revelation and channelling procedures thus emerge as an essential and integrant part of a larger and coupled process of self-organisation. In connection with this process we can ascertain the successive edification of an I-subject conceived as a progressively wrought work of abstraction, unification, and emergence. The fixed points which manage to articulate themselves within this channel, at the level of the trajectories of neural dynamics, represent the real bases on which the "I" can reflect and progressively constitute itself. The I-subject can thus perceive to the extent in which the single visual perceptions are the end result of a coupled process which, through selection, finally leads the original Source to articulate and present itself as true invariance and as "harmony" within (and through) the architectures of reflection, imagination, computation and vision, at the level of the effective constitution of a body and "its" intelligence: the body of "my" mind. These perceptions are (partially) veridical, direct, and irreducible. They exist not in themselves, but on the contrary, for the "I", but simultaneously constitute the primary departure-point for every successive form of reasoning perpetrated by the observer. As an observer I shall thus witness Natura naturata since I have connected functional forms at the semantic level in accordance with a successful and coherent "score".

Vision as emergence aims first of all to grasp (and "play") the paths and the modalities that determine the selective action, the modalities specifically relative to the revelation of the afore-mentioned semantic apparatus at the surface level according to different and successive phases of generality. The afore-mentioned paths and modalities thus manage to "speak" through my own fibers. It is exactly through a similar self-organising process, characterised by the presence of a double-selection mechanism, that the mind can partially manage to perceive depth information in an objective way. The extent to which the simulation model succeeds, albeit partially, in encapsulating the secret cipher of this articulation through a specific chain of programs determines the model's ability to see with the eyes of the mind as well as the successive irruption of new patterns of creativity.

BIBLIOGRAPHY

[1] H. Atlan Self-organising networks: Weak, strong and intentional, the role of their underdetermination. In A. Carsetti (ed.), *Functional Models of Cognition. Self-Organising Dynamics and Semantic Structures in Cognitive Systems*, Kluwer, Dordrecht 1999.
[2] R. Carnap and R. Jeffrey. *Studies in Inductive Logic and Probability*. University of California Press, Berkeley 1971.
[3] A. Carsetti. Randomness, information and meaningful complexity: Some remarks about the emergence of biological structures. *La Nuova Critica*, 36: 47–109, 2000.

[4] A. Carsetti (ed.) *Functional Models of Cognition. Self-Organising Dynamics and Semantic Structures in Cognitive Systems.* Kluwer, Dordrecht 1999.
[5] A. Carsetti (ed.) *Seeing, Thinking and Knowing. Meaning and Self- Organisation in Visual Cognition and Thought.* Kluwer, Dordrecht 2004.
[6] G. Chaitin. *Algorithmic Information Theory.* Cambridge University Press, Cambridge 1987.
[7] G. Chaitin and C. Calude. Mathematics/randomness everywhere. *Nature*, 400: 319–320, 1999.
[8] W. Freeman. *Neurodynamics: an Exploration of Mesoscopic Brain Dynamics.* Springer, London 2000.
[9] S. Grossberg. Neural models of seeing and thinking. In A. Carsetti (ed.), *Seeing, Thinking and Knowing. Meaning and Self-Organisation in Visual Cognition and Thought*, Kluwer, Dordrecht 2004,
[10] J.J. Hoppeld, Neural networks and physical systems with emergent col- lective computational abilities. *Proc. of the Nat. Ac. Sci.* 79: 2254–2258, 1982.
[11] E. Husserl. *Erfahrung und Urteil.* Claasen, Hamburg 1954
[12] R. Jackendoff. *Semantic Structures.* MIT Press, Cambridge 1983.
[13] G. Kanizsa. *Organisation in Vision: Essays on Gestalt Perception.* Praeger, NewYork 1979.
[14] A.S. Kauffman. *The Origins of Order.* Oxford University Press, New York 1993.
[15] K. Koffka. *Principles of Gestalt Psychology.* Harcourt, New York 1935.
[16] W. Kohler and R. Held. The cortical correlate of pattern vision. *Science*, 110: 414–419, 1947.
[17] T. Kohonen. *Self-organisation and Associative Memories.* Springer, Berlin 1984.
[18] W.V. Quine. *Pursuit of Truth.* MIT Press, Cambridge, MA 1990.
[19] L. Talmy. *Toward a Cognitive Semantics.* MIT Press, Cambridge, MA 2000.

Consciousness and the problem of different viewpoints

KATJA CRONE

1 Introduction

In his book *Sweet Dreams* (2005), Daniel Dennett once again takes issue with fellow philosophers who claim that any theory of consciousness must take the perspective of subjective experience into account. Philosophers such as David Chalmers, John Searle, Thomas Nagel, and Colin McGinn are subjected to considerable criticism because they believe that analysis from a third-person perspective — from without, so to speak — cannot fully capture what is truly distinctive about mental states. From their point of view, for example, imaging techniques can show what neural networks become active in the brain when someone sees a dog, but cannot describe what it feels like to see the dog and, say, experience fear.

Dennett, on the other hand, claims the exact opposite: he not only suggests that consciousness can be represented and explained in its entirety from a third-person perspective (i.e. from the perspective of an external observer), but also that such an approach provides descriptions far superior to those made from a first-person perspective. He goes even further by claiming that subjective experience must be entirely excluded from scrutiny since the consideration of mental states from a first-person perspective is by its very nature incapable of providing objective insights into the nature of consciousness. This, he argues, is evident merely from the fact that a person who finds himself in a particular conscious state is frequently mistaken regarding the way in which, for example, a sensual perception presents itself to him. Accordingly, Dennett accuses these philosophical approaches of being nothing but sweet dreams, yet taken for reality by the dreamers.

In what follows, I will critically discuss Dennett's view. I will defend the thesis that a first-person perspective is indispensable when it comes to understanding states of consciousness, for I believe that subjective experience is an integral part of any state of consciousness. Excluding the perspective of subjective experience would mean fundamentally changing how we understand the explanandum "consciousness". We would, to adapt the words of John Searle, be studying consciousness without consciousness [13, p. 18].

The urge to exclude everything subjective a priori from our field of enquiry is the result of, among other things, the principle of scientific precision. I therefore begin by considering a certain understanding of objectivity that presents itself in several theoretical discourses of philosophy and neuroscience. After subjecting it to critical scrutiny, I turn to some of Dennett's epistemological remarks and show, with reference to them, what appraisals from a first-person perspective really have to offer. Finally, I put forward a suggestion for combining a variety of perspectives in a way that takes account of the normative character of the phenomenal aspect of mental states in shaping the explanandum "consciousness".

2 The objective ideal

The first target of Dennett's criticism is the general methodological context in which the philosophers under attack have positioned themselves: they understand consciousness as a feature of mental processes, which, at various levels of intensity and in various qualitative forms, become recognizable as phenomena, and they seek to describe these phenomena by making recourse to individual experience. Such an approach, Dennett claims, is unmistakably in line with Cartesian tradition: as practised by Descartes in his *Meditations*, the philosophers in question choose a standpoint from which one's own view is turned inwards, in order to identify general features of conscious experience. According to Dennett, however, this kind of approach rests on a problematic supposition that he refers to as the "first-person-plural presumption". It takes descriptions of structures of consciousness to be intersubjectively valid by assuming that all beings capable of higher-order consciousness have generally similar internal experiences. Dennett's line is that such an approach latches on to something that everyone discovers or can discover in her own stream of consciousness — and that as a result there is no guarantee of sufficient objectivity, the more so given the absence of a coherent, generally accepted method. Consequently, his argument goes, philosophers who adopt a phenomenological approach should expect to be reproached not only for being arbitrary but also for conspiring against truly scientific analysis [4, pp. 25–34].[1]

This criticism is not particularly surprising given Dennett's theoretical background. According to Dennett, who is indebted to strong functionalism[2] and borrows from behaviourism, the only mental events of analytical relevance are those that have a *demonstrable* causal influence on other events in the system. "Demonstrable" means that the influence of neural and mental events on other neural and mental events must be measured and demonstrated empirically — with the help, for example, of forms of imaging that record certain metabolic reactions, changes in bloodflow, and electrical reactions. "Demonstrable", though, also means that participants in psychological studies, say, can supply information about whether they saw and identified a visual stimulus presented to them.[3] Internal experiences, on the other hand, such as qualia (the sensation of a rose's redness or of the characteristic sweetness of an apple), are, in so far as they have no demonstrable influence on

---

[1] See also [3, p. 67].

[2] The type of functionalism Dennett is devoted to is a non-teleological version of functionalism, often called "machine functionalism", where "function" is interpreted in the strictly mathematical sense: what is at stake is the mapping from, for instance, inputs and outputs of an information processing system. See [16, pp. 382–385].

[3] According to Dennett, this procedure, which he calls "heterophenomenology", guarantees neutrality, unlike positions with a phenomenological foundation. See especially [3, pp. 72–78].

other mental processes and behavioural patterns, treated as not real or nonexistent, and therefore have no analytical relevance. From this perspective, a sore tooth exists as a mental event only in so far as it results in my having certain beliefs: I may think, for example, that taking medicine will make me feel better, and this belief in turn leads to a certain course of action, such as taking the medicine or going to see a dentist. It therefore accords with the basic functionalist position that mental events be examined from a third-person perspective with the help of causal explanations.[4] Building on this, Dennett's maxim goes further to say: "Such a theory will have to be constructed from the third-person point of view, since all science is constructed from this perspective" [3, p. 71].[5]

To obtain a better general understanding of Dennett's criticism, however, we must go on to ask what it is exactly that lies behind this methodological stipulation. Dennett's position resembles many found in neuroscience in so far as its crucial defining factor is, alongside various tenets of the natural sciences such as the ability to test hypotheses by measurement and experimentation, an ideal according to which the world should be perceived in a stringently objective manner. "Objective" in this context describes the trait of a certain kind of stance on, attitude to, or perspective toward something: a point of view as free as possible from value judgements and bias. Distortions and arbitrariness should be excluded from the process of appraisal as much as possible. By definition, a person's individual standpoint can never satisfy this requirement completely, for any individual is always open to a variety of arbitrary influences stemming, say, from personal experiences and preferences. Being aware of this very fact is thus a necessary condition for a process of abstraction to be released. In this way, objectivity corresponds to an intellectual operation we take advantage of when we want that judgments are valid not only for us alone but rather for everybody.

Even so, Thomas Nagel pointed out some time ago that the ideal with which we are dealing here is questionable in theoretical terms and that objectivity, rather than being an absolute concept, should be treated as imperfect in the true sense of the word: objectivity is constantly open to further refinement. The objective view must be seen as a process of arriving gradually to a higher-order view: it is always developed on the basis of a subjective view from which we continuously distance ourselves through reflection [10, p. 77]. This, however, means that, because views are always held by individuals, the subjective view as such cannot be entirely circumvented. Thus, every description of the world implies the presence of subjective elements that cannot be removed [11, p. 5]. Again, this does not mean that a *general* understanding of what is at stake is a priori impossible.

---

[4]Needless to say that Dennett shares his discomfort towards what is often referred to as "introspection" with other paradigms, which made their way through in last century's philosophy and psychology. Eliminative materialism counts as the most radical view of all. See in detail for example [2].

[5]Dennett has no doubt whatsoever that "ongoing scientific research [...] *can* explain consciousness, just as deeply and completely as it can explain other natural phenomena: metabolism, reproduction, continental drift, light, gravity, and so on" [4, p. 25, emphasis in original].

The concept of imperfect objectivity can be used to show that, for several reasons, Dennett falls short of the mark in his criticism of theorists who take a phenomenological approach to the study of consciousness. Two objections in particular can be raised in the present context; the first concerns methodology, the second contents.

1. It is important to realize that a phenomenological consideration of internal experiences is not, as Dennett suggests, at the mercy of the judgements and arbitrariness of a particular *personal* point of view whose authority is no greater than that of any other. Instead, the aim is to articulate *general* structural features of consciousness that can be experienced in order to become intersubjectively accessible and conceptually revisable. This requires us to distance ourselves through reflection from our own immediate experiences and transfer them to a higher-order consciousness. Thus described, the procedure involved meets the understanding of objectivity outlined earlier, namely as a property of a rational process tied to a qualitative change of viewpoint.

In terms of objectivity, this is what the approach has to offer: the identifying of structures of consciousness, which can be accepted as *general* structures does meet an important characteristic of objectivity: intersubjectivity. Moreover, the result is an objective view of mental phenomena, namely in the sense of *relative* objectivity. The balancing act lies in ensuring that mental phenomena must be understood independently of the perspective from which they originate, without abandoning the fact that this very perspective is part of them. This is all the more important given that the search for an objective understanding of *mental states* is concerned with something that is by definition bound to first-person perspective: a mental state is always a state of someone, of a subject that experiences these states in its own particular way. This brings me to my second objection, which concerns the explanandum itself.

2. In logical terms, first-person perspective is the condition for mental states to exist. Furthermore, a subject will always be in such a state in some way or another. Thus, first-person perspective cannot be excluded from the scope of our attention without something being lost in the process. This means, however, that the phenomenon itself, as an experience, is not accessible to an observer from a third-person perspective. If I try to observe the mental states of another person, I am observing his conscious behaviour. I can draw conclusions about certain structures and identify both the causal relations between structure and behaviour and various forms of interaction between the person in question and his environment. The internal aspect of his mental experiences, though, is hidden from me. So, when it comes to explaining *consciousness*, the search for an objective appraisal runs up not only against the practical barriers described above, but also against barriers that cannot be crossed by an intelligence that observes and does nothing else.[6] If, however, we accept that the internal subjective aspect of mental states, their nature as phenomena, cannot be ignored, then the question arises: what can the first-person participant's perspective contribute, what special insights can it produce, and what does it actually mean to articulate immediate experience?

---

[6] See also [10, p. 83].

And it is precisely here that Dennett's voice makes itself heard again.

3 Epistemological objections

In a broader context, namely in his book *Consciousness Explained* (1991), Dennett seeks to dismiss the basic assumption of phenomenology as untenable for epistemological reasons. His remarks are concerned throughout with forms of sensual perception: with touch and smell, and with hearing and seeing. In discussing them, he is interested in how we, starting from our own perceptual perspectives, appraise and evaluate the structure of sensual perception. Here, Dennett aims to show that there are certain aspects of the functioning of sensual perception that cannot be illuminated from within the process of perception itself. When we eat a piece of chocolate, for example, we perceive its characteristic taste with our noses, where scent molecules stimulate special receptors. This is hardly something of which we are immediately aware when the chocolate provides us with pleasure that we think we feel on our tongues. The fact that we actually taste with our noses is something we cannot come to know from within, from the data that our senses themselves provide. Instead, we acquire such knowledge from external sources, specifically with the help of external analysis (in this case, scientific investigation).

The situation is similar with regard to the faculty of vision, for example in the way we appraise and evaluate the scope and quality of our field of vision. Although we recognize objects on the periphery of our vision not as well as the ones at its centre, that is to say, we would always assume on the basis of how we experience seeing that we are still able at least to distinguish different forms and colours from one another. Actually, it can be shown with simple tests that we are practically blind on the edge of our field of vision and can hardly even come close to identifying information there. The effect of having our misconception corrected is one of surprise and thus, according to Dennett, highly significant [3, p. 68].

Dennett's examples are not aimed at showing that she who has experienced seeing or tasting has no insight whatsoever into the microbiological processes involved — that would be somewhat trivial. Instead, his primary intention is to draw our attention to an epistemic shortcoming that necessarily threatens to affect anyone who *makes judgements* about the scope and functioning of perception from a first-person perspective. And how do I find out that my judgment's underlying assumptions were mistaken? Well, by looking at findings, which are, contrary to my inner experiences, empirically demonstrable. Dennett concludes that the first-person perspective should be stripped of its authority [3, p. 96].

The strategy Dennett pursues in his argumentation could now be challenged with the following question: given that our aim is to understand better what it is that characterizes consciousness as a distinctive feature of mental states, is scepticism of the kind Dennett expresses, with its *epistemological* motivation, really the appropriate form of criticism? As our aim is to analyse the explanandum, to bring its special features into view, Dennett's objections fail to be fully convincing. The point I am making can be illustrated paradigmatically with reference to one of Dennett's examples just mentioned, the experience of seeing and tasting. If I want to learn more about the nature of consciousness as a complex property, then it is certainly relevant that I be informed that my field of vision, described spatially, has a centre and a periphery. It could even be said that this situation pertains not only to the composition of one's field of vision but also to perceptual states in general, as I can ascertain on the basis of my own internal experience: I can, that is, direct my attention deliberately at something of which I am currently only peripherally aware (e.g. at the almost imperceptible feeling of pressure caused by the collar touching my neck). The precise *effectiveness* of this peripheral perception is indeed a legitimate and interesting object of analysis, but it is not necessarily the most important issue in the context of a structural appraisal. The crucial point for epistemology, namely the fact that my ability to see becomes unreliable in the field of peripheral vision, would seem to be of minor importance if I am concerned primarily with describing structural features of mental phenomena.

The example of tasting chocolate clarifies this point yet in a different way. It would be a sound and analytically relevant target of a theory of consciousness to examine the accompanying sensation of taste, according to its phenomenal quality, and to make its characteristic features explicit. Yet the basic information for this endeavour is not something I get by the documented evidence that I in fact do not taste with my tongue but more exactly with my nose. Rather, I try to describe and paraphrase my own sensation of taste, and by doing this I may, for instance, discover that the perception has a special temporal extension and that a slightly bitter sweetness unfolds little by little, which changes qualitatively. Ultimately, a theory of consciousness that takes the phenomenal aspect of mental states seriously does not have to ask whether our consciousness is made in such a way that it provides us with adequate information about the outside world. This is because such a theory is at base indifferent regarding the existence of the outside world. Thus, in the context of such a theory, it becomes necessary to agree on a method that has a different, perhaps less strong claim to unassailable validity. What does this mean? Well, since subjective experience, as a feature of mental phenomena, cannot be completely brought to light and represented empirically by scientific techniques, we must turn to *descriptive* means. It is precisely here that approaches belonging to the philosophy of mind, rejected by Dennett, have their strength: they strive to describe and categorize the diverse facets and varieties of mental states as precisely as possible, not just to explain them causally.[7] This objective, of course, along with many other methodological assumptions, traces back to German philosophers such as Franz Brentano and Edmund Husserl. For their central target is to make the complexity of mental phenomena explicit by means of a methodologically guided process of "self-clarification" in the strict sense of the word: the main focus lies on the defining trait of intentionality, approached explicitly in the first-person-perspective. In this theoretical framework one main and basic assumption amounts to the fact that mental acts are not only directed to an

---

[7]Wolfgang Künne makes this point when discussing the foundations of Husserl's phenomenological method [7, p. 173].

object but also contain an accompanying awareness of the whole psychological act as such. This allows for characterizing various forms of mental acts having objects precisely by the way we experience them. And it also gives way to an understanding of how meaning arises from experience conceived of as the phenomenal aspect of mental acts.

4 The internal aspect of mental states

Approaches to the philosophy of mind that follow a phenomenological method assume that we can, to some degree or another, make judgements about mental experience on the basis of evidence. However, attention must be paid to an important point in the process. I have already mentioned, on a variety of occasions, the crucial distinguishing feature of mental states and processes, namely the fact that they have an element of immediate subjective experience. If I am now to try to describe *this* particular aspect of consciousness, I must position myself at a distance from the immediate experience concerned, and in so doing I will change my perspective. This process is accompanied by a categorial shift: subjective experience considered from a reflective distance is not subjective experience any more.[8] Thomas Nagel, for example, stresses that consciousness has an unavoidable internal subjective aspect and that in this sense one always feels that one is in some way in a certain mental state [9]; the generalizing observation thus made is obviously not the same as the state in question itself. And if I try to provide a concrete description of this "feeling in some way" and represent it in language, the state of immediateness is inevitably lost. Franz Brentano makes a similar claim when he says that internal experience (usually translated as *perception*) can never become internal observation, since the latter would make the object in question unavoidably disappear. Internal observation requires well-directed attention, which Brentano interprets as a psychological act of a special kind, since it must be conceived of as somewhat "detached" from the immediate experience.[9] Put more generally, as soon as something direct and unmediated is articulated and reflected on, propositionality arises.

This categorial shift in epistemic status also has epistemological consequences. The status of immediate experience is by its very nature epistemologically characterized as not involving knowledge and not being open to error. Judgements made about experience from a reflective distance, however, do run the risk of being mistaken. Statements about the internal aspect of consciousness are embedded in a process of understanding and can always be

---

[8] See also [14, pp. 97–98].

[9] As an example, Brentano refers to anger as a mental phenomenon. According to Brentano, the particular feeling of anger must already be diminished whenever somebody wishes to "observe" anger as a mental phenomenon [1, p. 41, I, §2]. This view, however, seems to presuppose different mental acts, which has been criticized by Husserl (see for instance [6, p. 355, II/1, Section 5]. What Husserl is mainly challenging here is the problem of infinite regress. It should yet be mentioned that Brentano, being aware of this possible misinterpretation, makes a significant distinction between primary objects (the object perceived) and secondary objects (i.e. the act of perceiving), as two different aspects of the object belonging to one single mental act [1, p. 180, I, §8]. However, as long as a theory of perception, as in the case of Brentano's, holds on to the subject-object-relation as the only conceivable structure, it will be hard to do away with Husserl's general criticism.

corrected and refined further, above all when the process of understanding is an open and intersubjective one. It should be noted here, however, that descriptions of phenomena, that is, articulations about how phenomena are experienced, can have an effect on how the phenomena themselves are experienced. If for instance a wine connoisseur can manage to give a good and detailed description of his own qualitative experience when tasting a certain wine (say, by using precise analogies and metaphors) another person, taking note of the specialist's report, will most likely experience this wine in a certain way, maybe in a more sophisticated way. This point, of course, is deserving of further study and involves the question of the extent to which experiences and sensations themselves have a structure that can be conceptually captured.

As a whole, though, it is clear that a theory of consciousness based on a phenomenological approach will necessarily have to oscillate between the mode of (pre-reflective) immediate experience and various levels of reflective distance — and emphatically not, as Dennett assumes, become entrenched in purely private experience [4, p. 27]. Thus, we are no longer dealing with just a single perspective.

In the next and final part of my paper, I hope to demonstrate the analytical relevance of internal features of consciousness, drawing in the process on the typological description of features of consciousness put forward by John Searle.[10]

Searle's discussion is shaped by the basic insight that 'consciousness' (in the nominative singular) cannot be grasped as a single coherent concept. Accordingly, structural features of the mental domain, rather than being derived systematically, are described and taxonomically classified. The main question is not, which conditions must be satisfied to determine whether someone or a state is conscious. The question is rather, what difference it makes to be in a particular state of consciousness and how these particularities are best described and made explicit. The aim is to arrive in the process at as complete as possible a description of consciousness in all its complexity. Consciousness is characterized by many structural elements accessible as phenomena. In part, they point back to the way in which mental acts are directed at objects of experience or of thought (intentionality).[11] In part, they also involve states that are just qualitatively experienced purely as phenomenal states (lacking objects) — specific feelings such as pain, for example, or feelings bound to sensual perceptions.

The *presence of mood* is an example of a typical structural feature as identified by Searle; it refers to the fact that mental states always have some kind of affective colouring and are thus accompanied by moods. This is so because I am always in a particular state of mind, even if that state is low in prominence and I am unable to put an exact name to it. Finally, this becomes apparent when, for example, there is a sudden change in my state of mind — if I receive unpleasant news, say.

Another special feature of conscious states can be described as *situatedness*. All our conscious experiences are set against some spatiotemporal background.

---

[10]What follows is based on [14, 15].
[11]See [15, pp. 159–192] and [13].

Typically, this background does not explicitly lie at the centre of attention; instead, its presence accompanies whatever particular conscious state I am in. I generally have a latent basic awareness of where I am on the earth's surface at any given moment; I implicitly know roughly what time of day it is and know, equally implicitly, whether I had breakfast in the morning. The absence of this basic latent awareness of situatedness has a disconcerting effect, as when for example I wake up at night in unfamiliar surroundings and do not know where I am, why I am not at home and what time it is.

I would like to point to two further important particularities of conscious experience, which I will subsume under the notion of *coherence*. With reference to them, it can be shown quite clearly, to what extend the viewpoint of an external observer meets a boundary. An important structural feature of mental processes lies in their *unity*. This means that at a given point in time I do have not just one single perception but a multitude of perceptions that are, despite their diversity, drawn together as a unified whole. I have at this moment, for example, particular tactile, visual, olfactory, and auditory impressions. And I do not experience all these sense impressions individually one after the other; instead, I experience them as parts of a unified field of consciousness. This important structural feature of unity cannot be represented adequately from the perspective of an external observer. Imaging techniques, for example, can provide information about the brain activity that accompanies a particular sensual perception, but not about the extent to which it is experienced together with others as a unified event (which is known as the 'binding problem').

A somewhat similar peculiarity of our conscious mind is the fact that it *produces coherence* on the basis of sparse information we get through sense perception. Conscious experience does have in fact a precise and well-defined composition: in a normal visual perception we do not see unqualified and blurry fragments (of items), we rather put them automatically together to a coherent whole; we do not just see disconnected shapes and colours, we rather see chairs, tables and elephants. Closely related to this peculiarity of consciousness is the tendency to 'organize' the whole conscious field in a way that we can immediately identify spatial arrangement of objects when perceiving them. For instance, I see that the pencil is lying *on* the sheet of paper and that the sheet of paper in turn is situated *on* the desk.

This list of structural elements of consciousness could be extended further. My concern in the present context is to show that the description of structures of consciousness can be substantiated in so far as we can bring internal experiences to mind — we can focus deliberately on various kinds of mental phenomena and formulate them conceptually. Now, if one, like Dennett, summarily rejects this way of approaching the topic, the quality of the study can only suffer as a result, for there are certain dimensions of consciousness it will fail to analyse.

What, then, should we conclude from all this? Do these findings entail reductionism in the opposite direction — do we have to declare that consciousness must be described *solely* from a first-person perspective because subjective experience is a fundamental building block of consciousness? This would definitely be the wrong answer. Again, it is important to keep in mind the objective of phenomenological approaches, which is (at least) twofold: they certainly put great emphasis on the fact that mental states of consciousness do have a subjective phenomenal aspect, which itself can by definition not be captured from an external point of view. Yet these theories also make clear that, in taking this perspective as a necessary starting point, a description of general features of mental states and processes is nonetheless possible, which in turn means that the state of immediateness must be abandoned. Now, if the aim of a good overall theory of consciousness is to examine, as far as possible, the phenomenon under consideration in its entirety, then it would seem imperative that we combine a variety of methods and theoretical perspectives, even though each of them will, of course result incomplete when taken individually. Integrating the different methods would allow us to draw on descriptions of experiential states as well as causal and functional explanations based on scientific measurements and models.

In the past, some attempts have been made into this direction: not only philosophers but also scientists have discussed to what extend it is actually possible to relate phenomenological descriptions to explanatory frameworks in contemporary neuroscientific research. The opinions differ widely: some are rather pessimistic regarding the conceivability of such a methodological combination given the transcendental grounding of classical phenomenological claims (most notably made by Husserl).[12] According to others, however, there are no obstacles in principle to a methodological integration since one need not necessarily be committed to a transcendental reading, which would thus give way to a naturalistic approach toward experiential features of mental states.[13] I think that it is indeed unproblematic to move away from a classical methodological focus, which also seems to be consistent with some more recent accounts in the philosophy of mind that take the first-person perspective seriously. As already mentioned, those views primarily concentrate on a detailed description of structures and features of conscious mental states. Even if such a reading is granted, the true challenge remains: a theory is needed that establishes a link between the neurobiological processes, describable in physical terms, and experiential features of mental states the descriptions of which call for a first-person based view as well as intentional and phenomenal concepts. Most importantly, this requires a conceptualisation that is both adequate, in view of the phenomenology at issue, *and* manageable in empirical research.

The best way to accomplish this task is to transform descriptions in an empirically plausible way. For this purpose, Thomas Metzinger [8] has suggested a useful step-by-step strategy, the aim of which is to ultimately provide a conceptual framework for a reductive explanation of experiential features of mental states (note, however, that the broad picture of Nagel's conception of objectivity as a gradual detachment from the subjective standpoint may be found in here as well): one starts from a phenomenological interpretation of some aspect of subjective experience based on the first-person perspective,

---

[12]See for instance [17].
[13]Cf. the essays in [12].

say, the above mentioned 'situatedness' of conscious mental states. As a next step, the representational content of this experiential feature is to be specified, such that, to stay with the example, the content of some mental states has a spatiotemporal aspect. This gives way to the next step, a functional description of the property consisting in the identification of its causal role within the cognitive system. This may in turn permit to isolate a set of physical properties needed in order for the target property to occur.[14]

Clearly, the suggested method is above all an empirically supported *semantic* transformation. And this means that it necessarily comprises a loss of meaning and precision with respect to the phenomenological description started from. However, one should notice that the method is dynamic insofar as it yields a conceptual framework open to refinement once empirical findings allow for a richer account of phenomenological details. Given the prevailing programmatic character of contemporary attempts to naturalize conscious mental states, the method of empirically supported semantic transformation can indeed be said to accommodate the claim of integrating different perspectives: descriptions of phenomena made on the basis of internal experience thus not only delineate heuristics in the natural sciences; they also function as a necessary tool for shaping the explanandum.

BIBLIOGRAPHY

[1] F. Brentano *Psychologie vom empirischen Standpunkt* I. Meiner, Hamburg 1955 (reprint of the edition of 1924).
[2] P. Churchland *Neurophilosophy: Toward a Unified Science of the Mind/Brain*. MIT Press, Cambridge, MA 1986.
[3] D.C. Dennett. *Consciousness Explained*. Penguin, London/New York 1991.
[4] D.C. Dennett. *Sweet Dreams. Philosophical Obstacles to a Science of Consciousness*. MIT Press, Cambridge, MA 2005.
[5] V. Gallese Embodied simulation: from neurons to phenomenal experience. *Phenomenology and the Cognitive Sciences*, 4: 23–48, 2005.
[6] E. Husserl: *Logische Untersuchungen* II/1. Max Niemeyer Verlag, Tübingen 1995 (reprint of the edition of 1900).
[7] W. Künne. Edmund Husserl: Intentionalität. In J. Speck (ed.), *Grundprobleme großer Philosophen, Philosophie der Neuzeit IV*, Vandenhoeck & Ruprecht, Göttingen 1986, pages 165–215.
[8] T. Metzinger. The Subjectivity of subjective experience: a representationalist analysis of the first-person perspective. In T. Metzinger (ed.), *Neural Correlates of Consciousness: Empirical and Conceptual Questions*, MIT Press, Cambridge, MA 2000, pages 285–306.
[9] T. Nagel. What is it like to be a bat. *The Philosophical Review*, 83 (4): 435–450, 1974.
[10] T. Nagel.The limits of objectivity. In S. McMurrin (ed.) *The Tanner Lectures on Human Values*, Vol. 1. Utah University Press, Salt Lake City 1980.
[11] T. Nagel. *The View From Nowhere*. Oxford University Press, Oxford 1986.
[12] J. Petitot, F. Varela, B. Pachoud, J.-M. Roy (eds.) *Naturalizing Phenomenology. Issues In Contemporary Phenomenology and Cognitive Science*. Stanford University Press, Stanford 1999.
[13] J.R. Searle. *Intentionality. An Essay in the Philosophy of Mind*. Cambridge University Press, Cambridge, MA 1983.
[14] J.R. Searle. *The Rediscovery of Mind*. MIT Press, Cambridge, MA 1992.
[15] J.R. Searle. *Mind. A Brief Introduction*. Oxford University Press, Oxford 2004.

---

[14]In his study on the awareness of one's body acting in space Vittorio Gallese, for instance, makes similar semantic shifts when he identifies representational content (body-related knowledge), functional features (body knowledge enabling us to understand actions performed by others) and physical properties (especially parieto-premotor cortical circuit) [5].

[16] R. Van Gulick. Functionalism and Qualia. In M. Velmans and S. Schneider (eds.) *The Blackwell Companion to Consciousness*, Blackwell Publishing, Malden/Oxford/Carlton 2007, pages 381–395.

[17] D. Zahavi. Phenomenology and the project of naturalization. *Phenomenology and the Cognitive Sciences*, (3): 331–347, 2004.

The whys and hows of extended mind
GIULIA PIREDDA

1 Introduction

In this article I will address the question raised by the provocative proposal known as the extended mind hypothesis [9, 10]. In particular, I am interested in analyzing the relationship between this proposal and the issue of individualism in psychology and cognitive science. First, I will briefly present the main positions in these fields. Thus, I will distinguish two forms of individualism and two forms of externalism. It is important to keep separate different fields that often have been unduly confused and mixed up: the semantic thesis regarding the internalism or externalism of mental content; the taxonomic thesis concerning the psychological constraint of individualism; the locational thesis about the vehicles of our thought.

After having disentangled some of the questions raised, I will discuss some arguments that underlie my refusal of the concept of extended mind. At the end of the paper, I will compare two forms of "weak individualism" — infra-individualism [26, 27] and "biological individualism" [1] — both developed on a naturalistic and mechanistic framework. One reason we might want to retain the notion of the individual could be, among others, its connection with the concept of person, which we would not like to lose, even in a naturalistic approach [22].

Here are some questions that gave rise to this article: Is the extended mind proposal tenable? What would be its main consequences in the domain of cognitive science? Is there a connection between the form of externalism supported by the extended mind thesis — that is, locational or vehicular externalism — and traditional semantic or taxonomic externalism [5, 23]? Does individualism come as a monolithic thesis that can only be accepted or rejected? Or can we distinguish different forms (and degrees) of acceptance? To begin with, I will briefly introduce the problem of individualism in cognitive science and distinguish its different possible versions.

2 Individualisms and anti-individualisms: taxonomic vs locational

In its traditional theoretical architecture, cognitive science has been developed on some basic assumptions: functionalism, representationalism, computationalism, innatism, modularity. In brief, these assumptions constitute the Representational-Computational-Theory-of-Mind. According to this theory, the mind is characterized as a computational engine capable of representing to itself the external world and acting in response to these representations. Together with these assumptions, another principle endorsed by many cognitive scientists and philosophers was that of methodological individualism, which has been derived from the theoretical core described above (more specifically, from the conjunction of the representational thesis with the local character of computations [15, 16, 29]); the other theses are also somewhat connected to individualism, but in more indirect and empirical ways. For example, innatist and modular research programs also typically endorse individualistic constraint — see Chomsky's linguistics or Marr's theory of vision.[1]

According to the theory of methodological individualism, one should study cognition abstracting the individual from his or her corporeal and environmental context. Thus, psychology had to create its taxonomies that reflected only the intrinsic (i.e. not relational, not historical) properties of the states realized by the individual.[15][2]

It seems that one reason to endorse individualism is the idea according to which the relevant external factors are reflected in internal states, and that is why there would be no reason to include relational or historical factors in the determination of mental content. From an internalist point of view on mental content, what counts as relevant is, above all, the role mental states play in the generation of behaviour: the effect of mental states, as it were. From an externalist perspective, on the other hand, what is more important is the causal and referential history of that particular mental content [20, p. 93].

In some way, individualism can be conceived as a consequence of representationalism, because what representationalism intends to defend is precisely the idea that the mind acts following its internal representations, and not mechanically responding to what is in the external world (like behaviourist model seemed to claim). As William Bechtel comments, "Fodor's [15] account of the research strategy of methodological solipsism, according to which only representational states within the mind are viewed as playing causal roles in producing cognitive activity, is an extreme characterization of this approach. [...] [I]t amounts to reversing behaviourism by construing the mind as a white box in a black world" [1, p. 1].

Given this framework, we are now able to distinguish different ideas that lie behind individualism in general. In order to do that, we begin by tracing a distinction between the taxonomic and the locational realms. The first is

---

[1]There has been a lively debate between proponents of either the internalistic or the externalistic character of Marr's theory of vision. While the question remains open, I find the individualists' reasons and arguments more compelling. For this reason, I consider Marr's theory of vision as an example of individualistic research [5, 6, 7, 14, 21, 30, 31].

[2]This would be the definition of what Fodor [16] calls "methodological solipsism", which he distinguishes from the weaker view of "methodological individualism", defined as a constraint that applies to each scientific taxonomy, according to which these are to be constructed by only considering individual properties bearing causal powers. In this sense, methodological individualism, though more general, implies a weaker constraint on psychology, as it accepts that relational factors can enter the constitution of taxonomies, as long as they have influences on causal powers. See Wilson [30, 31] for a critique of the argument for individualism from causal powers and for a general epistemological analysis about the physicalistic roots of individualism in the fragile sciences. However, both methodological solipsism and individualism are endorsed by Fodor.

characterized by the question of how should psychology or mental content theory pick up its states and create its taxonomies. On the other hand, locational theories address questions regarding the limits on cognitive systems in reference to perception, cognition and action. Extended mind theory offers answers to such questions. According to the different questions posed, we can sketch two versions of individualism. To the first question, both semantic internalism and methodological individualism are possible answers. Semantic internalism could be defined as the thesis opposed to semantic (or taxonomic) externalism, whereas this is characterized as a metaphysical modal thesis that can be expressed in this way:

> Given two intentional agents who share the same internal (brain and computational) state, and given differences in their respective environments, they can instantiate different mental states [21].

Roughly, semantic externalism is the idea that content of a mental state is at least partially determined by the environment. According to semantic internalism, on the other hand, mental content is determined by intrinsic and computational properties of mental representations. As long as these properties are also the syntactic and formal ones, this view goes together with methodological individualism derived from (local and formal) computationalism and from considerations about causal powers [30, 31]. That is, even through different paths, semantic internalism and methodological individualism get to the same answer to the taxonomic (or metaphysical) question.

This is what concerns the taxonomic and the semantic realms. In terms of the so-called locational one, there are several positions to be distinguished, all of which must be kept distinct from the taxonomic and semantic versions of individualism or anti-individualism we discussed above. As noted, locational theories typically respond to questions of where we should locate the boundaries of our mind. Some of these theories even refuse to admit the presence or the relevance of such boundaries. The extended mind thesis that I discuss represents an extreme position in that it mantains that the internality or externality of thought vehicles does not make a difference in cognition. From a functionalist point of view, the distinction between internal and external would not be relevant. An external (dispositional[3]) belief counts as a regular belief as long as it plays a convenient (functional) role in the production of behaviour. Even if extended mind supporters — understandably — do not mean to give up the notion of the individual (only to suggest a revisable and mobile notion of it), they do in fact jeopardize this notion as it has been conceptualized in traditional ways. Obviously they do not want to give up the idea that there is — even for extended processes — a locus of control to be individuated somewhere in the brain or in the nervous system.[4] That is probably why they naively interpret individualism as a constraint which,

---

[3]The dispositional character of the belief on which the whole argument about extended mind rests should be deeply analyzed. I think Clark and Chalmers [10] take this point too much for granted. They do not define what can and what cannot count as dispositional belief, whereas it goes without saying that not every belief can. As far as I know, this point has not been analyzed as it should be in the literature about extended mind.

[4]On the notion of "locus of control" see [1, 4].

instead of posing a normative methodology on the study of mental life, only tells part of the story ("an individualistic psychology could only, at best, tell part of the story about cognitive processing: the inside story" [32]). What they instead explicitly intend to give up is the boundary interposed between agent and environment. In so doing, they are probably compelled to give up the individual as an entity to be held in the description of mental activities. Instead, holding the notion of the individual — against locational externalism — corresponds to what we could call "weak individualism". I think that there are good reasons to maintain a concept of the individual grounded on biological and epistemological considerations about the notion of mechanism and mechanistic explanation. This notion is not compatible with the form of externalism — active or locational — proposed by extended mind theorists. However, before presenting two possible forms of weak individualism, I will briefly going over some important points and critiques of the extended mind thesis.

3 What extended mind is not

The main point of the previous section was the difference between taxonomic and locational realms. This leads us to conclude one important point about extended mind: extended mind — like externalistic explanation [21] — is not an extreme form of semantic externalism, as its supporters are willing to claim [10, 31]. We should not concede any additional appeal to extended mind on the basis of this presumptive connection with semantic externalism.

Another negative point about extended mind — about what it is not — comes from mental architectural considerations. In distinction to the post-Chomskyan consensus in the cognitive sciences and more recent developmental psychological evidences, extended mind supporters insist on the external resources our mind can count on. They claim that these are exactly equivalent to internal ones. Among these external resources are, importantly, the so-called "cognitive artefacts", that is, formal or material structures that change and reorganize the nature of cognitive tasks. Examples range from pen and paper to pocket calculators, computers and, above all, language. Even if referring to external resources has evident explicative advantages in evolutionary terms, this viewpoint, supported by Andy Clark, Daniel Dennett and Robert Wilson, has been criticized from two different vantage points: one considers architectural issues, the other appeals to the evolution of language.

From an architectural point of view, by appealing to the importance of external resources, locational externalists seem to suggest an idea of mind characterized by poor internal resources (mnemonic, linguistic). Locational externalists seem to support some form of what we could call the "inverse proportionality thesis" between internal and external resources, and this reasoning is mirrored in the scarce interest for architectural issues in the literature about extended mind. However, could we possibly think that the richer the external resources the poorer the internal? Is this really a viable way of thinking about the mind? I think not, for two main reasons.

First, this line of thinking would put us at odds with the evidence of developmental psychology, which has increasingly demonstrated the richness of the human mind even in the first months of life. Second, even ignoring the previous consideration, the inverse proportionality thesis is not plausible for another reason: contrary to what locational externalists seem to think, in order to be able to exploit and coordinate external resources we need to display strong internal and well developed resources (among others: [17, 27]). In brief, extended theories of mind or cognition sometimes tend to underestimate the importance of psychological theory and constraints, falling in the trap of the inverse proportionality thesis. Overlooking this risk would be a fatal error for extended theories. This leads us to the point about language.

Many critics have considered the understanding of language as a typical external resource to be mistaken. According to Clark, language is a cognitive artefact *par excellence*, and its existence would manifestly show our being "natural born cyborgs" [9]. This seems to be a brave, if not dangerous, claim, given the findings of generative linguistics and universal grammar studies. In fact, it is now widely accepted that language does not come for free, and that, as a result, it cannot be considered a given, something just lying there outside, waiting for us to use it for cognitive purposes. As many authors unfailingly outlined, in order to account for language development we need to postulate important internal resources already there in the individual mind. More specifically, Origgi and Sperber [19] argue that in order to account for the evolution of coded communication, and therefore of natural languages, we have to assume a pre-existent form of inferential communication. Hence, it follows that the possibility of the extended mind is to be found in an already developed human mind, endowed with the cognitive abilities that enable communication with others. To sum up, an extended mind cannot be an impoverished mind. We are now ready to present and discuss some of the critiques that were articulated against the extended mind thesis.

4 Some (almost) conclusive arguments against extended mind

First, a concise outline. As Clark and Chalmers [10] put it, the idea of extended mind is based on a thought experiment consisting of a comparison between two characters, Otto and Inga, dealing with their belief about a museum address. Content belief being the same, what is different in the two cases concerns belief location: that is, Inga has her belief stored in her biological memory, whereas, because he suffers of Alzheimer's disease, Otto's belief is physically located outside his brain, in a notebook which he constantly holds, and which he regularly consults before undertaking any action. The claim is: "insofar as beliefs and desires are characterized by their explanatory roles, Otto's and Inga's cases seem to be on a par: the essential causal dynamics of the two cases mirror each other precisely" [10]. This quotation introduces the next point, that is: extended mind theory rests on a hard assumption of functionalism, which implies a strong acceptance of the multiple realizability argument [23]. This is the necessary background for claims like that above.

Now, we know that cognitive science states the possibility of a functional analysis of mental activities, and that functionalism was an ingenious way of escaping both identity theory regarding mental and physical states, and the behaviourist claim that reduced mental states to behavioural dispositions. We also know that strong functionalism, as articulated in classic Artificial Intelligence theory, has been critiqued and revised, in particular the validity of the multiple realizability argument, on which functionalism is strongly based. In fact, even if it is true that mental states can efficaciously be described in terms of functional states (namely connections between typical causes and typical effects) it is always important to remember that another principle on which the whole cognitive scientific enterprise rests is after all the possibility of investigating mental states and mechanisms[5]. An important role in this debate has been played by Bechtel and Mundale's paper "Multiple Realizability Revisited" [3]. In this paper, the authors analyze the use of the very concept of multiple realizability in cognitive science literature. They consequently argue that most of the concept's success is due to "a tacit adoption of some sort of double standard" in thinking about mental and physical. Bechtel and Mundale conclude, "in thinking about psychological capacities, it is common to describe them coarsely — as the capacity for vision, or for short term memory. By contrast, realizing neural structures and their "immediate" functions receive comparatively fine-grained description".

Moreover, Larry Shapiro has contributed to this debate by proposing two conditions necessary for an effective case of multiple realizability [24, 25]). On the one hand, the same psychological function has to be individuated; on the other hand, its neural realizations have to be significantly different. Now, putting together these two last considerations, it could be said that "given that a coarse grain of description facilitates the view that two cases are instances of the same (psychological) kind, and that a fine grain of description does the same for the view that two cases are instances of different (neurological) kinds, the bias that this double-standard introduces is one that creates the impression that such cases satisfy both of Shapiro's constraints" [3]. Thus, the strong interpretation of the multiple realizability argument has been revised towards an increasing acceptance of some constraints deriving from our being embodied and "embrained". In this way, limits have been placed on the extreme version of functionalism.

A cautious attitude toward functionalism is also justified on some other grounds. As noted above, the extended mind hypothesis holds up on a strong functionalist account about mental states, particularly beliefs. The risk of this assumption, however, is an involuntary return to a behaviouristic account of mind. Though a functionalist approach should, to some extent, be maintained in order to legitimate a genuine cognitive viewpoint, functionalism that goes too far — like the functionalism presupposed by extended mind theory — is not desirable. In fact, it really seems that functionalism as it is endorsed by the extended mind thesis is so extreme that the constraints necessary to speak of belief (or mental states in general) are practically non-existent. This loose characterization of a mental state — with no reference at all to mental structure or mechanisms which instantiate it — runs the risk of bringing us

---

[5]With the exception of eliminativists, both (of course!) intentional realists (like Fodor) and interpretativists (like Dennett) admit some kind of reality to mental states and show some interest in mental architecture and mechanisms.

back to the conception of mind as an unanalyzable monolithic black box. Once again, weakening the interface between the mind and the environment seems to cost us more than we earn.

Another kind of critique comes from some epistemological considerations about the issue of levels of description in science. This critique concerns the difference between personal, sub-personal, and non-personal levels. The question is strictly related to a central point about extended mind: the problem of criteria. Clark and Chalmers propose to reform the traditional ways of speaking about the mind, weakening the boundaries of head and skull. These authors have offered some criteria that serve to outline the condition under which we can individuate an extended mind. Here they are:

1. Otto never acts without consulting the notebook (constant presence and automatic use);

2. Information in the notebook is directly available without difficulty;

3. Otto automatically endorses information retrieved in the notebook;

4. Information in the notebook has been endorsed in the past and it is there because of this.

There are several problems with these criteria. First of all, they are too strict. That is, even for biological cognitive resources — like our memory — they sometimes do not apply. We do not always endorse our own memories, but even when we are distrustful toward our memory, we are not dissociated from it. Moreover, the presumed paradigmatic example of cognitive external resource — Otto's notebook — does not fulfill the criteria at all. In order to work as Clark and Chalmers describe it, the notebook would have to contain billions of data items, which surely could not be easily retrieved, as the second criterion requires.

Also, these criteria drive us back to the point about the distinction between personal, sub-personal, and non-personal. As Di Francesco rightly noted, these criteria make strong reference to the idea of personal mind, a notion we have had to dispense with according to extended mind theory [12, 13]. So, the whole argument necessitates what it claims to refuse. Clark and Chalmers need the notion of personal mind in order to describe the characteristics of a transparent interface. But relying on personal mind, they contradict their own hypothesis. The attention shown in cognitive science to sub-personal states and mechanisms does not authorize per se extended mind supporters to shift the interest toward non-personal ones, like notebooks or pocket calculators. That is, we at least need some criteria to distinguish non-personal from sub-personal states, and Clark and Chalmers give us none.

The relationship to personal mind is also involved in the critique about the incompleteness of extended mind theory. Even accepting it as an interesting theory about how the mind works (and what it really is), extended mind would be at best incomplete, as it rules out any reference to some important aspects of mind, such as experiential and phenomenal ones [11]. However, this last point is relevant not only to extended mind theory; it plays a role in any functionalist approach to mind which individuates mental states by their causal roles.

Finally, there is one last criticism to be addressed regarding the extended mind, which I find convincing and probably conclusive, for it does not rest on any particular previous theoretical bond. The argument consists in adopting the extended mind idea, applying it to a situation and finally showing how this leads to big problems in understanding the scenario [18]. Emma and Anna are two lazy girls who have to translate a text from Latin into Italian for homework. Emma, the rich one, has a brand new automatic translator she constantly uses for this kind of task. Anna, on the other hand, cannot afford such new technology. Nevertheless, her father is a real Latin expert, and he regularly agrees to help his daughter do her Latin homework. While the case of Emma and her translator would fit the framework of extended mind, problems arise in analyzing Anna's case. If in some sense we can apply the extended mind framework to Anna and her father translating Latin, something exceeds this representation of things. While translating, Anna's mind feels bored and desires to go out with her friends; her father's mind feels also bored and desires to go fishing. Thus, how many minds should we count in this scenario? Accepting extended mind framework, we seem to not even be able to solve this simple but crucial ontological problem.

5 Biological individualism and infra-individualism

Having reviewed these critical points, we are now ready for a theoretical assessment. Extended mind thesis is problematic. Nevertheless, it still entails some attractive points connected to its evolutionary explicative potential. Whereas the relationship between mind, body, and environment had been somehow ignored by previous cognitive science, this relationship is taken into account by extended mind approaches. An acknowledgement of the strict connection (and co-evolution?) between our environment and ourselves offers an evolution-friendly perspective that we ought to maintain. The alteration of environment taken into account in the extended approach to cognition is an important aspect of our evolutionary history. But does this necessarily mean that we have to give up the boundary between our environment and ourselves? In light of what we have seen, I propose to argue in favour of a form of weak individualism. It seems in fact reasonable to maintain an idea of mind linked with the notion of the individual. In the last part of this paper, I would like to compare two different versions of weak individualism, which substantially differ in the entity they individuate as locus of control. They are the infra-individualism advocated by Dan Sperber [26, 27] and biological mechanistic individualism advocated by William Bechtel [1]. Neither accepts extended mind extremisms; their comparison can be useful in order to construe an alternative to strong individualism (à la Fodor) that eschews the problems raised by the extended mind theory.

Infra-individualism, inspired by an epidemiological explicative model, claims that "an adequate explanation of social phenomena should invoke the behaviour or properties of infra-individual entities" (like modules). This choice is functional to the adoption of a strong naturalistic approach, where reasons are interesting only qua causes and "only organisms, and not persons, or actors, are manifestly natural entities". The departure question of this thought would be: "What are the individuals of methodological individualism? Are they human organisms as studied by biology and naturalistic psychology? Are they persons, conscious subjects, the characters common sense psychology is about? From the point of view of a naturalistic psychologist or philosopher [...], personhood is not a given" [26].

Bechtel's point of view is built on the notion of mechanism. Construing the individual as a cognitive mechanism, and positing there the locus of control, Bechtel preserves the boundary between individual and environment and criticizes the infra-individualistic, modular, sub-personal position. Thus, Bechtel supports a conception of unity of mind. He argues that, in relying on a mechanistic explanation there is neither the need to lose the boundary between individual and environment nor to give up the unity of the mind. "What serves to explain a phenomenon is an account of the mechanism responsible for producing it. [... T]he parts of a mechanism are often highly interactive in the production of any phenomenon. Yet, they also have an identity of their own and there are good explanatory reasons to differentiate them from their environmental context" [1, p. 5]. Mechanisms are then defined as bounded systems that are selectively open to their environment and often interact with and depend upon their environment in giving rise to the phenomenon for which they are responsible (p. 2). Cognitive mechanisms, like biological ones, "are always situated and dependent on their environments while also being in a critical sense distinct from them" [1, p. 2]. "A mechanism is a structure performing a function in virtue of its components parts, component operations, and their organization. The orchestrated functioning of the mechanism is responsible for one or more phenomena" [2, p. 6]. If we intend a mechanism in this way, the entire individual can be considered as a unified mechanism composed of different parts in interaction with each other. The only difference between mechanism parts and modules appear to be the strong interactive character of the first ones.

Though Bechtel's position is attractive, it also presents some difficulties. First of all, it considers modularity only in the Fodorian sense. In fact, even as he presents the various versions of modularity theory and explains their differences, Bechtel then rejects these theories appealing to an aspect which is certainly characteristic of Fodorian modularity, but not of the evolutionary one, namely informational encapsulation. This alone does not justify the rejection of central modularity, in which "the defining mark of modules [is] that they operate on specific domains of inputs" [1, p. 4], which then remains an open possibility. Secondly, and maybe more importantly, Bechtel's theory neatly distinguishes two separate fields: mental (psychological, behavioural) phenomena, "for which it is appropriate to treat the mind/brain as the locus of the responsible mechanism and to emphasize the boundary between the mind/brain and the rest of the body and between the cognitive agent and its environment" [1, p. 1] and social phenomena, in which "the agent is so intertwined with entities outside itself that the responsible system includes one or more cognitive agents and their environment" [1, p. 1]. In the case of an organism, it is the cognitive agent who maintains himself through his activities, including the ones that modify the world around him. Likewise in larger systems, such as a social network, it is the network itself that "becomes the locus of control for certain phenomena — those that are carried out by the social network in the service of it" [1, p. 14]. The reason for shifting the locus of control to the social network is that "it is the network itself that it is being maintained by the operations being performed" [1, p. 14]. In this sense, situated cognition is said to refer "to the cognitive activities of agents situated in an environment, and the locus of control for these cognitive activities remains the individual cognitive agent" [1, p. 14]. It stops being so in social activity. But it is not at all easy to individuate the boundary between cognitive and social activity. This kind of strict separation between mental and social, typical of classic individualistic cognitive science, can be risky because it seems to suggest an artificial separation between individual and social activities, completely ignoring social components in individual behaviour and its potential social and public effects.[6]

Sperber's infraindividualism is linked to his particular naturalistic and mechanistic approach to individual and social cognition, inspired by epidemiological methodology. In applying epidemiological explanation to social causation "one has a complex causal chain linking a variety of causal processes, some internal to individual organisms, others taking place in the environment. Among internal causes, one has mental causes. Among mental causes, one has beliefs, desires, and practical syllogisms leading to actions" [26]. These causal chains are said to extend beyond the individual and to represent social and cultural facts in a naturalistic framework [27]. Thus, Sperber's view gives a much more plausible interpretation of the boundary between cognitive and social: "[e]very stabilized social phenomenon, be it described as a social practice, or as a cultural representation, or as an institution, is the outcome of [...] processes of distribution" of representations of all kinds (beliefs, values, techniques, projects, intentions and so on) [26]. Sperber concludes:

> A human population is inhabited by a much wider population of mental representations of all kinds [...]. These mental representations are distributed in the brains of individuals. Behaviours are caused by mental representations. The behaviour of an individual, for instance walking or speaking, may be perceptible to other individuals, or it may leave perceptible traces, for instance footsteps or writing. I will call such perceptible behaviours and traces "public productions". The public productions of an individual may provide an input to the mental processes of other individuals, causing them to construct their own mental representations. These representations can in turn result in public productions, which can trigger the construction of yet other mental representations in other individuals, and so on. A human group is thus crisscrossed by a mesh of causal chains where mental and environmental links alternate. Everything social, I would argue, is caught in that mesh. A description of social facts from this epidemiological point of view is both mechanistic and naturalistic. Complex processes are decomposed into chainings of elementary processes. Some of these elementary processes are to be studied by ecology, other elementary processes are to be studied by cognitive psychology [26].

---

[6] Similar observations are made by Sterelny [28] in assessing Clark's examples in favour of extended mind.

6 Conclusion: In defence of a reasonable individualism

Having analyzed and rejected extended mind hypothesis, I would like to argue for a notion of the individual which takes into account our being embedded in an environment we continuously change and which in turn has some effect on the ways we perceive, cognize and interact. Two possible starting points for a weak (and reasonable) individualism are Bechtel's and Sperber's views, both compatible with extended cognition. Both Bechtel's and Sperber's views have some aspects I would like to support. On one side, Bechtel defends what we could call a reasonable individualism, grounded on biological and mechanistic considerations. On the other hand, Sperber's naturalistic approach has the advantage of not separating neatly individual and social realms, explaining the latter from individual behaviours. Although Bechtel refuses modularity altogether, I think one can find a way to keep together the main advantages of each theory, in order to develop an effective form of weak individualism as an alternative to extreme locational externalism.

BIBLIOGRAPHY

[1] W. Bechtel. Explanation: mechanism, modularity, and situated cognition. In M. Aydede and P. Robbins, *Cambridge Handbook of Situated Cognition*, Cambridge University Press, Cambridge 2007.
[2] W. Bechtel and A. Abrahamsen. Explanation: A mechanist alternative. *Studies in History and Philosophy of Biological and Biomedical Sciences*, 36: 421–441, 2005.
[3] W. Bechtel and J. Mundale. Multiple realizability revisited: Linking cognitive and neural states. *Philosophy of Science*, 66: 175–207, 1999.
[4] W. Bechtel and R.C. Richardson. *Discovering Complexity: Decomposition and Localization as Strategies in Scientifc Research*. Princeton University Press, Princeton 1993.
[5] T. Burge. Individualism and psychology. *The Philosophical Review*, 95: 3–45, 1986.
[6] N. Chomsky. Language and nature. *Mind*, 104: 1–61, 1995.
[7] N. Chomsky. *New Horizons in the Study of Mind and Language*. Cambridge University Press, Cambridge, MA 2000.
[8] A. Clark. *Being There: Putting Brain, Body and World Together Again*. MIT Press, Cambridge, MA 1997.
[9] A. Clark. *Natural Born Cyborgs: Technologies and the Future of Human Intelligence*. Oxford University Press, Oxford 2003.
[10] A. Clark and D. Chalmers. The extended mind. *Analysis*, 58 (1): 7–19, 1998.
[11] M. Di Francesco. Mi ritorni in mente. Mente distribuita e unità del soggetto. *Networks*, 3-4: 115–139, 2004.
[12] M. Di Francesco. Extended cognition and the unity of mind. Why we are not spread into the world. In M. Marraffa, M. De Caro and F. Ferretti (eds.), *Cartographies of the Mind. Philosophy and Psychology in Intersection*, Springer, Berlin 2007.
[13] M. Di Francesco. Soggettività e trasparenza. Clark, Marconi e la mente estesa. *Rivista di estetica*, 34 (1): 233–250, 2007.
[14] F. Egan. In defence of narrowmindedness. *Mind and Language*, 14 (2): 177–194, 1999.
[15] J.A. Fodor. Methodological solipsism considered as a research strategy in cognitive psychology. *Behavioral and Brain Sciences*, 3: 63–109, 1980.
[16] J.A. Fodor. *Psychosemantics: The Problem of Meaning in the Philosophy of Mind*. MIT Press, Cambridge, MA 1987.
[17] P. Jacob. Review of The Body in Mind: Understanding Cognitive Processes by Mark Rowlands. *Mind and Language*, 17 (3): 325–331, 2002.
[18] D. Marconi,Contro la mente estesa. *Sistemi Intelligenti*, XVII (3): 389–398, 2005.
[19] G. Origgi and D. Sperber. Evolution, communication and the proper function of language: A discussion of Millikan in the light of pragmatics and of the psychology of mindreading. In P. Carruthers and A. Chamberlain (eds.), *Evolution and the Human Mind: Language, Modularity and Social Cognition*, Cambridge University Press, Cambridge 2000, pages. 140–169.
[20] A. Paternoster. *Introduzione alla filosofia della mente*. Laterza, Roma-Bari 2002.
[21] A. Paternoster. Esternismo e spiegazione psicologica. *Epistemologia*, 1: 45–78, 2003.

[22] A. Paternoster. La naturalizzazione del concetto di persona. *Fenomenologia e società*, 1: 85–95, 2007.
[23] H. Putnam. The meaning of meaning. In *Mind, Language and Reality: Philosophical Papers*, vol. 2, Cambridge University Press, Cambridge 1975, pages 215–271.
[24] L. Shapiro. Multiple realizations. *Journal of Philosophy*, 97: 635-654, 2000.
[25] L. Shapiro. *The Mind Incarnate*. MIT Press, Cambridge, MA 2004.
[26] D. Sperber. Methodological Individualism and cognitivism in the social sciences. Unpublished english version of *Individualisme méthodologique et cognitivisme*. In R. Boudon, F. Chazel and A. Bouvier (eds.), *Cognition et sciences sociales*, Presse Universitaires de France, Paris 1997, pages. 123–136.
[27] D. Sperber. Why a deep understanding of cultural evolution is incompatible with shallow psychology. In Nick Enfield and Stephen Levinson (eds.), *Roots of Human Sociality*, Berg Publishers, Oxford 2006, pages 431–449.
[28] K. Sterelny. Externalism, epistemic artefacts and the extended mind. In R. Schantz (ed.), *The Externalist Challenge. New Studies on Cognition and Intentionality*, de Gruyter, Berlin and New York 2004.
[29] S. Stich. *From Folk Psychology to Cognitive Science*. MIT Press, Cambridge, MA 1983.
[30] R.A. Wilson. *Cartesian Psychology and Physical Minds. Individualism and the Sciences of the Mind*. Cambridge University Press, Cambridge 1995.
[31] R.A. Wilson, *Boundaries of the Mind: The individual in the Fragile Sciences: Cognition*. Cambridge University Press, Cambridge 2004.
[32] R.A. Wilson and A. Clark. How to situate cognition: Letting nature take its course. In M. Aydede and P. Robbins, *Cambridge Handbook of Situated Cognition*, Cambridge University Press, Cambridge 2007.
[33] R.A. Wilson and C.F. Craver. Realization. In P. Thagard (ed.) *Handbook of the Philosophy of Psychology and Cognitive Science*, Kluwer, Dodrecht 2006.

Movement in the philosophy of mind: traces of the motor model of mind in the history of science

CARMELA MORABITO

1 The motor model of mind

A new model of mind, the so called "motor model", is gaining the scene within contemporary neurosciences, raising from a fertile "triangulation" [36] of data and from the acquisitions — theoretical, experimental and clinical — of different disciplines, from experimental psychology of cognitive processes to neuropsychology, from cognitive neurosciences "systemic" or "holistic" (in the sense these terms are used by Kandel) [1] to mathematical modelling and to the most recent philosophy of mind. It is a model of the "incarnate" or "embodied" mind, rooted at the very intersection of these different disciplines (each endowed with specific conceptual and methodological tools, as well as with a specific level of complexity in the explanation of behaviour) and, departing exactly from their convergence, this model aims at imposing a new concept of mind. A mind whose genetic roots are located far "below" and much "before" consciousness and will, in the organism's vital drives and in kinesthesia. As a consequence, it is in the body and in the brain that the basic premises of the study of the cognitive functions are to be identified. The brain, within this theoretical framework, is specifically intended as an organ whose development was principally aimed at predicting the consequences of action rather than, in a classical fashion, as a generator of responses to stimuli coming to the organism from the more or less external environment. This new, action-based approach to the mind, in fact, attributes to body movement a basic and fundamental role in the development of consciousness and cognition [19, 6, 10, 7, 12].

Thus, with the aim of preserving the fertile epistemological interaction of a phenomenology of behaviour with the models of its underlying causal mechanisms, the new philosophy of mind aims at a philosophical foundation of the so called physiology of action. Choosing action as a cornerstone, as a theoretical lens through which to observe the behaviour in its wholeness, and therefore mind, naturally implies a stronger emphasis on the specificity of the organism, on its being intrinsically goal-oriented and in an active and constructive interaction with its environment. The organism is, in fact, conceived as a sort of constant generator of hypotheses, that selects sensory information depending upon the aims of the action. In this theoretical perspective — biological, dynamic and integrated —, rather than as a bare motor expression of sensory computation, action is conceived as an active and goal-oriented "kinetic melody",[1] a structured ensemble of co-ordinated movements in function of a specific aim.

Attributing to body movement a basic and fundamental role in the development of consciousness and cognition, allows a peculiar conceptual inversion, through which mind is interpreted as "moulded" by movements (which it traditionally plans and directs), and movement is no more the means to satisfy the needs of higher cerebral centres (mind): to the contrary, it is mind to be the tool to perform actions; thinking equals to decide what movement to perform next. Mind is intrinsically a motor system: thought, memory, cognition, perception, consciousness, motivation, meaning, in short, all that is mental, is a product of constructive motor capacities. Of course, strongly stressing the biological matrix of mental phenomena implies the overcoming of the Cartesian and universal epistemic subject, on which modern philosophy was based (a subject non-biologically conceived, thus separated from "external reality" that he aims at understand); it implies also the grounding of cognitive functions in evolution and history, in personal and interpersonal experience.

Hence it derives a model of the living being, of the environment and of the mind, aimed at finally overcoming the limitations of mechanism and of the metaphysical watershed that has kept body and mind separated for centuries. From the study of movement and form cognitive neurosciences, a new way to the embodiment of mind is thus taking form, based on a bodily and non-propositional concept of representation; in this sense the philosophy of action proposes itself as a theoretical route to the overcoming of the dichotomic contraposition between bodily mechanism and mental representation, between subject and object, mind and world. For an authentic understanding of cognitive functions it is in fact considered indispensable the fundamental relation between organism (with its aims, its needs, its history, etc.) and environment, between observer and phenomenon, within the scope of a concept basically grounded on an interactive constructivism. It incorporates the co-evolution of species and environment and the complex interaction between the subject and the world in a theoretical frame characterized by a complex and dynamic interaction: of the organism with the environment (intended as *Umwelt*), of the body with the brain and of the "bodybrain" with the mind.

The tight intertwinement of motricity and thought is by now evident at a phylogenetic as well as at an ontogenetic level.[2] The incarnation of *cogito*

---

[1] Pierre Janet (1859–1947) — in open contrast with the reductionist, molecular approach adopted in those very years by the American behavorist psychology — developed the concept of "conduct", intended as "global behaviour, intentional and intrinsically meaningful" [18] opposing it to the conception of behaviour in terms of mere Stimulus-Response associations.

[2] Developmental psychology and contemporary neurosciences have clearly demonstrated that the embryo is primarily a motor organism, before than a sensory one; in the embryonic phase, in the phoetal one and in early infancy action precedes sensation, reflex movements are performed before any concept of them is developed (already Bain, in the mid-Nineteenth Century, had clearly expressed such a concept, conjugating philosophical reflection, coming from Anglo-Saxon associationism of empiricist tradition, with Darwinian intuitions and with the experimental acquisitions by the physiologists of the "Berlin Circle", Helmholtz among them). Movement is a basic factor in infant development: it is through observation

emerges from neurosciences as the recognition of the capacity of the body to anticipate, imagine, mimic and forecast the actual body movement. The fundamental theoretical assumption, the unifying frame, is the constitutively temporal and material dimension of experience, the mutually formative interaction between organism and environment. Experience is conceived as an anticipatory construction, insofar as it is considered the adaptive outcome of the essentially active nature of a subject that determines by itself the object of possible experience. It is an important step, maybe nearly the goal, of a process of naturalisation of mind that from Darwin to contemporary neurosciences has aimed at arriving at symbols starting from matter, rather than looking at the latter, in our perception of reality, in terms of hypotheses and calculations, languages and symbols to decipher. Contrary to the 20th Century functionalistic approach, brain is not conceived as a computer, nor as any machine resembling an AI device, rather it is an original biological construction, the product of evolution, history and culture. Looking at the brain as a "proactive" rather than a "reactive machine", perception and consciousness are fundamentally predictive functions, insofar as they allow anticipation of the consequences of actual or potential actions.

In the progress of psychological research on perception, its projective character is testified by many experimental data on the capacity to "fill in the gaps", integrating the missing information.[3] These data are made intelligible

---

and motor action that the child operates a series of concrete learning actions that progressively develop into abstract concepts. The development of human mind unfolds along stages that are based on the concreteness of motor actions and sensations, instead than on the abstraction of language and logico-symbolic thought: we adjust to reality through forms of learning and generalisation. It will at this point be useful to recall the words of Piaget (1896–1980), to whom — as it is well known — consciousness is based on the concrete activity of the entire organism, in the sensory-motor coupling of mind, body and environment; cognitive structures emerge from recurrent schemes of sensory-motor activity, mostly unconscious basic capacities. According to Piaget [26] biology and evolution, constructivism and history, have to lead research on the mind. Every kind of knowledge is linked to an action, and knowing an object or an event means using them, assimilating them to schemes of action. Knowing does not mean, in fact, copying reality, rather acting upon it and transforming it (apparently and actually), so as to understand it as a function of the systems of transformation to which those activities are linked. Sensory-motor intelligence consists in directly co-ordinating actions, without going through representation or thought. Perception has a meaning only inasmuch as it is linked to actions.

[3] The obvious reference is here to the "revolutionary" acquisitions by Gestalt Psychology in the early 1900s. Considering the epistemological standards derived from mechanical physics and empiricist epistemology inadequate to the interpretation of some important mind-related facts, they stressed the necessity of keeping in mind the fundamental value of the experimental method, upholding at the same time the priority of a phenomenal dimension and the need for a holistic approach, aiming at the overcoming of the mind-nature dualism and at the eliminating of the distinction between sensation and perception, experience and "external" reality. The critique to the notion of alterity of environment with respect to mind is based on the fact that to each organism a behavioural environment inheres, and each organism is the centre of its own environment. To Wolfgang Kohler (1887–1967), effects depend not only on given causes, but also on the characteristics of the system in which they come into being. And, according to a metaphor by Kurt Koffka (1886–1941),the builder puts his own bricks together and builds the house: he forgets to have piled them within a gravitational field, without which no house could be built, just as it could not be built without bricks; but bricks are so much more tangible than gravity, that he only cares about them; so his concept of reality is forged.

by the hypothesis that the brain operates as a simulator, constantly inventing models to project on a constantly changing outside world. In this perspective, emphasizing the plastic, flexible and adaptive character of biological mechanisms, in the context of an ecological approach to mind and behaviour,[4] the nervous system is conceived as a complex and dynamic generator of hypotheses and, consequently, the brain does not limit itself to produce responses to stimuli, to passively combine sensations and to organise perceptions in view of successive transformations. Instead, it bases itself on an internal repertoire of actions, that make it a simulator capable of evaluating the interaction among goal-directed actions and their consequences.

As is evident, contemporary researches have produced a lot of hypotheses and data, clearly hinting at the possibility to isolate a single explicatory principle in the motor model of cognitive functions. A reference frame comes into vision, unitary enough for the study and explanation of phenomena, but within a plurality of approaches, theoretical assumptions and research perspectives. It is only fair to remember, however, that, on the one hand, science itself is often subject to fashion (and the contemporary emphasis on the motor component of mind certainly runs the risk of becoming one); on the other hand, that stressing the complex, integrated and dynamic dimension of living being always runs the risk of being perceived as a "mystic permeation" of organism and environment, and that the "top-down" approach runs the risk of being assimilated to a holism that has had in the past strong anti-scientific tones (the Gestaltists themselves, as it is well known, were in some sense accused of this by critics). This would rather seem an instance of the developmental dynamics of scientific knowledge, characterized by the re-surfacing — this time in an evidence-based fashion, at the experimental as well as the clinical level — of a theoretical frame and approach to the living being that in the course of history, with varying fortune, has importantly contributed to the scientific understanding of mind, starting from mid-1800s.

2   Movement as a cognitive factor in a historical perspective: from reflex to action

The historical and interdisciplinary dimension of the motor theory of mind stems clearly from the analysis of different aspects of scientific and philosoph-

---

[4]In Gibson's (1904–1979) "ecological perspective", the world we perceive is not the world of physics, or of geometry, in which space is an abstraction and the position of an object is specified by the co-ordinates of given axes in an isotropic space. The world, or, better, the environment, is the eco-system in which the organism is immerged, in a dialectical complementary relation. From this holistic, dynamic and integrated theoretical assumption, Gibson derives a critique to classical analysis of perception, which distinguishes sensory data from the meaning they would receive by means of an intellective act. Perception is, instead, an active process depending on the organism/environment interaction and it is always fundamentally gained in relation to the percipient body's position and to its activities. The ecological theory of perception therefore postulates that the act of perception directly gathers information, without implying any involvement of conscience or any mechanisms for the elaboration of stimuli in a sort of "internal theatre". Gibson proposes instead to critically re-consider perception and cognition in the light of direct realism and affordance. An affordance transversally cuts the subjective/objective dichotomy; it is directed in both directions, towards environment as well as towards the observer. The idea of an interface between us and the world is useless and unintelligible and, with regard to this relational aspect of affordance, Gibson recognises the Gestaltic origin of the term.

ical thought in the 1800s and 1900s: it hints hypotheses and models which have been more or less abandoned or included trough re-definition by the contemporary cognitive sciences. The above sketched concept of "perception-action-cognition" is based on the idea that all the organism's resources, used in action as well as in perception, substantially share the character of motor anticipation, and that the understanding of actions rests on "a sort of [species-specific] vocabulary of actions related to prehension" [30, p. 220].

> What characterizes action and differentiate it from a movement is the presence of a goal. Action is accompanied by the creation of an expectation that the goal will be met. Thus, an individual performing an action is able to predict its consequences. He knows what to expect. Objects, as pictorially described by visual areas are devoid of meaning. They gain meaning because of an association between their pictorial description (meaningless) and the motor behaviour (meaningful). The starting process is motor and is based on the expectations about the final outcome of progressively more and more complex actions. The neurophysiological data provide a new insight about the neural mechanisms that might subserve the process of object categorization and action understanding. Both these processes in our perspective seem to be deeply grounded in the bi-directional relationship between agent and environment. This relationship is basically dependent upon action execution. Action appears to represent the founding principle of our knowledge of the world [30, pp. 221–227].

In relation to this concept, the historical perspective emphasizes how, in the course of time, through different theoretical routes, the development of both philosophy and scientific knowledge has led to a process of naturalization and progressive embodiment of mind, deeply changing the traditional concept of cognitive functions and rooting them in the organism's development and in its interaction with environment. In the historical development of the knowledge on mind and behaviour, produced in the last two Centuries, it would be possible to choose several different case studies, in order to reconstruct a sort of map, to facilitate orientation within the complex theoretical landscape of the progressive naturalization of mind. Here I will only consider one single 'chapter' of this dense and stimulating theoretical route,[5] promising in terms of heuristic value and developments, the so-called "Physiology of Activity" developed within the Russian physiological community in the second half of the 1900s as a deepening, a critique and, finally, an overcoming of the reflex concept. The analysis will be especially focussed on Bernstein's theory and on the complex motor model of mind he develops exactly as an attempt to theoretically overcome the simple S/R account of behaviour. The deepening of the reflex concept — initially conceived as an arc, a linear and sequential connection between sensation and movement — has led Bernstein to question the neat distinction between stimulus and response, posture and voluntary activity (traditionally conceived as a sum of complex motor sequences made up of simple reflex "building blocks").

By the end of the 1800s, already Dewey (1859–1952), thinking about the reflex arc as a possible key to understand motor behaviour, states the inadequacy of an elementary approach in psychological investigation and, more at large, for a biological understanding of the organism, whose activities, of whatever nature, are always global and continuous processes. Dewey pointed

---

[5]With Kuhn, historical reconstruction becomes an essentually selective and interpretative activity, but data can retroact back on expectations.

out that the very distinction between sensation and movement, sensory stimulus and motor response, is but an abstraction if applied to behaviours other than simple automatisms. The distinction has of course been of great importance as a heuristic principle to investigate the functioning of the nervous system, but it overlooks the bare fact that in the organism's actual behaviour there always is a fundamental circular connection, so that response actually acts back on stimulus. This lets the observer appreciate some aspects previously not adequately evaluated, to produce, as a consequence, a new, more effective response that will in turn trigger a new circular process, and so on. In Dewey's own words, it would be more appropriate to look at the reflex arc as a "reflex circle": "The circle is a coordination. It is the coordination which unifies that which the reflex arc concept gives us only in disjointed fragments. It is the motor response which assists in discovering and constituting the stimulus. It is the holding of the movement at a certain stages which creates the sensation, which throws it into relief" [11, p. 370].

Few years later, Sherrington (1857–1952) conceived the reflex not as a simple reaction elicited by a specific organ, better as an already co-ordinated movement, depending on the excitement of a given region of the organism, whose effects are also determined by the organism's global state. "A simple reflex arc is probably a pure abstract conception, because all parts of the nervous system are connected together and no part of it is probably ever capable of reaction without affecting and being affected by various other parts, and it is a system certainly never absolutely at rest". In other words, the reflex movement, even in its most simple, analytical aspects, is a form of behaviour; it is the reaction of an organic whole to a change in its relation with environment [31, pp.7–8]. Beyond these important changes in perspective produced, on one hand, by the functionalist and pragmatic American philosophy/psychology and, on the other hand, by British neurophysiology, I consider the development of Soviet "Physiology of Activity" as a paradigmatic example of the production of a drastically different conception of mind, still from within an undoubtedly reductionistic and experimental theoretical framework which originally conceived the reflex as a constitutive "building block" of nervous activity, the minimal unit to account for mind and behaviour in neurophysiological terms.

In the mid-1800s, Secenov (1829–1905)[34] had first tried to trace the whole behaviour back to reflex, and to reduce mental processes to physiological mechanisms. He demonstrated that the brain can produce inhibitory influences on the reflex activity (developing an intuition already put forward by Weber in the 1840s), so he employed the concept of "inhibitory action" in the nervous system as a means to overcome the clear limits of any attempt to account for behaviour exclusively in terms of reflexes. In his thought we can recognise the premise of the whole theoretical horizon of Soviet Reflexology, which has in Pavlov and Bechterev its most outstanding representatives. At the dawn of 1900s, Pavlov (1849–1936), who recovered and developed

Secenov's intuitions, was among the founders of the so-called "Reflexologic School" and proposed a more dynamic conception of reflex, enriched by the effects of experience (conditioning).

> The inborn reflexes by themselves are inadequate to ensure the continued existence of the organism, especially of the more highly organized animals. The complex conditions of everyday existence require a much more detailed and specialized correlation between the animal and its environment than is afforded by the inborn reflexes alone. This more precise correlation can be established only through the medium of the cerebral hemispheres; and we have found that a great number of all sorts of stimuli always act through the medium of the hemispheres as temporary and interchangeable signals for the comparatively small number of agencies of a general character which determine the inborn reflexes, and that this is the only means by which a most delicate adjustment of the organism to the environment can be established. I have termed this new group of reflexes conditioned reflexes to distinguish them from the inborn or unconditioned reflexes. Compared with the inborn reflexes, these new reflexes actually do depend on very many conditions, both in their formation and in the maintenance of their physiological activity. We might retain the term 'inborn reflexes", and call the new type "acquired reflexes"; or call the former "species reflexes" since they are characteristic of the species, and the latter "individual reflexes" since they vary from animal to animal in a species, and even in the same animal at different times and under different conditions [24, p. 17].

The reflex concept retains therefore its validity in accounting for the complex and dynamic way in which the animal's behaviour adapts to the environment. Pavlov's conditioning shows the reflex to be plastic and modifiable by experience, thus plausibly conceivable as the basic neurophysiological mechanism of learning and of all the "higher functions" of the nervous system. In the same years, Bechterev (1857–1927) viewes these functions in terms of coupling of reflexes, or progressively more complex integrations thereof, the so-called "associative reflexes". To Bechterev's opinion, Reflexology consists in examining from a rigorously objective standpoint not only the most elementary, but also all the higher human functions that in everyday language are called psychic activity. Thus, the investigation has to be limited to the external features of human actions and it is necessary to undertake a naturalistic observation of the subject in its social environment, with the aim of defining the relations between man and the surrounding physical, biological and, especially, social world [3]. Around the half of the XXth Century it is Anochin (1908–1989), the most famous pupil of Pavlov, to call the attention of neurophysiologists on the need to finally overcome the reflex bottleneck, in order to concentrate on the complexity and on the integrated and unitary dimension of action. Studying conditioned reflexes under Pavlov's guidance, Anochin came to a radical critique of the traditional physiological culture: to his opinion, in fact, one of the most meaningful aspects of the history of brain research has been the complete exclusion of the results of action from the physiological concepts. This obviously has been a serious methodological limit in the study of the integrated activity of the brain, since it is the very results of action that constitute the final goal of behaviour. The reflex arc concept holds nervous processes as linear by nature, leading the physiologists' attention on the accomplished fact, lying thus down an impenetrable barrier between the act itself and the evaluation of the obtained results, which are an intrinsic consequence of action. "The behavioural act (conceived as a functional system) has a harmonious structure, an integral unity. The behavioural act constitutes the link between neurophysiology, higher nervous activity and psychology" [2]. Thus, from within experimental neurophysiology, deeply rooted in the quest for the simplest elements, the presumed minimal units of behaviour, the necessity had grown to acknowledge the integrative, goal-oriented, dynamic and unitary nature of behaviour. Anochin is, with Bernstein, one of the great representatives of the Physiology of Activity, the Soviet "School" that between the 1930s and the 1960s implemented a qualitative shift in physiological and psychophysiological investigation, from the acknowledgment of the bare fact of integration to a real systemic perspective. Once the mechanism has been abandoned, to embrace the concept of "process", it is not sufficient to just assume the integration among reflexes: one must recognise the specific organisation of the system itself. Bernstein's theory brings to completion the critique of the reflex arc, as well as of the traditional rigid concept of the relation between stimuli and responses; conversely, the fundamental value of the motor component for the development of mind and the organisation of behaviour is emphasized.

3 Bernstein on action and perception: movement and mind

"Reading Bernstein is somewhat like reading the Bible" [32, p. 22]. These words clearly express how Bernstein's work on motor control in the last decades of XXth Century was recognised as the starting point of contemporary movement sciences, providing a new understanding of the organization of movements. Nikolaj A. Bernstein (1896–1966) is actually considered "the father of motor control in humans" [20], with special reference to natural, voluntary, non-automatic (naturally occurring) movements. It must be underlined, however, that his contribution is mostly well known within the "human movement sciences" community (rehabilitation, sports training, sport medicine), whilst almost unmentioned by scholars interested in behaviour, mind and mind-brain relations. Even Lurija (one of the "fathers" of contemporary neuropsychology) defines him "a rare case of a scientist who practically devoted his whole life to one problem: the physiological mechanisms of human movements and motor actions", just overlooking Bernstein's interest in brain and mind, in the integrated models of behaviour and their epistemological value. It is instead of the utmost importance to underscore how Bernstein actually aimed at understanding the brain through the study of movement and, vice versa, how he used his knowledge of the brain to improve and develop knowledge on movement. By integrating different theoretical approaches and methodologies in his own research,[6] he tried to correlate all the different levels of organisation of movement, with the aim of defining a new, ecological and integrated, concept of mind and behaviour. It is exactly this emphasis on the

---

[6] Starting from a mechanistic position, Bernstein adopted in the 1930s and 1940s a global dynamic approach; he went through a renovated mechanism and cybernetics in the 1950s, to finally reach in the 1960s an "ecological" and again dynamic conception, that will allow a completely naturalistic account of behavioural planning, without recurring to any dualism whatever.

interaction among brain, motor system, natural and cultural environment, that should be acknowledged as his most relevant contribution. It is my opinion that the great heuristic value of his interdisciplinary approach and of his peculiar theoretical progression extend the relevance of his contribution well beyond contemporary movement sciences, making it a theoretical articulation of crucial importance to the development of a motor model of mind.

Starting with his works of the 1930s, and then with the collections of his most relevant works, appeared in the 1960s,[7] Bernstein accomplished a powerful synthesis of neurophysiology, psychology and cybernetics, introducing in the study of motor system physiology new methods and concepts: action-perception cycle, "motor synergies", posture as "keeping oneself ready to action". The starting point of his experimental work are his researches on biomechanics and on the physiology of movement, within a clearly neuropsychological theoretical frame since the beginning.

With the aim of extending the knowledge of the brain through the study of movement, in fact, since 1924 he started a collaboration with Kornilov's Moscow Institute of Experimental Psychology. There Vygotskij and Lurija worked, who will become (together with Leontijev), the highest representatives of the "Psychology of Activity", a psychological model that emphasizes the role of action and experience in the development of mental functions, and the social dimension of human behaviour, conceived as a complex of essentially cultural "higher functions" intrinsically different from the lower, "natural" ones.[8] Without explicitly referring to this psychological approach, in 1962 Bernstein will name his theoretical system "Physiology of Activity" to highlight his attempt to provide a non-idealistic alternative to Pavlovian Reflexology, based on a small number of basic pillars: movement, brain and mind, organism and environment. Developing a hierarchical conception[9] of nervous control of movement, based on evolutionism and clinical neurosciences, Bernstein proposes two basic concepts: 1) movement as structure; 2) motor regulation and control (hierarchically organised co-ordination). Movements are not to be seen as chains of details, rather as structures articulated into details; they are structural wholes, characterised at the same time by a high degree of differentiation of the elements and by differences in the relations among the parts.

Thus, he comes to underline the importance of an organisation in which the same goal is reachable by different paths, i.e., the "functional non-univocality between impulses and effects: Changes in muscle tension bring about a movement and the movement affects the condition of the muscles by shortening or stretching them causing further changes in their tension. Consequently, this form of interaction does not presuppose a one-to-one correspondence between force and movement, that is, one and the same sequence of changes in forces

---

[7]Almost all the following quotes are taken from [6], a selection (and English translation) of Bernstein's works, made by the author shortly before his death.

[8]In a game of reciprocal acknowledgements, Leontjev himself, in 1959, underlines the importance of Bernstein's theory of the multilevel and hierarchical motor co-ordination, and of the fundamental role it attributed to the relation between 'moving organism' and environment, as a theoretical input towards the development of his own theory of activity.

[9]Clear, in this respect, is the influence of Jackson [17].

may produce different movements on successive repetitions" [4, p. 62]. This precludes the mechanistic idea of a central signal "just striking a piano key" [4]. In motor control there is a circular flux of information, aimed at assuring the overall co-ordination of movemement organs, conceived as complex systems. Such a position implies a shift from purely descriptive biomechanics to the problems of central control and regulation of movement, starting exactly from a critique of the reflex concept, elaborated — it must be emphasized — from within a materialistic perspective.

Thus Bernstein succeeds in deeply penetrating the structure, organisation and programming of goal-oriented motor acts and comes to focussing on the crucial concept of "co-ordination" as "overcoming excessive degrees of freedom of our movement organs, that is, turning the movement organs into controllable systems" [5, p. 41]. "The reflex arc cannot exist — he claims — and the organization of movement requires reflex rings" [4]: "The period of struggle towards the recognition of the biological importance, the reality and the generality of the principle of cyclical regulation of life processes is now behind us" [4]. The organisation of motor apparatus control, typical of biological systems, implies afference as well as efference, perception as well as action. In the action, "a whole sequence of movements that together solve the motor problem, all the movements are related to each other by the meaning of the problem" [5, p. 146].

Mastering the very many degrees of freedom involved in a particular movement, reducing the number of independent variables to be controlled, the organization of movement, coordination, emerges as the reciprocal attunement of several simultaneous kinetic and informational processes. An interdisciplinary and integrated approach, and a new concept of movement, call then for a new theory, both of behaviour and of brain organisation. Conceiving co-ordination as a patterning of body and limbs motions relative to the patterning of environmental objects and events, Bernstein views it both as a process and as a structure showing itself in the "motor field", i.e. the space in which movements take shape.[10] Hence he develops his notion of localization in the brain, in clear accordance with what will be Lurjia's theory of diffused localisation in a functional system:[11] the brain is the centre of diffused and parallel processses, the central signal is written in terms of the overall structure of movement and not in terms of its spatial details. Thus, from the study of motor co-ordination Bernstein obtains an insight into the "true categories" [4] of the organization of movement and of the brain itself.

This conception implies a harsh critique to the Reflexology of Pavlov, who "failed to understand the brain because he failed to understand its most important function, that is, the organization of movement" [5].[12] In this theoret-

---

[10] Bernstein stresses that the motor field has a global topology rather than a specific metrics; here he explicitly refers to Kurt Lewin for a "non-Euclidean, non-rectilinear geometry" [4].

[11] Lurija 1962, 1973.

[12] The clash between politics and science in the Soviet Union is one of the most important factors in Bernstein's biography: his idea that motor behaviour never replicates itself identically is in fact incompatible with the neo-pavlovian theory of conditioned reflexes. Bernstein is therefore considered a public enemy, and is fired from his job on the grounds

ical position the influence is clear of the XXth Century German thought and of its search for alternatives to the mechanism/vitalism counterposition. So, by developing a comparative and evolutionary approach (based on what he calls "interphyletic awareness"), in the early 1900s Bernstein proposes himself as "an exception to the overall distinction between the domain of neuronal control and that of motor behaviour [...] His research integrated concepts deriving from the behavioural field with neurophysiological, neuromuscular and biomechanical data, especially in the study of locomotion [...]. While the two domains (behavioural/neuronal study of movements) were progressively integrating in Russia, this was not happening in the USA or Great Britain, where most of the studies on movement were taking place" [32]. Self-determining goals and trying to find ways to solve motor problems are functional properties of the cortex; however — Bernstein holds — neurophysiology is by itself not sufficient to explain these higher phenomena; it is necessary to develop a sort of motor model of mind, in-between neurosciences and psychology, adopting action as a theoretical framework to the study of mind: "every skill arises in answer to a particular motor problem" [4].

> Motor problems arise out of the external environment, upon which the organism actively operates and from which it receives sensory feedback. Biological activity implies the cognition of the surrounding world through action and the regulation of action within it. Each meaningful motor directive demands not an arbitrarily coded, but an objective, quantitatively and qualitatively reliable representation of the surrounding environment in the brain. This also leads to knowledge through action and revision through practice which is the cornerstone of the entire dialectical-materialistic theory of knowledge and serves as a sort of biological context for Lenin's theory of reflection [4, pp. 114–120].

"Physiology of Activity" aims to be a non-metaphysical, naturalistic understanding of life: animals pursue aims which must have a natural origin. If movements are goal-directed, they must be controlled by something "as yet unrealized", i.e. a sort of "model of the future" [4]. In their interaction with the environment, organisms must "plan action through an active sampling incorporating a measure of uncertainty into their motor acts. By the way of a probabilistic extrapolation they predict the course of events in the environment" [4].[13] Since the 1930s.[14], then, Bernstein identifies the key to understand movement of organisms in the goal of action and in the formulation of the motor program. He then considers Cybernetics insufficient for a convincing account of the essential features of life: "the honeymoon of this union between automatic processes and physiology is over. Cybernetics may capture self-programming automata that are able to estimate what will happen but cannot model what has to happen" [4]

---

of his "displayed adoration of foreign scientists [he] neglected the importance of the work of Pavlov" (cited in [13]). Only after Stalin's death (1953) he will be gradually "rehabilitated" [8].

[13] See the "proactive" model of the brain recently formulated by A. Berthoz: "The brain is above all a biological machine for moving quickly while anticipating. Evolution obviously selected receptors capable of predicting the future" [6].

[14] "The problem of the relation between co-ordination and localization" is published by Bernstein in 1935, at least twelve years before it was focussed upon by Wiener and the Cyberneticists.

Integrating, through an accurate philosophical and psychological elaboration, the laws of control and regulation of the whole organism's movements into a wider concept of "living being's activity", based on biological, cultural and social factors with cybernetic principles, Bernstein formulates in the 1960s a fully naturalistic — neurobiological and psychological — account of goal-oriented behaviour. Such a conception clearly shows many important common features with the ecological psychology being developed more or less at the same time by James Gibson, centered on the basic tenet that one must move in order to perceive, but also perceive in order to move, its ground assumption being the mutuality of an animal and its environment.

In conclusion, a historical reconstruction of Bernstein's thought shows how, through a conception of organism as a self-regulating system, that actively accomplishes the genetically-and environmentally-determined goals of its action, a decisive qualitative shift is produced, from within the materialistic-dialectical analysis of the relation between organism and environment. So, the limits set by classical physiology and reflex theory (the Pavlovian concepts, as well as the S/R model of the behaviourists, which were dominant, as it is well known, in the mid-1900s) are overcome. Bernstein's "poor orthodoxy",[15] his daring and pragmatic theoretical and methodological eclecticism, are thus determinant factors which led him, who studied movement with an eye on brain and mind, to develop hints, intuitions and suggestions that represent important premises to, and meaningful theoretical elements of, the contemporary motor model of mind.

In the most recent studies on the physiology of movement, and in the discovery of mirror neurons, it is then possible to dig out the neurophysiological evidence, the experimental grounding, of a model that has appeared, disappeared and re-emerged over and again in the development of behavioural and mind sciences. And, without constraining historical analysis within silly quests for precursors [9], recognising instead resemblances and "family likeness" (à la Wittgenstein) among concepts and hypotheses developed over time, with the aim to find the solution to a specific problem, can help perceiving the actual historical dimension of the development of knowledge. It can help to grasp the ways in which in the course of time a process of naturalization of the mind has taken place on the basis of a functionally integrated approach to the organism-environment system. The minimal unit of analysis is the perception-action cycle in intentional contexts, and the unifying theoretical frame is the continuous dialectic relation between man and its physical, biological, historical and cultural environment.

BIBLIOGRAPHY

[1] T.D. Albright, T.M. Jessell, E.R. Kandel and M.I.Posner. Neuroscience. *Cell*, 100: 1–55, 2000.
[2] P.K. Anochin. *Voprosy psichologii*, n. 3. Transl. in L. Mecacci. *Neurofisiologia e cibernetica*, Ubaldini Editore, Roma 1973, pages 50–79.
[3] V.M. Bechterev. *Obscie osnovy refleksologii celoveka*. Tranls. in L. Mecacci. *La psicologia sovietica 1917–1936*. Editori Riuniti, Roma 1976.

---

[15]The reason why he had no official position and many limits to his scientific activity during all his life.

[4] N.A. Bernstein. *The Coordination and Regulation of Movement*. Pergamon Press, Oxford 1967.
[5] N.A. Bernstein. On dexterity and its development. In M.L. Latash and M.T. Turvey (eds.), *Dexterity and its Development*, N.J. Erlbaum, Mahwah 1996.
[6] A. Berthoz. *Le Sens du Mouvement*. Odile Jacob, Paris 1997. Engl. Transl. *The Brain's Sense of Movement*, Harvard University Press, Cambridge, MA 2000.
[7] A. Berthoz and J.-L. Petit. *Phénoménologie et physiologie de l'action*. Odile Jacob, Paris 2006.
[8] R. Bongaardt and O.G. Meijer. Bernstein's theory of movement behaviour: Historical development and contemporary relevance. *Journal of Motor Behaviour*, 32 (1): 57–71, 2000.
[9] G. Canguilhem. *Idéologie et rationalité dans l'histoire des sciences de la vie*. Librairie Philosophique J. Vrin, Paris 1997.
[10] A. Damasio. *The Feeling of What Happens: Body and Emotion in the Making of Conscousness*. Harcourt Brace, New York 1999.
[11] J. Dewey. The reflex arc concept in psychology. *Psychological Review*, 3: 357–370, 1896.
[12] G.M. Edelman. *Second Nature: Brain Science and Human Knowledge*, Yale University Press, New Haven and London 2006.
[13] I.M. Feigenberg. Chronologisches Verzeichnis aller Publikationen N.A. Bernstein. In L. Pickenhain & G. Schnabel (eds.), *Bewegungsphysiologie von N.A. Bernstein*, 2nd ed., Barth, Leipzig 1988, pages 255–263.
[14] W. Freeman. *How Brains Make Up Their Minds*. Columbia University Press, New York 2000.
[15] V. Gallese and G. Lakoff. The brain's concepts: The role of the sensory-motor system in reason and language. *Cognitive Neuropsychology*, 22: 455–479, 2005.
[16] S.R. Hurley. *Consciuosness in Action*. Harvard University Press, Cambdrige, MA 1998.
[17] J.H. Jackson. The Croonian Lectures on the evolution and dissolution of the nervous system. In *Selected Writings of John Hughlings Jackson*. Hodder and Stoughton, London 1932
[18] P. Janet. Autobiography. In C. Murchison (ed.) *History of Psychology in Autobiography*. Vol. 1, 1830, , pages 123–133.
[19] M. Jeannerod, *De la physiologie mentale. Histoire des relations entre biologie et psychologie*. Odile Jacob, Paris 1996.
[20] L.P. Latash and M.L. Latash. A new book by N.A. Bernstein: "On Dexterity and Its Development". *Journal of Motor Behavior*, 26: 56–62, 1994.
[21] L. Mecacci. *Neurofisiologia e cibernetica*. Ubaldini Editore, Roma 1973.
[22] L. Mecacci. *La psicologia sovietica 1917–1936*. Editori Riuniti, Roma 1976.
[23] R. Nunez and W. Freeman. Reclaiming cognition: The primacy of action, intention and emotion. *Journal of Consciousness Studies*, 6: 11–12, 1999.
[24] I.P. Pavlov. *Conditioned Reflexes: An Investigation of the Physiological Activity of the Cerebral Cortex*. Oxford University Press, London 1927.
[25] J.-L. Petit (ed.) *Les neurosciences et la philosophie de l'action*. Vrin, Paris 1997.
[26] J. Piaget. *Biologie et Connaissance*. Gallimard, Paris 1967.
[27] R. Port and T. van Gelder. *Mind as Motion: Explorations in the Dynamics of Cognition*. MIT Press, Cambridge, MA 1995.
[28] G. Rizzolatti et al. Understanding motor events: a neurophysiological study. *Expermental Brain Research*, 91: 176–180, 1992.
[29] G. Rizzolatti and L. Craighero. The mirror neuron system. *Annual Review of Neuroscience*, 27: 169–192, 2004.
[30] G. Rizzolatti and V. Gallese. From action to meaning. A neurophysiological perspective. in J.-L. Petit (ed.), *Les neurosciences et la philosophie de l'action*, Vrin, Paris 1997, pages 217–229.
[31] C.S. Sherrington. *The Integrative Action of the Nervous System*. Constable and Co., London 1906.
[32] R.A. Schmidt. *Motor Control and Motor Learning. A Behavioral Emphasis*. Human Kinetics, Champaign, IL, 2nd ed. 1988.
[33] R.A. Schmidt. Motor and action perspectives on motor behaviour. In O.G. Meijer and K. Roth (eds.), *Complex Movement behaviour: The Motor-Action Controversy*, North-Holland, Amsterdam 1988, pages 3–44.
[34] I.M. Secenov. *Refleksy golovnogo mozg*. Engl. transl. *The Reflexes of the Brain*, the MIT Press, Cambridge, MA 1965.
[35] O. Sporns and G.M.Edelman. Bernstein's dynamic view of the brain. The current problems of modern neurophysiology. *Motor Control*, 2: 283–305, 1998.

[36] S.L. Star. Triangulating clinical and basic research: British localizationists 1870–1906. *History of Science*, 24: 29–48, 1986.
[37] D.G. Stuart. Integration of posture and movement: Contributions of Sherrington, Hess and Bernstein. *Human Movement Science*, 24: 621–643, 2005.
[38] M.T. Turvey. Coordination. *American Psychologist*, 45: 938–953, 1990.
[39] T. Van Gelder. Dynamic approaches to cognition. In R. Wilson and F. Keil (eds.), *The MIT Encyclopedia of the Cognitive Sciences*, MIT Press, Cambridge, MA 1999.
[40] H.T.A. Whiting (ed.) *Human Motor Actions: Bernstein Reassessed*. North-Holland, Amsterdam 1984.

The brain, the person and the world
JEAN-LUC PETIT

It is not unusual to hear researchers in the neurosciences and the cognitive sciences saying that "the brain acts", "the brain decides", "the brain anticipates", "the brain simulates (or emulates) the real". However enigmatic or indeed senseless such expressions might seem to be at first sight — they require interpreting by the philosopher as the translation of a physiological thinking looking for the right way to view things and who strikes out in the direction of a type of description which still makes sense in a context where ordinary language is no longer relevant. Since all our usual ways of talking about practical or cognitive activities relate to the whole person, we still lack a language capable of reaching back to the point at which the organisms "strives to make sense of" — a form of words which is still too heavily marked by a vitalist teleology remote from the computational mechanicism dominant in the neurosciences. For all that, just such an effort at making sense does actually find expression across biopsychological values (hedonic, affective, pragmatic and not just cognitive values) progressively superimposed upon the activation patterns of cerebral circuits as they gradually get enriched. To the extent lthat the living system including not just the brain and the body of the individual but also its socio-ecological environment — functions in a normal or pathological mode and that the conditions of this functioning are integrated into intra- or extra-cerebral regions that are ever more varied, ever more extended and remote from each other.

1 Reductionism and anti-reductionism: A Dichotomy?

It is always surprising for a philosopher to note that the same scientists who in their laboratory research are extremely cautious in establishing the facts of the matter display a audacity bordering on ingenuity when it comes to giving public expression to their understanding of living organisms, particularly in what concerns their mental activity and the cerebral functioning that underpins it. Freely extrapolating from a pre-critical ontological thinking, they make conjectures about each level in the analysis of the biological substrate of human experience to make of it the direct support of mental states or the agent responsible for actions. "The brain, this or that cerebral circuit, the neuron wants or decides this or that". Without worrying about being at odds with their own professed functionalism they even seem to want to propose a teleological conception : "The brain, this or that cerebral circuit or again the neuron exists essentially (or is there for) this or that [...]". However opposed they may be officially to any dualism of the mind and brain they are unable to do without a substantialized mind that haunts the cerebral material, referring to the brain as a sort of demiurge capable of perceiving or of doing anything that the person is capable of perceiving and doing. But since this demiurge could not possibly possess this power as simple material in the brain, they are obliged to surreptitiously confer upon it command of the body and access to the enviroment, which thereby gives rise to the illusion of a brain that contains everything! The paradox is that those who are most prone to this temptation to confer upon the brain a maximum of properties, by precipitating upon the material substrate aptitudes stemming from the whole person, are those who are normally classified as reductionists: the neurophysiologists

From another angle, when the anti-reductionists protest by saying that man can not be reduced to his brain, they do it by presupposing a conception of the brain that is itself seriously reductionist, if not physicalist:

> After all, neural activation, be it here or there in the cortex, is simply neural activation. Something more is needed to explain why a particular neural activation activates a particular learning-like quality, and another activates a particular seeing-like quality. How could different neural activations possibly give rise to different feelings? [17, p. 379]

> From the point of view of the brain, there is nothing that differentiates nervous influxes coming from the retinal, haptic, proprioceptive, olfactory, and the other senses, and there is nothing to discriminate motor neurons that are connected to extraocular muscles, skeletal muscles, or any other structures. Even if the size, the shape, the firing patterns, or the places where the neurons are localized in the cortex differ, this does not in itself confer upon them any particular visual, olfactory, motor or other perceptual quality. [16, p. 941]

> You can no more explain mind in terms of the cell than you can explain dance in terms of the muscles.[...] we need to turn our attention away from individual neurons. [...] we need to widen our gaze to encompass large-scale populations of neurons and their dynamic activity over time. But why stop there? [...] Perhaps the proper scale at which to make sense of neural functions is that of the living, environmentally situated animal itself? If this seems like a far-fetched proposal, it may be because tradition teaches that the skull is the crucial boundary marking off what is inside from what is merely outside; and crucially, we are inside [12, p. 24].

Descriptions of this kind fail to take account of the θαυμάζω: the astonishment of the researcher confronted on a daily basis with the astonishing performances of the brain.[1] A brain that displays aptitudes one has difficulty in attibuting seriously and not in a merely metaphorical manner to anything less than the complete person. When the researcher feels obliged to talk about "the brain acting, deciding, anticipating, simulating or emulating reality" it is important, if one is a philosopher — more a philosopher following up the discoveries made in the neurosciences — to consider whether formulations as enigmatic if not plain senseless as these appear to be at first sight might not be the provisional substitutes for alternative ways of ushering in a new type of physiological thinking. A new kind of physiological thinking that is still trying to come to terms with itself and which is stumbling in the direction of more adequate descriptions in a context where ordinary language ceases to

---

[1] [Prehistoric man painting animals in a Lascaux cave] must have been amazed by what he saw, even though his brain was recreating these shapes from various cues, just as I am present as a spectator at my own lectures, an impression produced by a brain whose expression I listen to with astonishment [4, p. 135].

be relevant. In fact, since our habitual ways of describing cognitive activities refer to the whole person, we simply do not possess the language needed to render the "effort" made by the organism to make sense of ... Just such an effort is recognisable in the biopsychological values (hedonic, affective, pragmatic and not just cognitive values) with which the activation patterns of the cerebral circuits are invested as they move forwards gradually from the primary receptive areas to the association areas and from there to the motor areas; or again from the sub-cortical circuits of motivation to the cortical circuits of perception, of cognition and of action. But, in the debate between reductionism and anti-reductionism, no account is taken of this progression from lower to higher orders of meaningfulness.

"*This* (the mind) is nothing more than that (the brain as a cerebral tissue)". Against the reductionist who holds this true but terribly elliptical view, the anti-reductionist holds that the cerebral tissue is only what it is and that mind is not reducible to that. A new claim that is both true and terribly inadequate. What the the anti-reductionist does not see or pretends to ignore in his defence of the irreducible character of the mind or the person is that the conception of the brain he himself has uncritically taken over is itself extremely physicalist. Without knowing it, he is the product of a philosophical tradition going back to Descartes, refusing to see himself as a thinking subject in this "machine composed of flesh and bones that one encounters in a corpse" and finishing up with Hegel who responded to Gall's assimilation of mind (*Geist*) to a bone (the cranium) that slapping such a hollow head would only make it resonate, not get it thinking! Without always making their position clear, neurobiologist today are trying to get away from any physicalism of this kind. And they are trying to do this by moving towards a description of the brain that is not just anatomical and structural but also functional and dynamic: a description of the "brain in act" (Stanislas Dehaene), even of a "mental cinema" (Semir Zeki).

One finds this reductionism implicit in anti-reductionism in a recent development in the cognitive sciences: Embodied-Embedded Cognition. Initiated by Francisco Varela and now represented by the philosophers Andy Clark, Shaun Gallagher and Alva Noë, the psychologist Kevin O'Regan and others, this movement is trying to react against a neuroscience "looking exclusively at what goes on in the head or in the brain". Instead, what is upheld is an emphasis upon the contribution made to cognition by the body, by intersubjectivity as also upon the interaction of the organism with its environment. The contrast underlined in the passages cited above by Alva Noë, between lived experience, feelings or states of consciousness, on one hand, and the action potentials of neurons or electrical activity patterns in brain tissue, on the other, are to be read in the frame of this way of thinking. But this hidden reductionism is also to be found in major schools of contemporary philosophy. In phenomenology, the description of lived experience just as it is lived out from the point of view of the subject itself and without presupposing any underlying explanatory causes is a description that it ought, in principle, to be possible to develop on the plane of phenomenal appearances alone, just as they are apprehended within the horizon of the *Lebenswelt*. Here the reduction applies to the biological substrate of lived experience, and more particularly to the brain.

"Even though my body is at the centre of my experience, my brain, Paul Ricoeur observed, plays no part in my experience. It's an object for science" [6, p. 64]. Running parallel to this reductionism motivated by the defence of the irreducible character of lived experience, one also encounters in contemporary philosophy a linguistically motivated reductionism. Philosophy practised as logico-grammatical analysis claims that any speaker is in possession of a vocabulary of mental concepts sufficent to enable him to attribute mental properties to persons, to describe persons in mental terms, and to explain their behaviour in these same terms. Given that logicians have no doubts about the universality of language, the resources of this vocabulary should be enough to make it possible for the ordinary speaker to decide in a satisfactory manner any question concerning the mind of the other [20]. To be sure, ever since scientific psychology got started, there has been a tendency for ordinary language to borrow terms from the physiology of the brain. But the assimilation of such borrowed expressions can only lead to category mistakes which threaten to cloud ordinary language in obscurity and confusion. The systematic pursuit of the study of ordinary language aimed at forbidding any transgression of the "limits of sense" finishes up by enclosing he neurosciences in a physiology of the reflex and refusing to let them enter in the sphere of cognition [3].

2 The Homunculus in the brain: how to get rid of it?

As soon as expressions normally employed in reference to persons are recycled in the context of infra-personal sub-structures (brain, cerebral areas, sub-cortical centres, neuronal networks, cells assemblies or individual neurons) this very usage automatically introduces a reference to a fictive agent responsible for the thoughts and actions that would normally be imputed to the whole person. In his Preface to *New Essays on Human Understanding*, Leibniz warned us against the danger of falling back into a

> barbaric philosophy, like that of certain philosophical scholastics and doctors of the past, who, crippled by the barbaric character of their century, and today rightly disregarded, saved appearances by concocting occult qualities and faculties envisaged as little demons capable of doing unwittingly what one wanted, as if our pocket watches marked the time in virtue of a certain chronological faculty without needing wheels, or as if mills crushed grain by virtue of a fractional faculty, without needing mill stones.

The lesson Leibniz drew from this, to stick to mechanisms for the explanation of corporeal movements and to limit recourse to internal faculties for the living individual and its mental activity, would not be of much help in the cognitive sciences, where what is sought are the cerebral mechanisms correlated with mental acts.

So that it does become possible to accept the claim advanced by Ryle, Bennett and Hacker that certain speakers misuse language when they attribute personal properties to parts of the brain. But this doesn't prevent other speakers from abusing language in the same way by simply doing their job as neurobiologists. More precisely, in so doing they are simply testing Horace

Barlow's hypothesis linking mental concepts to the responses of individual neurons:

> The firing of one neuron would be important enough to trigger a major decision, such as stopping at a traffic light [...] I am suggesting that one cell would be enough, and the following psychophysical linking hypothesis expresses this claim: Whenever two stimuli can be distinguished, in normal life or in a psychophysical experiment, the proper analysis of the impulses occurring in *a single neuron* would enable them to be distinguished with equal or greater reliability. One can argue for the correctness of this hypothesis along the following lines. Nerve cells are the only means we know about whereby items of information occurring in different parts of the brain can be combined; sensory discriminations require the combination of information from different parts of the brain; therefore this operation must be performed by a cell, and if one could record from the cell that did this, one would obtain results at least as good as those of the whole animal [1, page 133–134].

But exactly how is this kind of selectivity of the information carried by the activity of an individual neuron possible? In fact this selectivity is already written into the presuppositions of the electroencephalographic record, since the linear and hierarchic organisation of the nerve pathways carrying the cognitive information has simply been assumed. In the frame of such a linear and hierarchic organisation, higher order neurons collect, combine and synthesize the information transmitted to these neurons by numerous neurons of a lower order. From level to level, an ever more important body of information is concentrated in an ever more limited number of neurons. And to such a degree that, if only two cells remained to be activated at the penultimate stage in the hierarchy dealing with the processing of visual information, it might be difficult to avoid admitting the logic of Barlow's position when he states:

> I don't see how the information from the two or more essential cells could be combined, except by *another cell*.

This is how the concept of the "grandmother cell" was devised, a hypothesis claiming to have identified a neuron without which it would be impossible for you to recognise your grandmother, even if she were to present herself to you in person. As grotesque as such a hypothesis might appear to be, it shows what can be done with a neuron once its psychological performance is placed on a par with that of the individual. The theory of the neuronal encoding of cognitive information without wanting to minimize the importance of the improvements brought to this theory by introducing computational procedures — is fatally committed to this paradox. It should be added that the grandmother neuron is nothing more than a modern version of Leibniz' homunculus. Leibniz talked of little demons capable of accomplishing, unwittingly, what one wants. Unwittingly, that is, without employing known means. But this is exactly what we don't know about Barlow's cardinal cell, what he doesn't even claim that we know: "It is also true that we do not know *the means* by which such a cell is able to make the discrimination".

The most disconcerting thing is that the paradoxical character of the concept in question — the irrationality of recurring to humunculi in the brain — has not prevented the neurosciences from getting closer to an empirical verification of Barlow's hypothesis. We know that the organisation of the principal visual nerve pathways, including, in this order, the retinae, the retinal gangliae, the optic nerve, the lateral geniculate bodies of thalamus, and the striate and extra-striate cortices of occipital brain areas, is globally linear and hierarchic. Everything seems to happen as if the entire functioning of these pathways was organised in such a way as to lead from the sensorial captors towards the perceptual representation that gets constructed in the polar temporal regions (especially). Towards the peak of this hierarchy in the perceptual processing of visual information, in the superior bank of the superior temporal sulcus, neurons have been recorded which respond to the presentation of features of the face or of the face in profile. But also, neurons, which seem to be responsive to the individual character of the face whether or not it is presented in profile. They are activated by the face of one experimenter but not by the face of another, even though the latter may be as familiar as the former to the monkey. One is tempted to attribute to these neurons the capacity to recognize the individuality of the observed face, a capacity one would have wanted to reserve for the person of the observer. But as to knowing how these neurons succeed in such a performance, all that can be said is that they do it by synthesizing the information supplied by cells of a lower order. Not forgetting that the computational approach makes it possible to arrive at a more detailed answer to the question. But the inventor of these facial neurons, David Perrett, is forced to admit:

> the details of the next stage of processing after the visual cortex but before the structural encoding that has been studied in the temporal cortex are to a large extent unknown [15, p. 92].

In other words, we are once again confronted with these little demons capable of accomplishing, unwittingly, what actually gets done.

3 Affective and motor resonance or the alienation to the internal homunculus.

What, in the end, could a homunculus in the brain possibly be, if not an alien in me who does everything for me without my personal participation? An alien all the more mysterious and worrisome for doing all this without knowing anything about my relation to the world and to other persons, since he is radically solipsistic and acosmic. Whether it is cerebral or numerical, a computer is always in fact shut up in itself. If it is indeed this absence of any personal participation on the part of the subject in its own mental life that is responsible for this feeling of alienation that one quite reasonably experiences when confronted with neuroscientific explanations presupposing a homunculus in the brain, a solution begins to dawn. A new current of research in the neurosciences, the neurosciences of emotion and action, is bringing to light the neural foundations for our being directly involved in the operations of perception and cognition.

The idea that is going the rounds is that emotion and action are not as one might have thought, accessory or peripheral functions with regard to a central core of cognition based upon representation and computation which, for their part, remain affectively neutral and kinaesthetically inert. On the one hand, emotion and affectivity in general is once again recognised as lying at the root of the pulsional core of mental life. On the other hand, we are learning to re-discover the fact that the motor capacities of our body disclose the practical resources of the environment and render us sensitive to the motor intentions of other agents. If we take this evolution seriously and draw whatever consequences follow therefrom for our problem the phantom of the homunculus in the brain should gradually give way to a better knowledge of the incarnation of cognition in a sensible and acting body. The foundations of our being involved in events and in action are being investigated in the new neurosciences of emotion and action under the head of phenomenona of *resonance*. Some are more interested in the affective repercussions of the predictable consequences of our decisions, repercussions which normally accompany and guide the taking of a decision, and which would be handicapped by any emotional deficit [2, p. 336]. Others are more interested in the resonance of the observed movements of another agent in the repertory of action and the motor memory of the observer, a resonance which makes it possible for him to immediately understand the meaning of the actions undertaken by the other. The more we know about these phenomena of resonance the closer we get to promoting resonance into a new paradigm, a paradigm which might even replace the paradigm of the computation of information developed with respect to internal representations. If the further development of the neurosciences makes it possible to confirm this prediction, we shall be able to get rid of the homunculus, this fiction of an abstract calculator with whom it is impossible to identify because he feels nothing and does everything effortlessly.

4   The contribution of the neurosciences of emotion

The neurobiology of the emotions is not limited to a conception of emotional experience modelled on the processing of visual information, that is, to an analysis of external stimuli, the statistical extraction of invariants and their interpretation from the standpoint of cognition in general. At the root of the most basic emotions, let us say emotions common to man and other mammals, the existence of specific nerve circuits is presupposed, circuits including a collection of sub-cortical centres in the brain[13]. Formed earlier on in the evolution of the sensorial and motor cortical regions responsible for cognition, the cerebral amygdala, the hypothalamus and the periaqueductal grey nucleus exert an excitatory and modifying influence on the former regions, which make it possible for behaviour to be adjusted to the emotional state of mind. The activation of these circuits, whose electrical and chemical conditions are beginning to be understood, provokes impulses releasing emotional behaviours at the same time that they invest the stimuli that prompt these comportments with positive and negative values. The entire wealth of the emotional experience of man is rooted in the diversity of these subjectively experienced impulses. Even if we still do not know the processes by means of which the subjective emotional experiences are engendered on the basis of the electro-chemical activation of these circuits, the relation in question is no longer thought of in the cold and arbitrary terms of an encoding of mental representations in the neuronal action potentials. And this because the association between the felt emotion, the stimulus by which it is released, and the behaviour is always motivated.

For the subject as a living organism, being moved is a matter of feeling, in its very being, the absolute seriousness of an episode of (potential) importance for its life. Emotions are written into the individual as an imprint of an ancestral history where survival depended upon bringing into play the behaviour in question, and rapidly mobilizing the energy needed to do so. Emotions are fundamental modalities deposited in the genetic memory of the individual, recording its active engagement in a situation of vital significance. Sedimented in our being, though constantly available for reactivation, they are the possible forms of our presence in the world. It is in this sense that Jaak Panksepp distinguishes (1) a "seeking" circuit responsible for directing our search for an object of interest, (2) a "rage" circuit directed against those who represent a frustration, (3) a "fear" circuit anticipating an imminent danger or a vague feeling of insecurity, (4) a "panic" circuit expressing attachment and distress at being separated from the object of attachment, etc. The direct activation of these circuits through intra-cerebral electrical stimulation evokes complex and complete comportments, together with their intentional orientation and their affective tonality. A cat jumps towards the face of the experimenter, its claws unsheathed; a rat lies prone or takes flight; a patient thinks he is being followed or in an obscure tunnel or that he has fallen into the sea. Rarely seen in their most basic state in the adult, these instinctive tendencies are filtered in daily life by culturally dictated learning patterns and by higher cognitive activities. But even across these modifications, emotions never stop saturating our mental life and orienting our behaviour in such a way as to ensure that the individual will be able to come to terms with the existential situations it is confronted with. And so uphold its readiness to get involved with the event, which is a contribution made by the living being to the sense of its life. This continual emotional saturation of human experience, taken together with its impulsional underpinnings, is enough to render futile and gratuitous the objection that the activation of a sub-cortical circuit is, in the end, nothing but a flux of chemical molecules or electrical potentials bearing no obvious relation to our passionate interests, our rage or our fear.

Having said this, we should be warned against any excessive hermeneutical optimism. However obvious it might be to the philosopher, this solution to the problem of the homunculus via affective resonance it unlikely to be as obvious to the scientist himself. And of course the problem is not nearly as serious for the one as it is for the other. Jaak Panksepp's hostility to the domination of computationalism in cognitive neuroscience ought to have pushed him in the direction of an incarnate and situated neuroscience, that is, a neuroscience that has rid itself of humunculi. For all that, wanting to promote his hypothesis concerning the biological foundations of consciousness against the competing hypothesis of Antonio Damasio, he finishes up characterizing the sub-cortical circuits of the emotions as if they were controled by a humunculus exercising sovereign control over the totality of mental life, even including the perceptual and the cognitive.

> The only reasonably well-developed alternative to that view is the possibility that emotional command systems can establish various distinct types of resonances in the neuro-symbolic representation of a primordial body (the "SELF") situated largely [...] within deep and ancient mesencephalic areas, such as the periaqueductal grey

nucleus and surrounding tectal and tegmental systems. [...] The SELF is capitalized to highlight that this is a postulate concerning some type of primordial organization of the brain a coherent neuro-symbolic humuncular schema of the organism, a virtual body heavily weighted toward the representation of the basic motor-orientational and visceral processes emotional and motivational processes control the attentional and information-processing capacities of the somatic-exteroceptive (i.e., sensory thalamic-neo-cortical) nervous systems [14, pp. 153–154].

The dominant current in cognitive science explains cognition through representational functions of the mind whose materialization is effected by cartographic properties of the homunculi lodged in the brain centres, thereby making it possible for the feeling and acting body to be brought under the control of the brain as if this body were reducible to an aggregate of external information captors and muscular movement effectors. A new tendency consists in emphasizing the role of the feeling and acting body as a major factor in high order cognition considered not just as an unconscious infra-personal mechanism but also as a dynamic process responsible for the emergence into full conscious awareness of psychical formations (affects, percepts, intentions). Antonio Damasio [7] bases the consciousness of self and, in addition to self-consciousness, the representational capacities of the subject (augmented by learning, language and culture), on the infrastructure of a proto-Self which he identifies with an intimate sense of the homeostatic control process of the internal milieu of the body. Nevertheless, his conception, inherited from Cannon's homeostatis, remains a non-dynamic point of view, closed in on the internal milieu. The feeling of the own body goes much further than the subject who experiences it. It is also a window open on the own body of the other as another subjective centre with its own world. An opening on an other I know something about from within as a result of a resonance going far beyond the purely intellectual cognitive capacities of a (solipsistic) subject. For the bearing of any such intellectual cognition is definitely limited to my ability to infer, whether syllogistically or analogically, on the basis of my representations alone. If the discovery of resonant systems in the brain and the determining role of such systems in the understanding of actions and emotions does not seem to have enabled neuroscience to make much progress in the direction of the recognition of the role of the body as an organ of cognition and not simply the effector of actions, this is undoubtedly due to the fact that the discovery of mirror neurons has been taken over by a cognitivist ideology, which rejects the incarnation of meaning and which refuses any somatological hermeneutics.

5   The contribution of the neurosciences of action

What the neurosciences of action presuppose is that the possession by the organism of a repertory of actions makes it possible for the latter not merely to choose the action adapted to the circumstances but also to project its own categories and practical values (affordances in James Gibson's sense of that word) upon the surrounding world and, amongst other things, to directly recognize the actions of others, without inference or computation but through a phenomenon of resonance. The concept of resonant system is a generalization of the concept of mirror neuron. Mirror neuron: that is, a nerve cell operating in a dual visuo-motor field linking the observation with the execution of an action. Resonant system: a functional loop integrating the nerve centres distributed about distant cortical areas (or sub-cortical centres) and linking the observation with the execution of an action or observation with an emotional experience. Example: a resonant system of manual prehension with all its modalities integrating a collection of premotor mirror neurons (in the monkey, homologous with the Broca area) with somato-sensorial neurons (in the parietal area). Reduced to its most elementary expression, the fact is the following: in electrophysiology, based upon the unitary recording of electrodes implanted in the monkey, the manual actions of the experimenter activate neurons in the frontal area 6/F5 of the monkey by either a positive or negative modulation of the frequency of the discharge, a discharge profile very similar to that spontaneously associated with the execution by the monkey of actions of the same type. These "actions" are different sequences of a complete chain running from attentive but passive observation to the execution of actions oriented toward the taking hold and manual ingestion of food. The general hypothesis is that any automatic mimetic comportment solicits the activation of a parallel resonant system in the brain [11, pp. 176–180].

The discovery of mirror neurons is therefore due to the recording of individual cells, an approach dedicated to the validation of Barlow's hypothesis, or to saving it in some improved form (population encoding, temporal encoding, etc.). Let's show this. Under what conditions are mirror neurons activated? We have just explained: under two conditions: 1) when the monkey executes manual gestures oriented towards the ingestion of food; 2) when it observes the experimenter (or a fellow monkey) in the process of executing one of the manual gestures belonging to its own motor repertory. Classically, the function of mirror neurons has been interpreted as that of matching an observed alien action with the corresponding action belonging to the repertory of the observer. However, operating between the mental act of recognizing the identity of an action and the simple and objectively verifiable fact of the similarity in this discharge of the neuron, the notion of matching seems poorly determined. This, despite the fact that there is no possibility of confusing two things: one, the similarity between the activation curves traced on a histogram, which, for its evaluation, requires examination by an expert; the other, the act of the perceiving subject engaged in recognizing an action he knows how to accomplish himself in one he sees being accomplished by an other agent. It is this confusion between these two things that introduces a homunculus into the brain: a fictive interior observer capable of recognizing the identity of the action on the basis of the intra-cerebral observation of the frequency curves of the neurons activated in the two sets of circumstances. In line with our Cartesian heritage we tend to associate the capacity to recognize an identity, or to grasp a thought, with an agent capable of bringing a local diversity into an integrative unity:

> Car on peut bien concevoir qu'une machine soit tellement faite qu'elle profère quelques paroles *à propos des actions corporelles qui causeront quelques changements en ses organes*; comme si on la touche en quelque endroit, qu'elle demande ce qu'on veut lui dire, si en un autre, qu'elle crie qu'on lui fait mal, et choses semblables; mais non pas *qu'elle les arrange diversement pour répondre au sens de tout ce qui se dira en sa présence*, ainsi que les hommes les plus hébétés peuvent faire (*Discours de la Méthode*, Cinquième Partie).

Let me cite myself. In one of the first philosophical articles to draw attention to the discovery made by the Giacomo Rizzolatti group, a citation repeated with approval in his recent book with Corrado Sinigaglia:

> Everything happens as if the neurons reacted not to the stimulus as such, that is to its form, its sensorial aspect, but to its meaning for the animal. But reacting to a meaning is what is meant by understanding. Should we not then be talking about understanding rather than about a simple stimulation? [18, p. 306], quoted in [19, p. 49]

A question that applies equally to man, with regard to the Broca area being postulated as the support of "the understanding of the same act of communication". Ever since, in cerebral imagery, this area has displayed a similar activation profile in cases of the production and of the simple observation of silent speech. And this applies yet again in man with regard to the cerebral amygdala or the insular cortex, which display similar activation profiles when the subject experiences an emotion and when it observes someone else experiencing the same emotion. Etc.

Situated in the classical rationalist tradition for which the mind of the other is not initially given in a fundamentally intersubjective experience but is the conclusion of a piece of reasoning on the part of a solitary subject, a neo-cognitivist tendency interprets the function of mirror neurons to be that of underpinning a strategy of attributing mental states to alien bodies whose behaviour we want to be able to predict. The resources we already possess for planning our own actions furnish us with an analogue for a theory of the mind of the other. Another neo-behaviourist tendency relies on the directly immediate, and necessarily unconscious, character of the synchronization of the agents resonant systems with that of the observer to advance the view that the motor repertories can, through their synchronization, explain not just motor control but also communication and social cognition. An outcome of the collaboration between the neurophysiologist, Vittorio Gallese, and the analytical philosopher, Alvin Goldman, a notion of simulation floating between resonance and analytical inference is not going to be enough to resolve the tension between these opposing tendencies [9, pp. 493–501].

All the more so given that the notion of resonance stemming from work on mirror neurons remains largely metaphorical. The subject of the verb "resonate" is still so poorly defined when one speaks of resonant systems that one hesitates between too many alternative applications: 1) that it is the person of the agent and the person of the observer of the same action or emotion that can be said to resonate; 2) that it is the brains of these persons that resonate; 3) that different resonant systems dedicated to the recognition and the execution of actions mobilize different areas or brain centres; 4) that individual mirror neurons, and by virtue of the duality of their modes of activation, directly link the visual stimuli of the observed movements with the motor programmes of the observer or his emotional systems. This ambiguity probably results from the fact that, split between the multi-unitary recording of neurons by micro-electrodes implanted in the brain of the monkey and functional cerebral imagery in the case of human beings, research on mirror neurons using the two technologies still lacks any common interface.

But resonant systems, hypothesized as neuronal groups distributed across distant regions in mutual interaction, play an active role in the neurodynamics of the whole brain. In order for the nature of this contribution to be specified more exactly, individual cellular activities will have to be compared with local field potentials and with whole brain cerebral rhythms by procuring EEG recordings simultaneously at all three levels. A seductive hypothesis [10, pp. 1578–1579],[8, pp. 474–480] is that resonance has to be attributed to a neuronal mode of communication based upon the agreement between oscillation phases of different anatomically connected regions of the brain, all of which are mobilized by the activation of one and the same resonant system. For individual neurons in distant, but synchronously oscillating regions, an effective channel of communication would open up, one that would be closed down by the failure to synchronize of the respective oscillation patterns. This synchronization-desynchronization mechanism should make it possible for us to offer a causal (albeit holistic and not localistic) account of the intentional sequence: emotion-motivation-intention-preparation-action. Except that we are still very far from realizing this ideal, if only because work on mirror neurons and eletroencephalographic measurement of the inter-regional coherence of the brain are carried out by quite different communities of researchers.

6 Kinaesthetic constitution: An extrapolation from the neurosciences

Faced with this deceptive ambiguity concerning the philosophical significance of neuroscientific evidence, the philosopher might be tempted to attempt an extrapolation. What follows should be taken as a fable by appeal to which the dilemma with which cognitive neuroscientists presently find themselves confronted might be resolved, and this without reference to the ongoing course of empirical research. Might it not be possible to account for mental acts in terms of underlying physiological processes without recreating, within the subject whose acts are now in question, a second subject responsible for the acts of the first?

Our point of departure in the philosophical tradition is the kinaesthetic theory of transcendental constitution[5], a theory developed by Husserl in manuscript material stemming from the thirties and from a point of view quite close to the intuitions of Helmholtz and Poincar on the origin of geometric space in the sensation of bodily movements. The idea is that any object of interest, any perceptual form, any unitary entity which might present itself in experience as endowed with the meaning of being something for a subject — that is to say, for myself — must have been engendered by the activity of this same subject in the course of its interaction with this object. By retracing, in an uninterrupted succession, the complete sequence of acts responsible for conferring meaning upon objects fully constituted in human experience, the theory of transcendental constitution should ideally be capable of dispelling the phantom of the homunculus, which latter only appears as a result of the gap that has been allowed to develop between the meanings finally constituted in and through the process of sense formation and the subjective acts responsible for this process of formation itself.

Moreover, this transcendental constitution does not presuppose any transcendental subject overseeing human experience and constituting its sense formations from above. On the contrary, here the operative subjectivity is incoporated in the intimate sense of my being able to activate ("I move myself") my organs of sense and my body, the body of a concrete human being. For such an essentially kinaesthetic subjectivity, "real" objects only make sense as invariants in a continual variation of profiles in the perceptual fields of the organism (binocular visual field, cutaneous tactile field, sonorous space), a variation correlated with the kinaesthetic series of bodily movements performed by the perceiving subject in the course of its exploration of its surrounding world. No permanent object without a kinaesthetic lived experience advising the agent interacting with this object of the recurrence of a series of perceptual profiles associated with the movement of the eyes, of the hands or the entire body as the inverse correlate of a previous movement. The thing is not constituted prior to the subjective experience of the thing but is dynamically constituted in and through the latter. The thing emerges from this process of constitution endowed with all its layers of meaning: as a simple thing in space, as a materially resisting thing, as a tool, as a work of art, etc. and this emergence of the thing will be strictly simultaneous with the act through which the acting subject gets hold of the thing in the course of an action. The connection between "meaning something for ..." and "giving meaning to ...", this formerly broken connection because of the common-sense or scientific objectivations, will now be re-established. Finally, any kinaesthetically embodied experience is, in addition, intersubjectively situated, to the extent that our kinaesthetic experience is duplicated, or rather get deepened through, our awareness of others, and this because we also have empathic access to the kinaesthetic experience of the other. Thanks to all this, our world can not be conceived as initially solipsist, only to become later a social world through some fictive convention, but is to be seen as the world of several persons from the outset. Objects in this common world do not just exist for me but always equally for others: the intersubjectivity of the operations responsible for conferring meaning also endow the objets with an absolute objectivity, the kind of objectivity one only normally concedes to the theoretical objects of the mathematician. And so it is that the closed world of every day life gets opened up upon the infinite world of the idealities of science.

7 Conclusion

- Anti-reductionism tends to favour a purely physical description of the functioning of the brain and in such a way as to highlight the irreducibility of mental life.

- A hypercritical philosophy leads us to condemn as absurd any attribution of the mental activities of a person to the brain or to parts of the brain.

- The residue of an ancient philosophical tradition, the humunculus argument is not so easy to dismantle, no matter what the approach attempted in the neurosciences, because the methodology employed lends itself to the introduction of humunculi.

- Prompted by its very method to relaunch Barlow's hypothesis concerning the grandmother neuron, research on mirror neurons runs the risk of conferring upon individual cells (or the resonant systems in which they are lodged) a personal capacity to understand the meaning of actions.

- Anchoring our philosophical interpretation in th effort made by the organism to make sense of, and in the ability of the neurosciences to elucidate the mechanisms at the root of such tendency, we focus our attention on an intermediary phase where one notes an interesting friction between mechanically oriented explanations and a teleologically oriented common intuition of the essence of the living being.

- For his own personal satisfaction, the philosopher can always claim the right to extrapolate, on the basis of empirical evidence, in a direction that brings about a subjective synthesis of his sympathy for a certain tradition of thought with the progress made in a science, just as long as he pays close attention to the development of this science. This is what we have tried to do by bringing Husserl's transcendental theory of kinaesthetic constitution to bear upon the work done in the neurosciences.

BIBLIOGRAPHY

[1] H. Barlow. The twelth Bartlett memorial lecture: The role of single neurons in the psychology of perception. *The Quarterly J. of Experimental Psychology*, 37A: 121–145, 1985.
[2] A. Bechara, A.R. Damasio. The somatic marker hypothesis: a neural theory of economic decision. *Games and Economic Behavior*, 52: 336–372, 2005.
[3] M.R. Bennett and P.M.S. Hacker. *Philosophical Foundations of Neuroscience*. Blackwell, Oxford 2003.
[4] A. Berthoz. *The Brain's Sense of Movement*. Harvard University Press, Cambridge, MA 1997/2000.
[5] A. Berthoz et J.-L. Petit. *Physiologie de l'action et phénoménologie*. Odile Jacob, Paris 2006.
[6] J.-P. Changeux and P. Ricœur. *Ce qui nous fait penser. La nature et la règle*. Odile Jacob, Paris 1998.
[7] A. Damasio. *The Feeling of What Happens. Body and Emotion in the Making of Consciousness*. Harcourt Brace, New York 1999.
[8] P. Fries. A mechanism for cognitive dynamics: neuronal communication through neuronal coherence. *Trends in Cognitive Sciences*, 9: 474–480, 2005.
[9] V. Gallese and A. Goldman. Mirror neurons and the simulation theory of mind-reading. *Trends in Cognitive Sciences*, 12: 493–501, 1998.
[10] R.T. Knight. Neural networks debunk phrenology. *Science*, 316: 1578–1579, 2007.
[11] G. di Pellegrino, L. Fadiga, L. Fogassi, V. Gallese and G. Rizzolatti. Understanding motor events: a neurophysiological study. *Experimental Brain Research*: 176–180, 1992.
[12] A. Noë. Product of the senses. *TLS*, 15: 24, 2007.
[13] J. Panksepp. *Affective Neuroscience*. Oxford University Press, Oxford 1998.
[14] J. Panksepp. The neuroevolutionary cusp between emotions and cognitions. Implications for understanding consciousness and the emergences of a unified mind science. *Evolution and Cognition*, 7 (2): 153–154, 2001.
[15] D.I. Perrett, M.H. Harries, R. Bevan, S. Thomas, P.J. Benson, A.J. Mistlin, A.J. Chitty, J.K. Hietanen and J.E. Ortega. Frameworks of analysis for the neural representation of animate objects and actions. *J. Experimental Biology*. 146: 87–113, 1989.
[16] K.O'Regan and A.Noë, A sensory account of vision and visual consciousness. *Behavioral and Brain Sciences*, 24(5), 2001.

[17] K. O' Regan, E. Myin, A. Noë, Sensory consciousness explained (better) in terms of "corporality" and "alerting capacity". *Phenomenology and the Cognitive Sciences*, 4: 369–387, 2005.

[18] J.-L. Petit, Constitution by movement: Husserl in light of recent neurobiological findings. In J. Petitot, F.J. Varela, B. Pachoud and J.-M. Roy (eds.), *Naturalizing Phenomenology. Issues in Contemporary Phenomenology and Cognitive Science*, Stanford University Press, Stanford, California 1999, pages 220–244.

[19] G. Rizzolatti and C. Sinigaglia. *So quel che fai. Il cervello che agisce e i neuroni specchio*. Raffaello Cortina, MIlano 2006.

[20] G. Ryle. *The Concept of Mind*. Hutchinson, London 1945.

PART VI

GENERAL PHILOSOPHY OF SCIENCE

The whole truth about Linda: probability, verisimilitude, and a paradox of conjunction

GUSTAVO CEVOLANI, VINCENZO CRUPI, ROBERTO FESTA

1 Linda's story and the paradox of conjunction

In a seminal work on the psychology of reasoning and judgment under uncertainty, Tversky and Kahneman [42] presented the following description of a fictitious character, Linda, which would then become famous:

> Linda is 31 years old, single, outspoken and very bright. She majored in philosophy. As a student, she was deeply concerned with issues of discrimination and social justice, and also participated in anti-nuclear demonstrations.

In a series of experimental inquiries, Tversky and Kahneman asked several samples of participants (both statistically naïve and sophisticated subjects) to judge the probability of some hypotheses about Linda, including the isolated statement "Linda is a bank teller" (b from now on) and the conjunctive statement "Linda is a bank teller and is active in the feminist movement" ($b \wedge f$). The results showed a strong tendency to judge $b \wedge f$ as more probable than b. In a particularly neat demonstration of the phenomenon, 142 university students were simply asked to choose the more probable state of affairs between b and $b \wedge f$: 85% of them chose the latter.

This pattern of judgments is puzzling in that it conflicts with a basic and uncontroversial principle of probability theory, known as the "conjunction rule", prescribing that a conjunction of statements can not be more probable than any of its conjuncts. This "paradox of conjunction" (our preferred label in what follows) is widely known in the literature as the "conjunction fallacy" or the "conjunction effect". Tversky and Kahneman themselves, along with many others in subsequent investigations, obtained similar results on a variety of experimental scenarios, showing that the phenomenon can hardly be got rid of as a curiosity. Their "medical prognosis" example is a case in point: when given the description of a 55-old woman with a pulmonary embolism documented angiographically 10 days after a cholecystectomy, a large majority of physicians (internists) judged that the patient would be more likely to experience "emiparesis and dyspnea" than "emiparesis" [42, p. 301].

The paradox of conjunction has become a key topic in debates on the rationality of human reasoning and its limitations (see [39], [21], [15] and [36]). However, the attempt of providing a satisfactory account of the phenomenon has proved rather challenging. If only roughly, alternative approaches can be classified depending on their reliance on a mainly psychological *vs* epistemological conceptual background.

2 Psychological perspectives

One reaction to the paradox of conjunction has been the claim that the experimental evidence has not demonstrated the occurrence of a reasoning error after all. As instantiated in the psychological literature, this line of argument has been inspired by recurrent concerns about the pragmatics of communication in experimental settings: in the Linda problem, participants might have in fact interpreted the isolated conjunct b as $b \wedge \neg f$ (see, for instance, [34] and [9]), or they might have read the ordinary-language conjunction "and" as a disjunction [27]. The results of recent experiments devised to investigate these possible sources of confound suggest that the first one of them might have contributed to the size of the effect in earlier documentations of the phenomenon [38, 2, 40]. However, these studies have also shown that the phenomenon persists despite such "conversational implicatures" [16] being strongly discouraged or otherwise controlled for.

The most widely known attempts to *explain* (as contrasted to *question*) the Linda paradox as a reasoning error have been grounded on Tversky and Kahneman's hypothesis of a "representativeness heuristic" for human judgment under uncertainty [41]. Elaborating on this hypothesis, Shafir, Smith, and Osherson [37] have collected typicality ratings of Linda's character relative to the single category "bank teller" and the conjoint category "feminist bank teller" and interpreted such ratings as reflecting intuitive assessments of the probability of the correctness of Linda's description (d for short) given b and $b \wedge f$, respectively. In the Linda problem, and in a set of similar cases, such typicality ratings have proven reliable predictors of the occurrence of the conjunction effect. One limitation of this "inverse probability" account — i.e., the explanatory hypothesis of people's misguided assessment of posteriors $p(b|d)$ and $p(b \wedge f|d)$ as reflecting evaluations of the likelihoods $p(d|b)$ and $p(d|b \wedge f)$ — is that it is not easily extended to the medical prognosis case above, as well as to other documented results [7]. In fact, this would imply the rather cumbersome judgmental strategy of focussing on the probability of the *known* clinical frame *conditional on future (hypothetical) events*, such as the manifestation of certain symptoms.

3 Epistemological analyses

Interestingly, ever since Levi's 1985 insightful review [25] of Kahneman, Slovic and Tversky's [20] influential work, the paradox of conjunction has attracted the attention of a number of epistemology scholars. An epistemologically-oriented case for the thesis that "there need not be anything fallacious or otherwise irrational about the conjunction effect" [18, p. 30] has been independently made by Bovens and Hartmann [3, pp. 85–88] and Hintikka [18]. Briefly put, the proposal is the following. Suppose "Linda is a bank teller" and "Linda is a feminist bank teller" are reports of two distinct sources of information s_1 and s_2 which are not perfectly reliable. Linda's description d may well suggest that source s_1 is *less reliable* than s_2. But then, probability theory *is* consistent with the statement that the probability of b *conditional on the relatively low reliability* of s_1 is lower than the probability of $b \wedge f$ *conditional on the relatively high reliability* of s_2. It is submitted that *this* is what participants' responses express. It has been observed, however, that standard experimental stimuli are completely silent about b and $b \wedge f$ being reports of two distinct sources of information (see [26, p. 37]; [32, p. 292]). And the plausibility of the above reconstruction is shown even more problematic by the conjunction effect occurring in problems (such as the medical prognosis example) involving hypotheses about *future* events. For one has to make the additional assumption that in such cases participants interpret the task as concerning *forecasts* ("emiparesis" and "dyspnea and emiparesis") as made by two distinct predictors, which again are never mentioned in the experimental scenario.

A different approach has been taken by Crupi, Fitelson and Tentori [7]. While recognising that the paradox of conjunction documents a genuine error in probabilistic judgment, these authors have outlined an explanatory framework based on the notion of *confirmation*, meant in terms of Bayesian confirmation theory [14, 10, 8, 6]). By a close analysis of previous empirical results [33, 24], they argued that the participants' fallacious probability judgments might reflect the assessment of confirmation relations among the evidence provided and the hypotheses at issue in the experimental scenario. Moreover, extending an earlier result by Sides et al. [38], they showed that in a class of cases including both the Linda and the medical example above, Bayesian quantitative models of inductive confirmation imply that the evidence provided does support the conjunctive statement more than the single conjunct. Roughly, this class of cases is identified by the evidence provided (e.g., Linda's description) confirming the *added* conjunct ("feminist") but *not* the isolated one ("bank teller"). (Further developments of this line of analysis can be found in [1].)

The latter confirmation-theoretic reading of the Linda paradox is one way to flesh out the otherwise esoteric statement by Tversky and Kahneman themselves that *"feminist bank teller* is a better hypothesis about Linda than *bank teller"* [42, p. 311]. In what follows, we will explore a different strategy to fill in the blanks of this noteworthy remark by providing a *verisimilitudinarian* analysis of the problem. In a nutshell, we will show that "feminist bank teller", while less likely to be true than "bank teller", may well be more likely to be close to the *whole* truth about Linda.

4 Verisimilitude and probability

The concept of *verisimilitude* or *truthlikeness* was introduced by Popper [35] in 1963 with respect to scientific theories and hypotheses.[1] Popper claimed that the main epistemic goal of science is truth-approximation and that scientific progress consists in devising new theories which are closer to the truth than preceding ones. In an effort to ground this theoretical framework, Popper advocated a neat conceptual distinction between verisimilitude and probability. In his own terms:

---

[1]In this paper, we use as synonymous terms like "verisimilitude", "truthlikeness" and "approximation or closeness or similarity to the truth", which have been however carefully distinguished and analyzed in the literature (see, for instance, [28]). An excellent survey of the modern history of theories of verisimilitude is provided by Niiniluoto [29].

> The differentiation between these two ideas [verisimilitude and probability] is the more important as they have become confused; because both are closely related to the idea of truth, and both introduce the idea of an approach to the truth by degrees. [...] Logical probability [...] represents the idea of approaching logical certainty, or tautological truth, through a gradual diminuition of informative content. Verisimilitude, on the other hand, represents the idea of approaching comprehensive truth. It thus combines truth and content [35, p. 236].

Popper's focus on "logical" probability (as it was conceived by other influential sholars of his time, such as Carnap [4]) rather than "epistemic" or "subjective" probability is immaterial for our present concerns. Under both kinds of interpretation, probability is a *decreasing* function of logical strength (and, in this sense at least, of content). On the contrary, a measure of verisimilitude must be *positively* associated to high content. This is simply because "nothing is as close as the truth as the whole truth itself" [30, p. 11], the latter clearly being a uniquely accurate *and exhaustive* description of a given matter of interest.

In general terms, a hypothesis or theory is highly verisimilar if it says many things about the domain under investigation and if many of those things are true. Thus, an appropriate measure of the verisimilitude of a theory must depend on both its content (how much the theory says) and its accuracy (how much of what the theory says is in fact true). Intuitively, it is easy to see that neither content nor accuracy alone is sufficient to define verisimilitude. In fact, suppose that $p \land q \land r$ is the maximally informative true description of a certain domain of inquiry. Then hypotheses p and $\neg q$ are equally informative in that both make a single claim about the domain at issue — still only the former is true and hence more verisimilar than the latter. On the other hand, p and $p \land q$ are equally accurate to the extent that both are true — still the latter is more informative and hence more verisimilar than the former.

Verisimilitude theorists did not fail to notice the obvious fact that in most interesting cases it is *not* known which is the complete true description of a domain of inquiry, so that the *estimated* verisimilitude of alternative hypotheses is the crucial point of interest. Accordingly, the theory of verisimilitude has been traditionally seen as including a *logical* and an *epistemic* problem.[2] The logical problem of verisimilitude amounts to the preliminary definition of an appropriate notion of verisimilitude, allowing for a comparison of any two hypotheses with regards to their closeness to the truth. The epistemic problem of verisimilitude, on the other hand, amounts to the definition of an appropriate notion of *expected* verisimilitude by which the estimated closeness to the truth of any two hypotheses could be compared on the basis of the available data.

In the following sections (5 and 6) we will outline the formal background to briefly address both problems in turn, introducing the basic traits of a theory of verisimilitude and expected verisimilitude for hypotheses expressed in a propositional language. (A more extensive treatment of the theory is presented in Cevolani, Crupi and Festa [5] as satisfying a number of epistemologically relevant adequacy requirements arising from the literature. See

---

[2]See in particular Oddie [30], Niiniluoto [28], Kuipers [22] and, for a recent survey, Oddie [31].

also Festa [11, 12, 13].) Then, in section 7, we will come back to the conjunction paradox and provide a novel verisimilitudinarian analysis of the Linda scenario.

5 Propositional hypotheses: formal background

The definition and application of our account of verisimilitude and expected verisimilitude will preliminarily require a certain amount of formal machinery.

Basic propositions. Consider a propositional language $\mathcal{L}$ with n atomic propositions denoted by the statement letters $a_1, \ldots, a_n$. Given an atomic proposition a_i we will say that the propositions $\alpha_i^1 \equiv a_i$ and $\alpha_i^2 \equiv \neg a_i$ are the *basic propositions* (or *b-propositions*) associated to the statement letter a_i. We will denote as **A** and **B**, respectively, the set $\{a_1, \ldots, a_n\}$ of the n statement letters and the set $\{\alpha_1^1, \alpha_1^2, \ldots, \alpha_n^1, \alpha_n^2\}$ of the $2n$ b-propositions of $\mathcal{L}$.

Constituents. The most informative propositions of $\mathcal{L}$ will be called *constituents*. A constituent C of $\mathcal{L}$ tells, for any atomic proposition a_i, if either a_i or $\neg a_i$ is true. A constituent C can then be written in the following form:

$$\pm a_1 \wedge \cdots \wedge \pm a_n \tag{1}$$

where "$\pm$" is either empty or the negation symbol "$\neg$". Alternatively, C can be written as follows:

$$\alpha_1^{j_1} \wedge \cdots \wedge \alpha_n^{j_n} \text{ where } j_1, \ldots, j_n \in \{1, 2\} \tag{2}$$

Any b-proposition occurring in (2) will be called a *basic claim* (or *b-claim*) of the constituent concerned. A constituent C can be seen as *the most complete description* of a possible world by means of the expressive resources of $\mathcal{L}$. Accordingly, it can be said that any b-claim α of C (α being a variable over **B**) is true in the possible world described by C or, for short, that α is *true in C*. Let us call $C^+ \equiv \{\alpha \in \mathbf{B} : C \models \alpha\}$ the set of all b-claims of C.

One can easily check that the constituents of $\mathcal{L}$ form a set of exactly 2^n elements, hereafter labelled $\mathbf{C} \equiv \{C_1, \ldots, C_{2^n}\}$. Also notice that there will be an *unique true constituent* of $\mathcal{L}$, which can be seen as "the (whole) truth" about the investigated domain. This (usually unknown) true constituent will be labelled $C_\star$ from now on.

Quasi-constituents and c-hypotheses. While a constituent C identifies a complete list of the allegedly true b-propositions in $\mathcal{L}$ (i.e., the elements of C^+), a *quasi-constituent* (or *q-constituent*) H identifies a (possibly) incomplete list of such b-propositions. A q-constituent H can be written in one of the following forms:

$$\pm a_{1_H} \wedge \cdots \wedge \pm a_{k_H} \tag{3}$$

$$\alpha_{1_H}^{j_{1_H}} \wedge \cdots \wedge \alpha_{k_H}^{j_{k_H}} \text{ where } k_H \leq n \text{ and } j_{1_H}, \ldots, j_{k_H} \in \{1, 2\} \tag{4}$$

Any b-proposition occurring in (4) will be called a b-claim of the q-constituent concerned. A q-constituent H can be seen as a possibly *incomplete* description of the domain under inquiry by means of the expressive resources of $\mathcal{L}$. Given the conjunctive form of q-constituents, we will also call them *conjunctive (propositional) hypotheses* or, for short, *c-hypotheses*.

Let us call $H^+ \equiv \{\alpha \in \mathbf{B} : H \models \alpha\}$ the set of all b-claims of H. Constituents themselves are nothing but a special kind of q-constituents, i.e., such that $k_H = n$. Another notable special case of q-constituent is represented by the *tautology*, denoted as $H_\top$ and corresponding to the case $k_H = 0$, i.e., $H^+ = \varnothing$.

Note that c-hypotheses and constituents are related in the following straightforward way: a non-tautological c-hypothesis H is *true in C* iff any b-claim of H is true in C, i.e., iff $H^+ \subseteq C^+$; otherwise, H is *false in C*. Moreover, H is *completely false in C* iff none of H's b-claims is true in C, i.e., iff $H^+ \cap C^+ = \varnothing$.

EXAMPLE 1. Consider Linda's description according to the following features: "Linda is a bank teller" (b), "Linda is active in the feminist movement" (f) and "Linda takes yoga classes" (y). Let us consider the simple language $\mathcal{L}$ containing only three statement letters a_1, a_2, a_3, denoting the three atomic propositions b, f, y respectively. Thus, $\mathbf{A} = \{b, f, y\}$, $\mathbf{B} = \{b, \neg b, f, \neg f, y, \neg y\}$ and $\mathbf{C} = \{C_1, \ldots, C_8\}$.

Each constituent of $\mathcal{L}$ gives a complete description of Linda, specifying which elements of $\mathbf{B}$ are true: for instance, $C_1 \equiv b \wedge f \wedge y$ claims that Linda is a feminist bank teller who takes yoga classes; thus, $C_1^+ = \{b, f, y\}$ is the set of the three b-claims of C_1. Let us consider the c-hypothesis $H \equiv b \wedge \neg f$, with $H^+ = \{b, \neg f\}$. H claims that Linda is a bank teller but not a feminist, and it is silent on whether Linda takes yoga classes or not. Clearly, H is false in C_1, since only one of H's b-claims (i.e., b) is true in C_1, whereas the other (i.e., $\neg f$) is not.

6 Expected verisimilitude of propositional hypotheses

Given a measure $\mathrm{s}(H, C)$ of the *similarity* (or *closeness*) of a c-hypothesis H to a constituent C, the *verisimilitude* $\mathrm{Vs}(H)$ of H can be identified with the similarity (closeness) of H to the (usually unknown) *true constituent* $C_\star$, i.e., $\mathrm{Vs}(H) \equiv \mathrm{s}(H, C_\star)$. For this reason we will first define a similarity measure $\mathrm{s}(H, C)$ over all pairs of c-hypotheses H and constituents C.

Similarity of c-hypotheses to constituents. From an intuitive point of view, the more truths H tells about C, the more similar H is to C; thus, $\mathrm{s}(H, C)$ is maximal when H tells exactly n truths about C (recall that n is the number of C's b-claims). Consequently, the definition of $\mathrm{s}(H, C)$ obeys the following strategy. In order to evaluate the similarity of H to C, we assign a "prize" or a "penalty" to each b-claim α of H, depending on whether α is true or false in C. We will denote as τ and ϕ, respectively, the "weight" of truths and of falsehoods, with $0 < \tau, \phi < 1$. Thus, $\frac{\tau}{n}$ will be the prize for each of H's truths, while $\frac{\phi}{n}$ will be the penalty for each of H's falsehoods.

Formally, this amounts to define, for each constituent C, a *payoff function* which assigns to each $\alpha \in \mathbf{B}$ the following payoff $\pi_C(\alpha)$ depending on whether

$\alpha \in C^+$ or $\neg \alpha \in C^+$:

$$\text{For any } \alpha \in \mathbf{B}, \, \pi_C(\alpha) = \begin{cases} \frac{\tau}{n} & \text{if } \alpha \in C^+ \\ -\frac{\phi}{n} & \text{if } \neg \alpha \in C^+ \end{cases} \quad (5)$$

From now on, it will be convenient to posit $\phi = 1 - \tau$, thus having:

$$\text{For any } \alpha \in \mathbf{B}, \, \pi_C(\alpha) = \begin{cases} \frac{\tau}{n} & \text{if } \alpha \in C^+ \\ \frac{\tau - 1}{n} & \text{if } \neg \alpha \in C^+ \end{cases} \quad (6)$$

Intuitively, different values of τ (and of ϕ) reflect the relative weight of truth and falsity in inquiry. If $\tau = 0.5$, and then also $\phi = 0.5$, an inquirer will equally value the prize obtained by endorsing a truth and the penalty obtained by endorsing a falsity. In all other cases, if $\tau > \phi$ then the inquirer will care more endorsing a truth than he suffers from endorsing a falsity, and viceversa if $\tau < \phi$.

Given the payoff function, the similarity of H to C can be defined as the sum of the prizes and penalties assigned to H's b-claims:

$$s(H, C) = \sum_{\alpha \in H^+} \pi_C(\alpha) \quad (7)$$

Definition (7) immediately implies that the similarity $s(\alpha, C)$ of a "singleton" c-hypothesis α to C equals the payoff $\pi_C(\alpha)$, i.e.:

$$s(\alpha, C) = \pi_C(\alpha) \quad (8)$$

Moreover, (7) and (8) imply that:

$$s(H, C) = \sum_{\alpha \in H^+} s(\alpha, C) \quad (9)$$

i.e., that the similarity of a c-hypothesis H to C amounts to the sum of the similarities of H's b-claims to C.

Verisimilitude of c-hypotheses. As anticipated, once the similarity function $s(H, C)$ has been defined, the verisimilitude of H can be equated to its similarity to the (usually unknown) true constituent $C_\star$:

$$Vs(H) = s(H, C_\star) = \sum_{\alpha \in H^+} s(\alpha, C_\star) \quad (10)$$

$Vs(H)$ is thus the sum of the prizes attributed to the truths of H and of the penalties attributed to the falsehoods of H. Since $C_\star$ is the maximally informative true description of the domain of concern, the verisimilitude of H expresses the similarity or closeness of H to the whole truth about that domain.

EXAMPLE 2. Consider again the c-hypothesis $H \equiv b \wedge \neg f$, which claims that Linda is a bank teller but is not a feminist, and assume that $C_\star = C_1 \equiv b \wedge f \wedge y$ is the true constituent (recall that we are considering a language with only

3 atomic propositions). In order to evaluate the verisimilitude of H, i.e., its similarity w.r.t. $C_\star$, we consider the payoff assigned to each of its b-claims, i.e., from (6):

$$\pi_{C_\star}(b) = \frac{\tau}{3} \quad \text{since } b \in C_\star^+$$
$$\pi_{C_\star}(\neg f) = \frac{\tau - 1}{3} \quad \text{since } f \in C_\star^+$$

Thus we have:

$$\text{Vs}(H) = \text{s}(H, C_\star) = \sum_{\alpha \in H^+} \pi_{C_\star}(\alpha) = \frac{\tau}{3} + \frac{\tau - 1}{3} = \frac{2\tau - 1}{3}$$

If, for instance, $\tau = 0.5$ then $\text{Vs}(H) = 0$. In other words, if the weight of truths equals the weight of falsehoods, and H tells exactly one truth and one falsehood, then H's verisimilitude is 0.

Expected verisimilitude of c-hypotheses. The true constituent $C_\star$ being typically unknown, actual values of $\text{Vs}(H) = \text{s}(H, C_\star)$ are also usually unknown. However, given a probability distribution p over $\mathbf{C}$, *expected* verisimilitude values can be computed as follows:

$$\text{EVs}(H) = \sum_{C \in \mathbf{C}} \text{s}(H, C) p(C) \tag{11}$$

The expected verisimilitude $\text{EVs}(H)$ expresses the probability of H being similar to the whole truth, given that we are uncertain about which is the true constituent $C_\star$.

Let $\pi(\alpha)$ denote the (usually unknown) *actual* payoff of α — i.e., $\pi_{C_\star}(\alpha)$ — and let $\text{E}\pi(\alpha)$ be its *expected value*. It follows from (11), along with (8) and (9) — see the Appendix for a proof — that the expected verisimilitude of H can be expressed in terms of the value of the expected payoff $\text{E}\pi(\alpha)$:

THEOREM 1. *For any H*, $\text{EVs}(H) = \sum_{\alpha \in H^+} \text{E}\pi(\alpha) = \sum_{\alpha \in H^+} \frac{p(\alpha) - \phi}{n}$

Thus, the expected verisimilitude of a c-hypothesis H amounts to the sum of the expected payoffs of H's b-claims.

7 A verisimilitudinarian account of the Linda paradox

Let us come back to the Linda paradox, i.e., to the fact that most people, when confronted with Linda's description (see section 1), rank the conjunction "Linda is a feminist bank teller" as more probable than "Linda is a bank teller", so departing from the relevant probabilistic relationship according to which a conjunction can not be more probable than any of its conjuncts.

The Linda problem can be reformulated in terms of c-hypotheses. The relevant b-propositions involved are "Linda is active in the feminist movement" (f) and "Linda is a bank teller" (b). The two c-hypotheses at issues are: $b \wedge f$, i.e., "Linda is a feminist bank teller" and b, i.e., "Linda is a bank teller". By the conjunction rule, $p(b \wedge f) \leq p(b)$ necessarily holds. Thus, "feminist bank teller" can never be more probable than "bank teller". However, the following theorem shows that the (expected) verisimilitude of $b \wedge f$ may well be higher than the (expected) verisimilitude of b (see the Appendix for a proof):

THEOREM 2. $\text{EVs}(b \wedge f) > \text{EVs}(b)$ iff $p(f) > \phi$

where $\phi = 1 - \tau$ is the weight of falsehoods. Thus, the expected verisimilitude of "feminist bank teller" is higher than the expected verisimilitude of "bank teller" if the probability of "feminist" is sufficiently high, i.e., higher than the threshold value ϕ. As far as the expected verisimilitude of $b \wedge f$ is concerned, ϕ may be intuitively read as the threshold above which the expected prize guaranteed by the greater content of $b \wedge f$ w.r.t. b outweighs the risk of obtaining a penalty due to the falsity of f.

This means that, if one believes that the probability that Linda is a feminist is higher than ϕ, then one should rank $\text{EVs}(b \wedge f)$ as higher than $\text{EVs}(b)$. In particular, in case that $\tau = 0.5$ (and thus $\phi = 0.5$), if Linda is more likely than not to be active in the feminist movement, then "Linda is a feminist bank teller" has an higher expected verisimilitude than "Linda is a bank teller". In other words, "feminist bank teller", although less probable than "bank teller", may well be a better approximation to the whole truth about Linda.

8   Concluding remarks

Presumably, the only undisputed fact about the Linda paradox is that people's responses can not be accounted for by assuming *both* (i) that participants indeed mean to provide judgments about the simple probabilities of b and $b \wedge f$, *and* (ii) that they are elaborating their judgments in a rational fashion. From here on, agreement gives way to open controversy.

According to several spirited critics, assumption (i) is the only culprit: based on the experimental stimuli, participants typically mean to judge something else other than $p(b)$ and $p(b \wedge f)$ — e.g., $p(b \wedge \neg f)$ and $p(b \wedge f)$ — and, in doing so, they are perfectly rational. In this perspective, it is argued that the "conjunction fallacy" reflects nothing else than "intelligent inferences" which only "*look* like reasoning errors" (see [17]). On the other hand, Tversky and Kahneman, along with many other investigators, made an articulated case that assumption (ii) can not be retained at the expenses of (i). Interestingly, they themselves referred to alternative notions (other than mathematical probability) as explaining peoples' behavior (e.g., representativeness or typicality), but interpret them as "heuristic attributes" on which human reasoners rely precisely to make intuitive judgments of chance and probability. Being of only limited value, it is then argued, such heuristic attributes may act as biasing factors and lead to outcomes conflicting with compelling standards of rationality, as in the Linda case. Indeed, it has been suggested that the whole "heuristics and biases" research program can be reframed as the study of the limited validity of intuitive judgment by common processes of heuristic attribute substitution [19]. Briefly put, "the answer to a question can be biased by the availability of an answer to a cognate question — even when the respondent is well aware of the distinction between them" [42, p. 312].

Notably, the divide outlined above is not limited to the psychological literature on the issue, but cuts across both the psychological and epistemological field. This is illustrated by a comparison between the account based on the "reliability of different sources" as presented by Bovens and Hartmann [3] and Hintikka [18] and that based on confirmation relations outlined in Crupi, Fitelson and Tentori [7]. In fact, the former analysis explicitly questions assumption (i) above while aiming at preserving (ii), and thus the full rationality of human judgment as far as the conjunction paradox is concerned. The latter proposal, on the contrary, follows the opposite strategy by presenting confirmation relations as defining a novel kind of relevant heuristic attributes, much along the general lines of Tversky and Kahneman's "cognate question" quote.

In order to draw some conclusions from our preceding analysis, the theoretical landscape on the conjunction paradox can thus be conveniently mapped in terms of two distinct questions:

(1) Are experimental procedures which are typically employed suitable to elicit judgments concerning the simple probabilities of a conjunction *vs* an isolated conjunct?
(2) Which attributes (other than the simple probabilities above) are guiding participants' prevailing responses?

Although related, questions (1) and (2) are largely independent. A positive answer to (1) establishes the conjunction rule as a relevant norm of rationality for the experimental task, thus fostering the diagnosis of a cognitive bias, whereas a negative answer hinders such application of the rule, thus leading to the rejection of that diagnosis. Notably, in our verisimilitudinarian analysis of Linda paradox, we did not tackle directly question (1), on which we would like to keep a non-committal attitude here. Suffice it to say that, following Popper's remarks on the issue (see section 4), probability and verisimilitude can be seen as distinct formal *explicata* of a presystematic notion of "plausibility" (see also [23] and [28, Ch. 5]). Thus, it does not seem unreasonable to assume that in human intuitive judgment they may overlap in one way or another. Indeed, our main goal has been to show that "expected verisimilitude" is an interesting candidate answer to question (2). More precisely, that it is an independently motivated and formally definable epistemological notion relying on which many judges *would* rank "feminist bank teller" over "bank teller" in the Linda problem. This is of interest to the extent that researchers concerned with the conjunction paradox do not seem to have been fully aware of the fact, despite its potential relevance having been somewhat obscurely perceived, as illustrated by the following passage, again from the comprehensive discussion by Tversky and Kahneman [42, p. 312]:

> The expected value of a message can sometimes be improved by increasing its content, although its probability is thereby reduced. The statement "Inflation will be in the range of 6% to 9% by the end of the year" may be a more valuable forecast than "Inflation will be in the range of 3% to 12%", although the latter is more likely to be confirmed. A good forecast is a compromise between a point estimate, which is sure to be wrong, and a 99.9% credible interval, which is often too broad. The selection of hypotheses in science is subject to the same trade-off. [...] Consider the task of ranking possible answers to the question "What do you think Linda is up to these days?" The maxim of value could justify a preference for $b \wedge f$ over b in this task, because the added attribute *feminist* considerably enriches the description of Linda's current activities at an acceptable cost in probable truth.

Verisimilitudinarian ears cannot help hearing a subtle verisimilitudinarian tune.

Appendix

Theorem 1: For any H, $\text{EVs}(H) = \sum_{\alpha \in H^+} \text{E}\pi(\alpha) = \sum_{\alpha \in H^+} \dfrac{p(\alpha) - \phi}{n}$.

Proof. The first part of the theorem is proved as follows: according to (11), $\text{EVs}(H) = \sum_{C \in \mathbf{C}} s(H, C) p(C)$; given (9), this is equivalent to

$$\sum_{C \in \mathbf{C}} \sum_{\alpha \in H^+} s(\alpha, C) p(C),$$

i.e., by (8), to $\sum_{C \in \mathbf{C}} \sum_{\alpha \in H^+} \pi_C(\alpha) p(C)$, which can be expressed as

$$\sum_{\alpha \in H^+} \text{E}\pi(\alpha).$$

As far as the second part, i.e., the value of $\text{E}\pi(\alpha)$ is concerned, we have that:

$$\begin{aligned}
\text{E}\pi(\alpha) &= \sum_{C \in \mathbf{C}} p(C) \pi_C(\alpha) \\
&= \sum_{C \in \mathbf{C}: \alpha \in C^+} p(C) \frac{\tau}{n} + \sum_{C \in \mathbf{C}: \neg \alpha \in C^+} p(C) \frac{-\phi}{n} \\
&= p(\alpha) \frac{\tau}{n} - p(\neg \alpha) \frac{\phi}{n} \\
&= p(\alpha) \frac{1 - \phi}{n} - (1 - p(\alpha)) \frac{\phi}{n} \\
&= \frac{p(\alpha) - \phi}{n}.
\end{aligned}$$

It follows from this that $\text{EVs}(H) = \sum_{\alpha \in H^+} \dfrac{p(\alpha) - \phi}{n}$, which completes the proof. ∎

Theorem 2: $\text{EVs}(b \wedge f) > \text{EVs}(b)$ iff $p(f) > \phi$.

The theorem is an immediate consequence of the following more general proposition: For any H, $\text{EVs}(\alpha_i \wedge \alpha_j) > \text{EVs}(\alpha_i)$ iff $p(\alpha_j) > \phi$.

Proof. $\text{EVs}(\alpha_i \wedge \alpha_j) > \text{EVs}(\alpha_i)$ iff, according to Th. 1, $\text{E}\pi(\alpha_i) + \text{E}\pi(\alpha_j) > \text{E}\pi(\alpha_i)$ iff $\text{E}\pi(\alpha_j) > 0$ iff, again by Th. 1, $\dfrac{p(\alpha_j) - \phi}{n} > 0$ iff $p(\alpha_j) > \phi$. ∎

Acknowledgements

Research supported by PRIN 2006 grant *Probability and causal structures in the construction of knowledge and in decision processes* (Research unit: Department of Philosophy, University of Bologna), by a grant from the SMC/Cassa di risparmio di Trento e Rovereto for the CIMeC (University of Trento) research project on *Inductive reasoning* and by PRIN 2006 grant *Rational decisions, strategic interactions, complexity and evolution of social systems* (Research unit: Department of Philosophy, University of Trieste).

BIBLIOGRAPHY

[1] D. Atkinson, J. Peijnenburg, and T. Kuipers. How to confirm the conjunction of disconfirmed hypotheses. *Philosophy of Science*, 76: 1–21, 2009.

[2] N. Bonini, K. Tentori and D. Osherson. A different conjunction fallacy. *Mind and Language*, 19: 199–210, 2004.

[3] L. Bovens and S. Hartmann. *Bayesian Epistemology*. Oxford University Press, Oxford 2003.

[4] R. Carnap. *Logical Foundations of Probability*. University of Chicago Press, Chicago, 2 edition, 1962.

[5] G. Cevolani, V. Crupi, and R. Festa. Verisimilitude and belief change for conjunctive theories. Submitted for publication.

[6] V. Crupi, R. Festa, and C. Buttasi. Towards a grammar of confirmation. In M. Dorato, M. Rèdei, and M. Suárez (eds.), *EPSA Epistemology and Methodology of Science: Launch of the European Philosophy of Science Association*, Springer, Dordrecht 2010, chap. 7, pages 73–93.

[7] V. Crupi, B. Fitelson and K. Tentori. Probability, confirmation and the conjunction fallacy. *Thinking and Reasoning*, 14: 182–199, 2008.

[8] V. Crupi, K. Tentori and M. Gonzalez. On Bayesian measures of evidential support: Theoretical and empirical issues. *Philosophy of Science*, 74: 229–252, 2007.

[9] D.E. Dulany and D.J. Hilton. Conversational implicature, conscious representation and the conjunction fallacy. *Social Cognition*, 9: 85–110, 1991.

[10] R. Festa. Bayesian confirmation. In M.-C. Galavotti and Alessandro Pagnini (eds.), *Experience, Reality, and Scientific Explanation*. Kluwer Academic Publishers, Dordrecht 1999.

[11] R. Festa. The qualitative and statistical verisimilitude of qualitative theories. *La Nuova Critica*, 47–48: 91–114, 2007.

[12] R. Festa. Verisimilitude, cross classification, and prediction logic. Approaching the statistical truth by falsified qualitative theories. *Mind and Society*, 6: 37–62, 2007.

[13] R. Festa. Verisimilitude, qualitative theories, and statistical inferences. In S. Pihlström M. Sintonen and P. Raatikainen (eds.), *Approaching Truth: Essays in Honour of Ilkka Niiniluoto*, College Publications, London 2007, pages 143–178.

[14] B. Fitelson. The plurality of Bayesian measures of confirmation and the problem of measure sensitivity. *Philosophy of Science*, 66: S362–S378, 1999.

[15] G. Gigerenzer. On narrow norms and vague heuristics: a rebuttal to Kahneman and Tversky. *Psychological Review*, 103: 592–596, 1996.

[16] H.P. Grice. *Studies in the Way of Words*. Harvard University Press, Cambridge, MA 1989.

[17] R. Hertwig and G. Gigerenzer. The "conjunction fallacy" revised: how intelligent inferences look like reasoning errors. *Journal of Behavioral Decision Making*, 12: 275–305, 1999.

[18] J. Hintikka. A fallacious fallacy? *Synthese*, 140: 25–35, 2004.

[19] D. Kahneman and S. Frederick. Representativeness revised: attribute substitution in intuitive judgment. In T. Gilovich, D. Griffin and D. Kahnemann, editors, *Heuristics and Biases: the Psychology of Intuitive Judgment*, Cambridge University Press, New York 2002, pages 49–81.

[20] D. Kahneman, P. Slovic and A. Tversky. *Judgment under Uncertainty: Heuristics and Biases*. Cambridge University Press, Cambridge 1982.

[21] D. Kahneman and A. Tversky. On the reality of cognitive illusions. *Psychological Review*, 103: 582–591, 1996.

[22] T. Kuipers. *What is Closer-to-the-Truth?* Rodopi, Amsterdam 1987.

[23] T. Kuipers. *From Instrumentalism to Constructive Realism. On Some Relations between Confirmation, Empirical Progress, and Truth Approximation*. Kluwer Academic Publishers, Dordrecht 2000.

[24] D.A. Lagnado and D.R. Shanks. Probability judgment in hierarchical learning: a conflict between predictiveness and coherence. *Cognition*, 83: 81–112, 2002.

[25] I. Levi. Illusions about uncertainty. *British Journal for the Philosophy of Science*, 36: 331–340, 1985.

[26] I. Levi. Jaakko Hintikka. *Synthese*, 140: 37–41, 2004. Reply to [18].

[27] A. Mellers, R. Hertwig and D. Kahneman. Do frequency representations eliminate conjunction effects? An exercise in adversarial collaboration. *Psychological Science*, 12: 269–275, 2001.

[28] I. Niiniluoto. *Truthlikeness*. Reidel, Dordrecht 1987.

[29] I. Niiniluoto. Verisimilitude: the third period. *The British Journal for the Philosophy of Science*, 49 (1): 1–29, 1998.
[30] G. Oddie. *Likeness to Truth*. Reidel Publishing Company, Doredrecht 1986.
[31] G. Oddie. Truthlikeness. In E.N. Zalta (eds.). *The Stanford Encyclopedia of Philosophy*. 2007.
[32] E.J. Olsson. Review of L. Bovens and S. Hartmann, *Bayesian Epistemology*. *Studia Logica*, 81: 289–292, 2005.
[33] D.N. Osherson, E.E. Smith, O. Wilkie, A. Lopez and E. Shafir. Category-based induction. *Psychological Review*, 97: 185–200, 1990.
[34] G. Politzer and I.A. Noveck. Are conjunction rule violations the result of conversational rule violations? *Journal of Psycholinguistic Research*, 20: 83–103, 1991.
[35] K.R. Popper. *Conjectures and Refutations: the Growth of Scientific Knowledge*. Routledge and Kegan Paul, London, 3rd edition, 1969.
[36] R. Samuels, S. Stich and M. Bishop. Ending the rationality wars: how to make disputes about human rationality disappear. In R. Elio (ed.), *Common Sense, Reasoning and Rationality*, Oxford University Press, New York 2002, pages 236–268.
[37] E. Shafir, E.E. Smith and D. Osherson. Typicality and reasoning fallacies. *Memory and Cognition*, 18: 229–239, 1990.
[38] A. Sides, D. Osherson, N. Bonini and R. Viale. On the reality of the conjunction fallacy. *Memory and Cognition*, 30: 191–198, 2002.
[39] S. Stich. *The Fragmentation of Reason: Preface to a Pragmatic Theory of Cognitive Evaluation*. MIT Press, Cambridge, MA 1990.
[40] K. Tentori, N. Bonini and D. Osherson. The conjunction fallacy: a misunderstanding about conjunction? *Cognitive Science*, 28: 467–477, 2004.
[41] A. Tversky and D. Kahneman. Judgment under uncertainty: heuristics and biases. *Science*, 185: 1124–1131, 1974.
[42] A. Tversky and D. Kahneman. Extensional versus intuitive reasoning: the conjunction fallacy in probability judgment. *Psychological Review*, 90: 293–315, 1983.

Probabilistic graphical models and the logic of scientific discovery

ANTONINO FRENO

1 Introduction

One major question in the philosophy of science is whether there can be a normative theory of scientific discovery, as opposed to a normative theory of justification. In the twentieth century, two representative stands on this subject have been taken for example by Karl Popper [16] on the one hand, who denied the possibility of a logic of scientific discovery, and on the other hand by Herbert Simon [21], who strenuously advocated the plausibility of such a research program, based on preliminary results achieved in artificial intelligence. Although Simon regarded computational models of discovery as plausible models of *human* discovery [13], the significance he attaches on artificial implementations of discovery strategies is not dependent on Simon's views concerning human psychology. While the possibility of a logic of scientific discovery has been the subject of much debate among philosophers, it is highly instructive to consider how the basic insights of a normative theory of discovery have been turned into a thriving research program by computer scientists, namely into the research field which is now known as *machine learning*.

Probabilistic graphical models, such as Bayesian networks and Markov random fields, are among the most powerful machine learning methods developed in the last decades. While these models allow for efficient representation of joint probability distributions and automated inference over stochastic domains, one of their main limitations lies in the high computational cost of learning them from data, which makes it infeasible to learn them in domains involving relatively large numbers of variables.

The aim of this paper is twofold. On the one hand, I describe some current research on probabilistic graphical models, aimed at reducing the computational cost of learning them from data. In particular, I present a hybrid graphical model exploiting features of both Bayesian networks and Markov random fields. The hybrid model exploits some factorization properties of Markov random fields in order to merge a large number of (small) Bayesian networks into a compact model of a (large) stochastic domain, so as to allow for efficient learning in domains that are otherwise prohibitive for both Bayesian networks and Markov random fields. Section 3.1 reviews the theory of Bayesian networks, while Section 3.2 describes the hybrid Bayesian/Markov network model.

On the other hand, I stress those features of probabilistic graphical models (and of machine learning formalisms in general) that make them plausible as computational counterparts of scientific theories. To this aim, after noticing how these models allow for the main kinds of inference that have traditionally been regarded as a trademark of scientific theories (namely induction, deduction, and abduction), I focus on the problem of learning statistical models from data. The analysis will show how machine learning is casting new light on traditional problems in the philosophy of science. In particular, in Section 4 I will explore the implications of some results in statistical learning with respect to the role of simplicity in theory choice, while in Section 5 I will argue that one philosophical lesson we can draw from machine learning research is that scalability (as a formal property of theories) can play a fundamental role in making scientific discovery effective.

2 Machine learning as a logic of scientific discovery

First of all, let me state what I mean by "logic of scientific discovery". Based on Herbert Simon's usage of that phrase, the logic of scientific discovery is meant as *a normative investigation of the inference processes leading to the introduction of (novel) scientific theories*. Concerning this notion, it is very important not to confuse the logic of scientific discovery with the psychology of scientific discovery. While Simon believed that AI techniques provide plausible models of human cognition, the viability of the project of a normative theory of discovery does not depend at all on a psychologistic view of AI.

Of course, one may wonder whether such a normative investigation of scientific discovery is possible at all. A possible strategy for answering this question might be to provide a philosophical argument aimed at showing that the project at issue is indeed theoretically sound (and practically viable). However, this will not be my strategy. Rather than using philosophical arguments in order to advocate the plausibility of a logic of scientific discovery, I will exhibit a piece of real scientific research as evidence for the plausibility of that project. In particular, the philosophical side of my argument will simply consist in showing how machine learning can be interpreted broadly as a project in the logic of scientific discovery. Given such an interpretation of machine learning research, I will then describe some results delivered by research on statistical learning methods. The main motivation for introducing such results to a philosophical audience is that they show that the logic of scientific discovery is not merely a philosophical project, but it is instead a mature (and thriving) scientific discipline.

Machine learning is the theoretical and experimental study of computational systems whose performance at specific tasks improves with experience [10]. "Computational systems" means (more or less complex) combinations of algorithms, typically implemented in real computer programs. Performance is usually measured by evaluation metrics that depend on the considered tasks, such as classification accuracy for pattern recognition. Experience is given by a collection of data items. Such data points can take the form of vectors of features (i.e. variables), sequences, graphs, or other suitably formalized objects. In intuitive terms, the aim of a machine learning system is to acquire the capability of solving a certain class of problems by being trained on a set of solved problems (belonging to that class).

In a sense, machine learning algorithms deliver *models* of the data they are trained on. But how should we regard those models, from the point of view of the philosophy of science? My claim is that such models are nothing but computational counterparts of what we commonly regard as scientific *theories*. "Theory" is a much debated term in the philosophy of science, and several different views of its meaning have been advocated thus far. Philosophical notions of theory range, for example, from the logical-empiricist conception of a formal system, i.e. a set of sentences expressed in first-order logic, to the structuralist idea that a theory should be identified with the set of its models, in the strict (model-theoretic) sense of semantic structures.[1] Now, let us reflect on the following passage, drawn from Ian Witten and Eibe Frank's introduction to data mining:

> What is learned by a machine learning method is a kind of "theory" of the domain from which the examples are drawn, a theory that is predictive in that it is capable of generating new facts about the domain — in other words, the class of unseen instances. Theory is a rather grandiose term: we are using it here only in the sense of a predictive model. Thus theories might comprise decision trees or sets of rules — they don't have to be any more "theoretical" than that. [23, pp. 179–180]

The remarks just quoted contain a simple yet fruitful insight. In my view, a possible reason why no general consensus has been reached by philosophers in analyzing the notion of theory lies in the fact that the philosophical aim has generally been to explain *what a theory is*, rather than *what a theory is useful for*. Strictly speaking, the latter question does not even have a precise meaning until we answer the former. Nevertheless, while the former question has in fact no universally accepted answer as yet, it would be hard to deny that *inference* (hence explanation and prediction) is the main purpose of scientific theories. In fact, it is clear that any notion of theory should be consistent at least with the following fact: theories are tools for performing (different kinds of) inference. Although a loose notion of theory as an inferential device (or a "predictive model", as Witten and Frank put it) may fall short of the expectations of philosophers of science, that notion has the important effect of encouraging us to turn our interest from the reflection on the notion of theory to the identification of rational strategies for developing (extending, revising, etc.) scientific theories.

Viewing theories broadly as inferential devices allows us to realize how machine learning methods are nothing but methods for automating the construction of scientific theories, since any machine learning method is aimed at supporting some kind of inference. I think that the success of machine learning (and of AI in general) in opening new perspectives for the application of computational methods to challenging scientific problems makes a fairly strong case for the plausibility of viewing machine learning as a logic of scientific discovery.[2]

---

[1] [11] is an example of the former conception; for the latter, see e.g. [22]. An excellent overview, with special emphasis on the problem of intertheoretic reduction, can be found in [3].

[2] Automated theorem-proving and bioinformatics are two significant examples of scientific application areas where AI/machine learning techniques are changing our way of

3 Probabilistic graphical models

Probabilistic graphical models, such as Bayesian networks [15] and Markov random fields [9], are among the most flexible machine learning formalisms developed in the last decades. While these models allow for efficient representation of joint probability distributions and automated inference over stochastic domains, one of their main limitations lies in the high computational cost of learning them from data, which makes it infeasible to learn them in domains involving relatively large numbers of variables. Some current research is showing how the computational limitations (in terms of time complexity) of both Bayesian networks and Markov random fields can be overcome by a hybrid Bayesian/Markov network model, called "hybrid random field" [8]. I will now review some basic principles and results in the study of probabilistic graphical models, concerning on the one hand Bayesian networks, and on the other hand hybrid random fields. In order to keep the presentation as simple as possible, many technical details will be omitted, when such details are not strictly necessary for understanding the philosophical upshot of my argument.

3.1 Bayesian networks

Bayesian networks [15] are used to represent joint probability distributions over sets of random variables. A Bayesian network is made up of two components: a directed acyclic graph (DAG), and a set of conditional probability tables (CPTs). Each node in the graph represents a random variable, and for each node there is a probability table specifying the conditional distribution of the variable given (each possible combination of) the values of its parents in the DAG. A simple Bayesian network is exemplified in Figure 1.

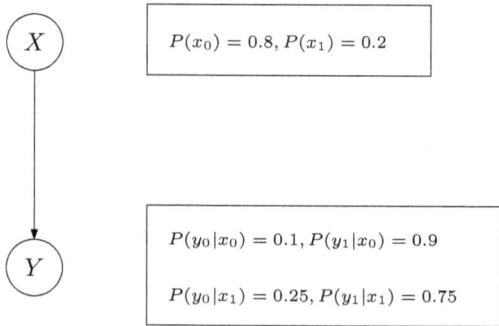

Figure 1. A Bayesian network for the binary variables X and Y. Given a variable X and a value x, the notation $P(x)$ is used as shorthand for $P(X = x)$.

In order to derive a joint probability distribution from a Bayesian network,

---

doing science, both in formal and in empirical research fields. See [18] and [4] for general overviews.

the *Markov assumption* is made according to which each variable is independent of its non-descendants in the DAG given the values of its parents. Consider the set **X** of random variables $X_1, \ldots, X_n$, and an arbitrary state $\mathbf{x} = x_1, \ldots, x_n$ of the variables in **X**. If $\mathcal{PA}(X_i)$ is the set of parents of X_i, let $pa(X_i)$ denote the state of $\mathcal{PA}(X_i)$, i.e. some specific configuration of the values of the variables in $\mathcal{PA}(X_i)$. Then, the Markov assumption entails the following equality:

$$P(\mathbf{X} = \mathbf{x}) = \prod_{i=1}^{n} P(X_i = x_i | pa(X_i)) \quad (1)$$

When X_i is a root node, $P(X_i = x_i | pa(X_i))$ refers to the absolute distribution of X_i, i.e. $P(X_i = x_i)$. Since the local distributions $P(X_i | pa(X_i))$ are provided by the CPTs, equation 1 specifies how to compute a joint probability distribution from a set of CPTs.

Let $I(X, Y | Z)$ mean that X is independent of Y given Z. Then, if **X** is a set of random variables $X_1, \ldots, X_n$, a *Markov blanket* $\mathcal{MB}(X_i)$ for X_i in **X** is any subset $\mathcal{S}$ of $\mathbf{X} \setminus \{X_i\}$ such that $I(X_i, (\mathbf{X} \setminus \mathcal{S}) \setminus \{X_i\} | \mathcal{S})$. In other words, the variables in $\mathcal{MB}(X_i)$ are such that $P(X_i | \mathbf{X} \setminus \{X_i\}) = P(X_i | \mathcal{MB}(X_i))$. An important property of Bayesian networks is that, for each variable X_i, the set containing the parents, the children, and the parents of the children of X_i is sufficient in order to form a Markov blanket of X_i within the Bayesian network [15]. This independence property will be exploited in the formulation of hybrid random fields.

The usefulness of Bayesian networks is given by the fact that they allow for efficient probabilistic inference [18, 12]. That is to say, Bayesian networks allow to efficiently automate the process of computing the probability distribution of any random variable within a certain set, given that the values of an aribitrary subset of the remaining variables are known. This process is nothing but (probabilistic) *deductive inference*. Moreover, algorithms are available for performing *abductive inference* by means of Bayesian networks [12]. In other words, the Bayes net formalism supports both deductive and abductive inference, which are the two basic ways of using scientific theories for explanation and prediction. Now, since the problem of learning a model from data is nothing but the AI counterpart of *inductive inference*, once we design an algorithm capable of learning Bayesian networks from specific datasets, we will have at our disposal a (mechanic) "theory construction procedure", so to speak. I now address this problem.

The strategy I describe for learning Bayesian networks from data resorts to hill-climbing search in a space of possible networks [7]. The search is aimed at finding the network that maximizes a certain heuristic function, inspired by the *minimum description length* (MDL) principle [17]. According to this principle, we seek the Bayesian network that maximizes the likelihood function while minimizing the length of the network description. The length of describing a Bayesian network is nothing but the length of encoding it in a specified language. The version of the MDL principle that we use takes the form of the heuristic function $mdl(h)$, defined as follows [23]:

$$mdl(h) = \log P(\mathbf{D}|h) - \frac{par(h)}{2} \log |\mathbf{D}| \qquad (2)$$

where $par(h)$ is the number of parameters specified in the Bayesian network h. $mdl(h)$ penalizes the likelihood of h to an extent that is proportional to the network complexity, where complexity is measured by $par(h)$. The aim of this heuristic is to encourage introducing parameters when the parameters really capture regularities in the data (and hence produce a strong increase of the network likelihood), and to discourage parameter introduction when this only captures the noise in the data (and hence increases the likelihood to a relatively small extent). In other words, the idea is to maximize the likelihood while keeping the DAG as sparse as possible. Under certain assumptions, it can be shown that the MDL evaluation function is asymptotically correct, i.e. that $mdl(h)$ converges to the true posterior probability of h [19, 12].

It is interesting to note that the score assigned to h by the MDL heuristic is tightly related to the posterior probability of h. This remark derives from the following argument [10]:

$$\begin{aligned}
\arg\max_h\ P(h|\mathbf{D}) &= \\
= \arg\max_h\ &\frac{P(\mathbf{D}|h) \cdot P(h)}{P(\mathbf{D})} \\
= \arg\max_h\ &P(\mathbf{D}|h) \cdot P(h) \\
= \arg\max_h\ &\log_2 P(\mathbf{D}|h) + \log_2 P(h) \\
= \arg\min_h\ &-\log_2 P(\mathbf{D}|h) - \log_2 P(h)
\end{aligned} \qquad (3)$$

As originally shown by [20], the result of derivation 3 can be interpreted as stating that a way of maximizing the posterior probability of h is by minimizing the sum of the length (in bits) of encoding the data given the information provided by h and the length (in bits) of encoding h. While the first quantity corresponds to the likelihood of h, the second quantity corresponds to the prior probability of h, which means that decreasing the length of the model description increases the prior probability of the model. Clearly, the number of parameters specified in a network is a measure of the network description length. Therefore, the prior probability of a network is inversely proportional to the number of its parameters. Of course, when we are not able to assess $P(h)$, we cannot be able to assess $\log_2 P(h)$ either. For this reason, we use a heuristic based on the MDL principle, since we cannot measure properly the description length of models. In other words, the MDL heuristic offers a well-grounded way to approximate the prior probability of different candidate models when no exact estimate of priors is available.

3.2 Hybrid random fields

Just like Bayesian networks (and Markov random fields), hybrid random fields are aimed at representing joint probability distributions underlying sets of random variables. Given the set $\mathbf{X}$ of variables $X_1, \ldots, X_n$, a hybrid random field for $X_1, \ldots, X_n$ is a set of Bayesian networks $BN_1, \ldots, BN_n$ with DAGs

$\mathcal{G}_1, \ldots, \mathcal{G}_n$. Each Bayesian network BN_i contains the variable X_i together with a subset $\mathcal{N}(X_i)$ of $\mathbf{X} \setminus \{X_i\}$. Within each Bayesian network BN_i, the node X_i will have a Markov blanket $\mathcal{MB}_i(X_i)$, given by the parents, the children, and the parents of the children of X_i within $\mathcal{G}_i$. In particular, we assume that the set $\mathcal{MB}_i(X_i)$ is a Markov blanket of X_i within the whole hybrid random field, i.e. that $\mathcal{MB}(X_i) = \mathcal{MB}_i(X_i)$. This assumption, which I call "modularity assumption", is needed in order to compute the joint probability distribution of $X_1, \ldots, X_n$ in the way I will explain. An example of hybrid random field is provided in Figure 2.

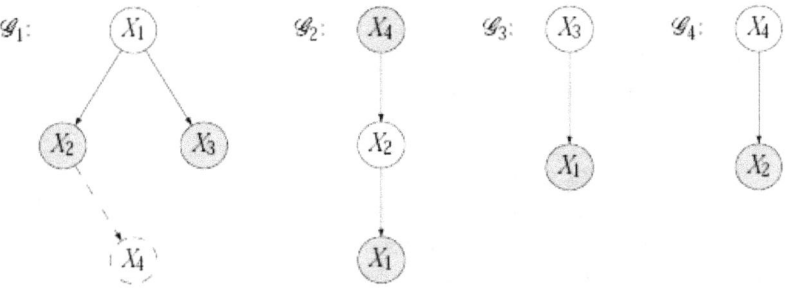

Figure 2. The graphical components of a hybrid random field for the variables $X_1, \ldots, X_4$. Since each node X_i has its own Bayesian network (where nodes in the Markov blanket of X_i are shaded), there are four different DAGs.

Given a vector $\mathbf{x} = x_1, \ldots, x_n$ of values of the variables in $\mathbf{X}$, the joint probability of $\mathbf{x}$ is represented as follows:

$$P^*(\mathbf{X} = \mathbf{x}) = \prod_{i=1}^{n} P(X_i = x_i | mb(X_i)) \qquad (4)$$

where $mb(X_i)$ is the state (determined by $\mathbf{x}$) of the Markov blanket of X_i. In other words, hybrid random fields employ the pseudo-likelihood measure introduced by Julian Besag [2] for Markov random fields. Since the local distribution of each variable X_i given its Markov blanket is modeled by a local Bayesian network BN_i, the conditional probability $P(X_i = x_i | mb(X_i))$ can be computed in a particularly simple way, as shown by [14].

The main advantage of hybrid random fields over Bayesian networks lies in the strong scalability properties of the learning algorithm. The procedure used to learn hybrid random fields from data, called "Markov blanket merging", resorts to an iterative strategy in order to identify an assignment of Markov blankets to the model variables that optimizes (up to a local maximum) the model pseudo-likelihood. As part of Markov blanket merging, the routine learning the local Bayesian networks within the global model employs the same hill-climbing algorithm described in Section 3.1, which makes the comparison between learning in Bayesian networks and learning in hybrid random fields particularly significant. Although the details of Markov blanket merging go far beyond the scope of this paper,[3] the most important result to consider in the present context is that, while the space searched when learning Bayesian networks grows exponentially with the number of variables contained in the model, the space explored by Markov blanket merging grows only linearly with that number. An empirical comparison of the computational burden of learning hybrid random fields to the burden of learning Bayes nets is illustrated in Figure 3. As the plot shows, the computational advantage of hybrid random fields over Bayesian networks is overwhelming.

At the same time, some experiments in pattern recognition and link prediction show that hybrid random fields are able to achieve accuracy levels analogous to (or even higher than) other probabilistic graphical models [8], including not only Bayesian and Markov networks, but also the naive Bayes classifier [5]. This means that the higher scalability of hybrid random fields with respect to Bayes nets does not entail a loss in the accuracy of the learned models.

4 Simplicity reconsidered

A common view in the philosophy of science maintains that, other things being equal, simpler theories should be preferred over more complex ones.[4] The puzzle concerning this view is that no account of simplicity has ever been able to reach universal consensus among philosophers. Indeed, several notions of simplicity should be kept distinct from one another, such as *syntactic* simplicity, relating to the form of scientific theories, and *ontological* simplicity, relating to the objects postulated by theories. In this section, I propose to reflect on the syntactic notion of simplicity, based on the theory presented in the previous sections. However, my aim is not to establish a philosophical account of simplicity. My strategy will be instead to focus on a notion of simplicity grounded in the theory of statistical learning, and to stress the importance of the role played by simplicity in making learning (i.e. induction) effective. While this may not answer many open questions in the philosophical debate about simplicity, it should nevertheless help to understand what part of the philosophical puzzle can be saved as a meaningful problem in the methodology of science. The theory related to the MDL principle (discussed in Section 3.1) is of fundamental importance to the problem we are going to address. In particular, the present philosophical reading of that theory is nothing but a way of reformulating it in less technical terms, which does not add anything new to what is regarded as "received wisdom" in the field of statistical learning. This means that, if the reader feels that something is wrong with the ideas expressed in this section, the proper way of refuting them would be by refuting the content of Section 3.1.

The reason for using the MDL heuristic function is that, when learning a Bayesian network, in order to avoid overfitting the data we need to introduce, in our evaluation function, a careful tradeoff between the likelihood and the complexity of the evaluated models (where complexity is measured by the number of parameters contained in a given model). In other words,

---

[3] See [8] for a technically detailed treatment of learning in hybrid random fields.
[4] For a general introduction to the philosophical problem of simplicity, see e.g. [1].

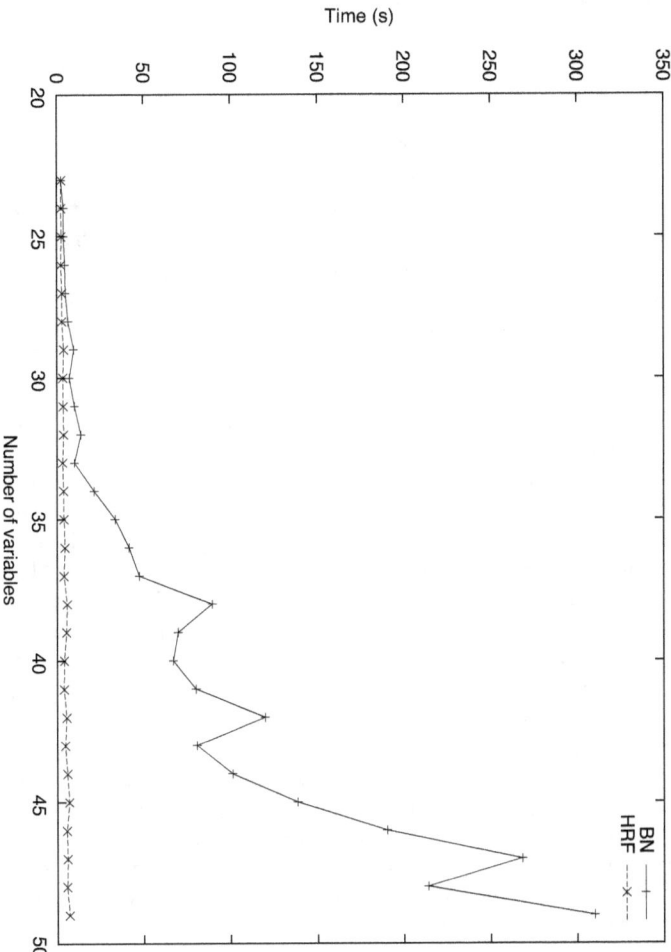

Figure 3. Time required for learning Bayesian networks (BN) and hybrid random fields (HRF) as the problem size increases. For each n such that $23 \leq n \leq 49$, the time (in seconds) is measured using a dataset containing 100 patterns, where each pattern is composed of n binary variables.

such a tradeoff within the scoring function is necessary for the learned model to *generalize well* to new data. Now, since the models being evaluated are hypotheses concerning some specified domain, and since the number of model parameters is nothing but a measure of the syntactic simplicity of those hypotheses, one general implication of the theory of Section 3.1 is that an appropriate weighting of the simplicity of hypotheses is necessary in order to find hypotheses that generalize well to future data. That is to say, when formulating hypotheses, *taking into account (syntactic) simplicity plays a precise role in making induction effective*. An interesting question to ask (and which was already answered in Section 3.1) is then: why is simplicity so effective? The simple moral to be drawn from derivation 3 is that minimizing the description length of a hypothesis h, i.e. maximizing its "simplicity", means nothing but maximizing the prior probability of h. That is to say, the intimate mathematical connection between simplicity (i.e. description length) and prior probability explains the contribution given by simplicity to the success of induction (i.e. learning): other things being equal, a simpler hypothesis will have a higher posterior probability than a more complex one.[5]

5 Scalability as an epistemic virtue

From the perspective of the philosophy of science, one interesting point emerging from research on hybrid random fields is the following. As we saw, the key advantage of hybrid random fields over Bayesian networks is the low computational cost of learning them from data, which makes it feasible to learn them even in high-dimensional domains. In other words, a preference for hybrid random fields over other probabilistic models is suggested by their strong *scalability* properties. The philosophical reader should not take this advantage to be a merely practical (and relatively unphilosophical) one. In fact, in many important cases (e.g. in bioinformatics or web mining) higher scalability means capability of allowing for induction, where less scalable techniques may defuse any attempt to induce a model from data. In other words, improving scalability means extending our attempt of inducing novel scientific theories to domains where induction was not even conceivable before. In this sense, one philosophical lesson we can draw from the effort of designing probabilistic graphical models that are suitable for application to high-dimensional tasks is that scalability should be regarded as an *epistemic virtue* in its own right, i.e. as one of the basic parameters we ought to consider when evaluating scientific theories, on a par with other virtues such as simplicity or consistency (and, of course, besides explanatory and predictive power).

6 Conclusions

The aim of this paper was on the one hand to introduce a novel probabilistic graphical model to a philosophical audience, and on the other hand to discuss some philosophical implications of machine learning research (with particular emphasis on statistical learning methods). Although many technical details

---

[5] The same philosophical point is also made in [6] by discussing Akaike's statistical results, which are very similar in their spirit to the meaning of the MDL principle underlying our evaluation function.

were omitted from the description of the hybrid Bayesian/Markov network model, and from the related theory of Bayesian networks, some philosophical points should have emerged clearly from the discussion. First, I argued that machine learning is establishing itself as a full-fledged implementation of the philosophical project that Herbert Simon called "logic of scientific discovery". In particular, such a realization of Simon's project is not only consistent and theoretically sound, as shown by the theory of statistical learning, but it is also practically effective, as the technological impact of machine learning research (and AI in general) is no longer a mere promise, but it is part of the world we are living in today. Second, I showed how an elegant and mathematically robust solution for the philosophical puzzle of simplicity can be drawn from statistical learning theory, in particular from an analysis of the MDL principle and its role in learning algorithms for probabilistic graphical models. Finally, I claimed that scalability, as a formal property of induction techniques for scientific theories, should be recognized to play a very important role in making induction effective, and hence in allowing science to develop more rapidly.

Acknowledgments

I am grateful to John Bickle, Sandro Nannini, Anca Ralescu, Edmondo Trentin, and Giuseppe Varnier for discussing with me many of the ideas explored in this work.

BIBLIOGRAPHY

[1] A. Baker. Simplicity. In E.N. Zalta (ed.), *The Stanford Encyclopedia of Philosophy*, 2004.
[2] J. Besag. Statistical analysis of non-lattice data. *The Statistician*, 24: 179–195, 1975.
[3] J.W. Bickle. *Psychoneural Reduction*. MIT Press, Cambridge, MA 1998.
[4] J. Cohen. Bioinformatics — An Introduction for computer scientists. *ACM Computing Surveys*, 36: 122–158, 2004.
[5] P. Domingos and M. Pazzani. On the optimality of the simple Bayesian classifier under zero-one loss. *Machine Learning*, 29: 103–130, 1997.
[6] M. Forster and E. Sober. How to tell when simpler, more unified, or less ad hoc theories will provide more accurate predictions. *British Journal for the Philosophy of Science*, 45: 1–35, 1994.
[7] A. Freno. Selecting features by learning Markov blankets. In B. Apolloni, R.J. Howlett and L.C. Jain (eds.), *Knowledge-Based Intelligent Information and Engineering Systems*, volume 4692 of *LNAI*, Springer, Berlin 2007, pages 69–76.
[8] A. Freno. *A Hybrid Bayesian/Markov Network Model for Scalable Statistical Learning*. PhD thesis, Università degli Studi di Siena 2008.
[9] R. Kindermann and J.L. Snell. *Markov Random Fields and Their Applications*. American Mathematical Society, Providence, RI 1980.
[10] T.M. Mitchell. *Machine Learning*. McGraw-Hill, London 1997.
[11] E. Nagel. *The Structure of Science*. Harcourt, Brace & World, New York 1961.
[12] R.E. Neapolitan. *Learning Bayesian Networks*. Prentice Hall, Upper Saddle River, NJ 2004.
[13] A. Newell and H.A. Simon. Computer science as empirical inquiry: symbols and search. *Communications of the ACM*, 19: 113–126, 1976.
[14] J. Pearl. Evidential reasoning using stochastic simulation of causal models. *Artificial Intelligence*, 32: 245–257, 1987.
[15] J. Pearl. *Probabilistic Reasoning in Intelligent Systems*. Morgan Kaufmann, San Francisco 1988.
[16] K.R. Popper. *The Logic of Scientific Discovery*. Hutchinson & Co., London 1959.
[17] J. Rissanen. Stochastic Complexity. *Journal of the Royal Statistical Society*. Series B, 49: 223–239, 1987.

[18] S. Russell and P. Norvig. *Artificial Intelligence*. Prentice Hall, Upper Saddle River, NJ, second edition 2003.
[19] G. Schwarz. Estimating the dimension of a model. *The Annals of Statistics*, 6: 461–464, 1978.
[20] C.E. Shannon and W. Weaver. *The Mathematical Theory of Communication*. University of Illinois Press, Urbana, ILL 1949.
[21] H.A. Simon. Does scientific discovery have a logic? *Philosophy of Science*, 40: 471–480, 1973.
[22] W. Stegmüller. *The Structuralist View of Theories*. Springer, New York 1979.
[23] I.H. Witten and E. Frank. *Data Mining*. Morgan Kaufmann, San Francisco, second edition 2005.

Perfected science and the knowability paradox

MASSIMILIANO CARRARA, DAVIDE FASSIO

1 Introduction

In *The Limits of Science* [5], N. Rescher embraces a logical argument known as the *Knowability Paradox*, according to which, if every true proposition is knowable, then every true proposition is known, i.e., if there are unknown truths, there are unknowable truths. Rescher argues that the paradox, providing evidence of a limit of our knowledge (the existence of unknowable truths), could be used for arguing against perfected science.

In this article, we present two criticisms of Rescher's argument. The first one points out that Rescher is ambiguous on the meaning of "impossibility of a perfected science": it could be interpreted in at least two different ways, one of which is plainly unproblematic compared with the *Knowability Paradox*. In the second criticism, we argue that the kind of unknowability involved in the paradox is semantic, rather than epistemic. Therefore, it is not a real problem for science. The final conclusion of the paper is, if our criticisms are correct, that the paradox leaves open the possibility of a perfected science.

The paper is divided into three parts. In the first one, we give an account of the paradox and our reading of Rescher's argument. In the second and third parts, we point out our criticisms. If our arguments are correct, Rescher's conclusion, according to which the *Knowability Paradox* constitutes a problem for perfected science, is mistaken.

2 The *Knowability Paradox* and Rescher's argument for the imperfectibility of science

N. Rescher, in *The Limits of Science*, argues that "perfected science is a mirage; complete knowledge a chimera" [5, p. 150]. The above thesis is a consequence of the *Knowability Paradox*, a logical argument published by F. Fitch in an article entitled *A Logical Analysis of Some Value Concepts*.[1] Fitch's argument, starting from the assumption that every true proposition is knowable, reaches the strong conclusion that every true proposition is known or, in different terms: if there are unknown truths, there are unknowable truths. *Prima facie*, this argument seems to seriously narrow our epistemic possibilities and to constitute a limit for knowledge in general, for scientific knowledge in particular. The argument runs as follows: take the epistemic

---

[1][2]. For an introduction to the literature about the argument, see [1] and [3].

operator K, where Kp stands for "someone knows that p" or "it is known that p",[2] and "p" is a proposition in a formal language.

Assume the following two properties of knowledge:

1. the distributive property over conjunction (Dist), i.e., if a conjunction is known, then also its conjuncts are, and

2. the factivity of knowledge (Fact), i.e., if a proposition is known, then it is true.

Formally:
$$K(p \wedge q) \vdash Kp \wedge Kq \qquad \text{(Dist)}$$
$$Kp \vdash p \qquad \text{(Fact)}$$

Assume the following two unremarkable modal claims, which can be formulated using the usual modal operators $\Diamond$ (which is read "it is possible that") and $\Box$ (which is read "it is necessary that"). The first is the *Rule of Necessitation*:
$$if \vdash p, then \Box p \qquad \text{(Nec)}$$

The second rule establishes the interdefinability of the modal concepts of necessity and possibility:
$$\Box \neg p \dashv\vdash \neg \Diamond p \qquad \text{(ER)}$$

Assume also the *Knowability Principle*, according to which every true proposition is knowable, formally:
$$\forall q (q \rightarrow \Diamond Kq) \qquad \text{(KP)}$$

Finally, assume that we are not omniscient, i.e., there is at least a truth that is not known:
$$\exists r (r \wedge \neg Kr) \qquad \text{(NO)}$$

an instantiation of (NO) is:
$$(p \wedge \neg Kp) \qquad (2)$$

Consider an example of (KP) resulting by the substitution of q with (2):
$$((p \wedge \neg Kp) \rightarrow \Diamond K(p \wedge \neg Kp)) \qquad (3)$$

By (2) and (3), we obtain:
$$\Diamond K(p \wedge \neg Kp) \qquad (4)$$

Consider the following argument "per absurdum" (independent from (2)–(4)):

(5)	$K(p \wedge \neg Kp)$	[assumption]
(6)	$Kp \wedge K\neg Kp$	[by (5) and (Dist)]
(7)	$Kp \wedge \neg Kp$	[applying (Fact) to (6)]
(8)	$\neg K(p \wedge \neg Kp)$	[by (5)–(7), refusing (5) for the inconsistency of (7)]
(9)	$\Box \neg K(p \wedge \neg Kp)$	[by (8) and (Nec)]
(10)	$\neg \Diamond K(p \wedge \neg Kp)$	[by (9) and (ER)]

---

[2] Kp is commonly generalized at every subject and time: "someone knows at some time that p". For the purposes of our paper, the chosen reading of Kp is irrelevant.

(10) is inconsistent with (4).³ If so, (NO) and (KP) are incompatible. One of the two assumptions must be abandoned. The advocate of the view that all truths are knowable must negate (NO):

$$\neg \exists r(r \wedge \neg Kr) \qquad \text{(Not-NO)}$$

according to (Not-NO), there are not unknown truths, i.e., every truth is known:

$$\forall r(r \rightarrow Kr) \qquad \text{(Not-NO*)}$$

Otherwise, one must negate (KP):

$$\neg \forall q(q \rightarrow \Diamond Kq) \qquad \text{(Not-KP)}$$

obtaining that there are unknowable truths:

$$\exists q(q \wedge \neg \Diamond Kq) \qquad \text{(Not-KP*)}$$

"This argumentation shows that in the presence of (relatively unproblematic) principles [(Dist)–(Fact)], the thesis that all truths are knowable [(KP)] entails that all truths are known, that is, [(Not-NO*)]. Since the latter thesis is clearly unacceptable, the former must be rejected. We must concede that some truths are unknowable: $\exists q(q \wedge \neg \Diamond Kq)$" [5, p. 150].⁴

Rescher points out that "No doubt this sort of argumentation for the incompleteness of knowledge is too abstract [...] to carry much conviction in itself. But it does provide some suggestive stage setting for the more concrete rationale of the imperfectibility of science" [5, p.150].⁵

Rescher's argument for the imperfectibility of science could be analyzed in the following way:

if the *Knowability Paradox* holds, then there are unknowable truths (Ass.) (R1)

the *Knowability Paradox* holds (Ass.) (R2)

there are unknowable truths (Conclusion I) (R3)

if there are unknowable truths, then perfected science is impossible (Ass.) (R4)

perfected science is impossible (Conclusion II) (R5)

Rescher's argument (as it is here reformulated) is based on three different premises, (R1), (R2), and (R4). Here we are not interested in how correct (R1)–(R3) is: we just assume that Fitch's argument is sound.⁶ Is Rescher's second part of the argument (R3)–(R5) correct?

---

³Here we have take the freedom of substituting the argument as it was originally proposed by Rescher with the equivalent, clearer, and commonly used formulation. See, for instance, [1].

⁴For a formal proof of $\exists q(q \wedge \neg \Diamond Kq)$, see [6].

⁵Routley in [6] considers the paradox in a more serious way, as an authentic limitation for knowledge in general. For articles related to Rescher's, see [7]and [10].

⁶There is a long list of criticisms of the paradox. For an introduction to the main literature, see [1].

3 First criticism: (R4) is ambiguous

The first problem of Rescher's argument is that (R4):

If there are unknowable truths, then perfected science is impossible (R4)

is ambiguous. In particular, there are at least two meanings of "imperfectibility of science" — where the expression is here considered equivalent to "impossibility of a perfected science" — and one of them is plainly unproblematic compared with the paradoxical conclusion.

Consider (R4): if the existence of unknowable truths is a problem for the perfectibility of science, it seems reasonable to think that a perfected science is equivalent or at least implies an omniscient science.[7] The following is Rescher's train of thought: if there are unknowable truths, scientific omniscience is impossible, and a perfected science is impossible, too.

Here an ambiguity rises. What does it mean that "omniscience is impossible"? We could read it as:

It is impossible that every true proposition is known (IO1)

Formally:

$$\neg \Diamond \forall q(q \rightarrow Kq) \qquad (IO1)$$

But we could also read "omniscience is impossible" as:

not every true proposition is knowable. (IO2)

Formally:

$$\neg \forall q(q \rightarrow \Diamond Kq) \qquad (IO2)$$

Are (IO1) and (IO2) both implied by the paradox? If there are unknown truths, the result of the paradox (according to Rescher) is the negation of the Knowability Principle: (Not-KP) $\neg \forall q(q \rightarrow \Diamond Kq)$) and (Not-KP) is (IO2). So, (IO2) is the proper conclusion of the paradox.

What about (IO1)? Let us first note that the result of the paradox is that (NO) and (KP) are incompatible. The result of the paradox can be summarized in the following theorem:

$$\vdash \exists q(q \land \neg Kq) \rightarrow \neg \forall q(q \rightarrow \Diamond Kq) \qquad (T1)$$

Furthermore, notice also that the converse of (T1) can be easily demonstrated; in fact, by the principle that what is actual is possible, we obtain:

$$\vdash \forall q(q \rightarrow Kq) \rightarrow \forall q(q \rightarrow \Diamond Kq). \qquad (T2)$$

which is provably equivalent to:

$$\vdash \neg \forall q(q \rightarrow \Diamond Kq) \rightarrow \exists q(q \land \neg Kq) \qquad (T3)$$

---

[7] Here with *omniscience* we do not mean a property of a subject, i.e. the property of possessing an effective knowledge of every truth. Rather, *omniscience* is specifically referred to science: an omniscient science is a science having the means to acquire the knowledge of every truth.

(T1) and (T3) validate the following theorem:

$$\vdash \exists q(q \wedge \neg Kq) \leftrightarrow \neg \forall q(q \to \Diamond Kq)) \tag{T}$$

If (T) is a theorem, by applying the *Rule of Necessitation* to (T), we obtain:

$$\vdash \Box(\exists q(q \wedge \neg Kq) \leftrightarrow \neg \forall q(q \to \Diamond Kq)) \tag{TN}$$

Now, notice that (NO) $\exists r(r \wedge \neg Kr)$ — the non-omniscience thesis — is the result of a commonsensical observation according to which, *de facto*, actually there are true propositions that we do not know. It is not a logical principle of the paradox, nor it is introduced through a logical argument.[8] If it is so, (NO) is contingently true, i.e., it is possibly false:

$$\Diamond \neg \exists r(r \wedge \neg Kr) \tag{CNO}$$

But, by (CNO), (TN), and the modal rule ($\Diamond A, \Box(A \leftrightarrow B) \vdash \Diamond B$), we obtain the following:

$$\Diamond \forall q(q \to \Diamond Kq) \tag{CIO2}$$

(IO2) is contingent, that is, it is possibly false. By (CIO2), (TN), and the modal rule ($\Diamond A, \Box(A \leftrightarrow B) \vdash \Diamond B$) it is easy to derive (Not-IO1):

$$\Diamond \forall q(q \to Kq) \tag{Not-IO1}$$

Summarizing: we assumed that (NO) is only contingently true (i.e., it is possible that it is false). If (NO) is only contingently true, (IO2) is contingently true, too. But if (IO2) is only contingently true, then (IO1) is false. So (IO1) is false. Accepting the contingency of (NO), Fitch's paradox does not imply (IO1), Fitch's paradox implies the negation of (IO1). With this result at hand, let us return to our considerations concerning (R4); given the two readings of "omniscience is impossible", there are two corresponding readings of "imperfectibility of science":

Assuming (IO1) $\neg \Diamond \forall q(q \to Kq)$ (it is logically impossible that every truth is known), the imperfectibility of science is equivalent to the logical impossibility of a perfected science.

Assuming (IO2) $\neg \forall q(q \to \Diamond Kq)$ (Not every truth is knowable), and the actual, contingent existence of unknown truths, the imperfectibility of science is equivalent to the actual unrealizability of a perfected science.

The *Knowability Paradox* is an argument only for the second reading. It is not an argument for the first one. If it is so, Rescher's premise (R4):

if there are unknowable truths, then perfected science is impossible (R4)

is ambiguous.

---

[8]As C. Wright writes in [9], (NO) says just that p is true and not actually known. Furthermore, if (NO) were necessarily true, (Not-KP) would be easily proved by it, without the necessity of introducing Fitch's argument.

Rescher does not seem to be aware of the above specified distinction, and for this reason he falls in the mentioned ambiguity: the *Knowability Paradox* is an argument for (IO2) and the actual unrealizability of a perfected science, but it is not an argument for (IO1) and the logical impossibility of a perfected science.

4   A second criticism: an incorrectness in Rescher's argument

The second problem of Rescher's argument is that, given

$$\text{there are unknowable truths (Conclusion I)} \tag{R3}$$

and

$$\text{if there are unknowable truths, then perfected science is impossible (Ass.)} \tag{R4}$$

the conclusion

$$\text{perfected science is impossible (Conclusion II)} \tag{R5}$$

is misleading.

The mistake is due to the fact that Rescher does not take into consideration the special *status* of the propositions that lead to the paradox. The propositions resulting unknowable by the paradox, as the *reductio* (5)–(8) in the paradox showed, are instances of (NO): e.g., (2) $(p \wedge \neg Kp)$. But why are those propositions unknowable?

First of all, let us distinguish between two different kinds of unknowability. A true proposition could be unknowable because of some epistemic limits. Take, for example, Heisenberg's indetermination principle: according to a certain interpretation, the principle seems to give an ineliminable epistemic limit to human knowledge. On the other hand, a different kind of unknowability is just based on semantic considerations: the unknowability of a proposition could result just from its meaning. In the last case, there are no effective limits to our (scientific) knowledge.

Consider the proposition:

$$\text{perfected science is unrealized} \tag{S}$$

(S) jeopardizes the realization of perfected science: if it is true, then it is false that perfected science is realized. But the reason for such unrealizability is not ascribable to an epistemic limit. It is simply a semantic consequence of the logical law that a proposition is incompatible with its negation (the *law of non-contradiction*): (S) is incompatible with:

$$\text{perfected science is realized} \tag{S*}$$

(S*) is false simply because (S) is true. Propositions that lead the paradox emerge, and have the logical form (2), present the same sort of problem. This semantic phenomenon has been studied in the literature, and some authors have called this kind of propositions "blindspots":[9] $(p \wedge \neg Kp)$ is unknowable just because it is a conjunction of two propositions, p and $\neg Kp$, and the knowledge of the first conjunct implies the falsity of the second one just for semantic reasons: "it is known that p" is trivially incompatible with "it is not known that p". Notice that the paradox does not concern the knowability of each conjunct in (2). Each one is independently knowable, whereas their contemporary knowledge is impossible, for the demonstrated semantic reason. If it is so, the problem of the paradox is strictly semantic, not epistemic: it does not concern any specific area of science or, more generally, human epistemic skills.

In light of the above explanation, Rescher's unproblematic acceptance of (R4):

if there are unknowable truths, then perfected science is impossible   (R4)

is mistaken. The unknowability problematic for science, referred to by Rescher in the antecedent of (R4), is an epistemic one: given our epistemic limits perfected science is impossible. On the other hand, the unknowability resulting in the paradox, assumed in (R3), is of the semantic kind. So, the conclusion of the paradox is not the intended premise of (R4), and from (R3) and (R4) we cannot infer (R5):

perfected science is impossible (Conclusion II)   (R5)

5 Conclusion

To conclude: if our criticisms are satisfactory, Rescher's argument according to which the *Knowability Paradox* constitutes a limit for perfected science, is ambiguous and mistaken. Specifically, the *Knowability Paradox* cannot be used — as Rescher has — as an argument for the imperfectibility of science. The final conclusion of our paper is that the paradox leaves open the possibility of a perfected science.

Acknowledgements

An early version of this paper was read at the "SILFS 2007 Conference". We are indebted to the participants at the conference for the stimulating discussion. Special thanks to Pierdaniele Giaretta for having improved previous versions of our paper. We also wish to thank Silvia Gaio, Enrico Martino, Vittorio Morato, Marzia Soavi and an anonymous reviewer for their detailed comments.

BIBLIOGRAPHY

[1] B. Brogaard and J. Salerno. Fitch's paradox of knowability. In E. Zalta (ed.), *The Stanford Encyclopedia of philosophy* (Summer 2004 edition), 2008.
[2] F. Fitch. A logical analysis of some value concepts. *The Journal of Symbolic Logic*, 28: 135–142, 1963.
[3] J. Kvanvig. *The Knowability Paradox*. Oxford University Press, Oxford 2006.
[4] B. Linsky. Factives, Blindspots and Some Paradoxes. *Analysis*, 46: 10–15, 1986.
[5] N. Rescher. *The Limits of Science*. University of California Press, Berkeley 1984.

---

[9]See, for instance, [8] and [4].

[6] R. Routley. Necessary limits to knowledge: unknown truths. In E. Morscher *et al.* (eds.), *Essays in Scientific Philosophy*, Bad Reichental, 1981, pp. 93–113 (reprinted in a special issue of *Synthese* edited by J. Salerno (forthcoming)).
[7] N.G. Schlesinger. On the limits of science. *Analysis*, 46: 24–26, 1986.
[8] R.A. Sorensen. *Blindspots*. Clarendon-Press, New York 1988.
[9] C. Wright. Truth as sort of epistemic: Putnam's peregrinations. *Journal of Philosophy*, 97 (6): 335–364, 2000.
[10] E.M. Zemach. Are there logical limits for science? *The British Journal for the Philosophy of Science*, 38 (4): 527–532, 1987.

Two problems for normative naturalism
LUCA TAMBOLO

1 The basic problem of meta-methodology

A method (from the ancient Greek *methodos*, "the pursuit of knowledge", or "a way of inquiry") is an orderly arrangement of parts or steps to accomplish an aim. The task of scientific methodology is to specify the steps that must be taken in order accomplish the cognitive aim(s) of science. These steps are stated in the form of methodological rules, which therefore can be defined as — allegedly — effective means for the achievement of the aim(s) of science.

Philosophers as well as scientists advocate conflicting views both on the aim(s) and on the correct method of scientific research. Here, however, we shall gloss over axiological controversies and focus on what we shall call "the basic problem of meta-methodology", which can be stated as follows: "How can the choice of a (set of) methodological rule(s) be justified?"; "By what means is one supposed to argue for the superiority of a certain methodological rule over its competitors?".

These questions arise because of the disagreement between the upholders of different methodologies. For instance, as Laudan and others [16] have shown, in the writings of such contemporary authors as Karl Popper, Thomas Kuhn, Imre Lakatos, Paul Feyerabend, and Larry Laudan, more than 250 methodological rules are defended. Of course, this disagreement stems from the fact that the above mentioned philosophers defend conflicting views on the aim(s) of science; however, it also partly depends on the fact that they propose different solutions to the problem of the justification of scientific method. Consequently, it seems not unreasonable to expect that, if it were possible to arrive at a shared solution to this problem, methodological disagreement would be significantly reduced.

In a number of writings, Larry Laudan [14, 15] has advocated the epistemological stance known as "normative naturalism", which is aimed, among other things, at solving the basic problem of meta-methodology, and which will be the focus of the present paper. Laudan's theory of the justification of methodological rules consists of the following theses: (*a*) methodological rules are disguised hypothetical imperatives, in which the antecedent concerns some cognitive aim, and the consequent suggests the means to be used for the achievement of the aim; (*b*) a methodological rule is justified if it is possible to establish, by means of a scrutiny (primarily) of the history of science, that the means-aims connection that it asserts actually holds; (*c*) all methodological rules are justified a posteriori.

In what follows, we shall raise two criticisms against this theory. First of all, we shall claim that Laudan has so far not provided us with convincing examples of methodological rules which are justified in the way envisaged by him. Secondly, we shall show that his solution to the problem of the justification of scientific method is belied by the fact that some interesting methodological rules are justified a priori. Therefore, we shall conclude that normative naturalism is an unsatisfactory solution to the basic problem of meta-methodology.

In section 2, Laudan's solution to the problem of the justification of scientific method will be introduced. In section 3, the so-called "circularity objection" to naturalism will be discussed. We shall argue that, although Laudan can plausibly claim that this objection is far from compelling, he still owes us an illustration of the potential fruitfulness of his meta-methodological research programme. In section 4, some methodological rules which are justified a priori will be discussed. We shall claim that, since Laudan maintains that all methodological rules are justified a posteriori, our discussion belies his theory of justification. Some concluding remarks will be made in section 5.

2 The justification of methodological rules in normative naturalism

The meta-methodological component of normative naturalism crucially revolves around Laudan's analysis of the implicit structure of methodological rules, which are typically stated as categorical imperatives (e.g., "Avoid *ad hoc* hypotheses", "Prefer simple theories to complex ones", etc.). According to Laudan, this way of stating methodological rules is misleading, since it may lead to neglect the fact that they do not emerge in an "axiological vacuum": a methodological rule is always advocated "because it is believed that following [it] will promote certain cognitive ends which one holds dear" [15, p. 132]. Methodological rules, Laudan maintains, are not categorical imperatives of the form:

(0) One ought to do x;

rather, they are hypothetical imperatives having the following form:

(1) *If* one's aim is y, *then* one ought to do x.

In imperatives of the sort of (1), the antecedent concerns some aim — end, goal, value — the consequent suggests the action to be performed in order to promote the aim: "every such rule presupposes that 'doing x' will, as a matter of fact, promote y or tend to promote y, or bring closer to the realization of y" [15, p. 133].[1]

There is another sense in which, according to Laudan, the usual way of stat-

---

[1] Clearly, underlying Laudan's discussion is the instrumental view of scientific rationality: theory-choices must be governed by means-ends rationality, where methodological rules are the means used by the scientists in theory-choice, and the aims of science are the ends that the rules are supposed to promote. However, it needs to be emphasized that Laudan is quite vague concerning the meaning of such expressions as "promoting y", "tending to promote y", and "bringing closer to the realization of y".

ing methodological rules is misleading. Lacking an analysis of their implicit structure, one will likely incline to think of them as commands: "they appear decidedly not to be the sort of utterance which could be true or false, but at best useful" [15, p. 132]. However, once their implicit structure is identified, the problem of their justification — of the warrant for accepting them — is immediately solved. In fact, in Laudan's view, it is obvious that they depend for their warrant on the truth of the statements concerning the connections between cognitive means and cognitive aims that they presuppose: "If I assert a rule of type (1), I am committed to believing that doing x has some prospect of promoting y" [15, p. 133]. As a consequence, the problem of the justification of a methodological rule boils down to that of the assessment of the evidence for the claim that the use of the rule promotes the desired aim: "Provided that we are reasonably clear about how low-level empirical claims (e.g., these alleged ends/means connections) are tested, we will know how to test rival methodologies" [15, p. 133].[2]

Laudan acknowledges that methodological rules "vary from the highly general ('formulate testable and simple hypotheses') to those of intermediate generality ('prefer the results of double-blind to single-blind experiments'), to those specific to a particular discipline ('make sure to calibrate instrument x against standard y')" [14, p. 25]. This classification may suggest, for instance, that different kinds of rules require different kinds of justification. However, Laudan advocates the thesis that meta-methodology is a wholly empirical discipline, whose only tool for ascertaining the effectiveness of a rule in promoting an aim is the empirical information gathered primarily from the history of science. More precisely, he claims that all the meta-methodologist has to do in order to justify any given methodological rule R is to empirically establish that, in the past, R has promoted the realization of the desired aim better than its rivals. As a consequence, he puts forward a meta-methodological principle for the choice between methodological rules which can be phrased as follows:

> (L) Two methodological rules R_1 and R_2 are said to be competing methodological rules if they suggest different means for the achievement of the same cognitive aim A — e.g., if R_1 suggests the means M_1, and R_2 suggests the means M_2. If, on the basis of a scrutiny of the available (mainly historical) evidence, the factual hypothesis that M_1 has so far promoted the achievement of A better than M_2 can be accepted as true (or as probably true), then one ought to infer that in the future M_1 will continue to be more effective than M_2 in promoting A; as a consequence, one ought to consider R_1 as the justified methodological rule.[3]

---

[2] Laudan maintains that "*a thoroughly 'scientific' and robustly 'descriptive' methodology will have normative consequences*" [15, p. 133]. Critics claim that a naturalistic philosophy of science can yield, at most, a description of scientific practice; see [9] for a useful introduction to this debate.

[3] For Laudan's wording of L, see his [15, p. 135]. The problem of the choice between competing methodological rules might be addressed by using the expected utility theory, but Laudan refrains from doing so, presumably because, in his view, the use of the con-

It needs to be emphasised that, according to Laudan, L is nothing but a starting point for the work of the meta-methodologist, since a systematic use of it will soon lead to "a body of complex methodological rules and procedures": Laudan promises that "simple inductive rules (like [L]) will quickly give way to more complex rules of evidential support, as soon as we have a body of methodological rules which has been picked out by these simple test procedures" [15, p. 136].[4]

Moreover, it is important to note that Laudan considers L as fairly uncontroversial for the purpose of choosing between competing methodological rules.[5] This depends on his — controversial — view on the aim of science. According to Laudan, "the aim of science is to secure theories with a high degree of problem-solving effectiveness" [15, p. 78], where "problem-solving effectiveness" indicates, in the first place, the capability of a theory to make successful predictions in the middle run.[6] Since this capability is an empirically ascertainable feature of theories, L seems to suggest itself as an obvious solution to the basic problem of meta-methodology. However, the upholders of alternative axiological views may well feel dissatisfied with L, which is incompatible, for instance, with scientific realism. In fact, scientific realists maintain that the scientists ought to be concerned with the truth-value of the claims that their theories make concerning observable as well as unobservable (or theoretical) entities and processes. But the capability of a theory to make true (or approximately true) claims concerning unobservable entities and processes cannot be directly ascertained in the way recommended by L. As a consequence, L is far from obvious as a solution to the problem of the justification of scientific method.

The above remarks highlight that Laudan's view on the aim of science has a strong influence on his meta-methodological theses. Here, however, we shall not deal with this issue. Rather, by discussing the so-called "circularity objection" to naturalism, we shall argue that Laudan's solution to the basic problem of meta-methodology is unsatisfactory by Laudan's own lights, since so far he has failed to provide us with convincing examples of methodological rules justified by appeal to the factual information gathered (primarily) from the history of science.[7]

3 Normative naturalism: a degenerating research programme

In a nutshell, the circularity objection runs as follows: if a philosopher engages in a scientific study of science, then the results of his endeavours will be

---

ceptual tools of this theory would undermine the naturalistic-empirical character of his meta-methodology.

[4]Laudan is never explicit concerning this body of complex methodological rules and procedures — he seems to rest content with the principled claim that repeated applications of L will yield it.

[5]However, see [15, pp. 135–136] and [19, p. 178] for a defence of L against various criticisms.

[6]For a presentation of Laudan's theory on the aim of science, see his [12].

[7][8] and [1], among others, offer useful discussions of other problems facing normative naturalism.

circular, thereby undermining the justificatory project he pursues. In fact, in order to study science scientifically, one needs to know, from the outset, how one must conduct scientific research. But this is exactly what the philosopher is supposed to discover through his inquiries; therefore, the naturalist — who sponsors a scientific approach to scientific knowledge — reasons in a viciously circular way.

Among the authors who have pressed this objection against Laudan's meta-methodological stance there is Colin Howson [5], who has forcefully argued that Laudan must face the following dilemma. The justification of a methodological rule can be either an a priori or an a posteriori issue. Laudan cannot choose the first horn of the dilemma, because going for the a priori solution to the problem of the justification of scientific method would amount to a rejection of a key premise of his whole philosophical outlook, that is, the claim that any plausible theory of science must take into account — indeed, must be firmly rooted in — the history of science. As a consequence, Laudan must choose the a posteriori solution. However, Howson claims, it is clear that, in this case, Laudan's enterprise cannot even commence. In fact, any claim concerning the justification of a methodological rule made by a researcher engaged in an empirical scrutiny of the history of science requires an appeal to a methodological principle which warrants the justification of the claim under consideration. But the appeal to such a methodological principle requires, in turn, the appeal to another methodological principle, and so on. Therefore, the choice of an a posteriori approach leads to a regress of justifications, which Laudan can stop only at the price of invoking some self-justifying principle of epistemic appraisal — that is, at the price of presupposing the results that he is supposed to demonstrate via a naturalistic inquiry into the workings of science.

Laudan claims that this criticism cannot be a cause for concern, since it rests on the assumption that the meta-methodologist can somehow gain a special position — conquered by "stepping out" of the practice of science — which would allow him to get at a set of unassailable principles of epistemic assessment. But the very idea of "armchair methodology", Laudan maintains, is no less absurd than that of armchair chemistry or armchair physics [14, p. 40]. The systematic failure of generations of philosophers who tried to devise, on purely a priori grounds, unassailable norms for the assessment of theories, suggests that the construction of an a priori methodology is hopeless. More generally, from the point of view of a naturalist, the circularity objection cannot be a problem because science and philosophy are continuous. This implies that no epistemological inquiry can start from a position independent of the body of our currently accepted theories: the meta-methodologist must do his best with what he has, that is, with his informed prejudices concerning the way in which, in our physical world, certain cognitive means are conductive to certain cognitive aims. Of course, as the research goes on, it may turn out that certain supposedly effective means are not conductive to certain aims, but this comes as no surprise to the naturalist, since to him it is obvious that "factual beliefs [...] shape methodological attitudes" [14, p. 39].

The disagreement between Laudan and Howson concerns, in the first place, the very existence of a neutral yardstick against which to measure the merits of their respective views. In any case, it seems to us that even those who are sympathetic to Laudan's defence of naturalism have every right to feel dissatisfied with the results that his meta-methodological research programme has so far produced. Recall that, according to Laudan, "simple inductive rules (like [L]) will quickly give way to more complex rules of evidential support" [15, p. 136]. Doubtless, this promise has not been kept: so far, not a single rule of evidential support has been validated thanks to the justification procedure envisaged by Laudan. To use Lakatos' apt phrase, normative naturalism is a degenerating research programme. Of course, it may one day stage an impressive comeback. However, this seems very unlikely, especially when one takes into account Laudan's discussion of the methodological import of the discovery of the placebo effect, which he considers as a clear-cut example of how a methodological rule can be justified by appeal to the empirical information gathered (primarily) from the history of science.

Due to the placebo effect, it often happens that patients report an improvement even in cases in which they are given pharmacologically inert medications: their expectations of betterment significantly affect the reliability of their reports. According to Laudan's historical reconstruction [12, pp. 38–39], when the placebo effect was discovered, it became clear that simple controlled experiments cannot be considered as sufficient tests of therapeutic effectiveness; as a consequence, scientists started to resort to single-blind experiments, characterized by the fact that the subjects do not know whether they are being given a real drug or a placebo. However, when it was discovered that the physicians administering drug tests often transmit their therapeutic expectations to the patients, it turned out that also single-blind experiments are inadequate as tests of therapeutic effectiveness; as a consequence, scientists started to resort to double-blind experiments, characterized by the fact that neither the patients nor the physicians administering drug tests know whether the subjects are being given a real drug or a placebo.

The improvement in the experimental techniques which followed the discovery of the placebo effect, Laudan claims, could not possibly have taken place as a consequence of a priori reflection: it took a change in factual beliefs to realize that simple controlled experiments and single-blind experiments are unreliable as indicators of the pharmacological effectiveness of drugs and therapies. The discovery of the placebo effect, Laudan maintains, justifies the following methodological rule:

> (R.DB) If one wants to learn whether a drug or therapy is effective, then one ought to prefer double-blind experiments to single-blind experiments and to simple controlled experiments.

Moreover, the case of the placebo effect would show, according to Laudan, that not only the theories, but also the methods of science change over time: since we continuously learn new facts concerning the world as well as human beings as observers of the world, it comes as no surprise that our standards for the assessment of theories are subject to revision and improvement.[8]

---

[8] See [20, 21] for a criticism of this claim.

For our present purposes, it is important to note that Laudan's discussion of this episode does *not* support his claim that a systematic scrutiny of the history of science — as that recommended by L — is required in order to justify such rules as R.DB. Rather, the conclusion that suggests itself is that it is our factual knowledge of the world which justifies the use of double-blind experiments. Consequently, Laudan's claim that the case of the discovery of the placebo effect is a clear-cut supporting instance for his theory of the justification of methodological rules must be rejected.

Consider, moreover, the case of the rule of predesignation, listed by Laudan [15, p. 131] among the methodological rules that his theory of justification is supposed to account for:

> R.P "Prefer theories that make successful surprising predictions over theories which explain only what is already known".

Laudan acknowledges that it is "difficult, and in certain cases patently impossible, to exhibit that a particular set of rules is the best possible way for realizing a certain set of values" [14, p. 35]. This is hardly surprising, he argues. In fact, if it were possible to demonstrate that there is only one set of rules which leads to the realization of a certain aim, then a widespread irrationality in the scientific and philosophical community would be the only sensible explanation of certain chronic methodological disputes:

> Consider, for instance, the 150-year-long (and still ongoing) controversy about the so-called rule of predesignation. [...] A host of prominent thinkers have been arrayed on each side of this issue [...]. All parties to the controversy would, I believe, subscribe to substantially the same cognitive aims. They seek theories that are true, general, simple, and explanatory. Yet no one has been able to show whether the rule of predesignation is the best, *or even an appropriate*, means for reaching those ends. *This failure is entirely typical.* [14, pp. 35–36, italics added]

In the above passage, Laudan implies that the issue has not yet been settled because both parties to the debate have been trying to make their case via some a priori method. According to him, as soon as we start to think of such rules as R.P in naturalistic terms, it becomes clear that there is "an empirical way to settle the issue" [15, p. 178], consisting in ascertaining whether, as a matter of fact, the theories that have been successful — that is, that have shown to possess the capability to stand up to subsequent testing — satisfied the requirement of making surprising successful predictions.

This way of defending normative naturalism has an unpleasant programmatic flavour. Over the past twenty years, Laudan has relentlessly repeated that the scrutiny of the history of science will put an end to the methodological bickering among philosophers. Still, surprisingly enough, he has never done the empirical work which would be necessary in order to keep this promise.

Consider the collection of essays entitled *Scrutinizing Science*, in which a systematic empirical assessment of the post-positivistic philosophies of science is attempted. The editors openly admit that they are far from having a "theoretically integrated and empirically solid" [2, p. 41] theory of science. Nevertheless, they pride themselves on the fact that the picture of the scientific enterprise which is gradually emerging thanks to the empirical scrutiny of its history "arguably represents a dramatic improvement" [2, p.41] over the positivistic and post-positivistic caricatures. Unfortunately, however, nothing in *Scrutinizing Science* seems to support this claim. For instance, three of the case-studies therein collected [3, 4, 23] concern, among other things, the thesis — related to R.P — that the acceptability of a set of guiding assumptions (i.e., of a paradigm, or of a research programme, or of a research tradition) is judged on the basis of the success of its associated theories at making successful surprising predictions. The editors reject this thesis on the ground that the capability of making successful surprising predictions played no significant role in the historical episodes studied by the contributors to the collection. We do not want to challenge the findings of the contributors, but it must be pointed out that the editors' conclusion does not follow from the premises. Nobody who seriously embraces a thoroughly historical approach to the philosophy of science can really believe that three case-studies, no matter how accurate, are enough to establish the claim that making successful surprising predictions is not a significant factor for the acceptability of a set of guiding assumptions.

Although Laudan pleads for a systematic scrutiny of the historical record, as a matter of fact he rests content with a principled defence of normative naturalism. What is worse, he has so far failed to provide us with convincing examples of methodological rules justified by appeal to the factual information gathered (primarily) from the history of science. Moreover, even if Laudan could persuasively argue that the history of science is the key to the solution to the problem of the justification of such rules as R.P, there remains the fact that other methodological rules are justified a priori, via conceptual analysis.

4 Methodological rules justified via conceptual analysis

We shall start our discussion in this section by considering the Popperian rule which recommends to avoid *ad hoc* hypotheses, phrased by Laudan as follows:

> (R.A) "If one wants to develop theories which are very risky, then one ought to avoid *ad hoc* hypotheses". [15, p. 133]

As Kaiser [6, p. 427] and Worrall [22, pp. 353–354], among others, have pointed out, there is no need of an empirical inquiry to ascertain the connection between the antecedent and the consequent stated in R.A, which can be established a priori. This becomes clear as soon as one considers what *ad hoc* hypotheses are.

In the epistemological literature one finds a lack of formal definitions of the notion of an *ad hoc* hypothesis. However, it seems to us that the following passage, in which Popper treats *ad hoc* hypotheses as special types of auxiliary hypotheses, clearly illustrates the idea behind R.A:

> As regards auxiliary hypotheses [and *ad hoc* hypotheses] we propose to lay down the rule that only those are acceptable [that is, non-*ad hoc*] whose introduction does not diminish the degree of falsifiability or testability of [a theoretical system], but, on the contrary, increases it. [...] If the degree of falsifiability is increased, then introducing the hypothesis has actually strengthened the theory [since the hypothesis is non-*ad hoc*]: the system now rules out more than it did previously: it prohibits more [18, p. 62].

When a theory T has false observational consequences, which are deemed genuine counterexamples to T, the upholders of T can try to save it from the refutation by introducing an appropriate *ad hoc* hypothesis H, which — so to speak — transforms T into a new theory T_1 which is compatible with the available evidence. For instance, suppose that $T \equiv$ "All swans are white". If a black swan is observed — let us say, in a remote and previously unknown region of Australia — then the upholders of T can try to save T from the refutation by devising an *ad hoc* hypothesis $H \equiv$ "The swans that live in such and such region of Australia are black". By combining, as it were, H with T, T can be transformed into a new theory $T_1 \equiv$ "All swans, except those living in such and such region of Australia, are white". T_1 is compatible with the available evidence, but it is less risky than T. In fact, in Popperian terms, a theory is said to be risky or informative if it has a large empirical content. The empirical content of a theory is the class of its potential falsifiers, that is, the class of the basic statements forbidden by the theory — where "basic statement" means, roughly, "a statement of a singular fact" [18, p. 21]. Here we shall not enter into the details of Popper's discussion on how classes of potential falsifiers can be compared. For our present purposes, it suffices to say that T_1 has fewer potential falsifiers than T, since T_1 does not forbid the basic statements concerning black swans living in such and such region of Australia, while T forbids them. As a consequence, T_1 is less risky or informative than T. This depends on the fact that T_1 has been obtained from T by the use of an *ad hoc* hypothesis: if one introduces *ad hoc* hypotheses in order to save a theory from the refutation, one always gets a less risky theory.[9]

Before considering another example of a methodological rule which is justified via conceptual analysis, we want to draw the reader's attention to the following scheme of methodological rule, put forward by Ilkka Kieseppä within the context of a discussion on rationalism, naturalism, and methodological rules:

> If precisely the theories with the property Q have the property P, then, in order to choose a theory which has a property P, one should choose a theory which has the property Q [7, p. 251, note 28].[10]

It seems to us that, if a methodological rule R is such that: (*a*) it exemplifies the above scheme, and (*b*) the connection between the property Q and the property P can be established a priori, then R is a clear-cut counterexample to Laudan's thesis that all methodological rules are justified a priori. Clearly, R.A is such a counterexample; however, conceptual analysis enables us to justify a priori some even more interesting methodological rules.

It must be noted that, on one occasion, Laudan has acknowledged that his thesis that all methodological rules state contingent connections between means and aims is perhaps too strong: "One can imagine some means/ends

---

[9]Laudan discusses also another rule, which is related to R.A, and which, given his view on the aim of science, is particularly important to him: "If one is seeking reliable theories, then one should avoid *ad hoc* modifications of the theories under consideration" [15, p. 137], where "reliable" means "theories which more often stand up to subsequent testing".

[10]Kieseppä's wording has been slightly altered.

connections which are, in effect, analytic and whose truth or falsity can be established by conceptual analysis" [15, p. 261, note 28]. Unfortunately, however, Laudan never draws the consequences of what he says in this quotation: nowhere else in his writings does he address the issue of methodological rules which are justified a priori. What is worse, he stubbornly ignores some examples of a priori justified rules which have been devised in response to his "confutation" of convergent realism (see [13]). In the concluding paragraphs of this section, we shall briefly discuss one of these examples, that is, the rule of success proposed by Theo Kuipers.

Over the past thirty years, a number of authors have been working on the methodological research programme of the post-Popperian theories of verisimilitude (see, e.g., [17] and [11]). Within this research programme, it is assumed that a high degree of verisimilitude — or, equivalently, a high degree of truthlikeness, or approximation to the truth — is the main cognitive aim of science. According to the upholders of the programme, the verisimilitude of scientific theories plays the role of a regulative ideal of scientific research. As a consequence, starting with an explication of the notion of approximation to the truth, some criteria — some methodological rules — to regulate the theory-choices made by the scientists in the pursuit of the aim of science are offered. For instance, the rule of success advocated by Kuipers [11, p. 114] can be phrased as follows:

> (R.S) If a theory T_1 has so far been proven to be more successful than a theory T, then eliminate T in favour of T_1, at least for the time being.[11]

This rule is the core of what Kuipers calls the "HD-evaluation" of theories, that is, a sophisticated version of the hypothetico-deductive method which recommends "to take falsified theories seriously" [11, p. 95], since a theory which has already been conclusively falsified can still be the best at our disposal. Kuipers suggests that R.S "may even be considered as the fallible criterion and hallmark of scientific rationality" [11, p. 114], since it can be embraced both by the instrumentalist and by the realist as an effective means for the achievement of the aim of science.

In fact, the instrumentalist looks for theories which enable the derivation of as many true consequences as possible, and as few false consequences as possible — and certainly R.S governs theory-choices in such a way as to serve this purpose.[12] The realist, on the other hand, maintains that the scientists ought to be concerned not only with the observational consequences of their theories, but also with the truth-value of the claims that the theories make concerning unobservable (or theoretical) entities and processes. More specifically, the realist who works within the methodological programme of the post-Popperian theories of verisimilitude looks for theories which are closer and closer both to the observational and to the theoretical truth. Kuipers forcefully argues that R.S serves also this latter purpose.

---

[11] The more successful theory is the one which has got, so far, more empirical successes and less counterexamples. The rule of success was first introduced in [10].

[12] Given his axiological stance, Laudan ought to embrace R.S.

Although here it is impossible to account for all the details of Kuipers' discussion of R.S, it must be noted that the rule is justified a priori, by appeal to two families of theorems: the Success/Forward theorems and the Projection/Upward theorems.[13] For instance, by establishing appropriate links between empirical success and verisimilitude, the Success and the Forward theorems lead to the conclusion that R.S is functional for observational truth approximation — that is, approximate empirical adequacy; more generally, these two families of theorems lead to the conclusion that the HD-evaluation of theories, based on R.S, is functional for truth approximation *tout court*, that is, approximation to both observational *and* theoretical truth.

For our present purposes, R.S is especially interesting for two reasons. First of all, within the context of the HD-evaluation of theories, it exemplifies the above mentioned scheme of methodological rule: it connects a property Q of theories (empirical success) to a property P (truthlikeness), and recommends that, if one is seeking theories with a high degree of truthlikeness, then one ought to choose empirically successful theories. Secondly, this connection is established a priori.

Our discussion of such rules as R.A and R.S belies Laudan's claim that all methodological rules are justified a posteriori.

5 Concluding remarks

In this paper we have highlighted that Laudan's solution to the basic problem of meta-methodology is strongly influenced by his view on the aim of science: his proposed principle for the choice between competing methodological rules can seem to be an obvious solution to the basic problem of meta-methodology only under the — controversial — assumption that the achievement of the aim of science must be empirically ascertainable. Moreover, we have argued that normative naturalism is a degenerating research programme. In fact, the results that it has so far yielded are unsatisfactory by Laudan's own lights, since so far he has failed to provide us with convincing examples of methodological rules justified by appeal to the factual information gathered (primarily) from the history of science. Finally, we have discussed some methodological rules which are justified a priori, via conceptual analysis; our discussion shows that Laudan is wrong in his claim that all methodological rules are justified a posteriori. We therefore conclude that normative naturalism is an unsatisfactory solution to the problem of the justification of scientific method.

Acknowledgements

I would like to thank Roberto Festa and Gustavo Cevolani for their comments on a draft of this paper. Research supported by PRIN 2006 grant Rational Decisions, Strategic Interactions, Complexity and Evolution of Social Systems (Research Unit: Department of Philosophy, University of Trieste).

BIBLIOGRAPHY

[1] A. Diéguez-Lucena. Why does Laudan's confutation of convergent realism fail? *Journal for General Philosophy of Science*, 37: 393–403, 2006.

---

[13] For the Success/Forward theorems, see [11, pp. 158–161, 164, 231, 216, 260–262]; for the Projection/Upward theorems, see [11, pp. 213–214, 216–218, 276].

[2] A. Donovan, L. Laudan, and R. Laudan (eds.). *Scrutinizing Science*. Kluwer, Dordrecht 1988.
[3] M. Finocchiaro. Galileo's Copernicanism and the acceptability of guiding assumptions. In A. Donovan, L. Laudan, and R. Laudan (eds.). *Scrutinizing Science*, Kluwer, Dordrecht 1988, pages 49–68.
[4] J.R. Hoffman, Ampére's electrodynamics and the acceptability of guiding assumptions. In A. Donovan, L. Laudan, and R. Laudan (eds.). *Scrutinizing Science*, Kluwer, Dordrecht 1988, pages 201–218.
[5] C. Howson. The poverty of historicism. *Studies in History and Philosophy of Science*, 21: 173–179, 1988.
[6] M. Kaiser. Progress and rationality: Laudan's attempt to divorce a happy couple. *Inquiry*, 34: 433–455, 1991.
[7] I. Kieseppä. Rationalism, naturalism and methodological principles. *Erkenntnis*, 53: 337–352, 2000.
[8] J. Knowles. What is really wrong with normative naturalism. *International Studies in the Philosophy of Science*, 16: 171–186, 2002.
[9] H. Kornblith (ed.). *Naturalizing Epistemology*. MIT Press, Cambridge, MA 1985.
[10] T.A.F. Kuipers. Approaching the truth with the rule of success. *Philosophia Naturalis*, 21: 244–253, 1984.
[11] T.A.F. Kuipers, *From Instrumentalism to Constructive Realism. On Some Relations between Confirmation, Empirical Progress, and Truth Approximation*. Kluwer, Dordrecht 2000.
[12] L. Laudan. *Progress and Its Problems. Towards a Theory of Scientific Growth*. University of California Press, Berkeley 1977.
[13] L. Laudan, A confutation of convergent realism. *Philosophy of Science*, 48: 19–49, 1981.
[14] L. Laudan. *Science and Values*. University of California Press, Berkeley 1984.
[15] L. Laudan. *Beyond Positivism and Relativism*. Westview, Boulder, CO 1996.
[16] L. Laudan, A. Donovan, R. Laudan, P. Barker, H. Brown, J. Leplin, P. Thagard and S. Wykstra. Scientific change: philosophical models and historical research. *Synthese*, 69: 141–223, 1986.
[17] I. Niiniluoto. *Truthlikeness*. Reidel, Dordrecht 1987.
[18] K.R. Popper. *The Logic of Scientific Discovery*, revised Edition: 1972. Reprinted Routledge, London and New York 1992.
[19] H. Sankey. *Rationality, Relativism and Incommensurability*. Ashgate, Aldershot 1997.
[20] J. Worrall. The value of a fixed methodology. *The British Journal for the Philosophy of Science*, 39: 263–275, 1989.
[21] J. Worrall. Fix it and be damned: A reply to Laudan. *The British Journal for the Philosophy of Science*, 40: 376–388, 1990.
[22] J. Worrall. Two cheers for naturalised philosophy of science — or: Why naturalised philosophy of science is not the cat's whiskers. *Science and Education*, 8: 339–361, 1999.
[23] H. Zandvoort. Nuclear magnetic resonance and the acceptability of guiding assumptions. In A. Donovan, L. Laudan, and R. Laudan (eds.), *Scrutinizing Science*, pages 337–358. Kluwer, Dordrecht 1988.

Explaining the scientific success. A critique of an abductive defence of scientific realism

SILVANO ZIPOLI CAIANI

This paper aims to show the main limit of an abductive defense of scientific realism. My work illustrates how the well known *Inference to the Best Explanation* (I.B.E.), when adopted to explain the success of scientific theories, should be considered inadequate to sustain a *metaphysical* acceptation of scientific realism. The present analysis shows how an *epistemic* notion of truth is indissolubly involved by the adoption of an abductive argument, revealing how metaphysical realism, when conceived the best explanation for scientific success, has to be judged an unsatisfactory suggestion.

The present work is divided into two sections. In the first part Peirce's conception of abduction (section 1.1) will be introduced as a fundamental reference inspiring the contemporary employment of the abductive reasoning. Afterward, it will be analyzed how an instance of abductive reasoning has been used to explain the success of experimental science by contemporary realists (section 1.2). In the second section it will be shown how the occurrence of an abductive reasoning involves epistemic constraints related with its selective character (section 2.1). Finally, some comments concerning the explicative limits of metaphysical realism will be suggested.

1 The explanatory problem and the metaphysical argument

Introducing the *explanatory problem* concerning the success of the best available scientific theories is usual to make a reference to a famous germinal paper by Gilbert Harman,[10][1] this is the locus classicus where the notion of Inference to the Best Explanation (I.B.E.) has been introduced to solve questions regarding the epistemic justification of enumerative induction. As it will be shown in the following sections, more recently an occurrence of Harman's argument has been advanced to restore a realist statement after years of skeptical dominance in philosophy of science.

In recent times, the renewed exigency to contrast the anti-realist movement in philosophy of science has been influenced by Putnam's works on the justification of the empirical success of science. Putnam [28] and others (for example Smart [30] and Richard Boyd [4, 5, 6]) have tried to show how the

---
[1]Gilbert Harman was the first to introduce the concept of Inference to the Best Explanation within the contemporary debate. Other significant topics where Harman use this notion are [11, 12].

positivist conception leaves unexplained factual features characterizing the scientific knowledge such as its unity among different theoretical domains, its continuity between successively paradigms and the experimental success of our best scientific theories. The problem of explanation represents the main area of inquiry where a great deal of contributes committed to the question of scientific realism are concentrated.[2]

Core of the realist explicative thesis is the *non-epistemic notion of truth*, that is, a metaphysical interpretation of Tarski's schema that endorses the possibility to define a direct connection between scientific constructs and the ontological domain (see below section 1.2). The main reasoning adopted by scientific realists can be considered an occurrence of what the early father of pragmatism C.S. Peirce called first an *ampliative* inference. For this reason, next section will be dedicated to analyze some aspects of Peirce's influential conception about explanation.

1.1 Peirce on abduction

Within the modern debate, the philosophical interest for the abductive reasoning is mainly related to the work of C.S. Peirce. The term *abduction* is drawn by Peirce from Aristotle's Prior Analytics [22, 5.144], but the structure of this kind of reasoning is a recurring theme along the entire history of science (one for all the case of Kepler's abductive reasoning described by [1]).

To sum up Peirce's idea, abduction can be considered a form of *ampliative reasoning* that makes it possible to increase knowledge "guessing" the right explicative hypothesis [22, 7.219-20]. Using a famous example introduced by Peirce, it's possible to define the abductive reasoning in a syllogistic schema as follows [22, 2.623]:

1. Rule: All the beans from this bag are white

2. Result: These beans are white

3. Fact: These beans are from this bag

In a mature phase of his thought (see the Hardware Lectures of 1903), Peirce has concentrated part of his work into the research of an *epistemic justification* for the abductive line of reasoning. The question:

> what should an explanatory hypothesis be to be worthy to rank as a hypothesis? [22, 5.197].

was used by Peirce distinguishing two basic aspects, a *generative* and an *evaluative*,constituting an explicative inference.[3] Following Peirce, the generative aspect of an explicative inference corresponds to the starting reasoning following the discovery of new recalcitrant facts. This early phase of the explanatory process is identifiable with the attempt to detect plausible hypothesis, or a

---

[2]Many different topics constitute the realism/anti-realism debate. Here it will be examined a typical form of epistemic approach called "explanatory thesis". Another form of approach is, for example, the *logical* approach well represented by discussions related to Fitch's paradox (also known as paradox of knowability).

[3]For an analogous distinction see [17], where the creative character of abduction is associated with the evaluative function of the Inference to the best explanation.

set of them, that seem to be able to provide intelligibility to facts otherwise surprising. In Peirce's words:

> Abduction is the process of *forming* an explanatory hypothesis [22, 5.171].

Whereas the evaluative moment concerns, for Peirce, the specific selection of one of these previously detected hypotheses. In Peirce's words:

> Abduction is the process of *choosing* a hypothesis [22, 7.219].

For Peirce, every explanation starts with the ascertainment of an unexpected state of marvel. The importance of this condition is represented efficaciously by the image of a ship proceeding his sailing over a smooth sea, without any augury otherwise the monotony of such a voyage, when suddenly, an *unforeseen* impact with a rock occurs [22, 5.51]. In light of this conception, abduction is considered by Peirce the first step of scientific enquiry, it represents the main form of answer to the *undesirable* state of uncertainty related with the common experience of *recalcitrant* facts. [4]

Furthermore, Peirce pays special attention to underlying the state of *uncertainty* related with the condition of plurality typical of many generative moments where more than one plausible explanation is inferable as regard the same set of (surprising) data. This unsatisfactory condition is well described by one of most famous Peirce's examples:

> given a certain phenomena discovered by a physicist, it's always questionable how does he know but the conjuncts of the planets have something to do with it, or that it is not perhaps because the dowager express of China has at same time, a year ago, changed to pronounce some word of mystical power, or some invisible jinnee may be present [22, 5.172].

So, following Peirce, trillions of trillions of hypotheses might be advanced, even if we are looking just for one explanation. An undesirable condition of ambiguity that reveals the incompleteness of an explicative reasoning based only on what Peirce calls the generative moment, evidencing the central role played by the practice of evaluation.

Beside the power to confer intelligibility to surprising facts, the evaluative process constituting a complete abductive inference is characterized for Peirce by a proper internal epistemic structure.Two are for Peirce the main characteristics ascribable to an evaluative moment: the *economicity* and the *observational character* [22, 7.220]. The economicity of an explicative hypothesis consists, for Peirce, in the evaluation of three distinct epistemic parameters: i) its *methodological* value, that is, the rationality used to generate a certain explicative proposal; ii) its power to involve other aspects of knowledge, defining new interactions between different disciplines; iii) and finally its costs of experimentation. On the other hand, the observational character represents an essential form of epistemic appraisal consisting in to take at proof every explicative proposal, evaluating its adherence with available or expected data.

---

[4]Following Peirce, the generation of a new plausible hypothesis starts with an *appeal to instinct* and evolves in a gradual way [22, 1.630]. It moves from a *dark laboring*, bursting out the *startling conjecture* and showing at the end how an explicative hypothesis corresponds perfectly to the initial anomaly, as well as *a key opens a locked door* [22, 6.469].

Therefore, the selective process of explanation is defined by Peirce not only by rational criteria (such as represented by the properties of economicity), but also by the comparison between hypotheses and facts:

> any hypotheses therefore, may be admissible, in the absence of any *special reasons* to the contrary, provided it be capable of *experimental verification*, and only insofar as it is capable of such verification [22, 5.197; the emphasis is mine].

Resuming the critical points individuated in this short introduction to Peirce's theory of abductive reasoning, it is opportune to underline the following points:

1. Peirce distinguishes the internal structures of abductive reasoning in a *generative* moment and in a *evaluative* one. The former concerns conjecturing about a range of plausible explicative hypotheses, the latter concerns the selection of *the best* explicative solution;

2. Peirce evidences the importance of an evaluative moment to obtain the satisfactory condition represented by the achievement of an unambiguous explanation;

3. to make possible an explanatory evaluation, Peirce's conception suggests the adoption of epistemic criteria such as *economicity*, and *empirical verifiability*.

With these distinctions in mind, it is now possible to move forward analyzing how the contemporary debate on scientific realism has interpreted and employed the Peirceian notion of abductive reasoning.

1.2 The abductive realism

The recent debate on realism is mainly characterized by the problem of justification related to the empirical success of our best scientific theories. Philosophers such as Peter Lipton (2001) and Stathis Psillos (1999) have dedicated to the justification of the factual success of science large part of their works, supporting the idea that only a realistic interpretation of scientific theories represents the best candidate to explain the possibility of the experimental success.

First to propose this solution, in the middle of the last century, was J.C. Smart who opened the door to the formulation of an explicative conception of realism contrasting phenomenalism on theoretical entities [30]. Following the line traced by Smart, Hilary Putnam, in the early period of his thought, has developed his famous *no miracle argument* introducing the exigency to relate the problem of justification concerning scientific success to an *externalist* notion of explanation [28]. With Putnam and successively Richard Boyd [6], the realist proposal has become one of the most popular topic in philosophy of science, a view well resumed and analyzed in an exhaustive book edited by Stathis Psillos [23].[5]

---

[5] A separated analysis should be dedicated to a different form of abductive reasoning, the *common cause principle* mutated by W. Salmon from H. Reichenbach's thought. However, the *ontic* realism, developed in Salmon's work [29], with his special conception of causality, is not the object of the present analyses.

In this version scientific realism is a *metaphysical* thesis introduced as a consequence of an abductive reasoning and formulated to explain the otherwise surprising fact represented by the existence of empirically well confirmed scientific theories. In other words, scientific realism is formulated here as an explicative thesis concerning the existence of a direct relation between the linguistic domain of successful theories and an independent and well fixed ontological statement. Using the abductive schema developed by C.S. Peirce (see section 1.1), the general form of this explanatory suggestion can be presented as follows [22, 5.189]:[6]

1. The surprising fact C is observed;

2. but if A is true, C would be a matter of course;

3. Hence, there is reason to suspect that A is true;

Where: 1. represents the starting point of every explanatory problem, that is, the appraisal of recalcitrant data C as regard as the available theoretical context; 2. represents an explicative hypothesis, where the truth value of a certain statement A involves the occurrence of the previously surprising data C, and finally 3. represents the assumption of a certain explanatory hypothesis.

Substituting the fact C with the *success of scientific theories* and considering the *realist thesis* as the hypothesis A, we obtain the general structure of the abductive reasoning adopted by many recent scientific realists. In other words:

1. The fact that scientific theories have empirical success is observed;

2. If scientific realism is true, the empirical success of scientific theories is a matter of course;

3. Hence there is reason to suspect that scientific realism is true;

To understand what assumptions constitute scientific realism, it is possible to adopt three line of argumentation well defined by Stathis Psillos [26]: the *metaphysical* thesis, concerning what is reality; the *semantic* thesis, concerning the truth statement of successful theories and the *epistemic* thesis, regarding the nature of what we know with true theories.

The *metaphysical* thesis is the simplest and intuitive. Assuming a metaphysical conception of realism we accept that the world has a defined statement, independent from every condition of knowledge and human ability to know. In particular, the metaphysical stance distinguishes scientific realism from the anti-realist and the phenomenalist accounts of science, assuming an absolutely independent conception of reality from whatever epistemic statement [8, 24].

This idea, also known as the *mind-independent* condition of knowledge, should be considered a prerequisite for any defense of an explanatory realism,

---

[6]See also Ikka Niiniluoto [20].

where entities posited by well confirmed theories are assumed to map an *already existing* world [23].

With the assumption of the *semantic* thesis the observational success of well confirmed scientific theories should be taken at the *face value*. Following the realist reasoning, a semantic interpretation of scientific theories makes it possible to define the passage from a factual condition of empirical success to a meta-observational condition of truth, defining what Lipton has called the *truth argument*. This argument relates the explicative power of a hypothesis with its truth likeness condition, so that, if we want to explain the empirical success of science, we ought to assume that experimentally successful theories are, at less, approximately true [16].

Aim of the third stance constituting scientific realism, the *epistemic* thesis, is to posit the passage from an ascertained condition of truth (semantic thesis), to a description of a metaphysical reality (as established by the first thesis). This result is allowed assuming a so called *inflationist* conception of truth, obtained associating the notion of truth with an ontological statement, while, for a deflationist conception, truth consists in the acceptation of a schema of equivalence — Tarski's schema — on the formal ground [13]. Only assuming that truth likeness corresponds to a relation between language and world, that is, assuming a *substantive* interpretation of the Tarski's schema, truth takes its independence from the epistemic context, showing the explicative value required by scientific realism [19].

In other words, the epistemic thesis claims that theoretical assertions have a *non-epistemic* truth value [26]. In this way, if with the concept of *epistemic conditions* we denote conventions, hypothetical assumptions or simply information introduced by the subject of knowledge, the non-epistemic conception of truth aims to defend the independence of truth-values from every theorethical context, differently from what is posited by constructive conceptions (see for this [32]). With the assumption of a non-epistemic notion of truth, scientific realism takes the form of an abductive reasoning, passing from the factual success of a scientific theory, to the assumption of its correspondence with an epistemic independent domain: the metaphysical reality.

The assumption of these three thesis configures realism as an *a posteriori* argument, suggesting the presence of a link between the employment of an abductive methodology and the achievement of a not epistemic condition of truth. Thus, if we infer abductively an explanatory account of the empirical success of science, as scientific realists do, we obtain that this *surprising result* is explained assuming a direct relation between experimental confirmed theories and certain fixed aspects of an independent world [18].

Resuming this section, the three major assumptions constituting the realist hypothesis should be defined as follows:

1. Reality is a metaphysical domain, it pertains to a *mind independent* world, that is, to an objective sphere not related to assumptions or conditions constituting knowledge;

2. Successful theories are true, or in other words, terms and propositions involved in well accepted scientific theories have a semantic reference;

3. True is an inflationary ontological notion, it pertains not exclusively to an observational or instrumental domain, but to a metaphysical one.

As it was evidenced above, scientific realism is mainly an explicative thesis, it aims to solve the problem of justification posited by the empirical success of scientific theories. The explicative character of realism is conceived in conjunction with the favor for a *non-epistemic* notion of truth. Only in this way it is possible for a realist to relate the truth likeness of scientific descriptions to an absolutely independent domain, that is, without considering the epistemic role of assumptions such as conventions, or other not-factual stances involved by explicative processes.

2 Is metaphysical realism the best explanation?

Bass van Fraassen [32] and more recently Henk de Reght and Dennies Dieks [7] have showed that the understanding of an explicative process involves the definition of particular purposes and assumptions. Following this line, to understand something is a pragmatic process in the sense that it implies the possibility to choice between different explanations concerning the same fact, assuming different theoretical backgrounds with a different philosophical or social nature. This general condition of context-dependence is usually adopted as a refutation of metaphysical realism, but differently from another well known *confutation of realism* introduced by Laudan (1981), the present argument is not based on a factual analysis related with the historical development of science, but is an epistemic argument concerning the procedures by means of which any explanation is made possible.

2.1 The epistemic character of selective abduction

Yemina Ben-Menahem [3] has sustained that the rationality of an explicative process depends indissolubly on the set of evaluative standards we adopt to assign *explanatory power* to a certain hypothesis (for an analogous position see also [9]). In this way, no explanatory problems as well as, no explanations tout court, are possible in a beliefs vacuum (see also section 1.1). In other words something can been judged lacking an explanation only when a set of beliefs are previously assumed, as in the case a certain fact appears surprising in light of a well accepted theory.

Considering every explicative claim as contextually dependent involves that in case of variation of a background set of assumptions differences could be also induced in what may be considered *problematic*. In this way, what is an explicative problem in a certain historical moment can be considered an acceptable and unproblematic condition in another, forcing us to consider our explicative problems evolving with the rest of our knowledge [3].

For example, while within Aristotle's physics is relevant the exigency to explain how a grave, flanged in the air, carries on its movement even if apparently nothing is sustaining it, with the introduction of the concept of *inertia* we assist at a theoretical change involving the very content of the physical knowledge, as well as at variation of the range of questions we are inclined to advance. For example, after Newton's theory it become irrelevant the question *why the grave has a movement without apparent causes?*, but the problem became *how to explain irregularities of grave's motion?*

As it's clear from the previous section dedicated to C.S. Peirce (1.1), an explanatory process may encounter cases of equivalence as regard as the same empirical statement. For all explicative exigencies related to a certain set of facts (the surprising facts), it's possible to define more than one explicative solution, so that for every explanatory hypothesis could be individuated different alternatives entailing the same evidence. Therefore, the processes of explanation can suffer of the existence of just too many undercutting solutions [25]. This condition is represented by cases where two or more different theories show the same empirical under-determination, or in other words, when different theories propose different kinds of explanation for the same range of (surprising) facts.

Historically this case is well represented by the famous controversy between the Copernican heliocentric theory and the Brahe's geocentric conception, both aiming to explain the same facts related to the observation of some irregular movements of planets. Recently in physics, cases of observational equivalence are represented by the debate between Copenhagen's interpretation of quantum mechanics and the bohmian theory of hidden variables, as well as by the debate between standard particles model of matter and the elegant strings' theory.

As noted by Peirce, the possibility to found alternative explanatory hypotheses justifies the presence of an evaluative moment within the formulation of a general theory of abductive reasoning (section 1.1). With this assumption it is possible to constrain the acceptation of a valid explicative solution, introducing a discriminatory process that makes it possible to promote the choice of only one proposal within a certain set of plausible explicative solutions.

Reformulating the previously introduced abductive schema we have that (Psillos 2002):

1. C is a collection of data (facts, observations);

2. the hypothesis A explains C;

3. No other hypothesis can explain C as well as A does;

4. Therefore the hypothesis A is (probably) true

In general, if there are several candidates to explain the same evidence, one must be able to reject all such alternatives until only a single satisfactory explicative inference is obtained. It is important to note that, in cases such as these, the presence of contextual assumptions can play an important role. Indeed, pre-accepted explanatory criteria guide the explanatory inference, revealing what are the salient explanatory relations, determining the ranking of rival explanatory hypothesis (a condition that is also accepted by a realist such as [27]).

Following Lipton [16], we can conceive the evaluative moment constituting the abductive reasoning a research defining the *loveliest* explanation within a pool of potential candidates. What is important to note here is that, to define whichever set of loveliest explicative arguments, we have to define previously some epistemic filters or, in other words, some *standards of choice* that make it possible the selection of the best available explanatory solution.

Without intention of completeness, it's possible to individuate some of the most relevant epistemic filters that usually guide the process of hypothesis selection:

1. *Unification*: based on the assumption that a hypothesis is an explanation if it unifies the explanandum with other background knowledge;

2. *Parsimony*: based on the assumption that a hypothesis is an explanation if it implies fewer particular assumptions than another;

3. *Consilience*: based on the assumption that the best explanation is the hypothesis that covers the major classes of facts;

4. *Importance*: the best explanation covers the most salient phenomena;

5. *Refute of ad-hocness*: a hypothesis explains if it avoids dogmatic or tautological solutions;

6. *Analogy*: assume that a hypothesis explains if it shares properties with other just accepted explanations;

7. *Observability*: concerning the empirical character of an explicative hypothesis;

8. *Simplicity*: one of the most difficult properties to define in force of its contextual relativity, but one of the most important;

Notwithstanding the incompleteness of the list above, the definition of an ideally complete list of criteria is not enough to understand the entire process of evaluation. Insofar the process of choice is multi-factorial, that is, involves more than one criteria of evaluation, it is not possible to reduce it simply to a quantitative stance. Instead, every standard presents a proper qualitative value. Since that is so to be judged the best explicative hypothesis available is not enough to fit the major number of explicative properties, but it requires to posses the most important of them. In other words, to employ the explanatory virtues listed above, it requires the definition of a *hierarchical structure* that makes it possible to establish priorities among criteria.

It's possible to formulate this condition considering the selective criteria as *principles of explanation*, introduced with the aim to define properties that every processes of understanding should possess. These principles constitute the set of basic assumptions defining what can be called a *theory of explanation*, that is the theory adopted to *individuate* and *solve* explanatory problems. Variations concerning the elements of the list, as well as variations concerning the epistemic value attributed to each of them, generate divergences between different perspectives about what *needs* and what *admits* an explanation.

As in the case of Peirce's condition of economicity (see section 1.1), these principles are not strictly reducible to factual conditions, instead they conserve the aspect of a priori conjectures, such as conventions, and are subjected to different judgments within different contexts (theorethical,historical or social) (see [32] or [25]).

Moreover, the variability of selective standards admits some internal constraints, limiting combinations between different criteria. This means that not all possible features ascribable to an explanation are reciprocally compatible in all circumstances. As it is noted by Thagard [31], a typical case of contrast is well represented by the tension between the criterion of *consilience* and the criterion of *parsimony*. Following this line, making a hypothesis more consilient can render it less parsimonious, as in the case when extra assumptions are added to explain some additional facts.

This condition reveals a dependency of explicative criteria from experiential conditions. For this reason it appears not possible to define a universal and invariable hierarchy of explicative principles, rather they evolve in combination with new ascertained facts and with the structure of our accepted theories of knowledge [2].

2.2 Limits of the abductive defence of realism

Turning back to the main topic of this paper, in light of the previous analysis we have now many elements that make it possible to consider scientific realism a controversial explicative conception. Within the realist argument it is now clear that two epistemic conditions are not sound: a) the assumption of a non-epistemic conception of truth, and b) the presence of the epistemic criteria characterizing the evaluative moment proper of every explicative process.

In the previous section (1.2), it was possible to realize how, for a scientific realist, an ontological and non-epistemic conception of truth plays a fundamental role into furnishing an explicative justification for the scientific success. As we have seen, the explicative claim advanced by scientific realists is made possible by the preliminary acceptance of a condition of independence between truth-reference and epistemic assumptions involved by scientific knowledge. The request, advanced by scientific realists such as Psillos, concerning the epistemic independent status of truth values ascribable to our best theories, underestimates the proper role of the evaluative moment (that is the adoption of some explicative standards) required to complete an abductive reasoning.

In light of this, a question emerges: how could a realist justify the explicative significance of a *truth value*, independently of every epistemic assumption concerning what should be considered explicative? In other words: how can we consider a set of explicative standards, characterized by an objective and ontological value as always preferable above all others?

This question configures a sort of *transcendental* argument for realism.[7] What kind of criteria makes it possible to select scientific realism as best explanation of the factual success of science? The selection between a multitude of different explicative proposals requires to be justified and, if not in epistemic terms, how? Now the burden of the proof passes in the hands of scientific realists.

---

[7]An argument concerning the transcendental condition of truth in realism is analyzed by [15].

Resuming, the problem individuated within the explicative conception of scientific realism is defined by the contrast between two assumptions:

1. scientific realism, in its abductive acceptation, involves a substantial and non-epistemic conception of truth;

2. every unequivocal abductive reasoning involves an epistemic evaluative moment concerning the adoption of conventions, conjectures and postulates regarding the definition of what should be considered explicative.

The assumption of both these two statements opens the door to the subsequent question: how can we defend the explicative statement of a theoretical hypothesis independently of whichever epistemic condition? The deficiency in to furnish an answer to this question configures an important limit of the explicative claim of realism. The explicative supremacy of scientific realism may be also questioned in force of its metaphysical character and its related not-observational nature. Without the possibility to establish any empirical confirmation, the process of hypothesis selection involved by abductive reasoning, should be drawn only on the ground of the explanatory principles we choose to adopt. A condition that elude the second criterion introduced by Peirce concerning the observational character of any explanatory process.

Furthermore, the metaphysical acceptation of reality, beside to be an unobservable assumption, also appears unable to furnish more empirical predictions than a not-metaphysically committed conception do. The possibility of a scientific progress, that is the possibility to discover new theories more confirmed than the older, with a better structure, or with a largest empirical domain, seems to be independent from the assumption of a particular metaphysical view concerning truth values or ontological conditions.[8] For this reasons, if realism has to be evaluated as any other scientific explanation, it appears nothing but a very unsatisfactory suggestion [21].

3 Conclusions

In this paper it was possible to reveal the presence of an inconsistence within the abductive argument usually advanced to establish scientific realism.

Starting from the analysis of Peirce's notion of abductive reasoning, it was possible to make explicit the distinction between two internal moments, the *generative* and the *evaluative*, both characterizing a typical kind of not-ambiguous explicative process. Moreover, following Peirce, it was possible to underling how both *rational* and *observational* criteria are involved by the evaluative process aiming to select the best explicative solution available.

After this preliminary introduction, the analysis of the explicative thesis concerning scientific success, endorsed by philosophers such as Smart, Putnam, Boyd and more recently Psillos, has made explicit the underlying abductive structure of scientific realism, as well as the role of a non-epistemic

---

[8]This at least we choice to pass from an epistemological to a psychological context of analysis, where assumptions of this kind may influence the development of subjective motivations.

notion of truth, linking directly successfully theoretical descriptions with an independent ontological domain.

Finally, an examination of the epistemic properties characterizing abductive reasoning has evidenced the untenability of an explicative defense of realism based on both a) the assumption of an abductive argumentation and b) the adoption of a non-epistemic notion of truth.

This contrast is generated by the frequently underestimated role of the evaluative process constituting any occurrence of a satisfactory explanation, that is, the result of a not ambiguous abductive reasoning. As it was shown in the last part of the work, the choice of a set of selective criteria with the aim to single out the best explicative hypothesis available is a typical process involving epistemic assumptions, contextual contingencies and is mutable with the progress of knowledge.

Concluding, the analysis of the abductive defense of scientific realism has revealed the presence of remarkable a limit. The selective character of abductive reasoning invalidates any attempt to identify an explicative proposal independently from the function of a restricted set of epistemic assumptions concerning the adoption of a specific theory of explanation. Any attempt to identify a link between theoretic constructs and an ontological domain is the consequence of a typical epistemic choice concerning the relevance of a previously selected set of explanatory principles. This condition configures scientific realism as an epistemic choice between others, a circumstance that no metaphysical stance appears capable to elude.

BIBLIOGRAPHY

[1] A. Atocha. *Abductive Reasoning: Logical Investigation into Discovery and Explanation.* Dortrecht Springer 2006.
[2] P. Baumann. Theory of choice and the Intransitivity of "is a better theory than". *Philosophy of Science*, 72: 231–240, 2005.
[3] Y. Ben-Menahem. The inference to the best explanation. *Erkenntnis*, 33: 319–344, 1990.
[4] R. Boyd. Observations, explanatory power and simplicity. In R. Boyd, P. Gasper and D. Trout (eds.), *The Philosophy of Science*, the MIT Press, Cambridge 1983, pages 349–377.
[5] R. Boyd. On the current status of the issue of scientific realism. *Erkenntnis*, 19: 45–90, 1983.
[6] R. Boyd. What realism implies and what it does not. *Dialectica*, 43: 5–29, 1989.
[7] H. De Regt, Henk and Di Dieks. A contextual approach to scientific understanding. *Synthese*, 144: 137–170, 2005.
[8] M. Devitt, *Realism and Truth.* Blackwell, Cambridge and Oxford 1991.
[9] M. Friedman. Explanation and scientific understanding. *The Journal of Philosophy*, 71: 5–19, 1974.
[10] G. Harman. Inference to the best explanation. *The Philosophical Review*, 74: 88–95, 1965.
[11] G. Harman. Detachment, probability, and maximum likelihood. *Nous*, 1: 401–411, 1967.
[12] G. Harman. Knowledge, Inference, and explanation. *American Philosophical Quarterly*, 5 (3): 164–173, 1968.
[13] P. Horwich. *Truth.* Blackwell, Oxford 1990.
[14] L. Laudan. A confutation of convergent realism. *Philosophy of Science*, 48 (1): 19–49, 1981.
[15] J. Leplin. Truth and scientific progress. In *Scientific Realism*. University of California Press, Berkeley 1984.
[16] P. Lipton. *Inference to the Best Explanation.* Routledge, London 2001.
[17] L. Magnani. *Abduction, Reason and Science.* Kluwer Academic Publishers, New York 2001.

[18] E. Mc Mullin. A case for scientific realism. In J. Leplin (ed.), *Scientific Realism*, California University Press, Berkeley 1984.
[19] W.-H. Newton-Smith. The truth in realism. *Dialectica*, 43: 31–45 1989.
[20] I. Niiniluoto. Defending abduction. *Philosophy of Science*, 66: 437–451 1999.
[21] P. Parrini. *Knowledge and Reality. An Essay in Positive Philosophy*. Kluwer Academic Publishers, Dordrecht, Boston, London 1995.
[22] C. Peirce. *Collected Papers*. The Belknap Press of Harvard University Press, Cambridge 1933.
[23] S. Psillos. *Scientific Realism: How Science Tracks Truth*. Routledge, London 1999.
[24] S. Psillos. The present state of the scientific realism debate. *British Journal for the Philosophy of Science*, 51 (Special Supplement): 705–728, 2000.
[25] S. Psillos. Simply the best: A case for abduction. In A. C. Kakas and F. Sadri (eds.), *Computational Logic: From Logic Programming into the Future*, Springer-Verlag, Berlin and Heidelberg 2002.
[26] S. Psillos. Scientific realism and metaphysics. *Ratio*, 18: 385–404 2005.
[27] S. Psillos. The fine structure of Inference to the best xplanation. *Philosophy and Phenomenological Research*, 74 (2): 441–448 2007.
[28] H. Putnam. *Mind, Language and Reality*. Cambridge University Press, Cambridge 1975.
[29] W. Salmon. *Scientific Explanation and the Causal Structure of the World*. Princeton University Press, Princeton 1984.
[30] J.J. Smart. *Philosophy and Scientific Realism*. Routledge, London 1963.
[31] P. Thagard. The best explanation criteria for theory choice. *The Journal of Philosophy*, 75: 76–92 1978.
[32] B. van Fraassen. *The Scientific Image*. Clarendon Press, Oxford 1980.

Van Fraassen, observability and belief
MARIO ALAI

1 Observation, observability and the strict empiricist compromise

Most of us are empiricists, as we believe that observation is the primary source of factual beliefs and the primary basis of justification for them. A *fundamentalist* empiricist would hold that it is also the *only* legitimate source of justification for them: deduction is a valid form of justification, but it adds nothing to the results of observation, while induction, abduction, and any sort of ampliative reasoning, cannot be trusted at all. Thus, only beliefs about observed facts can be considered justified, and only *observed* fact can be known.

Perhaps there exists no such *fundamentalist* empiricist, for these conclusions are too implausible to be accepted: it is obvious that there can be good reasons to believe in a lot of things we have never observed, i.e., that ampliative reasoning may be reliable. For instance, nobody has observed the asteroid that 2600 millions years ago struck the Earth, producing the large crater we still see at Suvajärvi;[1] however, observation of the crater and ampliative reasoning together give us strong reasons to believe in the existence of the asteroid and in its impact, which in fact are accepted by all informed people. So, while observation is the primary justification for our factual beliefs, it is not the only one, and although all our factual beliefs are ultimately based on it, they can go well beyond it.

But some philosophers believe we should be, if not fundamentalist, at least *strict* empiricist: it is justifiable to believe in unobserved entities, but on condition they are at least *observable*; ampliative reasoning may be trusted sometimes, but only when it concerns observable entities (henceforth I shall use 'entity' as a generic term for objects, properties, relations, facts, events, processes, etc.). This sort of compromise underlies the philosophy of scientific antirealists like van Fraassen, who therefore hold that, even if we may *accept* scientific theories, we should not believe in what they claim concerning unobservable entities.

Here I wish to suggest that this compromise is ill founded. While observation is the strongest source of justification for factual beliefs, warranting, in ordinary cases, practical certainty, ampliative reasoning may also offer a wide range of reasons for belief, ranging from very strong in the best cases to rather weak and conjectural in the worst cases. Yet, what makes an ampliative inference stronger or weaker is not the fact of concerning observable rather

---
[1] In the Finnish Carelia, now part of the Russian Federation.

than unobservable entities; much less, this fact can mark a sharp dichotomy between having some justificatory force, and no force at all. So, there is no reason to limit the scope of ampliative reasoning to observable entities. While the difference between being observed and not observed has a great epistemic relevance, that between being observable and not observable is epistemically irrelevant.

2 How van Fraassen draws the observable-unobservable distinction

As is well known, a number of problems have been raised concerning the very possibility of drawing an observable-unobservable distinction. The first is the theory-ladeness of concepts and terms: since all language is pervaded by theory, it has been claimed, no term or statement is purely observative (see for instance [22, pp. 100–103], [9, ch. 1], [11, ch. 10], [8]). But van Fraassen, following the "ontologic turn" which in recent philosophy of science has reversed the "linguistic turn" of early XX century, is not interested in a distinction between terms, but between entities. Now, entities can be distinguished into observable and unobservable, no matter whether they are described in theoretical or non theoretical terms [27, pp. 14–15], [23].

But shifting the focus from terms to entities does not avoid a second problem: *in a sense* everything is observable, scientists commonly speak of observing things like viruses, atomic decay, fields, the centre of the Sun, etc. However, taking the readings of very sophisticated apparatuses (like electronic microscopes, particle accelerators or solar neutrinos observatories) as observations involves a commitment not only to the reliability of those instruments, but also to a particular interpretation of their readings (for instance, it involves assuming that the "click" of a Geiger counter is in fact produced by the decay of an atom); but this involves believing a complex body of theoretical claims about the interactions between observable apparatuses and unobservable entities, which is precisely what van Fraassen rejects. So, in his view, atomic decay and the centre of the Sun are not observable.

A third related problem is that, as pointed out by Grover Maxwell, there is continuity between ordinary observation and plainly unproblematic observation by simple instruments as eyglasses or magnifying lenses, through detection by very sophisticated apparatuses:

> There is in principle a continuous series beginning with looking through a vacuum and containing these as members: looking through a windowpane, looking through glasses, looking through binoculars, looking through a low-power microscope, looking through a high-power microscope, etc. [as a consequence] we are left without criteria which would enable us to draw a non-arbitrary line between "observation" and "theory" [15], quoted in [27, pp. 15–16].

But van Fraassen remarks that, although there may be some kind of continuity among observation acts, there is a natural distinction between observable and unobservable entities: by "observable" he chooses to mean what can be observed directly, or by unaided senses:

> X is observable if there are circumstances which are such that if X is present under those circumstances, then we observe it.[27, p. 16].

So, something is observable even if as a matter of fact it is observed through a window, and the moons of Jupiter are observable even if from the Earth they may be seen only through a telescope, for if we were close enough we could see them by the naked eye. Granted, when so defined the observable-unobservable distinction is not perfectly clear-cut, leaving a grey area of cases which do not distinctly belong to either set; but is it still useful and significant, as the greater majority of cases clearly falls under one or the other classification.

It might be objected that it is arbitrary to limit our beliefs to what may be observed directly, by unaided human senses: there are animals whose senses are keener than ours; for instance dogs hear sounds and distinguish smells we cannot distinguish, and bats hear ultrasounds. So, why should we not believe that there are ultrasonic frequencies, perceived by them? Again, we know that some instruments, built by ourselves, are more powerful than our own senses. So, why should we not believe in the existence of entities detected by those instruments? Thus, one might want to consider as observation any detection by instruments or apparatuses based on well confirmed theories, and as observable what can be so detected; or one might take an intermediate position, as suggested by Fano [7, pp. 161–163], for whom observable is anything that might be observed by some conceivable sentient being, no matter how its sensory organs differ from ours. In this sense, viruses, red globules and ultrasounds are observable; on the other hand, such essentially mathematical entities as electrons and electromagnetic fields, which may be detected by instruments, could not conceivably be observed by any living being, and so should not be considered observable. Both ways of considering observation seem to have a rationale: on the one hand, instrumental readings are generally considered as substantive evidence, yielding a qualitatively stronger support than just ampliative reasoning; on the other hand, perception by a living being seems to offer an even stronger evidence than mere instrumental detection.

Nevertheless, van Fraassen could reply, once again, that trusting detection by animals or instruments presupposes a complex body of theoretical beliefs. It is true that even our sense organs are very complex devices, and to account for their functioning and reliability we need a rather complex theory (whose details not even completely available, yet). But while we learned to rely on the detections by other sentient beings or by instruments through a complex theory, we need nothing like that to trust our own senses: we just rely on them instinctively, from the very beginning; nay, being perceived by our own senses may be considered the very paradigm of what it means to be real, or to deserve belief. Of course, this reliance on our senses might be questioned, but then *a fortiori* the reliability of other animals or instruments should be questioned, and in fact, nothing would be certain anymore. On the other hand, even if the reliability of our senses is granted, that of other animals or instruments can still be doubted, for it is based on theories, i.e., on the testimony of our senses *plus* ampliative reasoning, which is fallible; moreover, we cannot conclusively test those theories by direct comparison with sensory experience: that there are entities unobservable to us but observable to other animals or detectable by instruments is something we cannot *ever* observe. So, there are conceivable circumstances in which we might doubt the reliability of instruments without *ipso facto* doubting that of our sense, but not *vice versa*.

In my opinion this is not to say we cannot reasonably believe that other animals or instruments can have access to a wider field of data than us (since I think that the ampliative reasoning on which we base such a belief is reliable); but it is to admit that from at least one point of view the testimony of our own senses is in principle a bit more certain than any other sort of data. And this means that not only van Fraassen succeeds in drawing a viable observable-unobservable distinction, but he has a rationale for drawing it exactly at the point where he draws it: in this sense, his distinction is not arbitrary.

Nevertheless, the observable-unobservable distinction can also be drawn differently: as we just saw, by "observable" one could understand *observable by some conceivable sentient being*, or *detectable by instruments*; or, in a more guarded way, *perceivable by means of instruments* (like magnifying lenses, microphones, etc.) *whose reliability dos not presuppose theoretical beliefs*, since we may check it in the range of directly observable entities; or, again, in a solipsistic vein, *observable by myself*. All of these borderlines (and one could even think of further ones) correspond to some distinguishable difference. So, van Fraassen's distinction is an admissible one, but not the only one. The question is rather if it really marks the limits of what we can believe, and my answer is no. In fact, it might be argued (but room is not enough here) that the none of tje possible observable-unobservable distinctions can mark those limits: first, because there is no dichotomy between warranted and unwarranted beliefs, but a continuum of more or less warranted beliefs; second, because actual observation is highly relevant to the justification of beliefs, but mere observability is not.

3 The legitimacy of inferences to unobservable entities

To begin with, let us consider the *dodo*, a species of birds discovered on the island of Mauritius in 1598 and later observed many times; but after being hunted very heavily, it became extinct by 1681. On the other hand, let us imagine a new species of butterflies living in the Amazonic forest, never observed so far; let us fully define it by all the features that may identify a species of butterflies, and even give it a name, say *"Papilia Silfsvestris"* (in honour of this congress). Now, we all believe in the existence of the dodo, but none believes in the existence of the *Papilia Silfsvestris*, for the good reason that we have no grounds — neither observation nor reasoning — to accept it. Since there was no reasoning evidence in favour of the dodo either, the only difference is that it was observed: hence, observation is epistemically relevant, it may decide between believing and not believing.

But this is not the case with observability: for instance, all informed people rightly believe in the existence of the HIV virus, just as all those who know enough about it believe in the Suvajärvi asteroid impact: neither has ever been observed, but our beliefs are grounded in ampliative reasoning. (By the way, this shows that, contrary to *fundamentalist* empiricism, ampliative reasoning is epistemically relevant). Now, while the Suvajärvi asteroid is observable, the HIV virus is not; yet we correctly believe in both: and this shows that observability is epistemically irrelevant, it does not make a difference for belief.

Van Fraassen might reply that I am begging the question here: no matter how many people believe in unobservable entities like the HIV virus on the basis of ampliative reasoning, the point in discussion is whether it is epistemically right to do so, and van Fraassen denies this. Of course, most of us have the strong intuition that, unfortunately, it is all too certain that the HIV exists, and so van Fraassen's position is highly counterintuitive. But philosophers have often clashed with commonsense intuitions, and sometimes their counterintuitive views have been proven correct in the long run. So, we should find a way to adjudicate this clash of intuitions beyond a mere evaluation of their psychological strength. We should be able to find out which reasons, if any, underlie these contrasting intuitions, and which one is rationally correct.

So, let us consider another imaginary natural kind: a new virus, well defined in all its properties (call it the "*VHD virus*", for *Very Highly Dangerous* virus), which could possibly exist, since there is nothing incoherent in its description. But luckily enough, so far we have neither observed it (nor could we, for it is too small to be observable), nor have we any inferential evidence for it, so we do not believe it exists. Now, let us compare it with the *Papilia Silfsvestris*, another kind in which we do not believe (notice that Van Fraassen's intuitions, here, agree with ours: we shouldn't believe in either kind). Yet, there is a difference: the *Papilia Silfsvestris* is observable, while the VHD virus is not. This shows that observability does not make a difference, it does not yield grounds for belief; for otherwise we should believe in the *Papilia Silfsvestris*.

It might be replied that although observability as such does not supply any positive ground to believe, it is a necessary condition for belief: while its presence does not warrant belief, its absence undermines it. This is what happens, for instance, with consistency: mere consistency does not make a story credible, but no story can be believed unless it is coherent. However, there is an obvious reason why consistency is necessary for belief: it is necessary for truth! Once we see that a story is inconsistent, we understand that it is not true, so we cannot believe it. But there is no similar reason why observability should be necessary for belief; in particular, there is no reason why hypotheses concerning unobservable entities could not be true. On the contrary, the realist has a good positive reason to hold that we may believe in unobservable entities: those beliefs are grounded on ampliative reasoning patterns, whose correctness is granted even by the antirealists: just as from the observation of the crater we infer the existence of the unobserved asteroid, so from the observation of a number of observable symptoms we infer the existence of the unobservable HIV virus.

Thus van Fraassen's position (the strict empiricist compromise according to which we may believe in unobserved entities, but only if they are observable) presupposes that ampliative reasoning is correct when inferring from observed to observable entities, but not from observed to unobservable entities (and as a consequence, since neither observation nor deduction can ground beliefs in unobservable entities, such beliefs cannot be grounded at all). But why should it be so?

The reason, one might claim, is that in the past the correctness of many ampliative inferences from observed entities A to observable entities B has been checked *by observation*, i.e. by actually observing at a later moment the entity B which we were not in a position to observe at the time of the inference. So, by a sort of meta-induction, we may assume that ampliative inferences from observed to observable entities are reliable. On the other hand, by hypothesis no such observative check is available for inferences to unobservable entities; so, we do not have the same meta-inductive reason to assume that the latter kind of inferences are reliable. (Notice that Hume's problem is not a issue here, as the principle of induction is taken for granted by scientific realists and anti-realists alike. Hence, the above reasoning does not commit the fallacy of circularity involved in trying to justify induction by induction. Rather, it is supposed to exploit the general principle of induction to show that while we have reasons to rely on the correctness of inferences to observable entities, we do not have reasons to rely on the correctness of inferences to unobservable entities, and so we should avoid them).

This argument, however, may be countered by observing that the reliability of inference patterns is a formal question: if an inference pattern (say, *modus ponens*, *modus tollens*, simple induction, enumerative induction, abduction, etc.) is accepted as reliable, then all its instantiations are reliable, regardless of their subject topic. So, whichever patterns of ampliative reasoning we accept for observable entities, we should accept also for unobservable entities. The following simple-minded imaginary example may clarify this point: suppose astronauts land on one of Jupiter's moons, and they find that the landing area is constantly hit by a rain of meteorites of any size, from barely perceptible up to approximately 10 cm. of diameter. They also notice that each new meteorite produces in the ground a crater whose diameter is on the average about 30 times the meteorite's diameter. So, when they observe an old crater 3 m. wide, they correctly infer that it was produced by a meteorites of a 10 cm. diameter; and when they observe (by the naked eye) an old crater 0.03 cm. wide, they equally correctly infer that it was produced by a meteorite of 0.001 cm. Van Fraassen would object to this inference, because the 0.001 cm. meteorite is too small to be observable by the naked eye; but his objection runs against the principle of induction, or of the uniformity of nature, stating that similar things behave in similar ways, or that for similar causes we should expect similar effects, and vice versa. Since this principle is inevitably presupposed by whoever trusts induction, as van Fraassen does, his position is contradictory.

Might he reply that one cannot infer from the reliability of inferences to observable entities to that of inferences to unobservable entities, because observable and unobservable entities, as such, are different, hence they are not similar? For instance, could he deny that from the fact that a crater of 3 m. is caused by a meteorite 30 times smaller we can infer that a 0.03 cm. crater is caused by a meteorite 30 times smaller, because these two craters are different? No, because as is well known, every thing is different from any other thing in so many ways, but also similar to it in so many other ways; hence, even if two items are different, they may be similar, and it may be correct to infer from the one to the other.

Of course, this means that the principle of the uniformity of nature is empty, unless it is specified which similarities and differences are relevant, which similarities allow inductive inferences and which differences prevent them. Admittedly, this is not always clear *a priori*: often an answer is drawn from our background knowledge, and sometimes it is discovered only *a posteriori*. But it is clear that, in general, only intrinsic properties of the involved entities are relevant, not their position or direction in space or time, or their relationships to other entities (unless, of course, the inductive inference concerns precisely those relationships). Thus, position in space and time cannot be inductively projected, nor prevent inductive projection of other properties: even if we have always seen that all free bodies fall toward the centre of the Earth, we should not infer that all free bodies will move toward the same point; and even if all the dogs we have seen live before A.D. 2010, we should not infer that all dogs live before that date. Equally, the fact all observations of water samples boiling at 100^c at 1 atm. were located on the Earth and before A.D. 2010, does not prevent one from inferring that the same would happen even on the Moon, or after A.D. 2010. In particular, it is even clearer that the relationships of entities to the observers are in general irrelevant to induction: future samples of water are different from those we have observed in the past, in that they have not been observed by us; but this should not prevent us from inferring that they too will boil at 100^c at 1 atm. Equally, we should not refrain from inferring that the same relationship we have observed between craters and meteorites holds between the 0.03 cm. crater and a corresponding meteorite, just because the other meteorites were observable and this one is not.

It might be noticed that the difference between being observed and not observed is purely extrinsic and relational, having to do with the contingent space-time location of the entity in question and the observers, while the difference between being observable and not observable is at least grounded in the intrinsic properties of the entity itself: the 0.001 cm. meteorite of our example is unobservable because it is too small; so, it is different from the meteorites observed by the astronauts in being smaller than them. But is this difference relevant? It would seem not, for, in the hypothesis, the astronauts had observed that the 1/30 ratio between meteorites and craters did not vary with size. And if we turn to background knowledge, nothing suggests that the dynamics of these impacts is influenced by the difference in size between barely observable and barely unobservable meteorites (say, between diameters of 0.01 cm. and 0.001 cm., although perhaps it might vary at much smaller scales).

Of course it is true that the inference to the existence of a 0.001 cm. meteorite takes us outside the quantitative limits of the observed data (as we have only observed meteorites between 10 and 0.01 cm.). So, our inference may be somewhat less confident than an inference to something included within those limits (say, to the existence of a 0.5 cm meteorite), and we might wish some further confirmation. Yet it is credible, enough to warrant belief: for in general we do not limit our inferences to the quantitative limits of the observed samples, unless there are positive indications that those limits are relevant to the inferences in question. For instance, suppose that nobody ever observed a sample of boiling water larger than 200 l. or smaller than 0.01 l.: still we would be entitled to believe that all samples of water, even outside those limits, boil at 100^c at 1 atm.

One could object that further research might show that, in spite of our background presuppositions, the difference between diameters of 0.01 cm. and 0.001 cm. after all is relevant to being the cause of a 30 times larger crater: say, that only meteorites of 0.01 cm. diameter or larger are massive enough to produce craters, and smaller craters are due to a different cause. Of course, in this case, it would be correct to refrain from the inference, *because of the size difference*, not because the lack of observability as such. And we certainly could not conclude that this particular difference in size (i.e., precisely the difference responsible for making entities observable rather than unobservable) is relevant to *all* inductive inferences. The same obviously holds for unobservability due to other differences (such as reflecting light outside the visible spectrum, etc.).

Summing up, the observable-unobservable distinction is not, *per se*, relevant to induction; hence, there is no reason to reject inductive inferences to unobservable entities as such. Van Fraassen's stricture is inconsistent with his own (and our) acceptance of the principle of induction, and his strict empiricist compromise is not viable: he should either reject induction and take the fundamentalist empiricist stand that we can only believe in observed entities, or grant that it may be warranted to believe in unobservable entities.

4 Strictly inductive vs. abductive inferences

Perhaps, when denying that ampliative inferences to non-observble entities are reliable, van Fraassen has in mind not strictly inductive inferences, such as those of my example, but abductive inferences, i.e., procedures of the following structure: we must explain a wide field of data, we conjecture a whole theory including various existential and lawlike assumptions on unobservable entities, we find out that it accounts nicely for our data, hence we assume it is true or approximately true. This is how we tend to picture the adoption of complex theories of modern and contemporary science: not only relativity theory or quantum theory, but even Newton's gravitation theory, Darwin's evolutionary theory, etc. In this way the assumption of unobservable entities is seen as a largely free creation of a genial mind, as Einstein has often stressed (see [3, 4], [5, 221–226, 272], [6, 10]). Now, the confirmation of any hypothesis introduced in this way meets serious obstacles: to begin with, it is fallible, for it is an instance of the affirmation of the consequent, a deductive fallacy. Moreover, by the principle of empirical underdetermination, there may always be a number of alternative theories globally coherent with the same body of data, hence we cannot ever be certain that our hypothesis is the true one. Again, although hypotheses of this kind are often confirmed by striking predictive successes [10, 25, 26, 1, 2, 24, 20, 18, 19, 13], important doubts have been raised on the strength of this kind of confirmation (see for instance [12, 14]) and some may feel that it is not enough to outweigh the uncertainty deriving from empirical underdetermination. So, one may think that inferences to unobservable entities are radically inconclusive.

But claims on unobservable entities are not always introduced top-down, in this abductive way. Often they are introduced bottom-up, by a strictly inductive inference, precisely as the claim on the 0.001 cm. meteorite in our previous example: i.e., just by an extrapolation from observed to unobservable items which relies simply on the principle of the uniformity of nature, as we often do in extrapolating from observed to unobserved but observable items. In fact, in the history of science inferences of this kind often are the way in which (a) new kinds of unobservable objects are discovered, or (b) unobservable properties of observable objects are discovered, or (c) unobservable properties of unobservable objects introduced by the abductive method are measured.

An example of case (a) is that of red globules, spermatozoa, protozoa and bacteria (all unobservable in van Fraassen's sense). Their existence was immediately accepted once van Leeuwenhoeck (1632–1723) after using magnifying lenses to check the fine details of the textile fabrics he traded, was able to cut lenses of grater magnifying power, which showed him entities not perceptible by the naked eye. Since it had been previously observed that variations in the lenses' curvature or in the device on which they were mounted (the "microscope") did not affect their reliability, but increased their magnifying power by a known proportion, nobody doubted that van Leeuwenhoeck's new lenses showed actual entities of proportional size: only the uniformity of nature was presupposed by this assumption.

An example of case (b) is spectrographic detection of the chemical components of the stars: given the physical conditions of the stars, there is no way in which their chemical composition might be observed in van Fraassen's sense; but since we may directly observe that the light emitted by a sample of a given element always produces the same spectrum, when observing the spectra of the light from the stars, by a plainly inductive inference we may tell which elements they contain.

Examples of kind (c) are the various experiments by which Perrin and others were able to measure Avogadro's number, and hence the volume and weight of molecules. One of the simplest procedures described by Perrin consists in dropping a droplet of a suitable solution of oil on a water surface covered by talc powder: the drop expands to the point that its thickness cannot be appreciated by the eye, although the shape of its surface can be easily distinguished, because while expanding it pushes the talc off to the edge. Thus, by measuring two observable properties (the volume of the droplet and its area once completely expanded on the water) and by a simple mathematical deduction, we can measure an unobservable property, the thickness of the oil layer, which can be of the order of 1 $\mu\mu$, and which gives us a maximum value for the diameter of molecules [21, § 32].

Another example of this kind is Millikan's measurement of the charge of an electron: when a droplet of oil is driven upwards by the attraction of an electric field, occasionally it may undergo a sudden acceleration; the experimental setup makes it plausible that it has attracted a new electron, thus increasing its own electric charge. This increase, corresponding to the electron's charge, may be computed from the difference in velocity and the mass of the droplet. So again, we can measure non-observable properties just by measuring two observable properties and by a mathematical deduction [17], [21, § 100].

The existence of atoms had been assumed as an explanatory hypothesis in a number of occasions since antiquity, and most notably in the XVIII and XIX century by scientists like Prout, Dalton, Avogadro and many others, up to Maxwell and Boltzmann. It is remarkable that, although in agreement with various empirical laws, it was surrounded by widespread scepticism until the end of XIX century, when Mach rejected the very notion of the atom as a scientific concept. Only when the aforementioned measurements were performed did scepticism suddenly give way to almost universal acceptance. I suggest that such a striking change of attitude is justified because such inferences to the unobservable entities are completely warranted by observation and by the principle of the uniformity of nature, that no empiricist would deny.

5 Are all observable entities described by science also observed?

One could think that the observable-unobservable distinction is epistemically relevant because in fact it *coincides* with the distinction between observed and unobserved. That is, all observable entities described by science are also *in a sense* observed: for no theory introduces a new *observable* kind of entity before some of its members have been observed, and no theory is seriously proposed for acceptance before its observative claims (i.e. its empirical laws) are empirically confirmed in at least some instances. That is to say, we should only accept the existence of natural kinds that have actually been observed (by observing some of their members), or empirical laws that have been directly tested (by testing some of their instances).

However, in the first place, sometimes scientists are not concerned with *kinds* of entities, but just with individual entities, as the in case of the Suvajärvi asteroid: in such cases, the entity in question may be observable but not actually observed. Secondly, in some circumstances an empirical law implied by a theory cannot be tested for lack of technology, funds, suitable circumstances, etc.: for instance, the bending of light pedicted by the Relativity Theory until Eddington's observation of a solar eclipse in 1919. Finally, since empirical laws are universal statements, strictly speaking we will never have observed that *everything* a law says is true: for instance, the law that water boils at 100^c at 1 atm. has been tested in the sense that we observed that water boiled at 100^c at 1 atm. yesterday, the day before, etc.; we observed that this happened in Milan, in Cesena, etc; but the law also says that the same will happen at any future time and at any place. So, we cannot justify our belief in it by saying that we have *observed* that everything the law claims will happen has indeed happened.

It may be replied that although we have not observed that *all* the entities the law is about behave according to it, we have at least observed that some entities *of that kind* behave according to it: for instance, we have observed that some samples of water in boil at 100^c at 1 atm.; from this, it is just a strictly inductive inference that all entities of the same kind behave in the same way: of course we cannot be *sure*, but we have good reasons to believe it; however this cannot happen with theoretical laws, referring to kinds of entities of which we cannot observe *even a single instance*. In other words, laws concerning observable entities may be supported just by observation and induction, while laws concerning unobservable entities require much more complex forms of reasoning, such as abduction, analogy, modelling, etc.

However, we have seen that sometimes even the existence and/or the properties of unobservable entities can be discovered just by observation and induction (or even observation and deduction). Thus, van Fraassen's observable-unobservable distinction does not coincide with this tentative distinction between a safer and a more precarious kind of belief. Moreover, this line of reasoning raises two problems: first, it assumes the notion of a natural kind and a precise understanding of the extension of a particular kind. Now, for some kinds, like water, the notion of the unity of the kind and the ability to identify its members may be very intuitive, and one could even think (as J.S. Mill did)[16, IV,2.2.] that they are based exclusively on observation; but the unity and the extension of other more complex kinds may be recognized only on the basis of beliefs concerning unobservable entities: for instance, mushrooms are so "observationally" different among themselves that we cannot assume that they constitute a unique kind, nor succeed in identifying all its various members, without relying on a good deal of theoretical considerations on unobservable entities. But if, as van Fraassen requires, we did not entertain beliefs concerning unobservable entities, in no way could we claim to have actually *observed* that *all* the claims made by the empirical law in question are true. For instance, although mushrooms are observable, and containing water is an observable property, the simple empirical law that all mushrooms contain water cannot be confirmed just by observation and induction, without presupposing other forms of ampliative reasoning as well.

Secondly, the more a kind is large and internally differentiated (say, as we go from mushrooms to the whole kingdom of *fungi*, to the class of all eukaryotic organisms, to all living beings), the more theoretical considerations on unobservable entities are involved in the assumption of its unity, the weaker the inductive reasons for extending to all members what we have observed concerning some of them. So, if one accepted this looser sense of 'observation' in which it may be claimed that the whole content of an empirical law has been observed, one could no longer hold that in general observation supplies very strong support or *practical certainty* to hypotheses, as opposed to a much weaker support offered by ampliative reasoning; rather, one should admit a continuum between cases in which observation bestows practical certainty and cases in which it offers merely some rather week reasons to believe. But in this way, since (as we saw) even ampliative inferences to unobservable entities may deserve some degree of belief, one would lose the very rationale for drawing a sharp observable-unobservable distinction: even maintaining the clear-cut definition of 'observable' as observability by unaided human senses, one could not hold that observed entities should definitely be believed, observable ones may deserve strong confidence, and unobservable ones cannot deserve any; in fact, observation itself, by presupposing more and more inductive generalization, analogical reasoning and theoretical assumptions, would gradually fade into theorization.

6 A pragmatically relevant distinction?

Finally, still another possible reason for maintaining an epistemic privilege of the hypotheses concerning unobserved but observable entities may be that when advancing one of them we know that at some future time we will be in a position to check it by direct observation, and so, possibly, to pass from the *reasonable belief* that may be warranted by any form of ampliative reasoning, to the *practical certainty* that may be warranted by actual direct observation.

But while this may be *pragmatically* relevant (for it may encourage the investment of time and money in the hypothesis at hand), it is not *epistemically* relevant, for it cannot add plausibility to our beliefs, neither at present, nor in future: not at present, for the mere possibility of confirmation is not yet confirmation; nor in the future, for *if* at some future time the entity in question is observed, the hypothesis would be confirmed by actual observation, not by mere observability. In fact, we have seen that observability, as such, neither confirms nor is a necessary condition for confirmation.

Moreover, there are observable things which we shall never be in the position to observe directly, such as the Suavjärvi asteroid. So, the distinction having some pragmatic relevance is not observable vs. unobservable, but *observable-in-practice* vs. *observable-just-in-principle-or-altogether-unobservable*.

Acknowledgements

I thank Lars Aagard-Mogensen, Alessandro Afriat, Howard G. Callaway, Vincenzo Fano and Carlo Rovelli for their useful comments to earlier drafts of this paper.

BIBLIOGRAPHY

[1] R. Boyd. Realism, underdetermination and the causal theory of evidence. *Nous* 7: 1–12, 1973.
[2] R. Boyd. The current status of scientific realism. In J. Leplin (ed.), *Scientific Realism*, pages 41–82. University of California Press, Berkeley 1984.
[3] A. Einstein. Reply to criticisms. In P.A. Schilpp (ed.), *Albert Einstein: Philosopher-Scientist*, pages 665–688. Library of Living Philosophers Evanston, Ill., 1949.
[4] A. Einstein. *Lettres à Maurice Solvine*. Gauthier-Villars, Paris, 1956, (lettre 7 mai, 1952).
[5] A. Einstein. *Ideas and Opinions by Albert Einstein*, edited by C. Seelig, Crown, New York 1954.
[6] A. Einstein. Autobiographische Skizze. In C. Seelig (hrsg.) *Helle Zeit — Dunkle Zeit*. pages 9–17. Europa Verlag, Zurich 1956.
[7] V. Fano. *Comprendere la scienza. Un'introduzione all'epistemologia delle scienze naturali*. Liguori, Napoli 2005.
[8] P.K. Feyerabend. *Against Method. Outline of an Anarchistic Theory of Knowledge*. New Left Books, London 1975.
[9] N.R. Hanson. *Patterns of Discovery*. Cambridge University Press, Cambridge 1972.
[10] G. Harman. Inference to the best explanation. *Philosophical Review* 74: 88–95, 1965.
[11] T.S. Kuhn *The Structure of Scientific Revolutions*. The University of Chicago Press, Chicago 1962.
[12] L. Laudan. A confutation of convergent realism. *Philosophy of Science* 48: 19–49, 1981.
[13] J. Leplin. *A Novel Defence of Scientific Realism*. Oxford University Press, Oxford 1997.
[14] T.D. Lyons. The pessimistic meta-modus tollens. In S. Clarke, T.D. Lyons (eds.), *Recent Themes in the Philosophy of Science. Scientific Realism and Commonsense*, pages 63–90. Kluwer, Dordrecht 2002.

[15] G.E. Maxwell. The ontological status of theoretical entities. In H. Feigl, G.E. Maxwell (eds.) *Scientific Explanation, Space and Time*, Minnesota Studies in the Philosophy of Science, Vol. 3, pages 3–27. University of Minnesota Press, Minneapolis 1962.
[16] J.S. Mill, *A System of Logic Ratiocinative and Inductive: Being a Connected View of the Principles of Evidence and the Methods of Scientific Investigation.* J.W. Parker, London 1843.
[17] R.A. Millikan. The isolation of an ion, a precision measurement of its charge, and the correction of Stokes's law. *Physical Review*, 32: 349–397, 1911.
[18] A. Musgrave. The ultimate argument for scientific realism. In R. Nola (ed.) *Realism and Relativism in Science*, pages 229–252. Kluwer, Dordrecht 1988.
[19] A, Musgrave.The 'miracle argument' for scientific realism. *The Rutherford Journal*, http://www.rutherfordjournal.org/current1.html
[20] W.H. Newton-Smith. Realism and inference to the best explanation. *Fundamenta Scientiae*, 7: 305–316, 1987.
[21] J. Perrin. *Les atomes*. Alcan, Paris 1913.
[22] K.R. Popper. *Objective Knowledge: An Evolutionary Approach.* Clarendon Press, Oxford 1972.
[23] H. Putnam. What theories are not. In E. Nagel, P. Suppes, A. Tarski (eds.) *Logic, Methodology and Philosophy of Science: Proceedings of the 1960 International Congress*, pages 240–251. University Press, Stanford (Ca); reprinted in H. Putnam, *Philosophical Papers, Vol. I: Mathematics, Matter and Method.* Cambridge University Press Cambridge (Mass.) 1975.
[24] H. Putnam. *Meaning and the Moral Sciences.* Routledge, London 1978.
[25] J.J.C. Smart. *Philosophy and Scientific Realism.* Routledge, London 1963 .
[26] J.J.C. Smart. *Between Science and Philosophy.* Random House, New York 1968.
[27] B.C. van Fraassen. *The Scientific Image.* Clarendon Press, Oxford 1980.

Reduction in dynamical systems: a representational view

MARCO GIUNTI

1 Introduction

Standard accounts have traditionally viewed reduction as a *deductive* relationship between two *formal* theories [23]. Schaffner's *General Reduction Paradigm* [25] was an early attempt to modify Nagel's classic account, so as to accomodate cases where the reduced theory is, strictly speaking, false. The most comprehensive and detailed deductivist account of reduction is Churchland and Hooker's *Imaging Approach* [9, 10, 13, 14, 16], which can be seen as a creative development of Nagel's basic insights, as well as a sensible departure from Nagel's explicit tenets [4, 6, 7, 19]. [19] has convincingly argued that Kim's *Functionalizing Approach* to reduction [17] is in fact a version of Nagel's account; such a version is essentially equivalent to the Imaging Approach.

This paper proposes an alternative view, according to which reduction is better conceived as a *representational* relationship between two mathematical models MS_1 and MS_2, which grants the retrieval, within the representing model MS_1, of an isomorphic image of MS_2.[1]

Bickle's *New Wave Reduction* [6, ch. 3] is a version of the Imaging Approach by Churchland and Hooker in which (i) theories are construed as sets of models (semantically), rather than sets of sentences (syntactically), and thus (ii) reduction is not a *deductive* relationship between *formal* theories, but a relationship between *semantic* theories (i.e. *sets* of models) that satisfies special conditions. Notwithstanding these differences, reduction is still analyzed by Bickle as a special relationship between *theories* (i.e. *sets* of models) and not as a representational relationship between *models*. Bickle shares his general view of reduction and theory structure with the *Structuralist Program* [27, 28, 21, 22, 3, 2].

The general representational theory of reduction that I advocate is in broad agreement with Suppes' *Reduction Paradigm* [29, 271],[2] and it is somehow consonant with some of the ideas of Hooker's *dynamically based* revision of the Imaging Approach [15].

Compared to traditional approaches to reduction (*deductivist* or, more generally, *theory-based* approaches), the representational one has several advantages, whose details will only be apparent later. For the moment, it suffice to

---

[1]The term "isomorphic image" is intended here in its rigorous mathematical sense. This is not the sense in which the Imaging Approach employs the same term.

[2]Section 4 (see case 2) will make clear that Schaffner's [25] "too weak to be adequate" [6, ch. 3] criticism of Suppes' Reduction Paradigm does not apply to my view.

say that the representational theory scores better as far as precision and depth of analysis are concerned. Also, this theory is fostering a unified, conceptually crisp, and formally developed account of *prima facie* conflicting aspects of reduction – total and exact reduction *vs.* partial, approximate and asymptotic one, on which traditional approaches hardly fare as well.

I will develop this general representational theory only for the special case of dynamical systems. As intended here [1, 30, 12], a *dynamical system* is a kind of mathematical model that captures the intuitive idea of an arbitrary deterministic system. Models of this kind allow us to study in a precise way typical features of complex systems. Among them, in recent years, the one of emulation has gained growing attention [33, 34, 35, 36, 37]. Intuitively, a dynamical system DS_1 *emulates* a second dynamical system DS_2 when the first one exactly reproduces the whole dynamics of the second one.

The emulation relationship can be defined in a precise way for any two arbitrary dynamical systems and it has been shown [12, ch. 1, th. 11] that, if DS_1 emulates DS_2, there is a third system DS_3 such that (i) DS_2 is isomorphic to DS_3; (ii) all states of DS_3 are states of DS_1; (iii) any state transition of DS_3 is constructed out of state transitions of DS_1. In this paper, I will focus on a more general version of this theorem [*Virtual System Theorem VST*], which is based on a weaker and simpler definition of emulation. I will then argue that this result allows us to claim: If DS_1 emulates DS_2, then DS_2 is *reduced* to DS_1.

The claim that emulation is sufficient for reduction (in force of [*VST*]) is a precise statement of the representational view of reduction for the special case of dynamical systems. Strictly speaking, this claim is intended to hold exclusively for dynamical systems as purely *mathematical models* with no empirical interpretation. In a different sense, however, dynamical systems typically function as *models of real phenomena*. In this second sense, a dynamical system is not a purely mathematical entity DS, but it is a pair (DS, I_H), where I_H is an empirical interpretation that links the purely mathematical model DS to a phenomenon H. This paper will also provide the main lines of an extension of the representational theory of reduction to *empirically interpreted* dynamical systems.

As said, the emulation relationship is the basis of a representational view of reduction for dynamical systems (either empirically interpreted or not). The simplest form of such relationship holds between two dynamical systems DS_1 and DS_2 when the *whole* dynamics of DS_2 is *exactly* reproduced by DS_1. This simple form may very well be the basis for a representational account of *total* and *exact* reduction, but we need a more sophisticated version of emulation for dealing with cases of asymptotic, partial and approximate reduction [15]. Such a version will be introduced in section 5, where it will then be employed for a treatment of partial and approximate reduction in empirically interpreted dynamical systems.

2 Dynamical systems and emulation

A dynamical system is a kind of mathematical model that formally expresses the notion of an arbitrary deterministic system, either reversible or irreversible, with discrete or continuous time or state space. Let Z be the integers, Z^+ the non-negative integers, R the reals and R^+ the non-negative reals; below is the exact definition of a dynamical system.

[1] *DS is a dynamical system* iff *DS* is a pair $(M, (g^t)_{t \in T})$ such that
1. M is a non-empty set; M represents all the possible states of the system, and it is called the *state space*;
2. T is either Z, Z^+, R, or R^+; T represents the time of the system, and it is called the *time set*; any $t \in T$ is called a *duration* of the system;
3. $(g^t)_{t \in T}$ is a family of functions from M to M; each function g^t is called a *state transition* of duration t, or a *t-advance*, of the system;
4. for any $t, v \in T$, for any $x \in M$, $g^0(x) = x$ and $g^{t+v}(x) = g^v(g^t(x))$.

[2] A *discrete dynamical system* is a dynamical system whose state space is finite or denumerable, and whose time set is either Z or Z^+; examples of discrete dynamical systems are Turing machines and cellular automata.[3]

[3] A *continuous dynamical system* is a dynamical system that is not discrete; examples of continuous dynamical systems are iterated mappings on R, and systems specified by ordinary differential equations.

[4] A *possible dynamical system* is a pair $(M, (g^t)_{t \in T})$ that satisfies the first three conditions of definition [1].

We can now define the concept of an isomorphism between two possible dynamical systems as follows.

[5] *r is an isomorphism of DS_1 in DS_2* iff $DS_1 = (M, (g^t)_{t \in T})$ and $DS_2 = (N, (h^v)_{v \in V})$ are possible dynamical systems, $T = V$, $r: M \to N$ is a bijection and, for any $t \in T$, for any $x \in M$, $r(g^t(x)) = h^t(r(x))$.

[6] *DS_1 is isomorphic to DS_2* iff there is r such that r is an isomorphism of DS_1 in DS_2.

It is easy to verify that the isomorphism relation is an equivalence relation on any given set of possible dynamical systems. (The concept of *set of all possible dynamical systems* is inconsistent, and we must then take as the basis of the theory of dynamical systems a specific, sufficiently large, set of possible dynamical systems.) It is also not difficult to prove that the relation of isomorphism is compatible with the property of being a dynamical system, that is to say: if DS_1 is isomorphic to DS_2 and DS_1 is a dynamical system, then DS_2 is a dynamical system. This allows us to speak of abstract dynamical systems in exactly the same sense we talk of abstract groups, fields, lattices, order structures, etc. We can thus define:

[7] an *abstract dynamical system* is any equivalence class of isomorphic dynamical systems.

It is easily shown that any two dynamical systems have exactly the same *structural properties* iff they are isomorphic.[4] Since *general dynamical sys-*

---
[3] The term "discrete dynamical system" is often used (see, for example, [18, 20, 24]) as a synonym for "dynamical system with discrete time", i.e., according to [30], a *cascade*. My use of the term "discrete dynamical system" is in accordance with [31].

[4] P is a *structural property of a dynamical system* (or a *dynamical property*) iff for any two mathematical models MS_1 and MS_2, (i) if MS_1 has P, MS_1 is a dynamical system and (ii) if MS_1 has P, and MS_1 is isomorphic to MS_2, then MS_2 has P. Thus, a dynamical property is a property *specific* to dynamical systems that is *preserved* by isomorphism.

tems theory[5] is exclusively interested in such properties, it regards any two isomorphic systems as identical.

Dynamical systems are appropriate models to study several interesting features of complex systems. The one of emulation is typical of computational systems [37], but it can in principle involve any two dynamical systems. The intuitive idea is that a dynamical system DS_1 emulates a second dynamical system DS_2 when the first one exactly reproduces the whole dynamics of the second one. Here are some examples. A universal Turing machine emulates any Turing machine; for any Turing machine TM there is a cellular automaton CA such that CA emulates TM [26, th. 3], and vice versa; the simple cellular automaton specified by Wolfram's rule 18 emulates the one specified by rule 90 (both CA are monodimensional, with 2 possible values for cell, and neighborhood of radius 1; see [34, p. 20]).

[12, ch. 1, def. 4] gave a formal definition of the emulation relationship that applies to any two arbitrary dynamical systems. Here, I will employ a weaker and simpler definition (see figure 1), which nevertheless suffices for the present purposes.

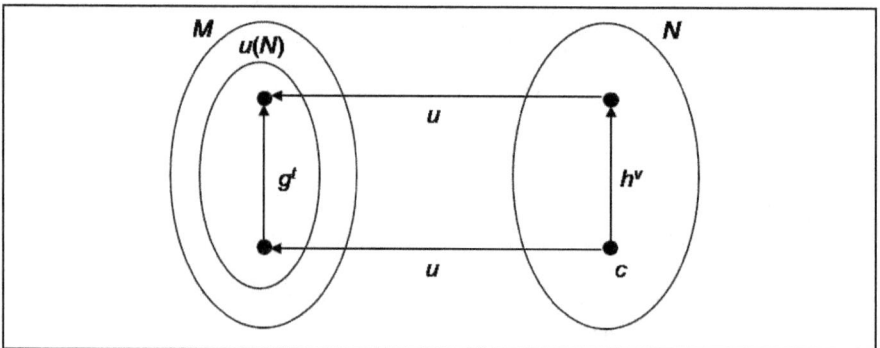

Figure 1. Emulation

[8] DS_1 emulates DS_2 iff $DS_1 = (M, (g^t)_{t \in T})$ and $DS_2 = (N, (h^v)_{v \in V})$ are dynamical systems, and there is an injective function $u: N \rightarrow M$ such that, for any $v \in V$, for any $c \in N$, there is $t \in T$ such that $u(h^v(c)) = g^t(u(c))$. Any function u that satisfies the previous condition is called an *emulation of DS_2 in DS_1*.

3 Emulation is sufficient for reduction

[12, ch 1, th. 11] proved that, if u is an emulation of DS_2 in DS_1, there is a third system DS_3 such that (i) u is an isomorphism of DS_2 in DS_3; (ii) all

---

The proof that any two isomorphic dynamical systems have exactly the same dynamical properties is immediate. Conversely, for any two non-isomorphic dynamical systems DS_1 and DS_2, there is a dynamical property they do not share; namely, the property of being isomorphic to DS_1.

[5] By *general dynamical systems theory* I mean the mathematical theory whose Suppes' style axiomatization [29, ch. 12] is given by def. [1].

states of DS_3 are states of DS_1; (iii) any state transition of DS_3 is constructed out of state transitions of DS_1. This result still holds for the weaker definition of emulation [8], as the following theorem shows.

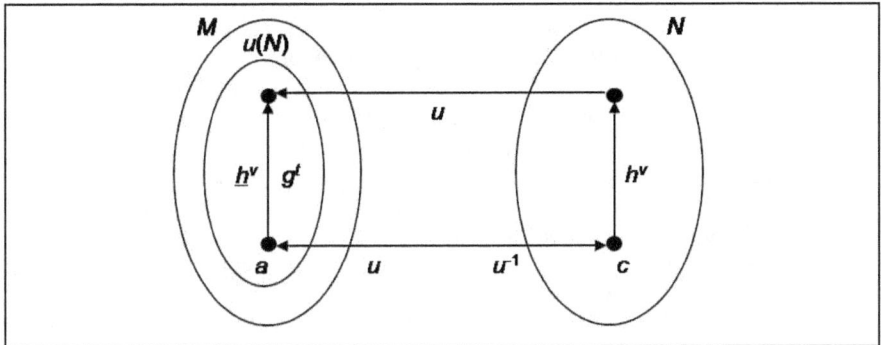

Figure 2. The u-virtual system DS_2 in DS_1

Virtual System Theorem [VST]
• Let $DS_1 = (M, (g^t)_{t \in T})$ and $DS_2 = (N, (h^v)_{v \in V})$ be dynamical systems, and u be an emulation of DS_2 in DS_1;
• let $DS_3 = (\underline{N}, (\underline{h}^v)_{v \in V})$, where $\underline{N} = u(N)$ and, for any $a \in \underline{N}$, for any $v \in V$, $\underline{h}^v(a) = u(h^v(u^{-1}(a)))$; the system DS_3 is called *the u-virtual system DS_2 in DS_1* (see figure 2);
then:
(i) u is an isomorphism of DS_2 in DS_3;
(ii) all states of DS_3 are states of DS_1;
(iii) for any state transition $\underline{h}^v$ of DS_3, for any $a \in \underline{N}$, there is a state transition g^t of DS_1 such that $\underline{h}^v(a) = g^t(a)$.
Proof of (i)
By the definition of DS_3, for any $c \in N$, $u(\underline{h}^v(c)) = u(h^v(u^{-1}(u(c)))) = h^v(u(c))$. Therefore, by the definition of isomorphism [5], u is an isomorphism of DS_2 in DS_3.
Proof of (ii)
Obvious, by the definition of DS_3.
Proof of (iii)
By the definition of DS_3, for any $v \in V$, for any $a \in \underline{N}$, $\underline{h}^v(a) = u(h^v(u^{-1}(a)))$. Let $c = u^{-1}(a)$. Since u is an emulation of DS_2 in DS_1, by definition [8], there is $t \in T$ such that $u(h^v(c)) = g^t(u(c))$. Therefore, $\underline{h}^v(a) = g^t(u(c)) = g^t(a)$. Q.E.D.

It is my contention that, if a dynamical system DS_1 emulates a second system DS_2, [VST] allows us to claim that DS_2 *is reduced to* DS_1. In other words, I maintain that, because of [VST], emulation is sufficient for *reduction*.

Before seeing the details of the supporting argument, it is important to make clear that dynamical systems, as intended here, are *purely mathematical* entities with no empirical interpretation; that is to say, at this level of analysis, a dynamical system is just a model of the mathematical theory whose Suppes' style axiomatization [29, ch. 12] is given by def. [1]. The claim that emulation is sufficient for reduction is thus exclusively limited to dynamical systems intended in this sense.

As just said, when I speak of a dynamical system as a *model*, I mean a model of a quite general *mathematical theory*, whose axiomatization is expressed by the definition, in set theory, of an appropriate set-theoretical predicate (def. [1]). It is important to sharply distinguish this sense of the term "model" from a different one, which also applies to dynamical systems, and is equally central to a complete understanding of their epistemological status. This second sense is the one intended when we say that a specific dynamical system is a model of a *real phenomenon*; however, this sense does not refer to a dynamical system as a purely mathematical entity (i.e., just a model of general dynamical system theory) but, rather, to such entity *together with* an empirical interpretation that links the mathematical model to the phenomenon which it is intended to describe.

A simple example will make the distinction clear. Let us consider the following system of two ordinary differential equations $\langle dy(v)/dv = \dot{y}(v)$, $d\dot{y}(v)/dv = -\boldsymbol{g}\rangle$, where $\boldsymbol{g}$ is a fixed real positive constant. The solutions of such equations uniquely determine the dynamical system $DS_e = (Y \times \dot{Y}, (h^v)_{v \in V})$, where $Y = \dot{Y} = V = R$ (the real numbers) and, for any $v, y, \dot{y} \in R$, $h^v(y, \dot{y}) = (-\boldsymbol{g}v^2/2 + \dot{y}v + y, -\boldsymbol{g}v + \dot{y})$. It is immediate to verify that DS_e satisfies def. [1], so that it is a model in the first sense.

On the other hand, let us consider the phenomenon of the free fall of a medium size body in the vicinity of the earth (henceforth, H_e), and let us interpret the first component Y of the state space of DS_e as the set of all possible values of the *vertical position* of an arbitrary free falling body, the second component $\dot{Y}$ as the set of all possible values of the *vertical velocity* of the falling body,[6] and the time set V of DS_e as the set of all possible values of *physical time*. Since all three of these magnitudes are measurable or detectable properties of the intended phenomenon H_e, the given interpretation is an *empirical* interpretation of the dynamical system DS_e on H_e. Let I_{H_e} be such an interpretation. Then, the pair $(DS_e, I_{H_e}) = \boldsymbol{DS}_e$ is an empirical model of H_e, i.e., such a *pair* is a model in the second sense. $\boldsymbol{DS}_e$ will be called the *falling body model*.

My claim that emulation is sufficient for reduction (in force of [VST]) is intended to hold exclusively for dynamical models in the first sense. This does not mean that such a claim does not have any bearing on the further question: What are the conditions for reduction of an empirically interpreted dynamical system (DS_2, I_2) to another one (DS_1, I_1)? I will return later (see sec. 4) to this question. For the moment, it suffice to say that, in my view, the conditions for reduction of the mathematical model DS_2 to the mathematical model DS_1 are a necessary component of the more complex conditions for

---

[6] For any falling body a, if p_a is the point where a is initially released, a's *vertical* position and velocity are taken with respect to an axis with origin in the earth center that passes through p_a; the positive direction of such axis is from the earth center to its surface.

reduction of (DS_2, I_2) to (DS_1, I_1).

I am now going to present a detailed argument to support the claim that emulation is sufficient for reduction. The complete argument relies on five premises, divided into three groups. The first premise (**A**) is the most general one, for it refers to systems of *any* kind. Specifically, **A** states a sufficient condition for reduction between two arbitrary systems. The premises of the second group (**B1** and **B2**) are at an itermediate level of generality, for they refer exclusively to *mathematical* systems of any kind, that is, systems that are models of *some* mathematical theory. **B1** explicitly states what it is to be intended for "constitutive entity of a mathematical model", while **B2** makes clear the meaning of "whole structure of a mathematical model". The premises of the third group (**C1** and **C2**) are the most specific, for they refer to *dynamical systems* (in the purely mathematical sense). In particular, **C1** states identity conditions for such systems, and **C2** makes explicit the exact meaning of "whole structure of a dynamical system". Below are the five premises. Each of them is followed by a brief elucidation, which is intended to pin point crucial features of the corresponding premise, as well as to provide an intuitive justification for its assumption.

A For a system S_2 to be reduced to a system S_1, it is sufficient that (a) all the constitutive entities of S_2 are constitutive entities of S_1 and (b) the whole structure of S_2 is a part of the whole structure of S_1. *Elucidation* – In general, a system S is characterized by a whole structure formed by a complex of interconnected elements; each of these structural elements is built out of a given stock of building blocks, which we call "the constitutive entities of S". Thus, if two systems S_1 and S_2 satisfy conditions (a) and (b) above, the system S_2 is in fact a subsystem of S_1; this allows us to claim that S_2 is reduced to S_1.

B1 The constitutive entities of a mathematical model are the entities in its domain. *Elucidation* – According to standard definition, a mathematical model MS is a set D together with a family $(\sigma_i)_{i \in I}$ of relations on D. For any $i \in I$, there is exactly one $n \geq 0$ such that σ_i has arity n, where relations of arity 0 are identified with members of D, and relations of arity $n > 0$ are identified with sets of n-tuples of members of D; the set D is called the *domain* of the model. A mathematical model can thus be thought as a special kind of system, whose structural elements are the relations in the family $(\sigma_i)_{i \in I}$, and whose constitutive entities are the members of D.

B2 The *whole* structure of a mathematical model $MS = (D, (\sigma_i)_{i \in I})$ is the union of all the relations in the family $(\sigma_i)_{i \in I}$;[7] accordingly, if the relata of "is a part of" are whole structures of mathematical models, "is a part of" is to be interpreted as set-inclusion. *Elucidation* – We have just seen that a mathematical model can be thought as a special kind of system, whose structural elements are the relations in the family $(\sigma_i)_{i \in I}$. Each of such relations is a set of n-tuples; thus, the union of these sets is the whole structure

---

[7] The condition in the text holds iff any relation σ_i has arity > 0. The general condition is as follows. Let $X = \{x : \text{for some } i \in I, x = \sigma_i \text{ and } \sigma_i \text{ is a relation of arity 0}\}$; then, the whole structure of $(D, (\sigma_i)_{i \in I})$ is the union of X and all relations σ_i of arity > 0. Obviously, this condition reduces to the one in the text when X is empty, i.e., when any relation σ_i has arity > 0.

formed by the complex of such relations. Given this interpretation of "whole structure of a mathematical model", it is then obvious that "is a part of" should be interpreted as set-inclusion.

C1 From the point of view of general dynamical systems theory, any two isomorphic dynamical systems are identical. *Elucidation* – General dynamical systems theory studies the structural properties (see notes 4 and 5) of dynamical systems, and any two dynamical systems have exactly the same structural properties iff they are isomorphic. Therefore, general dynamical systems theory does not distinguish between any two isomorphic dynamical systems.

C2 If a mathematical model is a dynamical system $DS = (M, (g^t)_{t \in T})$, the whole structure of the model is the set of all state pairs (x, y) such that, for some $t \in T$, $g^t(x) = y$. *Elucidation* – We should first of all notice that, by def. [1], a dynamical system is a mathematical model of a special kind, namely, such that any relation g^t is in fact a function from M to M. Then, **C2** is an immediate consequence of this observation and **B2**.[8]

SUFFICIENCY OF EMULATION FOR REDUCTION

1. For a mathematical model MS_2 to be reduced to a mathematical model MS_1, it is sufficient that (a) the domain of MS_2 is included in the domain of MS_1 and (b) the whole structure of MS_2 is included in the whole structure of MS_1; (logically follows from **A**, **B1** and **B2**;)
2. hence, if u is an emulation of DS_2 in DS_1, the u-virtual system DS_2 in DS_1 is reduced to DS_1; (logically follows from 1, **C2**, and theses (ii) and (iii) of $[VST]$;)
3. if u is an emulation of DS_2 in DS_1, DS_2 is isomorphic to the u-virtual system DS_2 in DS_1; (logically follows from thesis (i) of $[VST]$ and def. [6];)
4. consequently, if u is an emulation of DS_2 in DS_1, DS_2 is reduced to DS_1. (Logically follows from 2, 3, **C1** and the fact that dynamical systems, as intended here, are just models of general dynamical systems theory.)

4 Models of phenomena — sufficient conditions for total and exact reduction in empirically interpreted dynamical systems

Thus far, the *representational theory* of reduction has a precise formulation only if the models involved are dynamical systems in the purely mathematical sense. However, we have seen in sec. 3 that dynamical systems can also be intended as *models of real phenomena*. According to this second sense of the term "model", a dynamical system is not a purely mathematical entity DS; rather, it is a pair (DS, I_H), where I_H is an empirical interpretation that links the purely mathematical model DS to a phenomenon H. The representational theory should then be further developed to provide conditions for reduction of an empirically interpreted dynamical system (DS_2, I_{H_2}) to another one (DS_1, I_{H_1}). I will briefly sketch here the main lines of such development. The following exposition has no pretention to exhaustiveness. Its goal is just

---

[8]Thus, **C2** is not an *independent* premise of the argument, for it is entailed by def. [1], the standard definition of a mathematical model, and **B2**.

to trace a possible way along which an adequate representational theory of reduction for *empirically interpreted* dynamical systems might be worked out.

In general, a *phenomenon H* can be thought as a pair (F, B_F) of two distinct elements. The first one, F, is a *functional description* of (i) an abstract type of real system AS_F and (ii) a general spatio-temporal scheme CS_F of its causal interactions; in particular, the functional description of the abstract system AS_F specifies its structural elements (or functional parts) and their mutual relationships and organization, while the description of the causal scheme CS_F specifies the initial conditions of AS_F's evolution. The second element, B_F, is the set of all concrete systems of type AS_F that also satisfy the causal interaction scheme CS_F; B_F is called the *application domain*[9] of the phenomenon H.

For example, let $H_e = (F_e, B_{F_e})$ be the phenomenon of the free fall of a medium size body in the vicinity of the earth (from now on, I will refer to H_e just as *the phenomenon of free fall*). In this case, the functional description F_e is as follows. The abstract type of real system AS_{F_e} has just one structural element, namely, a medium size body in the vicinity of the earth; the causal interaction scheme CS_{F_e} consists in releasing the body at an arbitrary instant, and with a *purely vertical* velocity and position (relative to the earth surface) whose respective values are within appropriate boundaries. B_{F_e} is then the set of all concrete medium size bodies in the vicinity of the earth that satisfy the given scheme of causal interactions. Any such body will be called a *(free) falling body*.

Let $DS = (X_1 \times \ldots \times X_n, (g^t)_{t \in T})$ be a dynamical system whose state space $M = X_1 \times \ldots \times X_n$ has n components X_i ($1 \leq i \leq n$, where $i, n \in Z^+ =$ the non negative integers). An *interpretation I_H of DS on a phenomenon H* consists in identifying each component X_i with the set of all possible values of a magnitude M_i of the phenomenon H, and the time set T with the set of all possible instants of the time T of H itself. An interpretation I_H of DS on H is *empirical* if the time T and all the magnitudes M_i are measurable properties of the phenomenon H. A pair (DS, I_H), where DS is a dynamical system with n components and I_H is an interpretation of DS on H, is said to be a *model of the phenomenon H*. If the interpretation I_H is empirical, then (DS, I_H) is an *empirical model of H*. Such a model is said to be *empirically correct* if, for any i, all measurements of magnitude M_i are consistent with the corresponding values x_i determined by DS. An empirically correct model of H is also called a *Galilean model of H* [11], [12, ch. 3]. A *Galilean model* is then any empirically correct model of some phenomenon.

As an example, let us consider again the phenomenon of free fall H_e. Let DS_e be the dynamical system with two components specified in sec. 3, and I_{H_e} be its interpretation given in sec. 3; then, according to the previous definitions, I_{H_e} is an empirical interpretation of DS_e on H_e, and $(DS_e, I_{H_e}) = DS_e$ is an empirical model of H_e. For an appropriate value of the constant

---

[9] Since the functional description F typically contains several idealizations, no concrete or real system RS exactly satisfies F, but it rather fits F up to a certain degree. Thus, from a formal point of view, the application domain B_F of a phenomenon (F, B_F) might be better described as a fuzzy set.

g, such a model also turns out to be empirically correct.[10]

Let us now consider two phenomena $H_1 = (F_1, B_{F_1})$ and $H_2 = (F_2, B_{F_2})$, and two empirically interpreted dynamical systems $\boldsymbol{DS}_1 = (DS_1, I_{H_1})$ and $\boldsymbol{DS}_2 = (DS_2, I_{H_2})$ such that $\boldsymbol{DS}_1$ is an empirical model of H_1 and $\boldsymbol{DS}_2$ is an empirical model of H_2. What are the conditions for reduction of $\boldsymbol{DS}_2$ to $\boldsymbol{DS}_1$? I will divide the discussion into three distinct cases.

Case 1. Let us suppose that $B_{F_2} \subseteq B_{F_1}$. Under this hypothesis, it seems sensible to claim that, if DS_1 emulates DS_2, then $\boldsymbol{DS}_2$ *is reduced to* $\boldsymbol{DS}_1$. To see this point, let us notice, first, that the hypothesis $B_{F_2} \subseteq B_{F_1}$ ensures that any concrete system described by $\boldsymbol{DS}_2$ is also described by $\boldsymbol{DS}_1$. Second, let $u: Y_1 \times \ldots \times Y_n \to X_1 \times \ldots \times X_m$ be an emulation of $DS_2 = (Y_1 \times \ldots \times Y_n, (h^v)_{v \in V})$ in $DS_1 = (X_1 \times \ldots \times X_m, (g^t)_{t \in T})$. Thus, by def. [8], any state transition $h^v: (y_1, \ldots, y_n) \to (y_1', \ldots, y_n')$ corresponds to a state transition $g^t: (x_1, \ldots, x_m) \to (x_1', \ldots, x_m')$, where $u(y_1, \ldots, y_n) = (x_1, \ldots, x_m)$ and $u(y_1', \ldots, y_n') = (x_1', \ldots, x_m')$. In addition, since $\boldsymbol{DS}_2$ is an empirical model of H_2, for any j, y_j and y_j' are values of a measurable magnitude $\boldsymbol{M}_j$ of H_2, and v is a value of the time $\boldsymbol{T}_2$ of H_2; on the other hand, since $\boldsymbol{DS}_1$ is an empirical model of H_1, for any i, x_i and x_i' are values of a measurable magnitude $\boldsymbol{M}_i$ of H_1, and t is a value of the time $\boldsymbol{T}_1$ of H_1. For any concrete system $RS \in B_{F_2}$, both the $\boldsymbol{DS}_2$ and the $\boldsymbol{DS}_1$ descriptions apply to RS. But then, the emulation function u tells us exactly how the $\boldsymbol{DS}_2$ description of RS corresponds to the $\boldsymbol{DS}_1$ description.

As an example, let $\boldsymbol{DS}_e = (DS_e, I_{H_e})$ be the falling body model, where H_e = the phenomenon of free fall, and let $H_p = (F_p, B_{F_p})$ be *the phenomenon of projectile motion*, where its functional description F_p and its application domain B_{F_p} are specified as follows. The abstract type of real system AS_{F_p} is a medium size body in the vicinity of the earth, and it is thus identical to AS_{F_e}. However, the causal interaction scheme CS_{F_p} is more general than CS_{F_e}, for it consists in the body's being released at an arbitrary instant, and with *any velocity* and position (relative to the earth surface) whose respective values are within appropriate boundaries. B_{F_p} is then the set of all concrete medium size bodies in the vicinity of the earth that satisfy the given more general scheme of causal interactions.

Let us then consider the following system of four ordinary differential equations $\langle dx(t)/dt = \dot{x}(t), dy(t)/dt = \dot{y}(t), d\dot{x}(t)/dt = 0, d\dot{y}(t)/dt = -\boldsymbol{g} \rangle$, where $\boldsymbol{g}$ is a fixed real positive constant. The solutions of such equations uniquely determine the dynamical system $DS_p = (X \times Y \times \dot{X} \times \dot{Y}, (g^t)_{t \in T})$, where $X = Y = \dot{X} = \dot{Y} = T = R$ (the real numbers) and, for any $t, x, y, \dot{x}, \dot{y} \in R$, $g^t(x, y, \dot{x}, \dot{y}) = (\dot{x}t + x, -\boldsymbol{g}t^2/2 + \dot{y}t + y, \dot{x}, -\boldsymbol{g}t + \dot{y})$.

Let I_{H_p} be the following interpretation of DS_p on the phenomenon of projectile motion H_p. In the first place, for any projectile a, let p_a be the point where a is initially released; we then consider the plane that contains a's initial velocity vector and the earth center. On this plane, we fix both the x-axis and the y-axis of a Cartesian coordinate system, in such a way that its origin

---

[10] Quite obviously, if $\boldsymbol{g}$ = *standard gravity* ($\boldsymbol{g}$ = 9.80665 m/s^2), the model $(DS_e, I_{H_e}) = \boldsymbol{DS}_e$ turns out to be empirically correct within limits of precision sufficient for many practical purposes.

coincides with the earth center, and the y-axis passes through p_a. The positive direction of the y-axis is from the earth center to its surface; accordingly, we call the y-axis the *vertical axis*, and the x-axis the *horizontal axis*. We then interpret the first component X of the state space of DS_p as the set of all possible values of the *horizontal position* of the projectile a, the second component Y as the set of all possible values of its *vertical position*, the third component $\dot{X}$ as the set of all possible values of its *horizontal velocity*, the fourth component $\dot{Y}$ as the set of all possible values of its *vertical velocity*, and the time set T of DS_p as the set all possible values of *physical time*. Since all five of these magnitudes are measurable or detectable properties of the intended phenomenon H_p, I_{H_p} is an *empirical* interpretation of DS_p on H_p.

Let $\boldsymbol{DS}_p = (DS_p, I_{H_p})$; $\boldsymbol{DS}_p$ will be called the *projectile model*. By the respective definitions of B_{F_e} and B_{F_p}, $B_{F_e} \subset B_{F_p}$. Thus, by case 1, to show that the falling body model $\boldsymbol{DS}_e$ is reduced to the projectile model $\boldsymbol{DS}_p$, it suffice to exhibit an emulation u of DS_e in DS_p. Let $u\colon Y \times \dot{Y} \to X \times Y \times \dot{X} \times \dot{Y}$ and, for any $y, \dot{y} \in R$, $u(y, \dot{y}) = (0, y, 0, \dot{y})$; then, quite obviously, u is an emulation of DS_e in DS_p.

Case 2. Let us suppose next that $B_{F_2} \cap B_{F_1} = \varnothing$. In this case, no matter how DS_1 and DS_2 are related, $\boldsymbol{DS}_2$ *is not reduced to* $\boldsymbol{DS}_1$. For, even if DS_2 is identical to DS_1, *any* concrete system described by $\boldsymbol{DS}_2$ (that is to say, any concrete system $RS \in B_{F_2}$) is not a system also described by $\boldsymbol{DS}_1$.

Case 3. The case $B_{F_2} \cap B_{F_1} \neq \varnothing$ and $\neg(B_{F_2} \subseteq B_{F_1})$ is still left. This case is a combination of the previous two. In fact, for some concrete system $RS \in B_{F_2}$, both the $\boldsymbol{DS}_2$ and the $\boldsymbol{DS}_1$ descriptions apply to RS; however, if $RS \in B_{F_2}$ and $RS \notin B_{F_1}$, only the $\boldsymbol{DS}_2$ description applies to RS. Thus, in this case, if DS_1 emulates DS_2, $\boldsymbol{DS}_2$ *is incompletely reduced to* $\boldsymbol{DS}_1$.

We have just seen that case 3 only grants *incomplete* reduction of $\boldsymbol{DS}_2$ to $\boldsymbol{DS}_1$, provided that DS_1 emulates DS_2. However, $\boldsymbol{DS}_2$ may turn out to be *multiply* reduced to a *family* $(\boldsymbol{DS}_j)_{j \in J} = ((DS_j, I_{H_j}))_{j \in J}$ of empirically interpreted dynamical systems, each of which satisfies case 3 and emulates DS_2. This will be the case if the application domain B_{F_2} is included in the union of all application domains B_{F_j}. More precisely, for $\boldsymbol{DS}_2$ to be *multiply reduced to* $(\boldsymbol{DS}_j)_{j \in J}$, it is sufficient that, for any $j \in J$, $B_{F_2} \cap B_{F_j} \neq \varnothing$, $\neg(B_{F_2} \subseteq B_{F_j})$, DS_j emulates DS_2, and $B_{F_2} \subseteq \cup_{j \in J} B_{F_j}$.

A relationship between this condition for multiple reduction and the second order property version of the multiple realization concept [17, pp. 19–20, 103–104] is worth noticing. According to the latter, a property P *is multiply realized by properties of type* D just in case, for any x, x has P iff there is a property P_j of type D such that x has P_j. Any property P_j that satisfies the previous condition is said a *D-realizer* of the property P, and the property P itself is said a *second order property*.

Suppose now that $\boldsymbol{DS}_2$ is multiply reduced to $(\boldsymbol{DS}_j)_{j \in J}$ according to the previously stated sufficient condition. Let P_2 be the property that corresponds to functional description F_2 and, for any $j \in J$, P_j be the property that corresponds to functional description F_j. Let D be the property of being one of the properties P_j, for some $j \in J$. As $B_{F_2} \subseteq \cup_{j \in J} B_{F_j}$, it follows that, if x has P_2, then x has P_j, for some $j \in J$. Furthermore, if $B_{F_2} = \cup_{j \in J} B_{F_j}$, the converse holds as well, so that P_2 is multiply realized by properties of type D, and $(P_j)_{j \in J}$ is the family of its D-realizers.

From an intuitive point of view, the emulation relationship holds between two dynamical systems DS_1 and DS_2 when the *whole* dynamics of DS_2 is *exactly* reproduced by DS_1. I have argued so far that this relationship might be the basis for a new approach to reduction, which I have called *representational*. However, it is well known that, in many cases of inter-theoretic reduction, the relationship between the reduced theory S_2 and the reducing one S_1 is such that S_2 is only *partially* and *approximately* reduced to S_1. Furthermore, such a relationship typically is an *asymptotic* one, that is, it depends on some parameter p^* of either S_1 or S_2 in such a way that, for p^* tending to some fixed limiting value p, S_2 tends to be partially and approximately reduced to S_1, as established according to the limiting value p.[11]

The simple form of the emulation relationship considered so far may very well be the basis for a representational account of *total* and *exact* reduction (like, for example, the reduction of the falling body model $\boldsymbol{DS}_e$ to the projectile model $\boldsymbol{DS}_p$; see case 1 above). Nevertheless, we need a more sophisticated version of emulation for dealing with cases of asymptotic, partial and approximate reduction. In the next section, I suggest how this might be accomplished and provide (i) a formal definition of *partial* and *approximate* emulation, (ii) a simple example that shows how this relationship may turn out to be asymptotic, and (iii) sufficient conditions for partial and approximate reduction in empirically interpreted dynamical systems.

5 Partial and approximate emulation—sufficient conditions for partial and approximate reduction in empirically interpreted dynamical systems

Intuitively, a dynamical system $DS_1 = (M, (g^t)_{t \in T})$ partially emulates a second dynamical system $DS_2 = (N, (h^v)_{v \in V})$ if DS_1 exactly reproduces the dynamics of DS_2, limited to a fixed non-empty subset C of DS_2's state space N. This concept is thus a straightforward relativization of def. [8]. Let $C \neq \varnothing$, $C \subseteq N$, and define:

[9] DS_1 *C-emulates* DS_2 iff there is an injective function $u: C \to M$ such that for any $v \in V$, for any $c \in C$, there is $t \in T$ such that $u(h^v(c)) = g^t(u(c))$. Any function u that satisfies the previous condition is called a *C-emulation of DS_2 in DS_1*, and C is called its *emulation domain*.

Intuitively, DS_1 approximately emulates DS_2 if each state transition $h^v: y \to y'$ of DS_2 approximately corresponds to a state transition $g^t: x \to x'$ of

---

[11] Hooker ([15, p. 436]) maintains that "asymptotics provides the ground on which claims about inter-theoretic explanation, reduction and emergence must ultimately rest". According to him, "in physics, we find that the most famous theory pairs are all asymptotically related" ([15, p. 437]. Among such pairs, he explicitly mentions: (i) special relativity and Newtonian mechanics; (ii) optics and ray optics; (iii) quantum mechanics and Newtonian mechanics; (iv) statistical mechanics and thermodynamics (where, in each pair, the first element is the *reducing* theory and the second element is the *reduced* theory). According to Hooker, an analogous relationship may also hold between two different models of the *same* theory; an example is the following pair of models of Newtonian mechanics: a non linear classic pendulum model and a harmonic oscillator model ([15, p. 438]).

DS_1. This idea can be made precise by requiring that, for some injective function u, $u(y) = x$, and $u(y')$ be *sufficiently close* to x', where the two states $u(y')$, $x' \in M$ are sufficiently close to each other if their *distance* does not exceed a fixed non-negative real δ. Thus, the concept of approximate emulation in fact presupposes that M (i.e., the state space of DS_1) be equipped with a metric. Let $d: M \times M \to R^+$ be a metric on M, let $\delta \in R^+$. We then define:

[10] DS_1 δ-*emulates* DS_2 iff there is an injective function $u: N \to M$ such that, for any $v \in V$, for any $c \in N$, there is $t \in T$ such that $d(u(h^v(c)), g^t(u(c))) \leq \delta$. Any function u that satisfies the previous condition is called a δ-*emulation of* DS_2 *in* DS_1, and δ is called its *approximation degree*. If, for some δ, u is a δ-emulation of DS_2 in DS_1, the minimum of all such δ must exist, for R satisfies the least upper bound property.[12] Let δ^{min} be such a minimum; δ^{min} is then called u's *best approximation degree*. Thus, obviously, if, for some δ, u is a δ-emulation of DS_2 in DS_1, then u is a δ^{min}-emulation of DS_2 in DS_1.

Finally, by combining definitions [9] and [10], we get a definition of the intuitive idea of *partial* and *approximate* emulation. Let $C \neq \varnothing$, $C \subseteq N$, $d: M \times M \to R^+$ be a metric on M, and $\delta \in R^+$;

[11] DS_1 C-δ-*emulates* DS_2 iff there is an injective function $u: C \to M$ such that for any $v \in V$, for any $c \in C$, there is $t \in T$ such that $d(u(h^v(c)), g^t(u(c))) \leq \delta$. Any function u that satisfies the previous condition is called a C-δ-*emulation of* DS_2 *in* DS_1, C is called its *emulation domain*, and δ is called its *approximation degree*. If, for some δ, u is a C-δ-emulation of DS_2 in DS_1, the minimum of all such δ, indicated by δ^{min}, is called u's *best approximation degree*.[13] Thus, obviously, if, for some δ, u is a C-δ-emulation of DS_2 in DS_1, then u is a C-δ^{min}-emulation of DS_2 in DS_1.

5.1 An example

Let $X = \dot{X} = T = V = R$ (the real numbers), and $DS_n = (X \times \dot{X}, (g^t)_{t \in T})$ be the dynamical system that is determined by the solutions of the following non-linear system of ordinary differential equations $\langle dx(t)/dt = \dot{x}(t), d\dot{x}(t)/dt = -g\sin(x(t)/l)\rangle$, where g is a fixed real positive constant, and l is an arbitrary real positive parameter; note that this system is in fact a *non-linear classic pendulum*.[14] On the other hand, let $DS_o = (X \times \dot{X}, (h^v)_{v \in V})$ be the dynamical system that is determined by the solutions of the following linear system of ordinary differential equations $\langle dx(v)/dv = \dot{x}(v), d\dot{x}(v)/dv = -gx(v)/l\rangle$, where g and l are as above; this second system is a *harmonic oscillator*.

Let $0 \leq \theta \leq \pi$, and $C_\theta = \{c$ such that, for some $x \in R$, $c = (x, 0)$, and $-\theta \leq x/l \leq \theta\}$. As it can be visually verified by means of any dynamical systems software, for an appropriately chosen $\delta_\theta > 0$, for any $v \in V$, for any $c \in C_\theta$, $d(h^v(c), g^v(c)) \leq \delta_\theta$, where d is the usual Euclidean distance on $X \times \dot{X}$

---

[12] According to the least upper bound property, for any non-empty subset A of R, if A has an upper bound, then the minimum of all upper bounds of A exists. Also recall that, for any $B \subseteq R$, m *is the minimum of* B iff $m \in B$ and, for any $b \in B$, $m \leq b$; u *is an upper bound of* B iff $u \in R$ and, for any $b \in B$, $b \leq u$.
[13] Such a minimum exists (see def. [10]).
[14] If $c = (m\pi l, 0)$ for some $m \in Z$, then $g^t(c) = c$ for any $t \in T$; that is to say, c is a fixed point of DS_n.

$= R^2$. Let u be the identity function on C_θ. By def. [11], it thus follows that u is a C_θ-δ_θ-emulation of DS_o in DS_n. Let δ_θ^{min} be the minimum of all such δ_θ. Then, by def. [11], u is a C_θ-δ_θ^{min}-emulation of DS_o in DS_n as well, and so DS_n C_θ-δ_θ^{min}-emulates DS_o.

It is important to keep in mind that δ_θ^{min} represents the best approximation degree to which DS_n partially emulates DS_o with respect to emulation domain C_θ. Besides, δ_θ^{min} is a function of $\theta \in [0, \pi]$. Therefore, we can study the behavior of δ_θ^{min} for θ tending to 0 from the right, and it is not difficult to verify that $\lim_{\theta \to 0^+} \delta_\theta^{min} = 0 = \delta_0^{min}$.

That is to say, for θ tending to 0 from the right, the best approximation degree to which DS_n partially emulates DS_o with respect to emulation domain C_θ tends to the best approximation degree to which DS_n partially emulates DS_o with respect to emulation domain C_0. In this precise sense, then, the relationship of partial and approximate emulation of DS_n by DS_o (with respect to emulation domain C_θ, and to the best approximation degree δ_θ^{min}) turns out to be asymptotic (see sec. 4, penultimate paragraph).

5.2 Empirical interpretations of the two dynamical systems of the previous example

Both dynamical systems $DS_n = (X \times \dot{X}, (g^t)_{t \in T})$ and $DS_o = (X \times \dot{X}, (h^v)_{v \in V})$ can be given natural empirical interpretations on corresponding phenomena. As regards the first system, let $H_n = (F_n, B_{F_n})$ be the phenomenon of the unrestricted swing of a pendulum or, more briefly, *the phenomenon of (unrestricted) pendulum swings*, where its functional description F_n and its application domain B_{F_n} are specified as follows . The abstract type of real system AS_{F_n} (called *simple* or *classic pendulum*) is made up of two structural elements, namely, a light rigid arm of length l, with a much heavier "bob" on one of its ends; the arm is pivoted on the other end, so that the bob can frictionlessly swing along a circular path of radius l in a vertical plane. The causal interaction scheme CS_{F_n} consists in releasing the bob at an arbitrary instant and position on its swinging path, with an arbitrary tangent velocity. B_{F_n} is then the set of all concrete simple pendula that satisfy the given scheme of causal interaction. Any such device will be called an *(unrestricted) pendulum*.

Let I_{H_n} be the following interpretation of DS_n on the phenomenon of unrestricted pendulum swings H_n. The first component X of the state space of DS_n is the set of all possible values of the *bob position*[15] of an arbitrary unrestricted pendulum, the second component $\dot{X}$ is the set of all possible values of the *bob tangent velocity*, and the time set T of DS_n is the set of all possible values of *physical time*. Since all three of these magnitudes are measurable or detectable properties of the intended phenomenon H_n, I_{H_n} is an empirical interpretation of DS_n on H_n, and $\boldsymbol{DS_n} = (DS_n, I_{H_n})$ is thus a model of H_n. This model is empirically correct, for an appropriate value of the constant

---

[15] A positive (negative) bob position x is the distance (the opposite of the distance), along the positive (negative) direction of the circular swinging path, of the bob itself from the intersection O between the path and the vertical straight line passing through the pendulum pivot. We take the positive path direction to be anticlockwise.

g.[16] Henceforth, I will refer to $\boldsymbol{DS}_n$ as the *(unrestricted) pendulum model*.

As for the second system, let $H_{o\theta} = (F_{o\theta}, B_{F_{o\theta}})$ be the phenomenon of pendulum motion restricted to small swings or, more briefly, *the phenomenon of small pendulum swings*, where its functional description $F_{o\theta}$ and its application domain $B_{F_{o\theta}}$ are specified as follows. The abstract type of real system $AS_{F_{o\theta}}$ is a simple pendulum of length l, so it is identical to AS_{F_n}. However, the causal interaction scheme $CS_{F_{o\theta}}$ is more specific than CS_{F_n}, for it consists in releasing the pendulum's bob at an arbitrary instant, with zero tangent velocity, and in a position sufficiently close to the intersection O between the swinging path and the vertical straight line r passing through the pendulum pivot. This last clause can be put in the following form. Let θ ($0 \leq \theta \leq \pi$) be the measure, in radians, of the angle between r and a straight line s passing through the pivot; let x be the bob's releasing position on the swinging path (where the origin is O, and the positive path direction is anticlockwise); then, $-\theta \leq x/l \leq \theta$. Thus, for θ sufficiently close to 0, the pendulum only performs small swings, when its bob is released at an arbitrary instant, with zero tangent velocity and in position x. $B_{F_{o\theta}}$ is then the set of all concrete simple pendula that satisfy the given more specific scheme of causal interaction. Any such device will be called a *small swing pendulum*.

Let $I_{H_{o\theta}}$ be the following interpretation of DS_o on the phenomenon of small pendulum swings $H_{o\theta}$. The first component X of the state space of DS_o is the set of all possible values of the *bob position* of an arbitrary small swing pendulum, the second component $\dot{X}$ is the set of all possible values of the *bob tangent velocity*, and the time set V of DS_o is the set all possible instants of *physical time*. These three magnitudes are measurable properties of the intended phenomenon $H_{o\theta}$. Therefore, $I_{H_{o\theta}}$ is an empirical interpretation of DS_o on $H_{o\theta}$, and $\boldsymbol{DS}_{o\theta} = (DS_o, I_{H_{o\theta}})$ is a model of $H_{o\theta}$. Furthermore, if θ is sufficiently small, such a model turns out to be empirically correct (for an appropriate value of the constant g, see note 16). In what follows, $\boldsymbol{DS}_{o\theta}$ will be called the *small swing pendulum model*.

5.3 Sufficient conditions for partial and approximate reduction

Let us notice now that the unrestricted pendulum model $\boldsymbol{DS}_n$ and the small swing pendulum model $\boldsymbol{DS}_{o\theta}$ satisfy case 1 above (sec. 4), for $B_{F_{o\theta}} \subset B_{F_n}$ (by the definitions of the respective application domains B_{F_n} and $B_{F_{o\theta}}$). Moreover, we have seen (sec. 5.1, par. 2) that, for any $\theta \in [0, \pi]$, DS_n C_θ-δ_θ^{min}-emulates DS_o. The question then naturally arises whether this condition is sufficient for reduction of $\boldsymbol{DS}_{o\theta}$ to $\boldsymbol{DS}_n$.

Let us notice first that C_θ is, on the one hand, the emulation domain with respect to which dynamical system DS_n partially emulates DS_o and, on the other hand, C_θ is determined by the specific causal interaction scheme $CS_{F_{o\theta}}$ of the phenomenon of small pendulum swings. As a consequence, C_θ can be thought as singling out that part of the structure of DS_o that has an empirical interpretation according to $I_{H_{o\theta}}$. Let us call $E(C_\theta) = \{e: e = (c, h^v(c)),$ for some $c \in C_\theta$ and some $v \in V\}$ *the empirical substructure of DS_o relative to*

---

[16]$g = 9.80665$ m/s^2 (standard gravity) will be appropriate for many purposes.

interpretation $I_{H_{o\theta}}$.[17] Thus, by def. [11], the *whole* empirical substructure $E(C_\theta)$ is represented, through a partial emulation function u,[18] by corresponding structure of DS_n, within approximation degree δ_θ^{min}. Suppose now that $\Delta > 0$ is the desired approximation degree. Then, if $\delta_\theta^{min} \leq \Delta$, we can safely conclude that $\boldsymbol{DS}_{o\theta}$ is reduced to $\boldsymbol{DS}_n$.

In this connection, also recall that $\lim_{\theta \to 0^+} \delta_\theta^{min} = 0$, where $\theta \in [0, \pi]$ (sec. 5.1, par. 3). This means that the best approximation degree δ_θ^{min} to which the empirical substructure $E(C_\theta)$ is represented by corresponding structure of DS_n can be made as small as we please, by taking a sufficiently small value of θ. More precisely, for any desired approximation degree $\Delta > 0$, there is a sufficiently small θ_Δ such that, for any θ, if $0 < \theta < \theta_\Delta$, then $\delta_\theta^{min} < \Delta$. In addition, recall that $\delta_0^{min} = 0$ (sec. 5.1, par. 3); therefore, for any $\theta < \theta_\Delta$, $\delta_\theta^{min} < \Delta$. It thus follows that, for any $\theta < \theta_\Delta$, $\boldsymbol{DS}_{o\theta}$ is reduced to $\boldsymbol{DS}_n$.

In the general case, let $H = (F, B_F)$ be an arbitrary phenomenon, and $\boldsymbol{DS} = (DS, I_H)$ be any empirically interpreted dynamical system such that $\boldsymbol{DS}$ is an empirical model of H; let $DS = (M, (g^t)_{t \in T})$. Let us assume that, in force of interpretation I_H, there is a one-to-one correspondence between the *initial conditions* specified by the causal interaction scheme CS_F of phenomenon H (see sec. 4, par. 2) and a set of states of the dynamical system DS; let $C_F \subseteq M$ be such a set. Then, C_F is called *the empirical domain of DS relative to interpretation* I_H, and $E(C_F) = \{e : e = (c, g^t(c))$, for some $c \in C_F$ and some $t \in T\}$ is called *the empirical substructure of DS relative to* I_H.

Let $H_1 = (F_1, B_{F_1})$ and $H_2 = (F_2, B_{F_2})$ be two phenomena, and $\boldsymbol{DS}_1 = (DS_1, I_{H_1})$ and $\boldsymbol{DS}_2 = (DS_2, I_{H_2})$ be two empirically interpreted dynamical systems such that $\boldsymbol{DS}_1$ is an empirical model of H_1 and $\boldsymbol{DS}_2$ is an empirical model of H_2. Let C_{F_2} be the empirical domain of DS_2 relative to I_{H_2}, and $\Delta > 0$ be the desired approximation degree for DS_1 C_{F_2}-δ-emulating DS_2. The previous example thus suggests that case 1 (sec. 4) be supplemented with a weaker sufficient condition for reduction, as follows.

Case 1a. Let us suppose that $B_{F_2} \subseteq B_{F_1}$. If DS_1 C_{F_2}-δ-emulates DS_2 and $\delta \leq \Delta$, then $\boldsymbol{DS}_2$ is reduced to $\boldsymbol{DS}_1$.

A corresponding weaker condition can also be given for the case of *incomplete* reduction (case 3, sec. 4), as follows.

Case 3a. Suppose that $B_{F_2} \cap B_{F_1} \neq \emptyset$ and $\neg(B_{F_2} \subseteq B_{F_1})$. If DS_1 C_{F_2}-δ-emulates DS_2 and $\delta \leq \Delta$, then $\boldsymbol{DS}_2$ *is incompletely reduced to* $\boldsymbol{DS}_1$.

As for multiple reduction to a family $(\boldsymbol{DS}_j)_{j \in J} = ((DS_j, I_{H_j}))_{j \in J}$ of empirically interpreted dynamical systems, we get the following weaker condition. For $\boldsymbol{DS}_2$ to *be multiply reduced to* $(\boldsymbol{DS}_j)_{j \in J}$, it is sufficient that, for any $j \in J$, $B_{F_2} \cap B_{F_j} \neq \emptyset$, $\neg(B_{F_2} \subseteq B_{F_j})$, DS_j C_{F_2}-δ-emulates DS_2, $\delta_j \leq \Delta$, and $B_{F_2} \subseteq \bigcup_{j \in J} B_{F_j}$.

---

[17] See van Fraassen 1980 for a general discussion of the concept of an empirical substructure.

[18] Recall that, in this particular case, u is the identity function on $C_{\phi, \theta}$.

6 Concluding remarks

I have argued in this paper that reduction is better analyzed in terms of a *representational* relationship between *models*, rather than a *deductive* relationship between *theories*. Contrary to the received view, reduction has been conceived as a manifestation of an underlying representational relationship between mathematical models, namely, the one of *emulation*.

The representational theory of reduction has been developed so far only for the special case of dynamical systems (either empirically interpreted, or not). But, even in this special form, the theory is far from being complete.

Furthermore, even a complete representational theory for dynamical systems would not be sufficient to account for all relevant cases of reduction, for many models in real science are not of this kind. What we need is a *general* representational theory, as precise as the one restricted to dynamical systems, which apply to *arbitrary models*. The formulation of such a general theory, however, is not an easy matter, for it involves a preliminary investigation of fairly hard questions like: What is, *in general*, a purely mathematical model?[19] What is a structure preserving mapping between two *arbitrary* mathematical models? What is the relationship between two *arbitrary* mathematical models that generalizes the one of emulation between dynamical systems? What is, *in general*, an empirical interpretation of a mathematical model on a phenomenon?

BIBLIOGRAPHY

[1] V.I Arnold. *Ordinary Differential Equations*. The MIT Press, Cambridge, MA 1977.
[2] W. Balzer, C. U. Moulines, and J. D. Sneed. *An Architectonic for Science: The Structuralist Program*. D. Reidel Publishing, Dordrecht 1987.
[3] W. Balzer, D. A. Pearce, and H.-J. Schmidt (eds.). *Reduction in Science*. D. Reidel Publishing, Dordrecht 1984.
[4] A. Beckermann. Supervenience, emergence and reduction. In A. Beckermann, T. Toffoli and J. Kim (eds.). *Emergence or Reduction? Essays on the Prospects of Nonreductive Physicalism*, Walter de Gruyter, Berlin 1992, pages 94–118.
[5] A. Beckermann, T. Toffoli and J. Kim (eds.). *Emergence or Reduction? Essays on the Prospects of Nonreductive Physicalism*. Walter de Gruyter, Berlin 1992.
[6] J. Bickle. *Psychoneural Reduction: The New Wave*. The MIT Press, Cambridge, MA 1998.
[7] J. Bickle. *Philosophy and Neuroscience: A Ruthlessly Reductive Account*.Kluwer Academic Publishers, Dordrecht 2003.
[8] N. Bourbaki. *Theory of Sets*. Vol. I of the *Elements of Mathematics* series. Addison-Wesley Publishing Company, Reading, MA 1968.
[9] P.M. Churchland. *Scientific Realism and the Plasticity of Mind*. Cambridge University Press, Cambridge 1979.
[10] P.M. Churchland. Reduction, qualia, and the direct introspection of brain states. *Journal of Philosophy*, 82 (1): 8–28, 1985.

---

[19]In sec. 3 (see B1), I defined a mathematical model MS as a set D together with a family $(\sigma_i)_{i \in I}$ of relations on D. This definition is fine as far as *relational* models are concerned, but not all mathematical models are of this kind. For instance, a topological space (with the standard axiomatization in terms of open sets) is not a relational model. [8, ch. 4] contains a quite general treatment of mathematical structures. However, Bourbaki's general theory of structures is developed at the metamathematical level. What we need is a theory of models developed *within* set theory, and thus at the mathematical level, as general as Bourbaki's metamathematical theory of structures.

[11] M. Giunti. Dynamical models of cognition. in R.F. Port and T. van Gelder (eds.), *Mind as Motion: Explorations in the Dynamics of Cognition*, the MIT Press, Cambridge, MA 1995, pages 549–571.
[12] M. Giunti. *Computation, Dynamics, and Cognition*. Oxford University Press, New York 1997.
[13] A.C. Hooker. Critical notice: R. M. Yoshida's *Reduction in the Physical Sciences*. *Dialogue*, 18: 81–99, 1979.
[14] A.C. Hooker. Towards a general theory of reduction. *Dialogue*, 20: 38–59, 201–236, 496–529, 1981.
[15] A.C. Hooker. Asymptotics, reduction and emergence. *British Journal for the Philosophy of Science*, 55: 435–479, 2004.
[16] A.C. Hooker. Reduction as cognitive strategy. In B.L. Keeley (ed.), *Paul Churchland*, Cambridge University Press, New York 2005, pages 154-174.
[17] J. Kim. *Mind in a Physical World*. The MIT Press, Cambridge, MA 1998.
[18] M.S.R. Kulenovic and O. Merino. *Discrete Dynamical Systems and Difference Equations with Mathematica*. Chapman & Hall/CRC, Boca Raton 2002.
[19] A. Marras. Kim on reduction. *Erkenntnis*, 57: 231–257, 2002.
[20] M. Martelli. *Introduction to Discrete Dynamical Systems and Chaos*. Wiley, New York 1999.
[21] D. Mayr. Investigations of the concept of reduction, I. *Erkenntnis*, 10: 275–294, 1976.
[22] D. Mayr. Investigations of the concept of reduction, II. *Erkenntnis* , 16:109–129, 1981.
[23] E. Nagel. *The Structure of Science*. Harcourt, Brace & World, New York 1961.
[24] J.T. Sandefur. *Discrete Dynamical Systems: Theory and Applications*. Oxford University Press, New York 1990.
[25] K.F. Schaffner. Approaches to reduction, *Philosophy of Science*, 34, 2: 137–147, 1967.
[26] A.R.III Smith. Simple computation-universal cellular spaces. *Journal of the Association for Computing Machinery*, 18, 3: 339–353, 1971.
[27] J.D. Sneed. *The Logical Structure of Marthematical Physics*. Reidel, Dordrecht 1971.
[28] W. Stegmüller. *The Structure and Dynamics of Theories*. Springer-Verlag, New York 1976.
[29] P. Suppes. *Introduction to Logic*. D. Van Nostrand Company, New York 1957.
[30] W. Szlenk. *An Introduction to the Theory of Smooth Dynamical Systems*. John Wiley and Sons, Chichister, England 1984.
[31] A.M. Turing. Computing machinery and intelligence. *Mind*, 59: 433–460, 1950.
[32] B. van Fraassen. *The Scientific Image*. Clarendon Press, Oxford 1980.
[33] S. Wolfram. Statistical mechanics of cellular automata. *Reviews of Modern Physics*, 55, 3:601–644, 1983.
[34] S. Wolfram. Cellular automata. *Los Alamos Science*, 9: 2–21, 1983.
[35] S. Wolfram. Computer software in science and mathematics. *Scientific American*, 56: 188–203, 1984.
[36] S. Wolfram. Universality and complexity in cellular automata. In Farmer, Doyne, T. Toffoli and S. Wolfram (eds.). *Cellular Automata*, North Holland Publishing Company, Amsterdam 1984, pages 1–35.
[37] S. Wolfram. *A New Kind of Science*. Wolfram Media, Inc., Champaign, ILL 2002.

Duhem, Quine and the other dogma
ALEXANDER AFRIAT

1 Introduction

A resemblance[1] between positions held by Duhem and Quine has led to the conjunction of their names: one speaks of "Duhem-Quine". Whether the conjunction — amid differences[2] of period, provenance, profession, subject-matter, style and generality — is entirely justified is debatable, but not really the issue here. Quine's position is famously expressed in "Two dogmas of empiricism"; it was by disputing the second[3] (dogma2) that he came to be associated with Duhem. But there is also the first (dogma1), the "cleavage between analytic and synthetic truths".[4] Quine claims they are equivalent (dogma1 ⇔ dogma2), indeed "two sides of a single dubious coin", and contests both together. Duhem on the other hand attributes the impossibility (¬ dogma2) of crucial experiments to the "cleavage", as one might call it, between physics and mathematics. But surely the truths of physics are synthetic, those of mathematics (more or less) analytic. How then can the "Duhem-Quine thesis" (¬ dogma2) depend on the cleavage separating mathematics and physics (dogma1?), while a purportedly equivalent thesis (¬ dogma1) rejects the cleavage between analytic and synthetic? We appear to have something like

$$(\text{dogma1} \Leftrightarrow \text{dogma2}) \wedge (\text{dogma1} \stackrel{?}{\Rightarrow} \neg \text{dogma2}).$$
$$\text{Quine}\text{Duhem}$$

A kind of holism[5] — an entanglement of essences and accidents,[6] of essential experimental intention and accidental auxiliary assumptions — is the main obstacle to crucial experiments and (empirically grounded) meanings. Using notions hinted at by Duhem and Quine, formalised using the resources of set-theoretical axiomatisation, I argue that such holism and inextricability can

---

[1]On this resemblance, as recognised by Quine, see the footnote on p. 41 of [42], footnote 7 on p. 67 of [43] and the very beginning of [45].

[2]Krips [35], Ariew [7], Quine [45] and Vuillemin [49] have pointed out several. Too many according to Needham [39], who argues that Duhem and Quine share much common ground.

[3]In other words "*reductionism*: the belief that each meaningful statement is equivalent to some logical construct upon terms which refer to immediate experience", as Quine ([42, p. 20] puts it.

[4]Quine's rejection of it has met with much disapproval; see for instance [38], [46], [27], [33], [34], [6], [15].

[5]For a detailed analysis of various kinds of holism see [23].

[6]"Accident" and cognates will sometimes be used in a rather "Galilean" way. For Galileo an *accidente* deviates from or even interferes with the ideal purity of an object or scheme; hence air resistance and friction are *accidenti*, as is an imperfection spoiling a glass sphere or smooth plane.

be overcome[7] to an extent that's at least worth pointing out. Taking Quine's association — however questionable — of essence, meaning, synonymy and analyticity for granted, I also argue that analyticity is rehabilitated to the extent that the aforementioned entanglement of essences and accidents is undermined. If this recovery of the analytic completely dissociates it from the synthetic, giving it a distinct and separate identity, we arrive at the aforementioned paradox; for a rehabilitation of crucial experiments would appear to have the opposite effect on mathematics and physics, by consolidating the cleavage between them rather than undermining it. The matter is brought up, *not for resolution, but to shed light on the web of issues involved*, including relations between the arguments of Duhem and Quine. Trenchant conclusions or theorems should not be looked for; my purpose is exploration not proof.

I begin (§2) with a scheme for overcoming holism by disentangling essences from accidents, which leads (§3) to a new characterisation of the meaning and reference of sentences, involving "abstract tests". After noting (§4) that Duhem and Quine themselves already adumbrated such tests I show how they can be formalised in the language of model theory, in fact of set-theoretical axiomatisation. A quantum-mechanical example is looked at in §5. In §6 I consider how Quine relates meaning, essence and analyticity, in §7 how Duhem relates the cleavage between physics and mathematics to the impossibility of crucial experiments, and whether holism really does have conflicting implications for Duhem and for Quine.

Quine was renowned for bringing together logic, philosophy of language, epistemology and philosophy of science (*e.g.* [42], [43]); customary association with Duhem can introduce further elements to produce an even more varied mixture, reflected in the pages that follow.

2 Essences, accidents and holism

"The Aristotelian notion of essence", writes Quine ([42] p. 22), "was the forerunner, no doubt, of the modern notion of [...] meaning. [...] Things had essences, for Aristotle, but only linguistic forms have meanings. Meaning is what essence becomes when it is divorced from the object of reference and wedded to the word". Much here[8] turns on the fact that there is more to the object[9] of reference than just the essence intended — for if the essence exhausted the object why speak of an essence at all. Since there *is* more to it, we can distinguish between essence and the rest or *accident*. Such conceptual distinguishability is undermined, however, by physical inextricability, which produces a measure of logical entanglement too.

---

[7]Similar claims abound in the literature, *e.g.* "A naive holism that supposes theory to confront experience as an unstructured, blockish whole will inevitably be perplexed by the power of scientific argument to distribute praise and to distribute blame among our belief" [25]. See also [28], [29] — Quine replies in [44], Laudan defends Duhem in [36], claiming that Grünbaum has attacked too strong a version of "the Duhemian argument" — [26] and [6].

[8]This section serves only to introduce the next, and a notation, without any pretence of contributing to the abundant literature on the subject.

[9]Here an 'object' will be a *physical* object — even if mathematical objects have been considered ever since the early days of the meaning-reference distinction; see [24, p. 26] for instance.

A given meaning, then, breaks an object up into essential and accidental features, the latter being unintended and *dispensable*, in the sense that without them the object would remain 'what it is' and not be ontologically compromised.

Suppose a word W refers to an object O characterised by certain features $F = \{F_1, F_2, \ldots\}$. Whereas reference catches all the features, essential and accidental, only the essential ones $\bar{F}$ are meant by W. Even if we know that $\bar{F}$ is a proper subset of F, it may be less clear exactly which elements make it up. Hence the following test: remove the features one by one, and see what happens; if F_1 is removed and the object with features $\{F_2, F_3, \ldots\}$ is still intended, F_1 was not essential, and so on. Of course the test cannot be conclusive since the essences $\bar{F}$ are never found on their own, without accidents, some of which will necessarily be tangled up with essences (which could otherwise exhaust the object). Suppose a physical constraint prevents F_m from being separated from F_n. We notice that W still applies when both are present, but that it no longer does once they have been removed. What then? We cannot tell the three cases (1. $F_m \in \bar{F}, F_n \notin \bar{F}$; 2. $F_m \notin \bar{F}, F_n \in \bar{F}$; 3. $F_m \in \bar{F}, F_n \in \bar{F}$) apart and must therefore wonder about *dispensability*; for if a feature F_m cannot be removed without taking something essential with it, in what sense was that feature dispensable and hence accidental? F_m and F_n may be *conceptually* separable, just by thought, but *physical* separation can be considered more trustworthy and "empirical". This entanglement of essence and accident already adumbrates the holisms of Duhem and Quine.

For proper names and single objects the problem is insurmountable. But even if the essential features $\bar{F}$ cannot exist on their own, without accidental ones of some kind or other, they may be found with *different* sets of accidental features: W could refer to *various* objects (which perhaps constitute a 'natural kind'). In other words $\bar{F}$ may be accompanied by the accidents $\{F_1^1, F_2^1, \ldots\}$ or by $\{F_1^2, F_2^2, \ldots\}$ or $\{F_1^3, F_2^3, \ldots\}$ etc., in which case W, while *meaning* $\bar{F}$, would *refer to* object O^1 with features $F^1 = \{\bar{F}, F_1^1, F_2^1, \ldots\}$ or to object O^2 with features $F^1 = \{\bar{F}, F_1^2, F_2^2, \ldots\}$ and so on. Even without knowing the exact makeup of $\bar{F}$ beforehand, it is clearly a subset of $\hat{F} = \bigcap F^i$; and if the family of objects $O^1, O^2, \ldots$ is sufficiently large and the accidental features sufficiently varied, one can reasonably *identify* $\bar{F}$ with $\hat{F}$. The extension of W, if large and varied enough, therefore allows us to determine the intended essence. The idea being that even if that essence cannot be physically abstracted from the bearing object, with all its accidents, it can be abstracted from particular accidents (rather than others); for the distinguished features $\bar{F}$ emerge as the ones belonging to all the objects.

But of course not all linguistic forms are words. Quine seems to have been chiefly concerned with sentences, to whose meaning and reference we now turn.

3 The meaning and reference of sentences

For the empiricists an empirical procedure O was needed to give meaning to an (observation[10]) sentence W. But Quine wonders whether even that will

---

[10]Classification of sentences is not the issue here. Or rather it presupposes distinctions (analytic/synthetic *etc.*) that *are* the issue, and are best approached directly as such, rather

work; for such an O cannot help entangling W with the world in a messy, complicated way, involving all sorts of *unintended* sentences, or rather "collateral"[11] experimental features corresponding to assumptions one might even call "accidental". So, we again have a holistic problem of entanglement: an inextricability of ideal experimental *essence* or *intention* and unavoidable experimental "accidents" needed to implement that intention in the world. This is already reminiscent of the meaning and reference of words, and indeed I will propose a parallel characterisation for sentences, emphasised by a similar notation. Whereas sets and their intersections were enough to separate essences from accidents in my treatment of words, resources from elementary model theory will be used to effect the separation for sentences and the experiments used to test them.

Frege extended his *Sinn-Bedeutung* distinction from words to statements:

> Wir fragen nun nach Sinn und Bedeutung eines ganzen Behauptungssatzes. Ein solcher Satz enthält einen Gedanken. Ist dieser Gedanke nun als dessen Sinn oder als dessen Bedeutung anzusehen?[12]

A few lines on:

> Der Gedanke kann also nicht die Bedeutung des Satzes sein, vielmehr werden wir ihn als den Sinn aufzufassen haben. Wie ist es nun aber mit der Bedeutung? Dürfen wir überhaupt danach fragen? Hat vielleicht ein Satz als Ganzes nur einen Sinn, aber keine Bedeutung?[13]

In due course he answers:

> So werden wir dahin gedrängt, den *Wahrheitswert* eines Satzes als seine Bedeutung anzuerkennen. Ich verstehe unter dem Wahrheitswerte eines Satzes den Umstand, dass es wahr oder dass er falsch ist.[14]

But since the leap from an object to a truth-value is considerable, this seems an unnatural extension — however justified within his scheme — of the nomenclature first adopted for words. Attempting, then, a natural extension of the meaning-reference distinction from words to sentences, I suggest that a single experiment O provides not the *meaning* of a sentence — for the reasons urged by Quine — but something more like its "reference". With the analogy between experiments and physical objects in mind I propose, then, to say that a sentence W *refers* to a specific experiment — to experiment O^1 with features $F^1 = \{\bar{F}, F_1^1, F_2^1, \ldots\}$ or to O^2 with features $F^2 = \{\bar{F}, F_1^2, F_2^2, \ldots\}$ or to O^3 etc. — and that its *meaning* is given by the subset $\bar{F}$ of $\hat{F} = \bigcap F^i$ that

---

than indirectly in a derivative attempt at classifying sentences.

[11] Indeed one is reminded of the "collateral information" of [43], esp. §§9,10.

[12] [24, p. 32]. Quine may be in question, but not the indeterminacy of translation ([43], esp. §§12–16), in acceptance of which quotations have been left in the original. Translation: "We now wonder about the meaning and reference of a whole affirmative sentence. Such a sentence contains a thought. Is this thought to be viewed as its meaning or as its reference?" (The translations are mine.)

[13] Translation: "So the thought cannot be the reference of the sentence, rather we will have to take it as the meaning. What about the reference? Should we wonder about it at all? Does an entire sentence only have a meaning, but no reference?"

[14] [24, p. 34]. Translation: "We will thus be obliged to recognise the *truth-value* of a sentence as its reference. By the truth-value of a sentence I mean the circumstance, that it is true or that it is false".

corresponds to W by expressing an ideal experimental intention, an abstract logical core. It is up to the ingenuity of the experimenters to reduce $\hat{F}$ to $\bar{F}$ by producing enough experiments, with sufficiently varied auxiliary assumptions. Or rather the experimenters begin with the experimental intention $\bar{F}$ expressing W, and then go about finding many different ways to implement it physically. The trouble is that $\bar{F}$ is a tenuous, ideal object, which cannot be performed on its own; auxiliary features[15] of some kind or other are needed to realise it, to bring it about. Quine's point is roughly that W cannot be determined empirically because its counterpart $\bar{F}$ cannot be carried out alone, without accidental auxiliary features, which then confuse the logic of the experiment by unavoidable entanglement with $\bar{F}$.

The various experiments could agree or disagree. Disagreement complicates matters; for then which are to be trusted? Would the majority, or perhaps some privileged experiment or subclass of them, necessarily be right? To avoid such complications unanimity will be required: the experiments must all yield the same verdict.[16] It will then be claimed that, taken together, they are crucial. Such "cruciality" rests on the variety and prior plausibility of the auxiliary assumptions. Variety guarantees independence — for if the assumptions resemble each other too much, agreement will be no surprise[17] — and prior plausibility is inherited from other contexts. So, it will be assumed that the validity of every auxiliary assumption F_b^a made in each experiment O^a was established in several other experimental contexts $\{O_{b1}, O_{b2}, \ldots\}$; and furthermore that validity so established is maintained in the particular experiment O^a; $a, b = 1, 2, \ldots$. The unanimity of the verdict cannot then be reasonably attributed to a conspiracy of the auxiliary assumptions $\{F_1^1, F_2^1, \ldots\}$, $\{F_1^2, F_2^2, \ldots\}$, $\ldots$; it must be due to the experimental intention $\bar{F}$.

Another approach, adopted by Grice and Strawson ([27] p. 156) in response to Quine, is to deal with the troublesome auxiliary statements F_b^a by making "certain assumptions about the[ir] truth-values":

> [...] two statements are synonymous if and only if any experiences which, *on certain assumptions about the truth-values of other statements*, confirm or disconfirm one of the pair, also, *on the same assumptions*, confirm or disconfirm the other to the same degree.

But surely the truth-values of statements are subject to the same holistic entanglement as their meanings. Why should truth-values be less empirical, less susceptible to the intricacies of empirical determination, than meanings? Of course one could, while obliging meanings to maintain the *empirical* grounding that's causing all the trouble, arbitrarily adopt an "ontological" notion of truth and truth-values, admitting the very kind of purely conceptual disentanglement that holism precludes for empirical meanings — but surely the problems at issue here would thereby be left untouched. So one can wonder about the legitimacy of a fine-grained, detailed (ontological) assignment of

---

[15] Auxiliary *features* and *assumptions* seem related closely enough to justify conflation.

[16] Perhaps disagreement is more common or likely than agreement; but unanimous agreement remains possible nonetheless.

[17] As has been pointed out to me by John Earman and John Norton. The standard resources of confirmation theory, such as *probabilities*, have been deliberately avoided here.

individual *truth-values*, within and alongside the messy tangle of (empirical) *meanings*, when the "atoms" of meaning are so much larger than the "atoms" of truth.

In the approach I propose, the unanimity of the verdict provides *a posteriori* support for the prior plausibility of the auxiliary assumptions.

4 Abstract tests

Before attempting a characterisation of abstract tests we note that a similar idea can already be found in *La théorie physique*:

> Pour apprécier la variation de la force électromotrice, il pourra employer successivement tous les types connus d'électromètres, de galvanomètres, d'électrodynamomètres, de voltmètres [...]. *Cependant, toutes ces manipulations, si diverses qu'un profane n'apercevrait entre elles aucune analogie, ne sont pas vraiment des expériences différentes ; ce sont seulement des formes différentes d'une meme expérience ;* les faits qui se sont réellement produits ont été aussi dissemblables que possible ; cependant la constatation de ces faits s'exprime par cet unique énoncé : La force électromotrice de telle pile augmente de tant de volts lorsque la pression augmente de tant d'atmosphères.[18]

An *expérience* here is not a particular real experiment, subject to the difficulties Duhem will raise later, in Ch.VI §§II,III, but a class of equivalent experiments that all test or measure the same thing. Such an abstract experiment can be associated with the class of its *formes différentes* in the same way a *theory* (in the logical, Tarskian sense) can be identified with all its *models*. The accidental and logically confusing peculiarities of particular implementation are thus transcended.

There is something similar in *Word and object* too: "We may begin by defining the affirmative stimulus meaning of a sentence [...] as the class of all the stimulations [...] that would prompt [...] assent" [43, p. 32]. A couple of pages on:

> [...] a stimulation must be conceived for these purposes not as a dated particular event but as a universal, a repeatable event form. We are to say not that two like stimulations have occurred, but that the same stimulation has recurred. Such an attitude is implied the moment we speak of sameness of stimulus meaning for two speakers[43, p. 34][19].

Here the models are the 'repetitions' of the "repeatable event form."

So both Duhem and Quine have in mind an abstract test — an abstract *expérience*, a universal, a repeatable form — with many particular realisations. It is in such tests that the desired cruciality will be sought.

---

[18][21, p. 224]; emphasis mine. Translation: "To appreciate the variation of electromotive force, he can employ in succession all the known kinds of eletrometers, galvanometers, eletrodynamometers, voltmeters [...]. *However, all these manipulations, so different that a layman would see no analogy among them, are not really different experiments; they are only different forms of a single experiment*; the facts that really occurred were as different as possible, but can nonetheless be expressed in the same way: The electromotive force of such and such a battery increases by so many volts when the pressure increases by so many atmospheres".

[19]Quine argues, especially in [43] §§11,12, that stimulus meaning does not fix meaning well enough for all purposes and criteria. But his reservations, which regard behavioural linguistics, need not concern us here, especially as his characterisation of stimulus meaning is being taken only as a hint or rough ancestor.

One can wonder about appropriate formalisation, for the notion is nebulous and of little use as it stands. What the various realisations of an abstract test have in common is *structure*[20] of some sort; it is in that sense that they all test the same thing. But there remains the matter of what exactly "structure" is. The ordinary connotations of the word will hardly do; Duhem and Quine, who speak of *form*, provide little help. Specification of a means of description can clarify: of the many available ways of characterizing structure, the resources of set-theoretical axiomatisation, associated chiefly with Patrick Suppes (*e.g.* [48]) seem appropriate and will be used. In his language a set-theoretical predicate defines a *theory*, satisfied by *models*, whereas here the predicate will characterise an *abstract test*, again satisfied by *models*. It is the abstract test, rather than any particular model, that represents a crucial experiment. Auxiliary assumptions have admittedly to be made in each individual implementation, but again, they can be required to vary widely over the class, and to have a plausibility derived from other contexts.

The idea can be formalised by spelling out a set-theoretical predicate, after the manner of Suppes: a string $(A, B, \dots)$ of primitive notions "is an X", for instance, if certain axioms, say

1. A is a nonempty finite set.

2. The function $B : A \to \mathbb{R}^+$ is differentiable and ...

3. ...

$$\vdots$$

are satisfied. Any such particular $O^a = (A^a, B^a, \dots)$ satisfying the axioms is a *model*. The extension of the predicate 'is an X' is the set $\{O^1, O^2, \dots\}$ of models.

We are again dealing with essences and accidents, in the sense that a set-theoretical predicate defines the "essence" $\bar{F}$ common to all the models. Essential and accidental features are entangled, and indeed can be hard to tell apart, in any particular model O^a, which has its own contingent peculiarities $\{F_1^a, F_2^a, \dots\}$ in addition to the common, essential core $\bar{F}$ determined by the axioms. But once that model is considered alongside others, essences can be made out as what is common to all of them. The abstract test $\bar{F}$, in other words the set-theoretical predicate "is an $\bar{F}$-test", therefore gives the *meaning* of the sentence W, which *refers* to any model O^a of the test.

The cleavage between mathematics and physics (dwelt on in [21, Ch.VI §III]) is largely overcome by such abstract tests, which, being mathematical objects in themselves — despite having physical models — give physics much of the rigid necessity of mathematics. In §7 I consider the differences Duhem attributes to mathematics and physics, in §6 the way Quine links analyticity and "reductionist" meanings, after a much-needed example.

---

[20] In the logical literature "structure" is often a synonym of "model", whereas here its meaning is closer to that of "theory".

5 Example: Bell's inequality

If ever a scientific controversy stood sorely in need of experimental arbitration, the dispute over the foundations of quantum mechanics that developed around the positions of Einstein (*e.g.* [22]) and Bohr (*e.g.* [16] or [17]) certainly did (and still does). There have been celebrated efforts to satisfy the need; experiments to test Bell's inequality ([12], [14], and also [5]) by Alain Aspect and others (*e.g.* [8], [9], [10], [11], [18], [40], [50]) have been among the most spectacular and controversial attempts at empirical discrimination. But far from settling the debate they have given it new life and vigour ...

The hope at any rate was this: "Supposez [to follow Duhem] que deux hypothèses seulement soient en présence; [local realism is either valid or not] cherchez des conditions expérimentales telles que l'une des hypothèses annonce la production d'un phénomène et l'autre la production d'un phénomène tout différent; [Bell's inequality is either satisfied or violated] réalisez ces conditions et observez ce qui se passe; selon que vous observerez le premier des phénomènes prévu ou le second, vous condamnerez la seconde hypothèse ou la première; celle qui ne sera pas condamnée sera désormais incontestable; le débat sera tranché, une vérité nouvelle sera acquise à la Science."[21] Of course such conclusions are unwarranted, resting on assumptions that may be no less questionable than the principles supposedly refuted. Bell [13] for instance "always emphasize[d] that the Aspect experiment is too far from the ideal in many ways — counter efficiency is only one of them", and "that there is therefore a big extrapolation from practical present-day experiments to the conclusion that nonlocality holds".

Most attempts to test Bell's inequality, such as those of Aspect *et al.*, have involved photons, but these are seldom detected; this is the issue of "counter efficiency" referred to by Bell. To violate a Bell inequality with photons, assumptions (*i.e.* accidental features F_b^a) like

> Given a pair of photons emerging from two regions of space where two polarizers can be located, the probability of their joint detection from two photomultipliers [...] does not depend on the presence and the orientation of the polarizers. [19]

or

> The set of detected pairs with a given orientation of the polarizers is an undistorted representative sample of the set of pairs emitted by the source. [8]

have to be made. For our purposes they are equivalent, and give rise to the same consequences: they multiply the interval figuring in the inequality by the product of the efficiencies of the counters. The assumptions turn an interval running from -1 to 1, for instance, into one running from $-\eta_1\eta_2$ to $\eta_1\eta_2$ where η_1 and η_2 are the efficiencies. If the counters are relatively efficient, and each detect, say, a photon in four, the assumptions make the inequality

---

[21][21, p. 286]. Translation: "Suppose only two hypotheses are at issue; seek experimental conditions such that one of the hypotheses leads to the production of one phenomenon and the other to the production of a completely different phenomenon; realise these conditions and observe what happens; according to whether you observe the first of the predicted phenomena or the second, you will condemn the second hypothesis or the first; the one that will not be condemned will be incontestable; the issue will be settled, and Science will have a new truth".

sixteen times easier to violate.[22] This is the idea: Averaging involves adding up N terms, then dividing by N. But what if most of the terms are "duds", and do not contribute to the sum? Surely dividing by N is excessive; does it not make more sense to divide by the number of valid terms instead? In other words only a small fraction of the pairs get detected, so why not take that same fraction of the interval? After all, why should the sample not be representative of the whole population? Surely the photomultipliers act randomly and indiscriminately...

A sample that is almost the size of the whole population will clearly be very representative, whereas a much smaller sample may or may not be. Consider the assumption:

> For every photon in the state λ the probability of detection with a polarizer placed on its trajectory is less than or equal to the detection probability with the polarizer removed. [18]

The trouble is that the polarizer might *increase* the probability of detection, especially if that probability depends on the state λ, which could be altered by the polarizer. Suppose "detector" denotes both a vertically aligned polarizer π and a photomultiplier φ behind it. So a 'detection' involves both objects that make up the detector $\pi + \varphi$: a photon is detected when it gets through π *and* makes φ click. As horizontally polarized light will never get detected by $\pi + \varphi$ — its probability of detection vanishes — an oblique polarizer placed in front of π *increases* the probability of detection.

So if the experiment produces a number lying outside the narrow interval running from $-\eta_1\eta_2$ to $\eta_1\eta_2$, what is to be concluded?

Uncertainties concerning the particular additional assumptions made vitiate comprehensive statements an experiment may inspire, like "Bell's inequality is violated in nature". Who knows if the outcome really means that — and not the unfoundedness of this or that additional assumption instead. If kaons are used rather than photons, probability of detection, being very high, is no longer the issue; but their instability leads to other assumptions (see [1], [2]) of a completely different sort; and so on. Hence the abstract test, and the corresponding class of structurally equivalent experiments, with a whole range of different auxiliary assumptions: surely they cannot *all* be wrong.

Turning to the abstract test (*cf.* [3], [4]) itself, a *Bell test* will be a scheme

$$(\Xi, \Omega^s(k), \underline{\sigma}_n^s(k), \underline{B}; |\sigma\rangle, \sigma_n^s, B)$$

satisfying the following axioms:

1. $\Xi = \{(\Omega^1(1), \Omega^2(1)), \ldots, (\Omega^1(N), \Omega^2(N))\}$ is a large ensemble of pairs of objects.

2. Object $\Omega^s(k)$ has an intrinsic property $\underline{\sigma}_n^s(k) = \pm 1$ for every value of $n \in \mathbb{R}$.

3. $\underline{B} = \sum_{k=1}^{N} \{\underline{\sigma}_\alpha^1(k)\,\underline{\sigma}_\beta^2(k) - \underline{\sigma}_\alpha^1(k)\,\underline{\sigma}_{\beta'}^2(k) + \underline{\sigma}_{\alpha'}^1(k)\,\underline{\sigma}_\beta^2(k) + \underline{\sigma}_{\alpha'}^1(k)\,\underline{\sigma}_{\beta'}^2(k)\}/N$.

---

[22] Franco Selleri expresses this by distinguishing between *strong* and *weak* inequalities, described in [37], [5] and [2].

4. Ξ is accurately described by the quantum state vector[23]

$$|\sigma\rangle = \frac{1}{\sqrt{2}}(|+-\rangle - |-+\rangle) \in \mathbb{C}^{2(1)} \otimes \mathbb{C}^{2(2)},$$

where the $|\pm\rangle$ are orthonormal, and both Hilbert spaces $\mathbb{C}^{2(s)}$ are two-dimensional.

5. $B = \sigma_\alpha^1 \otimes \sigma_\beta^2 - \sigma_\alpha^1 \otimes \sigma_{\beta'}^2 + \sigma_{\alpha'}^1 \otimes \sigma_{\beta'}^2 + \sigma_{\alpha'}^1 \otimes \sigma_{\beta'}^1$, where $\sigma_n^s : \mathbb{C}^{2(s)} \to \mathbb{C}^{2(s)}$ is self-adjoint and unitary, with vanishing trace.

6. Measurement of σ_n^s faithfully reveals property $\underline{\sigma}_n^s(k)$, for all k, n (and both values of s).

The models of the axioms make up the extension of the predicate 'is a Bell test.' Here the *essence*, the *experimental intention* $\bar{F}$ is the abstract Bell test, and a fair sampling assumption like "The set of detected pairs with a given *etc.*" above would be one of the accidents $\{F_1^a, F_2^a, \ldots\}$ of a model O^a.

Leaving aside other difficulties — like the precarious counterfactual thinking required by axiom 6 — which would lead us too far astray, the axioms are inconsistent. The notation adopted in axioms 2 and 3, with just a single subscript, tacitly expresses a further axiom, say 7, by suggesting that property $\underline{\sigma}_n^s(k)$ only depends (once k and s have been fixed) on its subscript n, *and not on the subscript of the neighbouring factor*. This allows us to write

$$\underline{B} = \frac{1}{N} \sum_{k=1}^{N} [\underline{\sigma}_\alpha^1(k)\{\underline{\sigma}_\beta^2(k) - \underline{\sigma}_{\beta'}^2(k)\} + \underline{\sigma}_{\alpha'}^1(k)\{\underline{\sigma}_\beta^2(k) + \underline{\sigma}_{\beta'}^2(k)\}],$$

whose modulus cannot exceed 2, for purely arithmetical reasons. But it follows from axioms 4 and 5 that $\max(\langle\sigma|B|\sigma\rangle) = 2\sqrt{2}$; from axioms 3, 5, 6 (& 1, 2, 4) that $\langle\sigma|B|\sigma\rangle = \underline{B}$; from 4, 5, 6 (& 1, 2, 3) that $\max(\underline{B}) = 2\sqrt{2}$; and from 3, 5, 6, 7 (& 1, 2) that $-2 \leq \langle\sigma|B|\sigma\rangle \leq 2$. So we have all sorts of contradictions.

One approach would be to view the inconsistency as expressing the tension at issue, perhaps as representing a corresponding 'inconsistency' of nature itself. Of course if a model is a scheme *satisfying* the axioms, both 'model' and 'satisfaction' have to be understood in appropriately weakened, generalised senses.

The contradictory set has the advantage of allowing us to choose which axiom(s) — 2, 4, 6 or 7 — to blame, but it nevertheless remains simplest to make the axioms consistent by abandoning an axiom, say 4 or 6. Once consistent the axioms admit normal, classical models, in fact quite a variety of them, involving angles, polarizers and photons; or times and precessions generated by appropriate fields; or kaons and strangeness; and so forth — each with its own peculiar additional assumptions.

---

[23]The phase difference of π, which may seem an unduly strong requirement, is not the point here.

6 Quine on meaning, synonymy and analyticity

Let us now return to Quine, who by linking meaning, synonymy and analyticity argues that holism undermines analyticity along with meaning. We have already seen what holism has to do with meaning, and will now consider, with little comment, how Quine associates meaning, synonymy and analyticity. In "Two dogmas" [42, p. 22] he explicitly connects all three:

> Once the theory of meaning is sharply separated from the theory of reference, it is a short step to recognizing as the primary business of the theory of meaning simply the synonymy of linguistic forms and the analyticity of statements [...].

Fifteen pages on: "The verification theory of meaning [...] is that the meaning of a statement is the method of empirically confirming or infirming it", so that "[...] what the verification theory says is that statements are synonymous if and only if they are alike in point of method of empirical confirmation or infirmation"; meaning and synonymy are thus brought together through verificationist "reductionism". Reductionism also yields analyticity: "So, if the verification theory can be accepted as an adequate account of statement synonymy, the notion of analyticity is saved after all" [42, p. 38]. Analyticity and synonymy are again linked in *Word and object* [43, p. 65]:

> [...] synonymy [...] is interdefinable with another elusive notion of intuitive philosophical semantics: that of an *analytic* sentence. [...] The interdefinitions run thus: sentences are synonymous if and only if their biconditional (formed by joining them with "if and only if") is analytic, and a sentence is analytic if and only if synonymous with self-conditionals ("If p then p").

But again, this is not the place to dispute Quine's association of meaning, synonymy and analyticity, which will be taken for granted.

To understand whether holism really has conflicting implications for Duhem and for Quine, let us now see how Duhem relates the impossibility of crucial experiments to the 'cleavage' separating mathematics and physics.

7 Duhem on mathematics, physics and crucial experiments

Whereas Quine rejects the "cleavage between analytic and synthetic truths" (dogma1) along with "reductionism" (dogma2), Duhem's argument (against dogma2) *turns* (dogma1 $\Rightarrow$ ¬dogma2 ?) on a similar cleavage (dogma1?): over and over he returns to the troublesome "synthetic" character of physics by contrasting it with the clean necessity of mathematics (*cf.* [39, pp.109–11]) — in which analytic truths can be claimed to figure conspicuously, indeed paradigmatically.[24] Experimental refutation is often taken to be just like *reductio ad absurdum*:

---

[24]Until the difficulties and paradoxes that arose around the beginning of the twentieth century, mathematics was a paradigm of necessity. See [31], for instance, on the certainties of geometry: "Unter allen Zweigen menschlicher Wissenschaft gibt es keine [...] von deren vernichtender Aegis Widerspruch und Zweifel so wenig ihre Augen aufzuschlagen wagten. Dabei fällt ihr in keiner Weise die mühsame und langwierige Aufgabe zu, Erfahrungsthatsachen sammeln zu müssen, wie es die Naturwissenschaften im engeren Sinne zu thun haben, sondern die ausschliessliche Form ihres wissenschaftlichen Verfahrens ist die Deduktion. Schluss wird aus Schluss entwickelt ..."

> La réduction à l'absurde, qui semble n'être qu'un moyen de réfutation, peut devenir une méthode de démonstration ; pour démontrer qu'une proposition est vraie, il suffit d'acculer à une conséquence absurde celui qui admettrait la proposition contradictoire de celle-là ; on sait quel parti les géomètres grecs ont tiré de ce mode de démonstration. Ceux qui assimilent la contradiction expérimentale à la réduction à l'absurde pensent qu'on peut, en Physique, user d'un argument semblable à celui dont Euclide a fait un si fréquent usage en Géométrie.[25]

A few pages on Duhem points out that — quite apart from the rôles and validity of other assumptions — the *tertium non datur* usually assumed in mathematics does not hold in physics, where statements can be negated in many different ways:

> Mais admettons, pour un instant, que, dans chacun de ces systèmes, tout soit forcé, tout soit nécessaire de nécessité logique, sauf une seule hypothèse ; admettons, par conséquent, que les faits, en condamnant l'un des deux systèmes, condamnent à coup sûr la seule supposition douteuse qu'il renferme. En résulte-t-il qu'on puisse trouver dans l'*experimentum crucis* un procédé irréfutable pour démontrer en vérité démontrée l'une des deux hypothèses en présence, de même que la réduction à l'absurde d'une proposition géométrique confère la certitude à la proposition contradictoire ? Entre deux théorèmes de Géométrie qui sont contradictoires entre eux, il n'y a pas place pour un troisième jugement ; si l'un est faux, l'autre est nécessairement vrai. Deux hypothèses de Physique constituent-elles jamais un dilemme aussi rigoureux ? Oserons-nous jamais affirmer qu'aucune autre hypothèse n'est imaginable ?[26]

Not only does *tertium non datur* not hold in physics, the possibilities of negation are limitless: the negation $\neg H$ of hypothesis H can suggest, say, another hypothesis $H' = \neg H$; but why not some other $H'' = \neg H$ or $H''' = \neg H$ or who knows what else. So even if it were possible to refute a hypothesis in physics, its refutation would certainly not lead to the confirmation of another hypothesis — whereas the rejection of a hypothesis in mathematics typically allows a single, definite conclusion to be reached.

> La contradiction expérimentale n'a pas, comme la réduction à l'absurde employée par les géomètres, le pouvoir de transformer une hypothèse physique en une vérité incontestable ; pour le lui conférer, il faudrait énumerer complètement les diverses hypothèses auxquelles un groupe déterminé de phénomènes peut donner lieu ; or, le

---

[25] [21, p. 285]. Translation: "*Reductio ad absurdum*, which only appears to be a way of refuting, can become a method of demonstration; to demonstrate that a proposition is true, it is enough to push him who would assume the contrary proposition back to an absurd consequence; one knows what use the Greek geometers made of this mode of demonstration. Those who associate experimental contradiction with *reductio ad absurdum* think that one can, in physics, use an argument similar to the one Euclid used so often in geometry". Also [21, p. 280]: "Un pareil mode de démonstration semble aussi convaincant, aussi irréfutable que la réduction à l'absurde usuelle aux géomètres ; c'est, du reste, sur la réduction à l'absurde que cette démonstration est calquée, la contradiction expérimentale jouant dans l'une le rôle que la contradiction logique joue dans l'autre".

[26] [21, p. 288]. Translation : "But let us assume, for a moment, that, in each of these systems, all is forced, all is necessary of logical necessity, except a single hypothesis; let us assume, as a consequence, that the facts, by condemning one of the two systems, condemn with certainty the only doubtful supposition it contains. Does it follow that one can find in the *experimentum crucis* an irrefutable procedure to transform one of the two hypotheses at issue into a demonstrated truth, in the same way that the *reductio ad absurdum* of a geometrical proposition confers certainty on the contradictory proposition? Between two theorems of geometry that contradict one another, there is no room for a third judgement; if one is false, the other is necessarily true. Do two hypotheses of physics ever constitute so rigorous a dilemma? Would we ever dare to claim that no other hypothesis can be imagined?"

physicien n'est jamais sur d'avoir épuisé toutes les suppositions imaginables ; la vérité d'une théorie physique ne se décide pas à croix ou pile.

So Duhem's rejection ($\neg$ dogma2) of crucial experiments turns on a 'cleavage' which resembles the one (dogma1) repudiated in "Two dogmas", where it is claimed the dogmas are "two sides of a single dubious coin" (dogma1 $\Leftrightarrow$ dogma2).

Since the holism Duhem dwells on in Ch. VI §II (*Qu'une expérience en Physique ne peut jamais condamner une hypothèse isolée, mais seulement tout un ensemble théorique*) appears to be largely responsible for the cleavage invoked repeatedly in the following section, §III (*L'experimentum crucis est impossible en physique*), it could seem that overcoming holism would undermine that cleavage. This brings us to the difficulty raised at the beginning: that holism appears to have conflicting implications for Duhem and for Quine. In this connection let us briefly consider relations between Duhem's §II and §III (Ch. VI).

One relation is immediate succession — §III comes right after §II; another is that both are about crucial experiments. §II explains how holism prevents experiments from being crucial, the next section directly relates the impossibility of crucial experiments to the cleavage dividing physics and mathematics; one almost sees a simple syllogism:

II *Holism prevents experiments from being crucial.*

III *The impossibility of crucial experiments makes physics unlike mathematics.*

∴ *Holism makes physics unlike mathematics.*

The trouble is that the differences between physics and mathematics are only partly due to holism; single-valued, invertible negation,[27] for instance, which holds in mathematics but not in physics according to Duhem, has little to do with holism. Holism, which for Quine undermines meaning and hence analyticity, is therefore not entirely responsible for the cleavage repeatedly invoked by Duhem in his rejection of crucial experiments.

It must also be said that mathematics may not be as analytic as I have taken it to be; Kant and others have regarded much of it as synthetic. Kant [32, B190] defines the synthetic in terms of the principle of contradiction:

Der Satz nun: Keinem Dinge kommt ein Prädikat zu, welches ihm widerspricht, heit der Satz des Widerspruchs [...]. Denn, wenn *das Urteil analytisch ist*, es mag nun verneinend oder bejahend sein, so mu dessen Wahrheit jederzeit nach dem Satze des Widerspruchs hinreichend können erkannt werden.[28]

And Poincaré [41, Ch. 1] writes that the "règle du raisonnement par récurrence" — which he considers the raisonnement mathématique par excellence — is

---

[27]One can write $\neg(\neg H) = H$.

[28]Translation: "Now the statement that nothing can have a predicate which contradicts it, is called the principle of contradiction [...]. For if *the judgement is analytic* — be it negative or affirmative — its truth must always be adequately recognised by means of the principle of contradiction".

"irréductible [involving infinitely many syllogisms] au principe de contradiction", and hence is the "véritable type du jugement synthétique a priori". Crowe [20] argues that mathematics shares many of the difficulties attributed to physics in *La théorie physique*, and that Duhem attaches such weight to the distinctions of §III out of ignorance that mathematics is not so certain and 'analytic' after all. Physics, by becoming more and more detached from the world, seems moreover to be losing its synthetic character, and may have begun decades ago. The association of mathematics with the analytic, physics with the synthetic, could therefore be less straightforward than I have made it out to be. But again, definite resolution has not been my purpose; I have rather tried to explore the web of issues involved, and view in a fresh — perhaps questionable — light.

8 Final remarks

Troubling shades of grey have prevailed in these pages over the reassuring certainties of black and white. I have often spoken of degree and nuance, of more and less, rather than of *sic et non*, of true and false: holism is undermined but not completely eradicated, meaning acquires much definiteness, analyticity is recovered to the extent that holism is overcome and so on. But isn't Quine's point that analytic and synthetic differ in degree and not in kind?

The gains in cruciality and analyticity with respect to the concerns of Duhem and Quine may be a matter of degree, but that degree seems considerable, perhaps considerable enough to warrant representation as promotions "in kind". Indeed it can be misleading not to view certain differences in degree as differences in kind — and hence, for instance, not to call the unlikeliest events "impossible", to distinguish clearly from those that are only moderately unlikely. Nuances within a small enough "margin of discrimination" can be safely ignored.

As the distances from the ideal Platonic limits of absolute cruciality and analyticity can be made logically, conceptually negligible by a proper extrication of essences from accidents, the contested notions can reasonably be countenanced.

Acknowledgements

I thank Mario Alai, Mark Colyvan, Gabriele De Anna, John Earman, Michael Esfeld, Janet Kourany, Federico Laudisa, Rossella Lupacchini, John Norton, Ofra Rechter and Nino Zanghì for many fruitful discussions; an anonymous referee, for various suggestions; and the Center for Philosophy of Science, University of Pittsburgh, where as Visiting Fellow I began work on this project.

BIBLIOGRAPHY

[1] A. Afriat. Kaons and Bell's inequality. In C. Garola and A. Rossi (eds.) *The Foundations of Quantum Mechanics: Historical Analysis and Open Questions*. World Scientific, Singapore 2000.
[2] A. Afriat. *Waves, Particles and Configuration Space*. Accademia Nazionale delle Scienze, Lettere ed Arti, Modena 2001.
[3] A. Afriat. Bell's inequality with times rather than angles. *Journal of Modern Optics*, 50: 1063–9, 2003

[4] A. Afriat. Altering the remote past. *Foundations of Physics Letters*, 16: 293–301, 2003
[5] A. Afriat and F. Selleri. *The Einstein, Podolsky and Rosen Paradox*. Plenum, New York 1998.
[6] M. Alai. *Modi di conoscere il mondo*. Franco Angeli, Milano 1994.
[7] R. Ariew. The Duhem thesis. *British Journal for the Philosophy of Science*, 35: 313–25, 1984.
[8] A. Aspect. *Trois tests expérimentaux des inégalités de Bell par mesures de correlation de polarisation de photons*. Thèse, Université de Paris-Sud, Centre d'Orsay 1983.
[9] A. Aspect, G. Grangier and G. Roger. Experimental tests of realistic local theories via Bell's theorem. *Physical Review Letters*, 47: 460–3, 1981.
[10] A. Aspect, Grangier and G. Roger. Experimental realization of Einstein-Podolsky-Rosen-Bohm Gedankenexperiment: a new violation of Bell's inequalities. *Physical Review Letters*, 49: 91–4, 1982.
[11] A. Aspect, J. Dalibard and G. Roger Experimental test of Bell's inequalities using time-varying analyzers. *Physical Review Letters*, 49: 1804–7, 1982.
[12] J.S. Bell. On the Einstein-Podolsky-Rosen paradox. *Physics*, 1: 195–200, 1965.
[13] J.S. Bell. Letter to E. Santos dated 22 December 1986.
[14] J.S. Bell. *Speakable and Unspeakable in Quantum Mechanics*. Cambridge University Press, Cambridge 1987.
[15] A. Boghossian. Analyticity reconsidered. *Noûs*, 30: 360–91, 1996.
[16] N. Bohr. Quantum mechanics and physical reality. *Nature*, 136: 65, 1935.
[17] N. Bohr. Can quantum-mechanical description of physical reality be considered complete? *Physical Review*, 48: 696–702, 1935.
[18] J.F. Clauser and M.A. Horne. Experimental consequences of objective local theories. *Physical Review D*, 10: 526–35, 1974.
[19] J.F. Clauser and M.A. Horne, A. Shimony and R.A. Holt. Proposed experiment to test hidden variable theories. *Physical Review Letters*, 23: 880–4, 1969.
[20] M.J. Crowe. Duhem and history and philosophy of mathematics. *Synthese*, 83: 431–7, 1990.
[21] P. Duhem. *La théorie physique*. Vrin, Paris 1989.
[22] A. Einstein, B. Podolsky, and N. Rosen. Can quantum-mechanical description of reality be considered complete? *Physical Review*, 47: 777–80, 1935.
[23] M. Esfeld. *Holism in Philosophy of Mind and Philosophy of Physics*. Kluwer, Dordrecht 2001.
[24] G. Frege. Über Sinn und Bedeutung. *Zeitschrift für Philosophie und philosophische Kritik*, 100: 25–50, 1892.
[25] C. Glymour. Relevant evidence *The Journal of Philosophy*, 72: 403–26, 1975.
[26] C. Glymour. *Theory and Evidence*. Princeton University Press, Princeton 1980
[27] H. Grice and F. Strawson. In defense of a dogma *Philosophical Review*, 65: 141–58, 1956.
[28] A. Grünbaum. The Duhemian argument. *Philosophy of Science*, 27: 75–87, 1960.
[29] A. Grünbaum. The falsifiability of theories: total or partial? A contemporary evaluation of the Quine-Duhem thesis. *Synthese*, 14: 17–34, 1962.
[30] L. E. Hahn and A. Schilpp (eds.). *The philosophy of W.O. Quine*. Open Court, La Salle 1986.
[31] H. von Helmholtz. *Über den Ursprung und die Bedeutung der geometrischen Axiome*. Vortrag gehalten im Dozentenverein zu Heidelberg, 1870
[32] I. Kant. *Kritik der reinen Vernunft*. Hartknoch, Riga 1787.
[33] J. Katz. Some remarks on Quine on analyticity. *Journal of Philosophy*, 64: 36–52, 1967.
[34] J. Katz. Where things stand on the analytic-synthetic distinction. *Synthese*, 28: 283–319, 1974.
[35] H. Krips. Epistemological holism: Duhem or Quine? *Studies in History and Philosophy of Science*, 13: 251–64, 1982.
[36] L. Laudan. Grünbaum on 'The Duhemian argument'. *Philosophy of Science*, 32: 295–9, 1965.
[37] V.L. Lepore and F. Selleri. Do performed optical tests refute local realism? *Foundations of Physics Letters*, 3: 203–20, 1990.
[38] B. Mates. Analytic sentences. *Philosophical Review*, 60: 525–34, 1951.
[39] Needham. Duhem and Quine. *Dialectica*, 54: 109–32, 2000.
[40] W. Perrie, A.J. Duncan, H.J. Beyer and H. Kleinpoppen. Polarization correlation of the two photons emitted by metastable atomic deuterium: a test of Bell's inequality. *Physical Review Letters*, 54: 1790–3, 1985
[41] J.-H. Poincaré. *La science et l'hypothèse*. Flammarion, Paris 1902.

[42] W.V.O. Quine. Two dogmas of empiricism. In *From a Logical Point of View*. Harvard University Press, Cambridge MA, 1953.
[43] W.V.O. Quine. *Word and Object*. Wiley, New York 1960.
[44] W.V.O. Quine. A comment on Grünbaum's claim. In S. G. Harding (ed.). *Can Theories Be Refuted?*. Reidel, Dordrecht 1976.
[45] W.V.O. Quine. Reply to Jules Vuillemin. In L.E. Hahn and A. Schilpp (eds.), *The philosophy of W.O. Quine*, pages 619–22. Open Court, La Salle 1986.
[46] F. Strawson. A logician's landscape. *Philosophy*, 30: 229–37, 1955.
[47] F. Strawson. Paradoxes, posits and propositions. *The Philosophical Review*, 76: 214–9, 1967.
[48] P. Suppes. *Representation and Invariance of Scientific Structures*. CSLI, Stanford 2002.
[49] J. Vuillemin. On Duhem's and Quine's theses. In n L. E. Hahn and A. Schilpp (eds.), *The philosophy of W.O. Quine*, pages 595–618. Open Court, La Salle 1986.
[50] T. Walther and E.S. Fry. A Bell inequality experiment based on molecular dissociation – extension of the Lo-Shimony proposal to 199Hg (nuclear spin-$\frac{1}{2}$) dimers. In R.S. Cohen, M. Horne and J. Stachel (eds.). *Experimental metaphysics*. Kluwer, Dordrecht 1997.

Bionic simulation of biological systems: a methodological analysis

EDOARDO DATTERI

1 Introduction

Robotic simulations are extensively used to discover and test hypotheses on the mechanisms of biological behaviours (examples include studies on sensory-motor behaviours of rats [3], crickets [23], lobsters [14], locusts [2], and human beings [6]). In these investigations, comparisons between robotic and biological behaviours are taken as empirical basis for corroborating or rejecting biological hypotheses. The experimental methodology followed there is a robotic variant of more traditional computer-based simulation approaches which do not involve electro-mechanical devices [1].

Other machine simulation strategies, which have been adopted in recent studies on lamprey sensory-motor behaviours, are distinctively based on systems obtained by connecting biological and artificial devices.[1] Systems of this kind are often called *bionic systems* (BSs from now on). Research on BSs is primarily devoted to therapeutical purposes, insofar as bionic technologies can be fruitfully deployed to restore lost or injured sensory-motor capabilities in humans [15]. It has been claimed, however, that BS technologies may provide novel and powerful experimental tools for modelling neural systems. According to John Chapin [5, p. 669], "the general strategy ... of using brain-derived signals to control external devices may provide a unique new tool for investigating information processing within particular brain regions". Similarly, Miguel Nicolelis [19, p. 417] claims that brain-machine interfaces "can become the core of a new experimental approach with which to investigate the operation of neural systems in behaving animals".

Claims on the experimental potentialities of machine simulations in the theoretical modelling of biological systems have been often advanced since the beginning of the XX century [8, 24, 26]. However, these claims are only rarely supported by an analysis of the methodological difficulties and the related *auxiliary assumptions* that are needed to draw theoretical conclusions on the basis of simulation behaviours. Corroborating these assumptions, which are often neglected in simulation studies, is crucial in the machine-based theorizing on biological behaviours[10]. Methodological analyses of this kind are conceptually akin to philosophical reflections on experiments in other areas of scientific research (including physics [12]).

A methodological analysis of bionic experiments will be pursued here, with reference to classes of BS-based experiments whose distinguishing feature is

---
[1] This work focuses on *invasive* connections [18].

the use of BSs obtained by connecting parts of the target biological system and machine (robot and computer-based) *simulations* of other parts of the target system.[2] An analysis of the methodological difficulties that arise in the setting-up and performing of BS-based simulation experiments will reveal a number of background assumptions that are crucial to the BS-based theorizing on biological behaviours. This analysis may also contribute to understanding what kind of theoretical conclusions on the mechanisms of adaptive biological behaviours may flow from BS-based simulation experiments, and to adding significant methodological caveats on the aforementioned Chapin's and Nicolelis' claims on the potentialities of bionic experiments in neuroscientific research.

These issues will be addressed here with reference to simulation BS-based inquiries on the mechanisms of body stabilization in lampreys. The *explanandum* and a mechanistic hypothesis formulated there is illustrated in section 2. Sections 3 and 4 discuss the role of two classes of simulation BS-based experiments in testing the mechanistic hypothesis at stake. These sections will also address a number of methodological issues, mainly concerning the relationship between the biological hypothesis and the structure of the BS involved in the experiments. Such issues become integral part of background assumptions that are needed to draw theoretical conclusions on the target biological system on the basis of an analysis of BS behaviours. Significant methodological commonalities between BS-based experiments and other simulation approaches involving fully artificial systems will be revealed in the ensuing analysis. Addressing these issues may contribute to developing regulative methodological principles for "good" simulation BS-based experiments and, at large, to understanding the potentialities of BS-based experiments in biological theorizing.

2 The explanandum: body stabilization in lampreys

During navigation, lampreys are able to maintain fixed roll and pitch angles, against external deviations, by means of tail, dorsal fin, and other movements of the body. This regular behaviour is hypothesized to be generated by a mechanism involving parts of the lamprey sensory-motor system, notably the following: left and right vestibuli, which detect changes in orientation; neurons in the left and right vestibular nuclei (including the Intermediate Octavomotor nuclei, or nOMI, and Posterior Octavomotor nuclei, or nOMP), which receive vestibular input; neurons in various reticular nuclei of the brainstem, including the posterior rhombencephalic reticular nucleus or PRRN, whose activity is elicited by vestibular neurons; neurons in the spinal cord, which receive reticular stimulation and modulate the behaviour of the motor organs of the animal (see fig. 1).

In 1992, Orlovsky and collaborators [20] mentioned the following open issues concerning this mechanism:

> We wanted to understand (1) What signals are coming from the vestibular sensory

---

[2]Bionic technologies enable a variety of non-simulative experiments, that are invasive variants of more traditional techniques for analyzing and manipulating neural activity. Some of these experimental possibilities are discussed in [19].

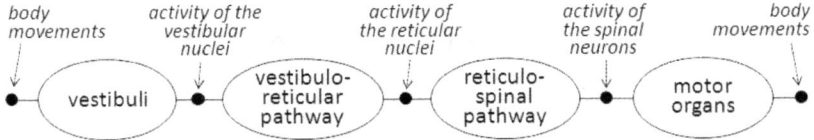

Figure 1. Components of the mechanism of body stabilization in lampreys

organs when the orientation of the animal in the gravity field is changing; (2) How this information is processed in the brainstem and what commands the brainstem sends to the spinal cord; (3) How the spinal motor mechanisms respond to commands coming from the brain, and what motor pattern is used for correcting the body orientation.[20, p. 479]

Bionic systems have been deployed to address some of these issues. In particular, vestibular processing in the brainstem has been investigated in the BS-based inquiries reported in [22, 17], while the causal relationship between reticular neurons and overt behaviours, through the intermediary of spinal neurons, is studied in [27]. These case-studies illustrate two classes of simulation BSs. Systems belonging to one of these classes (called ArB here for short) are obtained by replacing a component of the biological target system with an artificial component, while systems in the other class (called BrA here) are obtained by replacing part of an artificial system, such as a robot, with a biological component. The ensuing sections illustrate these experimental inquiries, analyze the contribution of the BSs involved in corroborating or rejecting the target mechanistic hypothesis, and discuss related methodological problems.

3 Artificial replacement of biological components

Consider the diagram represented in Figure 2, in which M_B represents schematically a purely notional mechanism description formulated for explaining some sensory-motor capacity of biological system B. Consistently with many models of functional/mechanistic explanation [13, 9, 7], this mechanism description isolates parts of B whose behaviour is hypothesized to be relevant to the generation of the behaviour of interest. Suppose, furthermore, that M_B specifies the behavioural regularities governing these parts (e.g.: the firing rate of neuron m is proportional to the firing rate of neuron n).[3] As shown in the figure, theoretical model M_H is obtained by substituting reference to artificial component a_1 for reference to biological component b_1. Implementing system H on the basis of M_H amounts to building up an *hybrid simulation* of B with respect to theoretical model M_B, insofar as H includes biological components and an artificial component which matches the behavioural specification obtained from M_B. H may be obtained by removing or inhibiting component

---

[3]Analyzing the character of the regularities identified in neuroscientific models is out of the scope of this work, which introduces general methodological problems concerning ArB and BrA experiments. Looking closer at the character of these regularities, as it is made in the aforementioned studies on the structure of functional/mechanistic explanations, may provide further elements for dealing with the methodological problems sketched here.

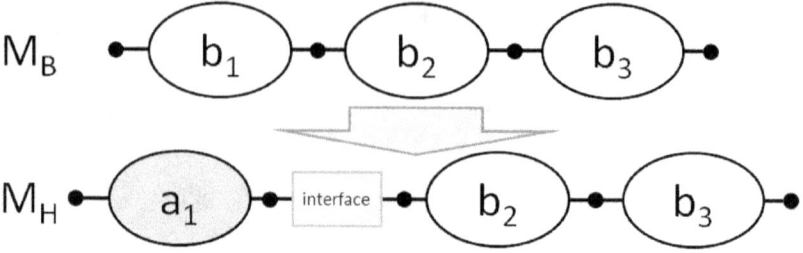

Figure 2. The ArB strategy

b_1 from B, and by establishing a connection between components b_2 and a_1, as specified by the model. This connection usually requires an interface, illustrated in the figure, such as an electrode implanted in the vicinity of some b_2 cell, together with devices for signal acquisition and pre-processing. The resulting hybrid system is called an ArB system here (standing for "artificial replacement of a biological component").

An ArB system has been used by Zelenin and collaborators [27] to address a specific hypothesis on the regularity governing the relationship between reticular neurons and overt lamprey behaviours, namely that each reticular neuron causes an ipsilateral roll tilt of the lamprey, and the resulting corrective motor response is proportional to the activity of the left and the right reticular neurons (see also fig. 1). The experimental setting used to test this hypothesis is as follows. A lamprey is fixed on a platform which allows one to control the roll angle of the lamprey, thus stimulating its vestibuli, while preventing the lamprey to rotate autonomously by movements of its own effector organs. Two recording electrodes are inserted in the vicinity of the reticular neurons (one on each side of the brain), and the acquired signal is delivered, after some processing step, to an electrical motor which can impose roll movements to the platform, hence to the lamprey. The signal processing module has been designed to reproduce the same mapping between reticular neurons and roll movements that the reticulo-spinal pathway of the lamprey is hypothesized to perform: the platform is rotated proportionally to the difference between the signal acquired from the left and the right reticular neurons. As a result (see fig. 3, where dashed lines indicate the replaced biological components), roll movements are caused by the electromechanical device whose behaviour is driven by the activity of reticular neurons, rather than by the biological mechanism connecting reticular neurons, spinal cord neurons, and effector organs.

What role can this system play in testing the addressed hypothesis? Consider the notional case illustrated in fig. 2, and suppose that H matches B's behaviours in response to similar inputs and background conditions. Assume the following:

(**ArB1**) no perturbation is introduced by the interface and by the acquisition and signal processing modules.

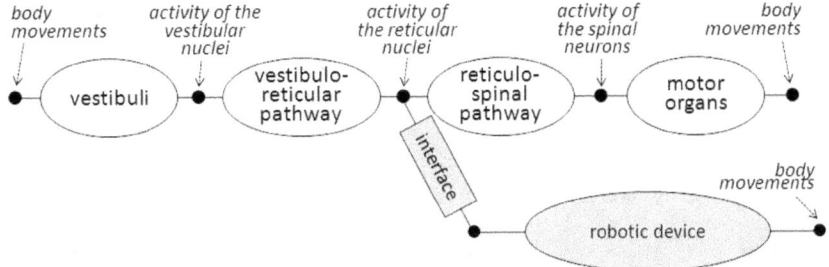

Figure 3. Schema of the ArB system used to study lamprey navigation

ArB1 is needed to exclude that BS behaviours result from artifacts introduced by the interface (due, for example, to electrode biocompatibility problems, or to deteriorating connections between electrodes and neural tissues). Assume also the following:

(**ArB2**) the non-replaced part of B (including b_2 and b_3) has undergone no internal change after bionic implantation.

If one endorses this assumption too (which is more extensively discussed below), behavioural similarities between B and H may be taken to support the hypothesis that a_1 matches b_1's behaviours. Indeed, if the bionic system reproduces the behaviour of the biological system, and the rest of the biological system has not changed after the implantation, one may reasonably conclude that the artificial device reproduces the behaviour of the replaced component. Now assume also that:

(**ArB3**) a_1 is governed by a "machine-system version" of the behavioural regularity assigned by M_B to b_1.

According to the presently examined lamprey hypothesis, the regularity r_b governing the behaviour of the replaced component is as follows: corrective motor responses are proportional to the difference between the activity of the left and the right reticular neurons. For the artificial device to reproduce this regularity, as prescribed by ArB3, one needs first to "translate" r_b (which mentions *biological* components and their properties) into a behavioural specification for an artificial device. Call this the *artificial system design* stage. The resulting behavioural specification r_a serves as blueprint for the *implementation* of the artificial device. While r_b refers to biological components and their properties, r_a may impose, for example, that the output signal of the artificial device be a binary encoding of $a(l-r)$, where l and r are left and right reticular signals (after proper filtering and binary encoding) and a is a proportionality coefficient.

If one assumes ArB3, to assert that a_1 matches b_1's behaviours amounts to asserting that b_1 meets the behavioural specifications expressed in M_B. Similarity between H's and B's behaviours is then brought to bear on the relationship between B and mechanistic hypothesis M_B, thus supporting theoretical conclusions on the behaviour of b_1.

Assumptions ArB1-ArB3 are needed to draw theoretical conclusions on the replaced component of an ArB system. It is worth noting, however, that the conditions expressed in these assumptions may fail to hold in practical circumstances, for a variety of technological and theoretical factors.

ArB1. Inserting electrodes in the neural tissues is likely to produce biocompatibility problems which may deteriorate, in the long run, the quality of the signal [21]. In the design of brain-machine interfaces one has also to deal with relative motion of the brain and the skull (on which the pedestal of the electrode is fixed) which can alter unpredictably the connection [11]. These problems, which are not easily detected and solved in bionic implants, may cause significant signal perturbations.

ArB2. Many studies show that bionic implantation is likely to produce plastic changes in the biological system. In prosthetic systems, plastic changes occurring in the brain and in the peripheral nervous system are hypothesized to be needed to let users learn progressively to control the external actuator [4]. Plastic changes are also supposed to compensate for deteriorations due to the factors examined in connection with ArB1. The occurrence of plastic processes clearly violates assumption ArB2, insofar it determines changes in the biological component before bionic implantation.

ArB3. Methodological issues connected to the requirements set by ArB3 may arise in the artificial system design and implementation stages. First, note that the behaviour of artificial systems may be perturbed by environmental or internal factors. The occurrence of these factors is to be carefully monitored in experiments, insofar as it can determine violations of ArB3 (perturbing conditions may alter a_1's capacity of reproducing the desired regularity).[4]

Second, in some cases there are no clear criteria for evaluating the relationship between a_1 and biological regularity r_b. The presently examined inquiry is a case in point. Indeed, regularity r_b is not fully specified in this case (it does not constrain, for example, the value of the proportionality coefficient).[5] Thus, it does not allow one to obtain precise values of motor responses to reticular activity. Implementing the artificial device requires one to cope with this underspecification, insofar as one has to build up an artificial device which generates a particular motor response on the basis of reticular signals. At least two approaches may be pursued: (a) in the artificial system design stage, one can work out a device blueprint in which underspecified elements of M_B are fixed; or (b) one may leave underspecified elements of M_B unfixed in the artificial system blueprint, and provide the device with adjustable elements. For example, one may let the proportionality coefficient correspond to a variable resistors in the implemented system; by manipulating the resistor, after implantation, one observes the behaviour of the system in response to different proportionality coefficients.

---

[4]Methodological issues concerning boundary conditions in biorobotic studies have been addressed in [10, 25].

[5]Underspecified regularities are called *fragile generalizations* in [7].

Clearly, M_B does not constrain the choice of the coefficient in (a), nor the choice of the value of the variable resistor in (b). However, different decisions taken at these steps may result in different behaviours of a_1, thus affecting similarities between bionic and biological behaviours. In this case, it is not reasonable to bring bionic behaviours to bear on the biological hypothesis, insofar as there are no clear criteria for assessing whether the artificial device actually reproduces a "machine-system" version of the biological regularity assigned by the hypothesis to the replaced component. Poor performances of the bionic system may be explained by reference to improper choices in (a) or (b), and good system performances might be due to *ad hoc* adjustments: as shown by Hopkins and Leipold [16], it may happen that similar behaviours, with respect to some range of input conditions, are produced by simulations of radically different models, after suitable fixing of system parameters. In this circumstance, one may reasonably doubt that good system performances provide an adequate empirical basis for claiming corroboration of the biological hypothesis.

In the presently examined lamprey case-study, comparisons between bionic and biological behaviours are taken to support the initial hypothesis concerning the replaced component. Experiments consist in observing the behaviour of the platform in response to manually applied roll tilts. As extensively discussed in [27], any postural disturbance is rapidly compensated by the system: in response to tilts, the reticular neurons (whose activity is elicited by vestibular stimulations as the animal rolls) issue stabilizing motor commands to the electrical motor. Thus, the bionic system replicates the behaviour of the intact lamprey, as far as stabilization in roll axis is concerned. Consistently with the ArB approach sketched above, the authors take this result as support for concluding that the "replaced" part of the system (which is referred in the following quotation as the "output stage" of the mechanism) behaves according to the regularity obtained by the electro-mechanical device:

> these results provide a strong support to the initial idea ... that interaction of the two descending commands (their subtraction) occurs in the output stage of the system [27, p. 2886].

The methodological issues discussed before are by and large tractable in this case-study. One may hold on to the hypothesis that system calibration gives rise to no significant concern related to ArB3, insofar as the adjustable parameters mentioned in [27] are, in fact, proportionality coefficients associated to the left and right input: any assignment of values to system parameters will result in a device whose output is proportional to the difference between the left and the right input, consistently with the underspecified hypothesis. In addition to this, one may resort to already available and well supported models of the lamprey nervous system to facilitate detection and prediction of plastic adaptation processes. Nevertheless, these issues may become far more problematic in connection with more intricate artificial systems, whose behaviour may turn out to be poorly consistent with the initial underspecified hypothesis for particular arrangements of system parameters, as well as in connection with other biological systems, whose plastic adaptations may be hardly detected and predicted due to the lack of adequate theoretical models.

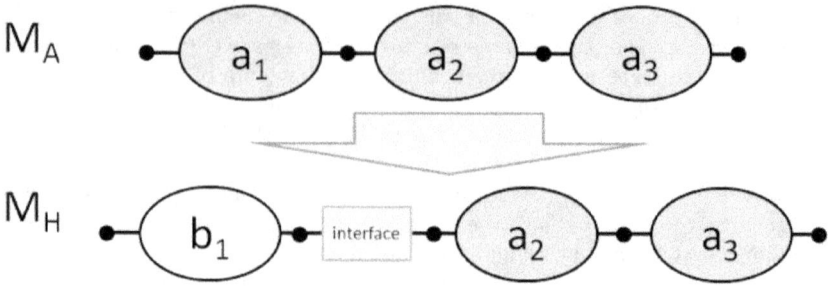

Figure 4. The BrA strategy

In these circumstances, one may reasonably doubt that the analysis of hybrid system performances can play a significant role in the scientific modelling of the target biological system, thus suggesting the need for adding crucial qualifications to Chapin's and Nicolelis's claims. Variants of these methodological problems, related to a different class of simulation bionic systems, will be discussed in the next section with reference to other BS-based studies on lamprey body stabilization.

4 Biological replacement of artificial components

While in the ArB methodology one investigates the behaviour of component b_1 by *replacing it* with an artificial component, the BrA strategy involves *including* the target component in an artificial system. With reference to fig. 4, suppose that in the implementation of artificial system A one leaves out component a_1 (say, a visual processing component). And suppose that knowledge of the behaviour r_a that a_1 should manifest for A to produce the desired sensory-motor behaviour (e.g. seeking light) is available on the basis of a previously formulated blueprint M_A of A. The BrA — "biological replacement of an artificial component" — approach proceeds, from this point on, by filling the mechanism gap with a biological component, represented as b_1 in the figure (for example, with a sensory-related portion of the brain of some biological system). As usual, electrodes and signal acquisition/filtering devices are needed to realize a working interface between b_1 and the rest of the system.

This approach can be exemplified by reference to a BrA system implemented to study the vestibulo-reticular segment of the mechanism of body stabilization represented in fig. 1 [22, 17]. The procedure used to build up this system, whose structure is schematically illustrated in fig. 5, is as follows. A lamprey brain is dissected and kept alive in a proper chemical solution. Four electrodes are inserted in the brain in order to record and stimulate extracellular activity. Two recording electrodes are inserted in the vicinity of the left and right PRRN axons respectively, and two stimulation electrodes are inserted, on each side of the brain, where the axons of the nOMI and nOMP (which receive vestibular inputs; see section 2) cross each other. While the

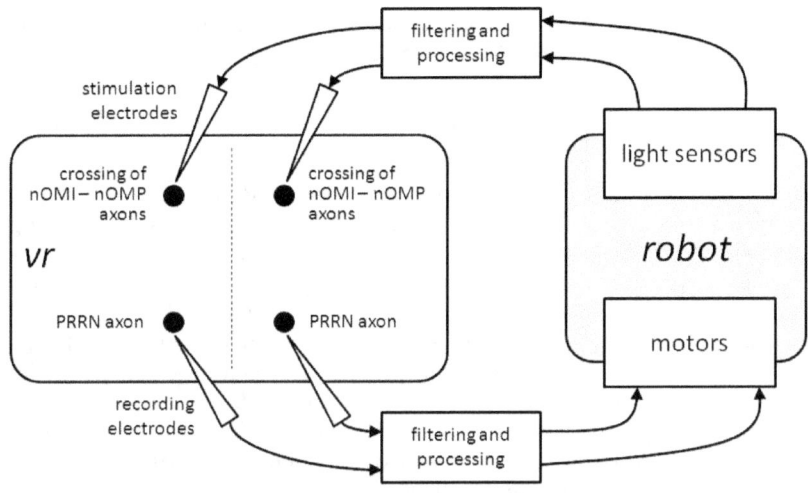

Figure 5. Schema of the BrA system described in section 4

axons of nOMI neurons remain ipsilateral, the axons of nOMP neurons cross the midline forming synapses with reticular neurons at the opposite side of the brain. As a consequence, stimulation of each nOMI-nOMP electrode elicits activity at both ipsilateral and controlateral PRRN. These electrodes are used to connect this portion of lamprey brain (*vr* from now on) with artificial sensors and motors located on a robotic wheeled platform. Light sensors located on the two sides of the robot are connected to the ipsilateral stimulating electrodes through the intermediary of filtering and processing modules, so that each electrode receives the maximum electrical signal when light comes from a 45° angle with respect to the corresponding side of the robot. As a result, light perceived by the sensors on the robot elicits activity of the nOMI-nOMP neurons. The "output" side of the neural circuitry (i.e. the set of recording electrodes) is connected to the motors of the robotic platform. Spike trains recorded by the two PRRN electrodes are filtered so as to avoid electrical artifacts created by the interface, and are subsequently thresholded, rectified, and averaged. The output signals are then used to control the motor velocities. As a result, *vr* "closes the loop" between artificial sensors and motors, by providing a control signal for the motors on the basis of the light data delivered to the stimulation electrodes.[6]

This BrA system has been used in experiments to study the relationship between the activity of vestibular and reticular neurons. To illustrate this strategy with reference to the general case represented in fig. 4, suppose

---

[6]For *H* to react to light, component *vr* should perform some sort of sensory input processing. Note that no specific hypothesis on the signal processing performed by *vr* is made: according to the strategy adopted by the authors, which will be discussed below, the space of possible hypotheses is to be restricted on the basis of an analysis of *H*'s behaviours (which may range from light attraction to light avoidance behaviours).

that H succeeds in generating the behaviour which A (i.e. the fully artificial system) would have generated. Analogously to the ArB case, assume the following:

(**BrA1**) no perturbation is introduced by the interfaces and by the acquisition and signal processing modules.

If one excludes that H's behaviour is deteriorated by effect of interfacing problems, H's ability of reproducing A's behaviours may be taken to support the hypothesis that the "biological filler" b_1 succeeds in reproducing the behaviour r_a which is needed to achieve good system working. This conclusion states that, in response to the inputs and background conditions to which b_1 is now subjected, b_1's behaviour is regimented by behavioural regularity r_a. On the basis of this result one cannot reasonably conclude that b_1 performs r_a in the framework of the intact biological system B from which it has been extracted, unless one endorses the additional assumption that these input and background conditions are similar to those to which b_1 is subjected in B. This assumption, which is discussed below, is called BrA2 here:

(**BrA2**) the input and background conditions to which b_1 is subjected in H are similar to those to which b_1 is subjected in B.

BrA2 is needed to conclude that the behaviour r_a which b_1 reproduces in the hybrid sensory-motor mechanism H corresponds to the behaviour that b_1 generates in the framework of B. This assumption needs more discussion.

BrA2. One may reasonably doubt that H's behaviours can be brought to bear on the behaviour of b_1 in B, insofar H involves de-contextualizing b_1 with respect to its original containing system. In the presently examined lamprey case-study, this is achieved by surgically removing the brain and putting it in a laboratory environment (a chemical solution). The behaviour of the biological tissue may well be significantly affected by the occurrence of completely novel background conditions, let alone damages due to surgery intervention. Moreover, component b_1 may react in peculiar ways when stimulated differently with respect to the stimulation it receives from adjacent neurons in the intact system. These stimulations may exceed the range of tolerance of the biological tissue, thus provoking damages or plastic adaptations which would not have occurred in the biological system. And the behaviour of b_1 in B may be modulated by concurrent mechanisms that become deteriorated or inactive by effect of the surgical intervention. In these circumstances, de-contextualizing b_1 may make b_1's behaviours poorly diagnostic of its actual behaviours in the context of B.

Thus, as a general regulative principle, in order to make sensible experimental use of H in the study of b_1, one should look carefully at the relationship between the behaviour of the artificial components of H (a_2 and a_3 in the case illustrated in fig. 4) and the behaviour of the biological components interacting with b_1 in the intact system. Ideally, the artificial components in H should reproduce the behaviour of the other biological components which are supposed to play a relevant role (together with b_1) with respect to the mechanism under investigation. Thus, theoretical conclusions on b_1 flowing from analyses of H's behaviours rest on additional assumptions (a) on the role of the other biological components interacting with b_1 in the framework of the target containing system capacity, and (b) on the capability of the artificial components of H of reproducing accurately the behaviour of those biological components. Understanding and reproducing artificially the relevant input and background conditions to which b_1 is subjected in B may be extremely problematic in many circumstances. These methodological problems cast serious doubts on the feasibility of the BrA approach for the modelling of biological behaviours.[7]

In the experiments reported in [22, 17], the bionic lamprey is put in the middle of a circular arena surrounded by walls on which some lights are placed. The lights are selectively turned on and the trajectories followed by the robot in response to light stimuli are monitored by a camera mounted on the top of the arena. The experimental results obtained there are not homogeneous, however, insofar as examples of light attraction and light avoidance have been observed in different preparations, in addition to a number of "intermediate" behaviours. The behavioural variability recorded in these experiments has been imputed to implementation problems, rather than to mechanisms at work in vr, insofar as its source has been identified in "the placement of the electrodes in the actual neural tissue" [22, p. 313]. Indeed, due to technological difficulties related to interface set up and electrode insertion, vestibular neuron stimulation changes from preparation to preparation, thus violating BrA1. Accordingly, the relationship between the electrical signal delivered by the electrodes and the activity of the vestibular neurons changes from case to case in a fairly unpredictable fashion, thus determining different behaviours in response to light.

These performances are an impressive result in the field of bionic interfacing, insofar as biological component vr succeeds in closing the loop between artificial sensors and motors and, as a matter of fact, drives robot movements by virtue of some kind of sensory processing. However, the methodological issues analyzed here (in addition to those addressed in the previous section) suggest the opportunity of adding significant caveats to Chapin's and Nicolelis' claims on the experimental potentialities of bionic experiments, as far as bionic simulations are concerned. The feasibity of BrA experiments is seriously undermined by the methodological and theoretical difficulties that have to be addressed in order to achieve the conditions imposed by BrA2. In the specific case of the lamprey-based simulation, due to interfacing problems which determine violations of assumption BrA1, the lamprey-machine BrA system fails to provide a sufficiently reliable experimental tool for the modelling of vr's behaviour, as far as the characterization of the positive or negative feedback processing performed by vr is concerned.

---

[7]Note that no precise qualification of the relationship of *similarity* between inputs and background conditions of the two systems, which is mentioned in the above formulation of BrA2, is provided here. Reflecting on the relationship between the artificial components included in H and the biological components interacting with b_1 in the intact system B may enable one to work out a more precise formulation of BrA2.

5 Conclusions

Bionic technologies are paving the way to new simulation approaches, thus stimulating epistemological and methodological reflections on the role of machine experiments in biological theorizing. Two classes of simulation BS-based experimental approaches have been considered here, which may support the corroboration or refutation of hypotheses on the behaviour of biological components and their role with respect to sensory-motor capacities of the containing system. Methodological issues arising in the framework of ArB and BrA experiments, which become integral part of auxiliary assumptions that are needed to bring bionic behaviours to bear on the biological hypothesis, have been discussed with reference to concrete case studies on lamprey navigation. Corroborating these auxiliary hypotheses, which may turn out to be highly problematical in some circumstances, is crucial to the BS-based theorizing on biological systems.

It is worth remarking, in this connection, that bionic technologies enable one to perform a variety of non-simulative experiments [19]. Addressing methodological aspects of these experiments is out of the scope of this work, which is distinctively concerned with *hybrid simulations* of biological systems. A variety of questions concerning simulation bionic experiments need further analysis. Namely, can BS-based experiments license other kinds of theoretical results, in addition to corroborating or refuting hypotheses on the behaviour of a biological component in the framework of a sensory-motor capacity of the target system? It is worth noting that the BrA system analyzed in section 4 has been used not only to investigate the role of vr with respect to lamprey body-stabilization capacities, but also to study the *internal mechanism* enabling vr to generate its characteristic behaviour [22, 17]. This experimental approach, whose methodological aspects are not analyzed here, may enable one to develop BS-supported multi-level mechanistic analyses of biological systems.[8] Reflecting on these experimental possibilities and to the related methodological issues may contribute to identifying regulative principles for "good" bionic simulation experiments, and to further our understanding of the role of machines in biological theorizing.

BIBLIOGRAPHY

[1] D.J. Amit. Simulation in neurobiology: theory or experiment? *Trends in Neurosciences*, 21 (6): 231–237, 1998.
[2] M. Blanchard, and F.C. Rind and P.F.M.J. Verschure. Collision avoidance using a model of the locust LGMD neuron. *Robotics and Autonomous Systems*, 30: 17–38, 2000.
[3] N. Burgess, A. Jackson, T. Hartley, and J. O'Keefe. Predictions derived from modelling the hippocampal role in navigation. *Biological Cybernetics*, 83: 301–312, 2000.
[4] J.M. Carmena, M.A. Lebedev, R.E. Crist, J.E. O'Doherty, D.M. Santucci, D.F. Dimitrov, P.G. Patil, C.S. Henriquez, M.A.L. Nicolelis. Learning to control a brain-machine interface for reaching and grasping by primates. *PLoS Biology*, 1,2: 193–208, 2003.
[5] J.K. Chapin, K.A. Moxon, R.S. Markowitz, and M.A.L. Nicolelis. Real-time control of a robot arm using simultaneously recorded neurons in the motor cortex. *Nature Neuroscience*, 2(7): 664–670, 1999

---

[8]See [7, chapters 4 and 5] for an analysis of multi-level explanations and inter-level experiments in neuroscience.

[6] C.-P. Chou and B. Hannaford. Study of human forearm posture maintenance with a physiologically based robotic arm and spinal level neural controller. *Biological Cybernetics*, 76: 285–298, 1997.
[7] C.F. Craver (ed.) *Explaining the Brain. Mechanisms and the Mosaic Unity of Neuroscience*. Oxford University Press, Oxford 2007.
[8] R. Cordeschi *The Discovery of the Artificial. Behavior, Mind and Machines Before and Beyond Cybernetics*. Kluwer Academic Publishers, Dordrecht 2002.
[9] R. Cummins. Functional analysis. *The Journal of Philosophy*, 72, 20: 741–765, 1975.
[10] E. Datteri and G. Tamburrini. *Bio-robotic experiments and scientific method*. College Publications, London 2006, pages 397-41.
[11] J.P. Donoghue, A. Nurmikko, M. Black, L.R. Hochberg. Assistive technology and robotic control using motor cortex ensemble-based neural interface systems in humans with tetraplegia. *Journal of Physiology*, 579 (3): 603–611, 2007.
[12] A. Franklin (ed.). *Can that be right? Essays on Experiment, Evidence and Science*. Kluwer Academic Publisher, Dordrecht 1999.
[13] S. Glennan. Modeling mechanisms. *Studies in History and Philosophy of Biological and Biomedical Sciences*, 36: 443–464, 2005.
[14] F.W. Grasso, T.R. Consi, D.C. Mountain and J. Atema. Biomimetic robot lobster performs chemo-orientation in turbulence using a pair of spatially separated sensors: Progress and challenges. *Robotics and Autonomous Systems*, 30: 115–131, 2000.
[15] L.R. Hochberg, M.D. Serruya, G.M. Friehs, J.A. Mukand, M. Saleh, A.H. Caplan, A. Branner, D. Chen, R. Penn, J.P. Donoghue. Neuronal ensemble control of prosthetic devices by a human with tetraplegia. *Nature*, 442: 164–171, 2006.
[16] J.C. Hopkins, R.J. Leipold. On the dangers of adjusting the parameters values of mechanism-based mathematical models. *Journal of Theoretical Biology*, 183 (4): 417–427, 1996.
[17] A. Karniel, M. Kositsky, K.M. Fleming, M. Chiappalone, V. Sanguineti, S. Alford, F.A. Mussa-Ivaldi. Computational analysis in vitro: dynamics and plasticity of a neurorobotic system. *Journal of Neural Engineering*, 2: 250–265, 2005.
[18] M.A. Lebedev and M.A.L. Nicolelis. Brain-machine interfaces: past, present and future. *Trends in Neurosciences*, 29 (9): 536–546. 2006.
[19] M.A.L. Nicolelis. Brain machine interfaces to restore motor function and probe neural circuits, *Nature Review Neurosciences*, 4: 417–422, 2003.
[20] G.N. Orlovsky, T.G. Deliagina and P. Walln. Vestibular control of swimming in lamprey. I. Responses of reticulospinal neurons to roll and pitch. *Experimental Brain Research*, 90: 479–488, 1992.
[21] V.S. Polikov, P.A. Tresco, and W.M. Reichert. Response of brain tissue to chronically implanted neural electrodes. *Journal of Neuroscience Methods*, 148: 1–18, 2005.
[22] B.D. Reger, K.M. Fleming, V. Sanguineti, S. Alford, F.A. Mussa-Ivaldi. Connecting brains to robots: an artificial body for studying the computational properties of neural tissues. *Artificial Life*, 6 (4): 307–24, 2000.
[23] R. Reeve, B. Webb, A. Horchler, G. Indiveri, R. Quinn. New technologies for testing a model of cricket phonotaxis on an outdoor robot. *Robotics and Autonomous Systems*. 51: 41–54, 2005.
[24] A. Rosenblueth and N. Wiener N. The role of models in science. *Philosophy of Science*, 12: 316–321, 1945.
[25] G. Tamburrini, E. Datteri. Machine Experiments and Theoretical Modelling: from Cybernetic Methodology to Neuro-Robotics. *Minds and Machines*, 15 (3–4): 335–358, 2005.
[26] B. Webb and T.R. Consi. *Biorobotics. Methods and Applications*. AAAI Press/The MIT Press, Cambridge, MA 2001.
[27] P.V. Zelenin, T.G. Deliagina, S. Grillner, G.N. Orlovsky. Postural control in the lamprey: A study with a neuro-mechanical model. *Journal of Neurophysiology*, 84: 2880–2887, 2000.

Some remarks on a heuristic point of view about the role of experiment in the physical sciences

Luca Guzzardi

1 Introduction

One major contribution of the 20th century epistemology is the focus on a non-naïve concept of experiment. We have learnt experiment does not provide either a "verification" or an immediate "falsification" of a theory; physical observations are theory-laden and the experimental praxis follows the theory and is at the expenses of a silent, long and slow theoretical work. Otherwise the experimental praxis wouldn't simply exist, and the work of experimenters is by far and in the first place a theoretical one. We can trace this attitude back to Pierre Duhem's *Théorie physique* (1906) and Karl R. Popper's *Logik der Forschung* (1934; *Logic of Scientific Discovery*, 1959). Associated with this point of view, which seems in itself very difficult to question, one can often find the unexpressed assumption that experiments are control means of theory, so they have no other purpose than to test and control the correctness of a system of empirical statements. According to this, any other role experimentation could play has so little importance in the eyes of epistemologists that it is at best a collection of interesting peculiarities (and a waste of time in the worst case). This point of view is appropriately summarized in the *incipit* of Popper's *Logic of Scientific Discovery*:

> A scientist, whether theorist or experimenter, puts forward statements, or systems of statements, and tests them step by step. In the field of the empirical sciences, more particularly, he constructs hypotheses, or systems of theories, and tests them against experience by observation and experiment [15, p. 3], [16, p. 27].[1]

In this paper I will try to show how this view, that emphasizes the prominence and prevalence of theory against experiment, leads to overlook a major use of experiments, provided with strong historical evidence and supported by epistemological arguments. In the picture I will try to give, experiments do not play the *main* role of "empirical control" of theory, though they can be *used* with that function. But I shall argue that their main role is a different one.

---

[1] Note that the words "whether theorist or experimenter" were added in the English edition 1958–1959 (and were not reported in the German edition 1966), as if Popper felt a need of specification or emphasis: the "experimenter" must do the same work, follow the same procedure and — more importantly — apply the same methods as the "theorist" colleague

A cautionary argument is to be made in advance: I do not claim my suggestions will apply to any field of empirical sciences, though I think this would became arguable, provided that appropriate changes in my picture are introduced. However, I don't try here to embark on this enterprise. This is also the reason why historical examples in this paper are only taken from the history of physics and my suggestions are restricted to the physical sciences.

2 Proof: a legal concept and its epistemological *pendant*

In which sense does an experiment become an experimental *test*? How did this idea that a certain experience is actually a proof against or in favour of a certain belief, conviction or system of hypotheses acquired relevance till to the point it has become obvious? An answer to this could be find in a brief account of the history of the concept of *proof*.

The term "proof" traces most probably back to legal jargon. Generally speaking, to describe something as a proof means to establish a criterion for deciding about something: proof is what provides the basis of our decision and *justify* it. So, for instance, the verdict of "not proven", typical of the Scots Law, means an acquittal for insufficient proof, because evidence is so inadequate and defective that a judge would have no reason to convict a defendant. This legal background, which transpires from the humean (and than kantian) image of a "court of justice" for the reason,[2] involves sometimes pure scientific work as well: this is the case of William Thomson's (Lord Kelvin) assessment of a particular and very sofisticated theory of aether he himself contributed to developed some years before: "I am thus driven to admit, in conclusion, that the most favourable verdict I can ask for the propagation of laminar waves through a turbulently moving inviscid liquid [i.e. the ether] is the Scottish verdict of *not proven*"[19, p. 352]. According to Kelvin, the problem with this theory was not the refined mathematics developed by him and others, but merely the experience, which didn't suffice to support the theory.

The legal imprint of the term *proof* still sticks on this concept even if we move away from legal jargon.[3] In an arithmetical textbook dated around 1430

---

[2]See [11, p. 3], [12, A11, eng. tr. p. 101]. Not to mention what Kant argues about the legal sense of the concept of (transcendental) deduction in §13 of the Critique of Pure Reason. A good understanding of the concept of proof, its development and background also in legal terms is provided by [6] (see more in particular about Kant and Hume [6, pp. 15–23 it. tr.]; about proof and the legal tradition [6, pp. 35–43 it. tr.]).

[3]Very interestingly the concept of proof seems originally have been affected by religious tradition too. One of the first occurrences of the term "proof" (intertwined with its "legal" meaning) you can find in English is contained in a Middle-age text known as *Ancren Riwle* or *Ancrene Wisse*, a Regula for Anchoresses written around 1200 by an anonymous English churchman for the instruction of a small community of three women about to become religious recluses. In this book we can find one of the first occurrences of the word in English. What in the eyes of the anonymous author justifies rules (therefore providing proofs) are mostly stories from the Bible. So in a couple of passage the speaker who gives the rule to the three young ladies argues: "That this is true [...] here is the proof [*preoue*]" [1, pp. 52, 53] and further: "Because I said that we find this both in the Old Testament and also in the New, I will, out of both, show an example and proof" [1, pp. 154, 155]. The term "proof" is used here in the sense of what makes good — that is proves — a statement:

and called *The art of nombryng* (a translation of a Latin textbook *De arte numerandi*, written in the 13th century and attributed to John of Holywood), proof [*prouffe*] takes explicitely the meaning of a test or a trial to check the correctness of an arithmetical calculation: "The subtraccioun is none other but a prouffe of the addicioun, and the contrarye in like wise"[2, p. 6]. In fact, this sentence, despite the triviality by which it is only seemingly affected, brings to bear what I called the legal origin of the concept of proof. Inverse operations in Mathematics are treated as proofs because they allow to decide — to *judge* — about the correctness of calculi, providing a justification of them. Nevertheless, by this way the arithmetical proof achieves somehow a new feature, namely the feature of a *trial* — a term involving an obvious legal background. Like in a lawsuit, the fact to have overcome a trial (the subtraction proof for the addiction, for example) puts de iure the calculation into the realm of what is legitimate, so to speak, beyond reasonable doubt.

As in Hume's and Kant's metaphor of the court of justice for human reason, on this theoretical level "proof" can indicate both the individual evidences used to convince the mind and the entire process of convincing someone — a process sometimes called *demonstration*, that is the ability to deducing something from certain, definite assumptions produced before the attentive and watchful *eye* of understanding. The relationship between proof and demonstration was developed by John Locke (in Book IV of his *Essay Concerning Human Understanding*, 1690). He put emphasis on the first one, with an important reference to the process of vision:

> Those intervening *Ideas*, which serve to shew the Agreement of any two others, are called *Proofs*; and where the Agreement or Disagreement is by this means plainly and clearly perceived, it is called *Demonstration*, it being shewn to the Understanding, and the Mind made see that it is so [13, Book IV/ii, 3, p. 532].

A proof builds up a demonstration in so far as it brings to evidence what, as beeing *seen*, takes its own place within positive and true knowledge. Something that has achieved this status of a demonstrated and established truth can never be moved away from that place by any kind of contrary judgement. Indeed, the understanding *see* it and "the Mind made *see* that it is so".

But Locke points out all this has no relation with experience, both crude or controlled in the form of experiment. According to him, the way of finding thruths by proofs and demonstrations applies only to ideas and even what he calls "the Art of finding Proofs" is one major breakthrough of an argumentation style which "is to be learned in the Schools of Mathematicians, who from very plain and easy beginnings, by gentle degrees, and a continued Chain of Reasonings, proceed to the discovery and demonstration of Truths" [13, Book IV/xii, 7, p. 643]. On the contrary, experiments would absolutely be not able to yield established knowledge. Though they have a different and very important role:

> *A Man accustomed to rational and regular Experiments* shall be able to see farther into the Nature of Bodies, and guess righter at their yet unknown Properties, then one, that is a stranger to them: But Yet, as I have said, this is but Judgement and Opinion, not Knowledge and Certainty [13, Book IV/xii, §10, p. 645; italics mine].

---

an evidence that is sufficient or contributes to establish anything; for instance, a rule for living.

3 A patient experimental culture

Locke remarkably reduced the significance of experiment in the sense of a proof: because of the weakness of achieving knowledge by means of experience, every "test" we undertake even to falsify (not to say to verify) our theories is not but "judgement and opinion". Above all, what is important in Locke's idea about experimenting nature is not that experiments are means to prove or test our knowledge, but that they are a substantial part of a heuristics, so they can lead and ultimately *help grow* our knowledge. Of course, we need to examine step by step any individual case of a phenomenon to be sure that "our principle will carry us quite through, and not be as inconsistent with one *Phoenomenon* of Nature"[13, Book IV/xii, §13, p. 648]. However, this is nothing but a side effect of a preliminary "be accustomed to rational and regular Experiments", that lead our opinion and compel us to some guesses. Therefore, experimentation is the source of the whole process of guessing. This aspect reaches far beyond Locke's empiricist perspective because it doesn't involve merely the *logic* of scientific discovery; it rather deals with both the *praxis* and the *practice* of science: such a specific praxis as we can by and large outline it.

As well known, Locke had close relationship with the environment of the Royal Society.[4] According to its Statute "frequent" meetings of the fellows — if possible once a week — should be hold. "The business of their weekly Meetings shall be, to order, take account, consider, and discourse of Philosophical Experiments, and Observation" [18, p. 145][5] — where the expression "philosophical experiments" means nothing but experiments in natural philosophy, i.e. scientific experiments as we are used to call them.

Remarkably, to make experiments — an activity that Bishop Thomas Sprat (who was himself a fellow and one of the first historians of the Royal Society) regarded as the substantial part of the meetings — was not the business of any Fellow. The Constitution of the Society stated that a person should be selected for this purpose and his exclusive duty had to be providing for experiments. The name of such "employee" was the *Curator of experiments*. While the Fellows came from the most different occupations and cultivated natural philosophy as their personal interest and passion, however serious these might be, curators were professional experimenters and they got by the Society — i.e. by the Fellows themselves — a salary for their work (the Statute established a maximum of two hundred pounds per year, but the first curator, Robert Hooke, got only thirty pounds in addition to an apartment in the buildings of the Royal Society).

---

[4]Locke was elected a fellow of the Royal Society in November 1668 (six years later its foundation). It is known he was quickly appointed to a committee for experiments and twice served on the council, but apparently he has little contributed to the work of the Society. Nevertheless, he dedicated the Essay, first published in 1690, to his friend Thomas Herbert, then President of the Society; throughout his life, either in England or in exile, Locke followed with interest the Society's activities. About the relations between Locke and the Royal Society with regard to the purposes of present paper see [3, vol. 1, pp. 245–249], [17, pp. 73–74], [21, pp. 52–58], [23, pp. 36–38], and [22, pp. 17–23]. About Locke's "experimental Knowledge" see [4, 204–209].

[5]Sprat's quoted text follows the edition London 1667.

The office of a curator must have been a delicate one, as the complicated appointment procedure points to: to begin with, unless an eminent person was known to the Fellows for his competence and worth, usually curators had to be examined very carefully by the Society before "election". They were first recruited for a trial period, which normally didn't take a full year, at the end of which "they shall be either elected for perpetuity, or for a longer time of probation, or wholly rejected" [18, p.147]. Specifically, their business was to

> take care of the managing of all Experiments, and Observations appointed by the Society, or Council, and report the same, and perform such other tasks, as the Society, or Council shall appoint: such as the examining of Sciences, Arts, and Inventions now in use, and the bringing in Histories of Natural and Artificial things, & c.[18, p.147].

The Statute provided also a brief sketch of a typical curator: his background and competence had to be appropriate to the office, he had to be "skilled in Philosophical, and Mathematical Learning, well versed in Observations, Inquiries, and Experiment of Nature and Art" [18, p.147]. In other words, curators were essentially technicians, who had of course to know the elements of natural philosophy and mathematics and, so to speak, feeling at home in applying them. However, this knowledge was first of all intended for their *job*, i.e. managing experiments before an audience consisting of the Fellows of the Royal Society.

According to Locke, the most valuable gift of a good philosopher trained at the School of Mathematicians, who proceed by proofs and demonstrations (and maybe refutations) of their own ideas is perhaps "sagacity", that is the ability to find out quickly and without delay a basis for the arguments he want to use, providing them with an indubitable certainty. The most important quality of a curator of experiments must have been, instead, the infinite patience he needed to vary every time his experiments and carefully examine a profusion of individual cases. After all, any experimenter knows that everything is nothing but an individual, unique case... *"Natur ist nur einmal da"*, as Ernst Mach put it in *The Science of Mechanics*: nature is but once there.

But after all varying experimental conditions of a phenomenon means to show, *ad excludendum*, what does not change in variation and how things generally are, so that one can illustrate through experiments the typical behaviour of things. Therefore, curators were required to have a great deal of imagination, both for observing nature in so many aspects as possible and for illustrative, teaching purposes. The aim of experiments in public demonstrations (at the Royal Society, for example) was not only to show nature, but also to illustrate theories in so many ways as possible.

Like in Locke's account, for learning science is needed to see and discuss experiments mainly through the ability of a good experimenter. Remember that according to Locke experiments had a very important role in teaching: people "accustomed to rational and regular Experiments shall be able to see farther into the Nature of Bodies, and guess", and so on. Experiments acted somehow as a propaedeutics to research, for young and less young natural philosophers (like the Fellow of the Royal Society) could learn the "art" of "seeing farther into the Nature of Bodies" and make guesses about their properties. But no illusions about that: in Locke's eyes experiments provide very restricted control, in most cases they say nor "yes" neither "no" and there is no assurance of an unquestionable verdict from them.

So experiments would play their major role in giving so to speak a sensible illustration of theories and training scientists about them. Often during 17th and 18th century experimenters were called *demonstrators* in scientific fields, meaning with that laboratory professionals who gave "public demonstrations" of an experiment for teaching purposes before a more or less large audience. Their aim was not the same of the theorists (or something like that the theorists could expect by experimenters), i.e. to test a set of hypotheses. They rather aimed to provide what I called "a sensible illustration" of a theory, so that people could learn and discuss *that* theory. In order to perform this, the experimenter-demonstrator, an institutional version of which is embodied by the curator of experiments at the Royal Society, did not have to take critical attitude against theories he was "demonstrating". He had to be a loyal supporter and a strenuous defender of these; he had to be convinced of these in order to convince his audience. In this regard, by no means an experimental demonstration, either in Popper's or in Locke's sense, would provide a proof for or against a theory. But a demonstration illustrates a theory[6] and in the meantime it can also give an *apology*, so the experimenter properly plays the role of an *apologist*.

One may think, *I* actually am trying here to make an apology of something namely of the inductive method, that is inferring theories from experiments. But to state an apology through experiments in the sense I have just described does not involve an application of inductivism. Experimenters as I have described them — scientists of a well recognizable kind in the history of science — don't bother stating a (new) general theory starting from empirical data. Scientists as Robert Hooke, Francis Hauksbee or Jean-Théophile Desaguliers — all of them Curators of experiments at the Royal Society — and many others, when they act as apologists (and uniquely in this case), only take care of defending theory against possible assaults by enemies and "infidels" and preventing potential "heretics" from proselytizing.[7]

To state an apology is also very different from performing an experimental test of a theory: in last case, even if we are expecting a positive result in favour of a certain theory, *in principle* an error could occur — that is the fallibilistic point of view. But apology does imply that there is no doubt that theory is fundamentally correct (though some details could vary).[8] Therefore, there is no need of any control at all. And this applies *in principle*: the possibility something might be wrong in the very fundament of the theory at issue is not even taken into account. Apology implies two actions are basi-

---

[6]This sense of scientific demonstration was probably relevant for Kant himself. On this issue see [14, pp. 95–97].

[7]Sometimes they are "driven" in doing this job by theorists or eminent scientists, as it happened in the case of Newton and "his" curator Francis Hauksbee. See [10, pp. 229–234].

[8]I don't need to point out that often experiments can be designed in order to extend a given theory. As well known, this feature has been emphasized by Bas van Fraassen, whose point of view I discuss in the third section of present paper.

cally needed: *defending* and *convincing*. A theory will not be confirmed nor "corroborated" in Popper's sense, which implies a reference to a (fallibilistic) degree of confidence that doesn't occur here. A theory will rather be literally "*confirmata*" in Latin meaning of the word: experiments *confirm* a theory in the sense that they make it firmer, more solid against possible assaults than it would be without them.

No matter of "corroboration degree"; no matter of psychology or convictions. This is something objective. As historians of science have recognized since some decades, the rapid success of Newtonianism and his general acceptance as the standard view during 18th century throughout the Continent was also due to an efficient apologetics. Amongst its major advocates are to be mentioned Dutch authors as Willem Jacob 'sGravesande and Pieter van Musschenbroek. Following Hauksbee's and Desaguliers' experimental tradition, 'sGravesande wrote in 1721 a physics textbook addressed to students that very quickly became (with its third edition, 1724) one of the most important and influent treatises of his age. Its Latin title remarkably was *Physices Elementa Mathematica, experimentis confirmata sive Introductio ad philosophiam Newtonianam* — i.e. *The Mathematical Elements of Physics Confirmed by Experiments*. What is really impressive about this book is the profusion of experiments compared with the small number of pages devoted to mathematical scholii. It's also not surprisingly that the first English translation of this textbook (1725) was made by a curator of the Royal Society, namely Jean-Théophile Desaguliers.[9]

4 Two cultures of experiment and a heuristic point of view on its role

I have contrasted the usual concept of experiment as a control, which I described using analogies from a legal background (but note how many and powerful analogies with originally legal concepts Popper uses in Chapter 5 of *The Logic of Scientific Discovery*, namely "The Problem of the Empirical Basis"),[10] with a broader concept, where experiments have at least a threefold role: first, they are *illustrations* of a theory and by this way can provide a good deal of examples for (in second place) *teaching* and (in third place) *defending* the theory itself. These different nuances about the role of experiment are of course related and interacting issues. They share at least one thing: up to this point experiments tightly remain in the hands of theorists, who need to illustrate and defend their creatures and perhaps to find and teach followers. And of course the theorist (not the experimentalist, pace Popper) feels the

---

[9] For further details see [8]. More in particular about 'sGravesande's and J.-T. Desaguliers' role in defending and "confirming" Newtonianism, see [7, pp. 96–97].

[10] So for example: "The *verdict* of the jury (*vere dictum* = spoken truly), like that of the experimenter, is an answer to a question of fact (*quid facti?*) wich must be put to the jury in the sharpest, the most definite form". Than, Popper makes clear that "what question is asked, and how it is put, will depend very largely on the legal situation, i.e. on the prevailing system of criminal law (corresponding to a system of theories)". And again: "In contrast to the verdict of the jury, the *judgement* of the judge is 'reasoned'; it needs, and contains, a justification. The judge tries to justify it by, or deduce it logically from, other stetements: the statements of the legal system, combined with the verdict that plays the role of initial conditions" [16, pp. 109–110].

need to test his theory and exploits the ability of skilled experimenters for this purpose.

In addition to this, following Locke I have argued that experiments can provide what we could properly call a heuristics. Though this last feature might be related with the three nuances I have mentioned above, I want to suggest that providing a heuristics means something very different. I will try to explain what the difference is.

At the origin of modern physics there might be a distinction both influent and elusive between two different kinds of experimental culture, which also shaped different research styles and different ways to consider science and scientific methods. *One culture* thinks of experimentation as if it would only be depending on theories and points out that it provides a control for theories. Following a suggestion by Ian Hacking, I shall call this *the theoretical approach to the experiment*.[11] Supporters of this culture mostly ignore other features of experimentation, which on the contrary are as crucial as the only one aspect they emphasize, though maybe less elevated (such as the illustrative-teaching-defending role of experiment). Therefore, they can regard an experiment as a proof in favour of or against a theory. Maybe this first approach to experiments, with the concept of proof as its legal pendant, arose in courtrooms, thanks to the job of brilliant orators, lawyers and judges; then it settled from the dusty reading stands of the universities on the aseptic writing desks of natural philosophers and finally reached the chairs of the fellows of titled institutions like The Royal Society.

Roughly in the meantime *another culture* of experiments, which I shall call *the experimental tradition*, was growing both in craftsman laboratories and in the first machine shops, amid dusty workbenches, scraps of unsuccessful experiments and instruments without any apparent utility, unawarely developed by scarcely educated people who have no fear to dirty their own hands. Of course we can find this tradition in the same rooms of the Royal Society; but their representatives did not sit in that educated and mixed audience. The experimental tradition rather passed through the skilled hands of clever professionals like the Curators of experiments: people with technical-practical background who had to perform experiments before that audience. For them experiment was synonymous with uncertainty and doubt, because we only faced with individual cases, and no proofs can be made from experiments. But in their hands experiments, if treated with care, could provide a useful heuristics, so that the experimenters could make (and let make to their audience) some guesses about the nature of bodies, to put it again in Locke's terms. An attentive intervention by a skilled and patient experimenter could and did open new ways to research, driving his guesses and those of his audience.

To do this one needs a very peculiar kind of knowledge. I argue that the design and implementation of an experiment requires more craftsmanship than pure, theoretical and objective knowledge: it involves an actual manipulation

---

[11] Hacking refers to "experimental and rational faculties" complaining "Popper and Lakatos", amongst others, because they "emphasize only the rational faculty" (that is the style I term here "the theoretical approach to the experiment") [9, pp. 260–261].

of the world, based not only on knowing something, but most of all on knowing how things are to be made and, last but not least, on personal experience. In other words experiments are a pragmatic matter, indeed *matter of praxis and practice*. In *The Aim and Structure of Physical Theory* Pierre Duhem gave a colourful and vivid idea of this sort of "knowledge", when he claimed what a researcher has to know to enter a laboratory. Duhem's suggestion is, maybe consistently with the author's will, often read as an example of theory-ladeness of observation or experiment. But the theory of which experiments would be "laden" with is of this very peculiar kind: it is, so to speak, a *theory-intended-for-the-experiment*. The theoretical work of an experimenter, if there is something like that, does not simply deal with abstract notions. This kind of knowledge is meaningful only in the context of an actual making experiments. It is a practical, applied knowledge of how instruments are to be used, how devices are to be manufactured, how they as experimenters have to dirty their own hands.

This pragmatic knowledge — this knowledge of a theory-intended-for-the-experiment — plays an extremely important role in the actual implementation of experiments. I have emphasized that one of the most important qualities of an experimenter is patience in varying experimental conditions to show what does not change and what is to be considered the typical behaviour of things, that is the way things generally are. That was for exemple the case of Francis Hauksbee, the curator of experiments of the Royal Society whose appointment in 1703 was heavily supported by Newton himself as the President of the Society: his ability in varying experimental conditions became legendary. In particular, it was this ability that led him to his remarkable results in examining some subtle electrical phenomena as the luminosity of phosphorus in a Torricelli vacuum.[12]

From this point of view making experiments is very different from simply testing (controlling) a theory, although experiments can be used for such a purpose. And yet, this doesn't suffice to claim for a heuristic role of experiments. According to a well-known witty remark à la Clausewitz, dued to Bas van Fraassen, experimentation is nothing but continuation of "theory construction" by other means. However, van Fraassen is not willing to admit that "experiments are [...] designed to test theories, to see if they should be admitted to the office of truth-bearers" [20, p. 73].[13] Experiments, he points out, rather aims to "fill in the blanks in a developing theory", so a whole "theory construction" can make advance. In fact, as Ian Hacking has observed, this *filling in the blanks* becomes the major issue of experimentation in van Fraassen's account of the "scientific image": indeed, "filling in the blanks [means] guiding the continuation of the construction, or the completion, of the thoery". On the other hand, the theories formulate the questions worth being answered and embody "a guiding factor in the design of the experiments to answer those questions" [20, p. 74].[14] Experiments are always theory-dependent and theories come first. By no means an experi-

---

[12] About Hauksbee's experiments see the quoted work by [10, p. 230].
[13] About his famous Clausewitz-styled statement see [20, p. 77].
[14] For a criticism see also [9, pp. 238–240].

menter could suggest to a theorist a new insight through his own work: in van Fraassen's account, even if experiments are not only tests "for empirical adequacy of the theory as developed so far", nevertheless they cannot be anything but useful "blank-fillers" *for* a theory could make progress. According to van Fraassen, this "being-for-a-theory" of any experiment (as opposed to the "theory-intended-for-the-experiment" outlined above) should entirely fulfil and exhaust the role of the experiment. Though experiments could help to construct or complete a theory, it's up to the theory to provide a heuristics.

However, this seems not to be consistent with historical evidences. Not only the case studies van Fraassen refers to are prone to a quite different interpretation, as shown by Hacking. In addition, his view can neither explain the relevance of professional experimenters as the Curators and the function the Statute of the Royal Society stated for them, nor account for the threefold role of experiment (illustrating-teaching-defending theories) I discussed above. Finally, van Fraassen's perspective doesn't take into adequate account that sometimes experimentation can be the *guiding factor* which come first in the design of a specific theory.

The double-slit experiment provides a good exemple of that. In XIX century this was supposed to be (in fact, it was) a crucial experiment between corpuscular and wave theory of light; but in XX century the double-slit experiment became a major issue for quantum mechanics; and here his role as a test is evanescent against his role as a heuristic tool. Look at the use Richard Feynman makes of it in his presentation of quantum mechanics: it is a heuristic one — a heuristics which indicates the most important features of quantum theory.

Note that it doesn't matter here if this experiment is actually a *Gedankenexperiment*, but only the role it plays. As Feynman states: "We are doing a "thought experiment", which we have chosen because it is easy to think about. We know the results that have been done, in which the scale and the properties have been chosen to show the effcts we shall describe" [5, vol. 3, pp. 1–6]. Just in the same way a skilled experimenter as Francis Hauksbee chose, so to speak, "the scale and the properties" of phenomena "to show the effcts" he *aimed* to describe: in other words he chose a heuristics. This is not filling in the blanks of a guiding theory behind the experiment; this is simply a good example of a "theory-intended-for-the-experiment" (or of an "experimental faculty", to put it in Hacking's terms).

According to Feynman, the double-slit experiment "has in it the heart of quantum mechanics. In reality, it contains the *only* mystery. We cannot make the mystery go away by "explaining" how it works. We will just *tell* you how it works. In telling you how it works we will have told you about the basic peculiarities of all quantum mechanics"[5, vol. 3, pp. 1–13]. So the experiment is a little bit more than a blank-filler in the theory constuction of quantum mechanics: in reality, it *embraces* the whole theory. Feynman even refuses any theoretical explanation behind the phenomenon: "One might still like to ask: "How does it work? What is the machinery behind the law?" No one has found any machinery behind the law. No one can "explain" any more than we have just "explained". No one will give you any deeper representation of the situation. We have no ideas about a more basic mechanism from which these results can be deduced" [5, vol. 3, pp. 1–13]. Explanation here simply ratifies in the unchangeable statement of the law that *typical*, paradigmatic behaviour of things the experiment has found, shown and demonstrated. To put it in Locke's terms, the double-slit experiment helps "guess righter at [...] yet unknown properties" of nature and Feynman still uses it to shape his path integrals formulation of quantum mechanics.

So, shall we maybe reverse van Fraassen's claim? Is really experimentation the continuation of theory construction by other means? Or shouldn't we admit, without burding this with any inductivist nuance, that *theory is often the prosecution of experiments with other means* (and experiments sometimes provide a first glimpse of a theory)? Maybe Popper is right to argue, quoting Novalis, one has to cast nets to catch fish — and hypotheses are our nets. We need audacious, open-minded hypotheses, "marvellously imaginative and bold conjectures", to put it in Popper's terms. I would like to suggest that one way to make such "marvellously imaginative and bold conjectures" is indeed *to make experiments*.

BIBLIOGRAPHY

[1] *The Ancren Riwle. A Treatise on the Rules and Duties of Monastic Life*. Edited and translated from a semi-saxon manuscript of the 13th century by J. Morton, printed for the Camden Society, London 1853.

[2] *The art of Nombryng*. A translation of John Holywood's De arte numerandi. In *The Earliest Arithmetics in English*. Published for the Early English Text Society (Extra Serie, n. 118), Oxford University Press, Oxford 1922.

[3] H.R.F. Bourne. *The Life of John Locke*. Scientia Verlag, Aalen 1969 (Reprint of the Edition London 1876).

[4] J. Colman. Lockes Theorie der empirischen Erkenntnis. In *John Locke, Essay über den menschlichen Verstand*. Herausgegeben von U. Thiel, pages 197–221. Akademie Verlag, Berlin 1997.

[5] *The Feynman Lectures on Physics*. Addison-Wesley, Reading 1964.

[6] F. Gil. *Provas*. INCM, Lisboa 1986. It. tr. *Prove. Attraverso la nozione di prova/dimostrazione*. Jaca Book, Milano 1990.

[7] G. Gori. *La fondazione dell'esperienza in 'sGravesande*. La Nuova Italia, Firenze 1972.

[8] W.J. 'sGravesande. *Physices Elementa Mathematica, experimentis confirmata sive Introductio ad philosophiam Newtonianam*. Leidae 1720–1721.

[9] I. Hacking. *Representing and Intervening. Introductory Topics in the Philosophy of Natural Science*. Cambridge University Press, Cambridge 1983.

[10] J.L. Heilbron. *Electricity in the 17th and 18th Centuries. A Study of Early Modern Physics*. University of California Press, Berkeley 1979.

[11] D. Hume. *A Treatise of Human Nature*. Edited by D. Fate Norton and M.J. Norton. Oxford University Press, Oxford/New York 2000.

[12] I. Kant. *Kritik der reinen Vernunft* [1st edition]. In Kant's *Gesammelte Schriften*, herausgegeben von der Königlich Preussischen Akademie der Wissenschaften, erste Abt., Bd. IV. Reimer, Berlin 1911. Eng. tr. *Critique of Pure Reason* [The Cambridge Edition of the Works of Immanuel Kant]. Cambridge University Press, Cambridge 1998.

[13] J. Locke. *An Essay Concerning Human Understanding*. Edited with an introduction, critical apparatus and glossary by P.H. Nidditch. Clarendon Press, Oxford 1975.

[14] F. Moiso. Morfologia e filosofia. In *Annuario Filosofico*. Mursia, Milano 1992.

[15] K.R. Popper. *Die Logik der Forschung*. J.C.B. Mohr (Paul Siebeck), Tübingen 1934, 1966.

[16] K.R. Popper. *The Logic of Scientific Discovery*. Hutchinson & Co., London et a. 1959, 1968.

[17] M. Purver. *The Royal Society: Concept and Creation*. Routledge and Kegan Paul, London 1967.

[18] T. Sprat, *The History of Royal Society of London For the Improving of Natural Knowledege*, edited with critical apparatus by J.I. Cope and H.W. Jones, Routledge and Kegan Paul, London 1959.
[19] W. Thomson. On the propagation of laminar motion through a turbulently moving inviscid liquid. In *The London, Edinburgh, and Dublin Philosophical Magazine and Journal of Science*, 24, 1887.
[20] B. van Fraassen. *The Scientific Image*. Clarendon Press, Oxford 1980.
[21] J.W. Yolton. *Locke and the Compass of Human Understanding. A Selective Commentary on the "Essay"*. Cambridge University Press, Cambridge 1970.
[22] P. Walmsley. *Locke's Essay and the Rethoric of Science*. Bucknell University Press, Lewisburg and Associated University Presses, London 2003.
[23] R.S. Woolhouse. *Locke*. The Harvester Press, Brighton 1983.